ORE PETROLOGY

ORE PETROLOGY

R. L. STANTON
Department of Geology
The University of New England
Australia

McGRAW-HILL BOOK COMPANY

New York   St. Louis   San Francisco   Düsseldorf   Johannesburg   Kuala Lumpur   London
Mexico   Montreal   New Delhi   Panama   Rio de Janeiro   Singapore   Sydney   Toronto

ORE PETROLOGY

Library of Congress Catalog Card Number 71–157486

07–060843–1

9–MAMM–76 5 4 3 2 1 0

Contents

Preface

This book is concerned with the study of ores as rocks—that is, with *ore petrology*.

For much of the last three hundred years—and certainly for much of the twentieth century—most students of ore genesis have regarded the majority of metalliferous deposits as exotic entities, derived from extraneous sources and superimposed on geological situations to which they bore little or no genetic relationship. Almost the only exceptions have been those, such as ores of chromium and nickel, that have apparently formed as part of the igneous rocks that enclose them. From time to time it has been suggested that at least some other ore types, such as disseminated sulfides in black shales, might be "rocks in their own right," but these views have usually been submerged by those involving introduction from elsewhere. Belief in a general exotic origin was natural when mining technologies were crude, since under these circumstances only rich vein fillings merited the term "ore." With the more recent refinement of mining and concentrating techniques, however, vein fillings have lost their preeminence, and a wide variety of materials now attract attention as potential sources of metal. While the minerals of veins are undoubtedly introduced, those of many of these other rock bodies are by no means of such clearly foreign origin. This has led to the suspicion—and, in some cases, the conviction—that in many instances the ore minerals concerned are an original, primary part of the rock formation in which they occur and hence that ore genesis and rock genesis are one and the same thing. This book is written with the latter possibility very much in mind: that ores as a class are divisible into *ore types*, that each ore type is a part of a particular *petrological association*, and that a fundamental consideration in genetic studies is geological *environment*.

This book is intended primarily for senior undergraduates and postgraduate students in economic geology. However, it is hoped that it will also be useful for geologists already involved in the search for ore and for those active in scientific fields adjacent to economic geology. The later parts of the book should be useful for such investigators as igneous petrologists, chemists, and physicists—people whose work impinges on or has application in some aspects of ore studies—who are interested in knowing something of the present state of the science of mineral deposits.

Like most books on petrology, this one is concerned only with primary, unweathered materials. No consideration is given to weathering processes, to the formation of "secondary" ores and gossans, or to ores in unconsolidated sediments (placers). The book is divided into two parts: "Principles" and "Important Associations."

Part One begins with a chapter on the historical development of theories of ore formation. This has been written not just as an interesting introduction but almost as a philosophical justification of the book. Its purpose is to show how views on ore deposits have evolved to the point where we may regard their study as ore petrology. To a lesser extent it is intended to show how repetitious ore-genesis theory has been and how old many of our "modern" ideas really are. This is followed by chapters on the nature and classification of ore minerals, the solubilities and stabilities of ore minerals at high temperatures and pressures, the segregational and mixing processes in ore formation, the mechanisms of crystal growth in "open space" and in solid, polycrystalline aggregates, and the processes and results of deformation. Chapters 3 and 4, which are principally concerned with crystal chemistry and ore mineral groups, may appear somewhat elementary in the present context, but this material has been included for those readers whose early training has not included basic ore mineralogy and for those among exploration geologists and others who may have lost some of their earlier familiarity with formal mineralogy. In the last three chapters of Part One—"Growth and Growth Structures in Open Space," "Growth and Growth Structures in Polycrystalline Aggregates," and "Structures due to Deformation and Annealing"—the features and behavior of ores are approached from the point of view of modern "materials science." Frequent reference is made to analogies with metals and ceramics.

While all this may seem a heavy emphasis on mineralogy, physics, and chemistry for a book on ore deposits, there is little doubt that students of ore formation will have to become increasingly familiar with these fields if any real progress is to be made in their own. Part One is intended to emphasize this and also—largely for the sake of those who are already practicing mining and exploration geologists—to collect into one book information that until now has been scattered widely through the scientific literature.

Part Two is concerned with the nature, mode of occurrence, and origin of the principal *ore types*. The ores are grouped under broad petrological headings, but any immediate genetic connotation is avoided. Chapters are devoted to "Ores in Igneous Rocks," "Ores in Sedimentary Rocks," "Ores in Metamorphic Rocks," and "Ores in Veins." This primary grouping is not, however, based on any preconceived ideas of genesis—its basis is simply the observable fact that any given ore occurs within an igneous, sedimentary, or metamorphic rock or as part of a vein. Within this framework each ore type is treated as part of a rock association. For each, the principal features of environmental association, form, mineralogy, and chemical constitution are given, and this is followed by a discussion of views on genesis. Part Two ends with a chapter on "Ore Type and the Tectonic Cycle," an account of the apparent parallelism between the evolutionary pattern of the continental crust and progressive changes in the ore types that each continent contains. This chapter is intended, in a brief way, to place the considerations of Part Two in perspective and to draw together the threads of the book as a whole.

As this book has progressed the author has been concerned to draw attention not only to what *is* known but also, most importantly, to what is *not* known. In doing this he may be accused of frequently leaving the reader "up in the air," but he would like to think that in pointing to present inadequacies in our knowledge he might stimulate the inquiring mind to look to the *future* of ore-genesis studies as well as to their past.

This book sets out to give an up-to-date picture of a very broad field. Such a task is not easy, particularly at a time when data is accumulating, and ideas changing, as rapidly as they are as the final quarter of the twentieth century is approached. Breadth of approach has inevitably meant that, here and there, one has had to tread near less familiar ground. The author is all too conscious that some inadequacies must inevitably have resulted from this and hopes, like Lindgren, that his shortcomings may be judged leniently. Similarly, the rapid proliferation of the literature makes it virtually impossible—in the space available and in the time it takes to complete and publish a book—to include all the very latest information. The author has regretfully been forced to omit reference to many excellent technical papers, and where, because of timing, it has been impossible to incorporate the most recent developments in the text, important papers have simply been included in "Recommended Reading" at the end of each chapter.

Acknowledgements. The author is greatly indebted to many helpers: first and foremost to Helen Willey—successively student, assistant, and colleague—for patient assistance in innumerable ways; to Bruce Whan for his meticulous work on illustrations; to his wife for much help in checking and indexing; to Heather Roan and Sandra O'Keefe for their tireless help in typing and checking; to the many authors and publishers who so kindly permitted reproduction of their figures and quotation of their texts; to Broken Hill South Ltd., the Broken Hill Mining Managers Association, the Consolidated Zinc Corporation of Broken Hill, the Electrolytic Zinc Company of Australasia Ltd., the International Nickel Company of Canada Ltd., North Broken Hill Ltd., Western Mining Corporation, and the Division of Mineral Chemistry, C.S.I.R.O., for their long continued moral and material support; and finally, and especially, to Haddon King and Donald Fairweather for their help and encouragement over many years.

R. L. STANTON

ORE PETROLOGY

1
Introduction

Minerals are the most important materials of industrial activity, and if past developments are a key to the future, there is little doubt that their importance will increase.
Even now there is hardly an avenue of human endeavor that is not in some way
dependent on substances that minerals produce.   Parallel with this increasing dependence on minerals in general has gone a great increase in the use and technology
of metals.   This has, of course, led to a corresponding increase in demand for
those minerals from which metals may be extracted, and so the present has become
an age of anxious search for ore.

     Because demand has become so immediate and so huge, the search for mineral
deposits can no longer be left to individuals.   One cannot help but be a little sad
that this is so; prospectors are an indomitable breed whose courage and tenacity
have not only supplied the world with treasure but have, through the ages, led to
the discovery of new lands and contributed to the development of new societies.
The change has, however, been inevitable.   The historical methods of search are
simply incapable of covering the ground with sufficient speed.   In addition, much
of the ore that has appeared at the surface has now been found.   "Ore-finding"
has now become a matter for large and thorough surveys, rather than isolated small
searches, and of detecting those deposits not yet bared by erosion—deposits that
are concealed by soil and rock and that could never have been found by simple

surface observations.   The discovery of hidden ore deposits now requires not only the efforts of many men but all the guile that science can supply.

Such an increase in thoroughness and dependence on technique has naturally led to problems of method and expense.   The cost of finding new deposits has risen enormously and now involves heavy responsibilities in determining the most effective methods and using these with maximum efficiency.   This emphasizes the necessity for reliable information on ore formation and likely patterns of occurrence —knowledge that in turn can be acquired only by the development of an extensive and rigorous theoretical background in ore-forming processes.

It is probably not unfair to say that the development of ore-genesis theory has been slow and rather haphazard.   It has also been repetitious, and the popularity of various well-known theories has often waxed and waned substantially as a matter of fashion.   This may have been due—at least in part—to a certain lack of co-herency in research.   In silicate petrology, for example, field observations, micros-copy, and mineral chemistry have proceeded more or less hand in hand and have continuously complemented each other.   In this way each technique or method of approach has received continuous help from the others, and many problems have been approached and solved in a relatively systematic manner.   While there have certainly been exceptions, this does not seem to have been the case with ores.   In the more extreme instances laboratory workers have tended to look upon their counterparts in the field and mine as a rather poorly informed band of empiricists, and the field geologists have reciprocated by accusing the mineralogists of "looking at mines through microscopes."   Although such exaggerated views are far from universal, there is no doubt that at least some lack of coordination between field and laboratory—and theory and practice—has been fairly general and that this has had a retarding influence on ore research.

With increasing pressures, such lack of coordination can no longer be afforded. Two things seem desirable: the development of very much better theory and the more serious application of such theory to the problems of practical geology. Other human activities abound with examples of close linkages between science and industry.   The use of mathematics in engineering is of ancient origin, and this continues in the application of highly sophisticated mathematics in such modern developments as aircraft design and satellite control.   The elegant discoveries of solid-state physics have led to much of the recent development in communications, and to the "pure" research of the physical metallurgist all our modern knowledge of alloys is due.   That there should be a similar relationship between those who study ores and those who find and mine them seems clear, and there are now many organizations attempting to carry the principle into practice.

What now seems to be needed is a substantial increase in our understanding of primary ores.   One possible approach in this endeavor is to accept, as a working hypothesis, that ores are no more or no less than rocks and, provided the relevant conditions for the stability of ore minerals are observed, that they may have formed in all the ways that "ordinary" rocks have formed.   The ores themselves are then seen quite simply as natural polycrystalline aggregates, conforming with the prin-ciples of physical metallurgy and "materials science."   If such an assumption is correct, their development, pattern of distribution, and physicochemical characteristics should conform quite systematically with that of the associated silicates and carbonates. The problem then becomes part of petrogenesis in its broadest sense.

part one

Principles

2
Historical Development
of Ore Petrology

INTRODUCTORY REMARKS

Some 40 years ago, Thomas Crook, in a very wise and balanced survey of the development of theories of ore genesis, drew attention to the sharp division between orthodox petrology and the science of ore deposits. He observed that although metalliferous-vein and other mineral deposits were essentially rocks in the best sense of the term, most petrologists ignored them, and he ventured the opinion that not until all mineral deposits came within the scope of rock studies would petrology become a fully fledged science.

In spite of this and the efforts of a small number of petrologists such as Fenner and Buddington in America and Niggli in Europe, the situation has changed little since the time of Crook's comments. Neither field of study seems to have had much impact on the broader aspects of the other. For the most part, "orthodox" petrology has been concerned with either the fine detail of silicates or the very broad aspects of igneous, sedimentary, and metamorphic rock formation, and any consideration of related economic minerals has been incidental. Students of ores, in contrast, have tended to neglect many petrological features and have treated their deposits as isolated entities related only incidentally to the rocks that enclose them. This has led to the development of a succession of ad hoc hypotheses of ore formation, few of which have been genuinely capable of prediction.

7

Such a situation, however, is by no means exclusively modern. If one looks back at the development of ideas on rocks and ores, it is clear that for most of the time since the sixteenth century the two fields of study have been regarded as fairly distinct. There have indeed been exceptions—periods of thought briefly influenced by men of unusual perception—but for most of the time each has followed essentially its own path. Any connection—as in the case of igneous rocks and hydrothermal ore deposits—has been of a very general and tenuous kind.

Such deep-rooted distinctions usually do not develop without good reason, and the reason in this instance is probably not difficult to see. In fact it seems very likely that it has had an economic basis in the early requirements of primitive technology. With relatively inefficient mining methods and crude concentrating techniques (often no more than picking by hand), mineral deposits had to be both rich and coarse to be regarded as ore. Since it was usually only in veins that the necessary richness and coarseness were achieved, the term "ore deposit" became virtually synonymous with "vein filling." Crook himself generally referred to ore deposits as "mineral veins," although mining and ore dressing had achieved a high level of sophistication at the time he wrote, and many occurrences other than veins had entered the category of ore deposits. Certainly, in the early days of modern European mining, when the metalliferous veins of the Harz began to be exploited intensively and the German students of the sixteenth and seventeenth centuries were pronouncing on their origin, most ores were necessarily of vein occurrence. It was in this setting that Agricola recognized that ores were generally of a different age from the rocks that enclosed them and in which he made his now famous pronouncement: "to say that lodes are of the same age as the earth itself is the opinion of the vulgar" ("De Re Metallica," 1556).

Twentieth-century technology, however, has completely changed this view of "ore." Methods of mining and concentrating have become so efficient that materials having extremely low metal content and fineness of grain are now worth exploiting. As a result, metalliferous rocks of a far wider range of type and occurrence have become of interest to the miner. The former necessary preoccupation with veins has gone, and the "economic geologist" is concerned with almost every material of the petrological kingdom. Naturally, this has had its effect on ideas on the origins of ore minerals.

AGRICOLA TO DE BEAUMONT: SIXTEENTH TO MID-NINETEENTH CENTURY

Neglecting the ancients, most of the early thinkers regarded ores as exotic and essentially unrelated to the environment in which the minerals had finally come to rest. This followed naturally from the then current definition of "ore" that almost required its occurrence as a fissure filling.

Agricola (1546) considered the materials of veins to have been derived from the crust. He thought that meteoric waters, seeping down into the earth from the surface, became heated and resurgent, dissolving metals at depth and redepositing them in fissures at higher, cooler levels. Descartes (1644), however, considered most ores to be additions from the subcrust. He regarded the earth as a cooled star having a chilled envelope that enclosed a still hot interior, and he suggested that beneath the crust was a shell of heavy material, some of which was expelled as exhalations. These, he thought, found their way upward and precipitated or sublimed in fissures in the cooler upper layers to form mineral deposits.

About 130 years later, the English geologist Pryce (1778) returned to Agricola's "lateral secretion" idea, observing:

> We may reasonably infer that water, in its passage through the earth to the principal fissures, imbibes, together with the natural salts and acids, the mineral and metallic particles with which the strata are impregnated . . . to form different ores more or less homogeneous, and more or less rich according to the different mixtures which the acid had held dissolved, and the voids in which it is deposited.

This view of ore deposits as being characteristically products of migration, superimposed on geological situations to which they were essentially unrelated, persisted in the work of Pryce's great contemporary, Hutton. Hutton was, however, rather more specific concerning source than the earlier writers and seems to have been the first to suggest a tangible link between "ores" on the one hand and "rocks" on the other. In his "Theory of the Earth" (1788) he maintained that there were two principal kinds of mineral matter within the earth—siliceous and sulfurous. Since silica and silicates were apparently insoluble in water, he considered that these must have been emplaced in a fused state, and he pointed to the intrusion of igneous rocks and the thermal metamorphism of the rocks they invaded as proof of this. By analogy he thought that metallic sulfides, which also appeared to be insoluble in water, could only have been transported in the fused state, and so he concluded that the sulfides of veins had been injected as molten material. He was thus an ore magmatist—perhaps the first—and regarded vein ores as igneous rocks.

Werner (1791), on the other hand, regarded the ores as chemical sediments, though of a rather special kind. His views were diametrically opposed to those of Hutton: Werner held that not only stratified rocks but also granite and basalt had been precipitated from aqueous solution. He extended this theme to mineral veins that he attributed to the filling of cavities by solutions percolating into them from above—the solutions from which the overlying sediments were being deposited concurrently.

While it is easy to criticize Hutton and Werner for holding views that may now appear to be extreme, it is not unlikely that, like some modern extremists, they were products of their environment and that, for the deposits they knew, their ideas were not entirely unreasonable, in spite of the ridicule later heaped upon some of their ideas. This may be particularly the case with Werner, who was doubtless influenced in his views by the German Kupferschiefer. The extensive bedded copper deposits have the appearance of chemical sediments deposited and preserved in marine euxinic sediments, and this view is accepted by many present-day geologists. The bedded ores are accompanied by minor veins filling cracks within the adjacent sedimentary rocks, and it would not have been difficult—or unreasonable—for anyone working in Werner's time to have attributed the vein sulfides to essentially the same source and processes as those responsible for the bedded materials. Whatever the validity of their views, however, in Hutton and Werner we have two early workers who saw the formation of mineral deposits not as particularly special events but as part of the broad scheme of petrogenesis.

While the views of these two men were of great importance in their connecting ores with specific processes of rock formation, their extremism led naturally to reaction. The views of Werner concerning the origin of igneous rocks (that they were marine precipitates) were soon abandoned, and his theories of ore formation fell with them—perhaps not entirely deservedly. Similarly the views of Hutton, whose experience with ores was far less than that of Werner, were soon found to be inconsistent with the bulk of observation. The development of delicate crustification, the thinning of many veins with depth, and the lack of thermal metamorphism of the wall rocks of many veins, quickly led to doubt of Hutton's thesis and the view that if the veins were indeed igneous, they had been deposited from vapors given off by intrusive rocks rather than as intrusive rocks themselves. This was supported by observations about volcanoes, where it quickly became clear that the production of lava was associated with the emission of much water vapor and sulfurous gas.

Breislak (1811), Boué (1822), and Scrope (1825) all began to emphasize the likely importance of water in inducing fluidity in magmas, and Boué pointed to the possibility that such igneous water was also important in the formation of mineral veins. Others, such as Necker (1832), stressed the importance of sublimation. Looking back, it is apparent that in spite of the prominence of the Hutton-Werner controversy, the views of Descartes had retained substantial adherence, and as the more extreme views declined in popularity the plutonic vapor theory reasserted itself, this time encompassing the activity of both "wet" and "dry" vapors and solutions and bringing to clearer focus the possibility that igneous magmas were their source.

The first beginnings of the genetic pattern favored by the American school of the first half of the twentieth century can be seen in the early writings of Fournet (1834), who recognized that igneous water and vapors might be responsible not only for veins but also for deposits formed along the margins of intrusive rocks by reaction with those they invaded—that group of ores now recognized as contact metamorphic. Although Fournet was later to revert to ore magma extremism, his earlier work appears to have been the first clear indication that igneous ores might constitute a lineage from "orthomagmatic," through contact metamorphic, to migrant types deposited from fugitive vapors and solutions.

As the middle of the nineteenth century approached, the study of ores was becoming an integral part of petrogenesis, and with this—some might say insidiously—was quietly growing the assumption of an almost exclusively igneous connection.

THE NINETEENTH CENTURY: 1840 TO 1870

The stage was now set for the work of de Beaumont, who at the July 1847 meeting of the Geological Society of France, presented his "Note sur les émanations volcaniques et métallifères," a paper regarded by Crook as perhaps "the most important and influential paper ever published on the theory of ore deposits" (1933, p. 60).

De Beaumont was probably the most eminent French authority of his time, and he coupled a very wide knowledge of geology with an outstanding sense of scientific perspective. He saw clearly that continents exhibited broad patterns of folding, often delineated by mountain chains, and that related to this folding were features such as metamorphism, intrusion of granitic rocks, and incidence of mineral veins. He proposed that the main structural lines were due to periodic collapse of the crust induced by shrinkage of the earth, leading to folding, the development of parallel trend lines within each folded province, and the formation of veins which, in conforming with these trend lines, were generally parallel for any single epoch. By analogy with volcanic processes, he considered that the materials of the veins were derived from the intrusive igneous rocks with which they were regionally coincident. In observing modern volcanicity he distinguished two kinds of volcanic products—lavas and sublimates—and he noted what appeared to be ubiquitous and abundant water vapor accompanying the deposition of the subli- mates. He also distinguished between "vapors" and thermal springs, the former giving rise to sulfur and a variety of metallic chlorides and the latter to ferruginous and calcareous deposits or, at higher "activities," deposits of silica and compounds of such elements as sulfur, arsenic, boron, and fluorine. It was to these latter solutions, which by analogy he considered to be evolved by the intrusive counter- parts of volcanic rocks, that he attributed the formation of ore veins.

It is interesting to note that de Beaumont did not consider all ores to be migrant with respect to their intrusive parent nor all ores of igneous derivation to have been necessarily deposited beneath the surface of the time. He observed segregations of chromite, magnetite, and platinum in basic rocks that he thought had formed by crystallization during the cooling of the intrusives concerned, and he noted a number of deposits of oxides and sulfides that he attributed to contact metamorphism. He also recognized that ore minerals such as copper sulfides, iron and manganese oxides, and various carbonates and sulfates now found within sedimentary beds might have been deposited as chemical precipitates *during sedi- mentation*. In view of the modern trend toward emphasizing the importance of sedimentation in the formation of ores, it is interesting to look back on de Beau- mont's speculations of 120 years ago. Unlike Werner, however, de Beaumont did not regard such ores as being derived from the body of the sea but from igneous solutions that happened—via fumaroles and thermal springs—to reach the sur- face of the earth, depositing their load on the floor of the ocean rather than on the walls of subterranean cavities.

The work of de Beaumont was very neatly—and appropriately—complemen- ted by that of his countryman Daubrée. Daubrée was essentially an experimental petrologist and may now clearly be seen to have been a pioneer of the experimental activity that was to become such an important part of the mid-twentieth century geological scene. Daubrée emphasized the possible importance of water in regional metamorphism and the likely place of mineralization as an event in the hydrothermal processes involved.   From the point of view of ore genesis his most important contribution was his demonstration of mineral transport by heterogeneous reaction.   Whereas others had referred to the *sublimation* of mineral matter, Daubrée showed that in at least some cases the substances were transported chemically as a result of reversible heterogeneous reactions.   This was also recognized by Bunsen (1852), who showed that the Fe_2O_3 of volcanic sublimates could be transported in a stream of HCl, and by Saint-Clair Deville (1861), who showed that SnO_2, TiO_2, and MgO could be made to migrate by similar means.   Daubrée had observed that minerals such as fluorite and tourmaline were commonly associated with ores, and by analogy with volcanic degassing he concluded that elements such as fluorine, chlorine, and boron were, together with all important water vapor, important activators of mineral movement.   His transportation of tin as $SnCl_4$ in a stream of water vapor and HCl gas is one of the best known of all geological experiments.

At about the same time, ore deposits were being looked at from a very different point of view by T. Sterry Hunt in America.   Sterry Hunt was another proponent of the metamorphic theory of granite formation, and he maintained that the concentration of metals into ores was an accompaniment of the conversion of metal-bearing sediments into granitic rocks.   His work is particularly interesting since it encompassed a wide range of processes and anticipated much of the explosion of ideas on sedimentation and ore genesis that was to take place in the middle of the twentieth century.   The basic premise of Sterry Hunt appears to have been that ore deposits represent the segregation of materials that originally occurred as disseminations.   He thus saw the problem of ore formation as one of *collection* from preexisting rocks and subsequent *concentration* by suitable localized precipitating agencies.   He maintained that water, with carbon dioxide, alkali carbonates, and other reactive substances in solution, was the universal collector, and this water picked up the normally diffuse metals and carried them to areas of sedimentation where they were precipitated and eventually built up into stable accumulations. Further, Sterry Hunt regarded organic agencies as being of prime importance in deposition: he pointed to the formation of pyrite in organic muds, the occurrence of traces of heavy metals in the bodies of many marine organisms, and the generation of H_2S by decaying organic matter in lagoons and on the sea floor.   He maintained that he could "hardly conceive of an accumulation of iron, copper, lead or gold in the production of which animal or vegetable life has not, either directly or indirectly, been necessary" (1861).

He considered that processes of this general kind led to the formation of greater or lesser accumulations of the metals that ultimately were buried with the enclosing sediments and then redissolved by circulating underground water, carried to fissures in adjacent rocks, and built up by accretion as mineral veins.   The conversion of sediments to granites was, in this context, simply a process leading to the expulsion of metalliferous solutions which redeposited their load in other rocks adjacent to the site of granitization.   The broad lines of Sterry Hunt's views on ore genesis are summarized well by his statements:

The metals seem to have been originally brought to the surface in watery solutions, from which we conceive them to have been separated by the reducing agency of organic matters in the form of sulphurets or in the native state, and mingled with the contemporaneous sediments, where they occur in beds, in disseminated grains . . ., or are the cementing material of conglomerates.   During the subsequent metamorphism of the strata, these metallic matters, being taken into solution by alkaline carbonates or sulphurets, have been redeposited in fissures in the metalliferous strata, forming veins, or, ascending to higher beds, have given rise to metalliferous veins in strata not themselves metalliferous (1861);

and

If the view which I hold, in common with many other geologists, that most, if not all, of our known eruptive rocks are but displaced and altered sediments, be true, then it may be fairly affirmed, not that eruptive rocks are the agents which impregnate sedimentary deposits with the metals, but on the contrary, that in such deposits is to be sought the origin of the metalliferous eruptive rocks, and that all our metallic ores are thus to be traced to aqueous solutions (1873).

It is interesting to pause at this point to reflect on the range of ideas that had been put forward by about 1870.   Agricola had recognized the epigenetic nature of vein deposits and had drawn attention to the possible importance of vadose water in collecting and transporting metallic compounds.   Descartes had pointed to the possibility that the ores were deposited from juvenile solutions and vapors and, in the language of his day, had anticipated those who were later to suggest a "mantle" origin for many deposits.   Hutton was the first important proponent of the "ore magma" hypothesis and played a leading role in popularizing the idea that granites and ores were causatively connected.   Werner had emphasized the role of marine sedimentation, anticipating by some 180 years those who were to espouse the sedimentary hypothesis in the mid-twentieth century.   De Beaumont, in masterly style, had encompassed the whole broad conspectus of tectonism and igneous activity and had shown how ores might be connected with all stages of the cooling history of igneous rocks and modified by all the circumstances the latter might encounter.   Not only did he consider the deposition of igneous materials beneath the surface, but he also saw clearly the possibility of fruitful interplay between volcanic and sedimentary processes at the surface.   He stated the magmatic theory of primary ore genesis with remarkable completeness and—with the perceptiveness that is the mark of the outstanding scientist—saw the ores he wrote of as a part of the geological rhythm and hence a part of their environment in the broadest sense.

While there were many others, such as Fournet, Scheerer, Bischof, von Cotta, and Belt, who were well known as protagonists of one current view or another, the scene really closes with Sterry Hunt.   Like de Beaumont he was a man of unusual perception and scientific balance.   He put in highly coherent form the importance of surface—particularly biological—processes in collecting and localizing metals and appears to have been the first sophisticated proponent of the process of "lateral secretion."   Sterry Hunt was clearly aware of the possible importance of "source beds" and anticipated by almost 100 years those who were to propose in modern geochemical terms a connection between granitization and the formation of ores.

Thus only shortly after mid-century the germ of all the principal theories of the origin and mode of formation of "mineral veins" had appeared. A pattern of ideas was set, and most of the work on ore deposits that was to follow would be concerned with the elaboration or clearer definition of these earlier hypotheses. Not only the pattern of ideas but also the pattern of thinking was established. Most concentrations of sulfides and related minerals were still regarded as almost unquestionably exotic—an attitude that persisted well on into the next century and about which many of the controversies of the 1950s and 1960s were to revolve. There was also a strong tendency to conformity: where a process appeared to be established for one set of deposits, there was often the suggestion that this might be universal—that it could also be made to account for other primary ores in spite of conspicuous differences in morphology and mineralogy.   There was, too, a widely held tacit assumption that the emplacement of most ores represented the last stage, or climax, of the geological history of the area in which they occurred.   Such tendencies were not, of course, universal, and there were certainly some who cautioned against them.   The Englishman Taylor was one; he had earlier (1833) cautioned in very clear terms:

> Our present state of knowledge as to the formation of veins should therefore, in my opinion, be allowed to admit that most of the causes which have been stated have operated at various periods and through a long succession of time, some prevailing at one epoch, and some at another, modified by circumstances which we can but imperfectly comprehend or explain.

Others, such as von Cotta and von Groddeck, called similarly for a balanced view, though as in many other fields of geology, there persisted a strong urge to favor one theory over another in a general way and to follow vogue almost as much as reason.

THE NINETEENTH CENTURY: THE LAST QUARTER

The remaining years of the nineteenth century saw few developments of note.   S. F. Emmons, who might be regarded as the first of the modern generation of economic geologists in America, was advocating a meteoric origin for some of the Colorado lead deposits—in this case the leaching of porphyries by descending ground waters, leading to the subsequent replacement of adjacent rocks.   Phillips, in England, was propounding views not dissimilar in principle from those of Agricola, Sterry Hunt, and Emmons, involving the collection of formerly disseminated metals by subsurface waters that then transported and deposited their load in appropriate fissures.   At about the same time (1882) Sandberger, in Germany, was attempting to place such lateral secretion hypotheses on a firmer factual basis by carefully determining the amounts of metals in various common rocks and minerals.   He showed that in fact the heavy metals were of wide occurrence in small amounts and that in many cases they occurred in rather specific associations, such as nickel and cobalt in ultramafic rocks.   Although, as always, there were dissidents, the prevailing atmosphere of thought in the 1880s seems to have been that most vein deposits were produced by the leaching of trace metals from the upper parts of the earth's crust and the redeposition of these in preexisting cavities. The stage was thus set for another reaction and this—the inevitable—came in the 1890s, with the work of Pošepný and Vogt.

Pošepný's work, which came to general notice with the publication of his "Genesis of Ore Deposits" in 1894, in essence represented a sharp return to the igneous hydrothermal ideas put forward in such sophisticated form by de Beaumont some 50 years earlier. Pošepný regarded ore-bearing faults as contributors rather than receivers of ore solutions. He considered that ores occurred in faults because the latter were able to tap the metalliferous "barysphere," not because they happened to be convenient—and passive—repositories for solutions circulating independently of them. While he allowed that one might conceive of two zones of circulation and ore deposition—a shallow one dominated by vadose waters and low-temperature leaching and a deeper one dominated by hot juvenile water—he considered the latter zone to be of dominating importance.

These somewhat extreme views were opposed by a number of geologists, notably Le Conte (1895) and also, with nice balance, Louis (see second edition of Phillips, "Treatise on Ore Deposits," 1896), who drew attention with quiet effectiveness to the subjectivity of the argument by saying:

> The manner in which the minerals were deposited in the fissures admits of various interpretations. Even the statements upon which the opposing theorists base their arguments are not by any means unchallenged by their opponents, and even if they were, most of them admit of diametrically opposite explanations. Thus, if it were definitely proved that the rocks within reasonable distance of a fissure vein contain minute proportions of the same metals as found in the vein, one side would see in this fact a proof that the metals in question were derived from the rocks, whilst the others would argue that the metals in the rocks were derived from the vein . . . (1896).

The final decade of the nineteenth century also saw the important contributions of J. H. L. Vogt, who, while following essentially the lines set down by Hutton, Fournet, de Beaumont, and their adherents, was able to refine the earlier ideas by applying to them the more modern principles of chemistry. He came to prominence with his studies of differentiation in basic magmas, showing that concentrations of nickel-iron sulfides in such rocks might form by immiscible liquid segregation and that ilmenite and titanomagnetite deposits might form by crystal settling. He also suggested that at least some vein deposits were formed by the injection of such orthomagmatic material following its segregation in the magma chamber. While this could be said to be simply a more sophisticated presentation of the ideas already put forward much earlier by Hutton and Fournet, Vogt did see magmatic differentiation as a more extended process and, like de Beaumont, regarded its final stage as that of hydrothermal deposition.

THE TWENTIETH CENTURY: 1900 TO 1919

The stage was now set for the development, and overwhelming ascendency, of the igneous theory—a theme that was to dominate the study of ore deposits for almost

half a century.   There was just time for a final warning from C. R. Van Hise, who
cautioned:

> The source of a metal for an individual district is not to be ascribed a priori to igneous,
> sedimentary or metamorphic rocks, but can be determined only after an inductive in-
> vestigation of the facts.   The metal of a district may be derived from the late igneous
> rocks, from ancient igneous rocks, from sediments, from metamorphosed rocks, or from
> any combination of the above.   When the important economic districts of the world
> are inductively studied, and certain knowledge obtained, I believe that it will be dis-
> covered that a great number, if not the majority, of ore deposits, are not the result of a
> single segregation, but are the accumulated fruits of a great interrupted process of
> segregation, a part of the metals for the deposits having been worked over many times by
> the metamorphic processes (1903).

Such a balanced view deserved greater attention than it got.   There was more
in Van Hise's statement, however, than just a balanced view.   Whether or not he
recognized it he sowed the seed of an idea—a seed that came tantalizingly close to
breaking its dormancy on several occasions but that in the end was to take almost
50 years to germinate.   The idea concerned was simply that the deposition of an
ore in its present position might not have been the final event in the geological
history of its surroundings: ores might be metamorphosed and hence they might
constitute an intrinsic part of their present environment rather than something
merely superimposed upon it.   Such a possibility was noted again on several
occasions in the ensuing years, though its potential significance seems to have
remained unrecognized.

The trend begun by Pošepný was established firmly and given further momentum by the famous American geologist J. F. Kemp.   In a series of papers beginning
in 1902, Kemp maintained that magmatic solutions were of overwhelming importance in the development of both vein and contact deposits.   He considered that
were ore deposits the result of meteoric activity, they should be of common occurrence, whereas in fact they were relatively rare.   Because of this he thought that
they must result from "some sharply localised, exceptional and briefly operative
cause . . ." (1907, p. 4) and that the phenomenon most likely to fulfill these
requirements was intrusive igneous activity.

The views of Kemp were reinforced by the essentially similar ones of Lindgren, a prodigious worker and an outstanding student of ore genesis whose influence persisted strongly for over 50 years.   Lindgren was a strong advocate of the
magmatic hydrothermal hypothesis—of this there was no question.   Yet he saw
ores as a broad geological group and did not question that while some might form
beneath the surface by both magmatic and meteoric activity, others formed on
the surface by weathering and sedimentation.   While his main interest appears to
have been in ores of igneous affiliation, he emphasized that many, such as those of
the limestone–lead-zinc and the sandstone–vanadium-uranium-copper red bed
provinces, were apparently quite unconnected with igneous activity.   He reaffirmed de Beaumont's orthomagmatic contact-metamorphic hydrothermal
lineage, but he clearly attempted to set these in their place, as igneous phenomena,
beside deposits formed by sedimentary, metamorphic, and weathering processes.

In 1911 he put forward a classification of ore deposits based essentially on genetic criteria. This not only took account of major processes but attempted to define certain categories by pressure-temperature conditions. Without doubt, igneous processes were uppermost in his mind as far as "primary" ores were concerned, though other processes had prominent positions in his classification, and he was very fair in noting deposits of which he was uncertain and for which he could not provide in his genetic tabulation. While Lindgren worked in a period in which igneous processes were in overwhelming vogue, and while he himself contributed substantially to the movement, it is quite apparent that for many ores he had his doubts and was in fact a man of moderate views. Certainly the reading of his work leads one to suspect that he was a good deal more moderate than some of his disciples might have us believe! By and large, however, Lindgren regarded ore minerals as additions to rocks rather than as parts of them.

The first decade of the new century closed with two interesting papers, both early stirrings of Van Hise's seed: Emmons' (1909) "Some regionally metamorphosed ore deposits and the so-called segregated veins" and Lindgren and Irving's (1911) "The origin of the Rammelsberg ore deposit."

Emmons examined a number of sulfide lenses in the Appalachians, particularly those of the Milan and Blue Hill mines in Maine, and concluded that they had suffered regional metamorphism along with the volcanic and sedimentary rocks that enclosed them. He noted that:

> It is no easy task to determine the character of these deposits before they were metamorphosed, but this may be inferred from their present condition and mineralogical character. The diagnostic value of the minerals is prejudiced, however, since nearly all of them may be produced under conditions of regional metamorphism where the necessary elements are at hand. Before metamorphism the pyritic copper deposits . . . were presumably fissure veins of replacement types and disseminated ores along fractured zones (1909, p. 761).

While he was clearly of the opinion that the initial deposition of the sulfides had followed that of the host rocks, Emmons very fairly quoted Hitchcock (1878), who had stated:

> Our theory as to the origin of the deposits has been that they were originally beds not fissure veins and that in later periods the copper has been segregated from the general metalliferous belt into several strings and veins making up the richest portions.

While Lindgren and Irving concurred with the then well-established view that the Rammelsberg sulfide lenses were for the most part concordant with the foliation of the sediments enclosing them, they noted local discordancies and gave weight to those in their consideration of genesis. They did not doubt, however, that the sulfides had been metamorphosed:

> The structural relations of the orebody indicate, according to our view, without any doubt, that the deposit is a bedded vein; that is, a fissure vein lying in part conformable to the surrounding slates. The distribution and the structure of the ore itself is inconsistent with the theory of sedimentary deposition. As far as our experience goes the

structure is unique in ore deposits, but as to its interpretation there can be no reasonable doubt.   The sulphides do not occur with their primary texture.   The structure is that of a dynamo-metamorphic rock, in which all of the constituents, except pyrite, have been drawn out into streaks which are intricately mingled.   The appearance . . . could easily be duplicated from any fine grained gneiss area resulting by pressure from an original granular rock.   The different constituents have acted under pressure as plastic materials and are thoroughly mashed and squeezed (1911, p. 312).

Thus while the belief that most ore deposits were introduced features was firmly held, it was coming to be recognized that *in essentially their present positions* some of them had shared at least a part of the history of the enclosing rocks.

The year 1914 saw the publication of an important paper in this vein by the Englishman Thomas Crook (*Mineral. Mag.*, **17**:55–85).   From the point of view of the philosophy of the study of ores this was a contribution of quite fundamental importance, but strangely it appears to have received little attention and to have been almost instantly forgotten.   In it Crook set forth his thesis—later to be presented in much greater detail in his book "The History of the Theory of Ore Deposits"—that ores were no more and no less than rocks and should be studied in this light:

Ore deposits are rock masses in which occur one or more metalliferous minerals in sufficient quantity to make them useful as sources of metals.   This definition can be extended to deposits of economic minerals generally, i.e. an economic mineral deposit is a rock mass in which occur one or more useful minerals in such quantity that it is practicable to extract them for use in the arts.   Any one who faces the facts fairly will find it impossible to escape from this petrological definition of an ore deposit.

In some instances an economic mineral in itself constitutes a rock.   Masses and beds of iron ore, chromium ore, manganese ore, . . . and numerous other instances could be quoted.   More commonly, however, the economic mineral is, from a quantitative standpoint, a comparatively insignificant constituent of the rock in which it occurs; but in all cases the deposit as a whole is essentially a rock.   Indeed, there are all possible gradations between instances in which economic minerals in themselves form the chief part of a rock mass, and those in which they are disseminated through a large amount of useless rock matrix.

An important end is attained by this way of looking at ore deposits: it shows the fundamental value of petrology in the study of these deposits.   If the recognition of this fact has been slow in growth, the petrographers are to blame; for had vein deposits and other types of economic mineral deposits received adequate scientific treatment at their hands, there would doubtless have been a quicker, and by this time a fuller recognition of the almost inseparable connexion that exists between rock genetics and ore genetics, and of the mutual aid that these studies can afford.

It is certainly unscientific to allow an igneous dyke to rank as a rock, and to deny that rank to a vein deposit, especially when we take into consideration the fact that a vein deposit is often of far greater dimensions than a dyke.   The recognition of this fact by petrologists has at most been extremely feeble . . . (1914, pp. 76–77);

and

Ore deposition is, indeed, even from the petrological standpoint, an important and not merely a trifling incident in the process of change that is continually affecting the rocks of the unstable earth's crust; and whether it has taken place *pari passu* with the formation of the more ordinary rock in which the deposit occurs, or subsequently, it is one of several causally related incidents that need to be considered together.

The origin of ore deposits is thus inseparably bound up with the origin of rocks. It almost follows as a matter of course that the best genetic grouping of ore deposits should correspond closely with that of rocks . . . (1914, pp. 77–78).

It is surprising that what seems so balanced and so sensible a view should be barely noticed. Certainly it anticipated by some 40 years the stirring and questioning of the 1950s and 1960s. Perhaps the best explanation of this neglect is the old one that "the man lived before his time." Perhaps, too, with the discovery and exploitation of the great vein deposits of the western United States, the tide of thought was already running too strongly in the other direction.

The igneous theory, and particularly the hydrothermal, or "ascensionist," aspect of it, established itself with increasing firmness during the next few years. This was an inevitable result of the great leap forward in the understanding of magmatic differentiation processes and of the detailed observations of Kemp, Lindgren, Weed, Graton, and others on the mineral vein occurrences and their igneous affiliates in the western United States. Subsurface deposition from magmatic waters, by cavity filling or metasomatic replacement, dominated ore-genesis theory.

The year 1919, however, saw the publication of two notable papers—one in Norway by C. W. Carstens and the other in Japan by Ohashi. Both men had been studying concordant, layered, sulfide deposits associated with volcanic rocks— Carstens, the pyritic copper mineralization of the Trondhjem area in the Caledonides of Norway, and Ohashi, the Kuroko (black ore) of the Kosaka mine in Tertiary rhyolitic tuff and breccia near Akita. Both observed the concordant, laminated nature of the sulfide-bearing beds and the close volcanic association. Each concluded that his respective deposits resulted from submarine volcanic activity and that the ores were deposited essentially as sinters about hot springs that had discharged acid metallic salt solutions onto the sea floor. Interestingly this idea had arisen almost 90 years earlier in connection with stratiform iron ores. In 1831 Henry de la Bêche had suggested that certain iron oxide ores in sedimentary rocks might have formed by precipitation from sea-floor volcanic hot springs— and in 1911 Van Hise and Leith, in their monumental United States Geological Survey Monograph "The Geology of the Lake Superior Region," concluded that this was the mechanism by which the great Lake Superior banded iron formations had principally formed. Although de Beaumont had mentioned a submarine volcanic group of ores, the work of Carstens and Ohashi appears to have been the first clear application of the idea to *specific sulfide ore deposits*. The idea did not receive wide acceptance, and while it lingered on in what became known as the "hot-spring school," it was virtually forgotten for the next 30 years. However, it was clearly the forerunner of those put forward in the early 1950s and grasped with such enthusiasm—as the "submarine-exhalative theory"—in the 1960s. It was thus considerably more important than it was judged to be at the time; in fact it laid down the new principles that orthomagmatic ores were not necessarily the only igneous ones deposited syngenetically and that igneous and sedimentary factors might contribute *simultaneously* in the development of primary sulfide concentrations.

THE TWENTIETH CENTURY: 1920 TO 1945

The following two decades saw few major developments.   For the most part it was a period of reexamination and refurbishing of earlier conceptions.   J. E. Spurr, in his book "The Ore Magmas" (1923), put forward the old idea of ore magmas in a somewhat extreme form.   Geologists in England and America continued their preoccupation with intrusive magmas and hydrothermal solutions.   Those of the newly developing countries of Africa, Australia, and India were beginning to express opinions based on the deposits of their own lands, though such opinions did not diverge far from those of England and America.   The Europeans, too, were heavily involved in the igneous hydrothermal (subsurface) theme, though in characteristic fashion they showed a certain resistance to conformity!   Schneiderhohn's introduction in 1932 of a genetic classification of ore deposits was perhaps the most notable event of the period.   It presented one of the more balanced views of ores achieved to that time and indeed probably showed a rather better sense of perspective than many since.   Schneiderhöhn did the simplest and most obvious thing— like Van Hise and Crook before him, he recognized ores as rocks and classified them as such.   Its elements were:

A. Magmatic rocks and ore deposits
     (*a*) Intrusive magmatic
            I. Intrusive rocks and liquid magmatic deposits
            I–II. Liquid magmatic-pneumatolytic
            II. Pneumatolytic
                   1. Pegmatite veins
                   2. Pneumatolytic veins and impregnations
                   3. Contact pneumatolytic
            II–III. Pneumatolytic-hydrothermal
            III. Hydrothermal
     (*b*) Extrusive magmatic
            I. Extrusive-hydrothermal
            II. Exhalation
B. Sedimentary deposits
     1, Weathered zone (oxidation and enrichment); 2, placers; 3, residual; 4, biochemical-organic; 5, salts; 6, fuels; 7, descending ground-water deposits
C. Metamorphic deposits
     1, Thermal contact metamorphism; 2, metamorphic rocks; 3, metamorphosed ore deposits; 4, rarely formed metamorphic deposits

The period immediately prior to the Second World War saw the magmatic theory almost completely in the ascendancy.   The majority of primary ores were ascribed to one stage or another of the cooling histories of intrusive igneous rocks —orthomagmatic and pegmatitic crystallization, contact metamorphism, and deposition, by accretion or replacement, from fugitive gases and solutions. The magmatic water theory had been established as far as most people were concerned, and it was simply a matter of working out the details.

The prevailing outlook—in the United States at least—was well illustrated by the emphasis on "hypogene" deposits in the Lindgren Volume—a remarkable gathering together of the then current American opinion and, from the historical viewpoint, an extremely valuable document. While, since the central theme was the heavily epigenetic mineralization of the Western states, it was only natural that igneous processes should be emphasized on the local plane, there is little doubt that the volume was compiled with the view that this area illustrated something of dominating importance. Weathering and sedimentary processes were indeed dealt with—albeit briefly—but magmatic processes received overwhelming attention. Fenner contributed a masterly paper on pneumatolytic processes; Bowen gave a résumé of igneous differentiation processes and their possible role in the generation of ore fluids; Lindgren himself wrote on differentiation and ore deposition on a broad scale; Graton wrote on hydrothermal depth zones; Emmons wrote on the deposition of lode systems associated with batholiths; and Buddington gave his now famous paper on the correlation of kinds of igneous rocks with kinds of mineralization. No group more competent to put the view of the day could surely have been assembled!

By an interesting coincidence, 1933 also saw the publication of Thomas Crook's "History of the Theory of Ore Deposits." In addition to tracing the development of ideas on ores during the previous 400 years, Crook examined the relationship between "ore" studies and "silicate" studies and reiterated his views of 1914. His principal themes were what he considered to be the then currently exaggerated view of the importance of "endogenous" processes in the formation of ores and the schism between petrology on the one hand and the study of ores on the other. Of ores and petrology he said:

> Early on in the second half of the [nineteenth] century petrologists ceased for the most part to interest themselves in the possibilities as regards the scope of their science. . . . Since that time petrologists have settled down to a study of the microscopical, physical and chemical aspects of silicate and sedimentary rocks, with no interest in metalliferous vein and other ore deposits as far as petrology goes. . . .
>
> Various writers have at different times in recent years suggested broad bases of classification that would bring the study of ore deposits within the scope of petrology; but in spite of the fact that metalliferous-vein and other mineral deposits are essentially rocks in the best sense of the term, petrologists continue to ignore them. As already pointed out, however, not until all mineral deposits come within the purview of rock studies will petrology become a fully-fledged science (1933, pp. 160–161).

Of the limited importance of endogenic processes in ore formation, he clearly had no doubt:

> From a geological point of view, indeed, the insistence on a deep-seated or baryspheric shell of metalliferous sulphides to explain the origin of metalliferous veins is very far-fetched; for we have no satisfactory evidence that, excepting basalts, etc., any of the rocks known to us by observation in the earth's crust can safely be assumed to have been formed, in the condition in which we now know them, at depths exceeding ten miles or

so. Basalts traverse the crust for comparatively long distances through fissures in the superincumbent cover of granitic, metamorphic and sedimentary rocks; but as we know very well, apart from the ordinary rock elements, the metalliferous content of these basalts and their near relatives is very meagre indeed. It is in association with rocks of comparatively shallow origin in the crust, namely, the sedimentary and metamorphic rocks and the granites intrusive in them, that we find metalliferous veins.

A consideration of mineral deposits as a whole, and the nature of the concentrations represented by deposits of economic importance, enables one to get the problem of the genesis of metalliferous veins in truer perspective than it can be got by dealing with these veins alone. If we limit our attention to the commoner oxides of the outer earth's crust, including silica, alumina, iron oxides, lime, magnesia, soda, potash, water, titanium dioxide, phosphorous pentoxide, and manganese oxide, which probably account for about $99\frac{1}{2}$ per cent of the outer solid crust, we find that the economically important products in which these oxides figure are almost entirely exogene concentrations, i.e., due to processes operating superficially on the earth's crust, and downwards into it.

If this be the case for the commoner elements, is it likely to be otherwise as regards the scarcer metallic and other elements making up the remaining 0.5 per cent of the outer crust?

In fact, if we make an overall estimate of the percentage of mineral concentrations due to exogene processes, we find that they account for not less than 85 or 90 per cent of the present annual value of the world's mineral output. It is doubtful if more than 1 per cent or so can be attributed to igneous segregations, while, in the writer's opinion, there is at least some uncertainty as to whether exogene processes have not played a prominent part even in the remaining 10 per cent or so of the concentrations to be accounted for.

We see, therefore, that as regards the commoner elements of the earth's crust, there is an overwhelming case for the importance of exogene as compared with endogene processes as agents of concentration in the formation of economic mineral deposits. If we adopt the safe policy of arguing from the known to the unknown, we shall be led to doubt the efficacy of igneous processes as agents of concentration in the large measure claimed by modern igneous theorists (1933, pp. 129–130).

Crook's views, however, were again to go largely unheeded, and the magmatists proceeded on to their zenith, which—for this cycle—they appeared to reach early in the 1940s with the publication of A. M. Bateman's "Economic Mineral Deposits."

Bateman was in a particularly good position to survey the scene of the day: in addition to wide experience as a practical geologist, he was active in that part of the American academic world that virtually led the movement of which he was a part, and he edited what was undoubtedly the most prominent journal in the field of ore deposits. No one could have been in a better position to follow the ebb and flow of ideas than he. It was therefore not surprising that he should produce the most comprehensive treatment of "hypogene" ores put forward up to that time. So far as sulfide and related "primary" deposits were concerned, almost no allowance was made for the operation of sedimentary or metamorphic processes. While acknowledging the views of other theorists, Bateman clearly favored the magmatic hydrothermal theory even for such as the Mansfield and Roan copper beds, copper red beds, and sandstone uranium deposits. He stated:

Magmas are the source of essentially all the ingredients of mineral deposits. A few constituents, such as oxygen, carbon dioxide, or water, are derived from the atmosphere or the oceans, but even these are in part of magmatic derivation (1950, p. 300).

Apart from obviously sedimentary types such as certain bedded iron and manganese ores, the only syngenetic ores recognized by Bateman were those of orthomagmatic formation.   This general view was supported by most economic geologists of the day, and most of the world's major ore bodies were regarded as products of migrant magmatic solutions.   Such a viewpoint was clearly the antithesis of Crook's, denying as it did the possibility of any real genetic relationship between most metallic ores and the rocks in which they occurred.   Having taken up this position, classification of ore bodies then resolved itself more or less entirely into "morphogenetic" division of the orthomagmatic and hydrothermal groupings. This Bateman did very neatly, providing a highly systematic and useful subdivision of ores that may—though some would say not exclusively—be igneous in their immediate origin.

THE TWENTIETH CENTURY: THE "POSTWAR" PERIOD

It may well be that Bateman's work saw the popularity of the magmatic theory at its peak for the twentieth century.   For a few years the hypogene and igneous nature of most of the world's great base metal sulfide deposits was hardly doubted, and the world of economic geology became preoccupied with "structural control." Since the ores had been deposited from solutions, something must have provided a means of movement from source to host and the impounding, or host-rock preparation, necessary to induce highly localized precipitation.   This factor was "structure," and it became the structural geologist's day in the search for ore.   The mere existence of ore began to necessitate the existence of a localizing structure, and so wherever ore occurred, an appropriate structure was looked for—and usually found.   Geologists seemed quite unperturbed that very similar structures might occur nearby—devoid of ore—and there seemed no necessity to explain why one structure should acquire all the sulfide while others—of similar nature and close at hand—received none.   However, while structural geology was frequently helpful in predicting the behavior of *known* ore bodies, particularly where the structure was a break rather than a fold, it was not very reliable in predicting the whereabouts of unknown ores.   And so geologists once again began to turn their minds to the genesis of ores and to the development of ideas that not only explained the occurrence of deposits that were known but predicted the occurrence of those that were not.

The new mood began to be felt not long after the Second World War.   In 1948 Sullivan, stimulated by a recurrence of the magmatic-metamorphic controversy on granite formation, considered again the possible place of granitization in the development of mineral concentrations—a problem already examined by Scheerer, Fournet, Sterry Hunt, and others in the previous century.   Sullivan added to the earlier work, however, by applying some principles of geochemistry that had been developed in the meantime.   He noted that the outpouring of basaltic volcanic rocks led to a geochemical imbalance in the continental crust and that granitization of such metal-rich additions would inevitably tend to eliminate the distinction between these and the relatively barren "sial." He suggested that:

> In the process of the formation of granitic rock, the valuable elements of ore deposits are concentrated in inverse ratio to the extent to which they are incorporated by isomorphous substitution in the common rock-forming minerals. This is held to explain, to a large extent, the association of specific ores with specific types of granitic rocks (1948, p. 472).

The earlier investigations had simply suggested that the heavy metals were driven from the site of granitization in the original pore-space solutions of the altering sediments; Sullivan attempted to give some reason why specific elements should be driven out while others were not.

Sullivan's paper was quickly followed by a number of others concerned with principles, and the ensuing 15 years saw much reexamination of old ideas, many of which were refurbished and put forward in more modern form. F. Hegeman (1950) again drew attention to the volcanic-sedimentary idea—a revival of the earlier suggestions by Carstens and Ohashi. Goodspeed put forward again the idea of lateral secretion related to granitization:

> The process of granitization envisages the transformation into granitic rocks of the material originally deposited in a geosynclinal prism. Such a process would bring about a regional redistribution of certain chemical elements. It should be emphasized that kaolinitic material so abundant in a sedimentary series contains about 14 per cent water. The change of kaolinitic material to felspars must drive off this water. Water so released by granitization can easily become the source of hydrothermal solutions that produce mineralisation (1952, p. 146).

Ramdohr (1951) suggested that the massive sulfide ore of Broken Hill, which had long been regarded as an example of hydrothermal replacement in its pristine state, had been metamorphosed. From a careful study of textures he concluded that all the sulfides and associated minerals enclosed and were included in each other and therefore any earlier depositional textures had been obliterated by some event following original emplacement. He suggested that the ore might have originally been an epithermal one that was later highly metamorphosed and coarsened along with the enclosing sedimentary silicates, carbonates, and others. While this work aroused a great deal of attention, it was simply an application of the principles already put forward so clearly by Van Hise and Crook.

Although it involved an idea already established for many years, Ramdohr's paper seems to have come at a critical time. It set the stage for a most important paper by H. F. King and B. P. Thompson (1953) that in turn was to have a profound influence on the development of ideas during the next decade—and indeed whose influence seems likely to continue well into the future.

The sulfides at Broken Hill had always been regarded as having been introduced, in one way or another, after the folding and metamorphism of the associated rocks. Following the work of Moore (1916) it seemed well established that they had migrated in hot solutions from some hidden granite and that they had been deposited by selective substitution of certain well-defined layers of the preexisting metasedimentary rocks. It was to this process of selective replacement that the remarkable stratigraphic disposition of the ores and their mineralogies had been attributed. However, the very extensive studies of King and Thompson and their colleagues at Broken Hill led them to suspect that the sulfides had been emplaced *prior* to folding:

> An understanding of the emplacement of the Broken Hill deposit is beset by two difficulties—first, the transport of the ore constituents to their present position, and second, their segregation to the extent described . . . .
> There is no doubt that the ore-bearing layers are stratigraphic horizons, and little doubt that in places there has been more or less substitution of new constituents for the original rock-forming minerals. There is, however, no evidence of the mode of transport of these new constituents. The boundaries of high grade ore against barren rock are commonly sharp; the sillimanite gneiss within inches of high grade ore is not recognisably different from sillimanite gneiss miles distant from ore and is equally devoid of lead. In short, except in the southern end of the field, where there is limited silicification and garnet development around and between the ore layers, there is no evidence of the passage of the tremendous quantity of solution postulated. Even if it were known that the transporting medium would leave little or no trace, its passage could nevertheless only be assumed. An alternative mode of emplacement not involving transport in dilute solution must therefore be considered, even sought. The persistence and the difference in metal ratio and gangue minerals in the various ore-bearing beds must be faced. Since there is no upward or downward extension of the Broken Hill ore deposit, it is presumed that all the ore-forming material available was retained by the favourable beds and structures. If the ore constituents were introduced into these structures as they now are, the differences in ore type would represent a sorting-out and in some respects a segregation of the available ore constituents among the recipient beds and structures. Not only therefore were particular beds able to accept and reject certain constituents according to their individual composition, but these selected constituents and proportions were able to extend to virtually all parts of the particular ore bed over miles in length. Even more remarkable, the total quantities and proportions available appear to have been those which the various beds would together accept.
> There is, therefore, doubt of the hydrothermal mode of emplacement. An introduction of the ore constituents into their present location as sulphides and silicates without benefit of a dilute solution would be more in accordance with the evidence, at least to the extent of not requiring an assumption of the mode of transport for which there is little observational support.
> The differences between the ore types, and the segregation implied, is equally incomprehensible under these conditions. On the one hand, the fact that the two principal ore types may be in contact and yet remain distinct argues that their deposition cannot have been simultaneous. On the other hand, to imagine that a $3\frac{1}{2}$-mile replacement of particular beds in a particular structure could have occurred twice in almost identical fashion appears absurd. The result is a strong suspicion that the formation of the different ore types did not take place simultaneously in their present environment (1953, pp. 569–570);

and

> Each successive study of the Broken Hill district and its lead-zinc deposits has revealed further complexity in the mode of ore occurrence. It has become necessary to conceive that a single event of mineralization may be of tremendous intensity and yet of the utmost selectivity; that the structural and the stratigraphic factors in ore localization may be

equally strong; that though the whole of certain structurally favorable portions of particular beds may be replaced by ore, adjoining beds were not replaced at all; that closely adjacent beds were replaced by similar constituents in significantly different proportions over distances measured in miles; that this replacement occurred with virtually no effect on the enclosing rocks other than the substitution of sulphides and silicates for the original constituents of certain beds; and that despite its considerable extent, there was no zoning of constituents such as is seen in some other large sulphide bodies. As the complexity is progressively revealed it becomes increasingly difficult to imagine that a single event of mineralization could simultaneously produce all the observed phenomena. . . .
   The sequence of events is therefore now visualized as:

1. Willyama sedimentation; mainly argillaceous sediments with some ferruginous horizons and possibly local calcareous and/or manganiferous sediments. Into these, and to a lesser extent into the ferruginous horizons, lead and zinc sulphides were early introduced in some manner at present unknown but probably influenced, at least, by syngenetic factors. . . .
2. Severe deformation, metamorphism, and migmatization of the sediments producing granite gneiss, aplite, and pegmatite in great abundance. . . . Both the iron-bearing layers and the lead-zinc deposits were recrystallized and structurally concentrated along particular fold axes . . . (1953, pp. 573–574).

While King and Thompson were careful not to present a dogmatic view, it is clear that they suspected the sulfides to be chemical sediments and that these had developed their present features through the whole of the geological history of the area concerned. Interestingly, and although unknown to King and Thompson in 1953, the great German (Freiberg) geologist A. W. Stelzner had, as early as 1894, suggested that the Broken Hill sulfides were in the form of a bed and hence might be primarily sedimentary. However, by 1953 this had been long forgotten and Broken Hill had come to represent "replacement" *par excellence*. King and Thompson had therefore thrown down a theoretical challenge of the most serious kind. Largely as a result of this paper the next 10 years became the most active that ore-genesis theory had probably ever seen.

At about the same time Schneiderhöhn (1953) was putting forward ideas on "secondary mineralization." He pointed out that there were large numbers of ore deposits in Europe and North Africa within Mesozoic and Tertiary terrains that had not been involved in orogenic folding. Any movement of these rocks had been essentially epeirogenic—their only deformation was warping and the development of joints related to older joint sets in the underlying basements. Since there did not appear to be any igneous rocks to which these ores might be connected, Schneiderhöhn suggested that the ore materials had been abstracted, by chlorinated thermal waters, from preexisting deposits in the underlying older rocks. These were then transported to the overlying cover rocks and deposited in appropriate joints. Such a history is supported by the low-temperature "gel" textures of the sulfides and by lead isotope data that indicate the leads of the galena-bearing ores to be considerably older than the rocks that contain them. Schneiderhöhn suggested that regenerative processes of this kind might be of quite general importance.

Concurrently, Ehrenberg, Kraume, and a group of colleagues were reinvestigating the deposits of Meggen and Rammelsberg in Germany. They concluded (1954, 1955) in no uncertain fashion that both were the result of submarine volcanic activity. They suggested—as had de Beaumont, Carstens, and Ohashi before them—that soluble heavy metal compounds might be contributed directly to the sea floor via hot springs, there to precipitate as sulfides and related minerals and build up as stratiform concentrations.

At about the same time Stanton (1955) suggested that some stratiform ores— particularly those composed largely of iron sulfide—might constitute a distinctive rock type of wide but characteristic occurrence. He pointed out that many of these appeared to occur in reduced, near-shore sediments of old volcanic sequences and that their iron sulfide at least seemed to be sedimentary. He proposed that the formation of these deposits began with the deposition of iron sulfide by bacterial sulfate reduction in highly organic off-reef or equivalent facies and that this had later acted as a precipitant of such metals as copper, zinc, and lead removed in solution as halides from associated beds of andesitic to rhyolitic tuff. Such deposits were thus regarded as resulting from an *interplay* between sedimentary facies and volcanic emission. Since the major part of marine andesitic volcanism occurs along the arcs of volcanic island festoons, it was suggested that deposits of this group should have formed here and there about individual islands along such arcs. Further, since numbers of ancient arcs now appear to be incorporated in the continents, Stanton suggested that there should be a systematic relationship among old eugeosynclines, individual volcanic centers within them, and the distribution of ores of this particular stratiform type. This was, of course, very close to the spatial relationship proposed by de Beaumont. Most old eugeosynclines become loci of folding, metamorphism, and granitic intrusion. However, whereas de Beaumont had emphasized the late, plutonic, stage of the tectonic cycle, Stanton stressed the likely importance of the early, volcanic, period. In this way stratiform ores of this type became linked with the "Steinmann Trinity"—serpentines, volcanic "greenstones," and ferruginous cherts. The stratiform sulfides, many localized iron and manganese oxide concentrations, and large thicknesses of andesitic and more siliceous tuffs, were suggested as additional members of the association.

Following this, Stanton suggested that many mineral veins (*discordant* bodies as distinct from the *concordant* stratiform ones) associated with granitic intrusions might represent *dissipation* rather than *concentration*; i.e., they might be manifestations of destruction rather than of formation, the generation and intrusion of granites causing the destruction of earlier formed volcanic deposits and their dissipation into discordant plutonic deposits. While the evolution of a tectonic belt is probably more or less continuous, Stanton suggested that it might be broadly and loosely divided into two main phases as far as the associated mineral deposits were concerned:

> We may . . . visualise the first phase—vulcanism and sedimentation, and the formation of a number of conformable orebodies—completed. Folding, already under way in the later stages of sedimentation, intensifies, and leads to metamorphic recrystallization and the formation of granites. The sediments, including the sulphide-rich ones, are squeezed, heated and recrystallized, and attacked by the intergranular solution expelled by compaction. Some of the sulphide-bearing layers may be folded with the rocks that enclose them. Others, particularly if the principal sulphide is galena . . . may be involved in

thrust faults, and squeezed bodily up the fault planes. Other concentrations, by solution, melting or volatilisation, may be destroyed and their components dispersed to form irregular and irregularly occurring veins throughout the folded belt and within, marginal to, or about the granites that have developed in it (1961, p. 54).

DIGRESSION ON THE RISE OF GEOCHEMISTRY

The middle years of the 1950s may now be seen to have been of unusual importance —as the age of the rapid rise of geochemistry in the study of ores. This had its beginnings in the work of Daubrée and Vogt and their contemporaries, but it was not until the period between the two great wars that geochemical studies really began to gather momentum. It was at this time that Goldschmidt, Strock, and others in Europe and Bowen, Fenner, Allan, Zies, and others in America began applying the knowledge and experimental method of chemistry to a wide range of geological problems, including those of the minor (ore) metals. Goldschmidt studied the distribution of elements in the crust and indicated, from their chemical properties, the processes each was involved in and the ways in which they might be separated and concentrated in nature. In connection with sulfides, perhaps his best known contribution was his demonstration (1937, with Strock) of the way in which sulfur and selenium were separated during erosion. Through this he showed that sedimentary sulfides were characteristically deficient in selenium, and he suggested that this might be used to distinguish between ores of sedimentary and nonsedimentary origin. Allen and Zies (1919) made valuable contributions on the chemistry of volcanic fumaroles and hot springs, and Fenner (1933) showed how their observations might be applied to the study of ores of igneous derivation. Bowen (1933) stated very explicitly—largely on the basis of his own work on silicate equilibria—how heavy metal compounds might be concentrated in the residual fractions of a differentiating magma and how they might be expelled. While his work was essentially an extension of Vogt's, it was, expectably, much more refined. He summarized:

> The differentiation of igneous magmas follows a course basic to salic largely as a result of fractional crystallization. In the salic types there is, therefore, an enrichment of all substances of the original magmas that share with the salic ingredients the property of forming low-melting combinations and associations. Some of these substances have this property to a superlative degree and may be called hyperfusibles. They are principally water, sulphur, chlorine, fluorine, boron, and a number of others. As the salic magmas themselves crystallize, an increased concentration of the hyperfusibles in the residual liquid results. The trend is toward alkaline aqueous solutions, but with slow crystallization under heavy load the materials of these residues may be completely consumed in the formation of crystalline compounds that carry water, chlorine, boron, sulphur, fluorine, etc. Under less deep-seated conditions the residual liquid of crystallizing salic magmas frequently boils and a fractional-distillation column is set up in the interstices and fractures of the intrusive and its surroundings. This action results in the formation of a distillate which is an acid aqueous solution and it is such acid solutions that are probably the principal ore bringers (1933, p. 128).

At about the same time Merwin and Lombard (1937) carried out their experimental study of the Cu–Fe–S system. During the same period Newhouse,

Bastin, Schwartz, Stillwell, and others were attacking the petrography of ores, using the newly developed reflecting microscope. This led to the discovery of many microstructures suggestive of exsolution and replacement, and considerable effort was directed to the experimental reproduction of these microstructures. In particular, exsolution textures were investigated as possible indicators of the temperatures of ore formation—i.e., "geothermometers."

All this was, unfortunately, interrupted by the events of 1939 and the following years, but it set the stage for an explosion of geochemical studies after the Second World War. By about 1955 these were beginning to have a substantial influence on methods of investigation of ores and ore-forming processes.

Tudge and Thode (1949) had shown that there were substantial differences in the ratios of the isotopes ^{32}S and ^{34}S in various natural sulfur compounds and that these were induced systematically by fractionation accompanying various oxidation and reduction reactions. It was, for example, found that the reduction of sulfate to sulfide by certain sulfur bacteria led to gross enrichment of the lighter isotope in the sulfide. Since most sedimentary sulfide was regarded *ipso facto* as "biogenic," a high $^{32}S/^{34}S$ ratio appeared to be a likely characteristic of sedimentary ore. Igneous rocks and related sulfide segregations, on the other hand, usually possessed ratios rather similar to those of meteorites and the "average" crust. Thus the isotopic composition of sulfur began to be investigated as a possible method of distinguishing between ore-forming processes where these were unclear, and there was great activity in this field after about 1955.

At about the same time Kullerud (1953) published his paper "The FeS–ZnS system: a geological thermometer." This reported an investigation that used the same techniques as, and in principle followed on from, Merwin and Lombard's classical 1937 work on the Cu–Fe–S system. Kullerud showed what appeared to be a systematic relationship between the capacity of sphalerite to accept iron in its structure and its temperature of formation. Very conveniently the sphalerite did not exsolve this iron on cooling so that the depositional relation was "quenched in." Application of the method required a pressure correction, the coexistence of FeS as pyrrhotite, and some limitations on the amounts of other impurities. These requirements were usually readily met, however, and it was not long before the "sphalerite geothermometer" was being applied to controversial deposits all over the world. Not only was the method itself quickly grasped and applied, but it soon became the starting point for a whole new wave of activity in sulfide phase equilibria. During the next 10 years almost every conceivable binary and ternary system of any possible geological relevance was investigated—particularly as possible geothermometers or as indicators of other conditions of ore deposition.

At much the same time, too, a group of "solution chemists" was becoming interested in low-temperature aqueous equilibria, particularly in connection with the great bedded iron formations of the Precambrian shields. This appears to have led from the work of M. J. N. Pourbaix, who, as Director of the Belgian Institute for the Study of Corrosion, had long been interested in low-temperature aqueous equilibria involving industrial metals. Some of his techniques were quickly taken up by R. M. Garrels in America, who, with a number of collaborators, examined first the solution chemistry of iron compounds and then that of a variety of ferrous and nonferrous metals.

Initially Garrels' principal concern was the delineation of stability fields of various components in terms of oxidation-reduction potential and hydrogen ion concentration (Eh–pH). This he applied with great success to the problem of iron sedimentation. The fusion of the work of James on the mineralogy and field occurrence of iron formations (James, 1954) and of Garrels and his coworkers on the solution chemistry of iron formations is an excellent example of the way in which the more basic sciences were coming to be applied to problems of ore formation. Following this, Garrels turned his attention not only to other elements but also to the effects of additional factors, such as the partial pressures of oxygen, carbon dioxide, and sulfur, and of temperature. Although this approach had its initial success in connection with ores formed as chemical sediments, i.e., at the low temperatures and pressures of the earth's surface, it was soon applied to subsurface vein deposits. This was done by extending experiments and calculations to include liquids and gases at high temperatures and pressures. Thermodynamics had now established itself as a powerful technique in the more detailed understanding of rock formation, and it was soon used to examine and refine ideas on ore deposition.

At this time the "endogenous" theories remained in a preeminent position, and their operation and importance were widely accepted without question; the attention of many investigators was now simply with the details of the postulated mechanisms of extraction, migration, and deposition. Bowen and Fenner had, of course, both made notable contributions on extraction and transport mechanisms in 1933. Both had concluded that the metals of ores had probably been extracted and transported principally as volatile halides. Now J. S. Brown (1948) suggested that the metals moved as volatilized sulfides; Sullivan (1954) suggested movement as volatilized metal; and Morey (1957) and Morey and Hesselgeser (1957) suggested movement as solutes in supercritical steam. In his "The heavy metal content of magmatic vapor at 600°C," Krauskopf set out to examine in a detached way the feasibility of these various hypotheses, and his treatment of the problem is a good example of the developing thermodynamic approach. In this he examined, by calculation, the behavior of some important metals and metallic compounds in a gas of postulated "igneous" composition. The temperature of 600°C was chosen as a reasonable approximation to that at which igneous intrusions might evolve their volatiles and at which high-grade metamorphism might take place. Among his conclusions Krauskopf noted:

> Metals may be present in the vapor state as sulfides, oxides, chlorides, fluorides, and the native elements. Maximum amounts of the metals in these various forms are determined by equilibrium with the most stable solid metal compound. The stable solid for most common ore metals at 600°C is the sulfide, but for tin it is the oxide, for manganese the silicate, and for gold the free metal.
>
> All the common metals for which data are available are most volatile as the chlorides, except gold (whose chloride is decomposed at 600°) and copper (most volatile as the sulfide). The maximum amounts of the chlorides present in magmatic vapor are in general less than the measured volatilities because of equilibrium with the solid sulfide and with H_2S and HCl in the vapor. Maximum amounts for many metals are in the

range 10^{-2} to 10^{-5} atm (equivalent to $10^{-1.3}$ to $10^{-4.3}$ grams/liter of solution), adequate to transport enough metal to form large deposits.

The low volatilities of most sulfides at 600° eliminate the possibility of transporting the metals of ore deposits as volatile sulfides.

The low vapor pressures of most free metals, and the negligible decomposition of most metal sulfides at 600°, indicate that transportation of metals as metal vapor cannot play a significant role in ore formation.

Metallic melting points show no clear relation to simple vapor pressures, to pressures of metal vapor in equilibrium with sulfides, or to maximum concentrations of metal compounds in magmatic vapor. Hence there is no justification for assuming an influence of metallic melting point on the formation of ore deposits.

Any hypothesis of vapor transport must be able to show that magmatic gases at 600° are capable of holding in the vapor state sufficient quantities of the common heavy metals to form an ore deposit. Regardless of the later history of the vapor, regardless of the mechanism by which metals are deposited from the vapor, still the metals must be present in significant amounts in the gas at high temperatures. Thermodynamic data show clearly that some *but not all* of the common metals would volatilize in substantial quantity at 600°. The very low volatilities of copper, silver and gold, together with anomalies in relative volatilities such as the greater volatility of lead than manganese, show that *no* hypothesis of *simple* vapor transport can satisfactorily account for the origin of ore deposits.

Gases must nevertheless play an important role in transporting metals wherever a magmatic gas phase forms in appreciable quantity, simply because many of the heavy metals *will necessarily vaporize* at 600° if given an opportunity to do so. The gas phase may actually be more important than this investigation indicates, either because of the physical transport of solid particles in the gas or absorbed on gas-liquid interfaces, or because of the solvent action of highly compressed water vapor. Such supplementary hypotheses cannot at present be tested against quantitative experimental data (1957, p. 806).

THE "POSTWAR" PERIOD RESUMED: 1957 TO 1967

At the same time C. L. Knight was expressing dissatisfaction with the igneous theory and proposing sedimentary–lateral-secretion principles similar to those put forward earlier by Sterry Hunt. Knight took a very practical stand:

> . . . the author had become dissatisfied with the popular theory—that the majority of orebodies had a magmatic origin and developed by a process of fractionation from the magma—largely because the theory was of no use in the practical business of finding new orebodies. This dissatisfaction deepened as it became apparent that the theory did not even fit the facts in many important instances. In several fields in Australia, where granite outcropped and the orebodies were of the shear type, genetic linkage of granite and ore did seem a reasonable assumption. However, in the four major lead-zinc fields of Australia, namely Broken Hill, Mount Isa, Rosebery and Lake George, the linkage of ore with intrusives was not at all obvious and it was necessary to suppose that the parent magma in each case lay at considerable depth. On the other hand, in each of these fields, the importance of a particular sedimentary bed in localizing ore was obvious . . . (1957, p. 809);

and

> If the ore-bearing bed were invariably the one rock type it might be argued that chemical properties peculiar to the bed were responsible for its favorability to ore replacement.

This however is not the case.   Each of the common sedimentary rocks with the exception of graywacke . . . comprises the ore-bearing bed in different fields. . . .   On the other hand in some of the fields listed—for example Broken Hill, Mount Isa and Morocco—some measure of control of the shape of ore shoots by structure has been proved beyond reasonable doubt, and a simple syngenetic origin is therefore untenable.   If the sulfides of these orebodies were deposited originally as primary sedimentary components of the bed, they must have migrated appreciable distances within the bed at some subsequent time. . . .   The author proposes . . . that sulfide orebodies in the great majority of mining fields are, or were derived from, sulfide accumulations that were deposited contemporaneously with other sedimentary components at one particular horizon in the sedimentary basin which constitutes the field, and that the sulfides subsequently migrated in varying degree under the influence of rise in temperature of the rock environment (1957, pp. 815–816).

Knight voiced the rising suspicions of many geologists concerned with the discovery of new ore deposits.   Attitudes in ore genesis were changing with changing ideas on what constituted ore and with the pressures to find it.   Up to the Second World War most of the known large deposits had been found by prospectors and were exploited—and investigated geologically—as isolated, individual occurrences.   With the restricted view that this engendered, most deposits could, for one reason or another, be fitted to whatever the prevailing view on ore genesis happened to be.   The serious application of geological method, however, quickly led to regional studies and the viewing of ores in a wider setting.   This soon showed the existence of broad patterns of mineralization in many areas.   These were not, however—as might have been expected from the then prevailing view—related to patterns of plutonic intrusion but rather to those of sedimentation.   Many of the ores consisted simply of sulfides interstratified with the containing sedimentary or metasedimentary rocks, and they usually occurred in particular facies within particular horizons.   Others, as Knight stated, were discordant, but these still frequently exhibited stratigraphical preferences.   Such patterns soon became conspicuous in southern Africa and Australia, both of whose geologists began to play leading parts in developing—or redeveloping—theories of sedimentary ore formation.   Garlick, Davis, Brummer, and others in Africa and King and his associates in Australia soon became leading advocates of the sedimentary idea.   In general the metals were regarded as derived from sea water, and their precipitation as sulfides was ascribed to S^{--} generated by marine sulfate-reducing bacteria. Since the latter would have been most active in areas of bottom reduction, the frequent occurrence of ores of this type in near-shore, highly carbonaceous sequences was readily accounted for.   These ideas were accepted quite readily in Europe, where a school of "sedimentationists" had been in continuous existence anyway since the time of Werner.   (Whatever the fashions of other countries, the Germans always had the stratiform deposits of Mansfeld, Rammelsberg, and Meggen to remind them of the possibilities of sedimentary processes!)   North America was rather more conservative and continued to favor a hydrothermal replacement mechanism for the formation of stratiform ores.

By the end of the 1950s the "volcanic-sedimentary," or "exhalative-sedimentary," idea—already revived and set in motion earlier in the decade—suddenly began to gain remarkable popularity.   Oftedahl (1958) in Norway, Kinkel (1962) in America, and Williams (1960) in England became notable proponents, and the great Rio Tinto copper deposit in Spain—long regarded as a prime example of deep subsurface hydrothermal activity—soon became one of the centers of resurgence of the volcanic hypothesis.   The change was stated cautiously, but clearly, by Williams:

> Recent discussions on exhalative-sedimentary ores, containing references to the porphyries of Rio Tinto, Spain, prompt the author to recant some of his opinions of a generation ago concerning the nature of these porphyries.   At that time he subscribed to the prevalent belief that the porphyries had been intruded more or less parallel to the steeply-inclined cleavage of the Lower Carboniferous slates and that the pyritic orebodies had been formed mainly by the hydrothermal replacement of porphyry and slate along or close to their mutual contact.   No contemporaneous volcanic rocks had then been identified in the field, although the author had recognized in thin-section a rhyolitic vitric tuff. . . .   Repeated visits to Rio Tinto during the past decade, however, have convinced him that the porphyries consist mostly of rhyolitic flows and pyroclastic deposits which were extruded before the onset of folding and the impress of cleavage. . . .
>
> The localization of sulphide ore near the pyroclastic top of a rhyolitic suite of volcanics, overlain by argillaceous sediments, is repeated so often throughout the pyritic field of Huelva as to imply a close genetic relationship between the volcanics and the orebodies. . . .   It must be emphasized, however, that there is unequivocal evidence of the widespread replacement of porphyry and slate by sulphides, though the relative extent to which this is due to the activity of hypogene solutions and to the mobilization of sulphides during folding and metamorphism is still unknown (1962, pp. 265–266).

At this time it was also becoming established that many of the bedded manganese oxide deposits occurring about the marginal parts of old eugeosynclines were of the exhalative-sedimentary type.   Examples were described from the West Indies, South America, Japan, Europe, and elsewhere.   The volcanic-sedimentary hypothesis was thus being applied in one form or another to many bedded deposits of iron and manganese oxide, iron carbonate and silicate, and iron and other sulfides.   It was therefore only reasonable that these should soon be seen to be related.   James (1954, p. 235) had already shown that sedimentary zones of hematite, magnetite, iron silicates, iron carbonate, and pyrite might develop in conformity with facies patterns on sloping shelves.   Many stratiform base metal sulfide deposits were now being found to be associated with similarly stratiform iron oxide and iron silicate concentrations, and this led naturally to the suggestion that sulfides of this type might be analogous to the pyrite zones of bedded iron formations.   By the early 1960s the volcanic hypothesis had become well established in England and Europe and in Australia and Japan, and the latter— the scene of Ohashi's observations some 40 years earlier—was becoming a center of attention as an example of volcanic ore genesis *in vivo*.   North America was less enthusiastic, but it was not long before much of the base metal sulfide mineralization of the northern Appalachians, and of the Superior Province of Canada, was being regarded as of volcanic affiliation—a movement to which Kinkel (1962, 1966) in America and Goodwin (1965) in Canada have made particularly notable contributions.

As the final quarter of the twentieth century approached, three developments stood out: increasing awareness that sulfide concentrations were currently forming about modern volcanic cones in Japan; observation of substantial deposition of sulfides in pipes discharging geothermal bore waters from beneath the Salton Sea area in California and recognition of these waters as potential "ore solutions"; and—most spectacularly—the discovery of very large quantities of rich iron, zinc, copper, and lead sulfides being deposited from springs on the floor of the Red Sea. The stage had been reached where problems of ore deposition might be solved not only by studying ores that had formed in the past but also by studying those that were forming at the present day.

CONCLUDING REMARKS

While any historical review of this kind must inevitably suffer faults of omission and bias, it at least serves to show quite clearly that many "modern" ideas are really quite old and that, for some of them, current popularity is by no means the first.

It is quite apparent that the germ of most twentieth-century ideas on ore genesis had appeared by about 1860 to 1870. Agricola (1546) seems to have been the first "lateral secretionist," though the idea was put in a much more comprehensive form by Sterry Hunt in the 1860s. Hutton (1788) was probably the first advocate of dry ore magmas. Werner (1791) was the outstanding early proponent of sedimentary sulfide formation. Elie de Beaumont (1847) and his contemporary Scheerer were probably the first to combine the ideas of igneous origin and aqueous transport, and their ideas were, basically, those of the twentieth-century "igneous hydrothermalists."

Perhaps a surprising aspect of ore-genesis theory is the lack of a clear pattern of development. The various ideas have not really developed from each other. Rather they have tended to develop in parallel and in a mutually exclusive way, and there has been little evolutionary connection between them. There has in fact been a remarkable preservation of a "rival" relationship between theories, each of which has shown remarkable durability and resistance to elimination by the others. Each has waxed and waned in popularity—often as a result of the eloquence of a particular individual working on a particular group of ores—but no matter how strongly ascendent any one has been, the others have persisted in the background to rise again at some appropriate moment.

The remarkable persistence of this state of affairs leads one to the suspicion that *perhaps all these theories are correct:* that no single idea has been universally established simply because there is no universal—or even dominating—mechanism or ore formation. Perhaps, as Crook suggested, ores should not be viewed as particularly special things. Perhaps we should accept—at least as a working hypothesis—that primary ores are no more and no less than rocks and, provided that conditions for the stability of their minerals are preserved, that they form in all the ways that "ordinary" rocks do. We thus allow the possibility that different ores may form as igneous rocks, as sedimentary rocks, or as veins and, like other rocks, all may suffer metamorphism. If this were so, there is no call for a separate classification of ores. Having classified rocks, we have also classified those that happen to be called ores. Our "rival" theories no longer compete. Ore genesis becomes part of petrogenesis in the broadest sense. Perhaps, therefore, it is not inappropriate to see the study of ores and their habits as *ore petrology*.

RECOMMENDED READING

Adams, F. D.: "The Birth and Development of the Geological Sciences," Dover Publications Inc., New York, 1954.

Bateman, A. M.: "Economic Mineral Deposits," 2d ed., John Wiley & Sons, New York, 1950.

Crook, Thomas: "History of the Theory of Ore Deposits," Thomas Murby & Co., London, 1933.

3
Ore Mineralogy:
General Principles

If ores are to be considered as rocks it is appropriate to look first at their mineralogy. This involves two main aspects: the properties of minerals in general and the characteristics of ore minerals in particular. This chapter is concerned with the former.

The nature of a mineral is determined by its chemical composition and its crystal structure. The two are therefore combined to identify—or define—individual minerals and to classify groups of them. The chemical composition of a mineral is, clearly, dictated by the capacity of the relevant elements to react to form a stable compound of the composition concerned. Its structure is determined by the bonding forces between its constituent particles and the geometrical configuration in which the particles are most economically packed together—i.e., their coordination.

Most mineral crystals have an element of beauty, and this lies largely in their apparent perfection. They are, however, never more than *nearly perfect*. They depart from ideal composition through the presence of impurity or lack of stoichiometry and from structural perfection through the occurrence of dislocations and holes in the lattice. In addition to the broader considerations of composition and structure, the occurrence and movement of impurities and the ways in which these are related to structural imperfection and change are also of concern. Thus bonding, coordination, and structure type, as well as the processes of solid solution and diffusion, are the main concerns of this chapter.

BONDING IN MINERALS

Chemical bonding in minerals, as in all crystalline substances, is of fundamental importance in determining many of their characteristics. Not only does it influence the geometry of the structure but it is important in determining hardness, melting point, cleavage, conductivity, and other properties. The four principal bond types—ionic, covalent, metallic, and van der Waals—are all represented among minerals. In many cases bonds are hybrid, as with the widespread ionic-covalent transition, and in addition a number of minerals owe their distinctive properties to their possession of more than one discrete bond type.

Ionic-bonded substances—those formed by the chemical interaction of oppositely charged ions, such as Na^+ and Cl^-—are held together electrostatically. Because each charge is distributed evenly over the whole surface of the atom concerned, the bond is not highly directional, and the symmetry of the resulting crystal structure is usually high. Such substances are generally rather soluble, are of moderate hardness and melting point, and are poor conductors of heat and electricity.

Substances characterized by covalent bonding are held together by the sharing of electrons and as a result are very firmly bound; covalent bonds are in fact the strongest of the four types, and minerals possessing them are characterized by relative insolubility, considerable hardness, high melting and boiling points, and very low conductivity. In contrast to the ionic variety, covalent bonds appear to be sharply localized and hence highly directional, leading to a lowering of symmetry. Minerals having a high degree of covalent bonding therefore tend to show lower symmetries than those dominated by ionic bonds.

In metals there appears to be a lack of specific allegiance between electrons and nuclei, and the rather exclusive donation or sharing of electrons characteristic of ionic and covalent bonding is absent. Instead the electrons are highly mobile: the nuclei are held together in their packed arrangement by the aggregate charge surrounding each nucleus rather than by the action of particular electrons. This leads to the well-marked properties characteristic of the metallic bond. Outstanding among these are high thermal and electrical conductivities, high malleability, low hardness, and low melting point. As with ionic bonding there is a general lack of directionality, and symmetries therefore tend to be high: most metals have cubic or hexagonal close-packed structures.

The very weak van der Waals, or "stray-field," bond is not common among minerals, but where it does occur it usually manifests itself in easy cleavages and planes of low hardness.

Bonds are rarely "pure," and it is now well established that in most cases there is some electrostatic component in covalent-bonded substances and some electron sharing in ionic-bonding substances. In most sulfides too there is some metallic component in the bonding. Sphalerite shows characteristics of both covalent and ionic bonding, and galena shows those of ionic and metallic bonding. In addition to such "hybrid" bonds, most minerals exhibit more than one discrete bond type. Each tends to develop in particular crystallographic planes, and this may impart a strong directionality to a crystal structure. The nature of the bonding may have a pronounced effect on habit and also, clearly, may affect such properties as cleavage and the hardnesses of specific crystal faces and other planes. Two outstanding examples of the effect of differences in bond type are the micas and graphite. In the former, bonding parallel to the sheets is substantially covalent and strong, whereas that between sheets is ionic and rather weak. The combination leads to the excellent cleavage—a highly directional property—of the micas. In graphite, bonding within the sheets is covalent and strong, whereas that between sheets is of the van der Waals type and hence very weak. The result is an extraordinarily good cleavage—so good that graphite has the properties of a lubricant.

Ionic and covalent bonds are the most important among minerals, and there are several features pertaining to them that are important in the development and stability of crystal structures.

The first is the size of the ions or atoms, with the corresponding length of spacing between their centers. In ionic crystals the distance between centers of adjacent ions approximates to the sum of the two individual ionic radii. For individual elements the radius increases with decrease in the valence state and is smaller in positive ionization states than in the negative. Sizes of ions in the same group of the periodic table (e.g., the alkali and alkaline earth metals) increase with increase in atomic number:

Alkali metals			Alkaline earth metals		
Ion	Atomic number	Radius, Å	Ion	Atomic number	Radius, Å
Li	3	0.68	Be	4	0.35
Na	11	0.97	Mg	12	0.66
K	19	1.33	Ca	20	0.99
Rb	37	1.47	Sr	38	1.12
Cs	55	1.67	Ba	56	1.34

In covalent crystals the distance between the centers of adjacent atoms approximates to the arithmetic mean of the interatomic distances in crystals of the elemental substances.

The second consideration, which pertains particularly to ionic crystals or those having a large ionic component, is the effect of the surroundings of a given ion on its size. Since the electron cloud about each ion is not rigid, it may be compressed or dilated by adjacent ions, and in fact a large number of surrounding ions will cause expansion, whereas a smaller number may allow contraction. Ionic radii are normally given for the case of six "nearest neighbors"—what is commonly referred to as *standard six-fold coordination*. Asymmetry of the environment may, in addition, cause distortion of ions.

The third important consideration is the proportion of ionic and covalent character exhibited by a given bond. As already pointed out, covalent bonding is strongly localized about the shared electron, yielding a highly polarized arrangement. Increase in the covalent component of a hybrid bond therefore leads to increasing polarization. Compounds involving elements from widely separated groups in the periodic table are dominantly ionic, whereas those involving adjacent groups or the same group are essentially covalent. The alkali halides are almost completely ionic, the Al—O bond is 37 per cent covalent, the Si—O bond is 50 per cent covalent, and the Si—C bond is dominantly covalent.

COORDINATION

The basic process in the construction of a crystal is the grouping, or coordinating, of one set of ions about another: each individual ion gathers about itself as many oppositely charged ions as its size allows. In ionic-bonded substances the particles approximate to spheres, yielding fairly simple geometrical relationships; where the degree of covalency is substantial, polarization leads to more complex structures.

In ionic-bonded crystals the coordinated ions cluster about the central coordinating ions in a regular way such that their centers form the apices of regular polyhedra. Each cation thus lies at the center of a coordination polyhedron of anions and vice versa. In the galena structure each lead ion is immediately surrounded by six nearest-neighbor sulfur ions, and the lead is therefore in six-fold coordination with sulfur. Similarly the zinc of sphalerite (and wurtzite) is surrounded by sulfur in tetrahedral arrangement, or four-fold coordination, just as in the case of silicon and oxygen in the well-known SiO_4 tetrahedron.

Clearly the capacity of ions to coordinate is dependent on relative size. Where the coordinating particle is small relative to those surrounding it, only a small number of the latter can touch the former, and so the *coordination number* is necessarily small. As the relative size of the coordinating particle increases, a larger number of oppositely charged ions can gather in contact with it, assuming always that the requirements of valence are satisfied. This leads to a consideration of *radius ratio*. The radius ratio for any pair of ions is usually expressed as R_A/R_X, where R_A is the ionic radius of the coordinating cation and R_X is that of the coordinated anion.

Where $R_A/R_X = 1$, a central spherical unit may coordinate 12 other particles; i.e., it may have 12 "neighbors" all in contact with it. Such coordination may be achieved in two ways—by cubic or hexagonal closest packing. Each of these involves a somewhat different layering sequence, as follows: Suppose a number of spheres of similar radius are laid down in contact, as shown in Fig. 3-1. In two dimensions each sphere "coordinates" six others, which form a hexagon about it. We may call this the A layer. Suppose then that a second layer—the B layer —is laid down on the first so as to nest between the A spheres. The A spheres are shown as circles in Fig. 3-1b, and the B spheres are in their nested positions between them. The B spheres could equally well be put in the C positions of Fig. 3-1b at this stage, since to this point the B and C positions are indistinguishable.

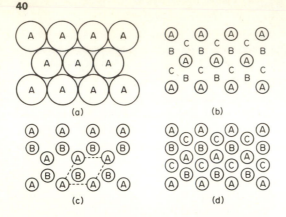

Fig. 3-1 Close packing of spheres of equal radius.
(*a*) A single close-packed layer; (*b*) a single close-packed layer showing the two alternative sets of nesting sites into which the second close-packed layer can fit; (*c*) hexagonal close packing, following the stacking sequence *ABABA*... (the outline of the hexagonal unit cell parallel to the basal plane is shown in dashed lines); (*d*) cubic close packing. (*Redrawn from Evans, "Crystal Chemistry," 2d ed., Cambridge University Press, New York, 1964, with the publisher's permission.*)

However, as soon as we come to lay down the *C* layer we are faced with two possibilities: the *C* spheres may be nested into the *B* layer either in positions immediately above the *A* spheres *or* into the new *C* positions of Fig. 3-1*b*. Whichever alternative is chosen we have 12 spheres in contact with the original central sphere—3 in the layer below, 6 in the same layer (as in Fig. 3-1*a*), and 3 in the layer above—and hence a coordination number of 12. However, two possible stacking sequences exist. In the first the layers of spheres lie alternately in the *A* and *B* positions, and the stacking sequence is *ABABABA*. . . . This yields a hexagonal structure and is referred to as *hexagonal close packing*. In the second the spheres nest in the *A*, *B*, and then *C* positions so that the stacking sequence *ABCABCA* . . . develops. This has a cubic structure—that of the face-centered cubic cell —and is referred to as *cubic close packing*. In the hexagonal sequence the vertical axis represents the *c* axis of the resulting crystal, and in the cubic sequence the vertical axis represents a body diagonal of the cube. The basal plane of the hexagonal and the {111} plane of the cubic are therefore common to both structures. Both types of stacking, and the existence of this common structural plane, are of great importance among the sulfides and oxides.

With smaller radius ratios the coordination number becomes progressively less. Where $0.732 < R_A/R_X < 1.0$, it is possible to coordinate eight ions, yielding eight-fold, or "cubic," coordination. For $0.414 < R_A/R_X < 0.732$, the possible coordination is six-fold, or "octahedral," coordination; for $0.225 < R_A/R_X < 0.414$, it is four-fold or tetrahedral; for $0.155 < R_A/R_X < 0.225$, it is three-fold, or "triangular"; and for $R_A/R_X < 0.155$, it is two-fold, or "linear." These are illustrated in Fig. 3-2. In each case the descriptive term describes the polyhedron formed by the hypothetical joining of the centers of the coordinated spheres. For eight-fold coordination this is a cube, for six-fold, an octohedron, and so on.

In crystals having a substantial covalent component in their bonding, the picture diverges from that of simple packed spheres. With increasing polarization, coordination numbers no longer reflect the simple factors of radius ratio and valence. The larger and more polarizable the anion and the smaller and more polarizing the cation, the greater the expectable divergence from the calculated radius-ratio limits.

Finally, in considering coordination, attention must be given to the way in which coordinated groups are themselves linked together. So far only the space-charge relationships between nearest neighbors have been considered. Although these are clearly dominant, others also affect the development of a crystal. In ionic substances, coordinated units tend to link in such a way that the cations are as far apart as possible, consistent with the space-charge requirements already considered. Thus the sharing of anions is arranged so that the distance between cations is a maximum, and cations tend to share as small a number of anions as possible. Thus linkage of polyhedra is more common through corners than edges and more common through edges than faces. Linkage between polyhedra is also affected by the strengths of bonds within them. Since the strengths of all bonds reaching an ion in an ionic-bonded crystal must sum to the valence of that ion, the relative strength of any bond in a structure can be determined by dividing the valence of the coordinating ion by the number of its nearest neighbors. The result is termed the *electrostatic valence*. In some cases the strengths of all bonds are the same, and the crystals concerned are termed *isodesmic*. In others, where small cations of high charge coordinate larger anions of lesser charge, the anions are bonded more strongly to their central coordinating ion than they can be to any other. This gives rise to a very strongly bonded unit, as in carbonates and sulfates

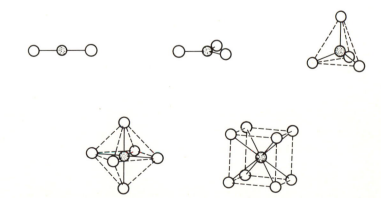

Fig. 3-2 Diagrammatic representation of coordination polyhedra, illustrating two-, three-, four-, six-, and eight-fold coordination.

in which the C—O and S—O bonds are stronger than the metal-CO_3 and metal-SO_4 bonds, which is said to be *anisodesmic*.   In other cases the strength of the bond between cation and anion is exactly half the total bonding energy of the anion. Here the anions in question are able to form bonds of like strength with another cation, which they may do, so that a given anion may form a part of more than one coordination polyhedron.   In this case the linkage of polyhedra is greatly facilitated, leading to polymerization and the formation of chains and sheets, as in carbon (organic) compounds and silicates.   Such crystals are termed *mesodesmic*.

IMPURITIES IN MINERALS

In spite of the remarkable constancy of many of the features of crystalline substances, these are rarely, if ever, pure.   This is particularly the case with minerals. It is safe to say that no mineral specimen is in reality a simple, "pure" compound composed only of its well-recognized major elements.   In addition to these elements there are always minor amounts of other elements, usually quite similar to the major ones in most of their properties, included in the crystal structure. These may range in amount from the minutest trace to substantial parts of the crystal concerned.

Such impurities may be "built into" the crystal as it grows, or they may move in by diffusion at some later stage.   They are accommodated in three principal ways:

1. *Mutual substitution.*   In this case the impurity atom simply occupies a lattice site normally occupied by an atom of the major element concerned.
2. *Coupled substitution.*   Here the impurity atom occupies a normal lattice site, but there is a valence difference between the substituted and substituting particles.   This is compensated for by an adjustment of the proportions of major elements of different valences or by the creation of vacancies in the structure—in each case a "coupling" with the charge on the new particle.
3. *Interstitial accommodation.*   In this case the foreign atom is fitted into spaces between those occupying normal lattice sites.   This requires that the impurity atom be relatively small.   Charge imbalances are neutralized as in 2.

The two substitutional mechanisms give rise to *substitutional solid solutions*. Insertion into interstitial spaces gives rise to *interstitial solid solutions* (Fig. 3-3). Such phenomena are extremely important, underlying virtually all compositional variation in minerals and, particularly, the tendency for many minerals to carry characteristic trace impurities.   It is therefore important to understand some of the principles involved.

SUBSTITUTIONAL SOLID SOLUTION

This includes substitution of both simple and coupled types.

Since the grouping of ions depends on size, charge, and bond type, ions of different elements may substitute for each other in certain groupings, provided the

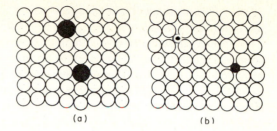

Fig. 3-3 (*a*) Substitutional and (*b*) interstitial solid solutions. (*Redrawn from Kingery, "Introduction to Ceramics," John Wiley & Sons, Inc., New York, 1965, with the publisher's permission.*)

(a)   (b)

size differences between substituted and substituting ions do not exceed certain limits. Consideration of size and charge led Goldschmidt (1937) to formulate three rules that he thought governed ionic substitution. These were based on the supposition that when two ions are capable of substituting for one another in a crystal structure, the ion that makes the greater contribution to the energy of the structure is preferentially incorporated:

1. For two ions to be able to replace one another in a crystal structure, the ionic radii must not differ by more than about 15 per cent.
2. When two ions that possess the same charge but different radii compete for a lattice site, the ion with the smaller radius is preferentially incorporated.
3. When two ions possessing similar radii but different charges compete for a lattice site, the ion with the higher charge is preferentially incorporated.

While Goldschmidt's rules are very useful in predicting in a general way which elements might substitute for each other, it has become apparent that such a simple view is not always adequate to explain observed substitution, or the lack of it. Neumann (1949) drew attention to the fact that magnesium and ferrous iron enter into zinc minerals to only a limited extent, though since the three elements have similar ionic radii ($Zn^{++} = 0.74$ Å; $Fe^{++} = 0.74$ Å; $Mg^{++} = 0.66$ Å) and charge, it might have been expected that they should substitute extensively. Fyfe (1951) suggested that this was probably because Goldschmidt's rules were a good fit for only simple ionic crystals, whereas most bonds in minerals are of the hybrid ionic–covalent type already discussed. Fyfe pointed out that even where the ionic component predominates, the partial covalent character of a bond may lead to a considerable shortening of the interatomic distance so that this no longer approximates to the sum of the radii of the atoms concerned. Where ionic bonding predominates, there are no special restrictions on bonds, and the structure therefore follows the radius-ratio rules and depends on the relative sizes of cations and anions. Where covalent bonding predominates, however, the structure is determined largely by the directional properties of the bonds. This is illustrated by carbon, in which the carbon atom always forms four bonds directed to the corners of a tetrahedron. From this, Fyfe suggested that Goldschmidt's rules required extending: as these rules predicted, two atoms will be mutually replaceable in *ionic compounds* if their ionic sizes are similar, but in more *covalent compounds* such mutual substitution will take place only if the number and directional properties of the bonds are similar.

These considerations of Neumann and Fyfe are concerned with possible complexities in the equal valence situation. However, as already noted—and as provided for in the Goldschmidt rules—substitution may also involve the introduction of ions of different valence from those whose place they are taking. Since the electrical neutrality of the crystal must be preserved, the potential charge imbalance must be nullified in some way. There appear to be two mechanisms:

1. *Valence adjustment.* In this case balance is preserved by changing the proportions of ions of similar sign but different valence. For example, suppose that the cation component of a compound involves two divalent elements, one of which is being partially displaced by a monovalent element. If uncompensated this would lead to the development of a negative charge on the crystal, which is not permissible. Compensation may be made, however, by the coupled incorporation of a trivalent element of suitable ionic radius. Kingery (1960, p. 164) points out, for example, that extensive substitution among Mg^{++}, Al^{3+}, and Fe^{++} is common in clay minerals of the montmorillonite structure. Charge deficiencies produced by replacing trivalent Al^{3+} by divalent Fe^{++} or Mg^{++} are made up by the adsorption of exchangeable ions on the surfaces of the minute particles of the clay. In a similar way, substitution of Si^{4+} by Al^{3+} in tetrahedral positions in kaolinite leads to a charge difference that is compensated for by appropriate surface adsorption of other ions. Perhaps the best example of coupled substitution among the rock-forming silicates is the plagioclase series, in which substitution of calcium for sodium is accompanied by a "coupled" substitution of aluminum for silicon so that the charge balance is preserved:

$$NaAlSi_3O_8 \qquad CaAl_2Si_2O_8$$
$$\text{16+} \qquad\qquad\;\; \text{16+}$$

This type of substitution is not common among the ore minerals, though it undoubtedly takes place among some of the oxides.

2. *Vacancy formation.* Another way in which charge imbalance may be avoided is by leaving some lattice sites vacant. Reverting to our example of the compound possessing two divalent cations, one of which is partially displaced by a monovalent one, rather than adjusting cation charges as before, we may compensate by omitting, say, one quadrivalent anion for every four cation substitutions that occur. There are many variants of this procedure. For example, extensive substitution occurs in the spinel $(MgO \cdot Al_2O_3)$–Al_2O_3 system (Kingery, 1960, p. 165). Clearly this involves the substitution of Al^{3+} for Mg^{++}. To maintain electrical neutrality, three Mg^{++} ions must be lost to accommodate two Al^{3+} ions, leading to the creation of a lattice vacancy. Similarly additions of Ca to ZrO_2 (cubic fluorite structure) involve the substitution of Ca^{++} for Zr^{4+} and hence the creation of an oxygen site vacancy for every calcium ion incorporated. In the same

way, additions of La^{3+} to CeO_2 and Zr^{4+} or Cd^{++} to Bi_2O_3 in ceramics lead to the development of corresponding numbers of vacancies in the anion (oxygen) arrays. In a parallel fashion the incorporation of additional Al^{3+} in $MgO\cdot Al_2O_3$, Mg^{++} in LiCl, Fe^{3+} in FeO, and so on, leads to the creation of vacancies in the cation array.

Among minerals, this kind of process is undoubtedly widespread in the oxides and silicates—certainly among the members of the large spinel group of oxides. While there do not seem to be any specifically documented examples among the sulfides and related minerals, perusal of trace element data suggests that the phenomenon is probably far from uncommon here.

INTERSTITIAL SOLID SOLUTION

Rather than substituting for major element ions in their established lattice sites, the foreign ions may fit into spaces *interstitial* to these. Clearly this requires that the incoming ions be relatively small. As would be expected those compounds with relatively "open" structures—such as the zeolites—form interstitial solid solutions with greatest ease.

The incorporation of ions in interstitial positions naturally requires a charge adjustment to preserve electrical neutrality. This may be achieved by vacancy formation or associated substitutional solid solution.

GENERAL REQUIREMENTS FOR SOLID SOLUTION

It is now clear that there are a number of general factors governing solid solution:

1. *Size.* The closer two ions are in size, the greater is the chance of substitutional solid solution. Extensive solid solution of this kind is possible up to a difference of about 15 per cent, after which it falls off rapidly. Size is by far the most important factor in substitution in ionic compounds. For interstitial solid solution the foreign ion must be small relative to the interatomic spacing in the host structure. Thus small ions and particularly open structures are those most frequently involved in this type of solid solution.
2. *Chemical affinity.* Any tendency to reactivity between host and substituting ions lessens the possibility of solid solution; in such cases a new phase is more stable than a solid solution.
3. *Valence.* Substitutional solid solution is favored by similarity in valence. Difference in valence between host and substituting ions does not preclude substitution; however, since it requires coupled substitution or vacancy formation—and hence structural changes—it does inhibit substitution. Interstitial solid solution always involves charge adjustment. This—and hence any structural modification—is least where the charge is low, and hence such solid solution should be favored by a low valence of the impurity ion.
4. *Crystal structure.* For complete or extensive solid solution the two end members must have very similar crystal structures. In minerals whose bonding possesses a substantial covalent component, bond angles must be essentially

similar. Solid solution is limited, though not precluded, by structural differences. Structure type, other than its openness, is of no concern in interstitial solid solution.

Where all these factors are highly favorable, substitutional solid solution is potentially unrestricted. Ions of the appropriate pair or group of elements may substitute for each other in all proportions, forming stable *solid-solution series*. Among the ore minerals the Co, Ni, Fe^{++} triarsenides (the skutterudite series), the tetrahedrite-tennantite series, and the spinels are well-known examples of extensive and stable solid solutions. The frequently associated rhombohedral Fe^{++}–Mg carbonates constitute another. In addition many minerals—and certainly many of those of ores—contain quite stably substituted "traces" of characteristic impurity elements. Pyrite not uncommonly contains small quantities of nickel and cobalt that substitute for the iron in its normal lattice position. Cadmium is a characteristic trace substitute for zinc in sphalerite—so characteristic in fact that sphalerite constitutes the world's major source of cadmium. Cobalt is a common constituent of pentlandite, and where pentlandite and pyrrhotite coexist, the cobalt is strongly concentrated in the pentlandite. Iron and manganese are also usually present in sphalerite, and Fe^{++} may substitute to an extent of more than 30 cation per cent. The substitution of Fe^{++} for zinc is dependent on temperature and pressure and is finally limited by structure (sphalerite and pyrrhotite being, respectively, cubic and hexagonal close packed).

On the other hand, solid solution commonly involves the incorporation of foreign ions of rather poor fit. In such cases the extent of "solution" is limited, but it tends to increase, and may be substantial, at high temperatures. This is due to the distention and partial disordering of the host structure that accompanies the rise in temperature and that renders easier the incorporation of particles of unlike size. Solutions formed at high temperatures in this way may be quenched and the high impurity content retained metastably at lower temperatures, or they may be cooled slowly, in which case the foreign ions are *exsolved*.

Temperature-dependent solid solutions are common in many mineral groups and are well known among the ore minerals. The sulfides may be used to illustrate the process. As the temperature of the crystal structure rises, two principal changes begin:

1. The radii of vibration of all atoms in the structure increase.
2. The cation portion of the structure breaks down, and the metal atoms become distributed statistically through the sulfur lattice, which, though distended, remains stable. (This is marked by an increase in symmetry, as will be discussed a little further on.)

The structure thus becomes distended and disordered in part, and under these conditions the incorporation of foreign atoms whose size would normally cause their exclusion is rendered easier. In this way sphalerite, for example, may accommodate quantities of tin, iron, and copper far in excess of those that could enter its structure at normal temperatures. With fall in temperature there is a return to the ordered state: the structure contracts and the metal lattice attempts to reorder itself. Foreign atoms become a "bad fit" and tend to diffuse out of the host. If they are present in sufficient quantity, they may aggregate to form small bodies of a new phase—a process termed *exsolution* or, less elegantly, *unmixing*.

ATOMIC MOBILITY AND DIFFUSION

Atoms are capable of moving about over comparatively long distances in solids. The process involved is referred to as *solid-state diffusion* and is essential for the performance of all chemical reactions and textural changes in crystalline and non-crystalline solids. The general nature of the process is illustrated in Fig. 3-4. If two substances A and B are placed in close contact, atoms may move, by a series of "atomic jumps," from each substance into the other. Such diffusion may result in the formation of a new crystal structure of composition AB or a random structured solid solution.

Diffusion of this kind is governed by Fick's law. We may consider a single phase in which the concentration of some substance is nonuniform. Suppose diffusion begins and takes place in one direction under conditions of constant temperature and pressure. Material moves in such a way as to reduce the initial concentration gradient, the quantity passing per unit time through unit area normal

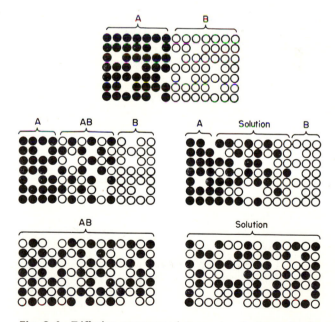

Fig. 3-4 Diffusion movement between two juxtaposed substances A and B, yielding a new compound AB or a random solid solution. (*Redrawn from Kingery, "Introduction to Ceramics," John Wiley & Sons, Inc., New York, 1965, with the publisher's permission.*)

to the direction of diffusion being proportional to the concentration gradient. The flux J is given by:

$$J = -D \frac{\partial c}{\partial x}$$

where D is the diffusion coefficient, c is the concentration per unit volume, and x is the distance moved. The relationship is analogous to that of Ohm's law, which states that an electrical current is proportional to the gradient of the electrical potential, and Fourier's law, which states that the rate of heat flow is proportional to the prevailing temperature gradient.

DIFFUSION IN CRYSTALLINE SOLIDS

In "pure" monatomic solids, such as metals, the atoms may exchange positions and move about extensively. This kind of atom movement is known as *self-diffusion*. In other cases the diffusing atoms are essentially impurities, in which case the process may be *interstitial* or *substitutional impurity diffusion*. Self-diffusion is important in the development of microstructure in single-phase solids. Impurity diffusion is, of course, important in all kinds of replacement, solid-solution, and exsolution processes.

Self-diffusion Here the like atoms move about by exchanging positions or by shunting into vacancies, and the overall structure is essentially the same after diffusion as it was before. The process may be studied by placing a small quantity of a radioactive isotope of the relevant element in a localized part of the crystal and then following its movement by very refined radioactive counting.

There appear to be four possible mechanisms of self-diffusion:

1. *Direct interchange.* Two adjacent atoms simply exchange positions, as in Fig. 3-5. Clearly this process must involve a large distortion of the surrounding

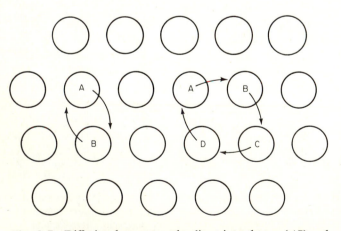

Fig. 3-5 Diffusional movement by direct interchange (AB) and by a ring movement ($ABCD$).

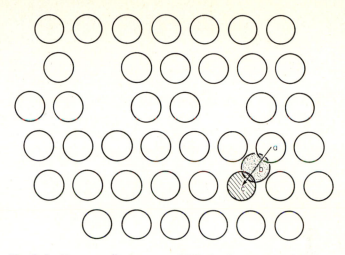

Fig. 3-6 Vacancy, divacancy, and diffusion by vacancy migration. A particle in position *a* moves through *b* to a vacancy at *c*; the vacancy is thus moved from position *c* to position *a*.

lattice and is not likely to occur unless the crystal structure is quite an open one.

2. *Ring mechanism.* The severe local distortion inherent in direct interchange may be avoided by the ring motion of Fig. 3-5. In this case four contiguous atoms move as a unit to produce a net displacement. While this involves less distortion, it requires the cooperative movement of four atoms instead of only two, with a corresponding lowering of probability. Which of the two mechanisms occurs therefore depends on the amount of energy required for the direct interchange and the probability of the cooperative movement necessary for ring motion. In general the ring movement is the more likely to occur. Even this, however, requires quite an open structure. It is probably important in sodium and may occur in chromium, but most metals appear to be too tightly packed for appreciable movement of this kind.

3. *Point defects: vacancy migration.* A point defect is an irregularity of atomic dimensions within a crystal structure. The most important kind of point defect is a vacant site, and such vacant sites may play an important part in diffusion. A single and a double vacancy are illustrated in Fig. 3-6. Atoms adjacent to a vacancy can slip into it, as shown, so that in effect the atom and the vacancy simply change places. In this way a vacancy can wander through a crystal, leading to a net transposition of matter. The same holds for a divacancy, though larger groups of omissions (trivacancies, etc.) are immobile and do not contribute to diffusion.

4. *Point defects: interstitials.* Another type of point defect is that in which an atom occurs interstitially to the surrounding "normal" sites. (Such atoms are not impurity atoms but merely additional atoms of the surrounding pure substance.) Interstitial atoms may move in either of two ways. The migrating

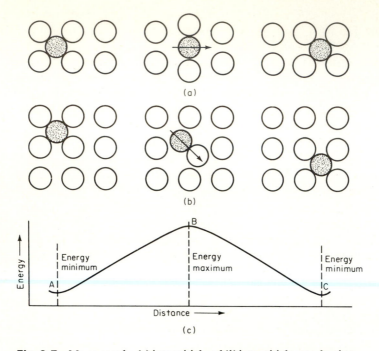

Fig. 3-7 Movement by (*a*) interstitial and (*b*) interstitialcy mechanisms. In each case the central position constitutes the *activated position*.

atom may simply squeeze past two of its neighbors, as shown in Fig. 3-7*a*. It forces its neighbors apart and then slips into a new interstitial position. Alternatively it may push an adjacent atom in a normal lattice site out of its position, taking over the site itself and shunting the other atom into an adjacent interstitial position.   This is shown in Fig. 3-7*b* and is termed an *interstitialcy mechanism*.

The advantage of the vacancy, interstitial, and interstitialcy mechanisms is that they do not involve the lattice distortion of the interchange and ring motions. Also, where there are abundant vacancies or interstitials in the crystal, the probability of movement may be high and extensive diffusion may therefore be possible.

Reference to the lesser distortion involved in the point-defect mechanisms is really another way of saying that the *activation energies* for these movements are less than those for direct interchange and ring motion.   Atomic movement of any kind in a crystal structure requires an increment of energy for its activation, and this is referred to as the *activation energy* for the particular movement concerned. It may be noted from Fig. 3-7*a* and *b* that movement of the diffusing atom requires the exertion of a repulsive force and the breaking and remaking of bonds.   This involves the expenditure of energy, and the potential energy states of a diffusing atom are illustrated diagrammatically in Fig. 3-7*c*.

Atoms forming a crystal structure vibrate about their average positions in that structure at all temperatures above absolute zero. The amplitude of these vibrations is minute, but it may increase if energy is added to the system. Thus if the crystal is heated, the amplitude of the vibrations of its constituent atoms increases. Any given atom may be visualized as sitting—and vibrating—in a cage composed of its neighbors. During each vibration it moves toward these neighbors until the resulting repulsive forces repel it back toward its mean position at the center of the cage. Only if sufficient energy is made available to it—by thermally agitated atoms on one side or the other, moving in concert and giving our atom a substantial "shove"—can these repulsive forces be exceeded and the cage walls penetrated. The central positions of Fig. 3-7*a* and *b* can thus be achieved only if the energy maximum of Fig. 3-7*c* is attained and just exceeded. The energy differences between the normal position *A* and that of *B* is *E*, the activation energy. When the atom is at *B* it is said to be in the *activated position* and the enclosing crystal in the *activated state*.

Clearly the activation energy for interchange and ring movement will generally be quite high, and for point-defect movement it will be comparatively low. This principle applies not only to self-diffusion but also to the diffusion of impurities.

Substitutional impurity diffusion in monatomic crystals This occurs when impurities at normal lattice sites, i.e., "substitutional solid-solution" atoms, become activated. The result is, of course, "exsolution." Since the foreign atoms occupy sites normally occupied by the major constituent atoms, their diffusion jumps follow similar paths to those of self-diffusion in the solvent. However, since such foreign atoms usually differ in size, and often differ in valence, from those of the solvent, there are some important differences between self-diffusion and substitutional impurity diffusion.

Suppose, for example, that the impurity atom is slightly larger than the solvent atoms. The repulsive force between the impurity and solvent atoms is greater than that between two solvent atoms, the local structure becomes distorted, and the surrounding solvent atoms are displaced slightly outward from their normal

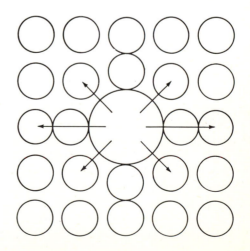

Fig. 3-8 Distortional effect of a large impurity atom.

positions (Fig. 3-8). This distortion may be partly relieved by the formation of a vacancy, and vacancies are therefore likely to form in this situation. The converse holds where the impurity atom is smaller than the solvent atom. Thus larger atoms will diffuse out (i.e., exsolve) from a crystal more rapidly than smaller ones. It has also been shown by electrostatic theory that the greater the valence of an impurity atom relative to those of the solvent, the more rapidly will it diffuse.

Interstitial impurity diffusion in monatomic crystals It has already been pointed out that interstitial impurities are most common in hosts having open structures. In dealing with their diffusion we are therefore chiefly concerned with crystals of wide spacing. Such spacing is essential not only for the accommodation of the impurity in the first place but also for its movement, since the diffusion process in this case is simply one of the impurities "squeezing past" the solvent atoms. The general nature of the mechanism is analogous to that shown in Fig. 3-7a.

A characteristically open structure of wide occurrence in the mineral kingdom is that of diamond—a configuration also possessed by Si, Ge, SiC, ZnS, $CuFeS_2$, and many others. The arrangement is such that each atom is at the center of a tetrahedron, the bonds being directed to the tetrahedron corners. As indicated in Fig. 4-2 this yields quite an open structure, and interstitial atoms can be accommodated in the lattice without unduly straining the crystal. The geometry of the diffusion jump appears to be quite simple and may be visualized from Fig. 4-2. The impurity atom passes through a kinked hexagon forming part of the {111} plane of the host atoms. This hexagon is so large that, as Girifalco (1964) points out, if the diffusing atom had a diameter equal to the interatomic distance in the host crystal, the hexagon would have to stretch its linear dimensions by only about 10 per cent to let the impurity pass. The open nature of this structure is therefore of utmost importance in diffusion.

Diffusion in ionic and covalent compounds To this point we have been concerned with monatomic crystals, of metallic nature for the most part. It is useful to consider these first since they provide the simplest picture of point defects and diffusion. However, most minerals are compounds and, as already noted, possess substantial ionic and covalent components in their bonding. The incidence of defects and the mechanisms of diffusion in such crystals are rather different from those in simple metals, though similar general principles hold. Ionic crystals are more complicated because they contain more than one kind of atom and because they possess charges. Covalent crystals frequently contain more than one kind of atom, and these are connected by bonds that are strongly directional.

In crystals having a substantial ionic component in their bonding, there are a number of possible kinds of point defects. These are similar to, or are readily recognized variants of, the defects found in monatomic metal crystals:

1. An ion in a wrong lattice position
2. A vacancy formed by a positive ion omission

3. A vacancy formed by a negative ion omission
4. A divacancy
5. A positive ion in an interstitial position
6. A negative ion in an interstitial position

If an ion occurs in an incorrect lattice position, it will be surrounded by other ions of like sign. This involves strong repulsive forces and hence a high potential energy. Such a situation is clearly unfavorable and is hence rare.

A positive ion vacancy is generated by removing a positive ion from its normal position and transferring it to the surface of the crystal. This does not result in a charge imbalance on a large scale (i.e., the crystal as a whole), but it does on a small scale since the charges are separated by a large distance in terms of the normal interionic spacings. The result is a small negative charge on the position formerly occupied by the positive ion. The reverse is the case for a negative ion vacancy.

In metals such imbalances may be substantially nullified by the mobility of their electrons. In ionic crystals, however, the electrons have rather specific allegiances to particular nuclei (as noted earlier in this chapter), and this imparts a rigidity of charge distribution, leading in turn to preservation of the vacancy and its local electrical imbalances. The formation of further vacancies of like sign involves a large increase in energy and is therefore not favored. However, the charge imbalance of an ion vacancy may be canceled by the nearby formation of another of opposite sign. Such an ion vacancy accompanied by a nearby compensating vacancy of opposite sign is termed a *Schottky defect*. Ion vacancies can also be neutralized by the insertion of an interstitial ion of the same sign in a closely adjacent position, and this is termed a *Frenkel defect* (Fig. 3-9).

Because the particles of ionic crystals are of different size and are electrically charged, relative movement by direct interchange and ring mechanisms would require prohibitively large activation energies. Diffusion is thus restricted to vacancy, interstitial, and interstitialcy mechanisms. Interstitial motion is shown in Fig. 3-10 and involves movement of the migrating ion between two others of opposite sign and hence through the field of the repulsive force between them. The movement of an interstitial ion involves a geometry similar to that of Fig. 3-7a but requires movement between two positively and two negatively charged ions. Initial, activated, and final positions are shown in Fig. 3-10. Interstitialcy movement also follows a geometrical path similar to that of metals, with the added complication of attractive and repulsive forces exerted by like and unlike charges on nearby ions.

The incidence of defects and the mechanisms of diffusion in covalent crystals are affected by a factor not encountered in metallic and ionic crystals. This is the directionality of the bonds. As already noted, crystals of the diamond structure combine this with a very open arrangement of the atoms, and the two yield a situation that is highly conducive to the acceptance and movement of interstitials. Where the latter have an atomic radius no greater than that of the atoms of the host crystal and do not tend to form directional bonds, the host structure is not

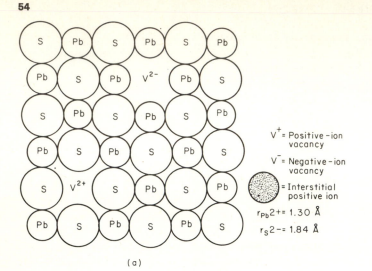

V^+ = Positive-ion vacancy

V^- = Negative-ion vacancy

= Interstitial positive ion

$r_{Pb^{2+}}$ = 1.30 Å

$r_{S^{2-}}$ = 1.84 Å

(a)

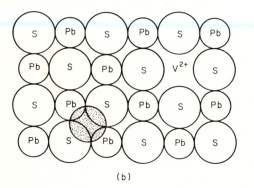

(b)

Fig. 3-9 (*a*) Schottky and (*b*) Frenkel defects. (While PbS is used as an example here, such defects are characteristically features of highly ionic substances such as the alkali and silver halides.)

distorted unduly and extensive interstitial solution is possible. Where the impurity does tend to form directional bonds, it is more likely to be accepted substitutionally.

Mechanisms of diffusion in covalent crystals appear to involve a number of factors whose interactions are complex and not yet clearly understood. In the

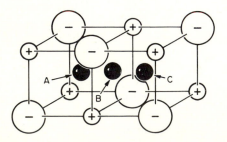

Fig. 3-10 Path of movement of a positive interstitial ion in a sodium chloride (that is, PbS) structure. Position *A* represents the initial interstitial position, *B* the ion in the activated position, and *C* the ion in the final interstitial position. (*Redrawn from Girifalco, "Atomic Migration in Crystals," Blaisdell Publishing Company, New York, 1964, with the publisher's permission.*)

diamond structure there seems no doubt that interstitial diffusion may occur readily. The interstitialcy mechanism, however, seems quite unlikely here, involving as it does the complete breaking of the four bonds. Ring movements seem highly unlikely for a similar reason. Diffusion by vacancy movement, however, may occur in some cases.

DIFFUSION IN NONCRYSTALLINE SOLIDS

When a substance cools and solidifies rapidly it frequently assumes a disordered structure, as in a silicate glass. Such glassy structures lack the ordered packing of crystals and characteristically contain many "holes." This poor packing and consequent holed structure is, of course, reflected in the lower densities of glasses as compared with their crystalline analogs. They are also reflected in substantial changes in other physical properties, including diffusion characteristics.

The effect of disorder on diffusion coefficients is not simple. However, it can be said that in general the diffusion coefficient in the random noncrystalline structure is greater by several orders of magnitude than it is for the corresponding crystalline solid. Where glass and crystal exist together it has been observed that diffusion tends to take place predominantly in the glassy phase. This principle is important in ceramics and is no doubt responsible for the preferential breakdown of the glassy matrix of many hypohyaline lavas during surface alteration.

DIFFUSION ALONG GRAIN BOUNDARIES

The nature of grain boundaries is discussed in some detail in Chap. 9. For present purposes it is sufficient to say that they are very thin zones (perhaps two or three atomic diameters wide) of poor packing that form the junction between two dissimilarly oriented crystalline grains. These grains are usually part of a polycrystalline aggregate composed of many grains.

Clearly such a boundary zone is less ordered than the crystal lattices on either side of it. Some of the atoms are closer than they would be in the normal lattice, and some are farther apart. Overall the atoms appear to be more loosely packed than in the adjacent crystals. The situation is thus somewhere between the high degree of ordering characteristic of crystals and the low degree characteristic of glasses. Diffusion should therefore be favored along the boundaries, and this is borne out by experiment. Diffusion of traceable radioactive atoms into a polycrystal and its boundaries is shown diagrammatically in Fig. 3-11. For this reason

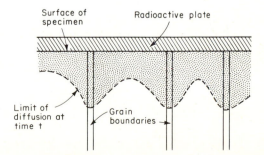

Fig. 3-11 Grain-boundary and crystal-lattice diffusion in a polycrystal; diffusion is favored by the grain boundaries.

the first stages of rock alteration, and initial replacements of early by later phases in ores, usually appear along grain boundaries in the host material.

STRUCTURE TYPE

Having considered the nature of bonding, coordination, and substitution, we may now appropriately look at the product they combine to make—the crystal structure. This may be done from the point of view of formal crystallography or, in a more restricted way, by considering some of the more important "structure types." For present purposes the latter is appropriate.

All crystals in which the centers of the constituent atoms occupy geometrically similar positions, regardless of the size of the atoms or the absolute dimensions of the structure as a whole, are said to belong to the same *structure type*. This concept of structure type is particularly important in the study of sulfides and oxides, among which a wide variety of mineral species have been generated by the substitution of different cations and anions in what is really quite a small group of basic structures. When this is kept in mind, the systematics of the ore minerals become much simpler, and many of the relationships among them become much more readily understood.

One of the simpler structure types among the sulfides is that of galena, in which there are equal numbers of cations and anions grouped in six-fold coordination. Other minerals possessing this arrangement are halite (NaCl), alabandite (MnS), cerargyrite (AgCl), sylvite (KCl), and periclase (MgO). These constitute an *isostructural group*, in this case the *halite*, or *rock-salt*, group. Other minerals contain equal numbers of cations and anions in tetrahedral coordination, having— as already noted—the same structure as diamond. Chief among them is sphalerite —hence the *blende* group. Other important configurations are the cubic *pyrite* group, the hexagonal *nickel arsenide* group, and the *spinel*, *hematite*, and other oxide groups. In each case there is a set ratio of cations to anions arranged about each other in constant coordinations. While the interatomic spacings and the size of unit cells may vary from one member to another, the structural pattern of cations and anions remains unchanged throughout the group, constituting a structure type.

An interesting and extremely useful development of this principle has been put forward by Ross (1957). Many native metals, sulfides, sulfosalts, and oxides are close packed in either the cubic or hexagonal manner. (In some cases, such as sphalerite and wurtzite, both types of stacking may occur within a single crystal, usually in a quite well-defined "stacking sequence.") The larger sulfur atoms are commonly close packed and, due to their greater size and to the inherent stability of the sulfur portion of the lattice, dominate the mineral structure. The metal atoms usually occupy the interstices between the sulfur atoms in the coordinating role discussed previously. Virtually unlimited variation of the detailed structure may be induced by substitution among metals of similar size and chemical properties. While such substitutions in the cation lattice often change the symmetry of the structure as a whole, they usually do not affect the sulfur portion, which continues to dominate the structure with its close-packed arrangement. In addition further modifications result from small deviations in stoichiometry, where structures are capable of sustaining minor omissions and additions without changing to another structure type.

Ross pointed out that structure types representing variations of cubic close packing are antifluorite (M_2S), sphalerite (MS), galena (MS), spinel (M_3S_4), and pyrite (MS_2). Variations of hexagonal close packing are found in the type structures of niccolite (NiAs), wurtzite (ZnS), molybdenite (MoS_2), melonite ($NiTe_2$), chalcocite (Cu_2S), stibnite (Sb_2S_3), and proustite-pyrargyrite [$Ag_2S \cdot (As \cdot Sb)_2S_3$]. Most of the structures of the sulfide and sulfosalt minerals (and native metals and oxides) can be related to these principal structure types, even though some are layered or chain-like. On this basis there are about a dozen basic structures, from which derivatives may be formed by substitution, defects of complete and omission type, and thermal disorder.

Examples of *substitutional derivatives* cited by Ross (1957, p. 759) are

Simple (basic) structure	Complex (derivative) structure
Sphalerite	Chalcopyrite
Zns	$CuFeS_2$
Cubic	Tetragonal
$a = 5.40$ Å	$a = 5.2$ Å
$Z = 4$	$c = 10.30$ Å
	$Z = 4$
Galena	β Matildite
PbS	$AgBiS_2$
Cubic	Orthorhombic
$a = 5.93$ Å	$a = 8.14$ Å
$Z = 4$	$b = 7.97$ Å
	$c = 5.69$ Å
	$Z = 4$
Pyrite	Cobaltite
FeS_2	CoAsS
Cubic	Cubic
$a = 5.405$ Å	$a = 5.58$ Å
$Z = 4$	$Z = 4$

Defect structures may be either complete or incomplete. *Complete defect structures* involve the unsystematic sharing of crystallographically equivalent lattice sites by atoms of two or more elements. Apparently the ideal condition where only one kind of atom fully occupies one set of positions seldom occurs; at normal temperatures most members of mixed-crystal series have metal lattices that are random distributions of more than one atomic species over the equivalent lattice sites concerned. *Incomplete defect structures* involve metal atom omissions or additions. Well-known examples are pyrrhotite ($Fe_{1-x}S$) and pentlandite [$(Fe,Ni)_9S_8$]. Linnaeite (Co_3S_4–Co_2S_3) apparently has a spinel structure which is able to support the substantial cobalt atom omissions involved in deviating from the formula M_3S_4.

Rise in temperatures and resulting thermal disorder of metal portions of mineral lattices may lead to the transformation of chemically complex minerals from lower to higher crystal symmetries. For example, metal atoms M and M' in a complex sulfide MM'S may, by thermal motion, assume a statistical distribution within the stable sulfur lattice; M and M' no longer occupy crystallographically distinctive sites, the basic structure MS or (MM')S is attained, and a higher symmetry results. Examples of this kind of transformation given by Ross (1957, p. 761) are

Low-temperature–low-symmetry structure		High-temperature–high-symmetry structure
Stannite Cu_2FeSnS_4 Tetragonal $a = 5.46$ Å $c = 10.73$ Å	600°C	"High" stannite (sphalerite structure) $4(Cu_{0.50}Fe_{0.25}Sn_{0.25})S$ Cubic $a = 5.40$ Å
β Matildite $AgBiS_2$ Orthorhombic $a = 8.14$ Å $b = 7.87$ Å $c = 6.69$ Å	Temperature not known accurately	"High" α matildite (galena structure) $2(Ag,Bi)S$ Cubic $a = 5.6$ Å
Chalcopyrite $CuFeS_2$ Tetragonal $a = 5.24$ Å $c = 10.34$ Å	550°C	"High" chalcopyrite (sphalerite structure) $(Cu,Fe)S$ Cubic $a = 5.40$ Å

Ross' classification is given in slightly abbreviated form in Table 3-1. It is developed in order of increasing anion-cation ratio and is divided throughout into three groups on the basis of cubic close packing, hexagonal close packing, and "miscellaneous" structures. *Simple* structures are taken as those containing one metal component. *Complex* structures are substitutional derivatives patterned after the simple types but containing two or more metals in essentially equivalent positions. Intermediate members of solid-solution series are involved here. *Defect* structures indicate those of the incomplete kind: those deviating from ideal composition through excess or deficiency of total metal. *Deformed* structures are also patterned after the basic types but with minor modifications of bond length and angle. These may be chemically simple or complex.

While this approach to the subdivision of sulfides and related minerals requires further work before it can be regarded as firmly established, even in its present incomplete state it provides a very systematic view of the minerals concerned and a very much clearer picture of the interrelationships between the simple

Table 3-1

Type	"Cubic close packed"			"Hexagonal close packed"			"Miscellaneous"		
	Group	Subtype	Members	Group	Subtype	Members	Group	Subtype	Members
M_2X	Argentite group	Simple	Argentite Ag_2S, Naumannite Ag_2Se, Hessite Ag_2Te	Chalcocite group		Chalcocite Cu_2S, Rickardite Cu_2Te	Covellite group		Covellite CuS, Klockmannite $CuSe$
		Defect	Digenite $Cu_{2-x}S$, Petzite Ag_3AuTe_2, Eucairite $CuAgSe$						
		Complex derivative							
MX	Galena group	Simple	Galena PbS, Clausthalite $PbSe$, Cinnabar HgS, Matildite $AgBiS_2$	Niccolite group	Simple	Niccolite $NiAs$, Breithauptite $NiSb$			
		Deformed / Complex-deformed derivative			Defect	Pyrrhotite $Fe_{1-x}S$			
	Sphalerite group	Simple	Sphalerite ZnS, Hawleyite CdS, Chalcopyrite $CuFeS_2$, Stannite Cu_2FeSnS_4	Wurtzite group	Simple	Wurtzite ZnS, Greenockite CdS			
		Complex derivative / Complex deformed	Tetrahedrite $(Cu,Ag,Fe)_{12}Sb_4S_{13}$		Complex derivative	Cubanite $CuFe_2S_3$, Emplectite $CuBiS_2$, Chalcostibite $CuSbS_2$			
M_3X_4	"Spinel" group	Simple	Linnaeite Co_3S_4, Polydymite Ni_3S_4				Chain structures		Galenobismutite $PbS \cdot Bi_2S_3$, Jamesonite $4PbS \cdot FeS \cdot 3Sb_2S_3$, Livingstonite $HgS \cdot 2Sb_2S_3$
		Defect / Complex derivative	Carrollite Co_2CuS_4, Violarite Ni_2FeS_4, Pentlandite $(Fe,Ni)_9S_8$						
M_2X_3	Tetradymite group	Simple	Tetradymite Bi_2Te_2S, Telluro-bismutite Bi_2Te_3				Chain structures		Stibnite Sb_2S_3, Bismuthinite Bi_2S_3, Bournonite $2PbS \cdot Cu_2S \cdot Sb_2S_3$, Berthierite $FeS \cdot Sb_2S_3$
		Complex derivative	Josite $Bi_{4-x}(Se,Te,S)_{3-x}$						
MS_2	Pyrite group	Simple	Pyrite FeS_2, Cattierite CoS_2, Sperrylite $PtAs_2$	Molybdenite group-layered structures		Molybdenite MoS_2, Tungstenite WS_2	Marcasite group	Simple	Marcasite FeS_2, Loellingite $FeAs_2$, Rammelsbergite $NiAs_2$, Safflorite $CoAs_2$
		Complex derivative	Cobaltite $CoAsS$, Gersdorffite $NiAsS$, Penroseite $(Ni,Cu,Pb)Se_2$					Complex	Arsenopyrite $(Fe,Co)(As,Sb)S$, Glaucodot $(Co,Fe)AsS$, Gudmundite $FeSbS$

and more complex members of the group as a whole.   It also shows very clearly
the fundamental importance of structure type.

RECOMMENDED READING

Hurlbut, C. S. (ed.): "Dana's Manual of Mineralogy," 17th ed., John Wiley & Sons, Inc., New
    York, 1959.
Girifalco, L. A.: "Atomic Migration in Crystals," Blaisdell Publishing Company, New York,
    1964.
Ross, V.: Geochemistry, crystal structure and mineralogy of the sulfides, *Econ. Geol.*, vol. 52,
    pp. 755–774, 1957.

4
Ore Mineralogy:
Principal Mineral Groups

Ores are usually regarded as being composed of two categories of material: *ore minerals*, from which the metals are extracted, and *gangue minerals*, which form the useless matrix. When one looks upon ores as rocks this distinction disappears: both groups of minerals are rock-forming constituents, and as such, each has been involved in and has affected the history of the other.

Having stated this petrological approach, it is necessary for reasons of space to undergo an immediate apparent reversal of outlook; while ores might indeed be better regarded as a whole—not as "ore minerals," which are interesting, and "the rest," which are not—this chapter is restricted to the ore minerals and their immediate relatives. This, however, is merely a matter of bowing to necessity. Since, from one ore type to another, ore minerals are associated with a wide variety of silicates, carbonates, and others, consideration of ores as a whole would obviously require the writing of yet another text on "mineralogy." In the present instance all that is necessary is to keep in mind that the "other" minerals are just as much a part of an ore as the ore minerals themselves and to regard sources of information on silicate, carbonate, and other mineral groups as complementary with this chapter.

Since the sulfides and related minerals occur in all major rock types, they are, as a result, found associated with all the more important nonsulfide mineral assemblages. Sulfides occur in a wide variety of igneous rocks and hence may occur with assemblages ranging from those of peridotites to those of granite pegmatites and with many textural variants. There seems to be a preferential association with more basic, intrusive, rocks, though there are numerous ores in granites and some notable occurrences in lavas. Ore minerals are abundant in sedimentary rocks, particularly shales, carbonate accumulations, and certain gray-wackes. Commonly they occur preferentially in certain facies of such rocks, or if they span more than one facies, sulfide and oxide mineralogy is often related to the facies pattern. The relation between bedded ferric and feroso-ferric oxide iron, iron silicates, iron carbonate, and iron sulfide with sedimentary facies of iron accumulation is an outstanding example. Similarly in some areas it is noteworthy that lead and zinc sulfides tend to occur in shales and copper sulfides in associated carbonate rocks—though galena and sphalerite may occur in abundance in lime-stone and dolomitic limestone of other environments. As might have been expected, ores also occur in metamorphic rocks of all grades—from the almost unmetamorphosed tuffs and lavas of the Kuroko deposits of Japan to the garnet sillimanite gneisses of Broken Hill, Australia, and the Grenville Province of North America. While there are certainly no obvious relationships between ore mineral assemblages and type and grade of metamorphism, textures often show a distinct tie. Contact-metamorphic sulfides are commonly very coarse grained, just as are their nonsulfide associates. Similarly sulfides dispersed in regionally metamorphosed rocks are usually fine grained where metamorphism has been slight and coarse grained where it has been intense. And finally, since ores also occur in veins, they are found here in association with all the other minerals of cavity fillings—quartz, calcite, fluorite, gypsum, barite, and many others—commonly exhibiting the same beautiful development of crustification and crystal faces so characteristic of these other minerals in this type of occurrence.

Just as it is possible to bring the common rock-forming minerals together into a relatively small number of groups—the feldspars, pyroxenes, amphiboles, micas, rhombohedral carbonates, and so on—so it is a fairly simple matter to systematize and classify the ore minerals. Although at first sight they may appear extremely diverse, many of them are closely related, and many show serial changes quite as ordered as any of those of silicate groups. When this is recognized their variations appear much more systematic and their diversity a great deal easier to understand.

CLASSIFICATION OF THE ORE MINERALS

Minerals may be classified according to their chemical composition, their crystal structure, or a combination of the two. In the well-known classification of Dana, division is into 12 classes on a chemical basis:

1. Elements
2. Sulfides
3. Sulfosalts
4. Oxides and hydroxides
5. Halides
6. Carbonates
7. Nitrates
8. Borates
9. Phosphates
10. Sulfates
11. Tungstates
12. Silicates

Each class is divided into families on the basis of chemical similarities, and families in turn are divided into groups as indicated by structural similarities. Each group is made up of species, which may in turn—generally on fine chemical distinctions—be divided into series and varieties.

A recent classification by Berry and Mason (1959, p. 273) is a modification of an earlier system devised by the Swedish chemist Berzelius and recognizes eight classes:

1. Elements
2. Sulfides (including sulfosalts)
3. Oxides and hydroxides
4. Halides
5. Carbonates, nitrates, borates, iodates
6. Sulfates, chromates, molybdates, tungstates
7. Phosphates, arsenates, vanadates
8. Silicates

As pointed out by Berry and Mason, this classification, though originally based on chemical differences, has an underlying structural pattern. The native elements include metals, with their metallic bonding, and semimetals and non-metals, with covalent bonding. The sulfides are hybrid: some have a notable metallic component, many are substantially covalent, and a few are dominantly ionic. The oxides and remaining classes, including the silicates, are dominantly ionic.

In the present instance we are concerned essentially with the first three classes of Berry and Mason. Their three-fold divison clearly includes almost all the ore minerals and provides a very convenient primary division of them. In the following pages the "primary" ore minerals are therefore divided into native metals, sulfides and sulfosalts, and oxides. They are then further subdivided on the basis of structure groups, and within these, representative members are considered in some detail.

PRINCIPAL CLASSES AND GROUPS

CLASS 1. NATIVE METALS AND SEMIMETALS

Gold group This includes metals of group Ib of the periodic table: gold, silver, and copper, all of which occur native in small quantity in crustal rocks. All are face-centered cubic, i.e., exhibit cubic close packing, and all show the properties associated with metallic bonding. The cell-edge lengths of gold, silver, and copper are 4.0786 Å, 4.0862 Å, and 3.6150 Å, respectively. Densities are high: approximately 19.3, 10.5, and 8.95, respectively, with some variation being induced by impurities.

Gold and silver and gold and copper are mutually soluble, but silver and copper are not, or are so only to a minute degree. Gold and silver appear to form a complete solid-solution series, but copper appears to be soluble only up to about 20 per cent in gold. In addition gold can contain trace quantities of palladium, rhodium, and bismuth; silver has been found to contain traces of arsenic, antimony, and mercury; and native copper may contain arsenic, antimony, bismuth, iron, mercury, and germanium as traces.

All three commonly occur in arborescent form, silver often occurs as "wire silver," and all on rare occasions are found as crystals. The most frequent form of subsurface occurrence is as scales on vein walls or as vein fillings. Silver exhibits spectacular cruciform and arborescent habit in certain native silver:cobalt-nickel arsenide ores, typified by those of the Cobalt area in Ontario.

Microscopically the outstanding feature of the group is their high reflectivity, causing them to render any adjacent sulfides quite dull. Silver has the highest reflectivity of any ore mineral—about 95—copper has a reflectivity of up to 81.2, and gold has one of 74.0. These may be compared with the mean reflectivity of pyrite, which is 54.5. In ordinary light each shows its expectable gold, silver, or copper color, though this is normally quickly modified by tarnish in air in the case of silver and copper. Silver turns pink and then brown, copper a brownish color. Silver is reported to be distinctly anisotropic when carrying substantial antimony, but all are normally isotropic. All tend to show scratches and the other polishing effects of soft minerals unless the greatest care is taken in the preparation of surfaces.

Platinum group The metals platinum and palladium and the natural alloy of iridium and osmium—osmiridium—are the principal members of this group.

Platinum and palladium are both cubic close-packed structures, with cell-edge lengths of 3.9158 Å and 3.8824 Å, respectively. The members of the osmiridium series are hexagonal close packed, with $a_0 = 2.714$ Å, $c_0 = 4.314$ Å, and $a_0/c_0 = 1:1.584$, these values varying slightly with variation in the proportions of the two metals in the alloy. Approximate compositional limits are Os/Ir = 23:77 to 80:20.

Each of the three may be alloyed with each other; in spite of their structural differences, iridium, for example, is known to occur in platinum at least up to 7.5 per cent. The most common impurity is Fe, though Au, Cu, Rh, Ru, and others may also occur in small quantities. The densities of pure Pt and Pd are 21.46 and 12.04, respectively, but owing to ubiquitous impurities the densities are always less than this in nature.

Under the microscope the three show the usual high reflectivity of native metals, though being harder than those of the gold group they take a fine polish. Pt and Pd are isotropic, but osmiridium, with its hexagonal structure, shows pleochroism and optical anisotropism.

These metals have a far more restricted occurrence than those of the gold group and are characteristically associated with ultrabasic (particularly) and basic igneous rocks.

Arsenic group The semimetals arsenic, antimony, and bismuth are far from uncommon in the native state. They are all hexagonal (trigonal) and of the same structure type. The group includes the natural "alloy" AsSb, allemontite. In

Table 4-1   Principal physical properties† of the arsenic group

Name	Formula	Crystal system	Cell dimensions, Å	Density	Hardness	Reflectivity
Arsenic	As	Trigonal	$a = 3.760$ $c = 10.548$	5.7	57–137	48–51
Antimony	Sb	Trigonal	$a = 4.307$ $c = 11.273$	6.7	45–101	47–58
Bismuth	Bi	Trigonal	$a = 4.546$ $c = 11.860$	9.7–9.8	9–19	67–68

† In Tables 4-1 to 4-14 the currently known ranges of density, hardness, and reflectivity are given wherever possible; where single figures are given, these are the only ones available.

contrast to the simple metal structures of the gold and platinum groups, the arsenic group shows clear anisotropism and hybridism in its bonding. Each atom is surrounded by six nearest neighbors but is more closely bonded to three of these than to the others. This yields double sheets of closely bonded atoms, each separated from the next double sheet through the longer bonds. The longer bonds are essentially metallic, but the shorter bonds are largely covalent in arsenic, becoming more metallic in antimony and then bismuth, leading to a progressive increase in overall metallic character. The short bonds are parallel to the basal plane, yielding the basal cleavage. The principal physical properties illustrate the serial relationship among these minerals and the increasingly metallic character developed in going from arsenic to antimony and bismuth.

Under the microscope they all have a rather metallic appearance, with relatively high reflectivities and the frequent poor surfaces characteristic of soft metals. All are white but tarnish quickly, though when familiar to the microscopist this tarnish may be quite useful in identification. Not infrequently the basal cleavage can be seen in polished section, particularly where this lies at a low angle to the surface of the section and has allowed the removal of portions of sheets during polishing. All are strongly doubly refracting, and antimony and bismuth show twinning in this way.

Native bismuth occurs associated with bismuthinite (Bi_2S_3) as a minor constituent of many base metal ores, and all are known as major constituents of veins in one place or another. In the veins they may form quite large, essentially pure, masses that on occasions are of several tons in weight. They are frequently associated with native silver:nickel-cobalt mineralization.

CLASS 2.   SULFIDES AND SULFOSALTS

These constitute a very large group, many members of which—as with numerous silicates—are quite rare and of little petrological significance. Consideration of a wide range of these would therefore be inappropriate in the present volume. We shall therefore concern ourselves with only the major groups, illustrating them by some of their more important members. Division of the groups follows essentially the lines suggested by Ross: primary division is substantially according to cation-anion ratios, with further subdivision based on structure type and its modifications. Type formulas are generalized, with M, M′ denoting cations and X the anions.

M$_2$X type: the argentite (antifluorite) group   This is quite a large group that Ross suggests may be regarded as being dominated by a cubic close-packed arrangement of antifluorite type.†   This, of course, refers only to the basic anionic structure; as a result of substitution and the development of defects, the symmetry —and hence the crystallographic nature—of many of its members is not, or is very doubtfully, cubic.   Silver and copper are the characteristic cations of the group, with Au, Fe, As, and Sb in minor roles and selenium, tellurium, and sulfur all prominent anions.

The more important members of the group are argentite (Ag$_2$S) itself, naumannite (Ag$_2$Se), hessite (Ag$_2$Te), digenite (Cu$_{2-x}$S), eucairite (CuAgSe), petzite (Ag$_3$AuTe$_2$), and aguilarite (Ag$_4$SSe).   Several undergo inversion at moderately elevated temperatures.   Argentite, for example, is cubic only above 179°C.   Below this it is the orthorhombic dimorph acanthite, and in fact Ag$_2$S showing external cubic form at ordinary temperatures is a pseudomorph having an internal structure of acanthite.   Naumannite and hessite also invert from cubic to less symmetrical forms on cooling below 133°C and 149.5°C, respectively.

Although most of the minerals of the group are constitutionally closely related as sulfides, tellurides, or selenides of copper, silver, and gold, it is noteworthy that there is little solid solution between the species; for example, eucairite is known to preserve quite constant Cu/Ag ratios, and the same is the case with the possible complex derivative α stromeyerite (CuAgS).   It is suggested by Ross that bornite (Cu$_5$FeS$_4$–Cu$_5$FeS$_6$) may be regarded as a complex-defect derivative member of the argentite group, though this mineral is generally regarded as a member of the MX type.

Microscopically the two outstanding features of the argentite group of minerals are softness and low to moderate reflectivity.   Argentite is the softest of all known naturally occurring sulfides (Vickers hardness = 20 to 91), and the various selenides and tellurides—and digenite—are very similar in this respect.   Reflectivities range from 22.0 (digenite) to about 45.0 (petzite), and colors are generally shades of gray, in some cases with a brownish tint.   Apart from hessite and eucairite, reflection pleochroism and double refraction are indistinct or absent.   Some members (e.g., hessite) may show inversion twinning.

Minerals of this group occur in a variety of settings.   Argentite is a normal constituent of argentiferous lead-zinc sulfide ores.   It also occurs in many native silver-silver sulfosalt veins and in various copper sulfide vein ores.   It is almost

---

† This configuration is identical, overall, with the fluorite (CaF$_2$) structure but differs from it in having the coordinating positions of cation and anion in reverse relationship.   In the fluorite structure each metal atom coordinates eight anions, and each anion coordinates four cations (see Fig. 4-6).   In the antifluorite structure the same geometry holds, but coordinating roles are reversed, yielding an M$_2$X compound in which each anion coordinates eight cations and each cation coordinates four anions.

Table 4-2 Chemical compositions and principal physical properties of minerals of the argentite group

Name	Formula	Crystal system	Density	Hardness	Reflectivity
Argentite	Ag_2S	Cubic 179°C Orthorhombic	7.0–7.4	20–91	29.0–36.0
Naumannite	Ag_2Se	Cubic 133°C Orthorhombic?	7.0	115–185	31.0–34.2
Hessite	Ag_2Te	Cubic 149.5°C Monoclinic	8.2–8.5	24–41	37.2–39.6
Digenite	$Cu_{2-x}S$	Cubic	5.546 ($x = 0.24$) 5.706 ($x = 0.16$)	56–95	22.0–25.0
Eucairite	$CuAgSe$	?Cubic at high temperatures	7.6–7.8	23–94	49.9
Petzite	Ag_3AuTe_2	?Cubic at high temperatures	8.7–9.0	43–74	45
Aguilarite	Ag_4SSe	Cubic at high temperatures	7.6	Soft	29.9

certainly the most important primary source of silver. Digenite is a frequent close associate of chalcocite and hence is more often secondary than primary. Naumannite, hessite, eucairite, petzite, and aguilarite usually occur as constituents of precious metal telluride-selenide veins, and are sources of gold and silver. Hessite has been found as a trace constituent of some of the silver-rich portions of the Sudbury nickel ore bodies.

M_2X type: the chalcocite group The only widespread and quantitatively important member of this group is chalcocite (Cu_2S) itself. Stromeyerite and α rickardite (Cu_2Te)† are possible members. As with argentite, it is only the high-temperature form of chalcocite that shows the "basic" symmetry (in this case hexagonal). The low-temperature form, β chalcocite, is orthorhombic, and this is stable to approximately 105°C.

Microscopically all three minerals are outstanding for their beautiful colors, particularly rickardite. In plain light chalcocite is the least spectacular, with a simple bluish white color. Stromeyerite, however, is a delicate pinkish gray and rickardite a bright purple or violet. All are pleochroic. Between crossed nicols chalcocite shows very delicate pink to light emerald green, stromeyerite is brilliant violet to brown, and rickardite shows brilliant orange, yellow, and blue colors. All have rather low mean reflectivities and all are quite soft; chalcocite appears to be the hardest of the group, with a mean Vickers hardness of 74.

Chalcocite is of wide occurrence but is of greatest importance as a secondary mineral. Surface waters charged with ferric sulfate and sulfuric acid derived from the weathering of iron sulfides of an ore may attack copper minerals such as chalcopyrite and bornite, producing $CuSO_4$ solutions. These solutions percolate

† Given as Cu_4Te_3 by Dana (1944, p. 198).

Table 4-3   Chemical compositions and principal physical properties of minerals of the chalcocite group

Name	Formula	Crystal system	Cell dimensions, Å	Density	Hardness	Reflectivity
Chalcocite	Cu_2S	Orthorhombic	$a = 11.82$ $b = 27.05$ $c = 13.43$	5.5–5.8	58–98	18.0–33.5
Stromeyerite	$CuAgS$	Orthorhombic	$a = 4.06$ $b = 6.66$ $c = 7.99$	6.2–6.3	27–62	25.5–32.3
Rickardite	Cu_2Te	Tetragonal	$a = 3.97$ $b = 6.11$	7.5	?	?

deeper into the ore body and further attack primary material, producing chalcocite and the commonly associated secondary sulfide covellite (CuS).   Both, however, occur in places as primary constituents—notably at Butte, Montana, and Kennecott, Alaska.

Rickardite usually occurs as a minor constituent of gold-silver telluride veins, and stromeyerite is a somewhat uncommon constituent of hydrothermal base metal sulfide veins.

MX type: the galena group   The MX category (or group of categories) is unquestionably the most important type among the sulfides, including as it does most of the common sulfides apart from pyrite.   Galena, sphalerite, chalcopyrite, and pyrrhotite are, of course, notable members, and the type includes numerous other sulfides and sulfosalts of frequent and widespread occurrence.

The galena group is somewhat heterogeneous from the chemical point of view, though it might be said to be dominated by Ag, Pb, Sb, and Bi among the cations. There are also, however, manganese, tin, and mercury sulfides included in the group, and although sulfur is the most important anion, selenium and tellurium also appear.

The principal isostructural members are galena itself, clausthalite (PbSe),

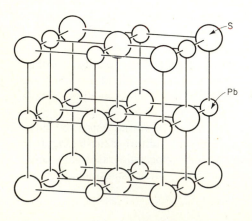

Fig. 4-1   The galena (or "rock-salt") structure.

Table 4-4   Chemical compositions and principal physical properties of minerals of the galena group

Name	Formula	Crystal system	Cell dimensions, Å	Density	Hardness	Reflectivity
Galena	PbS	Cubic	$a = 5.936$	7.58	56–116	42.4–43.2
Clausthalite	PbSe	Cubic	$a = 6.162$	7.8	43–63	43.1–50.4
Altaite	PbTe	Cubic	$a = 6.439$	8.15	39–60	63.2–65.5
Alabandite	MnS	Cubic	$a = 5.214$	4.0	150–266	23.4–23.9
Oldhamite	CaS	Cubic	$a = 5.686$	2.58	?	Translucent

altaite (PbTe), and alabandite (α MnS).   Oldhamite (CaS) is an interesting rarity. The atomic arrangement is the same as that of halite (Fig. 4-1), though PbS, PbSe, and PbTe possess semimetallic bonds instead of the ionic bonds of halite.   This is immediately indicated by the metallic luster of the minerals.   Alabandite presumably has a rather lesser metallic component in its bonding; its luster is submetallic and it has a low reflectivity.   In all members the anions are arranged in six-fold coordination about the cations and vice versa, as described in Chap. 3.

Probably the only physical characteristic common to all these minerals (other than packing arrangement) is the excellent cleavage parallel to the {001}—i.e., parallel to the planes of greatest spacing in the structure and perpendicular to the M—X bonds.   It is almost impossible to break galena other than along such a cleavage plane.

Under the microscope the lead members are characteristically white; alabandite, with its substantially ionic bonding, is dull gray.   Reflectivities and hardnesses also illustrate the bond differences between the lead and manganese members. Each of the minerals is of quite constant chemical composition.   Galena has been reported to contain impurities of Ag, Sb, As, and Bi, though it now seems well established that these occur largely as discrete, if extremely fine, particles of minerals such as the tetrahedrite-tennantite series, matildite, and so on.   Selenium and tellurium may possibly occur through very limited solid solution.   Cu, Zn, Cd, and Fe have also been reported, though it seems very likely that these are to be attributed to minute inclusions of chalcopyrite and sphalerite.   The small amounts of Hg, Ag, Au, Cu, Co, and Fe found in altaite and clausthalite are also thought to be present in discrete minerals rather than as foreign components of the lattice.

Galena is known to occur in greater or lesser amount—even if only as a trace —in almost all base metal ores.   Altaite and clausthalite are the "lead members" of many gold and silver telluride and selenide deposits.   Alabandite is an unusual constituent of some sulfide deposits containing other manganese minerals such as rhodonite and rhodochrosite.

A number of other minerals, including cinnabar and several sulfosalts, have structural affiliations with this group and are postulated by Ross (1957, p. 765) to be deformed and complex-deformed derivatives of it.

MX type: the sphalerite group   This is a most important group and is characterized by four-fold coordination in which the tetrahedra, joined through their

Table 4-5 Chemical compositions and principal physical properties of minerals of the sphalerite group

Name	Formula	Crystal system	Cell dimensions, Å	Density	Hardness	Reflectivity
"Simple" members						
Sphalerite	ZnS	Cubic	$a = 5.406$	4.096	128–276	16.1–18.8
Hawleyite	CdS	Cubic	$a = 5.818$	4.87	?	?
Metacinnabar	HgS	Cubic	$a = 5.854$	7.65	73–86	26.8
Tiemannite	$HgSe$	Cubic	$a = 6.069$	8.30–8.47	26–39	25.5–29.2
Coloradoite	$HgTe$	Cubic	$a = 6.444$	8.04	23–28	36.2–37.7
"Derivative" members						
Chalcopyrite	$CuFeS_2$	Tetragonal	$a = 5.28$ $c = 10.41$	4.1–4.3	174–245	42.5–44.0
Stannite	Cu_2FeSnS_4	Tetragonal	$a = 5.46$ $c = 10.725$	4.3–4.5	171–307	27.1–28.0
Tetrahedrite-tennantite	$(Cu,Ag,Fe)_{12}Sb_4S_{13}$-$(Cu,Ag,Fe)_{12}As_4S_{13}$	Cubic	$a = 10.21$	4.6–5.1†	291–464	28.8–31.2†
Famatinite	Cu_3SbS_4	Tetragonal	$a = 5.38$ $c = 10.76$	4.50–4.65	315–397	25.1–28.7
Enargite	Cu_3AsS_4	Orthorhombic	$a = 6.41$ $b = 7.42$ $c = 6.15$	4.45	133–358	24.7–28.1

† Increases with increase in antimony and silver content.

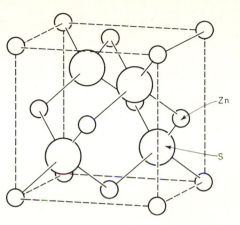

Fig. 4-2 The sphalerite structure (also known as the "diamond" or "blende" structure).

apices, are rotated through 60° with respect to each other. Each metal atom is surrounded by four anions and vice versa. The resulting structure, while very different from that of the galena group, has an overall cubic face-centered arrangement (Fig. 4-2). As already noted, minerals of this group are isostructural with diamond, silicon carbide, and one of the ice structures.

While this group is cubic by definition, some members—notably silicon carbide and zinc sulfide—have a pronounced capacity for changing their stacking pattern from cubic to hexagonal close packed and vice versa. As noted earlier the octahedral plane of the cubic structure and the basal plane of the hexagonal are common to each: the cubic structure is generated by the stacking sequence *ABCABC* . . . normal to the octahedral plane and the hexagonal structure by the sequence *ABABABA* . . . normal to what then becomes the basal plane. Very detailed work (Ramsdell, 1947) on synthetic SiC has shown the development of such *stacking faults* to be a common phenomenon and to yield a variety of extremely regular, but often complex, "stacking sequences." The general behavior of zinc sulfide seems very similar (Smith, 1955), and a number of stacking sequences have been determined in it. These yield a number of *polytypes*, each of which is characterized by a set proportion of sphalerite (cubic packing) and wurtzite (hexagonal packing). It must therefore be kept in mind that although the sphalerite group is classified as cubic, certain of its members may in fact be substantially hybrid.

The "simple" members are sphalerite (β ZnS), hawleyite (α CdS), metacinnabar [HgS $\rightarrow$ Hg(S_8Se_2)], the related mercuric selenide tiemannite (HgSe), and the mercuric telluride coloradoite (HgTe). There is some substitution between Zn and Cd in the sulfides and between S and Se in the mercury compounds. Iron is commonly a conspicuous impurity in sphalerite, and lesser quantities of Mn are common. The solubility of Fe in ZnS increases with temperature up to about 26 atom per cent and has been the subject of much experimental investigation in view of its promise as a geothermometer (see Chap. 5).

The mercury compounds all have a metallic luster, indicating a substantial metallic component in their bonding. Sphalerite, though, has almost a resinous luster, particularly in the low-iron varieties, and is known to be largely covalent (as, of course, are diamond and silicon carbide). Sphalerite has an excellent cleavage not present in the other compounds.

Under the microscope these minerals are all rather soft and of low reflectivity. Sphalerite has the lowest reflectivity (it is the dullest of the sulfides in polished section) but is rather harder than the mercury members, presumably reflecting the difference in bond type. Sphalerite also commonly exhibits internal reflections that change from white to pale amber in low-iron-manganese varieties through various shades of red to deep ruby as iron and/or manganese increase. At about 10 per cent Fe the mineral becomes opaque, and internal reflections can no longer be observed. Even lesser quantities of Mn seem to have the same effect. Twinning is conspicuous in sphalerite and may be revealed by etching. It is parallel to {111} and therefore related to the stacking phenomena already noted. Such twinning has been regarded generally as secondary and due to pressure—i.e., deformation twinning (Edwards, 1954, p. 36)—but recent observation indicates that this is certainly not always the case. Synthetic sphalerite grown in open space shows apparently similar twins. Recent work (Stanton and Gorman, 1968) indicates that some of them are of annealing type and that deformation twins, while of quite common occurrence, are of much lesser importance. Deformation is sometimes revealed by the bending of twins.

The derivative members of the sphalerite group include a number of important species, particularly if, as Ross suggests, the tetrahedrite-tennantite series $[(CuAgFe)_{12}Sb_4S_{13}–(CuAgFe)_{12}As_4S_{13}]$ can be included as "complex-defect" derivatives.

Chalcopyrite is clearly the outstanding derivative and provides a good illustration of the development of a substitutional member. Its packing arrangement is identical with that of sphalerite, but instead of the latter's simple rows of zinc atoms, the cation positions are occupied alternately by Cu and Fe atoms (Fig. 4-3). The result is that the unit cell of chalcopyrite corresponds to two unit cells of sphalerite stacked one on top of the other. Thus the cell-edge length of sphalerite is $a = 5.406$ Å and for chalcopyrite is $a = 5.28$ Å; $c = 10.41$ Å. Hence although the symmetry of chalcopyrite is tetragonal, the basic packing arrangement is still face-centered cubic. Similarly with stannite (Cu_2FeSnS_4) the zinc positions of the sphalerite structure are occupied by Cu, Fe, and Sn in the ratio 2:1:1, again yielding tetragonal symmetry—in this case with cell-edge lengths $a = 5.46$ Å; $c = 10.73$ Å. Sphalerite, chalcopyrite, and stannite, which are so unlike in composition, thus fall into a very simple and readily visualized group when viewed as a structural family. The famatinite-luzonite series ($Cu_3SbS_4–Cu_3AsS_4$) and a number of more complex copper-bearing compounds including tetrahedrite-tennantite also appear to possess the sphalerite type of packing.

Microscopically, chalcopyrite is distinguished by its yellow color. The others are gray or pinkish gray. Like sphalerite, chalcopyrite shows twinning on {111}, though with lesser frequency. Again this appears to be of two distinct types: sharp, evenly spaced twins apparently due to annealing and less regular, often curved and spindle-shaped twin lamellae that almost certainly result from deformation.

Next to the iron sulfides, sphalerite is the most abundant sulfide in the earth's crust. It occurs in all types of sulfide concentration and as disseminations in many

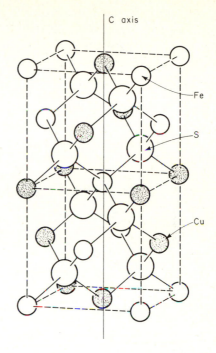

C axis

Fe

S

Cu

Fig. 4-3 The chalcopyrite structure; derivation from the sphalerite structure is clear and is emphasized by the central plane (shown in finely dashed lines) which divides the chalcopyrite unit cell into two equivalents of the sphalerite unit cell.

igneous, sedimentary, and metamorphic rocks. The mercury members are comparative rarities and normally appear in epithermal veins with selenides and tellurides of Au, Ag, Pb, and other metals. The derivative types are widespread. Chalcopyrite is found in all kinds of sulfide deposit, apparently representing the widest possible range of conditions for sulfide formation. However, although it is almost always present in galena-sphalerite concentrations, there is a strong tendency for segregation of Cu from Pb–Zn (see Chap. 15). Tetrahedrite-tennantite are rarely present in large quantity but are almost ubiquitous minor constituents of base metal sulfide deposits and are often associated with small amounts of arsenopyrite (FeAsS) in lead-zinc ores. Stannite is a rare mineral, and where it occurs it is generally intimately associated with sphalerite; this is the case in such widely contrasted occurrences as the copper-nickel ores of Sudbury, Ontario, and the complex hydrothermal vein ores of Tingha and Emmaville in New South Wales. In the latter ores, chalcopyrite, tetrahedrite-tennantite, and stannite all occur as complex exsolution intergrowths in sphalerite, reflecting the mutual compatibility of the members of this structure type. Tetrahedrite-tennantite, famatinite-luzonite, and enargite (Cu_3AsS_4), a derivative of the wurtzite group, are all common associates in copper ores containing high antimony and arsenic.

MX type: the wurtzite group The simple members of this group are wurtzite (α ZnS), which is the high-temperature form of zinc sulfide, and greenockite (β CdS), which is the low-temperature form of cadmium sulfide. As in the sphalerite group, M and S atoms are in four-fold coordination with respect to each other, forming

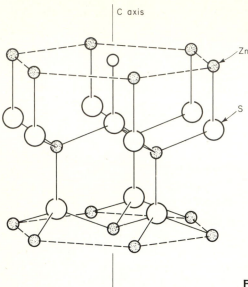

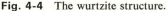

Fig. 4-4   The wurtzite structure.

tetrahedra that are linked through their apices.   Here, however, they are not rotated with respect to their neighbors, and the result is hexagonal, instead of cubic,
close packing (Fig. 4-4).

Several other species have related structures and are possible complex derivatives.   These are enargite, cubanite ($CuFe_2S_3$), emplectite ($CuBiS_2$), chalcostibite
($CuSbS_2$), sternbergite ($AgFeS_2$), and perhaps stannite.

Wurtzite and greenockite are very similar to sphalerite (and hawleyite) in their
grosser properties.   Both are of adamantine to resinous, rather than metallic,
luster and have a large covalent component in their bonding.   Wurtzite forms by
the inversion of sphalerite at approximately 1020°C and may be preserved metastably by rapid quenching.   The inversion point is lowered by the presence of Fe^{++}
in the crystal and is approximately 880°C for Fe^{++} = 17 per cent.   It may also be
precipitated in acid solutions above 250°C and appears to form at room temperature
from chloride solutions if small quantities of cadmium are present (unpublished
experiments by Stanton and Richards).   Presumably in this case the wurtzite is
nucleated epitaxially on slightly earlier formed particles of greenockite, and it is
possible that some naturally occurring low-temperature wurtzite has grown in this
way.

Microscopically both are, again, very similar to sphalerite, and the two are
very difficult to distinguish in reflected light.   Both are dark gray with correspondingly low reflectivity (sphalerite, 17.5; wurtzite, 17.4; greenockite, 19.0), and both
show internal reflections.   Reflection pleochroism and double refraction are not
perceptible in polished surfaces.   Wurtzite has not been reported to show twinning,
and this has been proposed as a method for distinguishing it from sphalerite.   However, since each of these minerals may exhibit a substantial amount of the stacking

of the other, such a distinction cannot be entirely valid. Wurtzite may show sheaf-like and radiating aggregates, and concentric zoning is more common here than in sphalerite.

A common mode of occurrence of wurtzite is as a constituent of *schalenblende*, an intimate intergrowth of wurtzite and sphalerite. Somewhat surprisingly, wurtz-ite commonly occurs—with marcasite—in ores that appear to have formed at relatively low temperatures, e.g., the limestone–lead-zinc ores of Joplin, Missouri. Greenockite is not an abundant mineral. Where it occurs it is generally in minor amount in association with the zinc sulfides.

The derivative group is somewhat heterogeneous and includes several sulfo-salts. All are distinctly more metallic than the simple members, and the dominant cation is clearly copper. Their more metallic nature gives a higher reflectivity, and in addition all are colored and clearly optically anisotropic. Although they are light colored, all have a distinct pinkish to pinkish-brown quality. All show quite distinct reflection pleochroism, and interference colors are generally strong and quite striking.

In general these minerals occur in vein ores—one associates them with the richer, more exotic, copper-silver veins of hydrothermal provinces rather than with, for example, stratiform types. However, enargite not infrequently appears as tiny specks here and there in massive base metal sulfide ores. Cubanite is the most common of the group and contributes a substantial part of the copper of some deposits. Outstanding examples are the Sudbury copper-nickel ores, in parts of

Table 4-6 Chemical compositions and principal physical properties of minerals of the wurtzite group

Name	Formula	Crystal system	Cell dimensions, Å	Density	Hardness	Reflectivity
"Simple" members						
Wurtzite	ZnS	Hexagonal	$a = 3.820$ $c = 6.260$	4.089	146–274	17.4–18.2
Greenockite	CdS	Hexagonal	$a = 4.142$ $c = 6.724$	4.9	52–91	19.0
"Derivative" members						
Cubanite	$CuFe_2S_3$	Orthorhombic	$a = 6.43$ $b = 11.04$ $c = 6.19$	4.03–4.18	150–260	39.2–42.5
Emplectite	$CuBiS_2$	Orthorhombic	$a = 6.13$ $b = 14.51$ $c = 3.89$	6.38	158–249	36.0–41.5
Chalcostibite	$CuSbS_2$	Orthorhombic	$a = 6.01$ $b = 14.46$ $c = 3.78$	4.95	193–285	37.1–43.0
Sternbergite	$AgFe_2S_3$	Orthorhombic	$a = 6.61$ $b = 11.64$ $c = 12.67$	4.101–4.215	40–74	32.0–40.0

which cubanite exceeds chalcopyrite in abundance and yields large quantities of copper.

MX type: the niccolite group   Members of this group are characterized by the *nickel arsenide* structure, a hexagonal close-packed arrangement in which cations and anions are in six-fold coordination with respect to each other.   They include niccolite (NiAs) itself, breithauptite (NiSb), and as a defect structure the very important mineral pyrrhotite ($Fe_{1-x}S$).   Although naturally occurring NiS, millerite, has quite a different structure, synthetic hexagonal NiS has a nickel arsenide structure.

In spite of their close structural similarity none of the three has been found to contain notable quantities of the relevant elements of the others.   Small amounts of Ni and Co have been reported in pyrrhotite, but careful microscopic examination usually shows that the nickel is present as very fine pentlandite and that the cobalt is within these pentlandite inclusions rather than in the pyrrhotite itself.   Niccolite sometimes contains some Sb, though this may be due to fine breithauptite inclusions. Small quantities of S, Fe, and Co in niccolite are also probably due to discrete inclusions rather than to lattice-bound impurities.

Distinct color with a red quality is a characteristic of the group.   Pyrrhotite is a distinct brownish pink, niccolite a light copper color, and breithauptite a red with a mauve tint.   All have a good metallic luster.   Pyrrhotite is variably magnetic; the others are nonmagnetic.

Microscopically the three minerals are all characterized by strong natural color, readily discerned reflection pleochroism, very strong optical anisotropism, and striking interference colors.   Natural colors are essentially those observed in hand specimen.   Pleochroism is strong enough to show up grain structure clearly. Interference colors are characteristic.   Pyrrhotite ranges from brown to greenish gray, niccolite from bright tan to brilliant apple green, and breithauptite from bluish green to violet-red.   Because of these pronounced optical features, the minerals of this group are among the easiest of all to identify in polished section.

Of the three, pyrrhotite is by far the most abundant, and after pyrite it is almost certainly the most abundant sulfide in the crust.   It occurs by itself or with other sulfides in many igneous, sedimentary, and metamorphic rocks and in a very wide variety of base and precious metal ore deposits.   It occurs in very large quantities in the ores of Sudbury and Sullivan in Canada, Mount Isa in Australia, and many others.   In spite of this very wide association it is notable, however, that among ores it has a clearly perceptible preference for those of copper, particularly where chalcopyrite and/or cubanite are prominent.   While there is no question that pyrrhotite does occur with lead and zinc sulfides (cf. Mount Isa and Sullivan), it conspicuously *tends to be* the iron sulfide of copper ores, while pyrite *tends to be* the iron sulfide of lead-zinc ores.

Niccolite may occur with nickel ores of noritic association as at Sudbury, but its principal habitat seems to be hydrothermal veins, particularly those carrying the native silver:Co–Ni arsenide assemblage typified by that of Cobalt, Ontario (see Chap. 17).   In massive nickel sulfide ores the niccolite usually occurs in simple

Table 4-7 Chemical compositions and principal physical properties of minerals of the niccolite group

Name	Formula	Crystal system	Cell dimensions, Å	Density	Hardness	Reflectivity
Niccolite	NiAs	Hexagonal	$a = 3.609$ $c = 5.019$	7.78	308–533	52.0–58.3
Breithauptite	NiSb	Hexagonal	$a = 3.93$ $c = 5.13$	8.23	527–644	45.3–54.6
Pyrrhotite	$Fe_{1-x}S$	Hexagonal	$a = 3.452$ $c = 5.762$	4.58–4.65†	212–363	34.0–45.6

† Variation due to variation in Fe/S ratio. Note also that the crystal structure of pyrrhotite is subject to some variation; cf. Chap. 5.

granular form, associated with such minerals as chalcopyrite, cubanite, gersdorffite ($NiAsS$), maucherite (Ni_3As_2), and perhaps, sperrylite ($PtAs_2$). In the hydro-thermal vein ores it may occur as massive bodies or as reniform, spheroidal, colum-nar, reticulated, or arborescent masses associated with a wide range of Co, Ni, and Fe arsenides and related species. This is also the principal association of breit-hauptite, which often occurs as veinlets within or closely associated with the niccolite.

M_3X_4 (spinel) type This is largely composed of cobalt and nickel sulfides, the more complex members containing some Fe, Cr, Cu, and perhaps, As, Bi, Pb, and Se. Although there are a number of possible derivatives, the more important members are those conforming to spinel structure and symmetry known as the linnaeite series. These are Co, Ni, Fe, and Cu sulfides, which may be written $M_2M'S_4$.

The series is characterized by extensive substitution, and properties such as hardness and specific gravity vary accordingly. Iron and copper are reported in most analyses but do not exceed one-third of the total metal atoms. A generalized formula for the series may be given as $(Co,Ni)_2(Co,Ni,Fe,Cu)S_4$. The identity of the mineral carrollite as described from Carroll County, Maryland, has been ques-tioned and attributed to a mixture of linnaeite and chalcocite, chalcopyrite, and bornite. However, the existence of a Cu member appears to be established in principle.

Table 4-8 Chemical compositions and principal physical properties of minerals of the (sulfide) spinel group

Name	Formula	Cell dimensions, Å	Density	Hardness	Reflectivity
Linnaeite	Co_3S_4	$a = 9.398$	4.85	450–613	46.4–47.5
Siegenite	$(Co,Ni)_3S_4$	$a = 9.41$	4.83	471–593	47.3–49.8
Carrollite	Co_2CuS_4	$a = 9.458$	4.83	351–566	44.0–44.9
Violarite	Ni_2FeS_4	$a = 9.40$	4.79	241–373	42.1
Polydymite	Ni_3S_4	$a = 9.405$	4.83	362–449	50.4

Microscopically the trio linnaeite-siegenite-polydymite are very similar.   All take an excellent polish.   Linnaeite is pale cream, polydymite is a pale but definite yellow, and siegenite is between the two.   All are isotropic.   Violarite is white with a pale, though distinct, violet tint.   It is rather softer than the others and may show a weak optical anisotropism.

An important sulfide usually included in the MX type but that Ross (1957, p. 765) suggests may be appropriately included in the spinel group is pentlandite $(Fe,Ni)_9S_8$.   In this case it is taken as a complex derivative characterized by an excess of metal over sulfur.   Like the linnaeite series, pentlandite belongs to the isometric hexoctahedral $4m/\bar{3}2/m$ crystal class, but it belongs to the space group $Fm3m$ as compared with $Fd3m$ for the linnaeite minerals.   Pentlandite is creamy yellow, and its other physical features are rather similar to those of linnaeite.   Its reflectivity is 52, it is isotropic, and its specific gravity is 4.6 to 5.0.   However, it is rather softer (215) than the linnaeite minerals and generally shows very well-developed octahedral parting planes, which the linnaeite minerals do not.   This parting may be so well developed as to make some specimens quite friable, with resulting difficulty in polishing.   Such parting planes are useful in identifying pentlandite microscopically and may provide pathways for replacement by later minerals (e.g., Hawley and Stanton, 1962, p. 139, fig. 63).

Minerals of the linnaeite series are comparatively rare and occur mainly in hydrothermal veins, particularly those carrying Co–Ni–Fe arsenide and sulfarsenide assemblages.   Pentlandite is the world's principal source of sulfide nickel but is quite restricted in its occurrence nonetheless.   It occurs as a trace constituent of ultrabasic rocks and serpentinites, but the major known part of it occurs associated with a very few basic intrusions.   Most of these are in Canada, the most famous being the group of deposits round the Sudbury intrusion.   There is also a well-known deposit at Petsamo, Finland, and minor amounts occur in the Bushveld igneous complex of the Transvaal.   It has also been found in quite a large group of deposits near Kambalda, Western Australia.   Pyrrhotite is a ubiquitous and generally abundant associate of pentlandite, and there are usually variable amounts of chalcopyrite and minor nickel minerals.

MX$_2$ type: general remarks   This includes the numerous members of the pyrite and marcasite groups and their structural affiliates and hence is another very important type.   It also includes the interesting molybdenite layered structure and the important krennerite group of gold and silver tellurides.

MX$_2$ type: the pyrite group   This is an important group if for no other reason than that it includes pyrite itself.   The pyrite structure is a cubic close-packed arrangement, the unit cell of which can be expressed as $M_4(X_2)_4$.   M may be represented by any one (in some cases two or more) of a large number of divalent metals, and X by S, As, Sb, Se, or Te.   The metal atoms occupy the cubic face-centered positions, and the X atoms lie along the trigonal axes in pairs (like "dumbbells") about each M atom.   In pyrite itself the sulfur atoms lie in pairs along the body diagonals, each at about three-eighths the length of the diagonal from the iron atom.

The sulfur atoms are in six-fold coordination about each iron atom, and iron is six-fold coordinated with respect to each sulfur pair. While the metallic luster of pyrite and many others of the group indicates metallic bonding, there is a substantial covalent component, and a number of these minerals are extremely hard.

The cobaltite group, although generally regarded as distinct, is in fact essentially isostructural with pyrite and can be regarded as a closely related derivative. Here, As or Sb is substituted for *one* of the paired X atoms of the simple structure so that the X pairs are no longer composed of a single atomic species. There is thus a polarity induced along the trigonal axes, and the symmetry, as distinct from the structure, is reduced.

As might be expected there is some substitution of both cations and anions, though this is often minor. Some members appear to adhere closely to their theoretical composition wherever analyzed. Pyrite itself may contain substantial Ni, and also Co, substituting in the Fe positions. Increase in nickel leads to the formation of the mineral bravoite (Fe,Ni)S, the unit cell size increasing sympathetically with nickel content. Cu, V, Mo, Au, and As have been reported frequently as spectroscopic impurities, though in some cases at least, these represent included mineral particles rather than solid solutions. Mutual substitution among the cobaltite minerals is general and may be extensive, as indicated in Table 4-9.

Microscopically the outstanding characteristic of the commoner members of

Table 4-9   Chemical compositions and principal physical properties of minerals of the pyrite group

Name	Formula	Crystal system	Cell dimensions, Å	Density	Hardness	Reflectivity
"Simple" members						
Pyrite	FeS_2	Cubic	$a = 5.417$	5.0	1027–1836	54.5
Cattierite	CoS_2	Cubic	?	?	Hard	?
Vaesite	NiS_2	Cubic	$a = 5.74$†	?	Hard	?
Hauerite	MnS_2	Cubic	$a = 6.11$	3.46	485–623	23.8–25.4
Sperrylite	$PtAs_2$	Cubic	$a = 5.967$	10.58	960–1277	51.5–54.2
Aurostibite	$AuSb_2$	Cubic	$a = 6.65$	9.98	Moderate	?
Laurite‡	RuS_2	Cubic	$a = 5.59$	6.23	1605–2167	42.7
"Derivative" members						
Bravoite	$(Ni,Fe,Co)S_2$¶	Cubic	$a = 5.5$–5.57	4.62	1003–1425	45.5–46.5
Penroseite	$(Ni,Cu,Pb)Se_2$	Cubic	$a = 6.01$	7.56	407–550	45.2
Cobaltite	$(Co,Fe)AsS$	Cubic	$a = 5.57$	6.33	948–1367	52.0–54.7
Gersdorffite	$(Ni,Fe,Co)AsS$	Cubic	$a = 5.719$	5.82	520–907	47.5–53.0
Ullmanite	$(Ni,Co,Fe)(Sb,As,Bi,)S$	Cubic	$a = 5.91$	6.65	460–560	46.9–48.0
Willyamite	$(NiCo)SbS$	Cubic	$a = 5.92$	6.87	Hard	?

† Artificial NiS_2.
‡ Reputedly the hardest sulfide known.
¶ In the case of a solid-solution series the first-named cation and/or anion is the essential element—i.e., gersdorffite is NiAsS, ullmanite is NiSbS, etc.—and the physical measurements given are those for the end-member compound.

the group is their strong tendency to develop crystal faces. It is a feature of innumerable pyritic ores that whereas minerals such as galena, sphalerite, and chalcopyrite typically occur as allotriomorphic grains and aggregates, pyrite stands out as sharply defined idiomorphic to subidiomorphic crystals. While it often shows concretionary forms as a constituent of sedimentary rocks, its development of fine cubic, octahedral, and pyritohedral crystals here, too, is well known. Sperrylite occurs as spectacular cubic crystals set in pentlandite and pyrrhotite at its type locality at Sudbury. Cobaltite and gersdorffite also show the strong family tendency to idiomorphism.

Members of the group are not strongly colored, and most of them show very pale—in some cases almost elusive—tints of white. Pyrite is a clear pale cream, sperrylite a bright white, cobaltite white with a very pale pink tint, and gersdorffite and ullmanite are white, but often with a faint blue quality. Hauerite is the exception and is a light brownish gray, with a distinctly lower reflectivity. Cobaltite has been reported to show weak "anomalous" reflection pleochroism. As cubic minerals, the group as a whole has been regarded as optically isotropic, though this is probably incorrect. Pyrite, cobaltite, and ullmanite have long been known to exhibit a form of double refraction, but this has been regarded as due to lattice strain (due to impurities, thermal changes, etc.) rather than being an intrinsic quality of the crystal. However, fairly extensive studies (Stanton, 1957) suggest that the anisotropism is real in the case of pyrite at least. Some 350 pyrite specimens representing a very wide range of natural depositional conditions were prepared with great care, and all were found to be doubly refracting on any plane other than the $\{111\}$. Spectrographic analysis indicated that this was not due to arsenic or other impurities, and heating *in vacuo* showed that the anisotropy was not irreversibly affected by temperatures up to 570°C. It seems likely from this that pyrite is in fact anisotropic, due to the low symmetry (class 23) of its crystal structure. Interference colors are pale blue-green to orange-red. It may, however, be rendered *apparently* isotropic by harsh polishing, as on a lead lap or a high-speed buff; the burnishing action inherent in such methods probably causes fine recrystallization of the surface layers, leading to elimination of polarization contrast and the appearance of optical isotropy.

Hauerite, with its low reflectivity, typically shows internal reflections. These range from red to brownish red, and are, as usual, emphasized by the use of an oil immersion lens.

Members of the pyrite group are found in a wide variety of occurrences. Pyrite is common in igneous, sedimentary, and metamorphic rocks and in veins. It is associated with many different mineral assemblages and clearly forms under a wide range of conditions. Sperrylite appears to be restricted to ultrabasic and basic rocks, its most notable occurrence being with the sulfides of the Sudbury norite. Hauerite occurs chiefly in volcanic areas associated with volcanic gypsum and sulfur; apparently it has been precipitated from solutions with an excess of sulfur. Members of the cobaltite subgroup are fairly widespread though far from abundant. They occur particularly in hydrothermal native silver:Co–Ni arsenide veins, contact metamorphic deposits, and some orthomagmatic ores. Gersdorffite is abundant in some of the arsenide-rich selvages of the Sudbury deposits and is an occasional member of arsenic-rich deposits generally.

MX_2 type: the marcasite group Members of this group diverge from the close-packed arrangements that have constituted a common thread running through all the other groups so far. Minerals of the marcasite group have orthorhombic structures, modified—by a minute change in the β angle—to monoclinic in some of the derivative members. Marcasite itself (FeS_2, the orthorhombic dimorph of pyrite) has its Fe atoms occupying the lattice points of a body-centered ortho-rhombic arrangement, with the S atoms in six-fold coordination about them. The various derivatives are produced by the substitution of Fe by Ni and Co and S by As and Sb. The principal subgroupings are marcasite itself, loellingite (Fe, Ni, Co diarsenide) group, and the arsenopyrite (Fe, Co sulfarsenide-sulfantimonide) series. As pointed out by Buerger (1937, p. 50) the crystallographic and structural dif-ferences of these three groups are due principally to slight differences in the nature of the coordination of the anions about the metal atoms.

Marcasite itself appears to be a simple iron disulfide and shows little variation in composition. However, substitution is quite extensive in the other two sub-groups. In loellingite, Co and Ni may substitute for Fe in part and Sb and S for As in part. In addition the Fe/As ratio is known to depart from the ideal 1:2. In safflorite ($CoAs_2$), iron may substitute up to about 16 per cent in natural material, and Ni, Bi, Cu, and S have been found in small amounts. Similarly rammelsbergite and pararammelsbergite ($NiAs_2$) may contain small quantities of Co, Fe, S, and Sb. Mutual substitution of essentially the same type takes place in the arsenopyrite-glaucodot-gudmundite subgroup.

With the members of the skutterudite series [$(Co,Ni)As_3$–$(Ni,Co)As_{3-x}$] the minerals of the loellingite and arsenopyrite subgroups are often referred to collec-tively as the *white arsenides*. There is good reason for this: all are indeed silver-white, and as a group they are the most highly reflecting of the ore minerals apart from the native metals and a few individual compounds. Marcasite itself is not quite so white; it has a very pale yellow tint which, however, is paler than that of pyrite, from which it may just be distinguished in reflected plain light. All take a brilliant polish and stand out against the less highly reflecting sulfides. All are weakly pleochroic; of the group, marcasite shows the most distinct reflection pleo-chroism, and it is through this that it is often first distinguished from pyrite where the two are intergrown. All show very strong double refraction, and striking inter-ference colors are a feature of the group. The brilliant green of marcasite is characteristic; arsenopyrite shows strong blue, green, and brownish yellow; loel-lingite shows bright orange-yellow to blue-green; and the other members exhibit similarly striking colors. A white arsenide assemblage is the ore mineralogist's delight and may far exceed in beauty any igneous rock except a peridotite when viewed between crossed nicols.

From the point of view of occurrence marcasite is clearly separated from the others. It is essentially a low-temperature form of FeS_2 and forms under acid conditions. At higher temperatures pyrite is the stable form. Marcasite usually

Table 4-10 Chemical compositions and principal physical properties of minerals of the marcasite group

Name	Formula	Crystal system	Cell dimensions, Å	Density	Hardness	Reflectivity
"Simple" members						
Marcasite	FeS_2	Orthorhombic	$a = 4.445$ $b = 5.425$ $c = 3.388$	4.89	824–1288	48.9–55.5
Loellingite	$FeAs_2$	Orthorhombic	$a = 5.25$ $b = 5.92$ $c = 2.85$	7.40	421–963	53.0–54.7
Safflorite	$(Co,Fe)As_2$	Orthorhombic	$a = 6.35$ $b = 4.86$ $c = 5.80$	7.20	606–988	50.5–56.9
Rammelsbergite	$NiAs_2$	Orthorhombic	$a = 6.35$ $b = 4.86$ $c = 5.80$	7.10	556–841	58.0–60.0
"Derivative" members						
Arsenopyrite	(Fe,Co) $(As,Sb)S$	Orthorhombic	$a = 6.43$ $b = 9.55$ $c = 5.69$	6.07	890–1283	51.7–55.7
Glaucodot	$(Co,Fe)AsS$	Orthorhombic	$a = 6.67$ $b = 9.62$ $c = 5.73$	5.90	841–1277	50.0–52.5
Gudmundite	$FeSbS$	Monoclinic	$a = 10.00$ $b = 5.95$ $c = 6.73$ $\beta = 90°$	6.91	588–1221	55.0

forms at or near the surface by precipitation in peat bogs and other suitably reducing environments or by deposition from ground waters as encrustations, concretions, or on occasions, as replacements of fossils.   It may result from the inversion of pyrite following the onset of more acid conditions, in which case pyrite and marcasite may be found intergrown.   It is also a common low-temperature-alteration product of pyrrhotite.   Such alteration, which usually proceeds from the margins, produces the microscopic "bird's eye" texture, with "eyes" of smooth pyrrhotite surrounded by finely cellular, pale cream marcasite.   The reaction is probably

$$2FeS \rightarrow FeS_2 + Fe$$

the iron being thrown down as fine magnetite dust.   Secondary pyrite may also be formed in this way, given the appropriate alkaline environment.

Occasionally marcasite constitutes an essential mineralogical part of a primary ore type.   The outstanding example of this is probably the low-temperature limestone–lead-zinc ore type, of which marcasite, though of no economic value, is a widespread constituent.   It occurs here as individual crystals or as coarse cockscomb groups associated with galena, sphalerite, and minor pyrite and chalcopyrite.

Just as pyrite is the most abundant sulfide, arsenopyrite is clearly the most abundant and widespread arsenic mineral. Its most frequent occurrence is as a constituent of veins, which are usually of a fairly high-temperature hydrothermal type. It also occurs within granitic rocks and as a constituent of contact-meta-morphic zones, and it is almost ubiquitous as a minor (trace) constituent of pyritic lead-zinc ores. In these it occurs as small crystals or patches of crystals of con-spicuously idiomorphic to subidiomorphic form, often associated with small particles of tetrahedrite-tennantite. Edwards (1943, p. 37) noted that small par-ticles of gold often occurred preferentially along such arsenopyrite and tennantite contacts. Glaucodot and gudmundite are uncommon minerals. They generally occur in hydrothermal sulfide-arsenide veins not necessarily accompanied by arsenopyrite. Minerals of the loellingite group are also of rather unusual occur-rence. Their major development, which is usually a spectacular one, is in the native silver: Co–Ni arsenide ores already referred to.

MX$_2$ type: the krennerite group This is a small but important group of gold-silver tellurides. There are three members: krennerite, calaverite, and sylvanite.

Although these three have not the simple structural affiliations possessed by most of the other groups, they are very similar in their chemical nature and in most of their physical properties. Dana (1944, p. 332) notes that in spite of their struc-tural differences certain crystallographic similarities become apparent when the three minerals are appropriately oriented. A high degree of metallic bonding throughout is indicated by the softness and the high reflectivity of each. Differ-ences in density clearly result from differences in Au/Ag ratios. Krennerite is essentially a ditelluride of gold, silver having been reported up to 3Ag/8Au. In calaverite the highest silver content reported is Ag/4Au. Although Au/Ag is usually close to 1:1 in sylvanite, Au is known to be in excess in some specimens.

Under the microscope all are characterized by high reflectivity and a creamy white color. Their softness, however, may preclude the production of a high

Table 4-11 Chemical compositions and principal physical properties of minerals of the krennerite group

Name	Formula	Crystal system	Cell dimensions, Å	Density	Hardness	Reflectivity
Krennerite	(Au,Ag)Te$_2$	Orthorhombic	$a = 16.54$ $b = 8.82$ $c = 4.46$	8.62	105–130	52.3–61.0
Calaverite	AuTe$_2$	Monoclinic	$a = 7.19$ $b = 4.41$ $c = 5.08$ $\beta = 90° \pm 30'$	9.24	198–237	63.0–66.5
Sylvanite	(Au,Ag)Te$_2$	Monoclinic	$a = 8.96$ $b = 4.49$ $c = 14.62$ $\beta = 145°26'$	8.16	102–203	48.0–60.0

polish, particularly where associated harder minerals tend to pluck.   All show weak to distinct reflection pleochroism, depending somewhat on orientation.   Optical anisotropism is quite strong.   Krennerite shows yellow to brown interference colors, sylvanite light gray to dark brown, and calaverite light to dark gray, with greenish and brownish tints.   All may develop twinning, sylvanite most frequently.

These three minerals are by far the most important natural compounds of gold and constitute a substantial source of this metal.   They are well known at Kalgoorlie in Western Australia, Kirkland Lake and Rouyn districts in Ontario and Quebec, respectively, Cripple Creek in Colorado, the Mother Lode in California, and the Nagyag area in Europe.   All occur as vein fillings, commonly in volcanic provinces. Such veins may be young (frequently Tertiary) epithermal types clearly associated with volcanic activity, or they may be of apparently deeper nature, in this case associated with much older (often Precambrian) volcanic rocks.   The older deposits may be of genuinely deep formation, though there is also the possibility that they originated as near-surface volcanic deposits similar to those of Tertiary age and were modified later by the various deformational processes undergone by the enclosing rocks.   In all these occurrences the krennerite minerals may be accompanied by other tellurides and selenides, pyrite and small quantities of other sulfides, and quartz, carbonate, and other introduced nonopaque minerals.

MX_3 type: the skutterudite group   The members of this group are often referred to as the *isometric triarsenides* of cobalt and nickel.   In fact they appear to be defect structures with arsenic omissions, and Ross (1957, p. 766) classifies them in the MX_2 type as defect members of the marcasite group.

The group constitutes a complete solid-solution series but for convenience is divided into three portions, namely, skutterudite $(Co,Ni)As_{3-x}$ approximately, smaltite $(Co,Ni)As_{3-x}$, and chloanthite $(Ni,Co)As_{3-x}$.   Dana (1944, p. 343) notes that for the skutterudite portion, x appears to range from 0 to 0.5, and for the smaltite-chloanthite portion it ranges from 0.5 to 1.0.   Some smaltites and chloanthites may therefore be of the form MAs_2.   Iron can substitute for Co and Ni up to about 12 per cent, but there is no member of the series containing iron as the major cation.   Similarly bismuth can substitute for Co and Ni in small amounts. Somewhat surprisingly there is little or no substitution of As by S.

The cell dimensions of the series ranges from $a = 8.19$ Å in skutterudite to 8.24 Å in smaltite.   Density is approximately 6.5.   All have a brilliant metallic luster that is modified on oxidation by a pinkish bloom in the case of cobaltian members and a greenish bloom where nickel is dominant.   Microscopically they are brilliant silver-white, with high reflectivity (skutterudite 55.8).   They are quite hard (skutterudite 589 to 724) and normally give a high polish.   The development of good crystal outlines is characteristic, and zoning is almost always present.

The skutterudite minerals are not common and are confined to sulfarsenide assemblages.   Their most notable habitat is the native silver:Co–Ni arsenide ore type, but they are also found in arsenic-rich patches and selvages in a variety of base metal sulfide ores.   Notable occurrences in native silver ores are at Cobalt, Ontario, and in Saxony.   They are well known in some of the arsenic-rich selvages of the Sudbury nickel deposits and have been found in small quantities—with other arsenic minerals—in such diverse ores as those of Cornwall, England; Broken Hill, New South Wales; and Franklin, New Jersey.

CLASS 3.   OXIDES

In a very general way it may be said that the native elements are the principal ore minerals of the precious metals, sulfides and sulfosalts the principal ore minerals of the nonferrous base metals, and oxides the principal ore minerals of the ferrous group.   While there are certainly exceptions, a moment's thought suffices to show that there is comparatively little overlap.   Gold and platinum occur as tellurides and arsenides, and copper and lead may be found in the native state, but the major parts of the two groups conform to the general chemical pattern described above. Similarly, iron occurs as the sulfide and zinc as the oxide (zincite, ZnO), but there is no question that the major concentrations of these metals occur as hematite-magnetite and sphalerite, respectively.

The principal elements of class 3 are thus iron, manganese, and chromium. Titanium, in the fourth period of the periodic table, is also important.   Vanadium, also in this period, is of less importance, as are niobium and tantalum, its comembers of group Va.   Tin is an exception; one might expect that tin would occur chiefly as a sulfide, such as stannite, but as it happens its principal occurrence is as the oxide cassiterite (SnO_2).

Hydroxides are understandably not common in primary ores, though primary goethite ($FeO \cdot OH$) has been found in copper ore of the Peko Mine (Northern Territory of Australia) by Edwards (1955), and there are no doubt other occurrences.

As with the sulfides and sulfosalts, the oxides may be conveniently grouped according to their cation-anion ratios.

MO type: the zincite group   In addition to zincite itself, this group embraces such compounds as CuO (tenorite), HgO (montroydite), and PbO (as litharge and massicot).   However, these are secondary products and hence are not of present concern.   Zincite is the only compound of this type found in primary ores.

The zincite structure is essentially similar to that of wurtzite.   Bonding is covalent-ionic, the lack of a significant metallic component being indicated by its adamantine rather than metallic luster.   Zincite can contain minor amounts of Fe and Mn and traces of other metals, but it is essentially ZnO.   Hardness is low but highly directional; mean values reported by Cameron (1961) are 154 normal to the cleavage, 305 parallel to it.   Reflectivity is very low: at 11.2 it has almost the lowest reflectivity of all "opaques" measured.

Microscopically, zincite is quite distinct.   As a result of its bright orange to red color and its low reflectivity, it shows abundant and characteristic internal reflections that obscure reflection pleochroism and optical anisotropism, though from the practical point of view the internal reflections are much more certain for identification.

Zincite is an uncommon mineral and is known in abundance only in the remarkable deposits of Franklin and Sterling Hill, New Jersey, where it is associated with willemite (Zn_2SiO_4), franklinite [$(Fe,Mn,Zn)O \cdot (Fe,Mn)_2O_3$], and coarsely recrystallized calcite.

MM_2O_4 type: the spinel group   This is a large group, of common and widespread occurrence, that derives its name from the naturally occurring cubic magnesium aluminum oxide, spinel ($MgFe_2O_4$).   Its members are usually regarded as double oxides $M^{++}O \cdot M_2^{3+}O_3$, where $M^{++}O/M_2^{3+}O_3$ is essentially $1:1$; M^{++} is one or more of the divalent metals Fe^{++}, Mg, Mn, Zn, Ni; and M^{3+} is one or more of the trivalent metals Al, Fe^{3+}, Cr, Mn. Ti^{4+} may enter the structure by coupled substitution and may occur in substantial amounts in the Fe-rich members.

It appears that most natural spinels fall into one or another of three series. In all these, $M^{++}O$ is essentially $(Mg,Fe)O$ (either Mg or Fe predominating), but $M_2^{3+}O_3$ varies from one series to another.   Where Al_2O_3 predominates, the spinel series proper results; dominant Fe_2O_3 yields a "magnetite series" and Cr_2O_3 a "chromite series."   In natural occurrences extensive solid solution is common within each series but is uncommon between them.   Stevens (1944) has constructed a triangular prism of composition showing a "principal zone of isomorphism" among the spinels (Fig. 4-5).   Ulmer and White (1966) and Ulmer (1969) have shown that at 1300°C there is complete solid solution between the end members:

Hercynite	$FeO \cdot Al_2O_3$
Spinel	$MgO \cdot Al_2O_3$
Magnetite	$FeO \cdot Fe_2O_3$
Magnesioferrite	$MgO \cdot Fe_2O_3$
Chromite	$FeO \cdot Cr_2O_3$
Magnesiochromite	$MgO \cdot Cr_2O_3$

However, they suggest that solid solutions developed at such temperatures may involve defect spinels containing trivalent cations in excess of the usual $1:2$ divalent-trivalent ratio.

Insertion of metals such as Mn, Zn, and Ni leads to the development of derivatives that also appear to have affinities for one or another of the three principal groups.   Zinc substituting for magnesium in the aluminum series gives the well-known zinc spinel gahnite, one of the accessory minerals of the Broken Hill lead-zinc lodes.

The structure of the spinels is comparatively complex in detail, but the cell is face-centered cubic, containing $8(M^{++}M_2^{3+}O_4)$ with the oxygen atoms in cubic close-packed arrangement.   The eight divalent metal atoms each coordinate four oxygen atoms in a tetrahedral arrangement similar to that of Zn and S atoms in sphalerite.   The 16 trivalent metal atoms each coordinate six oxygen atoms in an octahedral arrangement similar to that of Pb and S atoms in galena.   The whole

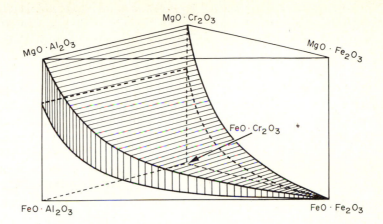

Fig. 4-5 Solid solution among the spinels. (*Redrawn from Stevens, Amer. Mineralogist, 1944.*)

structure is then linked together through the oxygen atoms, each of which is coordinated with one divalent atom and three trivalent atoms.

The true nature of the spinels, however, is not so well known as this statement of structure might imply. It is known that in artificial spinels there is a tendency for trivalent atoms to substitute for some of the divalent atoms and hence to be in excess, as noted by Ulmer and White (1966). Spinel itself can contain a considerable excess of Al_2O_3, and magnetite may contain excess Fe_2O_3 without apparent impairment of the crystal structure. Berry and Mason (1959, p. 353) point out, on the other hand, that this is in agreement with the fact that the γ modifications of Al_2O_3 and Fe_2O_3 (maghemite) both possess the spinel structure. Hausmannite ($MnO \cdot Mn_2O_3$) has a distorted spinel structure (tetragonal symmetry) that, however, is modified by heating and becomes cubic at about 1160°C.

Clearly a wide variety of compositions results from solid solution within each series, and appropriate modifying names are often desirable.

General properties of the whole group of spinel minerals are substantial hardness, low reflectivity, and optical isotropism. The last is often modified in the translucent members by "anomalous" double refraction. In hand specimen the spinel (Al_2O_3) series is characterized by bright colors—ruby red, yellow, brown, blue, green, and in hercynite and galaxite, black. The magnetite and chromite series are black. The Al_2O_3 series has an essentially vitreous luster, and the Fe_2O_3 and Cr_2O_3 series are more or less submetallic. The Fe_2O_3 series is magnetic to varying degrees. Both unit cell size and density vary according to proportions of the end members involved.

Microscopic observations on the Al_2O_3 series are naturally best carried out in transmitted light. They can usually be readily identified by their color, isotropism, and high refractive indices. The Fe_2O_3 and Cr_2O_3 series are opaque in all but the thinnest of thin sections. Chromite is commonly translucent in the inner parts of grains, where it is normally a deep cherry red, and opaque round its margins. This

Table 4-12 Chemical compositions and principal physical properties of minerals of the spinel group

Mineral	M^{++}	Cell dimensions, Å	Density	Hardness	Reflectivity
Spinel series ($MO \cdot Al_2O_3$)					
Spinel	Mg	$a = 8.080$	3.58	Hard	Low
Hercynite	Fe	$a = 8.119$	4.39	1378–1547	7.6
Gahnite	Zn	$a = 8.062$	4.62	861–1605	8.0
Galaxite	Mn	$a = 8.271$	4.03	Hard	Low
Magnetite series ($MO \cdot Fe_2O_3$)					
Magnesioferrite	Mg	$a = 8.366$	4.51	Hard	Low
Magnetite	Fe	$a = 8.391$	5.20	480–734	20.0–21.1
Franklinite	Zn	$a = 8.474$	5.07–5.22	720–824	18.5–20.0
Jacobsite	Mn	$a = 8.457$	4.03	724–870	18.5–19.7
Trevorite	Ni	$a = 8.41$	5.20	Hard	Low
Chromite series ($MO \cdot Cr_2O_3$)					
Chromite	Fe	$a = 8.36$	5.09	1036–1600	12.0–14.0
Magnesiochromite	Mg	$a = 8.305$	4.5–4.8	Hard	Low

appears to be due to a trend from a more magnesian to a more ferroan type of chromite with growth of the crystal. However, in spite of this pronounced change in opacity there is generally no obvious change in reflectivity, indicating that the increase in iron is not great (magnetite has a distinctly higher reflectivity than chromite) and that a small increase in iron may have a profound effect on translucency. In reflected light, chromite is gray and much darker than magnetite, which is pale gray and often misleadingly similar in appearance to sphalerite. When the latter two are in contact, however, magnetite is slightly paler and shows a faint brown quality—as well as its greater hardness and differences of habit. Magnetite less commonly shows a distinctly pinkish or pinkish-brown tint, and optical anisotropism, when it contains titanium in solid solution. This is quite pronounced in "titanomagnetite."

Chromite and magnetite show highly contrasting patterns of occurrence. Magnetite is perhaps the most widespread oxide mineral and occurs in many geological situations. It is an almost ubiquitous accessory mineral in igneous rocks and also occurs as substantial orthomagmatic concentrations. Because of its great stability magnetite is a common detrital constituent of sedimentary rocks, and it appears also as a chemical sediment in a number of large bedded iron formations. It is a common constituent of contact and regionally metamorphosed rocks and of veins. Chromite, however, is restricted to ultramafic rocks—either the dunites and peridotites (and derived serpentinites) of alpine-type intrusions or the ultramafic to basic layers of large differentiated sills. A minor amount of chromite is preserved in detrital sedimentary rocks. Jacobsite is found occasionally in metamorphosed manganese oxide deposits (with other manganese minerals, such as rhodonite, tephroite, etc.). Gahnite is sometimes found in rocks within and surrounding zinc sulfide concentrations; Broken Hill, New South Wales, is a notable

occurrence.   Franklinite is developed in abundance with zincite and willemite in
the metamorphosed zinc ores of Franklin and Sterling Hill, New Jersey.

M_2O_3 type: the hematite group   The principal members of this group are
hematite (Fe_2O_3) itself, ilmenite ($FeTiO_3$), and the nonopaque mineral corundum
(Al_2O_3).   There are in addition three quite rare minerals formed by substitution of
other divalent metals for Fe in the ilmenite structure.   These are pyrophanite
($MnTiO_3$), geikielite ($MgTiO_3$), and senaite [($Fe,Mn,Pb)TiO_3$].   Only hematite
and ilmenite are of present concern.

These are hexagonal (trigonal) and essentially isostructural.   The oxygen
atoms are in approximately hexagonal close-packed arrangement, with their layering parallel to the basal plane of the structure.   The metal atoms lie between these
layers, each coordinating six oxygen atoms.   Clearly if metal atoms filled all
available spaces, the cation-anion ratio would be 1:1 and the formula would be
MO.   However, only two-thirds of the spaces are filled, each metal atom coordinating three oxygen atoms from the layer below and three from above.   This
gives an octahedral arrangement in which any three oxygen atoms are shared by
two octahedra.   In the ilmenite structure half the Fe atoms of hematite are replaced by Ti, the two being arranged in ordered sequence.

As might have been expected, extensive substitution occurs and the two minerals are a well-known "solid-solution pair."   As early as 1926 Ramdohr established
that the two could be homogenized above about 600°C and that such solid solutions could be preserved metastably by quenching.   He reported that with slow
cooling, two products were formed: ilmenite containing about 4.2 per cent Fe^{3+}
in excess and hematite containing about 7.5 per cent Ti^{3+} in excess.   Substitution
by Mn and Mg in ilmenite is extensive, and it is likely that complete Fe–Mn and
Fe–Mg series can exist.

Microscopically both minerals are characterized by low reflectivity and gray
color, ilmenite being much the darker of the two.   Hematite is often difficult to
polish, whereas ilmenite normally gives a fine surface.   Reflection pleochroism is
barely discernible in hematite but is quite distinct in ilmenite.   Both show optical
anisotropism—hematite with gray-blue to yellow tints, ilmenite with greenish to
brownish grays.   Both are commonly twinned.   Hematite usually shows internal
reflections—the poorer the polish, the better the reflections—of intense blood red.

**Table 4-13   Chemical compositions and principal physical properties of minerals of the
hematite group**

Name	Formula	Crystal system	Cell dimensions, Å	Density	Hardness	Reflectivity
Hematite	Fe_2O_3	Trigonal	$a = 5.039$ $c = 13.76$	5.26	739–1097	24.0–30.6
Ilmenite	$FeTiO_3$	Trigonal	$a = 5.093$ $c = 14.06$	4.79	519–739	17.0–21.1
Corundum	Al_2O_3	Trigonal	$a = 4.758$ $c = 12.991$	3.98	Very hard	Translucent

Ilmenite, though, does not normally show internal reflections. When they do occur the ilmenite is usually a Mg variety, in which case they are dark brown.

Both minerals are of wide occurrence. Hematite is the principal commercial source of iron, the bulk of it being obtained from bedded accumulations of various kinds. Chief among these are the "bedded iron formations" of the type found in the Lake Superior area of North America, in which the hematite is associated with magnetite, iron carbonates and silicates, and much ferruginous chert. Oolitic hematite deposits are also extensive and are important in Paleozoic rocks of the eastern U.S.A. and Mesozoic rocks of Europe and England. Ilmenite commonly occurs as a constituent of basic rocks such as gabbros and anorthosites and may be concentrated into masses of some millions of tons (see Chap. 12). Much ilmenite occurs in beach sands produced by the breakdown of rocks containing disseminated ilmenite, which, following weathering and stream transport, is concentrated on beaches as placer deposits.

MO_2 type: the rutile group   This is an important "nonferrous" group and includes the isostructural, tetragonal minerals rutile (TiO_2), pyrolusite (MnO_2), cassiterite (SnO_2), and the rare lead mineral plattnerite (PbO_2). In addition the two mineral series mossite-tapiolite [$Fe(Nb,Ta)_2O_6$] and columbite-tantalite [$Fe,Mn(Nb,Ta)_2O_6$] can be regarded as derivative members. These have the same basic packing arrangement as the simple members, but as a result of the ordering—and hence larger repeat period—of the larger number of cations, the c axis is 3 times that of rutile itself. This is known as the *trirutile* structure. Artificial $FeTa_2O_6$ has a similar structure, but artificial $FeNb_2O_6$, $MnTa_2O_6$, and $MnNb_2O_6$ and a number of natural occurrences of tantalite-columbite possess orthorhombic symmetry.

Unlike many other mineral groups the simple members of the rutile series show little tendency to form solid-solution series. This is apparently due to the large differences in the sizes of the cations concerned, which for six-fold coordination are Ti = 0.68 Å, Mn = 0.80 Å, Sn = 0.71 Å, Pb = 1.30 Å. Where Fe, Ta, or Nb enter rutile to any appreciable extent, the structure changes to the trirutile arrangement. Although substantial quantities of these three elements have been reported in analyzed rutile specimens, such cases appear to be suspect and possibly due to very fine mixtures. Variation within each of the mossite-tapiolite and columbite-tantalite series is, on the other hand, extensive.

Cell dimensions of the various members are rather similar, though the a/c ratio is somewhat higher in rutile and pyrolusite than in cassiterite and plattnerite. The change in a/c going from rutile to trirutile structure is plainly illustrated in tapiolite. Density is clearly related to cation species. Hardness is variable and reflectivity is low; apart from pyrolusite the latter is in fact very low indeed.

With their low reflectivities these minerals are expectably rather dull microscopically. Pyrolusite is white, but the others are gray to dark gray. Rutile is similar in appearance to magnetite and ilmenite, but beside the latter it is distinguished by the absence of any brown tint. Cassiterite and the tantalite-columbite series are all dark gray and somewhat similar to chromite-magnetite. Pyrolusite, and to a lesser extent rutile, show fairly distinct reflection pleochroism; cassiterite

Table 4-14   Chemical compositions and principal physical properties of minerals of the rutile group

Name	Formula	Crystal system	Cell dimensions, Å	Density	Hardness	Reflectivity
"Simple" members						
Rutile	TiO_2	Tetragonal	$a = 4.594$ $c = 2.958$	4.25	978–1280	20.0–24.6
Pyrolusite	MnO_2	Tetragonal	$a = 4.39$ $c = 2.86$	5.06	225–405	30.0–41.5
Cassiterite	SnO_2	Tetragonal	$a = 4.738$ $c = 3.188$	6.99	992–1491	11.0–12.9
Plattnerite	PbO_2	Tetragonal	$a = 4.941$ $c = 3.374$	9.42	Hard	17
"Derivative" members						
Tapiolite	$FeTa_2O_6-$ $Fe(Nb,Ta)_2O_6$	Tetragonal	$a = 4.754$ $c = 9.228$	8.17	Hard	?
Columbite	$(Fe,Mn)(Nb,Ta)_2O_6$	Ortho-rhombic	$a = 5.74†$ $b = 14.27$ $c = 5.09$	5.48†	361–967	16.3–18.0

† Cell dimensions and density given for member containing 17 per cent Ta_2O_5. Density increases from 5.20 in the columbite end member to 7.95 in the tantalite end member.

and tapiolite show this only weakly, that of cassiterite being more readily detected where twinning occurs.   Between crossed nicols the various members are a little more distinctive.   Rutile, cassiterite, and tapiolite all show internal reflections that are normally deep red-brown in rutile, yellow- to red-brown in the tapiolite series, and pale yellow to yellow-brown in cassiterite.   All show distinct double refraction, though this tends to be masked in rutile and cassiterite by internal reflections, particularly when an oil immersion lens is used.

In spite of their similarities, members of this group show no great tendency to occur together in nature.   Rutile is a well-known accessory mineral in igneous rocks, but it is also common in veins (apparently of high-temperature type) and in high-grade schists, gneisses, and contact-metamorphic rocks.   In Norway there are notable occurrences of rutile in veins, associated with apatite.   Pyrolusite is a product of low-temperature, highly oxidizing aqueous environments and is often found as a lake, bog, or near-shore marine deposit associated with other manganese, and iron, oxides.   Cassiterite characteristically forms in highly felsic, often notably potassic, granites and in high-temperature pegmatitic-hydrothermal veins. Later erosion and stream transport yield placer deposits, and the major part of the world's cassiterite is obtained from concentrations of this kind.   Like cassiterite, the Ta–Nb minerals form initially as high-temperature igneous products, usually within granite pegmatites.

MO_2 type: the uraninite group   The principal members of this group are uraninite (UO_2) itself and thorianite (ThO_2).   The rare mineral cerianite (CeO_2) is of the same structure type, and baddeleyite (ZrO_2) is a slightly distorted analog.

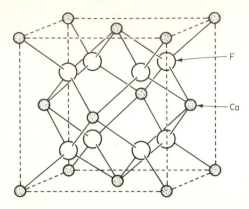

Fig. 4-6 The fluorite structure. In the antifluorite structure, cations and anions are in reverse relationship.

Apart from baddeleyite these minerals are cubic, with a fluorite structure, Fig. 4-6. In each case the metal atoms occupy face-centered positions (i.e., are cubic close packed) and coordinate eight oxygen atoms. Oxygen atoms occupy the body centers of each cubelet formed by four metal atoms and are hence in four-fold coordination with them. The unit cell thus contains $4(MO_2)$, each metal atom being in eight-fold coordination with oxygen atoms at the corners of a cube of edge length $a/2$ and each oxygen atom being in four-fold coordination with metal atoms at the apices of a tetrahedron.

Uraninite and thorianite form a complete solid-solution series. Substitution of Th to about 40 per cent ThO_2 has been found in nature, and the complete series has been synthesized in the laboratory. Cerium and other rare earth metals are also known to substitute. Change in the U/Th ratio is reflected in cell size: for pure UO_2, $a = 5.4682$ Å, and this increases progressively to $a = 5.5997$ Å in pure ThO_2. Density decreases from 10.95 for pure UO_2 to 9.87 for pure ThO_2. Both minerals are hard: uraninite has a Vickers hardness range of 280 to 839 (mean, 765) and thorianite one of 920 to 1235.

Microscopically these minerals are far from distinctive. Reflectivities are low (uraninite, 14.5 to 16.8; thorianite, 14.0 to 15.3), and both are dull gray in color and rather similar to magnetite. Both are isotropic but may show brown internal reflections.

Uraninite is found within intrusive rocks such as granites, syenites, and related pegmatites, in veins of various kinds, and in large stratiform disseminations. Where it occurs within igneous rocks it is often accompanied by monazite, tourmaline, and zircon and with Ti–Ta–Nb and rare earth compounds. Where it occurs in veins it may be associated with cassiterite, a variety of sulfides and, occasionally, with the Cobalt-type native silver : Co–Ni arsenide assemblage. Where it is a constituent of veins it commonly occurs as colloform masses, in which case it is referred to as pitchblende. From a quantitative point of view its most important mode of occurrence is as disseminated grains in Precambrian quartz-pebble conglomerates and associated quartzose sandstones. The two outstanding deposits of this kind are those of the Witwatersrand in the Transvaal and Blind River in Ontario (see Chap. 16).

Thorianite is found in intrusive igneous rocks, particularly pegmatites, but its principal occurrence is as a constituent of heavy mineral sands.   These occur as either stream or beach placers and usually contain monazite, zircon, rutile, ilmenite, and other minerals, such as garnet and spinels.

CONCLUDING REMARKS

This chapter has been devoted to showing that most of the quantitatively important ore minerals can be classified into quite a small number of groups and to considering the principal characteristics of the members of each of these groups.   It may now be seen that almost all the well-known sulfides and oxides, and many less common species, conform to one or another of about 20 structure types.   When viewed in this light, relationships between many "common" minerals on the one hand and "unusual" ones on the other are seen to be quite simple.

It must be noted, however, that attention has been confined to the more "basic" structures and that although these are of dominating importance there are others.   Many of the sulfosalts, for example, possess chain structures.   Stibnite (Sb_2S_3) and the isostructural bismuthinite (Bi_2S_3) both exhibit this kind of structure, as do proustite (Ag_3AsS_3)–pyrargyrite (Ag_3SbS_3) and others, such as jamesonite ($Pb_4FeSb_5S_{14}$), bournonite ($CuPbSbS_3$), and berthierite ($FeSb_2S_4$).   Other minerals have layered structures.   Molybdenite (MoS_2), for example, is hexagonal close packed, but its MoS_6 coordination octahedra are linked by the sharing of edges, forming sheets parallel to {0001}.   However, in both chain and layer structures the basic units are usually closely related to the simple structures of our classification; it is simply by variation in the linkage of these units that the strongly directional characteristics are created.

As might be expected, such variation in patterns of linkage does often greatly influence the physical nature of the minerals concerned.   Chain structures are often reflected in fibrous or acicular habit of resulting crystals.   Layer structures are usually reflected in excellent cleavages.   In molybdenite, for example, the sheets are linked by weak S—S bonds between sheets, giving the perfect basal cleavage —a very clear manifestation of the architecture of the mineral.

RECOMMENDED READING

Berry, L. G., and B. Mason: "Mineralogy: Concepts, Descriptions and Determinations," W. H. Freeman and Company, San Francisco, 1959.

Dana, J. W.: "The System of Mineralogy of James Dwight Dana and Edward Salisbury Dana, Yale University, 1837–1892," 7th edition, entirely rewritten and greatly enlarged by C. Palache, H. Berman, and C. Frondel, vol. I, John Wiley & Sons Inc., New York, 1944.

Ross, V.: Geochemistry, crystal structure and mineralogy of the sulfides, *Econ. Geol.*, vol. 52, pp. 755–774, 1957.

Uytenbogaard, W.: "Tables for the Microscopic Identification of Ore Minerals," Princeton University Press, Princeton, N.J., 1951.

5
Stabilities of
Ore Minerals

It is now apparent that the ore minerals are far from being a heterogeneous assortment of unconnected species. They are, like the minerals of silicate and carbonate rocks, largely separable into a relatively small number of groups whose members are closely related both in crystal structure and chemical composition.

Recognition of such distinctive groupings among silicates and related minerals has led to the determination of many serial relationships within each class. This has permitted the precise determination of the conditions of formation of the various members concerned and the physical and chemical conditions under which many of them are able to exist together in equilibrium. This in turn has led to a much better understanding of the conditions of formation of the rocks these minerals compose, and so mineralogy has had a natural—and dynamic—extension in petrology.

By analogy one might expect that the same could be done with the ore minerals, and in fact this has now been shown to be so. Work along these lines was begun by several earlier experimenters, including Bunsen and Daubrée, but it was not until early in the twentieth century that really systematic laboratory investigation of ore mineral stabilities was commenced. The first phase of this may be said to have culminated in the paper "The Cu–Fe–S system," by H. E. Merwin and R. H. Lombard, published in 1937. Since that time the investigation of sulfide and related mineral systems has been one of the most active parts of experimental geology, yielding much information on the composition and coexistence of ore minerals and the possible conditions of formation of the ores they compose.

Naturally most of the mineral systems examined are ones that appear to have some petrogenetic significance. Some of these are binary (involving two components), some ternary, and few are quaternary. The binary systems involve a variety of appropriate metals and, in each case, one of S, As, Se, or Te. The ternary systems involve one or two of the metals with, correspondingly, two or one of S, As, Sb, Te, and O. The quaternary systems generally comprise three metals and either sulfur or arsenic.

Among the binary systems, pairs such as Fe–S, Cu–S, and Ni–S are of obvious relevance. Of the ternary systems quite a number of important groupings immediately come to mind: Cu–Fe–S, pioneered by Merwin and Lombard, Fe–Ni–S, Zn–Fe–S, Ni–As–S, Au–Ag–Te, and others involving lead and silver sulfides, and iron, nickel, and cobalt arsenides. These in turn lead naturally to quaternary systems such as Co–Fe–Ni–As and Zn–Mn–Cd–S.

BASIC CONSIDERATIONS: EQUILIBRIUM AND THE PHASE RULE

The prime purposes of studying mineral systems are to determine the compositions of given minerals as functions of temperature and pressure and to determine which minerals are capable of coexistence under specified temperatures, pressures, and relative concentrations of the elements concerned. The principal investigations— and those most readily grasped—have been those involving studies of:

1. Variation of composition with temperature, at constant pressure
2. Mutual stability of two closely related substances (e.g., $FeS–FeS_2$, $CuS–Cu_2S$) as a function of temperature and pressure
3. Behavior of multicomponent systems as a function of temperature, usually at low, but not completely constant, vapor pressures

The underlying principle involved in such studies and in the application of their results is the assumption of the attainment of equilibrium and hence the application of the phase rule.

EQUILIBRIUM

A system may be said to be in a state of true equilibrium if the following apply:

1. It is sensitive to changes in external conditions.
2. The equilibrium concentrations are independent of the masses of the phases and of time.
3. The same equilibrium concentrations are attained when the equilibrium is approached from at least two directions.

If a system is uniform throughout its extent and hence has the same physical and chemical properties in all parts, it may clearly be said to be *homogeneous*. An equilibrium state developed in such a system—for example, a simple aqueous solution of some kind—is said to constitute *homogeneous equilibrium*. On the other hand, if the system consists of a number of distinct parts having distinctive physical properties, each being separated from the others by bounding surfaces, it may be said to be *heterogeneous*. A familiar example of such a system is that of ice, water, and water vapor, each of which is homogeneous within itself but clearly separated from the others. Equilibrium developed in such a system is said to constitute *heterogeneous equilibrium*.

The physically distinct portions of a system are called *phases*. Each such phase must, by definition, be physically and chemically homogeneous. It need not, however, be chemically simple; a mixture of gases or a solution constitutes a single phase; a mixture of different solid substances is composed of as many phases as there are substances present. In the system

$$3NiAs \rightleftharpoons Ni_3As_2 + As \text{ (vapor)}$$

there are two solid phases and one gaseous phase.

In any given system there can be only one gas phase, since all gases are completely miscible. There may, however, be any number (i.e., in conformity with the number of components—see below) of solid and liquid phases, since these are not necessarily completely miscible.

Those constituents of a system whose concentrations can undergo independent variation in the different phases are called *components*. The components of a system may be defined as *the smallest number of independently variable constituents by which the composition of each phase involved in the state of equilibrium can be expressed by a chemical equation*. It must be emphasized that the term "component" is not synonymous with "constituent." The case of ice-water-water vapor again provides a familiar example. Hydrogen and oxygen are elemental constituents, but it is only their chemical combination—water—that is a component. Hydrogen and oxygen are not components individually because they are not in a state of real equilibrium as that state has been earlier defined; they are combined in definite proportions in the compound water, and so their concentrations cannot be varied independently. They would, however, become separate components if the temperature were raised to 3000°C!

The same principle holds for more complex systems. For example, in the system pyrite-pyrrhotite-sulfur, there are three constituents in equilibrium. The compositions of all the phases present can be expressed by the following equations:

$$FeS_2 = FeS + S$$
$$FeS = FeS_2 - S$$
$$S = FeS_2 - FeS$$

Any two of the constituents are sufficient to define the composition of all phases present, and the system is therefore one of two components. This may be summarized as follows (see Findlay, 1951, p. 13):

1. The components of a system are chosen from among the constituents present when that system is in a state of true equilibrium, and they take part in this equilibrium.
2. Such components must be chosen as the smallest number of constituents necessary to express the composition of each phase participating in the equilibrium. Zero and negative quantities of the components are permissible.
3. In any given system the number of the components is fixed for any given set of conditions; this number can, however, alter with change in the conditions of the experiment.

The number of stable phases that can form from a set number of components is limited by the conditions of the experiment. This gives rise to a very important third number called the *variance*, or number of *degrees of freedom*, of the system. It is well known (from Boyle's and Charles' laws) that for an ideal gas

$$\frac{P_1 V_1}{T_1} = \frac{P_2 V_2}{T_2}$$

That is, specification of only pressure, or temperature, or volume (i.e., concentration) leaves the system undefined, but if two of the variables are arbitrarily fixed, the third is automatically determined. This applies to an entirely gaseous system.

Suppose, however, a system consists of coexisting liquid and vapor of the one composition. Such a system is immediately defined by the fixing of one of the other variables—temperature or pressure. If the temperature is fixed, pressure is determined and vice versa. Thus liquid and vapor can exist together under a given pressure only at a certain fixed temperature.

Further, suppose that such liquid and vapor be cooled until a solid begins to form. As soon as this third phase appears, both temperature and pressure are determined. Neither of the two variables can then be altered without causing the disappearance of one of the phases. We may therefore define variance as *the number of the variables temperature, pressure, and concentration of the components that must be arbitrarily fixed in order that the condition of the system is defined precisely.*

Thus different systems are capable of surviving the alteration of two or, in other cases, only one of the variables. Others are completely defined by the fact that they possess a certain number of phases and are destroyed by any change in temperature or pressure. They are therefore said to *differ in their degrees of freedom*, i.e., in the number of variables—temperature, pressure, and composition of the phases—that, when fixed, enable the condition of the system to be perfectly defined. It may be seen that for a one-component system:

1. A solid, liquid, gas, or vapor, each existing alone, has two degrees of freedom and hence constitutes a *bivariant* system.
2. A system solid-liquid, liquid-vapor, or solid-vapor has one degree of freedom and is hence *univariant*.
3. A system solid-liquid-vapor has no degrees of freedom and is hence *invariant*.

We may thus say that a given system has n degrees of freedom or that it is invariant, univariant, or bivariant as the number of its degrees of freedom is zero, one, or two.

THE PHASE RULE

The phase rule (set down by J. W. Gibbs in 1874) states:

$$F = C - P + 2$$

where F = variance
$\quad C$ = number of components
$\quad P$ = number of phases
of any given system at equilibrium.
$\quad$ Putting this in words:

1. A system of C components can exist in $C + 2$ phases only when $F = 0$, i.e., when total pressure, temperature, and the concentration of each phase have fixed and definite values.
2. A system of C components in $C + 1$ phases is defined when one of the variables is fixed.
3. A system of C components and C phases is defined when two of the variables are fixed.

$\quad$ Thus for a fixed number of components the phase rule imposes the condition that the greater the number of phases, the fewer the degrees of freedom, or the more numerous the phases, the more closely defined the system. Conversely, provided both the number of phases *and* the number of components vary appropriately, systems of widely differing complexities may have similar variances. Thus the system water-water vapor gives

$$F = 1 - 2 + 2 = 1$$

and the system FeS_2–FeS–S gives

$$F = 2 - 3 + 2 = 1$$

Both are similarly univariant, and both behave similarly in consequence. In this way widely different systems may be classified as invariant, univariant, or bivariant and, within each, be found to show analogous behavior.

MINERAL ASSEMBLAGES AND THE PHASE RULE

This is the basis of much of the work on mineral, particularly sulfide and oxide mineral, systems. Numerous multiphase systems have been studied isobarically (at constant pressure) to determine relationships between total composition and the identity of phases at different temperatures. Others have been investigated to determine compositions of coexisting solid phases as a function of temperature and the partial pressure of a vapor phase.

$\quad$ From the petrogenetic point of view the most interesting systems are the simple univariant ones, such as Fe–S and Cu–S, involving chemical variation of a continuous kind.   Where the precise compositions of the coexisting phases are known, fixing of pressure defines temperature and vice versa for the equilibrium concerned.   Where equilibria are not notably affected by pressure, compositions of coexisting phases can give an approximate indication of the temperature at which they formed together.   However, while these may indeed give a good *indication* of temperatures of formation of coexisting minerals—particularly in systems little affected by pressure—it is only at multiphase invariant points that a single pressure-temperature state is defined.   Clearly the chances of the natural achievement of such a point are rather remote, though a knowledge of their positions may sometimes be used to limit estimates of conditions under which the relevant assemblages formed.

It is sometimes maintained that such a simple picture is unlikely to hold in nature, where the environment of formation may well have been complicated by the presence of "mineralizers" and fluxes—substances that might have altered the position of equilibrium but have now disappeared.   We have seen that the phase rule applies to the equilibrium state and is concerned with neither the path nor the speed with which it is attained.   By acting as catalysts, various "fluxes" may indeed affect the *rate* at which the equilibrium state is approached, but they cannot affect the equilibrium itself.   They do not, therefore, diminish the relevance of laboratory studies.

EQUILIBRIUM AND QUENCHING IN NATURE

Whether or not a natural ore-mineral assemblage as deposited represents the equilibrium state under the conditions of its formation is never certain.   In many cases it probably does: the development of zoning in some natural minerals indicates that these minerals, at least, were very sensitive to changes in conditions and hence that their surfaces were close to being in equilibrium with their environment during deposition.   Whether or not this is the case in all ore-forming environments is not known.   It appears likely to hold where deposition has been from a liquid, vapor, or gas phase, but it is distinctly uncertain where solid-solid reactions have been involved.   Evidence for this lies in the very preservation of mineral zoning.

Quenching in nature is a much more difficult question and may constitute a serious weakness in the application of some of the synthetic phase-equilibrium studies.   This is considered in more detail a little later in this chapter, but its relevance must be appreciated whenever phase studies are applied.   It does not have to be emphasized that the preservation of the equilibrium state, or at least unambiguous evidence of it in spite of later retrograde effects, is vital to the whole argument.   If a high-temperature equilibrium relationship changes (retrogresses) with fall in temperature, all evidence of depositional conditions may be lost, and the application of any experimental results becomes irrelevant.   Unfortunately those systems that equilibrate most rapidly—and hence that are most certain to have attained equilibrium both in nature and in experiment—are those that retrogress most readily.   Some of these, such as the Cu–Fe–S system, are important, and so several systems that could be extremely useful through their widespread occurrences must be more or less discarded. Some minerals, such as the cubic form of Ag_2S (argentite), cannot be quenched under any conditions.

On the other hand, those minerals that resist retrogression are, expectably, those that equilibrate least rapidly. Among these are pyrite, loellingite, and sphalerite. These are useful because they are widespread and resistant to retrogression, but they have the limitation that they may lack sensitivity.

METHODS OF STUDYING PHASE EQUILIBRIA

Most of the experiments of present concern are carried out using a simple closed-tube method. In this the appropriate amounts of the relevant constituents are placed in a rigid glass tube (usually silica) that is then evacuated, sealed off, and placed in a furnace. While temperatures are readily controlled, pressures are not. These are, of course, the vapor pressures of the phases concerned, which may vary from a fraction of a millimeter of mercury up to several atmospheres.

The method has some disadvantages and involves some assumptions. These may or may not be serious but must be kept in mind:

1. *Lack of pressure measurement.* While sulfide systems themselves are comparatively insensitive to pressure, the partial pressure of sulfur is clearly very important in determining phase relations. For experimental work of highest quality it is necessary to measure and control this, and a review of methods has been given by Barton and Toulmin (1964).

2. *Attainment of equilibrium.* Where mineral systems are being studied over a range of temperatures, it is a prime assumption that the time allowed each "run" in the furnace has been sufficient for the attainment of equilibrium at the particular temperature concerned. Whether full equilibrium can or has been attained cannot, however, be completely demonstrated. If a number of experiments are carried out under the same conditions, it will be possible to say when equilibrium is approached but not when it is attained. It is known, however, that reaction rates in many sulfide systems are rapid, and it seems most unlikely that disequilibrium states are preserved metastably above about 150°C. As has already been pointed out, some of the more useful phases are rather slow to equilibrate. These are generally characterized by a very small component of metallic bonding, in contrast to a substantial component in those that equilibrate quickly.

3. *Quenching.* This is a problem in experiment just as it is in nature. Oxidation has to be avoided throughout, necessitating the rapid cooling not only of the specimen but also of the whole protective tube. Under these conditions it may be difficult or impossible to quench the more rapidly equilibrating systems.

4. *Identification.* The products of closed-tube experiments are generally very fine grained and may be difficult to identify. This applies not only to identification of the mineral species themselves but—perhaps more particularly—to subtle changes of composition *within* each mineral species. Identification of

phases is carried out by microscopy or x-ray methods and determination of composition within phases by the electron probe.

Where variation of vapor pressure has a significant effect on the compositions of the phases present or where the compositions of phases are being studied partly or wholly as functions of pressure, experimental procedures are modified to measure and control the pressure in the vessel.   The problems of attainment of equilibrium, quenching, and identification of phases are, of course, just the same as for the simple closed-tube method.

DIAGRAMMATIC PRESENTATION OF PHASE EQUILIBRIA

There are about a dozen types of diagrams† used to portray coexisting mineral groups, and compositions of minerals, as functions of temperature, pressure, and other variables.   Those most commonly used, and hence those with which we are now concerned, are the following.

TOTAL-PRESSURE–TEMPERATURE (P_t–T) DIAGRAM

This type of diagram is commonly used to depict phase relationships in univariant systems, such as that of Fe–S.   Here, for example, the effect of pressure on the temperature at which FeS_2 coexists with FeS and liquid S may be demonstrated. Since there is no compositional axis, the *compositions* of the phases in equilibrium at any given pressure and temperature do not appear.   Diagrams of this kind have particular application in systems of one component.   That of water is an excellent example and is shown in Fig. 5-1.   The curves OA, OB, and OC are the

† These are dealt with in detail by Darken and Gurry (1953), Garrels (1960), and Barton and Skinner (1967).

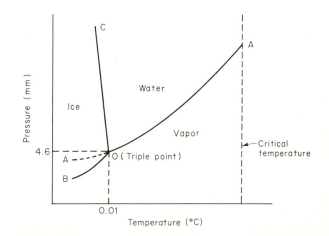

Fig. 5-1

vaporization, sublimation, and fusion curves, respectively, and each represents a univariant portion of the system as a whole. The point O is the invariant point:

$$F = C - P + 2$$
$$C = 1$$
$$P = 3$$
$$F = 0$$

Point A is the critical point (374°C and 218 atm pressure) at which liquid and vapor phases form a continuum.

Clearly in systems of one component a single phase yields bivariance, two phases yield univariance, and three phases fix the system in an invariant point. With two components, as in the FeS_2–FeS–S system, the presence of all three phases yields univariance.

TEMPERATURE–COMPOSITION (T–X) DIAGRAM

This is the most familiar form of presentation of mineral-stability studies.

The inclusion of composition in the consideration of stability implies the presence of more than one component. For a binary system $C = 2$, and applying the phase rule,

$$F = C - P + 2$$

it may be seen that four phases in equilibrium are required to define an invariant point. Clearly three phases are required to give a univariant, and two phases a bivariant system. With only one phase, the system becomes trivariant, and a third variable is required to describe it. This is the concentration of the components. Thus in two- and multicomponent systems we are concerned with variations in temperature, pressure, and concentration of the components in the different phases. This requires the use of three axes, as shown in Fig. 5-2. Such a construction may be "cut" to give sections normal to any one of these axes and hence at constant pressure, temperature, or composition as the case may be. With constant pressure such sections yield *isobars*, or T–X diagrams; with constant temperature, *isotherms*, or P–X diagrams; and with constant compositions, *isopleths*, or P–T diagrams. Present concern is with T–X relations.

These may take a number of forms, depending on the miscibility or immisci-

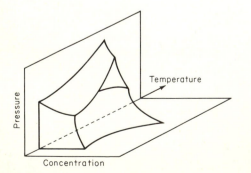

Fig. 5-2 (*Adapted from Findlay, "The Phase Rule," Dover Publications, Inc., New York, 1951, with the publisher's permission.*)

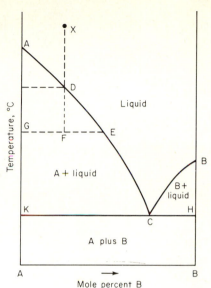

Fig. 5-3 (*Adapted from Findlay, "The Phase Rule," Dover Publications, Inc., New York, 1951, with the publisher's permission.*)

bility of the constituents, the nature of their melting processes, and their tendency to form compounds. We may consider three of the more common kinds of cooling history.

1. Consider first a system in which the only solid phases are the pure components, which when in the liquid state are miscible in all proportions. As shown in Fig. 5-3 we may begin with a liquid composition and temperature represented by the point X. On cooling to point D on the curve $ADEC$, component A will start to crystallize, crystals of A then being in equilibrium with liquid of composition D. With fall in temperature, further crystals of A are produced, and the composition of the liquid changes to that at E. At this stage crystals of A are in equilibrium with a liquid of composition E. The ratio of crystals to liquid will be EF/FG. With further cooling and separation of A the curve proceeds to C, at which point component B begins to precipitate. A and B then crystallize together in constant proportions ($A/B = HC/KC$). Point C is the *eutectic point*, and the composition of the mixture constitutes a *eutectic*. A corresponding path is followed when B preponderates and BC is the cooling curve.

2. The T–X diagram takes a somewhat different form when the two components A and B are capable of forming a compound. This may have a congruent or an incongruent melting point.†

† A compound is said to have a *congruent melting point* (i.e., melts "congruently") when it is capable of existing as a solid compound in equilibrium with a liquid of the same composition. It is said to have an *incongruent melting point* (i.e., melts "incongruently") when it melts to yield two or more substances of composition different from that of the solid; e.g., enstatite melts incongruently at 1557°C to give olivine and a liquid richer in silica than enstatite: $2MgSiO_3 \rightarrow Mg_2SiO_4 + SiO_2$-rich liquid.

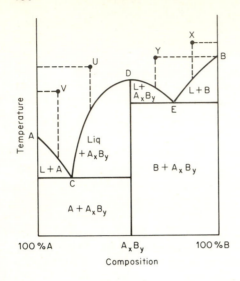

Fig. 5-4 (*Adapted from Findlay, "The Phase Rule," Dover Publications, Inc., New York, 1951, with the publisher's permission.*)

Where melting is congruent a third equilibrium curve must be added to those of Fig. 5-3, giving a series of relationships such as those of Fig. 5-4. The melting point of the compound A_xB_y is shown at D. With a preponderance of A, crystals of A are formed first, and the liquid descends along the path AC. A corresponding course is followed when component B is greatly in excess, cooling in this case leading to the formation of crystals of B and a trend of liquid compositions represented by BE. However, where the bulk composition falls between C and E, the compound A_xB_y crystallizes first, and

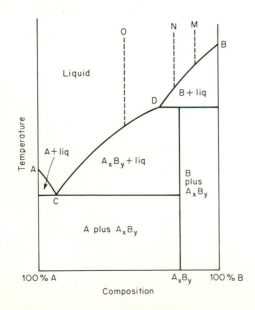

Fig. 5-5 (*Adapted from Findlay, "The Phase Rule," Dover Publications, Inc., New York, 1951, with the publisher's permission.*)

it proceeds to do so in equilibrium with liquids whose $T–X$ relations are given by the curves DC and DE, according as the original liquid contains more or less of component A than is required for the formation of A_xB_y. The points C and E are eutectic points, and the corresponding eutectics involve $A + A_xB_y$ and $B + A_xB_y$ in the proportions indicated by the diagram.

Where a compound is formed whose melting point is incongruent, solidification follows the possible courses indicated in Fig. 5-5. Where the initial liquid has, say, insufficient A to fully satisfy B in A_xB_y, the solidification process may be visualized as beginning at M. Crystals of B separate and the liquid proceeds down the curve BD to D. But at this point A_xB_y is the solid in equilibrium with the liquid. Solid B therefore reacts with the melt to form the compound. However, since there is insufficient A to react with all the B present, solidification goes to completion and there results a heterogeneous mixture of A_xB_y and B. The point D here is called the *incongruent melting point*, or *peritectic point*, and the reaction of B with the liquid is referred to as a *peritectic reaction*.

Where cooling commences from N to the curve BD, the amount of A is more than xA. Thus when the peritectic point D is reached, crystalline B reacts with the liquid to form A_xB_y, leaving some liquid. There is a temperature halt at D until the peritectic reaction is completed, and solidification then proceeds down DC to the eutectic point C. Cooling from O leads to crystallization along DC, yielding, as above, A_xB_y and liquid until C is reached, at which stage the eutectic $A + A_xB_y$ solidifies in the proportions indicated. There is no peritectic reaction involved in this case.

3. In some cases the components do not crystallize as pure substances or compounds but as solid solutions. Such solid solutions may, as was pointed out in Chaps. 3 and 4, be complete or only partial. We may therefore have complete solid-solid miscibility, as shown in Fig. 5-6, or a partial miscibility at lower temperatures, such as in Fig. 5-7.

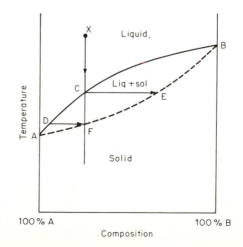

Fig. 5-6 *(Adapted from Findlay, "The Phase Rule," Dover Publications, Inc., New York, 1951, with the publisher's permission.)*

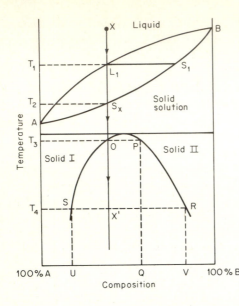

Fig. 5-7 (*Adapted from Findlay, "The Phase Rule," Dover Publications, Inc., New York, 1951, with the publisher's permission.*)

Suppose that a liquid containing the components A and B and represented by the point X of Fig. 5-6 is allowed to cool. With appropriate fall in temperature the isopleth meets the curve $ADCB$ in C. In this case, however, we do not have crystals of B forming in equilibrium with the liquid of composition C, as in Fig. 5-3. The crystals that form are of composition corresponding to E on the curve $AFEB$. With continued cooling, further crystals form and since these continue to be relatively enriched in B, the remaining liquid becomes impoverished in it, moving over toward D. When the liquid reaches D the crystals have the composition corresponding to F (and hence X) and solidification is complete. The curve of liquid composition is the *liquidus* and that for the coexisting solids is the *solidus*. In this case the two solid phases are miscible in all proportions, giving a complete range of solid solutions.

In the case of limited solid-solid miscibility a further equilibrium curve is added. This lies beneath the solidus and is referred to as the *solvus*. Suppose the liquid represented by X in Fig. 5-7 is cooled to T_1. Crystallization commences and proceeds as before, the initial products being the liquid L_1 and the solid S_1. At the moment of completion of crystallization there is a single homogeneous solid phase—the solid solution of S_x.

From this point the solid solution is unaffected by fall in temperature until T_3 is reached. Here the isopleth through X meets the region of partial solid miscibility defined by the solvus. Just as O is reached, a minute quantity of a new crystalline phase, of composition Q corresponding to the T–X point P, begins to form. As the temperature continues to fall this new phase follows the curve PR and hence becomes further enriched in component B. Correspondingly the complementary residual phase becomes enriched in A and follows the curve OS.   At the final temperature T_4 the system consists of two solid phases of composition U and V whose proportions are $U/V = RX'/SX'$.

"Condensed" T–X diagrams   In the T–X diagram proper, which includes all the types just described, pressure is, of course, kept constant, and vapor is stable only where specifically indicated on the diagram.   In the "condensed" diagram, on the other hand, there is a stable vapor phase throughout, and pressure varies with variation in the vapor pressures of the solid and liquid phases (i.e., the "condensed" phases) present.   Such variation in pressure is usually not great and is reckoned to range from insignificant to a few tens of atmospheres.   The nonvapor phases are essentially incompressible, and the effect of pressure is regarded as negligible.   These conditions apply in evacuated glass-tube experiments.

TERNARY COMPOSITION DIAGRAM

In most geological systems more than two components are involved.   Accordingly quite a large number of ternary sulfide and related systems have been investigated. In such cases the three axes of an equilateral triangle are taken as compositional axes, and temperature is measured along axes drawn perpendicularly to the apices of the triangle.   The result is a right prism (Fig. 5-8).   A typical solid-liquid equilibrium diagram resulting from this kind of construction is shown in Fig. 5-9.

The construction and reading of a diagram such as that of Fig. 5-9 is analogous to that of the two-component diagram of Fig. 5-3.   The curves drawn on each of the three sides of the prism are the same in principle as $ADEC$ and BC of Fig. 5-3.   Each begins at the melting point of a pure component.   Since the addition of B to A and vice versa lowers the melting point of each component, the curves AX_1 and BX_1 develop and meet at the eutectic point X_1.   Similarly AX_2 and CX_2, and BX_3 and CX_3, meet at their respective eutectic points X_2 and X_3.

If to a system represented by X_1 a small quantity of component C is added, the temperature at which the solid phases A and B can exist in equilibrium with the liquid is lowered further.   This continues to drop with progressive increase in the addition of C.   This generates the curve X_1E, which slopes downward into the interior of the prism and describes the change in composition of the ternary liquid

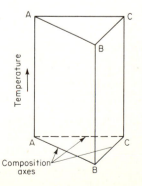

Fig. 5-8   (*Adapted from Findlay, "The Phase Rule," Dover Publications, Inc., New York, 1951, with the publisher's permission.*)

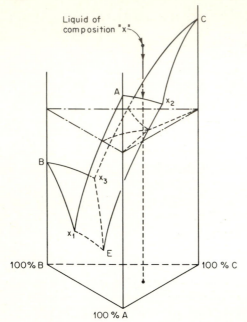

Liquid of
composition "x"

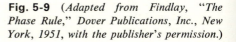

Fig. 5-9 (*Adapted from Findlay, "The Phase Rule," Dover Publications, Inc., New York, 1951, with the publisher's permission.*)

in equilibrium with both solid A and solid B as temperature falls. In the same way the curves X_2E and X_3E are generated by addition of components B and A to the binary eutectics X_2 and X_3, respectively. The three curves meet in E, which is the *ternary eutectic point*. This gives the composition of the *ternary eutectic* $A_xB_yC_z$ and indicates the maximum temperature at which it is stable.

It will be seen that these nine curves define three curved surfaces—AX_1EX_2, CX_2EX_3, and BX_3EX_1—within the prism. Any point on one of these surfaces indicates the composition of the ternary solution in equilibrium with one solid component at the temperature concerned. Any point on one of the curved lines of intersection indicates the composition of the liquid in equilibrium with two solid components. The point of intersection of the three surfaces and lines indicates the composition of the liquid in equilibrium with all three solid components simultaneously. The temperature here is fixed.

Recalling the phase rule

$$F = C - P + 2$$

and assuming the presence of vapor, it may be seen that:

1. For a point on one of the surfaces

    $$F = 3 - 3 + 2 = 2$$

    and the system is bivariant (both temperature and composition of the liquid plus vapor must be fixed to define the system).

2. For a point on one of the three curves of intersection

    $$F = 3 - 4 + 2 = 1$$

and the system is univariant (temperature *or* composition must be fixed to define the system).

3. At the ternary eutectic point

$$F = 3 - 5 + 2 = 0$$

and the system is invariant (the three solid phases can exist in equilibrium with liquid and vapor only at a definite temperature and composition).

Phase relations in such a system are usually shown in two dimensions by making one or more isothermal "cuts" through the prism. These are simply sections through the prism parallel to its base. One such cut is indicated in Fig. 5-9, and a series of them, corresponding to the various significant temperatures of Fig. 5-9, is shown in Fig. 5-10. The lines drawn from the apices of the triangles to the appropriate curves are known as *tie lines*, and they join the phases in equilibrium

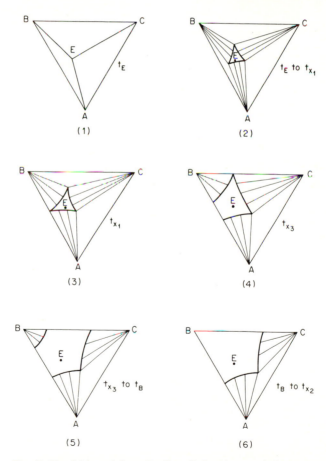

Fig. 5-10 (*Adapted from Findlay, "The Phase Rule," Dover Publications, Inc., New York, 1951, with the publisher's permission.*)

at the temperature concerned. Phase relations in ternary systems are also commonly shown by projections onto a plane, as in Fig. 5-31.

All cases of the T–X diagram have their counterparts in the ternary composition diagram. Systems possessing congruently or incongruently melting compounds, solid solutions, partial solid solutions, and so on, can all be represented by this form of diagram where an increased number of components requires it. All that must be remembered is that any isothermal cut is a "contour map" of what is really a three-dimensional construction. Such constructions, and sections of them, will, of course, become progressively more complex as compounds and solid solutions develop.

SOME IMPORTANT SYSTEMS†

BINARY SYSTEMS

The Fe–S system   This is an important system, if for no other reason than that it embraces the two most abundant sulfides, pyrite and pyrrhotite.

† Experimental ore mineralogy is currently developing so rapidly that any textbook treatment is inevitably slightly out of date as soon as it is written. This section is therefore intended to be somewhat historical. The reader should keep in mind that details of some of the systems dealt with here are always subject to at least slight modification and indeed that there have been several minor modifications since the time of writing.

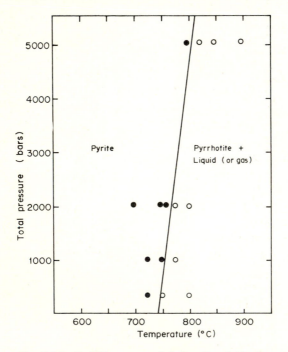

Fig. 5-11 Stability curve defining the fields for pyrite and pyrrhotite + liquid: $FeS_2 \rightleftharpoons Fe_{1-x}S + L$. (*Redrawn from Kullerud and Yoder, Econ. Geol., 1959.*)

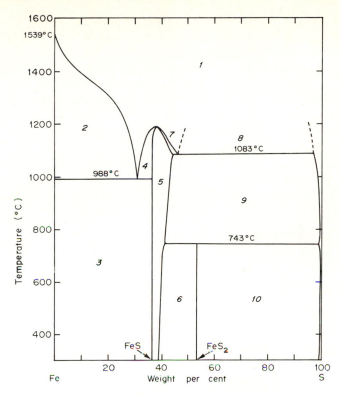

Fig. 5-12 Phase relations in the Fe–S system. (*Redrawn, after Kullerud and Yoder, Econ. Geol., 1959; Kullerud, Fortschr. Mineral., 1964.*)

Figure 5-11 shows the upper stability curve of pyrite.   This melts incongruently, to give pyrrhotite and sulfur, at 743°C at 10 bars and at progressively higher temperatures as pressure increases.   Figure 5-12 shows phase relations in the Fe–S system down to a temperature of 300°C.   Investigation below this is difficult due to very slow reaction rates.   However, the evidence of natural occurrence combined with limited experimental work in this region suggests that phase relations may be quite complicated at low temperatures.   Figure 5-13 shows detailed T–X relations for a portion of Fig. 5-12 according to Arnold (1962).   From this the composition of hexagonal pyrrhotite in equilibrium with pyrite appears to vary systematically with temperature between 325 and 743°C under the vapor pressure of the solid phases.   Confining pressures of up to 2000 bars do not affect the solvus, at least below 670°C.   Barton (Barton and Skinner, 1967) has now established that the solvus may be extended down as far as 200°C without interruption.

Such temperature dependence and pressure independence, coupled with a simple x-ray method for determining Fe/S ratios in hexagonal pyrrhotite, initially offered promise of a useful geothermometer if compositions corresponding to peak temperatures could be shown to be quenched in.   However, there is evidence that

Table 5-1 Temperature determinations on pyrrhotite from the Pennine area of northern England. (After Sawkins, Dunham, and Hirst, 1964)

Sample	Assemblage	Pyrrhotite dÅ natural	Pyrrhotite dÅ heated	Temperature, °C	Supporting temperature data
Rookhope borehole: 1083 feet	Quartz, pyrite, pyrrhotite, sphalerite; pyrite-pyrrhotite intergrowths; pyrrhotite-sphalerite intergrowths; mineralization in Smiddy limestone	2.0535† 2.088‡ 2.092‡	2.0603 2.0855	450 Low?	Sphalerite: $a = 5.409$ Å $< 1\%$ FeS; $<250°C$; no suitable fluid inclusions
Rookhope borehole: 1355 feet	Quartz, fluorite, pyrrhotite, calcite, pyrite; sulfides in center of vein, often lining vugs; mineralization in Weardale granite	2.053 2.063	2.0581	485	Fluorite: 160–174°C (8 readings) Quartz: $<220°C$; visual estimate
Rookhope borehole: 1562 feet	Fluorite, quartz, calcite, pyrrhotite, pyrite; sulfides in center vein, often lining vugs; mineralization in Weardale granite	2.0565 2.063	2.0586	475	Fluorite: 140–160°C (7 readings)
Great Sulphur Vein	Quartz, pyrrhotite, pyrite, chalcopyrite; pyrite veining pyrrhotite; from sheared zone in sediments in Sir John's Level	2.0565	2.0575	530¶	Quartz: 205–230°C (5 readings)

† From pyrrhotite-sphalerite intergrowth.
‡ From pyrrhotite-pyrite intergrowth.
¶ A correction of 35°C was added to the determined temperature to allow for the presence of chalcopyrite.

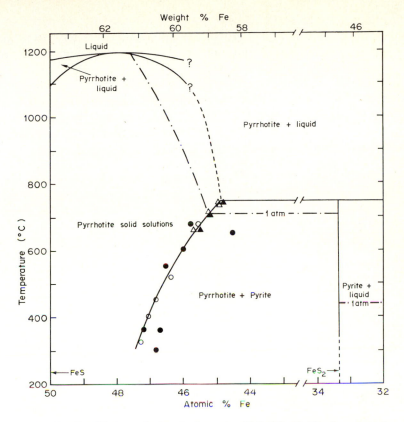

Fig. 5-13 Equilibrium relations in the system FeS–S for Fe/S ratios approximately between stoichiometric FeS and pyrite, FeS₂. (*Redrawn from Arnold, Econ. Geol., 1962.*) (Recent and current investigation is now adding to knowledge of the high-temperature region indicated by the question marks.)

the *T–X* relations are not quite so simple as Fig. 5-13 indicates. Monoclinic pyrrhotite is known to form only in the composition range 46.5 to 47.0 atom per cent iron—the composition range corresponding to 400 to 510°C in Fig. 5-13—and in some instances is known to have been deposited at low temperature. Erd et al. (1957) found highly magnetic (i.e., sulfur-rich) monoclinic pyrrhotite enclosed in calcite crystals for which fluid inclusion measurements indicated temperatures of formation in the 25 to 40°C range. Sawkins et al. (1964) studied monoclinic pyrrhotite from the Rookhope borehole and the Great Sulphur Vein of the Pennines of northern England and has found a similar state of affairs. These results are summarized in Table 5-1. From this Sawkins concludes that iron-deficient pyrrhotites can be deposited well below 250°C and that the pyrrhotite-pyrite solvus of Fig. 5-13 cannot, at least in some cases, be applied to monoclinic pyrrhotite even after it has been inverted to the hexagonal form. Barton and Skinner (1967) suggest that an as yet undetected feature of the phase diagram, presumably below 200°C, may permit the coexistence of pyrite with pyrrhotites having compositions and gross x-ray-diffraction characteristics similar to those of pyrrhotites formed experimentally in the 450 to 500°C range.   Clearly the system requires closer investigation below 250°C.

The Cu–S system   Minerals of this system do not commonly form primary assemblages, and it is therefore of rather limited interest.   However, there are a number of fairly well-defined inversions, and these have been used as temperature indicators in one or two notable instances.

The reaction

$$CuS \rightleftharpoons Cu_9S_5 + \text{liquid S}$$

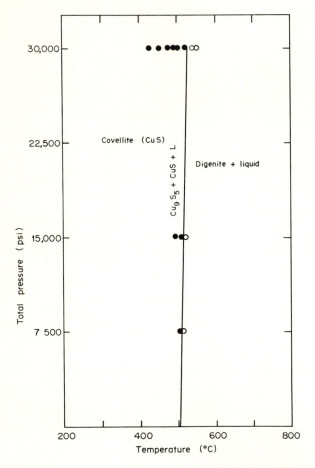

Fig. 5-14   Stability curve defining the fields for covellite and Cu$_9$S$_5$ plus liquid: CuS $\rightleftharpoons$ Cu$_9$S$_5$ + *L*.   (*Redrawn from Kullerud, Carnegie Inst. Wash., Ann. Dept. Dir. Geophys. Lab. Year Book, 1957.*)

is univariant, and Fig. 5-14 shows the upper stability curve of covellite, along which it melts incongruently. This is clearly almost insensitive to pressure, making the system a potentially useful geothermometer should quenchable changes of phase composition occur with change in temperature.

The compound Cu_2S, usually thought of as chalcocite, appears (Roseboom, 1966, p. 642) to include three minerals and, if the high-temperature forms of these are included, six phases.

1. *Digenite.* This shows a composition range of $Cu_{1.765}S$ to $Cu_{1.79}S$ at room temperature. The low-temperature form is uncertain but may be tetragonal, with $c \approx a$. The copper-rich end of the series inverts to a high-temperature cubic form ("high digenite"; $a = 5.56$ Å) at 76 to 83°C and changes to $Cu_{1.83}S$. Apparently the low-copper member does not undergo this inversion. The inversion of the high-copper member is readily reversed and high digenite cannot be quenched.

2. *Chalcocite.* This has a composition very close to Cu_2S and is orthorhombic at low temperatures. Cell-edge lengths are $a = 11.811 \pm 0.004$, $b = 27.323 \pm 0.008$, and $c = 13.49 \pm 0.004$ Å. Like digenite, this phase exhibits a range of composition from Cu_2S decreasing to about $Cu_{1.988}S$ at 105°C. At a little below this temperature (Roseboom, 1966: 103.5 ± 1.5 to $93.0 \pm 1.5°C$; Kullerud, 1964: $96 \pm 3°C$; inversion point is lowered with increase in sulfur content) it inverts to a hexagonal form of composition Cu_2S. The cell size of this has been determined at 112°C as $a = 3.89 \pm 0.04$ and $c = 6.68 \pm 0.07$ Å. Hexagonal chalcocite cannot be quenched experimentally and has not been observed as a mineral.

3. *Djurleite.* In 1958 Djurle reported three polymorphs at the composition $Cu_{1.96}S$—a low-temperature–low-symmetry form, a metastable tetragonal form, and a high-temperature cubic form. Roseboom (1966) has found the low-temperature form to be stable to $93 \pm 2°C$ and has named the mineral djurleite. Its symmetry is still uncertain but is apparently either rhombic or monoclinic, with cell edges $a = 26.95$, $b = 15.71$, and $c = 13.56$ Å. The metastable tetragonal phase coexists with the djurleite, complicating the study of both. The two together may have a composition of $Cu_{1.96}S$. Djurleite may approach $Cu_{1.97}S$ on the high-copper side. The tetragonal form may extend to $Cu_{1.95}S$ on the low-copper side but has been synthesized to Cu_2S at high pressures. The tetragonal phase has cell edges of $a = 4.008$ and $c = 11.268$ Å according to Djurle (1958). The structure is apparently a slightly deformed version of cubic close packing.

The high-temperature form noted by Djurle now appears to be simply high digenite of the appropriate $Cu_{1.96}S$ composition and is hence the stable polymorph above 83°C.

Phase relations in the Cu–S system as determined by Roseboom (1966) are given in Fig. 5-15.

Unfortunately the system does not offer great promise as a geological thermometer. Iron enters into several of the members and may alter relationships,

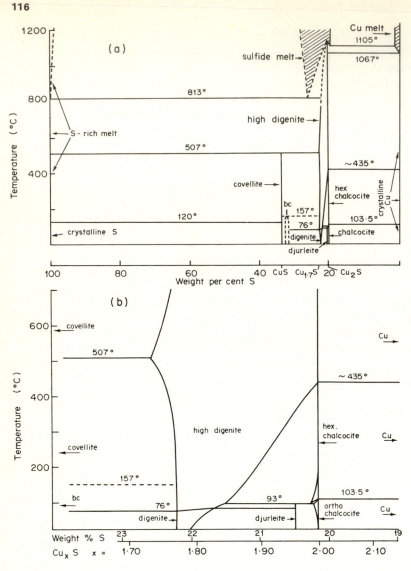

Fig. 5-15 (a) Phase diagram of the system Cu–S. All phases shown coexist with vapor. (b) An expanded portion of (a). There is no evidence at present that the Cu-rich limits of chalcocite, hexagonal-chalcocite, or high digenite (above 435°C) deviate from Cu_2S at the scale of the diagram. (*Redrawn from Roseboom, Econ. Geol., 1966.*)

though this difficulty may yield to future investigation. In general, reaction rates are very rapid, and higher-temperature phases and solid solutions cannot be quenched. The former existence of certain phases in ores might be inferred from twinning or pseudomorphs, though this is not always satisfactory. Roseboom (1966, p. 670) has suggested that the fact that the copper-rich limit of digenite is temperature-dependent up to 435°C might be useful, even though reequilibration is rapid even at room temperature. In some cases digenite-djurleite or djurleite-chalcocite mixtures might be found to show evidence of an original single cubic phase. If such samples were low in iron, a minimum temperature of formation

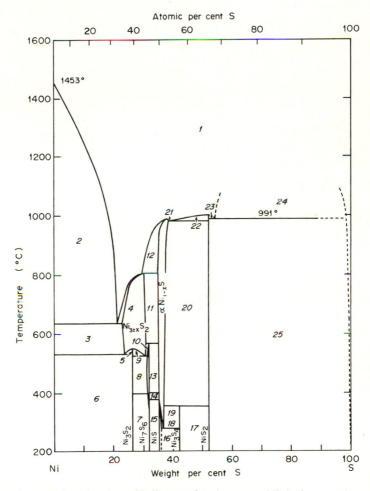

Fig. 5-16 "Condensed" diagram for the system Ni–S above 200°C. Phases present, in addition to vapor, in the numbered fields are as follows: 1, Liquid (L); 2, L + Ni s.s; 3, Ni s.s + $Ni_{3+x}S_2$; 4, L + $Ni_{3\pm x}S_2$; 5, Ni_3S_2 + $Ni_{3+x}S_2$; 6, Ni s.s + Ni_3S_2; 7, Ni_3S_2 + $\beta\,Ni_7S_6$; 8, Ni_3S_2 + $\alpha\,Ni_7S_6$; 9, Ni_3S_2 + $Ni_{3-x}S_2$; 10, $Ni_{3-x}S_2$ + Ni_7S_6; 11, $Ni_{3-x}S_2$ + $\alpha\,NiS$; 12, L + $\alpha\,Ni_{1-x}S$; 13, $\alpha\,Ni_7S_6$ + $\alpha\,NiS$; 14, $\beta\,Ni_7S_6$ + $\alpha\,NiS$; 15, $\beta\,Ni_7S_6$ + $\beta\,NiS$; 16, $\beta\,Ni_{1-x}S$ + Ni_3S_4; 17, Ni_3S_4 + NiS_2; 18, α + $\beta\,Ni_{1-x}S$; 19, $\alpha\,Ni_{1-x}S$ + Ni_3S_4; 20, $\alpha\,Ni_{1-x}S$ + NiS_2; 21, $\alpha\,Ni_{1-x}S$ + L; 22, L + NiS_2; 23, NiS_2 + L; 24, two liquids; 25, NiS_2 + L. (*Redrawn from Kullerud and Yund, J. Petrol., 1962.*)

(uncorrected for pressure) could be obtained by determining the bulk composition of the mixture and applying this to Fig. 5-15.

The Ni–S system   Like the Cu–S system, the Ni–S system is not of great geological relevance.   Both are completely overshadowed by the respective iron-bearing ternary systems.   However, the Ni–S system (Fig. 5-16) is a good example of a moderately complex binary system.   It possesses a large number of phases, several eutectics, and one quite clear dystectic.   Were the minerals involved of reasonably common occurrence in nickeliferous sulfide aggregates, the system might have provided some thermometric data, albeit of a rather approximate kind. However, the presence of iron almost always precludes this.

Figure 5-17 (from Kullerud and Yund, 1961) shows the upper stability curve of polydymite in the univariant system

$$Ni_3S_4 \rightleftharpoons 2\alpha\, Ni_{1-x}S + NiS_2$$

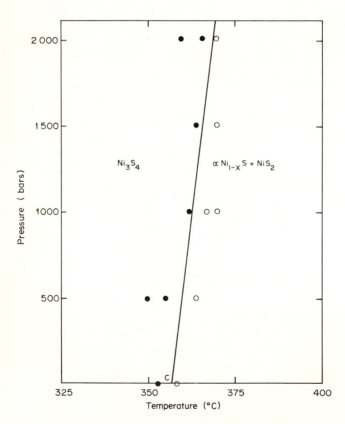

Fig. 5-17   Stability curve defining the fields for polydymite (Ni_3S_4) and $\alpha\, Ni_{1-x}S$ plus NiS_2: $Ni_3S_4 \rightleftharpoons 2\alpha\, Ni_{1-x}S + NiS_2$. *(Redrawn from Kullerud and Yund, J. Petrol., 1962.)*

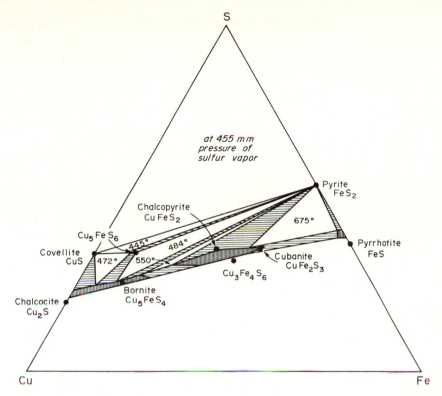

Fig. 5-18 The Cu–Fe–S system at 455-mm pressure of sulfur vapor; the historic diagram of Merwin and Lombard. (*Redrawn from Merwin and Lombard, Econ. Geol., 1937.*)

TERNARY SYSTEMS

The Cu–Fe–S system The early study of this system by Merwin and Lombard (1937) set the style for most of the sulfide-phase work done during the succeeding 30 years. Because of this early beginning and because of the ubiquity of its members and hence its potential mineralogical interest, it is the most thoroughly studied of the principal ternary sulfide groupings. The principal phases here are of common occurrence, and in fact few if any base metal sulfide deposits are found without at least one of them. It is not at all unusual to find deposits in which several of the members coexist in contact and in which equilibrium might reasonably be expected to have prevailed.

Figure 5-18 shows Merwin and Lombard's historic diagram of phase relations at variable temperatures under a constant sulfur vapor pressure of 455 mm of mercury. Figure 5-19 shows solidus relations (again in the presence of vapor) projected onto the temperature-Fe–Cu plane (the sulfur axis here is normal to the plane of the page) between 400 and 800°C. Figure 5-20 shows the system at a number of temperatures from 25 to 600°C.

From these diagrams it is apparent that there are a large number of reactions for which temperatures can be specified. However, the most interesting parts of the system are three solid-solution series of Fig. 5-20. The most extensive of these is that of the high-digenite field, which above about 290°C is continuous from digenite to beyond bornite. The second embraces both chalcopyrite and cubanite compositions above 550°C but splits at lower temperatures (see, for example, the 400°C cut of Fig. 5-20) into separate "chalcopyrite" and "cubanite" zones. The third is a field of hexagonal pyrrhotite. This is of very limited extent at low temperatures but becomes quite large at 600 to 700°C.

It has been pointed out by Barton and Skinner (1967) that all these solid solutions show marked variation in metal-sulfur ratios as well as in their cation contents. As already noted in connection with the Fe–S binary system, the pyrrhotite solid-solution field extends into highly metal-deficient material, and this extends to the ternary. Conversely the chalcopyrite-cubanite field extends out on the high-metal–sulfur side as well as showing substantial Cu/Fe variation. The high-digenite field extends out on both high-metal and high-sulfur sides of the digenite-bornite join. Since these very substantial variations are systematically tied to temperature, all seem at first sight to be potential indicators of temperatures of formation.

However, this promise is not fulfilled. With the possible exception of the pyrrhotite field of the Fe–S system, none of the solid solutions persists on cooling. In experimental runs they reequilibrate by exsolution and internal reaction to give compositions always reflecting the prevailing, lower, temperature. Clearly if they are able to reequilibrate in the time span of an experimental run, they are certainly

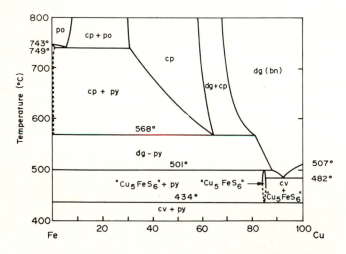

Fig. 5-19 Solidus relations in the Cu–Fe–S system projected onto the Cu–Fe-temperature plane. cp = chalcopyrite, po = pyrrhotite, py = pyrite, bn = bornite, cv = covellite, and dg = digenite (cf. Fig. 5-8). (*Redrawn from Roseboom and Kullerud, Carnegie Inst. Wash., Ann. Dept. Dir. Geophys. Lab. Year Book, 1958.*)

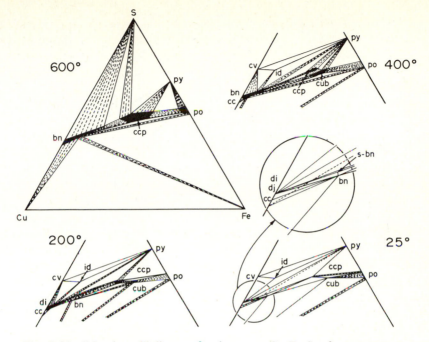

Fig. 5-20 "Condensed" diagram for the system Cu–Fe–S at four temperatures covering the major part of the range of ore-forming temperatures. [*Redrawn from Barton and Skinner, in Barnes (ed.), "Geochemistry of Hydrothermal Ore Deposits." Copyright © 1967 by Holt, Rinehart and Winston, Inc. Reprinted by permission of Holt, Rinehart and Winston, Inc., Publishers, New York.*]

able to do so in the presumably much longer period of the cooling of an ore body. They thus appear to hold no potential as geothermometers.

The Fe–S–O system Many ore deposits, and ore deposits of many different kinds, contain three or more of the assemblage pyrite-pyrrhotite-magnetite-hematite. Phase relations in this system between $T < 560°C$ and $T > 743°C$ have been determined by Kullerud (1957) and are shown in Fig. 5-21. Below 560°C the assemblages pyrite-pyrrhotite-magnetite and pyrite-magnetite-hematite are stable, and pyrite and hematite can coexist in the presence of SO_2. Above 560°C an FeO phase appears and yields the additional group $FeS–FeO–Fe_3O_4$. Above 675°C FeO disappears again, and the pyrite-magnetite tie is broken. There are then the stable assemblages pyrrhotite-magnetite-hematite, pyrite-pyrrhotite-hematite, and the continuing pyrite-hematite-SO_2 field. Stable groupings above 700°C are shown in Fig. 5-21e and f. At 743°C pyrite breaks down to FeS and S, and the coexisting phases are then pyrrhotite-magnetite-hematite and pyrrhotite-hematite-SO_2 gas.

Clearly these high-temperature relationships relate to orthomagmatic and kindred vein deposits and those of other kinds that have been subjected to quite high grades of metamorphism. However, many occurrences of assemblages

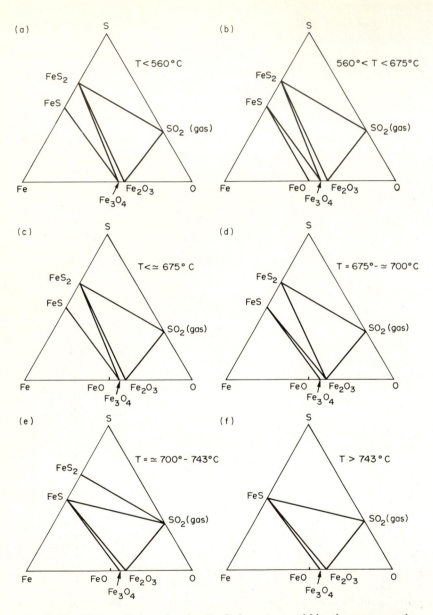

Fig. 5-21 Phase relations in the Fe–S–O systems within six representative temperature ranges. (*After Kullerud, Carnegie Inst. Wash., Ann. Dept. Dir. Geophys. Lab. Year Book, 1957; Fortschr. Mineral., 1964.*) (Most-recent investigation suggests that the appearance of $FeSO_4$ and possible appearance of other iron sulfate-type compounds may require substantial modification of these diagrams above about 700°C.)

belonging to this system develop under sedimentary and other very low-temperature conditions.   These are best studied in terms of low-temperature solution equilibria and are considered in this way in Chap. 6.

The Fe–Ni–S system   Phase relations in this system have been studied in great detail, due largely to interest in the great nickel deposits of Sudbury, Ontario. Since high concentrations of sulfide nickel are generally orthomagmatic, the Fe–Ni–S system was quickly recognized as one in which high-temperature anhydrous equilibrium studies might have particular relevance.   As a result this system is now one of the best understood.

Phase relations in the sulfur-rich portion of the system are shown in Fig. 5.22 and those for the disulfides, pyrite, and vaesite in Fig. 5-23.   Those in the Ni–S and Fe–S binaries have already been given.   The Ni–S system has little

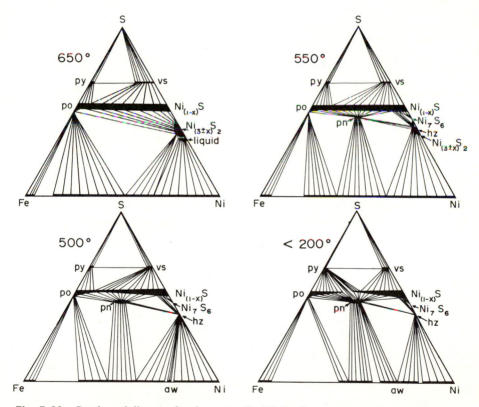

Fig. 5-22   Condensed diagram for the system Fe–Ni–S at four representative temperatures. All assemblages coexist with vapor: aw = awaruite, hz = heazlewoodite, pn = pentlandite, po = pyrrhotite, py = pyrite, and vs = vaesite.   [*Redrawn from Barton and Skinner, in Barnes (ed.), "Geochemistry of Hydrothermal Ore Deposits."   Copyright © 1967 by Holt, Rinehart and Winston, Inc.   Reprinted by permission of Holt, Rinehart and Winston, Inc., Publishers, New York.   After Kullerud, Fortschr. Mineral., 1964.*]

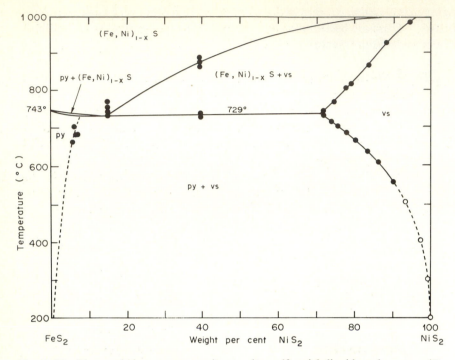

Fig. 5-23 Phase relations among pyrite, vaesite, sulfur-rich liquid, and vapor. (*Re-drawn from Clark and Kullerud, Econ. Geol., 1963.*)

application since the nickel sulfides are comparatively rare and because those portions of the binary that might have been useful do not quench. Clearly considerable solid solution occurs among the disulfides, but vaesite is rarely found in nature, and again the phases are metastable and tend to reequilibrate with cooling. The monosulfides form a complete nonstoichiometric solid-solution series $(Fe,Ni)_{1-x}S$ above about 500°C, the extent of which is indicated in Fig. 5-22. Below 500°C the field begins to "neck," and at about 450°C the tie line between the two areas of solid solution is broken. This is replaced by one between pyrite and pentlandite. If these assemblages were capable of quenching at their depositional temperatures, they would provide valuable geothermometers. However, it is rare to find natural pyrrhotite having more than 1.0 to 2.0 per cent nickel in solid solution (cf. Hawley and Stanton, 1962, p. 47), which strongly suggests that initial high-temperature compositions are not preserved. The Sudbury deposit is almost certainly of very high-temperature formation—perhaps about 1000°C—and were the natural solid solutions quenchable, they should certainly appear here. It seems, however, that slow cooling has been accompanied by almost complete exsolution to pyrrhotite and pentlandite.

In certain cases, such as Sudbury, it may be possible in spite of this to use the phase data for geothermometry. In some instances small pyrrhotite-pentlandite aggregates are completely enclosed in a mineral, such as chalcopyrite, that is not a nickel acceptor. If it be assumed that such aggregates represent the unmixing of a former Fe–Ni–S solid solution in a closed system, experimental homogenization may give a minimum temperature of formation.

The Zn–Fe–S system Although the Cu–Fe–S system may be regarded as the first to be thoroughly examined, the Zn–Fe–S system is undoubtedly the most famous. Through the demonstration of technique by its investigator, Kullerud, and the early promise of the system as a geothermometer, it first showed the potential power of phase studies in understanding the physical significance of sulfide assemblages. Although the early work has now been modified in detail, its value in emphasizing the importance of sulfide equilibria cannot be minimized. It stimulated the development of a whole new field of investigation, and there is no doubt of its place in the history of the theory of ore deposits.

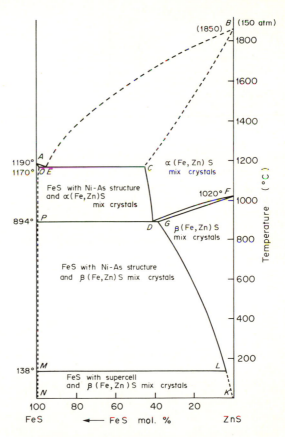

Fig. 5-24 ZnS–FeS-temperature relations in the system Zn–Fe–S system as worked out by Kullerud; this phase diagram is an important landmark in the history of sulfide equilibrium studies. (*Redrawn from Kullerud, Norsk Geol. Tidsskr., 1953.*)

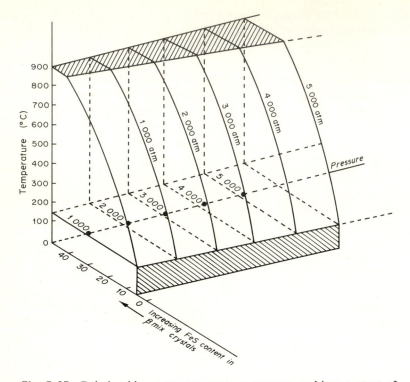

Fig. 5-25 Relationship among temperature, pressure, and iron content of sphalerite as originally deduced by Kullerud. (*Redrawn from Kullerud, Norsk Geol. Tidsskr., 1953.*)

At first sight the system is very promising indeed from the point of view of geothermometry. It is mineralogically simple, with no ternary phases. Sphalerite departs from a 1:1 metal-sulfur ratio by less than 0.1 atom per cent. Pyrrhotite is known to dissolve less than 0.1 per cent zinc, and pyrite dissolves even less. Sphalerite, however, is capable of incorporating substantial quantities of iron and is known to have done so in innumerable natural occurrences. Further, iron does not readily exsolve from sphalerite so that the system is "quenchable."

Figure 5-24 shows the FeS–ZnS join of the Zn–Fe–S system as determined by Kullerud (1953, 1964). The compositions of principal interest here are represented by the curve *LG*—that of (Zn,Fe)S mixed crystals in equilibrium with hexagonal FeS. FeS solubility is affected somewhat by pressure and the presence of such metals as cadmium and manganese. Solubility is inversely related to pressure, as indicated in Fig. 5-25, and a positive temperature correction of 25°C is required for each 1000 atm of added pressure. The presence of cadmium and manganese also inhibits solution, though according to Kullerud the quantities of these elements commonly found in sphalerite are insufficient to affect solubility significantly. According to Fig. 5-24 solubility-temperature relations at 150 atm range from about

6 mol per cent FeS at 200°C to about 38 mol per cent at 870°C.†   This looks to be
a very useful temperature-composition relationship.

    More recently, however, it has been shown (Barton and Toulmin, 1963, 1966)
that the solvus is greatly modified by variation in sulfur vapor pressure.   This was
not recognized by the earlier experimenters because they considered only a part of
the system—i.e., the ZnS–FeS portion—rather than the full Zn–Fe–S system.
This has been rectified by Barton and Toulmin (1966), whose results are shown in
Figs. 5-26 and 5-27.

    Figure 5-26 represents a cut parallel to the T axis and through the FeS–ZnS
joins of the ternary diagrams of Fig. 5-27.   It is thus normal to Fig. 5-27.   The
intersections of the lines A, B, and C of Fig. 5-26 are shown in each ternary diagram
of Fig. 5-27.   It may readily be seen from the latter that the nonstoichiometry of
the pyrrhotite permits variation in the compositions of the coexisting solid phases
if the amount of sulfur is allowed to vary.   Suppose a line representing a constant
Fe/Zn ratio (i.e., a line connecting some point on the Fe–Zn join to the sulfur apex)
intersects the bundle of pyrrhotite-sphalerite tie lines, say, just to the left of A.
Clearly for any given Fe/Zn ratio here a range of possible stable metal–sulfur
ratios exists.   It follows from this that for each value of the sulfur content there is a
different pair in equilibrium.   Thus for a given Zn/Fe ratio at a given temperature,
increase in sulfur causes the development of a pyrrhotite richer in sulfur and a

† At approximately 875°C and 1 atm pressure, sphalerite [β (Zn,Fe)S] containing 56 mol per cent
of FeS inverts to wurtzite [α (Zn,Fe)S] (Barton and Skinner, 1966, p. 295).

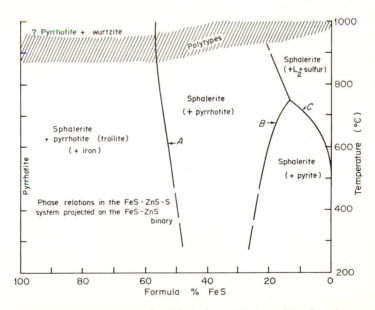

Fig. 5-26 Compositions of sphalerite in equilibrium with other phases
in the Zn–Fe–S system.   Curves A, B, and C appear as points in
Fig. 5.27.   (*Adapted from Barton and Toulmin, Econ. Geol., 1966.*)

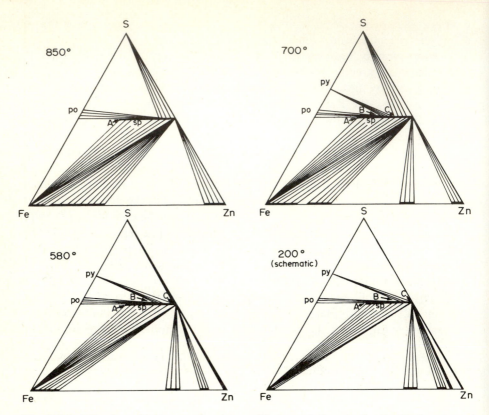

Fig. 5-27 "Condensed" diagram showing equilibrium relations in the Zn–Fe–S system at four representative temperatures. [*From Barton and Skinner, in Barnes (ed.), "Geochemistry of Hydrothermal Ore Deposits." Copyright © 1967 by Holt, Rinehart and Winston, Inc. Reprinted by permission of Holt, Rinehart and Winston, Inc., Publishers, New York. Also from Barton and Toulmin, Econ. Geol., 1963 and 1966.*]

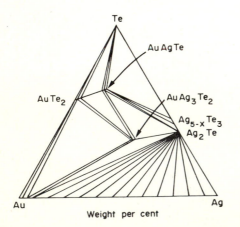

Fig. 5-28 The 300°C isothermal cut in the Ag–Au–Te system. (*Partly hypothetical view from Markham, 1960, redrawn from Markham by Kullerud, 1964, and by the author.*)

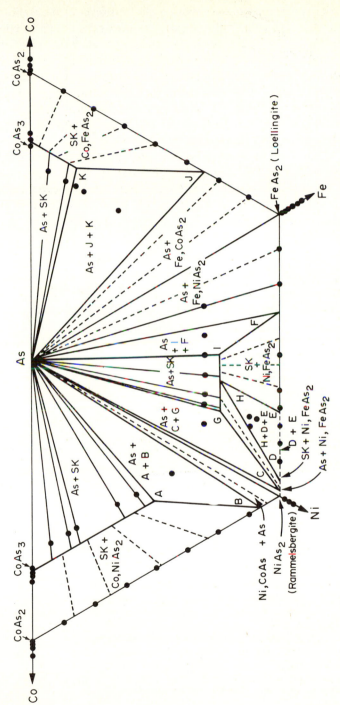

Fig. 5-29 Equilibria in the three ternary systems CoAs₂–NiAs₂–As, NiAs₂–FeAs₂–As, and FeAs₂–CoAs₂–As at 800°C and in the presence of a vapor phase. *A*, 65 per cent NiAs₃ 35 per cent CoAs₃; *B*, 92 per cent NiAs₃ 8 per cent CoAs₂; *C*, 98 per cent NiAs₂ 2 per cent FeAs₂; *D*, 94 per cent NiAs₂ 6 per cent FeAs₂; *E*, 70 per cent NiAs₂ 30 per cent FeAs₂; *F*, 35 per cent FeAs₂; *G*, 76 per cent NiAs₃ 24 per cent FeAs₃; *H*, 61½ per cent NiAs₃ 38½ per cent FeAs₃; *I*, 49½ per cent NiAs₃ 50½ per cent FeAs₃; *J*, 70 per cent FeAs₃ 30 per cent CoAs₂; *K*, 24 per cent FeAs₃ 76 per cent CoAs₃; all values to ±1 per cent.   SK = skutterudite (Co,Ni,Fe)As₃₋ₓ.   *(Redrawn from Roseboom, Carnegie Inst. Wash., Ann. Rept. Dir. Geophys. Lab. Year Book, 1957.)*

coexisting sphalerite poorer in iron.   Reduction in sulfur reverses these effects.   It is therefore necessary to know the sulfur vapor pressure and that this be constant.

It is the curve A of Fig. 5-26 that was thought to be represented by the curve LG in Fig. 5-24, but the latter is well away on the low-zinc side for true ZnS–FeS equilibrium.   This shift has apparently been induced by variation in sulfur pressure from one experimental run to the next in the earlier investigation.   That is, the original study was one of pseudo-$Fe_{1-x}S$–$(Zn,Fe)S$ relations rather than FeS–ZnS–S. The slope of A is clearly far steeper than that of LG so that the formerly supposed temperature-Zn/Fe relationship virtually disappears.   In fact it can almost be said now that the Zn/Fe ratio of sphalerite in equilibrium with *stoichiometric* pyrrhotite is independent of temperature of formation.   The curve C (Fig. 5-26), involving sphalerite-pyrite-liquid sulfur, is unlikely to have application in the great majority of ores.   Curve B, describing the composition of sphalerite coexisting with pyrrhotite *and* pyrite, may, however, be useful.   A limitation here will be the steepness of the slope near 500°C.

The Au–Ag–Te system   Most investigations of sulfide-type systems have naturally been concerned with base metal assemblages, though silver has been included in a few of them.   The Au–Ag–Te diagram is interesting as an example of a metal-tellu.ium assemblage and because it includes the principal elements of a very important class of silver-gold ores.   Figure 5-28 shows the 300°C isotherm as constructed by Markham (1960) and redrawn by Kullerud (1964).

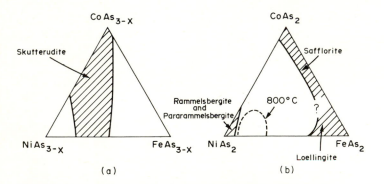

Fig. 5-30   (*a*) The extent of Co, Ni, and Fe substitution in synthetic skutterudite at 800°C.   Skutterudite solid solution is shown by the shaded area.   The extent of such substitution in natural skutterudite is probably about the same.   The compositions $NiAs_{3-x}$ and $FeAs_{3-x}$ are neither compounds nor minerals.   (*b*) The approximate extent of Co, Ni, and Fe substitutions among the natural diarsenides.   The natural solid solutions are shown by shading.   At 800°C the solid solutions cover the entire triangle except for the small area within the dashed lines toward the lower left.   (*Redrawn from Roseboom, Amer. Mineralogist, 1963.*)

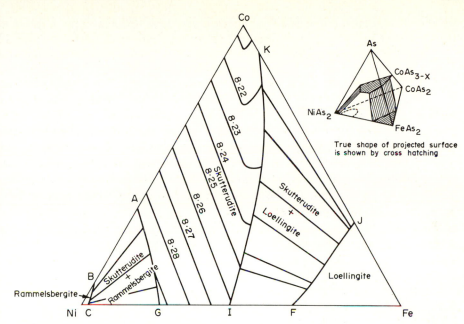

Fig. 5-31 Co, Ni, and Fe substitution in phases stable at 800°C in the presence of vapor (approximately 25 atm of vapor pressure) and crystalline arsenic. The diagram can be viewed as a projection from the arsenic corner of the tetrahedron (see inset) onto the $CoAs_2$–$NiAs_2$–$FeAs_2$ base. The fields of loellingite, *FJ*Fe, and rammelsbergite, *BC*Ni, lie on the base, and the other three fields are projected onto the base. The numbered contours in the field of skutterudite, *AGIK*Co, are lines of equal cell edge. In the field of skutterudite plus loellingite, *KIFJ*, and skutterudite plus rammelsbergite, *ABCG*, tie lines connect pairs of coexisting phases. Compositions are in molecular per cent of Co + Ni + Fe. (*From Roseboom, Amer. Mineralogist, 1962.*)

QUATERNARY SYSTEMS

The Co–Ni–Fe–As system Knowledge of quaternary systems is understandably less than that of binary and ternary systems. Some of them, such as Cu–Fe–Ni–S and Cu–Fe–Zn–S, are important, however, and will undoubtedly receive careful attention in due course.

The Co–Ni–Fe–As system stands out as one of potential geological interest. Its relevance to the exotic native silver:Co–Ni arsenide ore type has spurred interest, and there is now considerable information on it. Understandably much of this comes from various independent studies of constituent binary and ternary systems, though by far the major part is due to Roseboom (1957, 1962).

Figure 5-29 is a composite construction of the three constituent ternary groups in the high-arsenic region at 800°C. Figure 5-30 shows the degree of substitution found by Roseboom in synthetic skutterudite and in the natural diarsenides. Figure 5-31 depicts phase relations, in terms of Co–Ni–Fe ratios, in the presence of arsenic vapor at 800°C.   This is a very high temperature, though in at least some instances it may not be geologically unreal.   Arsenide selvages and patches are far from uncommon at Sudbury, and early stages of cobalt-type vein formation may well have involved quite high temperatures.

CONCLUDING REMARKS

There is now much experimental information on the stabilities of sulfide minerals and on relationships among temperature, pressure, and compositions.   The extent of many potential solid solutions has been determined, and there is now very precise information on a number of invariant points, univariant curves, eutectics, and dystectics.   Phase data for the ore minerals is now comparable in quantity and quality with that for the silicates.†

There is, however, a vital difference between the silicates and the sulfides that greatly affects the applicability of the phase information.   This is the capacity of the former to persist in a metastable state.   Kyanite, sillimanite, certain pyroxenes, and others characteristically form at high temperatures and pressures.   They do not, however, require a continuation of these conditions for continued existence— and coexistence.   Once formed they will, if allowed to do so, persist more or less indefinitely with little or no modification.   While retrogressive effects are well known among metamorphic silicates, these are usually limited and rarely completely obscure the original assemblage.   It is only because of such sluggishness, of course, that metamorphic conditions can be reconstructed.

Unfortunately most sulfide systems show no such reluctance to reequilibrate. The very features that make them so tractable in experiment—their high rates of lattice readjustment and diffusion—permit most of them to constantly reequilibrate in accordance with changing conditions.   With a few exceptions depositional relations are lost and the assemblages observed represent no more than the latest stage of their history—the present.   Thus while the sulfide-phase data are interesting in themselves, they have little geological relevance as yet.   This seems to call for further effort in a field that must surely yield useful results eventually.

RECOMMENDED READING

Barton, P. B., Jr., and B. J. Skinner: Sulfide mineral stabilities, in H. L. Barnes (ed.), "Geochemistry of Hydrothermal Ore Deposits," pp. 236–333, Holt, Rinehart and Winston, Inc., New York, 1967.

Findlay, A.: "The Phase Rule and Its Applications," 9th ed., revised and enlarged by A. N. Campbell and N. O. Smith, Dover Publications, Inc., New York, 1951.

† A situation for which Kullerud, in the first place, is largely responsible.

6
Ore Minerals
in Aqueous Systems

Many ores are formed at high temperatures and pressures, often in environments that are essentially anhydrous but that may be characterized by a high sulfur vapor pressure. Such ores are those formed as parts of igneous rocks (*orthomagmatic* ores), many of those formed in deep, high-temperature veins, and those produced or modified by higher grades of metamorphism.

Many other ores, however, appear to have formed in the first instance under quite low temperatures and pressures and in essentially aqueous media. Some of these ores have formed in low-temperature, near-surface vein systems. Examples are the *limestone–lead-zinc* ores and the *sandstone–uranium-vanadium-copper* ores, both of which are considered in Chap. 16. Others appear to have formed during sedimentation and hence to have been deposited under conditions prevailing at the mud-water interface in some area of marine or lacustrine sedimentation. Temperatures of formation among this class of deposit probably range from about 0 to 100°C, and pressures are those of a few to a few thousand meters of overlying water—i.e., from just over 1.0 to some 900 atm in deep marine troughs of around 10,000 meters. Sedimentary ores known to be forming at present on the floor of the Red Sea are doing so at temperatures around 50 to 60°C and at depths of about 2000 meters—i.e., under confining pressures of around 200 atm. It seems likely that many such ores have formed in the past at considerably lower temperatures and pressures than those currently observed in the Red Sea, and a few may have formed at higher ones.

All ores, precipitating as they do as solids from liquids and gases, involve *heterogeneous equilibria*. Our low-temperature category, with which we are concerned in this chapter, involves, particularly, aqueous solutions and the partial pressures of the various gases with which such solutions are associated. We shall now look briefly at some of the basic principles involved and then at some examples of natural ore solutions.

CHEMICAL PRINCIPLES GOVERNING ORE SOLUTIONS

CONCENTRATION

Clearly the concentration of a solute can vary from one solution to another. Equally clearly it is fundamental that such variations must be expressed precisely. Concentrations are therefore given as

1. Molar concentration, or *molarity*, M—the number of moles (gram molecules) of solute per liter of solution
2. Molal concentration, or *molality*, m—the number of moles per kilogram of solvent

HOMOGENEOUS EQUILIBRIA AND THE LAW OF MASS ACTION

It is well known that there are many reactions that, under a given set of conditions, proceed essentially to completion and are irreversible under those conditions. There are also many reactions that cease before all the reactants have been used up. The classical example of this is the reaction of acetic acid with ethyl alcohol, yielding ethyl acetate plus water. If equimolecular amounts of acetic acid and alcohol are mixed, the reaction proceeds only until about two-thirds of the reactants are consumed; no matter how long the substances are left to react, about one-third of both alcohol and acid remain. This is due to the fact that water reacts with ethyl acetate to produce alcohol and acetic acid. We thus have two reactions proceeding simultaneously:

$$C_2H_5(OH) + CH_3COOH \rightarrow CH_3COO \cdot C_2H_5 + H_2O \tag{6-1}$$

and

$$CH_3COO \cdot C_2H_5 + H_2O \rightarrow C_2H_5(OH) + CH_3COOH \tag{6-2}$$

i.e., we have the reversible reaction

$$C_2H_5OH + CH_3COOH \rightleftharpoons CH_3COO \cdot C_2H_5 + H_2O \tag{6-3}$$

the two sets of reactions eventually reaching an equilibrium dependent on concentration and temperature. The condition of equilibrium at a given temperature is given by the *law of mass action*, which states that the *velocity of a reaction at constant temperature is proportional to the product of the concentrations of the reacting substances*. Thus if we represent Eq. (6.3), and similar equations, by

$$A + B \rightleftharpoons C + D \tag{6-4}$$

the velocity at which A and B react is given by $v = k_1([A] \times [B])$, where brackets indicate molecular concentrations, and k_1 is a velocity coefficient. Similarly for the reaction of C with D, $v_2 = k_2([C] \times [D])$. At equilibrium $v_1 = v_2$, that is,

$$k_1([A] \times [B]) = k_2([C] \times [D])$$

and hence

$$\frac{[C] \times [D]}{[A] \times [B]} = \frac{k_1}{k_2} = K \qquad\qquad (6\text{-}5)$$

K being the *equilibrium constant* for the reaction at a given temperature. Generalizing, it is clear that for any reversible reaction of the form

$$x_1 A_1 + x_2 A_2 + x_3 A_3 \rightleftharpoons y_1 B_1 + y_2 B_2 + y_3 B_3$$

where x_1, x_2, x_3 and y_1, y_2, y_3 are the number of molecules of the reacting substance, the equilibrium constant is given by

$$K = \frac{[B_1]^{y_1} \times [B_2]^{y_2} \times [B]^{y_3}}{[A_1]^{x_1} \times [A_2]^{x_2} \times [A_3]^{x_3}} \qquad\qquad (6\text{-}6)$$

As originally formulated, the law of mass action was concerned not with simple *concentrations* as we have used but with what was visualized as the "active mass"—a quantity perhaps better referred to as *effective concentration* (Krauskopf, 1967, p. 73). *Concentration* and *effective concentration* are not necessarily the same thing, and this has to be allowed for in many cases. The two approach closely in very dilute solutions, but in most cases natural solutions are such that we are not concerned with concentrations; we are concerned with a precise measure of effective concentrations, or what are now referred to as *activities*. This is given by $a = \gamma c$, where a is the activity, c the concentration, and γ the *activity coefficient*. Activities are usually treated as "dimensionless numbers" but may be expressed as gram molecules per liter of solution or per kilogram of solvent, and it is these values that must be used when applying the law of "mass" action.

Consideration of liquid solutions in nature usually involves consideration of solids and gases in equilibrium with the liquid. Thus calculations involving activities of substances in solution also involve activities of solids and gases. For such solids and gases, the following apply:

1. By convention, the activities of solids and pure liquids (e.g., pure H_2O) are taken as unity.
2. The activity of a gas is taken to be numerically equal to the partial pressure of an ideal gas under the conditions concerned.

It may be remembered that Dalton's second law of partial pressures states that *when two or more gases are mixed, the total pressure is equal to the sum of the partial pressures of the constituents or is equal to the sum of the pressures that each constituent would exert if it alone occupied a volume equal to that of the mixture.* Thus for two constituents behaving as "perfect gases," $P = p_1 + p_2$, where P is the total pressure, and p_1 and p_2 are the partial pressures of the constituents. For three or more constituents, $P = p_1 + p_2 + p_3 \cdot \cdot \cdot$. If, for two gases, n_1 is the number of molecules of constituent 1, and n_2 is the number of molecules of constituent 2 in the mixture,

$$p_1 = P \frac{n_1}{n_1 + n_2} \tag{6-7}$$

where $n_1/(n_1 + n_2)$ is the *molar fraction* of constituent 1. In fact Boyle's law (which states that, for a given temperature, a gas conforms to the pressure-volume equation $pv = $ constant) does not hold at high pressures so that Eq. (6-7) is only an approximation at increased confining pressures. However, for surface and near-surface environments, pressures are sufficiently low for gases to be regarded as behaving "ideally," in which case the partial pressure and the activity of a gas can be regarded as identical. Partial pressure is expressed in atmospheres in most geological calculations.

THE LAW OF MASS ACTION AND SOLUTIONS OF ELECTROLYTES

The dissociation of an electrolyte in solution is, like the reversible reactions we have just been considering, dependent on temperature and concentration. For any given temperature there will develop a definite equilibrium between the un-ionized molecules of the electrolyte and the ions; this equilibrium depends on concentration, and the law of mass action may therefore be applied to it. Thus for a dilute solution of acetic acid,

$$CH_3COOH \rightleftharpoons CH_3COO^- + H^+ \tag{6-8}$$

so that, as in Eq. (6-5),

$$\frac{[CH_3COO^-] \times [H^+]}{[CH_3COOH]} = K \tag{6-9}$$

If 1 gram equivalent of an electrolyte is dissolved in a given volume v of solution, and if d is the degree of dissociation (ionization), the amount of each of the ions at equilibrium will be d gram equivalent, and the amount of un-ionized electrolyte will be $(1 - d)$. The concentration of each of the ions will then be d/v, and the concentration of un-ionized molecules will be $(1 - d)/v$. Substituting in Eq. (6-9),

$$\frac{d^2/v^2}{(1 - d)/v} = K$$

that is,

$$\frac{d^2}{(1 - d)v} = K \tag{6-10}$$

K here is the *dissociation*, or *ionization*, *constant*. Equation (6-10) expresses *Ostwald's dilution law*, which is applicable to weak, i.e., slightly ionized, electrolytes. For strong electrolytes Ostwald's law does not apply.

Hydrogen ion exponent: pH   For the ionization of water we have

$$\frac{[H^+] \times [OH^-]}{[H_2O]} = K \tag{6-11}$$

Since the concentration of the ions is extremely small, we may take it that $[H^+] \times [OH^-] =$ a constant $= K_w$.   The concentration of the hydrogen ion in neutral water has been determined from conductivity measurements to be 1.0×10^{-7} g equiv/liter at 25°C.   Since the concentration of hydroxide must be the same, K_w—the *ionic product* of water—is about 1.0×10^{-14}.

The hydrogen ion activity of a given solution is conveniently expressed as the pH, or *hydrogen ion exponent*, for that solution:†

$$pH = \log_{10} \frac{1}{[H^+]} \tag{6-12}$$

where $[H^+]$ represents the activity of the hydrogen ion.

In pure water $[H^+]$ is, as we have just noted, very close to 0.0000001 g equiv/liter; thus $\log_{10} 1/10^{-7} = 7$, so in neutral solutions pH $= 7$.

If $[H^+]$ is increased to 1 g equiv/liter, we may say that $[H^+] = 1 \times 10^{-0}$, in which case the pH of the solution is zero.   Thus pH decreases from 7 as $[H^+]$ becomes greater than that of pure water or a neutral solution.   In a similar way a solution containing 1 g equiv/liter of hydroxide ion has $[OH^-] = 1 \times 10^{-0}$. However, as in any aqueous solution, the product of $[H^+]$ and $[OH^-]$ is

$$[H^+] \times [OH^-] = 1 \times 10^{-14} \qquad \text{at 25°C}$$

if

$$[OH^-] = 1 \times 10^{-0}$$

The $[H^+]$ is given by

$$[H^+] = \frac{1 \times 10^{-14}}{1 \times 10^{-0}}$$

$$= 1 \times 10^{-14}$$

Therefore

$$pH = 14$$

Thus for solutions representing a range from $[H^+] = 1 \, N$ to $[OH^-] = 1 \, N$, pH ranges from 0 to 14, pH $= 7$ representing neutrality.

ENTHALPY, FREE ENERGY, AND FREE ENERGY OF FORMATION

Chemical reactions are accompanied by the liberation or the absorption of heat; i.e., they are exothermic or endothermic.   So that various reactions may be compared with one another in this respect, the amount of heat energy involved in a given

---

† The letter "p" here stands for exponent, i.e., "power."

reaction may be determined by calorimetry, the requisite calorimetric measurements being carried out under standard temperature-pressure conditions—usually 25°C at 1 atm pressure. The amount of heat thus found to be liberated or absorbed is the *heat of reaction* and is expressed as gram calories or kilogram calories (kg cal). Thus when metallic copper and rhombic sulfur are allowed to react,

$$Cu + S_{rhombic} \rightarrow CuS + 11.6 \text{ kg cal} \tag{6-13}$$

i.e., heat amounting to 11,600 calories is evolved for each gram molecule of CuS formed.

Since numerous reactions can take place in stages, one of the most important principles of thermochemistry is *Hess' law of constant heat summation*, which states that *the total heat of a reaction is constant whether such reactions are allowed to take place directly or in stages;* i.e., the heat of any given reaction depends only on the initial and final systems. This is illustrated by the formation of ammonium chloride solution from NH_3, HCl, and water, a process that can proceed via either of two routes.

First method:

$$NH_3 + H_2O \rightarrow NH_3 \text{ (aq)} + 8.4 \text{ kg cal}$$
$$HCl + H_2O \rightarrow HCl \text{ (aq)} + 17.3 \text{ kg cal}$$
$$NH_3 \text{ (aq)} + HCl \text{ (aq)} \rightarrow NH_4Cl \text{ (aq)} + 12.3 \text{ kg cal}$$

$$NH_3 + HCl + H_2O \rightarrow NH_4Cl \text{ (aq)} + 38.0 \text{ kg cal}$$

Second method:

$$NH_3 + HCl \rightarrow NH_4Cl + 42.1 \text{ kg cal}$$
$$NH_4Cl + H_2O \rightarrow NH_4Cl \text{ (aq)} - 3.9 \text{ kg cal}$$

$$NH_3 + HCl + H_2O \rightarrow NH_4Cl \text{ (aq)} + 38.2 \text{ kg cal}$$

Where, as in the case of Eq. (6-13), the reaction concerned is one involving the formation of a compound from its elements, we speak of the heat involved as the *heat of formation*. This may be defined as *the heat evolved or absorbed under standard conditions when 1 gram molecule of a compound is formed by the combination of the relevant elements when in their normal state.* The "state" of the elements is important; one sees intuitively that the "intrinsic energy" of an element is different in its different crystalline forms, and transition from one to the other involves a *heat of transition*, e.g., $S_{monoclinic} \rightarrow S_{rhombic} + 69$ cal. This leads to differences in heats of formation when different forms—amorphous, or belonging to different crystal systems—take part in a given reaction, e.g., the heat of formation of CO_2:

$$C_{diamond} + O_2 \rightarrow CO_2 + 94.42 \text{ kg cal} \tag{6-14}$$

$$C_{graphite} + O_2 \rightarrow CO_2 + 94.20 \text{ kg cal} \tag{6-15}$$

Such heats of formation (or reaction) represent changes in the heat content, or *enthalpy*, of the system concerned. If we represent enthalpy by *H*, we have for any reaction

$$H_{products} - H_{reactants} = \Delta H$$

where ΔH is the heat evolved or absorbed in a reaction taking place *at constant pressure*. No *absolute* quantities are involved; all we know is that ΔH—the heat of formation—represents the *change* in enthalpy of the system resulting from the reaction concerned. Where heat is evolved, ΔH is by convention taken as negative (heat is liberated and hence lost to the "system," enthalpy is reduced, and the enthalpy change is therefore regarded as negative) and vice versa; hence for the formation of CuS, as in Eq. (6-13) (an *exothermic reaction*), $\Delta H = -11.6$ kg cal. Two *endothermic* reactions are

$$H_2 + I_2 \rightarrow 2HI - 11.86 \text{ kg cal} \qquad \Delta H = 11.86 \text{ kg cal} \qquad (6\text{-}16)$$

$$C + 2S_{\text{rhombic}} \rightarrow CS_2 - 22.0 \text{ kg cal} \qquad \Delta H = 22.0 \text{ kg cal} \qquad (6\text{-}17)$$

Where we have compounds reacting with elements, or compounds with compounds, Hess' law of constant heat summation is observed, e.g., in the case of the oxidation, by burning, of H_2S (see Krauskopf, 1967, p. 207):

$$2H_2S + 3O_2 \rightarrow 2H_2O + 2SO_2 \qquad (6\text{-}18)$$

Clearly we are concerned with the heats of formation of H_2S, H_2O, and SO_2. Therefore we note that

$$H_2 + S \rightarrow H_2S + 4.8 \text{ kg cal} \qquad \Delta H = -4.8 \text{ kg cal}$$
$$2H_2 + O_2 \rightarrow 2H_2O + 136.6 \text{ kg cal} \qquad \Delta H = -136.6 \text{ kg cal}$$
$$S + O_2 \rightarrow SO_2 + 71.0 \text{ kg cal} \qquad \Delta H = -71.0 \text{ kg cal}$$

so that, remembering,

$$H_{\text{products}} - H_{\text{reactants}} = \Delta H$$
$$(-136.6) + 2(-71.0) - 2(-4.8) = \Delta H$$
$$-269 \text{ kg cal} = \Delta H$$

The *free energy* of a compound or of a reaction is closely related to the corresponding enthalpies but involves an additional consideration—*entropy*. Entropy involves the degree of order of a system and is a measure of what might be termed the *unavailable energy* S of that system. If a reversible reaction takes place isothermally, at an absolute temperature T, in a nonisolated system (i.e., one with access to heat energy) and absorbs an amount of heat q, no useful work is done and there is simply a transfer of unavailable energy TS. The increase in the unavailable energy will be $T \times \Delta S$, which is equal to q, the heat absorbed. Thus

$$T \times \Delta S = q$$

or

$$\Delta S = \frac{q}{T}$$

Since ΔS has the dimensions of energy divided by temperature, it is expressed as calories per degree.

We may then proceed to define free energy G as

$$G = H - TS$$

However, as with enthalpies, we are concerned not with absolute quantities but with *differences* in free energy. Thus

$$\Delta G = \Delta H - T \times \Delta S \tag{6-19}$$

at constant pressure-temperature.

Again in a manner analogous with enthalpies:

1. The standard free energy of formation of a compound ($\Delta G_f°$) is the *free-energy change* accompanying the formation of 1 gram molecule of the compound from the elements *in their standard states.*
2. Free energies are expressed as calories or kilogram calories (kg cal) and are given for standard temperature-pressure conditions, e.g., 25°C and 1 atm pressure.
3. Energy liberated is expressed as $-\Delta G$; energy absorbed as $+\Delta G$.
4. The law of constant summation applies, and equations can be multiplied, added, or subtracted as in ordinary algebra; thus:

$$H_2 + \tfrac{1}{2}O_2 \rightarrow H_2O \qquad F = -56.7 \text{ kg cal}$$
$$2H_2 + O_2 \rightarrow 2H_2O \qquad F = -113.4 \text{ kg cal}$$

FREE ENERGY AND EQUILIBRIUM

At first sight it would appear that the enthalpy change involved in a reaction might give a clue as to the readiness with which such a reaction takes place; i.e., if two substances are mixed, will they react and, if so, which way will the reaction go, and how energetically will it proceed? For example, it might be intuitively expected that reactions that are highly exothermic (large $-\Delta H$) would proceed more readily and with greater velocity than those with a lower, or positive, ΔH. To some extent these expectations are borne out by experiments, but in many cases they do not hold true. (All reactions are affected by Le Chatelier's principle—*whenever changes in the external conditions of a system in equilibrium are produced, changes occur* (*if possible*) *within the system that tend to counteract the effect of external changes*—and we therefore refer to reactions under *standard conditions.*) Many ionization reactions, for example, are endothermic, but the reactions are quite spontaneous and proceed quite rapidly. When added to water, common salt quickly cools the solution, the endothermic reaction

$$NaCl \rightleftharpoons Na^+ + Cl^-$$

proceeding rapidly and obtaining the necessary heat from the surrounding environment.

The measure of tendency to react we are looking for is, however, given by free energies or, more accurately, by the potential *change* in free energy of a system when the reaction(s) concerned takes place. As already noted, free energy is the sum of enthalpy and the entropy term, and it is the addition of the entropy term—the expression of *degree of order*—that enables us to give a measure of deviation from equilibrium. If we take *increase* in the degree of disorder as a positive change in entropy, a decrease in enthalpy coupled with an increase in entropy will combine to make a reaction go. Where a decrease in enthalpy is coupled with a decrease in entropy, the two, of course, work against each other, making a reaction less likely—another way of saying that the system concerned is closer to equilibrium. Thus the sum of enthalpy and the entropy term indicates divergence from equilibrium: whether a reaction will take place spontaneously, whether it will take place only if heat is supplied from outside, or whether the system is at equilibrium, i.e., if, under given pressure-temperature conditions,

1. $\Delta G < 0$, the reaction will proceed spontaneously, its velocity being greater at large negative values of ΔG and decreasing as $-\Delta G \to 0$.
2. $\Delta G = 0$, the reactants are at equilibrium under the P-T conditions involved.
3. $\Delta G > 0$, the reaction (e.g., the dissociation of water into its elements) will take place only if energy is supplied from the environment.

It will be remembered that the equilibrium constant K may be used to measure deviation from equilibrium. Since the magnitude of ΔG gives information of the same kind, it might be expected that K and ΔG had some systematic relationship. This is the case, and

$$G = -RT \log_e K + RT \log_e \frac{[B_1]^{y1} \times [B_2]^{y2}}{[A_1]^{x1} \times [A_2]^{x2}} \tag{6-20}$$

where $[A_1], [A_2], [B_1], [B_2], \ldots$ indicate activities, R is the universal gas constant (2 cal deg^{-1}/gram mol^{-1}), and T is the absolute temperature. Clearly where all relevant substances are present at unit activity,

$$\Delta G = -RT \log_e K = \Delta G° \tag{6-21}$$

$\Delta G°$ being the *standard free-energy change* for the reaction concerned, i.e., the free-energy change that would take place if all reacting substances were present at *unit activity*.

OXIDATION POTENTIAL

If a metal is placed in an aqueous solution of one of its salts, the metal may become negatively charged relative to the solution. The events involved are analogous to the case of a liquid passing into vapor: just as the vapor continues to form over the liquid until the vapor pressure achieves some definite value, so a metal when placed in a solution of one of its salts will begin to pass into solution as metal ions until the concentration of these ions in the immediately adjacent solution reaches some definite value. This tendency for the metal to go into solution as ions was noted by Nernst (1888), who termed it the *electrolytic solution pressure* of the element concerned. If, for example, a zinc bar is placed in a solution of zinc sulfate, zinc will pass into solution from the bar as Zn^{++}, leaving the bar negatively charged relative to the solution. Ionization will continue until the tendency of the metal to produce ions is balanced by the tendency of the ions to give up their charge. The equilibrium potential difference so developed between the metal bar and the solution is termed the *electrode potential* of the metal in the solution concerned.

Such electrode potentials are different for different metals and vary according to the nature and concentration of the solution involved. Zinc has a high "solution pressure"; it therefore passes ions into, and becomes negatively charged with respect to, solutions of all its salts. Copper, however, has an extremely low solution pressure and becomes positively charged even in the most dilute solutions.

Where a metal becomes negatively charged relative to the solution it may be said to have a *negative* electrode potential, and where it becomes positively charged it may be said to have a *positive* electrode potential. By convention, and for use in comparing the electrode potentials of different metals, the electrode potential of hydrogen (obtained by bubbling hydrogen at 1 atm pressure over a platinum electrode) is taken as zero. This brings us to the well-known electrochemical series of elements—a series indicating which elements will displace others from chemical combination in solution or, put another way, the relative tendencies of the elements to pass into the ionic state. A metal with a more negative potential will displace from solutions of its salts any other metal that comes after it in the series. Thus iron or zinc will displace copper from solutions of copper salts, and in turn copper will displace silver, and so on:

$$Zn + Cu^{++} \rightarrow Zn^{++} + Cu$$
$$Cu + 2Ag^+ \rightarrow Cu^{++} + 2Ag$$

Clearly in the first of these equations zinc loses two electrons and copper gains them, and in the second, copper loses two electrons and silver gains them. In the first case, zinc is oxidized and copper reduced; in the second, copper is oxidized and silver reduced. Such a reaction, involving transfer of electrons from one element to another, is termed an *oxidation-reduction reaction.*

Reactions of this kind may be carried out in such a way that the transfer of electrons from one element to the other takes place along a wire. In the zinc-copper reaction, bars of the two metals may be connected by a wire and their ends then placed in a copper sulfate solution. The zinc dissolves, copper from the solution deposits on the copper bar, and a current—which may be measured by a galvanometer—flows along the wire from the zinc to the copper. The process involved may be regarded as the sum of two parts: at the zinc electrode zinc atoms lose two electrons, $Zn \rightarrow Zn^{++} + 2e^-$, that flow to the copper electrode, where they reduce copper ions of the adjacent $CuSO_4$ solution, $Cu^{++} + 2e^- \rightarrow Cu$, yielding new copper metal. Reactions of this kind are appropriately termed *half reactions*, or *electrode reactions*, addition of one to the other giving the complete *oxidation-reduction reaction*

$$Zn + Cu^{++} \rightarrow Zn^{++} + Cu$$

Clearly the reduced form of a given element will contribute electrons to the oxidized form of any element below it in the electrochemical series. It cannot, however, contribute electrons to the oxidized form of an element above it. The farther apart any two elements are in the series, the more completely will the lower one be reduced and hence scavenged from the solution concerned. Where the two elements involved are close together in the series, the lower of the two is not completely eliminated from the solution, and ions of both elements may exist in substantial amounts.

By using the hydrogen electrode as one-half of the cell, it is possible to assign values, *relative to the standard hydrogen electrode*, to the various electrode potentials and potential differences, thereby giving quantitative expression to the electro-chemical series. Such potentials and potential differences are determined under standard conditions to give *standard potentials* or *standard potential differences*, designated E^0. Under standard conditions the electrode reactions take place at 25°C, with all substances at unit activity—dissolved substances at 1 mol/liter and gases at 1 atm pressure. Potentials obtained under any other conditions are denoted simply by E.

For the full oxidation-reduction reaction

$$Zn + 2H^+ = Zn^{++} + H_2 \qquad E^0 = 0.76 \text{ volt}$$

but by convention

$$\tfrac{1}{2}H_2 \rightarrow H^+ + e^- \qquad E^0 = 0.00 \text{ volt}$$

therefore

$$Zn \rightarrow Zn^{++} + 2e^- \qquad E^0 = 0.76 \text{ volt}$$

The capacity of a natural or experimental environment to induce oxidation or reduction processes is often extremely important geologically. We know that "oxidizing conditions" usually prevail in ground waters above the water table, leading to the development there of oxidation reactions in any sulfide concentra-tions (and any other rocks unstable under near-surface conditions) that happen to be present. "Reducing conditions" are normal below the water table so that any sulfides present there are preserved, and any oxidation products percolating down from above are reduced to form "secondary sulfides" and related minerals. Ancient coral and algal reefs, such as those that often act as traps for oil, frequently contain much decayed organic matter and constitute highly reducing environments. Similarly the waters of the open sea are "oxidizing," whereas those of the lower levels of confined seas (e.g., the Black Sea) and barred basins are often "reducing."

By measuring oxidation potentials in such "natural solutions" it is possible to assign values to the oxidizing or reducing capacity of the environment concerned. While such values must be regarded as semiquantitative rather than fully quantita-tive indicators, they may be extremely useful in understanding and predicting the general nature of chemical processes going on in the natural situation. The poten-tial difference usually measured is that between an inert electrode, such as platinum, and one of known potential, such as calomel. The two electrodes are immersed in the medium concerned—sea water, volcanic hot springs, lagoonal bottom water, etc., and the potential difference determined. The measurement obtained is termed the *oxidation-reduction potential*, or simply, the *redox potential*, and is denoted Eh. The range of Eh found in natural environments is indicated in Fig. 6-1.

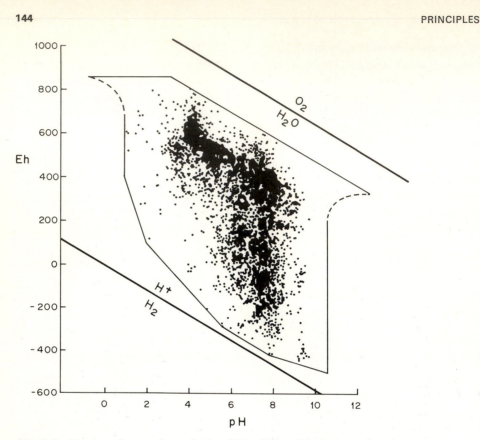

Fig. 6-1 Range and approximate limits of Eh–pH found in the natural environment. Points indicate individual measurements. (*From Baas Becking, Kaplan, and Moore, J. Geol., 1960.*)

The relation between any given redox potential and the standard potential for the electrodes used is given by the *Nernst equation*:

$$\text{Eh} = E^0 + \frac{RT}{nF} \log_e K \tag{6-22}$$

where R is the gas constant (1.9866 cal/deg in heat units), T is the absolute temperature, n is the number of electrons moved in the oxidation-reduction reaction, F is the Faraday constant (96,500 coulombs, or 23,066 cal/volt), and K is the equilibrium constant for the reaction concerned.

THE RELATION OF OXIDATION POTENTIAL AND FREE ENERGY

Under standard conditions the relationship is given by

$$E^0 = \frac{\varDelta G^\circ}{nF} \tag{6-23}$$

where again n is the number of electrons that move in the oxidation-reduction process, and F is the Faraday constant.

THE RELATION BETWEEN REDOX POTENTIAL AND pH

From Eqs. (6-22) and (6-23) it may be seen that redox potential and pH are connected through the standard free energy $G°$ of the reaction concerned.   Substituting in Eq. (6-22) the expression for E^0 given in Eq. (6-23), we have

$$\text{Eh} = \frac{\Delta G°}{nF} + \frac{RT}{nF} \log_e K \qquad\qquad (6\text{-}24)$$

In many of the reactions of interest, K involves the activities of two solids, water, and the hydrogen ion.   Since the activities of the solids and of water are unity by convention, K becomes equal to the hydrogen ion activity.   Further, since $\log_e K = 2.303 \log_{10} K$ and $\text{pH} = -\log_{10}[\text{H}^+]$, Eq. (6-24) becomes

$$\text{Eh} = x - y\,\text{pH} \qquad\qquad (6\text{-}25)$$

This equation is commonly referred to as the *Eh-pH equation* for the reaction concerned, and by its use it is possible to plot the stabilities, and the stability-field boundaries, of a wide variety of mineral compounds in terms of the Eh and pH of the aqueous medium concerned.

Eh-pH DIAGRAMS

The general nature of the Eh-pH diagram and the limits of Eh and pH found so far in the natural environment are shown in Fig. 6-1.

Although pH may reach extreme values in special cases—for example, less than zero in some acid volcanic waters and greater than 10 in some CO_2-free waters in contact with certain silicates—the limiting values for common near-surface environments are about 4.0 and 9.0.   The pH of the open sea is in the range of 7.0 to 8.0, but in confined basins this may drop to around 6.0.   Low values are also developed in and about sulfide ore deposits undergoing weathering, and values of around 3.0 to 6.0 are usually widespread in the soils of marshy and swampy terrains.   This is important in the development of "marsh ores" of iron and manganese, as we shall see in Chap. 13.

The limits of Eh in near-surface environments are set by the values at which water decomposes to form oxygen and hydrogen.   The strongest oxidizing agent common in nature is atmospheric oxygen.   Stronger oxidizing agents than this cannot persist, simply because they would react with water to liberate oxygen. The upper limit of Eh for most near-surface conditions is therefore defined by the reaction

$$\text{H}_2\text{O} \rightleftharpoons \tfrac{1}{2}\text{O}_2 + 2\text{H}^+ + 2e^- \qquad E^0 = 1.23 \text{ volts}$$

Remembering that

$$\text{Eh} = E^0 + \frac{RT}{nF} \log_e K$$

we have

$$\text{Eh} = +1.23 + 0.03 \log_{10}[\text{O}_2]^{\frac{1}{2}} \times [\text{H}^+]^2$$

Since oxygen constitutes about one-fifth of the atmosphere by volume, we may take the partial pressure of oxygen at the surface to be 0.2 atm. Thus

$$Eh = +1.23 + 0.03 \log_{10} (0.2)^{\frac{1}{2}} + 0.059 \log_{10} [H^+]$$
$$= 1.22 - 0.059 \text{ pH} \qquad (6\text{-}26)$$

is the equation giving the upper limit of Eh for most near-surface situations.

In the same way reducing agents are limited to those that do not react with water to produce hydrogen. The limiting potential is therefore that of the hydrogen electrode reaction

$$H_2 \rightleftharpoons 2H^+ + 2e^- \qquad E^0 = 0.00 \text{ volt}$$

For this

$$Eh = 0.00 + 0.03 (\log_{10} [H^+]^2 - \log_{10} [H_2])$$
$$= -0.059 \text{ pH} - 0.03 \log_{10} [H_2]$$

Since the partial pressure of hydrogen at and near the surface cannot exceed 1 atm (it is, of course, usually far less), the maximum possible potential in the presence of water would be

$$Eh = -0.059 \text{ pH} - 0.003 \log_{10} (1)$$
$$= -0.059 \text{ pH} \qquad (6\text{-}27)$$

Lower potentials than this may develop in natural environments, though these are of restricted extent and necessarily do not contain water. Such environments are for the most part those containing large quantities of organic matter.

The Eh-pH field in which the overwhelming majority of natural environments (as distinct from the approximate *limits* of the natural environments as given in Fig. 6-1) fall is therefore as shown in Fig. 6-2.

The construction of Eh-pH diagrams in general follows essentially the procedure we have just used to "define" the natural environment. The relevant electrode reaction is written down and the standard potential for it is obtained. This, together with appropriate activity values for the ions in solution, is then substituted in the equation

$$Eh = E^0 + \frac{RT}{nF} \log_e K$$

which is then converted to

$$Eh = E^0 + \frac{0.059}{n} \log_{10} K \qquad (6\text{-}28)$$

This then gives a pH value corresponding to any given value of Eh at given concentrations of the ions concerned.

An excellent example for illustrating the procedure is provided by the oxidation reactions of iron—reactions that are of great importance in nature, particu-

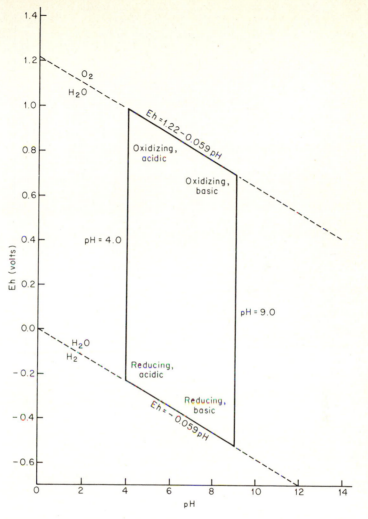

Fig. 6-2 Framework of Eh-pH diagrams; the parallelogram indicates the values more usually found in natural environments. (*From Krauskopf, "Introduction to Geochemistry."   Copyright McGraw-Hill Book Company, New York, 1967, with the publisher's permission.*)

larly in the formation of sedimentary iron ores as we shall see in Chap. 13.   The steps are particularly well set out by Krauskopf (1967, pp. 247–249), to whom the following is substantially due.

The first electrode reaction to consider is

$$Fe^{++} \rightleftharpoons Fe^{3+} + e^{-} \qquad E^0 = +0.77 \text{ volt}$$

Since this reaction is independent of pH (note that neither H^+ nor OH^- appears in this equation), the stability boundary can be plotted initially as a line

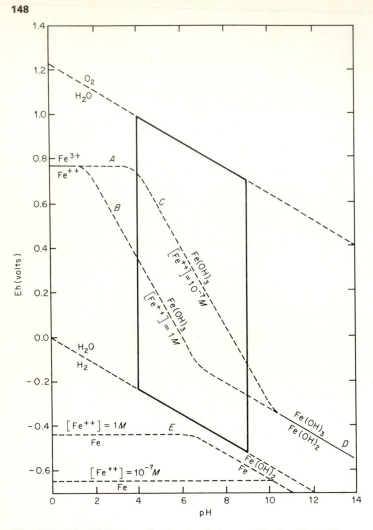

Fig. 6-3 Eh-pH diagram for the simple ions and hydroxides of iron at 25°C. (*From Krauskopf, "Introduction to Geochemistry." Copyright McGraw-Hill Book Company, New York, 1967, with the publisher's permission.*)

(*A*) parallel to the horizontal axis in Fig. 6-3 (from Krauskopf, 1967, p. 248, fig. 9-2). For Eh-pH values above this line $Fe^{3+}/Fe^{++} > 1$, and for values below it the ratio is < 1.

As the pH increases, the environment becomes one in which $Fe(OH)_3$ is stable, and this compound begins to precipitate. The precise pH at which it *does* begin to precipitate is, of course, dependent on the effective concentration of the iron in solution. The electrode reaction with which we are now concerned is

$$Fe^{++} + 3H_2O \rightleftharpoons Fe(OH)_3 + 3H^+ + e^- \qquad E^0 = +1.06 \text{ volts}$$

For this equation

$$Eh = E^0 + 0.059 \log_{10} \frac{[H^+]^3}{[Fe^{++}]}$$

the activities of H_2O and $Fe(OH)_3$ being taken, by convention, as unity.   Thus

$$Eh = 1.06 + 0.059(3 \log_{10} [H^+] - \log_{10} [Fe^{++}])$$

For Fe^{++} equals 1 M and 10^{-7} M, respectively,

$$Eh = 1.06 - 0.177 \text{ pH}$$
$$Eh = 1.47 - 0.177 \text{ pH}$$

The lines (stability-field boundaries) generated by these equations are shown as B and C, respectively, in Fig. 6-3.

If the solution is made progressively more basic, $Fe(OH)_2$ becomes a potentially stable compound at appropriately low Eh values, and the important electrode

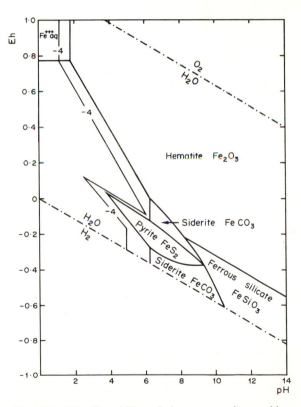

Fig. 6-4   Eh-pH stability relations among iron oxides, carbonate, sulfide, and silicate at 25°C, 1 atm total pressure, total $CO_2 = 10^0$ m, total sulfur $= 10^{-6}$ m, and in the presence of amorphous silica. [*Redrawn from Garrels, "Mineral Equilibria," after fig. 6.23 (p. 161), Harper & Brothers, New York, 1960.*]

reaction is

$$Fe(OH)_2 + (OH)^- \rightleftharpoons Fe(OH)_3 + e^- \qquad E^0 = -0.56 \text{ volt}$$

Substituting,

$$Eh = -0.56 + 0.059 \log_{10} \frac{1}{[OH^-]}$$

Again, the activities of the solids $Fe(OH)_2$ and $Fe(OH)_3$ are unity by convention, and hence

$$Eh = +0.27 - 0.059 \text{ pH}$$

and the field boundary concerned is shown at the lower right of Fig. 6-3.

Stability-field boundaries for Fe^{++} versus Fe for $[Fe^{++}]$ at $1 M$ and $10^{-7} M$

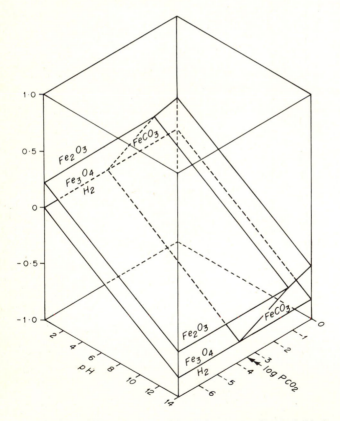

Fig. 6-5 Stability relations of hematite, magnetite, and siderite in terms of Eh, pH, and the partial pressure of CO_2 (given as log P_{CO_2}), at 25°C and 1 atm total pressure. Double arrow at log P_{CO_2} axis indicates the partial pressure of CO_2 in the present atmosphere. [*Redrawn from Garrels, "Mineral Equilibria," after fig. 6.7 (p. 130), Harper & Brothers, New York, 1960.*]

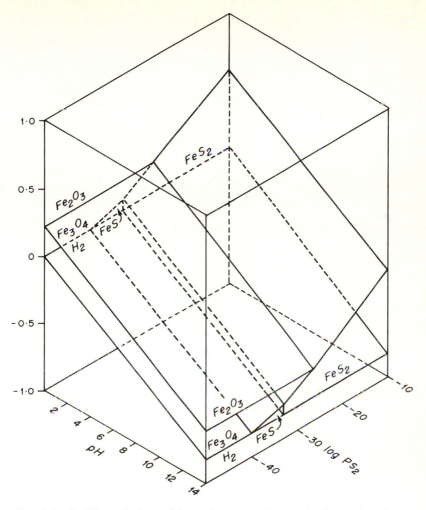

Fig. 6-6 Stability relations of hematite, magnetite, pyrrhotite, and pyrite in terms of Eh, pH, and partial pressure of sulfur (log P_{S_2}) at 1 atm total pressure in the presence of water. [*Redrawn from Garrels, "Mineral Equilibria," after fig. 6.15 (p. 145), Harper & Brothers, New York, 1960.*]

are shown in the lower part of Fig. 6-3. Clearly, metallic iron is stable *in the presence of aqueous solutions* only at Eh values lower than those normally found in near-surface environments. For this reason it is rarely if ever found in the upper crust, except where it occurs trapped as small particles in ultrabasic rocks.

Figure 6-4 (Garrels, 1960, fig. 6-23) demonstrates how stability relationships of a number of commonly occurring minerals can be shown together on an Eh-pH diagram. In this case oxides, carbonates, and sulfides are involved so that the diagram is valid only for specified quantities of dissolved carbonate, sulfur, etc. In this particular system the Eh-pH field for ferrous silicate ($FeSiO_3$) is virtually identical with that of magnetite, and progressive additions of silica to the solution lead to the elimination of magnetite and the substitution of silicate for it. This is shown in Chap. 13 (Fig. 13-5).

Eh-pH PARTIAL-PRESSURE DIAGRAMS

We have already noted that stabilities of compounds in aqueous media can be shown in terms of the partial pressures of important substances such as O_2, CO_2, and S. As would be expected from our Eh-pH equations, it is simple in principle to determine Eh-pH fields over a range of partial pressures of a particular component such as CO_2. The results are plotted on a three-axis diagram, which then gives stability in terms of the three variables.

Figures 6-5 and 6-6 show the aqueous stability relations of hematite-magnetite-siderite and hematite-magnetite-pyrite-pyrrhotite as functions of Eh, pH, and partial pressures of CO_2 and S, respectively.

COMPLEXES

Although the movement of metals in chloride, carbonate, sulfate, and a variety of other solutions is relatively easy under a variety of normal conditions in nature, transport of the "ore" metals in a sulfide-rich regime presents one of the great problems of ore-genesis theory. By far the largest part of all known nonferrous base metal deposits are in the form of sulfide. Clearly sulfide ion was present in large quantities at the site of deposition. But the sulfide ore minerals are highly insoluble; how, then, have enormous quantities of metallic and sulfide ions been transported together *in solution*? Perhaps, of course, they have not; it might be that the metals and sulfur have been derived from distinct sources, each set of ions moving in relatively concentrated solutions until the other solution was met, rapid precipitation then—and only then—ensuing. On the other hand, evidence obtained from many vein deposits suggests that in a great number of cases—though certainly not all—cation and anion have arrived in the same solution. This is supported by observations about modern volcanoes, where sulfur and metallic halide are often clearly being evolved together. The problem remains: how can such large quantities of sulfide be transported in aqueous solution when the "ore" sulfides are known to be so highly insoluble in water?

A possible explanation is that the metallic elements and sulfur move as *complex ions*, rather than as simple ones, such *complexes* having a much greater stability, i.e., capacity for remaining in solution in the ionized state, than the simple ions.

A *complex* may be defined as a (ionic) species formed by the association of two or more simpler species each of which, under the same or different conditions, is capable of independent existence. For example, $PbCl_2$ may dissociate into the simple ions Pb^{++} and Cl^-. Under appropriate conditions, however, it may go to complex ions such as $PbCl^+$, $PbCl_3^-$, or $PbCl_4^{--}$. Similarly Na_2S in solution may yield, in addition to the simple Na^+ and S^{--} ions, complex ions (i.e., *complexes*) such as $NaHS$, NaS^-, and so on. There is no restriction as far as the charges on the ions are concerned so that the term "complex" may be applied to associations of oppositely charged species, to molecular species, or to associations of similarly charged species that have gathered to form "higher order" complexes.

There are two particular groups of complexes that seem likely to be important in the movement of metals in aqueous sulfide-bearing regimes. These are the chloride and the bisulfide-polysulfide complexes.

CHLORIDE COMPLEXES

These develop in the presence of high concentrations of chloride ion, which are readily supplied by such compounds as HCl, NaCl, and KCl. In nature, the former is abundantly supplied in areas of fumarolic and volcanic hot-spring activity. NaCl and KCl are abundant in many deep artesian brines. Helgeson, perhaps the principal advocate of the importance of chloride complexes, considers (1964, p. 88; 1965, p. 1385) that in the temperature-pressure range 100 to 374°C and 1 to 2000 atm a number of chloride complexes have an even greater stability (i.e., capacity for remaining in solution as ions) than HCl, NaCl, and KCl. These are virtually completely dissociated at 25°C but achieve significant degrees of formation at near subcritical and supercritical temperatures. A number of complexes show a greater tendency to form, and a lesser tendency to combine and precipitate, at these higher temperatures.

Helgeson (the interested reader is especially referred to Helgeson's excellent monograph, "Complexing and Hydrothermal Ore Deposition," Pergamon, 1964) has paid particular attention to solution equilibria in the system $PbS-NaCl-HCl-H_2O$ at elevated temperatures (approximately 100 to 374°C), that is, to the solubility of lead in the presence of sulfur in highly chloride-rich brine solutions at high temperatures. Among his conclusions the following are particularly significant:

1. In $NaCl-HCl-H_2O$ solutions in equilibrium with galena at temperatures to 350°C, up to 600 ppm of Pb can be carried in solution—ample for the formation of ore deposits.
2. The formation of lead chloride complexes in the solutions is responsible for this high solubility of galena, the most important being $PbCl^+$ and $PbCl_4^{--}$.
3. $PbCl^+$ is principally responsible for lead in solution at low concentrations of NaCl, carrying 55 per cent of the total lead at 25°C and 95 per cent at 300°C. As the molal concentration (m_{NaCl}) is increased to 3.0, $PbCl_4^{--}$ becomes the principal lead species at low temperatures, carrying 80 per cent of the dissolved lead at 25°C. However, as the temperature is raised above 200°C, $PbCl_4^{--}$ begins to decrease in favor of $PbCl^+$, which is the dominant lead species at all m_{NaCl} up to 3 at all temperatures above 250°C. At 350°C, 95 per cent of dissolved lead occurs as $PbCl^+$ at all concentrations of NaCl.

The sulfides of the other important base metals also show relatively high solubilities, due to the formation of complexes, in high-chloride environments. According to Helgeson, the observed order of stability of chloride complexes for the principal sulfophile elements at 25°C is $Cu^{++} < Zn^{++} < Pb^{++} < Ag^+$.

Thermodynamic calculations indicate, however, that this order may be varied somewhat at higher temperatures.

While the capacity for chloride complexes to increase solubilities of sulfides is not without its doubters (e.g., Barnes, 1965, p. 1386; Czamanske, 1965, p. 1387), observations in nature seem to support Helgeson's ideas. As we shall see a little later in this chapter, natural hot brines in California have been found to carry substantial metal in the presence of sulfur. There is thus little doubt that significant quantities of sulfophile metals can be carried in solution in the presence of low concentrations of H_2S, HS^-, or S^{--} if conditions are appropriate for the formation of chloride complexes. The requirements are an abundance of the chloride ion, a pH a little on the acid side of neutral, and moderately elevated temperatures.

SULFIDE COMPLEXES

It has been suggested (see, e.g., Barnes and Czamanske, 1967) that the mineralogical constitution and the chemical features of the environment of many sulfide-bearing ore deposits indicate that transport cannot have been in acid solutions—and hence that chloride complexes cannot have been involved to any significant extent. It has therefore been put forward as an alternative that the metals are carried in the presence of abundant sulfide, in alkaline solutions, as *sulfide complexes*. Stabilities of the principal aqueous sulfur-bearing species in the H–S–O system as functions of Eh and pH at 25°C and $\Sigma S = 0.1$ are given in Fig. 6-7.

Information given by Barnes and Czamanske (1967) on the principal base metals includes:

1. *Zinc.* The solubility of ZnS in pure water appears to be less than 1.0 mg/liter up to 350°C. However, a complexing process appears to proceed in the HS^- region of Fig. 6-7:

   $$ZnS(s) + H_2S(g) + HS^- \rightleftharpoons Zn(HS)_3{}^-$$

   Barnes (1960) has found this to induce solubilities of up to 2700 mg/liter at 25°C, $P_{H_2S} = 7$ atm, and pH $\simeq 8.2$. If the solution is permitted to become even weakly acidic, the complexing, and hence solubility, decreases markedly.

2. *Lead.* PbS has a solubility of less than 0.1 mg/liter in pure water at all temperatures up to 350°C. At higher pH's, solubility can be raised by a factor of about 10, probably due to the formation of the complex $Pb(HS)_3{}^-$. Sulfide complexing, however, is not nearly so effective as chloride complexing in the case of Pb^{++}.

3. *Copper.* Experiments with chalcopyrite and covellite have shown both to have extremely low solubilities in pure water and in water saturated with H_2S. However, in bisulfide solution the following complexing process appears to occur in the case of covellite:

   $$CuS + H_2S(aq) + 2HS^- \rightleftharpoons Cu(HS)_4{}^{--}$$

   This has been found by Romberger (1967) to yield solubilities of up to 1300 mg/liter at 204°C, 24 atm pressure, and 4.1 m NaHS.

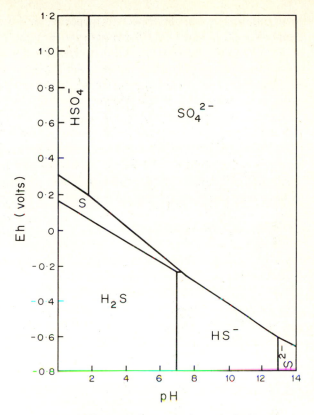

Fig. 6-7 Stability relations of the principal aqueous sulfur-bearing species in the H–S–O system in terms of Eh and pH at 25°C. [*Adapted from Barnes and Czamanske, in Barnes (ed.), "Geochemistry of Hydrothermal Ore Deposits." Copyright © 1967 by Holt, Rinehart and Winston, Inc. Reprinted by permission of Holt, Rinehart and Winston, Inc., Publishers, New York.*]

Extensive experimentation with these and other metals such as silver, gold, cadmium, and iron suggest that in weakly acidic solutions the development of sulfur-bearing complexes is no more than slight, and hence sulfide solubilities are low. The conditions required for the extensive formation of appropriate complexes seem to be high pH, low Eh, and an abundance of HS^-. Under these conditions stable bisulfide complexes appear to form, which in some cases (e.g., copper and zinc) leads to substantial increases in sulfide solubility.

"ORE SOLUTIONS" IN NATURE

These chemical principles we have just considered are most useful to the student of ore genesis. By their application we can determine the limiting conditions under which particular minerals persist or persist together and hence the kinds of chemical products that one might expect from particular kinds of solutions in particular kinds of environment. They also indicate the effects of temperature and pressure on the stabilities of minerals in aqueous environments and the significance of volatile components such as the common gases O_2, CO_2, and H_2S. With the multiplicity of variables involved in nature, it would, of course, be expecting too much to suppose that we could use such principles to *precisely* define an environment or to predict the *exact* nature of its solid products. Present knowledge is simply not sufficient for this, though doubtless the situation will improve with time. All that can be said at present is that application of the principles and techniques of physical chemistry—particularly those pertaining to Eh, pH, and concentration effects—are very helpful in understanding the significance of natural assemblages that appear to have developed in aqueous environments.

MODERN ORE SOLUTIONS

There appear to be four principal categories of modern "ore solution": sea water; "normal" ground waters occurring within a few hundred feet of the land surface; deep "geothermal" waters (including connate waters), such as those of artesian basins, principally of surface origin; and igneous, i.e., *juvenile* waters associated with both volcanic and plutonic activity. Apart from the second category—"normal" ground waters occurring within a few hundred feet of the land surface—these groupings cannot, of course, be regarded as entirely distinct. Sea water may be substantially modified in near-shore zones by additions from large river systems or from artesian outlet beds outcropping on the sea floor of the continental shelf. Sea water is also modified locally, though often substantially, by sea-floor volcanic activity. Deep "geothermal" waters, while generally of surface origin, may have an added volcanic component. Conversely, genuinely juvenile igneous waters may be mixed extensively with both deep and shallow subsurface waters and at the surface may mingle with the waters of rivers and of the sea floor.

With those points in mind we may look briefly at the four categories as potential carriers of ore materials.

1. Sea water A recent estimate of the composition of sea water by Krauskopf (1967) is given in Table 6-1. Krauskopf's estimate for the crust is given for comparison.

It may be noted that sea water does indeed carry small quantities of the metals common in sulfide, oxide, and related ores. The amounts, however, are minute. Total Fe is approximately 0.01 ppm; Mn, 0.002; Cu, 0.003; Zn, 0.01; Pb, 0.00003; and U, 0.003 ppm. In spite of the volume differences between ocean and crust, the amounts of these substances in sea water are very much smaller than in crustal materials. Of course, a large part of these metals eroded from the land surface may have remained incorporated within detritus and never really exposed to solution by the water of the sea. However, if notable quantities of eroded metals *have* initially gone into solution in the sea, much of this material must have been abstracted from the sea water by chemical means at some later stage. Through such reasoning it has been suggested that sea water is in fact a

Table 6-1   Estimated composition (in ppm) of the earth's crust and of sea water.   (From Krauskopf, 1967, p. 639)

Element	Crust	Sea water
O	46.4×10^4	857,000
Si	28.2×10^4	3.0
Al	8.2×10^4	0.01
Fe	5.6×10^4	0.01
Ca	4.1×10^4	400
Na	2.4×10^4	10,500
Mg	2.3×10^4	1350
K	2.1×10^4	380
Ti	5700	0.001
H	1400	108,000
P	1050	0.07
Mn	950	0.002
F	625	1.3
Ba	425	0.03
Sr	375	8.0
S	260	885
C	200	28
Zr	165	
V	135	0.002
Cl	130	19,000
Cr	100	0.00005
Rb	90	0.12
Ni	75	0.002
Zn	70	0.01
Ce	67	5.2×10^{-6}
Cu	55	0.003
Y	33	0.0003
Nd	28	9.2×10^{-6}
La	25	1.2×10^{-5}
Co	25	0.0001
Sc	22	0.00004
Li	20	0.17
N	20	0.5
Nb	20	0.00001
Ga	15	0.00003
Pb	12.5	0.00003
B	10	4.6
Th	9.6	0.00005
Sm	7.3	1.7×10^{-6}
Gd	7.3	2.4×10^{-6}
Pr	6.5	2.6×10^{-6}
Dy	5.2	2.9×10^{-6}
Yb	3	2.0×10^{-6}

highly dilute ore solution, and where it has found its way into appropriate environments, precipitation of ore minerals has taken place; the sea water supplied the cations, the environment the anions, and the two combined to form a compound whose solubility product is greatly exceeded in terms of the Eh, pH, temperature, etc., of the environment concerned.

The principal anions are oxygen and sulfide sulfur. Recent and Modern sea-floor oxidation of iron and manganese has led to the development of *manganese nodules* (composed of a variety of iron and manganese compounds, together with extensive assemblages of minor metals) over extensive tracts of the ocean floor, particularly in the Pacific. This is dealt with in some detail in Chap. 14. Paleozoic, and particularly Mesozoic, sea-floor oxidation of iron has led to the formation of extensive "ironstone" deposits, as we shall see in Chap. 13. Generation and provision of marine sulfide sulfur is thought to result principally from anaerobic sea-water sulfate reduction in areas of the sea floor characterized by a minimum of water movement and the accumulation of large quantities of organic matter. These stagnant, putrid, or *euxinic*, environments develop in landlocked seas (e.g., the Black Sea), in fjords, and in deep basins (e.g., the Santa Catalina and other faulted basins off the Californian coast). The H_2S content of some of these areas —both within the intergranular pore waters and for some distance up the overlying water column—may be quite high, and a variety of iron-sulfur compounds have been found in them.

However, we have no good evidence *so far* to suggest that sea water *by itself* has constituted an ore solution responsible for the formation of any large, highly concentrated ore deposits. While manganese nodules are widespread, no large *concentrations* of manganese and iron oxides have been found forming in simple situations on the ocean floor. Similarly in areas of sulfate reduction, only small quantities of metal sulfides have been found, and of these, virtually all are iron sulfide. Although in older marine sedimentary rocks that do contain large metal sulfide concentrations, and significant quantities of Cu, Pb, and Zn sulfides in addition to those of iron, the sulfides are often conspicuously confined to carbon-rich, "reduced" facies, the sulfide concentrations are rarely, if ever, as extensive as the facies in which they occur. This seems to indicate that the ores have not resulted from a simple interplay of sea water with a particular environment.

2. Ground water The chemical characteristics of ground water vary with respect to the water table and, quite often, the seasons. In general, ground water above the water table is oxidizing and on the acidic side of neutrality, whereas that below the water table has a low Eh and tends to be alkaline. Where a given climate is characterized by well-defined wet and dry seasons, the large addition to soil moisture during the wet season raises the water table and induces a higher Eh and lower pH in the upper parts of the zone of saturation. In tundra areas, covered by deep soil or glacial till and characterized by cold, moist conditions, marshes and swamps are widespread and the subsurface waters are normally strongly acidic. Since the water table in these areas is high, most of the subsurface water exhibits both a low pH and a low Eh.

Both oxide and sulfide—and carbonate—minerals are dissolved, transported, and deposited by ground waters. The sulfides are unimportant apart from those dissolved and redeposited during the weathering of an earlier formed ore body.†

A more important activity of ground water is the solution and transport of iron and manganese as the bicarbonates. This is especially important in tundra regions, where the ground waters, with their typical combination of low pH and low Eh, are able to leach iron and manganese from the soil and to transport these *stably* as $Fe(HCO_3)_2$ and $Mn(HCO_3)_2$. The two metals are precipitated as oxides, hydroxides, or carbonates when the bicarbonate solutions suddenly debouch into environments of appropriately higher Eh and pH. Ground waters acting in this way have contributed to the formation of substantial quantities of "marsh" or "lake" ores of Fe, Mn, and Fe–Mn. They thus constitute a particular kind of ore solution.

3. Deep geothermal water

Most artesian and subartesian waters are hot and saline. Where the relevant aquifer is a deep one, the water may reach very high temperatures ($T > 300°C$) as a result of the thermal gradient. Movement over long distances through the aquifer inevitably leads to the dissolving of soluble matter, and the waters may become quite highly saline as a result. Since many aquifers are sedimentary, and since in some cases evaporites form part of the sedimentary sequences concerned, it is not surprising that some deep geothermal waters become highly saline—so much so that they may best be referred to as *concentrated brines*.

Two outstanding examples of heavy metal-rich geothermal brines have been recognized recently: one tapped by deep bores in the Salton Sea area of California, the other detected by oceanographic sampling of the sea-floor and bottom waters of the central part of the Red Sea.

The *Salton Sea brine* is highly concentrated, containing some 25 to 30 per cent dissolved solids. It was discovered early in 1962 during drilling for geothermal power in the Imperial Valley, near the Salton Sea, and its high heavy metal content was recognized as a result of subsequent analysis of a siliceous scale deposited on the walls of the brine-discharge pipes. Two analyses of the brines are given in Table 6-2 (Skinner et al., 1967, p. 318) and three of the scale in Table 6-3 (Skinner et al., p. 320). The scale is composed of opaline silica in which are set extremely fine crystals, crystalline grains, and dendrites of sulfides and native silver. According to Skinner and his coauthors: ". . . between 5 and 8 tons of a scale containing approximately 20 per cent copper and precious metal values of several hundred ounces of silver plus one-tenth ounce of gold per ton accumulated during a 3-month well test in the summer of 1962" (1967, p. 316).

Sulfide minerals identified in the scale are digenite [Cu_9S_5 or $(Cu,Ag)_9S_5$], bornite [Cu_5FeS_4 or $(Cu,Ag)_5FeS_4$], chalcocite [Cu_2S or $(Cu,Ag)_2S$], stromeyerite, arsenopyrite, tetrahedrite, chalcopyrite, and pyrite. Table 6-4 (Skinner et al.,

---

† This leads to the *secondary (supergene) enrichment* of primary ores and in particular may be very important in the enrichment of the low-grade "porphyry copper" deposits of Chap. 12.

Table 6-2   Estimated composition of dissolved matter (in ppm) of brines produced from two bores in the Salton Sea area, California.   (After Skinner et al., 1967, p. 318)

Constituent	Bore No. 1 IID	Bore No. 2 IID
Sodium	50,400	53,000
Potassium	17,500	16,500
Lithium	215	210
Rubidium	137	70
Cesium	16	20
Ammonia (NH_4)	409	
Calcium	28,000	27,800
Magnesium	54	10
Barium	235	250
Strontium	609	440
Chlorine	155,000	155,000
Fluorine	15	Not reported
Bromine	120	Not reported
Iodine	18	Not reported
Sulfate (SO_4)	5.4	⎰ Total sulfur
Sulfide sulfur	16	⎱ = 30
Boron	390	390
Iron	2090	2000
Manganese	1560	1370
Silver	0.8	2
Copper	8	3
Lead	84	80
Zinc	790	500
Arsenic	12	Not reported
Antimony	0.4	Not reported
CO_2 as HCO_3	150	690
Silica	400	400
Total reported	258,360	258,765

1967, p. 322) gives analyses and sulfide mineralogies of samples of the more sulfide-rich portions of the scale.

To say that hot brines of this kind are potential ore solutions is putting it very mildly indeed.   Geological, chemical, and isotopic evidence has led White (see White, 1968; Skinner et al., 1967) to the view that the brines are principally of local surface origin, possibly with a very minor volcanic component.   White (1968, p. 313) suggests that rainwater falling in the Chocolate Mountains area, several miles to the east of the brine reservoir, seeps underground and descends perhaps 10,000 feet or more along front range faults.   As it percolates through the rocks it may collect salts from unexposed evaporites and, by leaching, from associated detrital material.   The water migrates westward and eventually upward in the sediments of the Colorado Delta area.

A remarkable feature of the brine is that, on an atom-for-atom basis, it contains much more of the heavy metals than of sulfide sulfur.   White (p. 313) notes that total sulfide ion is in the general range 15 to 30 ppm, or less than 5 per cent of that required to precipitate all available sulfophile metals as sulfides (Skinner and his colleagues had noted in 1967 that analysis of the brines gave sulfophile metal–sulfide sulfur ratios of 8:1 to 15:1). Since the overall pH is about 5.5 at 25°C (calculated as 4.6 at 300°C), almost all the dissolved sulfide must be present as H_2S and virtually none as S^{--}. The metals themselves are presumed by White to be present as chloride complexes. The problem does not appear to be the old one of rendering and keeping sulfides soluble in the presence of excess sulfide but rather of finding enough sulfur to precipitate the abundant metals as sulfides!

The *Red Sea brines* are also a recent discovery, and like those of the Salton Sea, they were discovered accidentally. Although the presence of warm brines in the median zone of the Red Sea had been known in a general way for some 80 years, it was not until 1964 that, more or less accidentally, the unusual nature of the brines was fully recognized. The Red Sea brines are now known to occur as three "pools," in adjacent depressions in the central portion of the median valley of the Red Sea (Fig. 6-8). Two of the pools are much larger than the third: the

Table 6-3   Estimated compositions (weight per cent) of three samples of sulfide-rich scale from pipes discharging Salton Sea brines at well No. 1 IID. (After Skinner et al., 1967, p. 320)

	W–769 (220 ± 30°C)	W–767 (170 ± 30°C)	W–768 (130 ± 20°C)
Si	M†	M	M
Cu	M	M	M
Fe	5.0	7.0	6.0
Ag	7.0	1.3	2.8
As	0.18	0.10	0.10
Sb	0.72	0.17	0.25
Bi	0.11	0.004	0.009
Mn	0.055	0.42	0.34
Co	0.0050	0.0004	0.0006
Al	3.	1.0	1.4
Ga	0.0016	0.016	0.012
Yb	0.0002	0.0002	0.0002
Be	0.0036	0.046	0.037
Mg	0.014	0.0085	0.0080
Ca	0.55	0.60	0.55
Sr	0.008	0.003	0.003
Ba	0.014	0.020	0.0090
B	0.019	0.11	0.080
Pb	0.012	0.011	0.007
Sn	0.002	0.002	0.002
Na	1.	1.	1.
K	1.5	1.	1.5
Ti	0.0007	0.0015	0.0007

†M = major constituent.

Table 6-4   Quantitative x-ray fluorescence analyses (weight per cent) and mineralogical
assemblages of sulfide-rich scale from No. 1 IID well, Salton Sea geothermal area.   (After
Skinner et al., 1967, p. 322; analyst H. J. Rose)†

Sample No.	Temp., °C	Cu	Ag	Fe	As	Sb	S	Mineral content
W–769 (1)	220 ± 30	10.0	1.2	3.8	0.23	1.05	6.6	bn, cp, Ag
W–769 (2)	220 ± 30	43.6	5.8	7.1	0.15	0.69	22.5	bn, cp, cc II, asp, Ag
W–767 (1)	170 ± 30	14.0	1.2	15.7	0.20	0.52	10.2	dg, py
W–767 (2)	170 ± 30	27.5	3.1	8.6	0.13	0.66	13.4	bn, cc II, Ag
W–767 (3)	170 ± 30	23.6	1.0	10.6	0.13	0.57	12.8	dg, bn
W–768 (1)	130 ± 20	10.1	1.0	25.4	0.30	0.55	9.9	dg, trace td
W–768 (2)	130 ± 20	13.4	1.2	18.8	0.23	0.53	10.8	dg, trace td
W–768 (3)	130 ± 20	19.2	1.6	11.3	0.15	0.56	12.2	dg, bn, py
W–768 (4)	130 ± 20	12.4	1.6	14.0	0.15	0.53	10.9	dg, bn
W–768 (5)	130 ± 20	11.5	2.1	12.0	0.14	0.45	11.6	dg, strm
W–768 (6)	130 ± 20	28.2	3.4	9.8	0.11	0.55	14.2	bn, cp, cc II, strm, Ag

† Individual bands in the scale are indicated by numbers in parentheses, band 1 being the
first band deposited in each case.   Mineralogical abbreviations: asp = arsenopyrite; bn =
bornite; cc II = dense Cu_2S; cp = chalcopyrite; dg = digenite; py = pyrite; strm = stromeyerite; td = tetrahedrite.

Atlantis Pool is reported to be about 12 × 5 km, the Discovery Pool about 4 ×
2.5 km, and the smaller Chain Deep about 3 × 0.66 km.   Principal features of
the brines of the three pools are illustrated in Table 6-5.   Relative to sea water, the
brines are of high Ca/Na, slightly lower K/Na, and very low in Mg/Ca, Br/Cl,
SO_4/Cl, and HCO_3/Cl.

Although the sediments accumulating on the floors of the Red Sea brine
pools do not attain concentrations of single metals as high as the 20 per cent of

Table 6-5   Principal physical and chemical features (approximate and
necessarily somewhat generalized) of the Red Sea brine pools.   (After
various authors in Degens and Ross, 1969)

	Atlantis	Discovery	Chain
Temp. (max. measured, °C)	56	44.9	34
Salinity (ppm)	255,000	255,000	74,000
Density	1.20	1.20	?
pH	≥6.5	?	?
Eh (mv)	≥ −0.1	?	?
Chloride (ppm)	156,200	155,200	42,000
Fe (g/kg)	8.1×10^{-2}	2.7×10^{-4}	?
Mn (g/kg)	8.2×10^{-2}	4.6×10^{-3}	5.0×10^{-3}
Cu (g/kg)	2.6×10^{-4}	7.5×10^{-5}	?
Zn (g/kg)	5.4×10^{-3}	7.7×10^{-4}	?
Pb (g/kg)	6.3×10^{-4}	1.7×10^{-5}	?

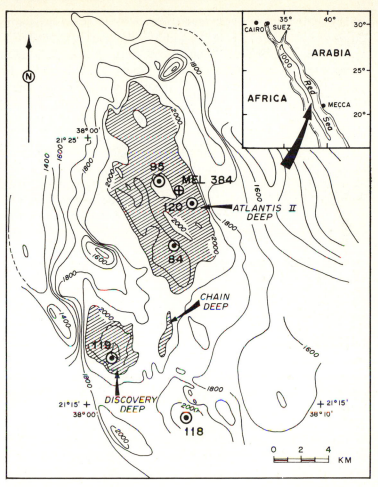

Fig. 6-8 The positions of the three hot brine pools in the median valley of the Red Sea.   Sampling localities are indicated by number.   [*Redrawn from Cooper and Richards, in Degens and Ross (eds.), "Hot Brines and Recent Heavy Metal Deposits in the Red Sea," Springer-Verlag New York Inc., New York, 1969.*]

copper in the Salton Sea pipe scales, their metal contents are nonetheless notable. The metal contents of some individual metal-rich samples recovered from the floors of the pools range up to some 21.0 per cent zinc expressed as ZnO, 4.0 per cent copper expressed as CuO, 0.8 per cent lead expressed as PbO, 85.0 per cent iron expressed as Fe_2O_3, and 5.7 per cent manganese expressed as Mn_3O_4.   In one analysis Mn expressed as Mn_2O_3 reached 45 per cent, although the MnO contents are generally below 5 per cent (Bischoff, 1969).

The sediments themselves are largely gel-like, and their principal nondetrital constituents are amorphous iron oxides and hydroxides, amorphous silica, goethite, hematite, dolomite, montmorillonite, anhydrite, siderite, rhodochrosite, sphalerite, pyrite, marcasite, and some amorphous manganese oxides-hydroxides. Copper and lead are present as both sulfides and carbonates.

These Red Sea sediments are notable in that they are the first chemical sedimentary ores to be found in the process of formation. Indeed it might be said that, apart from marsh and lake ores of iron and manganese (see Chaps. 13 and 14), they are the first extensive deposits of "ore grade" of *any kind* to have been found in the process of formation. There is therefore no doubt at all that the parent solutions can justifiably be called "ore solutions."

The precise nature of these solutions is, of course, not so easy to determine as in the case of the Salton Sea brines. However, by comparing the brine pools with the normal waters of the Red Sea elsewhere, it is possible to deduce many of their features. Present opinion is that they are deep geothermal brines of meteoric rather than volcanic origin and that they have acquired their high salinity by percolating through evaporite beds well known to be abundant in the Red Sea region. The very high Na content of the brines suggests that the evaporites were unusually rich in halite compared with the other common evaporite minerals. The chemical differences between the Salton Sea and Red Sea brines are also thought to suggest that the subsurface temperatures of the latter are considerably less than those of the Salton Sea waters. White (1968, p. 321) suggests that the subsurface temperatures of the Red Sea brines are probably of the order of 100 to 150°C. The salinities of the two brines are clearly very similar (slightly more than 25 per cent) and, judging from the presence of sulfophile elements as compounds other than sulfides in the Red Sea sediments, the brines of the latter are—like those of the Salton Sea area—low in sulfide ion relative to the available sulfophile metal ions.

4. Igneous solutions   That heavy metals are emitted in gases and vapors, and in aqueous solutions, from volcanic vents of various kinds has been known for a very long time. Extensive observations and descriptions of precipitates and sublimates deposited round the principal Mediterranean volcanoes were being made well over a century ago. More recently, deep drilling in volcanic areas for geothermal power has yielded much information on volcanic hydrothermal solutions. Observation of seaboard and particularly of sea-floor volcanic activity has, of course, been more difficult, but significant observations have been made over many years and improved facilities for oceanographic work are now permitting sampling of deep areas of the ocean bottom.

Studies of "igneous" solutions in volcanic areas are beset by a very major problem: how much of the water and of its contents is of immediate igneous origin, and how much has been gathered to it as it migrated from the magma chamber to the surface? That much normal ground water can be added to the igneous "juvenile" water has been firmly established by isotopic studies. How much of the sulfur, CO_2, halogens, heavy metal compounds, and others has been added from place to place we do not know, and to find out would clearly be very difficult indeed.

Probably because of frequent mixing, "volcanic" emanations *as a class* show

no clear chemical differences from "nonvolcanic" geothermal waters *as a class.*
In a general and somewhat subjective way it may be said that the products of volcanic areas tend to contain notably more sulfurous matter (H_2S, HS^-, S^{--},
SO_2^{--}, SO_3^{--}, polythionates, S) and more fluorine than the waters of nonvolcanic areas.   On the other hand, the nonvolcanic geothermal brines are far
more saline than are the volcanic products.   Despite this lack of distinctiveness it
can, however, be said without doubt that the gases, vapors, and derived aqueous
solutions of volcanic areas do often contain very significant amounts of Fe, Mn,
Cu, Zn, Pb, S, and other elements prominent in various types of ore deposits.
The accumulation of copper, lead, zinc, and silver compounds—particularly
oxides, carbonates, chlorides, and sulfides—about some of the Mediterranean
volcanoes has been recognized for centuries.   Perhaps the most spectacular and
best documented example of volcanic emission of heavy metal compounds is that
of the Valley of Ten Thousand Smokes, near Mount Katmai in Alaska.   Following
a violent eruption in 1919 a fumarole field developed in a valley filled with rhyolitic
pumice and ash—the Valley of Ten Thousand Smokes.   The fumaroles, hot
springs, and the mineral encrustations derived from them were examined in detail
by E. T. Allen, E. G. Zies, and C. N. Fenner (see Fenner, 1933) who found abundant chlorides and sulfides of lead, zinc, and copper and a great variety of other
metal compounds in lesser amounts.   A notable feature of the emanations was the
great abundance of HCl, H_2SO_4, and HF.   In general the gases and their associated heated waters seem to have had an extremely low pH, and an Eh sufficiently
low for the deposition of sulfides, but insufficient sulfide for the precipitation of all
available sulfophile metals as sulfide.

The contribution of volcanic solutions to the sea has been observed in a
qualitative way in a number of localities.   The precipitation of sulfides, particularly galena, in shallow waters about the volcanic island of Vulcano in the Mediterranean is a long-standing observation.   The formation of large quantities of ferric
hydroxide in the sea water overlying submarine hot springs near Santorin in the
Greek Islands, Ebeko volcano in the Kurile Islands, Rabaul Harbour in New
Britain, and about Simbo in the Solomon Islands is also well known.   A noteworthy semiquantitative investigation of the precipitation of heavy metals on the
deep sea floor off South America by Bostrom and Peterson (1966) has established
that quite considerable quantities of such elements as Mn, Cu, Pb, and Ba, in addition to iron, may be sedimented about areas of submarine volcanic effluent (see
also Chaps. 13, 14, and 15).   There seems no doubt in this case that the solutions
concerned *were* volcanic.   The sampling was carried out on the East Pacific Rise,
a zone of suspected volcanicity, and metal-rich sediments were obtained at distances
up to some 2000 miles from the nearest continental mass—South America.   There
is thus no chance that the solutions were geothermal brines of Salton Sea or Red
Sea type.   In addition the concentration of heavy elements in the sediments was
clearly related to a zone of anomalously high sea-floor heat flow—a strong indication of nearby igneous activity (see Fig. 14-6).   No chemical information was obtained on the solutions themselves, though the presence of abundant Ba suggests
that these did not contain significant sulfate.   If they did, the extremely low

solubility of $BaSO_4$ would have precluded the transport of barium in solution, though it may, of course, have been delivered to the sea floor in very finely particulate form.   Since the precipitation of such constituents as calcium carbonate and ferric and manganese oxides would have been substantially dictated by the sea water, nothing else can be said concerning the temperature, pH, or Eh of the volcanic solutions.

ANCIENT ORE SOLUTIONS

Very conveniently, small quantities of aqueous solutions that we know to have been present at the time of ore formation are sometimes found preserved.   These small samples are found in minute cavities in many ore minerals and their associates and are called *fluid inclusions*.   Such inclusions—or at least those termed *primary* fluid inclusions—consist of fluid trapped in imperfections that have developed in crystals as they have grown.   As we shall see in Chap. 8, many crystals growing in open space do so by the rapid initial proliferation of dendritic branches followed by the "filling in" of the spaces between them.   Imperfect filling of these spaces between dendrites results in the trapping of small bubbles of the "ore solution" present at the time, and the result is a fluid inclusion.   Inclusions of this kind were probably first noted by the English chemist Sorby in 1858 and provide us with our most accurate clues concerning the nature of the solutions from which the surrounding mineral was precipitated.   Care must, however, be taken to ensure that the inclusions chosen for examination are *primary*, i.e., *formed at the time of growth of the immediately surrounding portions of the crystal concerned*.   In some cases crystals may be subjected to stress long after their formation and suffer fine cracking as a result.   Such cracks may be filled with solutions much younger than —and quite unrelated to—the solutions from which the crystal precipitated; the cracks may then be sealed off by secondary growth and the intruding liquid trapped and preserved.   Because of this, extraction and analysis of fluid inclusions should always be preceded by the most careful microscopic examination of the openings concerned.

   Fluid inclusions usually occupy only a very small fraction of the total volume of a crystal.   Roedder, the outstanding modern investigator of fluid inclusions, estimates (1967, p. 517) that they seldom constitute more than a few tenths of 1 per cent of a crystal, in spite of the fact that they may be very numerous.   Many samples of common minerals owe their whiteness to the presence of very large numbers of inclusions—concentrations of the order of 10^9 per cubic centimeter. However, in spite of such enormous numbers, average volume is minute and total volume is usually only of the order of 0.1 per cent of the containing crystal. Instances of inclusions constituting 1.0 per cent of the crystal volume are uncommon but not rare; very occasionally a crystal, or a particular zone of a crystal, may contain as much as 5 per cent of inclusion.   As far as is known, significant leakage of inclusions does not occur unless the containing crystal becomes heavily deformed.

   The study of inclusions, while requiring great finesse and attention to detail (see Roedder, 1967, p. 515, for an excellent summary account), provides information on the composition and temperatures of ore-forming fluids and—though to a much lower degree of accuracy—pH. The determination of Eh naturally involves great problems in technique and has still to be successfully accomplished.

Inclusions may consist of liquid plus precipitated solid, liquid plus immiscible liquid, and liquid plus gas. The overwhelming majority of inclusions are composed of aqueous solutions together with a vapor bubble. Water is thus by far the most important constituent of inclusions *in toto*. In some environments, such as those of limestone–lead-zinc deposits (see Chap. 16), oil occurs as a second liquid. CO_2 is the commonest gas, though hydrocarbon and sulfur gases are also widespread. The aqueous solution component is almost always highly saline, with total salt concentrations ranging from near zero to over 40 weight per cent. According to Roedder (1967, p. 538) the salts include Na^+, K^+, Ca^{++}, Mg^{++}, Cl^-, and SO_4^{--} as major constituents and Li^+, Al^{3+}, BO_3^{3-}, PO_4^{3-}, SiO_3^{--}, HCO_3^-, and CO_3^{--} as minors. Many other elements and radicals occur, here and there, as traces. Usually, though not always, Na^+, K^+, and Cl^- are the principal ions, and crystals of NaCl and KCl are the most common solid components of inclusions. Some inclusions contain small quantities of sulfide precipitate— usually minute but reaching 1.0 per cent in some cases—indicating that the heavy metals may occasionally be present in significant quantities. The presence of such heavy metals in solution has now been established by Czamanske and his colleagues (1963), who, using neutron activation analysis on inclusions in quartz from Creede, Colorado, and fluorite from Illinois, obtained the results of Table 6-6.

Temperatures of ore formation (i.e., of the solution at the time of trapping) are determined by heating of the specimen and observing the temperatures at which

Table 6-6 Leachable copper, manganese, and zinc from inclusions in Creede, Colorado, quartz and southern Illinois fluorite. (From Czamanske, Roedder, and Burns, 1963, p. 402)†

Metal	Water leach (μg)	Concn. in inclusion fluid‡ ($\mu g/ml$)	Acid leach (μg)	Concn. in inclusion fluid ($\mu g/ml$)
Quartz sample (24.5 mg of fluid extracted)				
Cu	1.4	60	2.0	140
Mn	15.3	620	1.7	690
Zn	10.1	410	22.4	1330
Fluorite sample (4.87 mg of fluid extracted)				
Cu	0.71	150	43.8	9100
Mn	1.6	330	0.60	450
Zn	2.8	570	50.1	10,900

† Calculated on the basis of the combined water and acid leach data. For each calculated concentration it is assumed that all the metal has been present in solution in the fluid within the inclusions. The weight of the fluid extracted was calculated from the amount of water found by analysis (23.3 mg in the quartz sample and 3.65 mg in the fluorite sample), with a correction for dissolved salts known to be present (5 and 25 per cent by weight, respectively).

‡ Calculated on the basis of water leach data.

solid phases are redissolved and at which the solution and gas bubble become a single phase. Such observation is carried out microscopically, with the specimen on a carefully calibrated heating stage. A large number of formational temperatures have been estimated in this way, and as might have been expected the range is substantial. Limestone–lead-zinc deposits (see Chap. 16) show homogenization temperatures of less than 200°C and sometimes less than 100°C. Most are in the range 110 to 150°C. Other deposits, thought to be associated with subsurface igneous activity, give temperatures in the 150 to 400°C range. Pegmatites and associated deposits often give homogenization temperatures of over 500°C. Where a deposit appears to have formed at high pressures, an estimated pressure correction has to be applied to the temperature obtained from simple homogenization.

The determination of pH, though simpler than that of Eh, is difficult, and accurate figures have still to be obtained. On the basis of simple litmus tests, Roedder (1967, p. 553) estimates that most inclusions have a pH within about one unit of neutral. Values as low as $pH = 4.3$, however, have been reported by other investigators.

Evidence of fluid inclusions on ancient ore solutions thus indicates that many of the latter were highly saline, contained high concentrations of chloride ion and CO_2, and were of neutral to slightly acid pH. A range of temperatures of over 400°C is also indicated.

RECOMMENDED READING

Barnes, H. L., and G. K. Czamanske: Solubilities and transport of ore minerals, in H. L. Barnes (ed.), "Geochemistry of Hydrothermal Ore Deposits," pp. 334–381, Holt, Rinehart and Winston, Inc., New York, 1957.

Barton, P. B.: Some limitations on the possible composition of the ore-forming fluid, *Econ. Geol.*, vol. 52, pp. 333–353, 1957.

Barton, P. B.: The chemical environment of ore deposition and the problem of low-temperature ore transport, in P. H. Abelson (ed.), "Researches in Geochemistry," John Wiley & Sons, Inc., New York, 1959.

Degens, E. T., and D. A. Ross (eds.): "Hot Brines and Recent Heavy Metal Deposits in the Red Sea," Springer-Verlag New York Inc., New York, 1969.

Garrels, R. M.: "Mineral Equilibria at Low Temperatures and Pressures," Harper & Brothers, New York, 1960.

Helgeson, H. C.: "Complexing and Hydrothermal Ore Deposition," Pergamon Press, New York, 1964.

Holland, H. D.: Gangue minerals in hydrothermal deposits, in H. L. Barnes (ed.), "Geochemistry of Hydrothermal Ore Deposits," pp. 382–436, Holt, Rinehart and Winston, Inc., New York, 1967.

Krauskopf, K. B.: "Introduction to Geochemistry," McGraw-Hill Book Company, New York, 1967.

Roedder, E.: Fluid inclusions as samples of ore fluids, in H. L. Barnes (ed.), "Geochemistry of Hydrothermal Ore Deposits," pp. 515–574, Holt, Rinehart and Winston, Inc., New York, 1957.

Skinner, B. J., D. E. White, H. J. Rose, and R. E. Mays: Sulfides associated with the Salton Sea brine, *Econ. Geol.*, vol. 62, pp. 316–330, 1967.

7
Segregation and Mixing in Ore-forming Processes

Almost every geological process causes mixing of some elements and tends to induce segregation of others. In igneous processes, hybridism involves mixing and differentiation involves segregation. Erosion and sedimentation usually involve the production and mixing of detritus from highly heterogeneous terrains, and they are thus often great "averaging" processes. On the other hand, selective solution of certain elements such as sodium, potassium, calcium, and magnesium may eventually lead to the precipitation of beds consisting almost entirely of salts of these metals—an example of extreme segregation by sedimentary processes. Some homogenization of rocks is undoubtedly produced by regional metamorphism, though this process may also induce minor segregation—"metamorphic differentiation." Similarly subsurface solution activity, igneous and other, by selective solution and differential precipitation, may induce extensive mixing of some elements and extraordinarily efficient segregation of others.

Many cases of mixing and of segregation can be understood simply by a consideration of the chemical properties of the substances concerned. It follows that careful observation of relative abundances of certain substances may yield valuable information on the processes that led to the formation of the rocks containing these substances—information on source and, in some instances, on chemical and physical histories. Investigations based on this line of thought now constitute the principal field of geochemistry and, particularly, are finding substantial application in the elucidation of ore-forming processes.

There are three main lines of investigation involved: mixing and segregation of elements, of stable isotopes, and of radiogenic isotopes. The first is based on the chemical behavior of certain key elements and has led to a great proliferation of "trace element" studies since about 1930. The second is based on the fractionation and mixing of isotopes of stable substances such as oxygen, carbon (apart from ^{14}C), and sulfur by inorganic and by biological processes. The third is concerned with the radioactive generation of particular isotopes of certain substances—particularly lead and thorium—and the variable mixing of such isotopes by crustal and other processes. The two fields of isotope study have developed principally since about 1950.

All three types of investigation have been applied extensively to ore-genesis problems, and together they now involve a very large literature indeed—far too large to be examined comprehensively in a volume of the present kind. We shall therefore confine ourselves to three examples, keeping in mind that the principles involved have application in numerous other systems. Our examples are (1) the behavior of selenium and the elemental pair sulfur-selenium; (2) the stable isotopes ^{32}S and ^{34}S of sulfur; and (3) the radiogenic lead isotopes ^{206}Pb, ^{207}Pb, and ^{208}Pb generated by the decay of uranium and thorium.

SELENIUM AND SULFUR-SELENIUM

Sulfur and selenium are neighbors in the periodic table of elements and are members of the VIb subgroup. The radius of S^{--} is 1.74 Å and of Se^{--} is 1.91 Å (Goldschmidt, 1954). Like sulfur, the selenium atom has six electrons in its outer shell. It may form Se^{--} ions, through which it combines with metals such as Au, Ag, Fe, Cu, Pb, and Hg to form metallic selenides (Chap. 3). It can also occur as Se^{4+} and Se^{6+} ions that, analogously with sulfur, combine with oxygen to form SeO_2 and SeO_3. These lead to the formation of selenious and selenic acids and hence to selenites and selenates. Sulfide and selenide may develop as complete solid-solution series, as with galena-clausthalite, PbS–PbSe. In other cases substitution may occur, but only to a limited extent; according to Coleman and Delevaux about 4 molecular per cent $FeSe_2$ can enter the pyrite structure under low-temperature conditions, after which ferroselite, $FeSe_2$ with a rammelsbergite structure, is formed (this does not, however, preclude possible complete solid solution at higher temperatures). Overall the two elements are thus closely related in their chemistry and crystal chemistry. Small divergences in their geochemical behavior are hence likely to be very delicate indicators of chemical and physical conditions.

According to Goldschmidt (1954) the average content of Se in meteorites is about 7 ppm, and the average S/Se ratio is about 3000:1. He estimates an average content of 0.09 ppm Se in igneous rocks, with a corresponding S/Se ratio of about 6000:1. For shales, which of all sedimentary rocks are those generally richest in selenium, average Se is estimated at about 0.6 ppm, and S/Se is about 4000:1.

Table 7-1 Selenium and sulfur in natural sulfide and sulfur

Material	Locality	Se (ppm)	S/Se	Reference[†]
Ores in igneous rocks				
Ni–Cu ore in gabbro	Sudbury	160	?	1
Pyrrhotite	Sudbury	63	?	1
Chalcopyrite	Sudbury	97	?	1
Pyrrhotite in gabbro	Noril'sk	36–140	?	2
Chalcopyrite in gabbro	Noril'sk	50–200	?	2
Pentlandite in gabbro	Noril'sk	67	?	2
Ni–Cu ore in peridotite	Zeehan, Tasmania	42	7840	3
Volcanic sulfur				
Volcanic sulfur	Hawaii	5.18×10^3		2
Volcanic sulfur	Lipari	1.03×10^3		2
Volcanic sulfur	New Zealand	$19{-}30 \times 10^2$		2
Volcanic sulfur	Paramushir I.	$2.5{-}19 \times 10^2$		2
Sulfide in sedimentary rocks				
Pyrite, marine Tertiary	Torquay, Victoria	1	487,300	3
Pyrite, marine Cretaceous	Batavia Downs, Queensland	2	162,000	3
Pyrite, fresh water Jurassic	Inverlock, Victoria	4	81,130	3
Pyrite, marine Permian	Ravensworth, N.S.W.	2	83,630	3
Pyrite, marine Devonian	Kinglake, Victoria	3	111,670	3
Pyrite, marine Ordovician	Captain's Flat, N.S.W.	4	105,300	3
Pyrite, marine Cambrian	Broken Hill	4	103,000	3
Ores in sedimentary rocks				
Pyrite	Mt. Isa	39	11,680	3
Pyrrhotite	Mt. Isa	1–24	360,700	3
			–11,033	
Chalcopyrite	Mt. Isa	35–36	7630	3
Sphalerite	Mt. Isa	1	286,600	3
			–107,600	
Chalcopyrite	Captain's Flat, N.S.W.	29–33	10,880	3
			–10,200	
Sphalerite	Captain's Flat, N.S.W.	31	10,712	3
Galena	Captain's Flat, N.S.W.	2	99,200	3
Pyrite and pyrrhotite	Nairne, S. Australia	15	32,300	3
Pyrite	Michipicoten, Ontario	<15	?	1
Ores in veins				
Pyrite	Powell-Rouyn, Quebec	34	?	1
Pyrite	Fondulac, Saskatchewan	23	?	1
Pyrrhotite	Hollinger, Ontario	30	?	1
Pyrrhotite	Yellowknife, N.W. Terr.	26	?	1
Chalcopyrite	Hollinger, Ontario	23	?	1
Stannite	Conrad, N.S.W.	20	13,720	3
Stannite	Kangaroo Hills, Queensland	22	14,000	3
Tetrahedrite	Mt. Read, Tasmania	34	6470	3
Stibnite	Pt. Lookout, N.S.W.	9	30,200	3

[†] *References:*

1. Hawley and Nicol (1959).
2. Sindeeva (1964).
3. Edwards and Carlos (1954).

Sindeeva (1964) estimates that the average selenium content of "rocks of Russia" is about 0.14 ppm.   Selenium occurs in notable quantities in volcanic native sulfur (Table 7-1) and is also often abundant in mineral veins formed at shallow depth about volcanic centers.   Indeed, high levels of disseminated Se in volcanic rocks are in some notable cases responsible for the development of highly seleniferous soils.

River waters draining such seleniferous soils may contain quite high selenium. Perhaps the best known seleniferous province in the world is that stretching from the Gulf of Mexico up through the prairie regions of the United States to Saskatchewan in Canada—an area of some 300,000 square miles.   The selenium here is derived from Mesozoic—particularly Cretaceous—sedimentary rocks (also noted for their uranium, vanadium, and copper contents—cf. Chap. 16), and it has been suggested that certain tuff units may be the chief contributors.   Surface waters here have been found to show a range of about 0.001 to 2.7 ppm Se.   The Colorado River and its tributaries thus carry large amounts of selenium into the Gulf of California.   High levels of selenium are also found in Ireland, Israel, South Africa, northern Australia, and elsewhere, and it is being constantly transported by surface waters to the adjacent seas.   It is thus somewhat surprising—at least at first glance —that sea water has been found to contain only the minutest trace of selenium.

SEPARATION OF SULFUR AND SELENIUM IN THE CYCLE OF WEATHERING AND SEDIMENTATION

Clearly sulfur and selenium show many similarities of chemistry and general association.   However, one *dissimilarity* stands out: whereas sulfur is abundant in the waters of the ocean, selenium is, as we have just noted, conspicuously almost absent. Such a sharp divergence indicates some fundamental differences in behavior—a difference that is now well understood and that has been thought to be of substantial significance in ore geochemistry.

Selenium in igneous rocks and volcanic emanations is always a minor associate of sulfur, as sulfide-selenide or as a mixture of the elements.   In the more important case of sulfide-selenide, exposure to the atmosphere leads quickly to the oxidation of the sulfide and its removal as the sulfate.   The oxidation of selenide, however, is far less ready.   Figure 7-1 (after Delahaye et al., 1951; Coleman and Delevaux, 1957) shows that the oxidation of selenium requires a high oxidation potential and that metallic selenium is stable over a large part of the field of natural surface conditions.   Thus, as Coleman and Delevaux point out (1957, p. 524), selenites and selenates are not found near oxidized selenium-bearing sulfides, or selenides, but native selenium frequently is.   Because of this lesser mobility of selenium, sulfur and selenium begin to separate at the very onset of weathering.

This tendency to segregate is emphasized further during both transport and later deposition.   Any selenate that *is* formed and taken into solution at the point of weathering will clearly be reduced far more readily than the accompanying sulfate, and this leads to further segregation wherever the transporting medium encounters a slightly less oxidizing environment.   As a result, while some selenium unquestionably does reach the sea by surface processes, its original abundance relative to

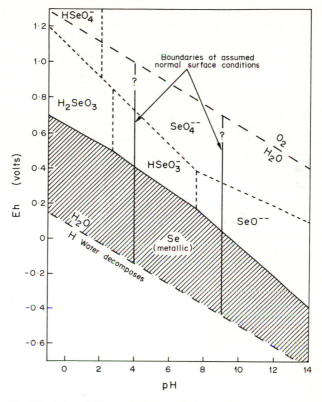

Fig. 7-1 Eh-pH stability relations of the principal aqueous selenium-bearing species in the system H–Se–O at 25°C, 1 atm total pressure, Se = 10^{-6} M. (*Redrawn from Delahaye, Pourbaix, and Van Rysselberghe, Comptes Rendus Troisieme Réunion Comite Thermodynam. Cinet Electrochim., Berne, 1952.*)

sulfur is enormously reduced. Having arrived, the small amount remaining in solution apparently is adsorbed rapidly on iron and manganese hydroxides and precipitated in organic (reduced) sediments. Thus oceanic sediments often contain small, but readily measurable, selenium: bottom sediments of the Bering Sea and North Atlantic contain 0.03 to 0.7 ppm; Atlantic Ocean, 0.2 to 2.0 ppm; Gulf of California, 0.1 to 5.0 ppm; and the Caribbean Sea (Bartlett Deep), 0.08 to 0.2 ppm (Rosenfeld and Beath, 1964, pp. 52–53). Conversely as a result of this abstraction of the selenium, the S/Se ratio of sea water is enormous—approximately 232,000:1 by weight, and for evaporite beds it is usually between 250,000:1 and 500,000:1. This is a huge increase over the average ratio for igneous rocks of approximately 6000:1.

From all this it is readily seen that if, as seems general, marine sedimentary sulfides are produced by reduction of sea-water sulfate, the amount of selenide incorporated in such sulfides will be minute and S/Se ratios will be orders of magnitude less than those in the sulfides "at source," i.e., "primary" or "igneous" sulfides.

SULFUR-SELENIUM RATIOS AS GEOCHEMICAL INDICATORS

Sulfur-selenium relations thus offered early promise of distinguishing between sulfides resulting from igneous processes (e.g., hydrothermal replacement) on the one hand and those resulting from sedimentary processes on the other. This possibility was first recognized by Goldschmidt and Strock, who in 1935 demonstrated that whereas a number of pyrites of apparent igneous-hydrothermal type had S/Se ratios of 10,000 to 20,000, others of clear sedimentary origin had ratios of the order of 200,000. Their work was supported by results obtained a little later by Carstens (1941) on a number of Norwegian pyrite deposits; those of apparent igneous-hydrothermal type contained some 20 to 30 ppm Se, whereas those of sedimentary type contained less than 1 ppm. At this point it seemed that the incidence of selenium might provide an objective method of settling the problem of the stratiform ores—the replacement versus sedimentation controversy discussed in Chaps. 2 and 15.

It was soon recognized, however, that not all "hydrothermal" ores were necessarily seleniferous; there were, in fact, selenium-rich and selenium-poor provinces. The Skellefte district of Sweden is a notable example of a province of selenium-rich vein mineralization; one occurrence—the Boliden deposit—contains sulfides with up to 6.4 per cent Se (Bergenfelt, 1953). At the other extreme Edwards and Carlos (1954) analyzed pyrite, obtained from *within a granite mass* in northern Australia (the Ennoggera granite), that gave S/Se = 170,000. This led Edwards and Carlos to suggest that while sedimentary sulfide should always show low selenium, igneous-hydrothermal ores might show high *or* low values, depending on the province. Further work has now demonstrated that it is not only igneous sulfides that vary—sedimentary sulfides have their selenium-rich provinces too! Table 7-1 gives examples of results obtained on a variety of occurrences.

The picture that has now emerged is not nearly so simple as the early work of Goldschmidt suggested it might be. Most sulfides of undoubted igneous affiliation show S/Se < 20,000, but some, such as the pyrite from the Ennoggera granite, show far larger ratios. Sulfur and sulfur compounds deposited as marine sediments usually show S/Se > 100,000, but there are notable exceptions—exceptions that might have been expected from the early recognized high-selenium contents of some black shales and manganese nodules. The sulfides of the Colorado Plateau uranium ores—apparently slightly modified sedimentary materials—are notably high in selenium. It appears that sedimentary rocks derived from seleniferous source areas—sediments such as those now being contributed to the Gulf of California by the Colorado River drainage system—are likely to be quite as seleniferous as many of primary igneous derivation.

We may therefore sum up by saying that the tendency for the separation of selenium from sulfur during weathering and erosion has a good theoretical basis and is borne out by observation. It is thus a good example of segregation of substances during ore-forming processes. Unfortunately, however, it is heavily influenced by local factors, and while it is a good example of a principle, it is not sufficiently reliable for confident diagnosis of sulfide origins. This is a weakness of all trace element techniques attempted so far.

THE SULFUR ISOTOPES ^{32}S AND ^{34}S IN SULFIDE ORES

Four stable isotopes of sulfur exist and their terrestrial abundances have been estimated by Bainbridge and Nier (1950):

Isotope	Per cent of terrestrial S
^{32}S	95.1
^{33}S	0.74
^{34}S	4.2
^{36}S	0.016

It is now well established that there are substantial deviations from these proportions in individual samples of sulfur-bearing materials. Variation in relative abundances is induced by slight differences in physical properties—particularly reaction rates—between the different isotopes. Certain reactions favor one isotope relative to another, thus increasing its relative amount in one of the resulting compounds. This tendency—and it must be emphasized that it is only a tendency —for the segregation of isotopes is termed *isotopic fractionation*. The effects of fractionation may, of course, be modified—either accentuated or diminished—by later mixing of sulfur compounds from different sources.

Because they are far more abundant than the other isotopes, and their amounts therefore are easiest to measure accurately, ^{32}S and ^{34}S have been studied most extensively. The $^{32}S/^{34}S$ ratio in the troilite phase of meteorites is very constant at 22.21, and, although difficult to estimate accurately, the ratios for the earth's crust and mantle, and for the earth as a whole, are probably also very close to 22.21. Deviations in $^{32}S/^{34}S$ from sample to sample, however, may be substantial, and a range of about 20.8 to 23.3—approximately 11 per cent—has been found in natural sulfides alone. The sulfur of present-day sea-water sulfate is very constant at 21.76.

$^{32}S/^{34}S$ ratios are often expressed relative to those of a well-known meteorite —the Cañon Diablo meteorite—and are given as "$\delta^{34}S$," where

$$\delta^{34}S = \frac{[^{34}S/^{32}S \text{ (sample)} - {}^{34}S/^{32}S \text{ (standard)}] \times 1000}{^{34}S/^{32}S \text{ (standard)}}$$

$\delta^{34}S$ for the Cañon Diablo troilite sulfur is thus zero. "Light" sulfur— with a $^{32}S/^{34}S$ ratio higher than that of meteoritic sulfur—has a negative $\delta^{34}S$ value, and "heavy" sulfur—$^{32}S/^{34}S$ less than that of meteorites—has a positive $\delta^{34}S$. $\delta^{34}S$ is given as ‰, or *per mil*, values.

FRACTIONATION OF ^{32}S AND ^{34}S

We have already noted that one of the basic phenomena of chemistry is the equilibrium reaction

$$aA + bB \rightleftharpoons xX + yY$$

where A and B are compounds of elements that can be rearranged to form compounds X and Y. K, the equilibrium constant for the reaction, is given by

$$K = \frac{[X]^x \times [Y]^y}{[A]^a \times [B]^b}$$

Isotopic equilibrium may be expressed in a similar way:

$$aA' + bB \rightleftharpoons aA + bB'$$

where A and B are compounds containing a particular element in common, the prime indicating the whereabouts of the heavier isotope of that element.   The equilibrium constant in this case is given by

$$K = \frac{[A]^a \times [B']^b}{[A']^a \times [B]^b}$$

Thus for the exchange reaction

$$H_2{}^{34}S + {}^{32}SO_4{}^{--} \rightleftharpoons H_2{}^{32}S + {}^{34}SO_4{}^{--} \tag{7-1}$$

$$\text{Gas} \qquad \text{Solution} \qquad \text{Gas} \qquad \text{Solution}$$

the equilibrium constant at 25°C is $K = 1.074$.   Hence at equilibrium at this temperature, ^{34}S concentrates in the sulfate by 7.4 per cent.   Values of K for the various ions and compounds involved in sulfur-isotope exchange reactions vary, but may be considerable (see Jacobs, Russell, and Wilson, 1959, p. 209).

In this general connection there are three important considerations:

1. The heavier isotope tends to be enriched in the molecules of higher oxidation state, the fractionation factor increasing with increasing difference in the oxidation state of the molecules concerned.
2. The fractionation factor is temperature-dependent and decreases with increase in temperature (Fig. 7-2).
3. The greater the fractionation factor, the greater its temperature dependence.

In addition Sakai (1968) has shown that ^{32}S/^{34}S ratios in sulfide minerals precipitated from aqueous solutions are heavily influenced by pH.   This stems from two factors:

1. H_2S in aqueous solution is in essential equilibrium with its dissociation products, HS^- and S^{--}, the relative concentrations of the three species being pH-dependent.
2. ^{32}S/^{34}S ratios in each of the three species are different, and the weighted average ^{32}S/^{34}S is therefore determined by their relative concentrations.

As a result, ^{32}S/^{34}S ratios are strongly dependent on pH, fractionation between gaseous H_2S and three coexisting aqueous species increasing with increasing

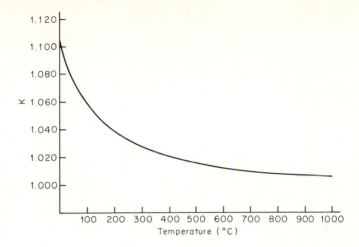

Fig. 7-2   Temperature dependence of K in Eq. (7-1).

alkalinity.   Thus at a temperature of 500°K (227°C) fractionation is zero at pH about 2.0, but at pH 8 the $\delta^{34}S$ of gaseous H_2S is approximately 3.5‰ greater than the weighted average $\delta^{34}S$ of the aqueous species, and at pH 9.5 the fractionation rises to about 5.05‰.   Once again the degree of fractionation is temperature-dependent, though for the $H_2S(g)-H_2S(aq)-HS^--S^{--}$ equilibrium the relation is not simple, and in the pH intervals 2.0 to 6.5 and about 8.7 to 11.4 the fractionation is, for example, *greater* at 500°K than at 300°K (Sakai, 1968, p. 41).

In an analogous manner sulfur-isotope fractionation between precipitating metallic sulfide species is pH-dependent and increases with increase in pH.   Thus, for example, Sakai has shown that the $\delta^{34}S$ of pyrite in isotopic equilibrium with a solution at 500°K and having a $\delta^{34}S$ of zero would be zero at pH 4, rising to 2‰ at pH 8 and 5‰ at pH 9.5.   The equilibrium values are again temperature-dependent, fractionation at 300°K being consistently lower here than at 500°K; viz., at 300°K the fractionation is approximately 3.5‰ at pH 4, 6.7‰ at pH 8, and 7.2‰ at pH 9.5.

The extent of fractionation between metal sulfide and coexisting sulfide species is different for the principal common ore minerals, and consideration of the various equilibrium constants indicate that where the sulfide minerals are in isotopic equilibrium, the heavier isotope ^{34}S is enriched in the order pyrite > sphalerite > chalcopyrite > galena.   The development of this order, i.e., pyrite sulfur being heavier than the coexisting sphalerite sulfur, and so on, has now been observed in numerous natural assemblages.

Thus in addition to the influence of oxidation state and temperature on the direction and degree of fractionation, pH of parental solutions and relative bond strengths in the precipitating minerals also exert considerable influence over abiological partitioning processes.

The above considerations pertain to simple inorganic exchange.   Fractionation of sulfur isotopes is also induced, however, by biological reactions, and indeed these may well be the most important fractionating agents in nature. That microorganisms could fractionate sulfur isotopes during metabolic function was first established in 1951 by Thode, Kleerekoper, and McElcheran, who showed that sulfide released during sulfate reduction by the bacteria *Desulfovibrio desulfuricans* was enriched in ^{32}S by 10 to 12‰. Since that time the fractionating capacities of sulfate-reducing and related bacteria have been investigated intensively, particularly by Thode, Kaplan, Jensen, and their respective coworkers. The principal findings are

1. Certain bacteria are capable of reducing oxidized sulfur compounds to sulfides; others are capable of oxidizing sulfide and other sulfur compounds to higher oxidation states.
2. Both processes induce fractionation of sulfur isotopes; reduction concentrates the lighter isotope in the sulfide and oxidation concentrates the heavier isotope in the newly formed oxidized compound.
3. Such fractionation may be substantial; for sulfate reduction, enrichment of ^{32}S to a value of $\delta^{34}S = -62$‰ has been found in natural sulfide, and experimental sulfate reduction by *Desulfovibrio desulfuricans* has produced a value of -46‰ (Kaplan and Rittenberg, 1964).
4. Degree of fractionation in nature is usually inversely proportional to rate of reduction, i.e., the more rapid the reduction, the lesser the degree of fractionation.
5. Temperature and sulfate concentration, within the normal ranges of physiological tolerance of the bacteria concerned, influence degree of fractionation insofar as they influence rate of reduction.

FRACTIONATION IN ORE-FORMING PROCESSES

It is thus established that fractionation of sulfur isotopes occurs (1) in both equilibrium and unidirectional processes and (2) through both inorganic and organic reactions. Oxidation, which enriches ^{34}S in the oxidized product, is induced by both inorganic and organic processes. Reduction, which enriches ^{32}S in the reduced product, is induced virtually entirely by organic processes. Since the production of true sedimentary sulfide has been thought to be almost entirely a product of bacterial reduction of sea-water sulfate, it follows that such sulfide sulfur should generally be lighter than that of sea-water sulfate—and, if conditions have been favorable, considerably lighter than "average crustal" ($\delta^{34}S = 0$) sulfur. Following this line of thought it was suggested that $^{32}S/^{34}S$ ratios might give valuable information concerning the geochemical history of the sulfide of ores and in particular might help to discriminate between sedimentary and subsurface hydrothermal-replacement processes in the formation of stratiform ores—the problem to which S/Se ratios had been particularly applied. Two characteristics of $^{32}S/^{34}S$ have been thought to be significant:

1. *Mean $^{32}S/^{34}S$ for a deposit as a whole.* $\delta^{34}S$ close to zero would suggest a deep, homogeneous source and hence subsurface replacement, whereas negative

δ^{34}S (i.e., high ^{32}S/^{34}S) would suggest bacteriological activity and hence sedimentary precipitation.

2. *Within-deposit variability of* $^{32}S/^{34}S$.   A high degree of uniformity of ^{32}S/^{34}S throughout a deposit or a steady, systematic change conforming to a sequence in mineral formation would suggest inorganic, subsurface deposition, whereas wide fluctuation in ^{32}S/^{34}S, or progressive change in conformity with stratigraphy, would suggest organic, sedimentary deposition.

Investigation of deposits whose nature is known with reasonable certainty has shown that in some cases behavior of ^{32}S/^{34}S ratios conforms quite well with these theoretical patterns but that in others it does not.   Sulfides of some *undoubted hydrothermal* conduits show great uniformity of δ^{34}S, though systematic changes of ratios with changes in temperature and pH have not been found. Some veins, though, show widely varying ratios conforming to no *apparent* systematic trend.   In sedimentary environments, semirandom fluctuations in ^{32}S/^{34}S from one bed to the next are well known and probably reflect changes in temperature, SO_4^{--} concentration, and the availability of nutrients, which in turn have induced substantial and sudden changes in degree of bacteriological fractionation.

Instances of essentially systematic unidirectional change related to stratigraphy may reflect reduction on the floor of a confined basin.   In this case initial SO_4^{--} might, for example, contain sulfur of δ^{34}S $= 0$.   The first sulfide produced by bacterial reduction would contain sulfur of negative δ^{34}S, thus leaving sulfur of positive δ^{34}S in the overlying reservoir.   With continued preferential abstraction of ^{32}S from the SO_4^{--}, the latter would become heavier and heavier so that eventually crops of sulfide of positive δ^{34}S would begin to form.   If in the end all the SO_4^{--} were used, the final sulfide would contain relatively heavy sulfur.   The mean δ^{34}S of the total sulfide of the sedimentary sequence would, of course, be that of the original SO_4^{--} reservoir, that is, δ^{34}S $= 0$; and ^{32}S/^{34}S in the bedded sulfide deposit would be systematically related to stratigraphy, being progressively heavier with increased height in the sequence.

In one case investigated—Rammelsberg—there is indeed a progressive "heavying" in sulfur going up stratigraphically.   In contrast, at Heath Steele, New Brunswick (Lusk and Crockett, 1969), the trend, though slight, is consistently the reverse!   This is discussed again a little further on.

^{32}S/^{34}S RATIOS IN NATURAL SULFIDES

We may now look at the isotopic constitution of sulfide in the principal classes of geological materials.

1. Igneous rocks   Sulfides occur in igneous rocks in amounts ranging from the sparsest disseminated traces to massive sulfide ores.   Table 7-2 lists δ^{34}S values for a number of intrusive and extrusive masses, including the nickel ore-bearing basic intrusions of Sudbury, Ontario, and Insizwa, South Africa.   Products of extreme differentiation, such as pegmatites, are not included here since their formation is likely to have induced substantial fractionation; such occurrences fall into the category of "vein deposits."

Data obtained so far indicate:

a. $^{32}S/^{34}S$ ratios in sulfides of igneous rocks are fairly uniform and average close to $\delta^{34}S = 0$.

b. Isotopic fractionation is therefore not great. It appears to be negligible in un-differentiated intrusive rocks, to have occurred to a small but measurable degree in differentiated intrusions such that the sulfur of the later differentiation is slightly enriched in ^{34}S, and to have occurred to a small degree in extrusive rocks, leading to a slight enrichment in ^{32}S.

2. Fumaroles and hydrothermal conduits associated with volcanism

Although we are concerned here with sulfides only, it should be mentioned that much work has been done on $^{32}S/^{34}S$ fractionation among a variety of sulfur compounds—sulfates, sulfites, polythionates, sulfides (and native S)—of volcanic

Table 7-2 $\delta^{34}S$ of sulfide sulfur in igneous rocks

Occurrence	No. of samples averaged	Mean $\delta^{34}S$	Reference†
Sudbury	23	+3.6	1, 2
Insizwa sill	3	−2.6	2
Duluth gabbro	6	−2.2	3, 4
Stillwater	3	+3.4	1
Newark Triassic diabases, eastern U.S.A.	19	+0.1	5
Newark lava flows	18	−4.7	5
Purcell gabbro sill, Idaho	1	−2.2	4
Peridotites, Maine	2	−0.7	5
Nipissing diabase, Cobalt, Ontario	1	+1.4	4
Paracutin lava, Mexico	1	+3.6	4
Noril'sk gabbro-diabase, Russia	13	+4.0	6
Kamchatka basalt	1	+2.3	6
Pyroxenite, Aldan	1	+0.4	6
Dunite, Raiiz	1	+0.4	6
Boulder batholith, Montana (diorite, quartz monzonite, and monzonite)	3	+4.5	4
Phonolite, Devil's Tower, Wyoming	1	+3.1	4
Syenite, Otto Township, Ontario	1	+2.3	7
Granodiorite, Urjala, Finland	1	−1.4	4

† *References:*
1. Macnamara, Fleming, Szabo, and Thode (1952).
2. Thode, Monster, and Dunford (1961).
3. Jensen (1959).
4. Ault and Kulp (1959).
5. Smitheringale and Jensen (1963).
6. Vinogradov (1957).
7. Thode (1949).

Table 7-3 $\delta^{34}S$ of fumarolic iron sulfides from New Zealand. (From Steiner and Rafter, 1966)

Locality	Drillhole No.	Depth, feet	Mineral	$\delta^{34}S$
Wairakei	12	1391	Pyrite	+7.1
	28	800–820	Pyrite	+5.0
	47	1621	Pyrite	+6.8
	71	1354.5	Pyrite	+7.2
	213	2633	Pyrite	+4.2
	215	1100–1105	Pyrite	+4.6
	218	2050–2052	Pyrite	+6.0
	220	2303	Pyrite	+6.8
	205	2350–2355	Pyrrhotite	+3.8
	205	2400–2405	Pyrrhotite	+3.0
	220	2400–2402	Pyrrhotite	+4.3
Waiotapu	3	707–714	Pyrite	+3.5
	4	2856	Pyrite	+5.5
	6	750–752	Pyrite	+4.6
	7	1348–1351	Pyrite	+5.6
	7	1551	Pyrite	+4.7
	7	1600	Pyrite	+4.3
	5	1050	Pyrrhotite	+4.7

products, particularly in New Zealand (Rafter and his coworkers) and in Japan (Sakai, Nagasawa, Tatsumi, and others). The isotopic constitution of sulfide from near-surface igneous (volcanic) hydrothermal deposits is given in Table 7-3. Most of the sulfides come from currently active conduits. Clearly the sulfur here is heavier than zero $\delta^{34}S$ and hence is distinctly heavier than that usually found *within* igneous rocks.

3. Sulfides in sediments and sedimentary rocks

There is a substantial amount of information on sulfide from both modern and ancient sedimentary materials.

a. Modern sediments. The small pyrite accumulations in the sediments of some of the southern Californian basins are good examples of sulfide in "normal" modern sediments. These have been investigated isotopically by Kaplan, Emery, and Rittenberg (1963), and their results are given in Table 7-4. $\delta^{34}S$ values are almost invariably negative but are clearly highly variable.

The isotopic constitution of the sulfide sulfur of the "abnormal" chemical sediments on the floors of the brine pools of the Red Sea (see Chap. 6) presents quite a different picture. The sulfur here occurs as sulfate, sulfide, and elemental sulfur, all of which have been investigated isotopically by Kaplan, Sweeney, and Nissenbaum (1969). Of the *sulfide sulfur*, all that in sediment cores obtained from beneath existing brine pools is heavier than meteoritic and of quite constant $\delta^{34}S$. For eight "total sulfide" analyses of

Table 7-4 $\delta^{34}S$ of pyrite sulfur in some modern sediments†

Basin	Depth below sediment surface, cm	$\delta^{34}S$
Southern California		
Santa Barbara basin	0–10	−26.6
	90–100	−17.3
	260–270	−18.6
	360–370	−19.9
Santa Monica basin	0–10	−10.7
	65–75	−15.1
	380–390	−10.7
Santa Catalina basin	0–10	−29.9
	120–125	−34.0
	360–370	−7.7
San Diego basin	0–10	+3.1
	90–100	−28.8
	240–250	−34.6
Newport marsh	0–5	−20.0
	30–35	−27.1
	60–65	−27.7
Red Sea		
Atlantis Deep (brine-bearing)		
Core 127P	610	+5.7
Core 120K	140–150	
	490–500	+3.6
Core 84K	200–210	+7.7
	310–320	+4.6
Nonbrine deep		
Core 118K	160–168	−31.7
	170–190	−31.6
	385	+10.6
	385–387	+10.0

† California analyses from Kaplan, Emery, and Rittenberg (1963); Red Sea analyses from Kaplan, Sweeney, and Nissenbaum (1969).

material from a single core of the Atlantis Deep brine sediment, the mean $\delta^{34}S = +5.7$, the range of $\delta^{34}S = +3.1$ to $+9.8$, and the standard deviation of $\delta^{34}S = 1.9$. Sulfide sulfur from cores outside the brine pools, however, i.e., from "normal" sediment, is, as might have been expected, much lighter than meteoritic, though still quite uniform; six "total sulfide" analyses representing the whole sampling from one such core in the Discovery Deep gave the mean $\delta^{34}S = -26.2$, the range of $\delta^{34}S = -23.3$ to -32.1, and the standard deviation of $\delta^{34}S = 2.8$. In several cases Kaplan, Sweeney, and Nissenbaum found that cores close to present brine pools, or in deeps adjacent to the pools, gave compound results; of samples representing different depths below the present mud-water interface in individual cores, some gave negative and others positive $\delta^{34}S$ values. This may indicate fluctuation in the size of the pools over a period of time as well as periodic overflow into adjacent deeps.

Examples (of pyrite sulfur only) are given in Table 7-4 (from Kaplan et al., 1969).

b. *Ancient sedimentary rocks.*   The number of analyses of sulfide (almost entirely pyrite) from old sedimentary rocks is very large indeed.   The range $\delta^{34}S$ for the class as a whole is extreme.   Most occurrences give ratios distinctly lighter than the meteoritic value, though a few are slightly heavier.   Very occasionally a sample yields sulfur much heavier than meteoritic (for example, $\delta^{34}S$ from $+15$ to $+30$), though this is rare.   Most sedimentary sulfide exhibits $\delta^{34}S$ values between about -5 and -35.   The sulfides of ores of the sandstone–uranium-vanadium-copper association (see Chap. 16), which occur in coarse fluviatile sediments and which may represent biologically reduced ground-water sulfates, are of conspicuously low $\delta^{34}S$.   Values between -10 and -45, with a substantial clustering in the -20 to -40 range, are almost characteristic.   Inevitably, however, there are exceptions; sulfides from the Woodrow mine, New Mexico, contain notably heavy sulfur, for the most part in the range $+10 < \delta^{34}S < +40$ (see Jensen, 1967).

4. Sulfides from sedimentary rocks of volcanic association   This group

includes sulfides both from sedimentary rocks containing notable pyroclastic components and from "normal-looking" sediments that, however, show a close spatial association with volcanic lavas and pyroclastic beds.   While the containing materials here are just as unquestionably sedimentary as those of group 3 above, the sulfides in them are usually conspicuously more abundant than in the sediments unassociated with volcanism.   Whereas the sulfides of nonvolcanic sequences are rarely more than sparse disseminations, those of volcanic association are often massive and amount to millions of tons in a small volume.   Also, whereas the sulfide of nonvolcanic sediments is almost wholly pyrite, sulfides of the present group include, in addition to pyrite, notable quantities of pyrrhotite, sphalerite, chalcopyrite, and galena.   It is, of course, sulfide accumulations of this type that constitute many important bedded ore deposits.

Because of the great economic importance of this class of materials, considerable attention has been paid to the isotopic constitution of their sulfur.   As in the case of sulfur-selenium, it was hoped initially that $^{32}S/^{34}S$ ratios might provide a neat and objective solution to the replacement-sedimentation controversy, thus leading to improved efficiency in exploration for ores of this kind.   Unfortunately results so far are not conclusive, but they do show the existence of a number of general, and quite conspicuous, characteristics:

a. The sulfur is *heavy*; mean $\delta^{34}S$ values are higher than meteoritic in all occurrences for which there are a reliable number of analyses.   Such means range from approximately $+1.0$ for Broken Hill to $+20.0$ for Zardu, Persia. Where mean $\delta^{34}S$ is close to meteoritic (e.g., Broken Hill), some *individual* analyses give negative $\delta^{34}S$ values, but these are minor.

b. In most cases the range of within-deposit $\delta^{34}S$ is not great and generally of the order of 7 to 8‰.   Mount Isa appears to be an exception to this, but most

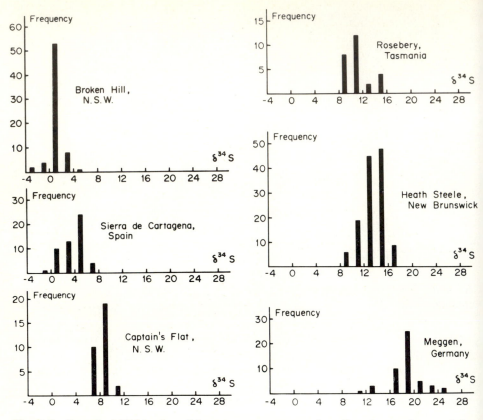

Fig. 7-3 Spread of $\delta^{34}S$ in six well-known ore occurrences in sedimentary rocks containing a volcanic component. (*From Stanton and Rafter, Miner. Deposita, 1966.*)

occurrences are remarkably uniform. Figure 7-3 shows spreads of $\delta^{34}S$ in six well-known deposits of this type. Thus in (*a*) and (*b*) the pattern of $\delta^{34}S$ for sulfide concentrations of this kind is similar to that of the Red Sea brine sediments.

c. Where it has been possible to make reliable stratigraphic samplings, $\delta^{34}S$ appears to be related to stratigraphy; that is, $\delta^{34}S$ tends to be uniform *within* beds and to vary *between* beds. Figure 7-4 gives results on samplings of two sulfide-rich beds at Mount Isa. These beds occurred 33 inches apart in the No. 6 ore body and could be traced continuously for about 1000 feet along the strike. Clearly there is a marked similarity *along* the bedding over a very large distance and a marked dissimilarity *across* the bedding over a short distance.

d. Where the requisite extensive samplings have been made, systematic change in $\delta^{34}S$ may occur proceeding stratigraphically from the base to the top of a deposit, as was mentioned a little earlier. At Rammelsberg, $\delta^{34}S$ in galena and sphalerite increases from about $+10$ at the base to about $+20\%_0$ at the

top (Anger et al., 1966).   This is as would be expected to result from bacteriological reduction of all or most of the sulfate present in a confined basin.   At
Heath Steele, New Brunswick, however, the trend is much finer and in the
opposite direction.   Lusk and Crockett (1969), in a very careful study of
δ^{34}S trends in three drill cores through the B-1 ore body, found a slight but
consistent *decrease* in δ^{34}S going from the base to the top of this deposit.

e. Partitioning of ^{32}S and ^{34}S between coexisting sulfide minerals is now known to
    be common and is probably general.   At the B-1 ore body at Heath Steele,
    Lusk and Crockett (1969) found that pyrites were consistently enriched in
    ^{34}S relative to all other sulfides and that sphalerite sulfur was consistently
    heavier than that of galena.   A similar state of affairs has been noted at
    Meggen, Germany, by Bushendorf and others.   At both Rammelsberg and
    Broken Hill, sphalerite shows consistently higher δ^{34}S than coexisting
    galena.   Figure 7-5 (after Stanton and Rafter, 1967) shows the fractionation
    in 16 coexisting sphalerite-galena pairs from Broken Hill.   This conforms well
    with the theoretical predictions of Sakai (1968) already referred to, and

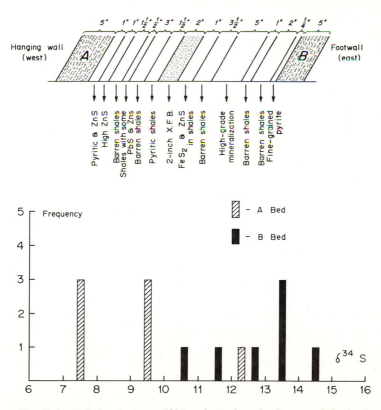

Fig. 7-4   Relation between δ^{34}S and stratigraphy in one of the lead-
zinc ore bodies of Mount Isa, Queensland.   (*From Stanton and Rafter,
Miner. Deposita, 1966.*)

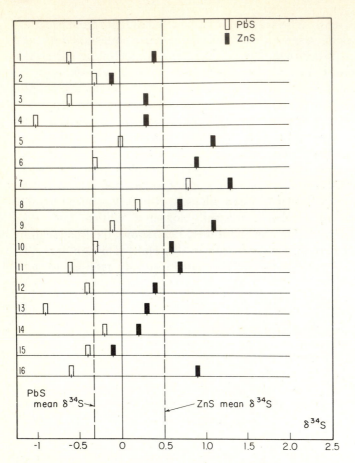

Fig. 7-5 Partitioning of ^{32}S and ^{34}S between galena and sphalerite at Broken Hill, New South Wales. (*From Stanton and Rafter, Econ. Geol., 1967.*)

presumably it is *primarily* due to sulfide-sulfide isotopic equilibria established during the original formation of the metallic sulfide minerals. The partitioning as now observed, however, may also reflect some *secondary* modification by metamorphism in certain cases.

5. Sulfides from metamorphic rocks Isotope ratios here seem essentially the same as those found in sedimentary and volcanic sedimentary rocks. There is some indication—though this is far from established—that the degree of ^{32}S/^{34}S partitioning as between PbS and ZnS decreases with increase in the metamorphic grade of the enclosing metasediments. Mean δ^{34}S (ZnS) minus mean δ^{34}S (PbS) in some unmetamorphosed low-temperature limestone–lead-zinc ores at Pine Point, N.W.T., Canada (Sasaki and Krouse, 1969), is about 3.5, and in some very young unmetamorphosed volcanic ores in Japan (Tatsumi, 1965) it is approximately 2.4‰; at Rammelsberg (Buschendorff et al., 1963) and Heath Steele (Lusk and

Crockett, 1969), which are both enclosed in low-greenschist-grade metasediments, $\delta^{34}S$ (ZnS) − $\delta^{34}S$ (PbS) is 1.5 and 1.9, respectively; and at Broken Hill (Stanton and Rafter, 1967), which is enclosed in metasediments of sillimanite grade, $\delta^{34}S$ (ZnS) − $\delta^{34}S$ (PbS) is 0.83.   Although not yet established, it is suspected that the partitioning as now observed represents an original primary depositional phenomenon that has later been progressively modified by increasing intensity of metamorphism.   This possible relation with associated metamorphic grade looks interesting but clearly requires much further investigation.

6. Sulfides from vein rocks   Since veins are an extremely heterogeneous category—their range of geological settings, morphologies, and mineralogies must be as wide as that of any set of crustal materials—it might have been expected that the isotopic constitution of their sulfur would cover a wide range too, and this is certainly so.   Veins in areas of plutonic intrusion sometimes show $\delta^{34}S \rightarrow 0$, but in other cases they deviate from this by large amounts.   Veins in limestone–lead-zinc deposits (one of the principal types of *stratabound* deposit, discussed in Chap. 16) usually—though not invariably—contain sulfur of high $\delta^{34}S$.   In addition to this very wide between-deposit variability, individual veins may show abundant and abrupt within-deposit variation.   Crustified ores often show fairly constant $\delta^{34}S$ *along* bands and sharp differences *across* them.   Figure 7-6 shows the striking variation in $^{32}S/^{34}S$ found by Ault (1959) within a single galena crystal from the Indian Creek mine in the limestone–lead-zinc belt of southeast Missouri.   Variations within and between veins are probably due to interplay of different and changing sources, variable migrational contamination, and variation, due to fluctuation in temperature and/or pH, in depositional fractionation.

Summing up on $^{32}S/^{34}S$ ratios in sulfide sulfur of different environments, it may be said that those of igneous rocks, older and modern sedimentary rocks, and

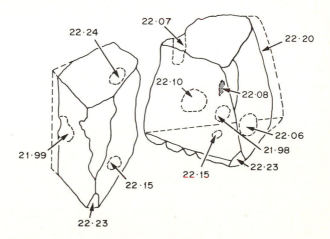

Fig. 7-6   Variation in $^{32}S/^{34}S$ in "single-crystal" galena from the Indian Creek mine, Missouri. (*From Ault and Kulp, Econ. Geol., 1960.*)

metamorphic rocks conform moderately well with the theoretical patterns predicted for them—predictions based on our knowledge of factors favoring *segregation* (i.e., fractionation) and *mixing* of the two isotopes in geological processes. However, the massive sulfides of sedimentary-volcanic association remain somewhat of an enigma; while they do show a number of well-defined, general characteristics, these are not attributable to any obvious theoretical pattern. Perhaps the processes responsible for forming these deposits are simple but as yet unrecognized; or it may be that there are sequences of segregation and mixing, or segregational processes, that we have yet to learn of.

ISOTOPIC CONSTITUTION OF LEAD

Recent studies of lead isotope abundances provide a good example of the use of radiogenic isotopes in investigations of geochemical histories of ores. Before discussing the subject it is necessary, however, to issue a word of caution: lead isotope abundances are difficult to interpret, and they generally permit a variety of interpretations. It is thus not surprising that their application has been highly controversial; indeed they have probably created more problems than they have solved, though this may well be an indication of the potential fertility of this particular field of study.

There are four principal isotopes of lead. One of these, ^{204}Pb, is nonradiogenic; that is, ^{204}Pb has always been present as ^{204}Pb, and its total terrestrial abundance may be presumed to have not changed with time. The other three isotopes are radiogenic and are produced by the radioactive disintegration of uranium and thorium via trains of intermediate daughter products:

$$^{208}\text{Pb} \leftarrow {}^{232}\text{Th}$$
$$^{207}\text{Pb} \leftarrow {}^{235}\text{U}$$
$$^{206}\text{Pb} \leftarrow {}^{238}\text{U}$$

This is illustrated diagrammatically in Fig. 7-7.

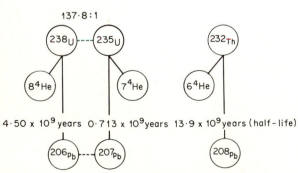

Fig. 7-7 Schematic diagram showing generation of ^{206}Pb, ^{207}Pb, and ^{208}Pb by radioactive decay of ^{238}U, ^{235}U, and ^{232}Th, respectively. (*From Stanton and Russell, Econ. Geol., 1959.*)

Table 7-5   Calculated atomic weights of common leads of widely different isotopic compositions.   (From Russell and Farquhar, 1960, p. 14)

Origin of sample	^{204}Pb, %	^{206}Pb, %	^{207}Pb, %	^{208}Pb, %	Atomic weight
Bathurst, New Brunswick	1.370	24.95	21.53	52.14	207.27
Chicobi Lake, Quebec	1.608	21.53	23.46	53.40	207.32
San Antonio mine, Mexico	1.343	25.28	21.21	52.16	207.27
New Dodgeville, Wisconsin	1.185	28.39	19.22	51.20	207.24
Joplin, Missouri	1.245	27.29	20.01	51.45	207.25
Ace Mine, Saskatchewan	1.044	41.87	19.45	37.64	206.93

As a result of different original U/Th/Pb ratios in source rocks and of differences in the times at which the various leads have been separated in nature from U and Th, the atomic weight of lead is—as would be expected—found to vary somewhat.   The calculated atomic weights for some leads of widely different isotopic compositions is given in Table 7-5.

It may be seen that the percentage mass differentials between the isotopes of lead, though present and readily measurable, are much less than those for sulfur. Consequently there is little if any *fractionation* of lead in nature.   Thus, whereas with sulfur we were concerned with *segregation* and *mixing* of stable isotopes, in the case of lead we are concerned with the *generation* and *mixing* of *radiogenic* isotopes.

GENERATION OF COMMON LEADS

In a simple disintegration process yielding a product whose atoms are stable, the relation between the number of parent atoms and their rate of decay is given by

$$\frac{dN}{dt} = -\lambda N \tag{7-1}$$

where N is the number of parent atoms present at time t, and λ is the *decay constant* for the radioactive process concerned:

$$\lambda = \frac{\log_e 2}{\text{half life}} \tag{7-2}$$

Integration of Eq. (7-1) gives the exponential relationship

$$N = N_0 e^{-\lambda t} \tag{7-3}$$

where N_0 is the original number of parent atoms, and N is the number remaining after a time t.

The number of daughter atoms D produced is clearly given by

$$D = N_0 - N$$

but from (7-3)

$$N_0 = N e^{\lambda t}$$

therefore

$$D = N(e^{\lambda t} - 1) \tag{7-4}$$

Thus, for example, in the production of ^{207}Pb from ^{235}U,

$$N(^{207}\text{Pb}) = N(^{235}\text{U})(e^{\lambda' t} - 1)$$

and of ^{206}Pb from ^{238}U,

$$N(^{206}\text{Pb}) = N(^{238}\text{U})(e^{\lambda t} - 1)$$

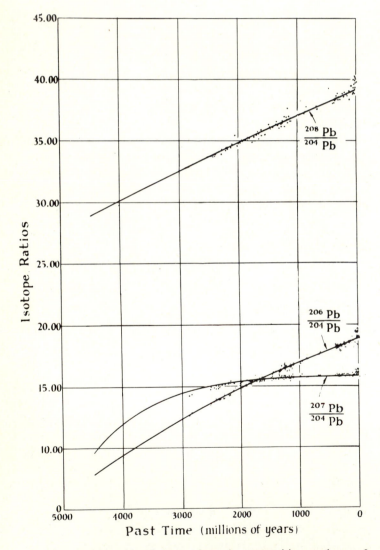

Fig. 7-8 Relationships between isotopic composition and age of ordinary common leads. (*From Russell and Farquhar, "Lead Isotopes in Geology," Interscience Publishers, Inc., New York, 1960.*)

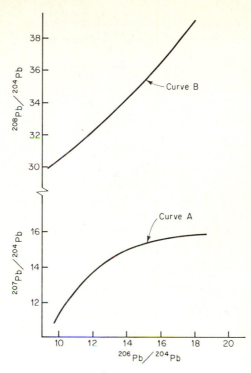

Fig. 7-9 Examples of curves developed by the generation of ordinary leads. The longer radioactive decay continues, the further upward and to the right does the resultant lead appear. Curve *A* depicts development of $^{207}Pb/^{204}Pb$ versus $^{206}Pb/^{204}Pb$, and curve *B* depicts development of $^{208}Pb/^{204}Pb$ versus $^{206}Pb/^{204}Pb$.

A basic assumption of all lead isotope interpretations is that at some very early time in the earth's history all lead present in the earth was of the one isotopic composition. Such lead is referred to as *primeval lead,* and the time of its existence is referred to as t_0. It is further assumed that all the lead now found at and near the surface of the earth has been produced by the addition of *radiogenic lead* to such *primeval lead.* These "mixed leads," which are no longer associated with uranium and thorium, are referred to as *common leads* and result from one or the other of two sequences of events, yielding either *ordinary* or *anomalous* leads.

1. Ordinary leads These are common leads of comparatively simple history. Their isotope ratios (L) can be accounted for mathematically as the result of the mixing of a primeval lead A with radiogenic lead B, produced by decay of uranium and thorium associated with the primeval lead in some ultimate source rock:

$$L = A + B$$

Such a lead is visualized as forming as follows. Trace quantities of primeval lead, and uranium and thorium, occur as disseminations in some fairly homogeneous, deep source rock. As long as such a source remains untapped its "total lead" content L is continually changed by the steady production of radiogenic lead from the associated uranium and thorium. If such a source is then momentarily broached (by deep faulting associated with an orogeny), there will be a local escape of lead ($L = A + B$) together, of course, with some uranium and thorium. If the lead is precipitated as an insoluble compound (generally PbS) in the upper crust or at the surface near a volcanic orifice, it is automatically chemically separated from the uranium and thorium that continue on as, for example, chlorides. Such lead thus suffers no further radiogenic addition and represents very closely the isotopic composition of our "source lead" at the time of its abstraction; i.e., it is a "single-stage" lead. It is lead of this isotopic type that we refer to as *ordinary lead*. Variation of the isotopic constitution of ordinary leads with age is depicted in Fig. 7-8. Equivalent changes in $^{207}Pb/^{204}Pb$ versus $^{206}Pb/^{204}Pb$ and of $^{208}Pb/^{204}Pb$ versus $^{206}Pb/^{204}Pb$ are shown in Fig. 7-9.

2. Anomalous leads These may be derived from *ordinary leads* by the occurrence of a further event. Suppose an *ordinary lead* is later taken into solution by subsurface water that percolates through sedimentary rocks containing uranium and thorium as two minor components. To our *ordinary* lead we would have added a new increment of radiogenic lead so that now

$$L = A + B + C$$

Whereas our *ordinary* lead is a "single-stage" lead, the new product is a "two-stage" lead—an *anomalous lead*. (Clearly our second stage may be repeated where geological histories become complex, and an anomalous lead may be the product of "*n*" stages. Usually, however, it is not possible to discriminate between a third and later stages.)

Figure 7-9 showed that the generation of *ordinary* leads yielded curved relationships between the relevant isotope ratios. In contrast the production of *anomalous* leads usually seems to involve the linear addition of radiogenic lead, leading to the development—within a single ore body or small mineralized province —of straight-line relationships, as shown in Fig. 7-10.†

This is, of course, a highly simplified picture of the development of the isotope ratios we observe in natural lead. In other cases a given lead may be produced by the mixing of two *ordinary* leads of different ages, yielding a product lying, for example, somewhere along the chord of Fig. 7-11. Others may be produced by the mixing of such an *ordinary-ordinary* mixture with a radiogenic lead, and so on. Clearly the possibilities are numerous and it is perhaps surprising that we find as much order as we appear to do in the isotopic constitution of ore lead.

† It should be pointed out that there is another usage of the term "anomalous lead." This is based on the relation between the ages of the lead and the containing rocks. If the age determined for the lead is less than that determined for the enclosing rocks the lead is referred to as *J-type anomalous* (J, after Joplin, Missouri, the original "type occurrence"); if the age determined for the lead is greater than that determined for the enclosing rocks the lead is referred to as *B-type anomalous* (B, after Bleiberg, Austria, the original type occurrence, though it appears now that the lead originally measured did not come from Bleiberg after all). There are, however, often considerable uncertainties involved in the calculations of these ages, and the author does not favor the usage. Keeping in mind the above uncertainties it appears that some *B-type* leads may be *ordinary* leads of our present usage and *vice versa*, that many *J-type* leads are *anomalous* leads of our present usage, and that in some cases *B-* and *J-type* leads may be closely related and differ only in the amount of radiogenic addition both have suffered.

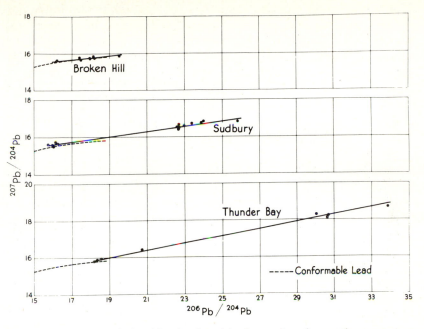

Fig. 7-10 Linear relationships developed in three suites of anomalous common leads; the plot also shows the relationship of each suite to the curve generated by the leads (apparently close to being "ordinary") of a number of stratiform ore bodies. (*From Stanton and Russell, Econ. Geol., 1959.*)

ORE LEADS

A plot of $^{207}Pb/^{204}Pb$ versus $^{206}Pb/^{204}Pb$ for a random selection of galenas from a variety of ores is shown in Fig. 7-12. Keeping in mind that significant *fractionation* of lead or of uranium isotopes is unlikely or insignificant, Fig. 7-12 indicates a wide variation in U/Pb ratios of the source materials. Examination of lead isotope abundances on the basis of *ore occurrence*, however, appears to simplify the picture a little.

In igneous rocks Little is known of the general behavior of ore leads formed orthomagmatically within large igneous intrusions, largely because igneous intrusions are not a common habitat of lead ores. The lead of very minor galena occurring in the massive nickel ores of Sudbury, Ontario (discussed in Chap. 11), which are found within and adjacent to a mafic intrusion, appears to be ordinary. This is not to suggest that all igneous ore leads are likely to have had simple histories; the characteristics of lead concentrations generated by igneous intrusions of different derivations seem likely to differ widely. The leads of some midoceanic basalts, for example, are very variable—though it must be emphasized that these are *rock leads*, not *ore leads*, and indeed that sulfide ore bodies *characteristically do not occur with basaltic rocks of this environment.*

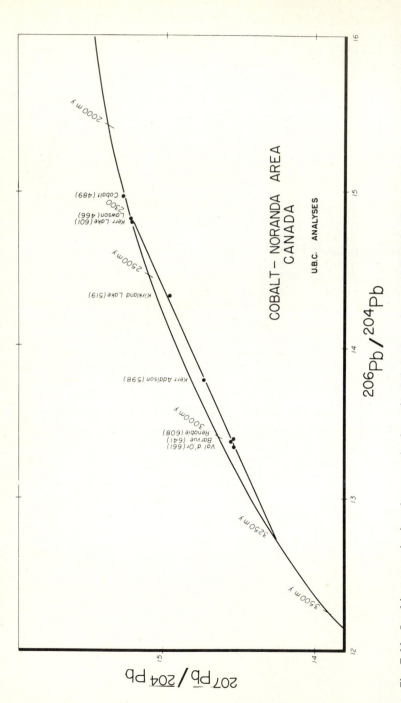

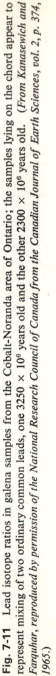

Fig. 7-11 Lead isotope ratios in galena samples from the Cobalt-Noranda area of Ontario; the samples lying on the chord appear to represent mixing of two ordinary common leads, one 3250×10^6 years old and the other 2300×10^6 years old. *(From Kanasewich and Farquhar, reproduced by permission of the National Research Council of Canada from the Canadian Journal of Earth Sciences, vol. 2, p. 374, 1965.)*

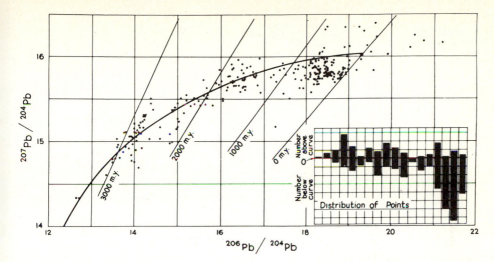

Fig. 7-12 Lead isotope plot for galenas of various ages and various geographical and geological origins. (The curve shown is *A* of Fig. 7-9, and the "isochrons" indicate the ages of the leads based on a single-stage model; the main purpose of this figure is, however, simply to show a representative range of isotope values such as would be obtained from a randomly chosen group of galena specimens.) (*From Stanton and Russell, Econ. Geol., 1959.*)

In sedimentary rocks (stratiform ores) The majority of these fall into the class considered as group 4 under *sulfur isotopes;* the deposits occur in sedimentary or metasedimentary rocks, are concordant with the latter, and are often well banded parallel to the associated bedding. A volcanic association is common. (Where such an association is not apparent it may be obscured by high-grade metamorphism or hidden as fine pyroclastic material; much detailed study is required in this connection.)   The leads of these occurrences show two outstanding features as a class:

1. For individual ore bodies, or groups of them occurring in close stratigraphic relation, within-deposit uniformity is very high. This applies to all of $^{207}Pb/^{204}Pb$, $^{206}Pb/^{204}Pb$, and $^{208}Pb/^{204}Pb$ and is often so extreme that even the most precise mass spectrometry can detect no differences between samples. For Broken Hill, New South Wales, for example, no detectable differences have been found in samples taken over a 4-mile length and a 2000-foot depth. Uniformity is not always quite so extreme as this, but it is invariably very high and far higher than in any other class of lead concentration.
2. For the whole class of deposits, covering a wide age, geographical, and time spread, there is a remarkable adherence to a single $^{207}Pb/^{204}Pb$-$^{206}Pb/^{204}Pb$ "growth curve" and, to a slightly lesser extent, to a $^{208}Pb/^{204}Pb$-$^{206}Pb/^{204}Pb$ curve. This is shown in Fig. 7-13*a* and *b* and suggests either a widespread, comparatively homogeneous source, or a series of remarkably

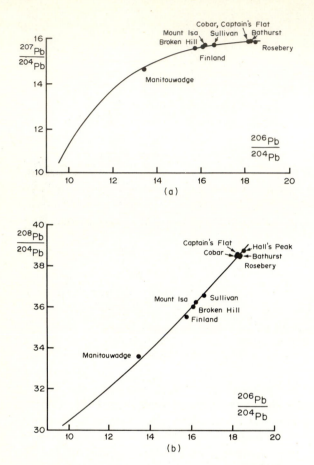

Fig. 7-13 (*a*) Plot of $^{207}Pb/^{204}Pb$ versus $^{206}Pb/^{204}Pb$ for galena leads of a number of stratiform ore bodies, the isotopic constitution of whose lead is very uniform; the curve is the best-fitting "single-stage" growth curve and is *A* of Fig. 7-9. (*b*) An analogous diagram for $^{208}Pb/^{204}Pb$ versus $^{206}Pb/^{204}Pb$. [*Adapted from Slawson and Russell, in Barnes* (*ed.*), *"Geochemistry of Hydrothermal Ore Deposits." Copyright © 1967 by Holt, Rinehart and Winston, Inc. Reprinted by permission of Holt, Rinehart and Winston, Inc., Publishers, New York.*]

similar sources, for leads of this class of ore occurrence. Recent more-refined work is showing that some ores of Paleozoic age and younger deviate from the growth curve slightly but significantly, usually on the low-^{207}Pb side. Such deviations represent only 1 or 2 per cent variation in original U/Pb ratios, however, and are an order of magnitude less than the variations found in leads from other ore types.

In metamorphic rocks   Many of the large class of stratiform ores are, of course, contained in *meta*sedimentary rocks and are thus problems with a metamorphic as well as a sedimentary component.   It has been suggested (cf. Richards, 1968) that the great within-deposit uniformity of lead isotope ratios in some of the older stratiform deposits (e.g., Broken Hill) of highly metamorphosed terrains might be due to metamorphic homogenization, i.e., thorough postdepositional *mixing*.   If this were correct the within-deposit pattern would reflect postdepositional events rather than features of source and migrational history.   However, most of the deposits concerned show conspicuous *elemental* heterogeneity, and the sulfur of the galenas has certainly not been homogenized isotopically (Fig. 7-5). Unless considerable solution activity is involved, it seems unlikely that metamorphism has any significant effect on lead isotope distributions within deposits and hence unlikely that there is any characteristic metamorphic pattern of lead isotope abundances.

Vein ores   These appear to contain three principal types of lead:

1. *Anomalous leads.*   A number of veins and vein provinces show conspicuously linear ^{207}Pb-^{206}Pb-^{204}Pb relations.   Vein galenas from Broken Hill, Sudbury, and Thunder Bay, Ontario, are shown in Fig. 7-10.   All appear to have their origin close to the growth curve of Fig. 7-13a and to be approximately tangent to it.
2. *Ordinary leads.*   Some veins—in particular those of very old Precambrian volcanic terrains such as those of the Canadian Superior province—contain leads that appear to be ordinary and that fall on, or very close to, the growth curves of Fig. 7-13.
3. *Hybrid ordinary leads.*   Some of the leads of the Canadian Superior province show linear relationships, suggesting that their isotopic constitutions have been produced by the mixing of two ordinary leads of different ages (Fig. 7-11 after Kanasewich and Farquhar, 1965).

CONCLUDING STATEMENT

We have seen through three examples how knowledge of segregational processes and the ability to recognize the products of mixing can be used to help in elucidating various geological processes responsible for the formation of ores.   *Segregation* of selenium from sulfur usually indicates that the two have been involved in solution processes operating on or near the earth's surface.   *Segregation* of sulfur isotopes may indicate a variety of biological or abiological processes and their operation under a variety of conditions.   The nature and degree of *mixing* of lead isotopes seem to indicate whether the source of a lead—and hence of the ore of which that lead is a part—was simple or compound and seem to indicate the relative ages of the events concerned.   Many other segregational and mixing processes —such as the concentration of nickel in olivine and hybridism in igneous rocks—

are known and understood and are of great assistance in tracing sequences of past geological events.

There are, however, a number of quite striking instances of segregation—several of them involving ores—whose mechanisms are not understood and indeed whose possible significance seems barely to have been recognized in some cases. The segregation of iron from manganese in the development of many sedimentary iron and manganese concentrations has been recognized for many years but remains one of the great enigmas of ore-genesis theory. Not so widely recognized is the extraordinary segregation of copper from lead and zinc in the development of stratiform and stratabound sulfide ores. In spite of their many chemical and crystal chemical similarities, and in spite of the fact that they occur in more or less equal amount in almost every terrestrial source rock one cares to consider, most lead-zinc-rich ores of sedimentary affiliation contain negligible copper and vice versa. High concentrations of lead and copper together in these ores are unknown. Antimonial (stibnite) ores in veins are virtually monometallic all over the world, and there are other examples of similar kind. An understanding of the processes that have led to such remarkable segregation must surely lead to a more accurate understanding of the ways in which the ores themselves have formed.

RECOMMENDED READING

Goldschmidt, V. M.: In A. Muir (ed.), "Geochemistry," Oxford University Press, London, 1954.

Jensen, M. L.: Sulfur isotopes and mineral genesis, in H. L. Barnes (ed.), "Geochemistry of Hydrothermal Ore Deposits," pp. 143–165, Holt, Rinehart and Winston, Inc., New York, 1967.

Kaplan, I. R., K. O. Emery, and S. C. Rittenberg: The distribution and isotopic abundance of sulfur in recent marine sediments off Southern California, *Geochim. Cosmochim. Acta*, vol. 27, pp. 297–331, 1963.

Russell, R. D., and R. M. Farquhar: "Lead Isotopes in Geology," Interscience Publishers, Inc., New York, 1960.

8
Growth and Growth Structures in Open Space

Since ores are composed of crystalline solids, an understanding of their growth structures requires an understanding of the mechanisms of crystal growth itself.

Crystals are solids exhibiting long-range order in the packing of their constituent atoms and molecules. The development of such ordered structures may proceed in four ways.

1. *Sublimation.* In this case the crystal is built up by deposition directly from the vapor. Familiar examples are the formation of snow from supercooled water vapor and the growth of crystals of sulfur and other substances on the walls of cracks leading from the throats of volcanic fumaroles.

2. *Precipitation from solution.* This is, of course, the most familiar method of crystal growth, and the one by which many of the finest natural and synthetic crystals have been produced. Deposition takes place when the solution in question becomes supersaturated either by evaporation of the solvent or by decrease in temperature. If a low degree of supersaturation is preserved over a long period and the site of crystal growth is kept undisturbed, large crystals of high quality may develop. Natural examples are the beautiful crystals of quartz, gypsum, fluorspar, and other substances formed on the walls of underground fissures through which subsurface solutions have

traveled, and the crystals and crystalline particles of salt deposits formed at the bottom of landlocked bodies of water.

3. *Solidification from a melt.*   The process here, of course, is freezing, solid particles separating out from the parent liquid as the latter reaches its freezing point (or, in the case of supercooling, where some nucleating agency manifests its presence).   The obvious natural example here is the freezing of lavas.

4. *Solid-solid transformation.*   This is perhaps the least familiar of the four mechanisms, though a widespread one that proceeds in the earth's crust continuously.   It involves the transformation of a given substance from one structural arrangement to another without chemical change.   Examples are the conversion of amorphous to crystalline matter and the various crystalline inversions.

All four processes take place because they are a means by which a lowering of the free energy of the substance concerned may be achieved under conditions prevailing at the time.   An ordered, densely packed, crystalline solid clearly possesses fewer unsatisfied "free" bonds than a vapor, solution, or melt, and when conditions permit the formation of such a solid, it will immediately begin to form as a means of reducing the free energy of the system.   While the media of growth are rather different, there do not seem to be any fundamental differences among the first three mechanisms.   Concentrations are, of course, much lower in vapors than they are in most solutions and in melts, and sublimation involves the release, at the growing surfaces, of comparatively large quantities of heat as a result of the simultaneous release of heats of fusion and of vaporization.   The effects of both of these differences are lessened, however, by the greater mobility of vapors, which permits the more rapid transport of material to the site of growth and the more rapid dissipation of heat from it.   Because of the very much lower mobilities involved, crystal growth in solids presents rather different problems; these are considered in Chap. 9.

Although the formation of crystals lowers the free energy of the system concerned, it does not eliminate the free energy—crystallization only reduces it, and there remains a certain amount of residual bond energy within the crystal, particularly at and near its surface.   The residual energy at the surface of such a crystal is termed its *surface energy* in the case of crystals directly exposed to the parent solution, vapor, or melt and its *interfacial energy* in the case of crystalline boundaries in polycrystalline aggregates.

It follows from this that just as the energy of the system is reduced by forming the crystal, the energy of the crystal—and hence, again, of the system—may be reduced by reducing the energy of its surface or interface.   This may be done in two ways:

1. By reducing the *total area* of the surface or interface, since the surface or interfacial free energy is the free energy per unit area of surface or interface

2. By reducing the area of any parts of the surface or interface characterized by comparatively high free energy per unit area, substituting a corresponding area of surface or interface characterized by lower free energy per unit area

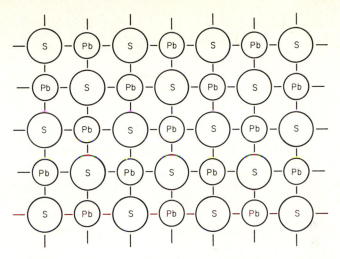

Fig. 8-1 Schematic two-dimensional diagram of galena, illustrating bonding of particles within the crystal, on the surfaces, and at the edges.

Where the crystal is perfectly isotropic, the first consideration is the only one; where the crystal is notably anisotropic, both are important. The case of an isotropic crystal may be illustrated simply, as follows. Suppose a freely growing single crystal of galena can be represented two-dimensionally as shown in Fig. 8-1. Each internal atom is bonded to four nearest neighbors (in three dimensions it will be six, as shown in Fig. 4-1). Each surface atom is exposed, and so "loses" one or more neighbors. In three dimensions it is quickly seen that, in the present case, atoms on a crystal face have lost one neighbor, those at edges two, and those at corners (coigns) three—i.e., half of their neighbors. The loss of a neighbor results in an unsatisfied bond and hence an added increment of surface energy. Therefore the smaller the area of face, the shorter the total edge length, and the fewer the number of corners, the less the total surface energy. The principle is shown simply in Fig. 8-2. Two crystals (e.g., of galena) contain the same number of atoms, similarly arranged and in similar spacing. Their surface areas are different, however. In crystal A there are 16 surface, 16 edge, and 4 corner atoms. In crystal B there are 14 surface, 18 edge, and 4 corner atoms. It is thus apparent that the surface free energy of A is slightly less than that of B; B is therefore a less

Fig. 8-2 Schematic two-dimensional diagrams of two crystals A and B of a single substance, showing the general direct relation between surface area, i.e., shape, and the sum of unsatisfied bonds. (*From Fyfe, "Geochemistry of Solids." Copyright McGraw-Hill Book Company, 1964, with the publisher's permission.*)

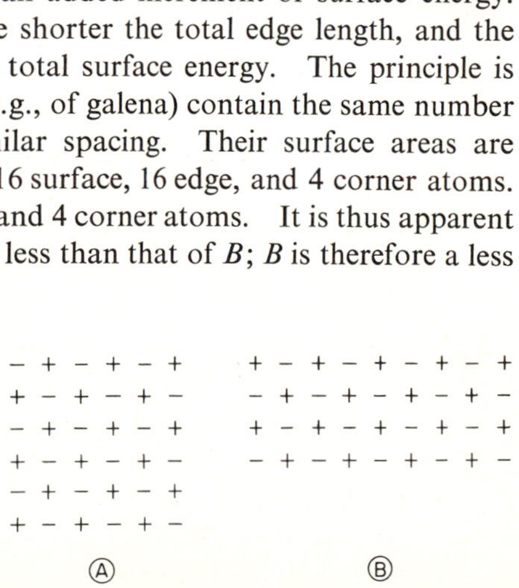

```
  − + − + − +        + − + − + − + − +
  + − + − + −        − + − + − + − + −
  − + − + − +        + − + − + − + − +
  + − + − + −        − + − + − + − + −
  − + − + − +
  + − + − + −

       Ⓐ                   Ⓑ
```

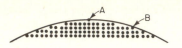

Fig. 8-3   Effect of the development of a curved surface on the number of surface particles on a crystal face. (*From Fyfe, "Geochemistry of Solids." Copyright McGraw-Hill Book Company, 1964, with the publisher's permission.*)

stable configuration than *A*, and if conditions permitted, it would tend to change its shape in the direction of that of *A*.

Where a crystalline substance is completely isotropic (which is probably rare) and change of *area* is the only factor influencing total surface energy, minimum energy will be achieved by the assumption of a spherical shape.   This does in fact appear to occur in some cases.   For example, much disseminated chromite is spherical, and "grape shot" and "pique" ore, consisting of rounded single crystals ranging from small pea size to about 1 inch across of chromite in serpentinite, are well known in several parts of the world.   However, when the crystal structure is considered on a fine scale it is apparent that the development of a curved surface may yield a large number of atoms in edge or corner positions—as in the case of *A* and *B* in Fig. 8-3—in which case planar surfaces of rather greater area may be energetically more favorable.   In a liquid, by contrast, there is no rigidity, and the spherical shape unquestionably yields the lowest surface energy.

Using similar two-dimensional models it will be apparent that in strongly anisotropic crystals the general tendency to reduce surface area will be accompanied by specific tendencies to reduce high-energy faces at speeds directly related to the order of the magnitudes of their energies.   This will be examined in more detail a little further on.

At this point it is appropriate to refer back briefly to the four principal mechanisms of crystal growth.   So far such growth has been considered in quite a general way, and no significance has been attached to the nature of the environment in which the crystals develop.   However, while all crystallization processes do indeed follow paths of energy reduction, there are some fundamental differences between them.   The grossest differences are between growth in a fluid phase and growth in a solid phase.   In the former, growth is essentially free: the developing rigid crystal simply grows into the enveloping nonrigid vapor, solution, or melt and, provided other conditions are favorable, may develop faces.   But where growth occurs "in the solid," the growing crystal is constrained by rigid surroundings, and the development of shape usually follows a rather different course.   In the case of free growth the interface between solid on the one hand and melt, solution, or gas on the other may be regarded as a "surface," and in many instances true surface tensions operate.   However, where growth occurs in the solid, we are concerned with the interaction of free bonds across the rigid interface between two solids, and instead of surface energies we are concerned with interfacial energies. While the results of the two phenomena are often surprisingly similar, the physical processes are not quite the same, and it is important to keep this in mind.   In view of this the present chapter is restricted to growth and growth structures in crystals formed in fluid environments—that is, in "open space."

It is self-evident that the ores in which free growth is important are those formed in open cavities. The crystals may form in suspension in a melt or, apparently much more importantly, from the walls of the cavity in a liquid or gaseous medium. While the crystals so formed grow, impinge, and are grown upon by later crystals, most are nucleated and grow initially as single crystals. We may therefore regard them in this way.

NUCLEATION

Before proceeding to examine the mechanisms of crystal growth it is appropriate to consider how such growth begins.

The instigation of the growth of a crystal is referred to as *nucleation*, a term arising from the fact that the structure necessarily develops from a nucleus of some kind. In some special cases such nuclei may be preexistent and are referred to as *seeds*. In the more general case nuclei are the initial products of precipitation from the medium concerned.

Nucleation in the latter case is substantially a matter of chance, and the growth of any given crystal is preceded by the formation and disintegration of large numbers of potential nuclei. Suppose that a solution is being slowly evaporated or a melt (such as water) is being slowly cooled. As soon as conditions are reached under which the solid can exist stably, crystallization can proceed. This, like all things, has to have its beginning, which in this case is the coming together of just sufficient atoms or molecules to form the initial structural pattern—i.e., crystal lattice arrangement—of the solid concerned. This is the nucleus.

However, such nuclei will necessarily have a very high ratio of surface area to volume and hence a very high free energy per unit volume. This may be illustrated by referring again to the galena crystal. Suppose that a single unit of PbS, as shown in Fig. 4-1, is formed in solution and persists momentarily as a separate and individual small unit. Of its 27 atoms, 26 are at the surface. Because all these possess unsatisfied bond energy, the free energy of such a nucleus is necessarily very high. This immediately implies a high solubility, and so there is a very strong tendency for such nuclei to go into solution again. It is here that chance plays an important role. The only way in which such a nucleus can avoid re-solution is by rapid growth to a size where the surface energy is a much smaller proportion of the total energy of the crystal. Stability is in fact reached only when the free energy of the crystal is reduced to a value just less than that of the free energy of the atoms remaining in the environment in which crystallization is taking place. A nucleus is said to reach its *critical size* when, through increase in volume, its energy per unit volume is reduced to that of its surroundings. At this point the nucleus can survive and grow. Such a point, however, is reached only when chance allows sufficient concentration of solute atoms at some particular spot for rapid accumulation and achievement of the critical size. Where concentration is insufficient for such rapid accumulation, the initial nuclei redissolve and are lost. It follows, and may be shown theoretically, that critical size depends on the degree of supersaturation in the case of solution; in the case of melts it is affected by the amount of undercooling, and with vapors by the size of the vapor droplets.

In other cases the nucleus is provided by a stable seed of some other substance whose lattice structure is similar in some critical way to that of the nascent crystals. The process involved has been termed *epitaxis* (by Royer, 1928), and the resultant growth is referred to as *epitaxial* growth. In this case the new crystal is essentially an oriented overgrowth. The fundamental minimum condition for such overgrowth appears to be the identity, or close similarity, of interatomic spacing and charge distribution of a row of atoms in the lattice structures of the two substances. The row of atoms may occur in the corresponding plane in the two structures, as is common in some closely related substances, or the atoms may be in quite different planes. It has been found that spacing and charge arrangement are the overriding factors in epitaxis. For example, similar substances, such as the alkali halides, will not exhibit epitaxis where the spacings exceed certain limits, but quite dissimilar substances will develop epitaxial relationships where spacing and charge distribution are similar in some direction of each. Similarity of structure type and of chemical and physical properties are of no avail unless the spacing requirements are met. However, where such requirements are not satisfied in isostructural, chemically similar substances by equivalent atom rows, they may sometimes be met by crystallographically nonequivalent rows. This leads to rotation of the overgrowth with respect to the seed.

Interesting examples of epitaxial relationships between chemically unlike substances are the nucleation of ice by covellite and silver iodide, galena by rock salt, and $NaNO_3$ by $CaCO_3$.

Clearly, whatever the medium of crystallization, the density of nucleation points will influence the nature of the final product. Where the solid forms from solution, heavy oversaturation leads statistically to the formation of many nuclei, in which case a colloid—frequently a gel—results. With lesser supersaturation, and hence fewer nuclei, a finely crystalline aggregate results, and with only slight supersaturation nucleation frequency may be reduced to the point where only a few large crystals develop.

GROWTH

An idealized picture of crystal growth is one in which the development of a stable nucleus is followed by the regular, ordered accretion of atoms, ions, or molecules that stack upon each other, layer by layer, to build up a perfect lattice. This is too simple a view, as will be seen shortly, but it provides a useful basis for some general considerations.

It is self-evident, and has been emphasized by Buerger and others, that crystal faces constitute a boundary between a crystal and the medium from which it is being deposited. Thus the manner of growth of the crystal, and the crystal faces displayed, might be expected to be determined both by the crystal itself and by the medium from which it forms.

Initial attempts to determine the role of the crystal were made quite early in the history of crystallography and were—naturally enough—concerned with geometrical considerations.   An early "rule" was that the faces of simplest indices were the most frequently developed and that frequency of occurrence and relative area of faces tended to decrease with increase in complexity of indices.   This has sometimes been referred to as the law of Hauy.   However, Buerger (1947) has pointed out that such a rule has obvious shortcomings; for instance, it implies that all cubic crystals should have similar form development, which they manifestly do not.   Bravais (1851) improved on the earlier approximation by postulating that the relative importance of the faces of a given crystal should be in the same order as their reticular density—i.e., their interplanar spacing parallel to the faces in question. Donnay and Harker (1937) improved on this again by showing that the Bravais principle had wider application when "interplanar spacings" were taken to include not only the simple spacings of the lattice but also those modifications of spacings required by space-group operations involving translation components.

Although such observations are purely empirical, and are concerned only with the recognition of systematic relationships between changes in one feature and differences in another, they must reflect the processes of crystal growth.

SIMPLE STACKING: THE KOSSEL-STRANSKI MECHANISM

The first theory of crystal growth relating crystalline form to surface energies was that put forward in 1885 by Pierre Curie.   While Curie could not at that time take into account the effects of the atomic arrangements of crystal surfaces, he did show in a general way that the rate of growth of any face should be directly related to its surface energy.

W. Kossel and J. N. Stranski, in separate papers published in 1928, seem to have been the first to put forward explanations of crystal growth that related the development of faces to the faces' atomic configurations.   The germ of their ideas had been suggested a little earlier by Volmer (1922) and Brandes (1927), but whereas the latter two simply put forward a general mechanism of crystal growth, Kossel and Stranski developed this in a quantitative way to explain the incidence of specific faces.

Earlier theories of crystal growth postulated the development, by adsorption, of a high concentration of atoms or molecules about the outermost lattice planes of the crystal.   It was suggested that this might be a few molecules deep and that it constituted an intermediate stage between the disordered environment and the ordered crystal.   Volmer and Brandes considered that crystals built up layer by layer, the "adsorption layer" continually moving ahead of the growing lattice and providing its source of atoms or molecules.   Kossel and Stranski also realized that local concentration and diffusion were important factors in growth.   However, they recognized in addition the importance of the fine structure of the face upon which accumulation was taking place.   Stranski stressed the importance of the "work of separation" of the unit particle from its position on the crystal surface. This was defined as the work required to move the particle (molecule, atom, or ion) from its position on the crystal to infinity.   Clearly the greater this value, the more firmly fixed the particle once it adhered to the surface and the more rapidly would others of its kind be attracted to positions beside it.

On this basis Kossel and Stranski postulated that crystals grew by the simple stacking of one lattice layer upon another.   Each layer began with the formation of an island nucleus standing up on the last-completed layer.   This presented an edge and, where the edge was irregular, kinks (Fig. 8-4) into which incoming particles could nest.   Any particle lodging in an edge would be subject to one-third of the bonding force experienced by an interior particle, and one lodging at a kink would be subject to one-half of that force.   Any isolated particle would, of course, be subject to only one-sixth of the bonding force and would be highly likely to lose adherence.   Growth was thus seen to proceed by the more stable lodgment of incoming particles at steps and kinks, the original nucleus thus spreading out in two dimensions over the face until that particular layer was complete.   Growth of the next layer then awaited the development of a new stable island nucleus (or series of nuclei) and then proceeded as before.   The crystal thus grew by the laying down of one lattice layer after another.

Such a simple mechanism required some modification for nonplanar faces, though this was fairly easily fitted to the theory.   The higher index faces of a cube, for example, are stepped, as shown diagrammatically in Fig. 8-5.   Such a face could not grow in the simple fashion illustrated in Fig. 8-4.   Kossel and Stranski considered that in these cases the steps in the face served as nuclei instead of the two-dimensional islands of the planar faces and that growth then followed the same pattern.   Figure 8-5 shows a (110) face made up of stepped increments of (100) and (010); growth here would thus, on the Kossel-Stranski model, take place by alternate growth on the (100) and (010) strips, each set of accumulations taking place in exactly equal proportions.

On the basis of layer growth of this general kind, it is quickly seen that the strength of bonding at the surface (Stranski's "work of separation") is likely to be related in some way to reticular density and the Bravais principle.   For simple cases the greater the reticular density, the stronger the bonding, and the stronger the bonding, the greater the speed of growth and hence the more rapid the elimination

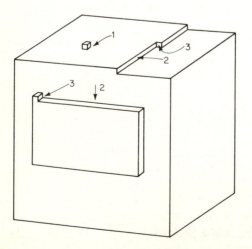

Fig. 8-4 Attachment of new particles on crystal surfaces (1) and at edges (2) and kinks (3), according to the Kossel-Stranski mechanism.

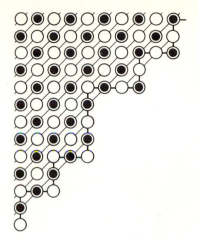

Fig. 8-5

of the face in question.   This has been examined in some detail by Buerger (1947), who has shown that the problem may be solved on a surface-energy basis.

Buerger first of all examined an ideal case where bonding was ionic and bond-strength energy was distributed evenly over the "molecule," which was taken to constitute the unit cell.   When such a molecule lands on a crystal face, the energy of the bond between the molecule and the crystal is approximately proportional to the area of the surface joining the molecule to the crystal.   If several alternative sites on the face are available, the incoming molecule will choose that providing maximum bond energy and hence—in terms of the present model—that offering maximum surface area for attachment.   The principle may be illustrated by the case of the three hypothetical pinacoid faces of Fig. 8-6 (after Buerger, 1947, p. 597, figs. 1 and 2).

Suppose such pinacoids represent three planes of different rectricular density within the crystal they bound.   Figure 8-6a shows the process of accretion where new layers have already been started.   Maximum area of attachment of our molecule is achieved by nesting in steps, as shown, and in each case this is half the total area of the parallelepiped molecule.   Thus if first coordinations only are considered, the bond strength offered by each pinacoid is the same; no plane has any energetic advantage over the others, and if there is an even supply of molecules, the crystal may be expected to grow at the same rate in all three directions.

However, this situation changes as soon as it becomes necessary to start a new layer.   Figure 8-6b shows single molecules in place on each of the pinacoids, as they might appear just as a new layer begins.   Clearly the surfaces of attachment are again the three sides of the cell taken one at a time.   Since the volume of our cell is the area of any such side multiplied by the corresponding interplanar spacing, it follows, because this volume is constant, that the area of any side is inversely proportional to the spacing and hence that the strength of bonding is inversely proportional to the spacing.   Thus the faces most likely to attract and hold incoming molecules are those with least interplanar spacing.   As the

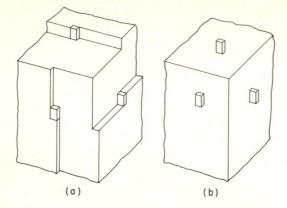

(a)   (b)

Fig. 8-6 (*a*) Deposition by nesting into edges and kinks of partially complete faces, and (*b*) the beginning of new pinacoidal layers. (*From Buerger, Amer. Mineralogist, 1947.*)

necessity to start a new plane continually recurs, the pinacoid of least spacing will be that of most rapid growth. However, where interfacial angles diverge from 90°, the most rapidly growing faces grow themselves out and eliminate themselves. In consequence the faces remaining—and thus ultimately those of greatest area and most frequent occurrence—are those that grow most slowly; those parallel to planes of greatest interplanar spacing. Thus in this simple situation Bravais' rule holds, and it may be said that crystals grow in such a way that increasing order of importance of the pinacoids is in the order of increasing spacing and vice versa.

Therefore surface energy continues to be important throughout the growth of the crystal. Bonding forces are not limited to the immediate surface, as assumed for the very simplified case above, but also include weaker energies associated with the layers immediately beneath. The sum of these may be referred to as the *total coordination*. If several sites are available to an incoming ion or other particle, those offering greatest coordination, and hence the greatest reduction in surface energy of both particle and crystal, will be favored. This appears to hold in a general way for all kinds of faces, provided there is a simple relation between the shape of the incoming particle and the geometry of the lattice as a whole. Thus this "ideal case" shows in simple fashion how the development of the crystal is an attempt to lower the energy of the system, and the development of particular faces is an attempt to lower the energy of the crystal.

This, however, begins to anticipate considerations of habit, which are examined in more detail a little further on.

SPIRAL GROWTH: THE INFLUENCE OF DISLOCATIONS

While the simple picture of nucleation of new layers by the attachment of single particles and the development of such layers by ordered stepwise growth is both neat and appealing, it fails to explain several features of growing crystals. The outstanding—and most embarrassing—of these is the rates at which crystals grow.

It is possible to calculate, for a given set of conditions, the rates at which crystals should grow by the simple stacking mechanism just discussed. However, the rates so obtained are often quite inadequate; crystals have been found to grow at speeds over 10^{10} faster than predicted by simple stacking theory. Such a discrepancy is not to be passed over lightly, and it is clear that something other than simple surface nucleation is operating. The explanation of the rapid growth of many crystals lies in a modern development of crystal physics—the recognition of *dislocations*.

Crystals have often been thought of as nature's closest approach to perfection, in the inanimate world at least. Nearly perfect as they are, they have, however—like most things—their imperfections, and it is perhaps largely because of these that they have been able to grow to a size where their beauty and apparent perfection have attracted the attention of man.

Crystals show several kinds of *imperfection*. A perfect crystal is, of course, one in which the particles form a regular pattern uninterrupted save by its bounding surfaces. By an imperfection is meant a small area where the regular pattern is interrupted so that some particles are not surrounded by neighbors in normal arrangement. By "small" is meant that the imperfection extends only a few atomic diameters in at least one dimension. Examples of imperfections are lattice vacancies, impurity atoms (both the "point defects" of Chap. 3), twin and stacking fault boundaries, and *dislocations*.

Dislocations were first recognized by physicists in the 1930s, and they have since become of central importance in the study of the physical properties of crystals. They are of two principal kinds: *edge dislocations* and *screw dislocations*.

Edge dislocations may be produced by slip, such as shown in Fig. 8-7. Here slip has occurred over the area $ABCD$, and the boundary AD of the slipped area within the crystal is a dislocation. By definition the dislocation AD is normal to the slip vector. The structural arrangement normal to AD is shown in Fig. 8-8.

Fig. 8-7 Slip that produces an edge dislocation. Unit slip has occurred over the area *ABCD*, and the boundary *AD* of the slipped area is a dislocation. By definition the edge dislocation *AD* is normal to the slip vector. (*From Read, "Dislocations in Crystals." Copyright McGraw-Hill Book Company, 1953, with the publisher's permission.*)

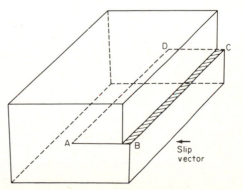

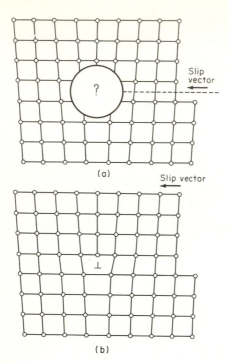

(a)

(b)

Fig. 8-8 A lattice plane normal to *AD* in Fig. 8-7. The dislocation is a region of severe atomic misfit, where atoms are not properly surrounded by their neighbors. (*a*) The elastic strain around the dislocation; the arrangement of atoms at the center of the dislocation is not known exactly. (*b*) An approximate arrangement. The symbol ⊥ denotes the dislocation. Note that the dislocation is the edge on an incomplete atomic plane; hence the name *edge dislocation*. (*From Read, "Dislocations in Crystals." Copyright McGraw-Hill Book Company, 1953, with the publisher's permission.*)

It can be seen here that the dislocation is the edge of an incomplete atomic plane, hence the term *edge dislocation*. An effect of large numbers of edge dislocations is to divide a crystal into blocks that are usually tilted with respect to each other by a few minutes of arc. This *substructure*, or *mosaic structure* (which may be induced by a variety of causes), within single crystals does not represent a complete disruption of the lattice—it is an *imperfection* in an otherwise nearly perfect structure. Its presence, and the degree of misorientation involved, may be neatly demonstrated by glancing x-rays at a low angle off a surface of such a crystal. The sharp, round Laue spots characteristic of a truly single crystal become elongated, and the x-ray photograph shows *asterism*. This is well illustrated by appropriately treated aluminum, as shown by Guinier (1953).

Whereas an edge dislocation is at right angles to the slip direction, a screw (or "*Burgers*") dislocation is by definition parallel to the slip vector. The kind of slip that produces a screw-type dislocation is shown in Fig. 8-9. In this case unit slip has occurred over *ABCD* in the direction indicated, yielding the screw dislocation *AD* parallel to the slip vector. Its more detailed nature (in a simple case) is shown in Fig. 8-10*a*. It is important to note, and it has been emphasized by Read (1953), that a crystal showing this type of dislocation is no longer composed of parallel atom planes stacked one above the other but has become a single atomic layer in the form of a helicoid, or spiral ramp. In connection with this figure it should be noted that the distortion about a screw dislocation is mostly pure shear,

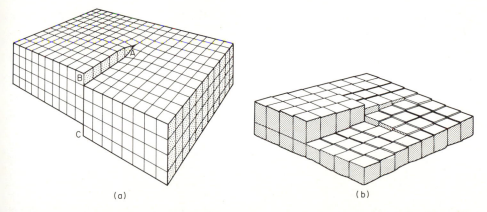

Fig. 8-9 Type of slip that produces a screw dislocation. Unit slip has occurred over *ABCD*. By definition the screw dislocation *AD* is parallel to the slip vector. (*From Read, "Dislocations in Crystals." Copyright McGraw-Hill Book Company, 1953, with the publisher's permission.*)

and so the situation close to *A* may be taken as that in Fig. 8-10*b*, in which undistorted cubes are displaced relative to each other, in simple fashion, in the direction of the slip vector.

The screw dislocation appears to be a most important factor in the growth of many crystals.

It has already been noted that observed growth rates are very much greater than might be expected from simple surface nucleation of new layers. In this case, just as with the formation of the original crystal nucleus, a certain critical supersaturation or vapor pressure is required before nuclei can form stably and grow. The probability that a nucleus of critical size will develop is very sensitive to supersaturation, and Frank (1952) has calculated that a supersaturation of 25 to 40 per cent is required for nuclei to form fast enough to account for crystal-growth rates

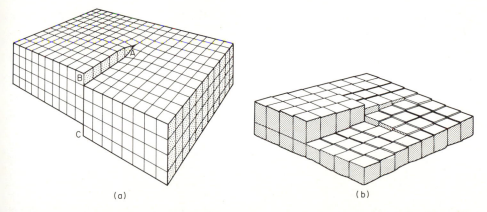

(a)   (b)

Fig. 8-10 Diagrammatic view of a screw dislocation at the surface. In (*a*) the dislocation *AD* (of which only the end *A* is visible) is parallel to the line *BC*, which is parallel to the slip vector. The atoms are represented by distorted cubes in (*a*) and by undistorted cubes in (*b*). Note that the crystal is a *single* atomic plane in the form of a helicoid, or spiral ramp. (*From Read, "Dislocations in Crystals." Copyright McGraw-Hill Book Company, 1953, with the publisher's permission.*)

commonly observed in the laboratory. In fact, however, substantial growth rates are observed at far lower supersaturations.

A very neat illustration of this was discovered by Volmer and Schultze (1931) in growing iodine crystals from vapor. Contrary to the theory based on recurrent nucleation, the growth rate of the iodine crystals was found by these investigators to increase almost in proportion to supersaturation from supersaturations just above 2 per cent. Even at 1 per cent the growth rate was clearly perceptible on their scale of millimeters per hour. Such a situation could develop only if nuclei were always present. It therefore appears that growth did not depend on the recurrent development of "island monolayers" for the nucleation of each new layer, as envisaged for the simple stacking mechanism. On the contrary the growth rate behaves as if steps were always present on the surface, never being eliminated by completion of layers as required by the simple stacking theory. This then raises the problem of how a surface can retain its steps no matter how far these steps advance, and it is here that the screw dislocation appears to provide an answer.

The significance of screw dislocations in crystal growth was first suggested by Frank (1952), and his theory of the growth of an imperfect crystal has a brilliant and elegant simplicity. It is based on the assumption that growing crystals are not perfect—as is more or less implicit in the simple stacking theory—but contain dislocations. This is supported by the less-than-theoretical strength of many crystals, which is predicted by dislocation theory and x-ray analysis. Figure 8-11 shows the mechanism proposed by Frank. The screw dislocation meets the surface at right angles, and the resultant step joins the end of the dislocation (*AD* of Fig. 8-9) to the edge of the crystal. As noted earlier, because the distortion about the screw dislocation is mostly pure shear, the atoms can be represented by undistorted cubes slipped over one another, as in Fig. 8-10. It can be seen that as the particles (ions, atoms, molecules) land on the surface they can immediately nest at the edge of the step, which constitutes a ready-made nucleation zone. The step thus advances, but it can never grow itself out by completing a layer as would our earlier island nucleus. It is simply the advancing front of the single-layer spiral ramp already referred to and, on a geometrical basis, has no alternative but to advance indefinitely as long as there are particles available to add to it. The crys-

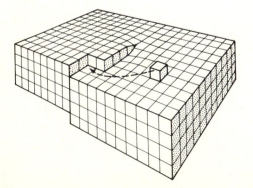

Fig. 8-11 Screw dislocation intersecting the surface of a crystal and providing a self-perpetuating edge for the nesting of incoming particles. The upper surface of the crystal has been converted into a spiral ramp by the appearance of the dislocation. The step cannot now be eliminated, and there is no further necessity for the nucleation of new layers. (*From Read, "Dislocations in Crystals." Copyright McGraw-Hill Book Company, 1953, with the publisher's permission.*)

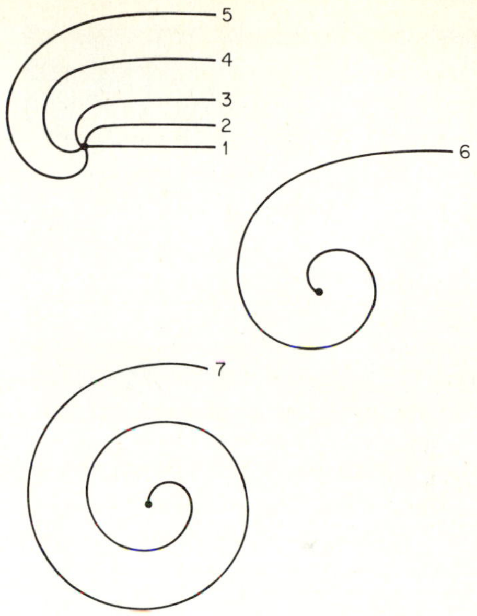

Fig. 8-12 Diagrammatic portrayal of seven successive stages in the development of a spiral. The slip plane is normal to the page and initially intersects the crystal surface along line 1. The spiral develops until it reaches the steady state of 7, at which stage it rotates about the central point without changing shape. (*From Read, "Dislocations in Crystals." Copyright McGraw-Hill Book Company, 1953, with the publisher's permission.*)

tal can therefore grow to any size, without the interruptions caused by delays in the nucleation of new layers.

Since the end point of the dislocation is fixed, the advancing step will tend to rotate about it, and so the ramp created instantaneously by the original formation of the dislocation is preserved and continued by growth on a spiral-shaped growth front. The geometry of the process is stated very clearly by Read:

> . . . a spiral develops; the outer sections have farther to go to make one complete revolution; so the inner sections make more revolutions in a given time. . . . The rate of advance of any section of the step is proportional to the excess of vapor pressure† over the local equilibrium vapor pressure, where the latter increases with curvature. As the spiral develops, the sections of step near the dislocation acquire a higher curvature. This increases the local equilibrium vapor pressure; now the inner sections advance more slowly than the outer sections. Finally a steady state is reached in which (rate of advance)/(distance from dislocation) = (angular velocity) = (revolutions per unit time) is the same for all points on the step. In the steady state, the form of the spiral remains constant; the whole spiral rotates uniformly about the dislocation. . . . Now the lines represent a step on a growth surface instead of a dislocation on a slip plane (1953, p. 145).

The development of such a spiral is illustrated in two dimensions in Fig. 8-12. Where more than one dislocation occurs, the resulting interaction of spirals produces a less simple picture, though this is generally quite ordered. Consideration of such complications is outside the scope of this book but is presented in detail by Read (1953).

† For solutions and melts, substitute concentration and solubility.

Fig. 8-13 Spiral growth pattern on the surface of a silicon carbide crystal. Sphalerite also commonly shows such structures. (*From Read, "Dislocations in Crystals." Copyright McGraw-Hill Book Company, 1953, with the publisher's permission.*)

To this point such a mechanism is quite theoretical and speculative, however. If crystal growth did in fact proceed on spiral fronts, the latter should be found on many crystal faces—"frozen" in the positions they had attained at the instant of cessation of growth. It might be expected, however, that such observations would not be easy, owing to the likely fineness of the structures concerned. Such fineness would require that the crystal surfaces were close to atomically plane, that the step height—i.e., the original lattice translation vector—was as large as possible, and that the growth steps were sufficiently far apart for microscopical resolution.

Very satisfactorily such growth spirals were soon found, and a number of beautiful examples have been photographed (see, particularly, Verma, 1951; figs. 15–18, 20–22, 30–32, 41–47). One is reproduced in Fig. 8-13 and shows the nature of the structures quite clearly. Although many other examples have since been photographed, Verma's fine photographs now have a historical quality since they showed without doubt the correctness of Frank's deductions.

It therefore appears that the earlier picture of crystal growth—the ordered, repetitive stacking of layer upon parallel layer to yield a perfect lattice structure —is in many cases an insufficient oversimplification. The surface-energy requirements of nucleus formation on one hand and observed rates of growth and relationships between these and degrees of supersaturation on the other indicate that crystals certainly do not always develop by the successive nucleation and growth of a series of parallel layers. Surface-energy requirements and observed rates of growth, however, are well explained by continuous nucleation along steps initiated by screw dislocations.

DENDRITIC GROWTH

Although many crystals grow as compact bodies, there are many that do not. Frequently crystals develop as branched structures that grow rapidly in a few directions, leaving quite large spaces between the individual branches. These are termed *dendritic crystals*, or *dendrites*. When carefully examined on a fine scale the extremities of the branches are found to be planar faces, and x-ray studies always reveal a crystalline structure in the dendrites themselves. A general form is shown in Fig. 8-14. There is no question that they are crystals. It is also clear that their development represents crystal growth of another kind. Such growth is highly favored by sublimation, but it also occurs by precipitation from solution and solidification from a melt.

Crystal surfaces may be classified into three types:

1. *Singular surfaces.* These surfaces are planar, as distinct from stepped, faces that yield cusped minima in plots of surface energy versus orientation.
2. *Vicinal surfaces.* These are surfaces whose orientations are comparatively close to those of singular planes and whose surface energies increase linearly with increasing departure from the orientation of the relevant singular plane. Vicinal faces are stepped ones, usually with broad bands of low index planes interrupted by widely separated monatomic or monomolecular steps.
3. *Diffuse surfaces.* These include all other orientations and are such that surface energy is more or less independent of orientation.

As pointed out by Cabrera and Coleman (1963) (and as is apparent from, for example, Fig. 8-5) singular and nonsingular surfaces differ in the number of their lattice layers that are parallel to the zone of transition from the crystal to its growth medium. With singular surfaces the transition occurs at one layer. For non-singular surfaces several layers are involved, so the interface becomes "diffuse."

If a given surface of a growing crystal is singular or vicinal, growth is of the layer type and requires a source of steps. These may be provided by the nucleation of island monolayers or by screw dislocations, as already discussed. But if

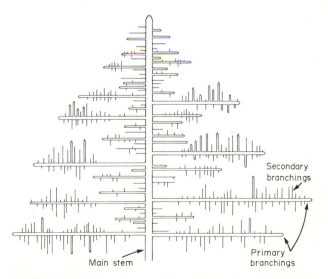

Fig. 8-14 Dendrite in two dimensions. (*From Buckley, "Crystal Growth," John Wiley & Sons, Inc., New York, 1951, with the publisher's permission.*)

the surface is diffuse no source of steps is required, since the latter are already present in abundance. Such surfaces are thus predisposed to dendritic growth.

Many crystal structures of low symmetry exhibit diffuse surfaces on which incoming particles can readily find stable attachment. No special nucleation is required, so the accumulation of particles is not delayed pending the formation of island nuclei or dislocations. Extremely rapid growth may therefore take place normal to the surface or surfaces concerned, leading to the formation of an open branching structure.

Such dendritic growth is favored by:

1. Lack of sources of steps on accompanying singular and vicinal faces.
2. A diffusion field in the surrounding medium favoring supply of material to nonsingular surfaces.
3. High supersaturation and high equilibrium concentration, both of which produce a high growth rate and enable the growing surface to move out and catch particles for itself rather than waiting for them to diffuse in.
4. Anisotropy, yielding a potentially high contrast between singular and non-singular surfaces (anisotropy is by no means essential, however; dendrites may develop normal to nonsingular surfaces of cubic crystals, such as the {111} in NaCl-type compounds).
5. Low thermal conductivity of the crystal. Accumulation of latent heat of crystallization will inhibit growth, and this will thus be favored at those points from which such heat is most readily dissipated. If the crystal has a low thermal conductivity, growth will be highly favored, and hence frequently highly localized, along edges and particularly at corners.

Dendrites are common in both natural and synthetic products. Perhaps the most familiar example is that of ice—as frost crystals and as snowflakes. Native silver frequently forms dendrites in veins where it has presumably been precipitated from solution. Magnetite often occurs as skeletal dendrites where it has solidified from molten lava. Dendrites are also extremely common in cast metals.

The frequency with which dendritic growth occurs may be rather greater than immediately apparent and is of considerable importance in the whole matter of crystal growth.

It has long been known that, in addition to the fine lattice imperfections in crystals, many crystals exhibit the blocky, or mosaic, structures already referred to. Such structures are found in many larger single crystals, and it has been suggested that there are probably few if any large crystals entirely devoid of them. Examination of many large galena crystals, for example, shows that the faces are not *quite* even but are made up of a rather irregular mosaic, each section of which is slightly misoriented with respect to its neighbors and is separated from them by quite easily discerned boundaries. Several explanations have been suggested.†

1. *Distortion about "Zwicky cracks."* It has been known for some time, and was noted particularly by Zwicky (1929), that the surface layers of some crystals

---

† We exclude here those involving deformation; these are considered in Chap. 10.

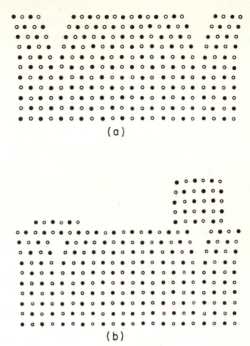

Fig. 8-15 (*a*) Development of and (*b*) keying in of "Zwicky cracks" in a galena-type structure. (*From Buerger, Amer. Mineralogist, 1932.*)

(notably those of the NaCl type) contract, leading to the formation of cracks which probably extend two or three lattice spacings down into the body of the crystals. Overgrowth of the kind shown in Fig. 8-15 (after Buerger, 1932, p. 184, figs. 7 and 8) would lead to the formation of new blocks "keying in" the cracks concerned. This would cause preservation of the crack and its accompanying distortion, yielding a slightly tilted block structure.

2. *Deflection by solid-solution particles.* Another possible source of structural distortion sufficient to initiate the development of a misoriented block is impurity particles. If these are incorporated on a developing crystal surface in such a way as to produce a feature capable of deflecting later growth, the local structure may then grow as a new, slightly misoriented block.

3. *Dendritic infilling.* Crystals may commence growth by the initial rapid development of dendritic branches and then at a later stage fill in. If for any reason a branch became slightly distorted, it would—as the nucleus for later filling in—give rise to a block slightly misoriented with respect to those "nucleated" by other branches. It is known that in some instances quite large droplets of liquid are keyed in by the growing together of adjacent dendritic limbs, and this sort of event might also cause sufficient deflection of growth for the formation of a misoriented block.

The first two mechanisms of mosaic formation have been considered in some detail by Buerger (1932), who pointed out that although neighboring blocks were slightly misoriented they could usually be traced back to a common parent, and for this reason he termed the structures *lineages*:

> The key to the explanation of the phenomenon is found in the fact that the orientation discrepancies between neighboring blocks are least near the center of the crystal and increase toward the edges, and, most important, grains which are discrete near the periphery of the crystal, when followed back to the center are found to join in one and the same parent grain . . . . In other words, the different "grains" are not actually different grains in the ordinary sense, but are really one and the same grain, connecting by way of the original nucleus . . . .
>
> The writer proposes the name *lineages* for all such "grains" whose orientations descend continuously from the same parent nucleus but whose mutual orientations may differ. Some such term is plainly called for since in the strictest sense all descendants from the same nucleus are the same grain and in the best sense they are also the same crystal. In this nomenclature, the discontinuities which mark the cleavage surfaces of a single crystal . . . are *lineage boundaries* (1932, pp. 180–181).

Whatever the influence of Zwicky cracks, solid-solution irregularities, and perhaps, dislocations, it seems not unlikely that many observed lineages owe their origin to the filling in of dendritic crystals. Buckley (1951, pp. 482–483) has in fact raised the question of whether all crystals might—at least in part—be filled in dendrites. The presence of inclusions in many crystals indicates that these crystals owe part of their bulk to such a process. This indicates the growth of dendrites at least to microscopic dimensions. While such evidence certainly does not establish a universality of dendritic nuclei, it does indicate that the very earliest stages of crystal growth might be—far more often than is obvious—of dendritic kind.

GROWTH SHAPES OF "FREE" CRYSTALS

SPHERICAL GROWTH FORMS

The simplest shape produced by free growth in liquid or vapor is the sphere. As already noted this is the least energetic form for a liquid or a completely isotropic solid, but since most solids are somewhat anisotropic (even where cubic), low-energy plane faces are usually energetically more favorable than the curved surface of a sphere. However, spherical crystals are not altogether uncommon in ores crystallized from melts. The case of chromite in peridotite has already been referred to, and spheres and ovoids of pyrrhotite in pentlandite, of pentlandite, pyrrhotite, and chalcopyrite in silicate, and, sometimes, of magnetite in pyrrhotite are quite well known in the Sudbury nickel ores (Hawley and Stanton, 1962).

POLYHEDRAL GROWTH FORMS

We have already seen how the development of faces may be influenced by the crystal itself. However, as emphasized by Buerger (1947), such faces constitute a boundary between the crystal *and its environment*. It is therefore to be expected that the environment should also exert considerable influence in determining shape, or *habit*. This indeed it does. Environment can induce the exaggerated growth of some faces with respect to others and can cause complete suppression of still other faces. Pyrite, for example, grows as cubes under some conditions and as octahedra or pyritohedra under others. Other crystals develop as dendrites—an extreme manifestation of habit—under some circumstances and as compact crystals under different environmental conditions. Two examples of environmental "habit modifiers" are impurity particles and concentration gradients.

1. Impurity particles The effect of these was probably first noted about 1780 by Rome de Lisle, who observed that common salt, which normally crystallizes from aqueous solutions as cubes, formed octahedra when crystallized from brines containing urea. At about the same time Leblanc produced alum (which normally crystallizes as octahedra) as cubes by introduction of alkali into the solution. Succeeding investigators found that the habit of alum could also be affected by inclusion of borax or HCl in the solution, that $NaClO_3$ could be modified from the cube to the tetrahedron by $NaClO_4$ or Na_2SO_4, and that NH_4Cl could be modified from octahedra to cubes by the presence of urea or of chlorides of such metals as Co, Ni, and Mn.

A major advance was made by Retgers, who in 1892 found that the habit of KNO_3 could be changed from needles to plates by the addition of the dye nigrosine to the solution. This led to the intensive work on the effects of dyes by Gaubert, Wenk, and others, which has continued throughout the twentieth century. Gaubert's first notable discovery was that lead and barium nitrates crystallize as cubes, instead of the usual octahedra or tetrahedra, when grown in the presence of the dye methylene blue (the well-known pH indicator). His explanation was that the habit change was caused by the deposition of crystalline layers of the dye preferentially on the {100} planes of the growing crystals. Growth of the {100} planes was thus inhibited, whereas that of the {111} planes, which remained free of the dye, continued. The {111} planes therefore grew themselves out, leaving the cubic faces in an unusual dominance.

In this particular case, habit modification could be induced only when the "impurity" was in high concentration in the parent solution—so high, in fact, that the dye itself was at the point of crystallization. This led Gaubert to the view that the impurity formed a *continuous crystalline layer* over the face concerned, more or less completely cloaking it from further growth. In addition he appeared to regard the crystal structures of dye and substrate to be epitaxially related, although at that time the nature of epitaxis had not been fully recognized. He was forced to modify this idea of a continuous layer, however, when in later experiments he found cases where the dye influenced growth at low saturations and without forming its own crystalline layers. He realized that here he was dealing only with scattered inclusions—a case of "parallel growth"—but accepted that these also were effective as growth inhibitors.

Bunn (1933) went on to emphasize the probable importance of coherent parallel growth (epitaxis). He considered the formation of mixed crystals and of oriented overgrowths, and the modification of crystal habit by impurities, to be related phenomena; if a substance *A* can grow in a crystallographically parallel position on a surface of crystal *B*, it should be possible conversely for substance *B* to grow on an appropriate face of crystal *A*, and in each case habit modification should follow.   Bunn did not regard the adsorbed layer as being necessarily continuous and considered that where crystals grew in solutions containing only small amounts of impurity the latter would be adsorbed as ". . . isolated molecules or ions of impurity here and there on the growing crystal face" (1933, p. 576).   However, isolated particles and continuous layers were regarded as having the same relationship with the underlying crystal.   The reason given why isolated particles could also inhibit growth was:

> A certain amount of [the] impurity is always included in those regions of the crystal built up by deposition on the affected faces, forming an unstable mixed crystal.   The fact that the mixed crystal is unstable gives the explanation.   On the growing face particles of impurity are strongly adsorbed, together with particles of the main substance; when more material is deposited on these, an unstable mixed crystal is formed, which, being unstable, tends to redissolve, the net result being that the rate of growth is lower than it would be in absence of the impurity (1933, pp. 576–577).

Buckley (1951) disagreed with Bunn's view of a high level of epitaxis and put forward an alternative theory.   This also was based on preferential adsorption, but it emphasized the possible importance of simple ionic substitution at the crystal surface rather than the necessity for longer range crystallographic similarities:

> If an ion of impurity has sufficient in common with a growing crystal surface as to adhere to it, it seems possible that, if it is not far removed in its dimensions from the alternative ion which it is replacing, it could gather to itself cations when present in sufficient quantity in the common solution and form its own crystals, and the size and position of such would naturally be conditioned by lattice dimensions and overall similarity in the two cases.   These, according to the degree of intimacy of the crystals, would be our mixed crystals or overgrowths or parallel growths.   But, when the ion is of such a shape and size that such a correspondence is quite out of the question and the only feature common to the growing surface ion and the impurity is some small portion of a group at one end of the latter, we are driven to conclude that, if such cases as the latter are found to exist, they provide better examples of pure habit modification than the others, for here would be an attachment of the impurity ion divorced from some spurious similarities.   Do such exist?   Undoubtedly.   There is a relatively huge series of organic compounds in the water-soluble dyes . . . (1951, p. 353).

Buckley acknowledged that where the impurity is a simple inorganic compound, or urea, lattice similarities exist, but he suggested that it is probably wrong to regard this as *generally* important.   He pointed out that great numbers of dyes, many of which possess large ions of complex configuration, may act as habit modifiers:

> It seems impossible to credit that, when practically every one of these configurations has effective power to modify several types of crystal surface, the conditions required by the theory adopted by Bunn and Royer should hold.   There is no evidence whatever that there is any lattice similarity in these cases—only analogy from the far simpler inorganic compounds (plus urea) (1951, p. 354).

He pointed out that the two groups of substances require sharply different levels of concentration to carry out their respective roles of inhibition. Where such an effect is produced by simple inorganic ions or urea, the required concentration of impurity is high and approaches that of the crystallizing substance. The range is of the order of 1:1000 to 1:1, values close to the latter being fairly general. With the larger, more complex ions, minimum concentrations are much lower— e.g., the dye Cotton Blue Conc. No. 1 is known to modify K_2CrO_4 crystals at a concentration of less than 1 part in 67,000. When the impurity and molecular weight of such a dye are taken into account, Buckley has calculated that it is capable of growth inhibition at a concentration of one molecule of dye to 330,000 molecules of potential crystal material. As he pointed out, there seems little prospect of several of these ions lying next to each other in the crystal surface and forming themselves into foreign nuclei. Rather they may attach themselves individually, in more or less statistical distribution on the adsorbing face, by ionic substitution. Such substitution, however, would involve only one end of what might be a large and complexly shaped ion; thus actual attachment might be through a small oxygen, sulfonate, or other grouping, the remainder of the ion standing up as a protuberance on the crystal face. Such protuberances would undoubtedly disrupt the further growth of faces concerned, so modifying habit.

If this is the case it appears that impurities may act in at least two ways as habit modifiers. In the case of high concentrations of "simple" impurities, the mechanism is primarily epitaxis and the result is growth zoning, variation of growth rate, and in some cases, habit modification. In the case of large ionic complexes the process is one of ionic substitution by an extremity of the complex, and the result is a diminution of growth and modification of habit.

2. Concentration gradients Concentration gradients, and hence currents, may be initiated by a variety of factors and may be important in influencing growth. Such currents may result simply from the large-scale movement of solution—as in the case of subsurface water or gas moving upward from a major igneous source—or they may be induced by quite local factors.

Under quiet conditions quite subtle factors may operate on a local scale. For example, in the growth of crystals from a liquid the subtraction of the solute from the parent solution may induce density changes sufficient to cause constant overturn. Rapid deposition of a lead compound would clearly leave the depleted layer much less dense than the remaining solution, and movement would result. Where the crystals were growing from the walls of a cavity, this would lead to upward movement close to the walls and downward movement near the center. This would, of course, be somewhat modified by the turbulence caused by the crystals themselves. In melts, density changes may be induced by heat contributed by crystallization (the heat of fusion) and by heat lost to the walls of the chamber. Again movement would be complicated by the interaction of the density streams and the developing crystals. Similar effects occur where crystals form from vapor, as in the case of snow.

Whatever the cause of a particular current, it is important in influencing

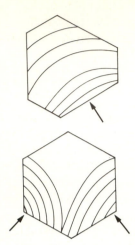

Fig. 8-16 Concentration steps developed on a (111) face of lead nitrate due to the development of currents indicated by the arrows. (*From Buckley, "Crystal Growth," John Wiley & Sons, Inc., 1951, with the publisher's permission.*)

the supply of material to the growing crystal, and hence it may have an important effect on its rate of growth and final form.

Keeping in mind that the crystal itself influences rates of growth in specific directions, growth rate generally is directly related to the supply of solute. Hence if the crystal in question is held in a single position and grows from a solution flowing in a more or less constant direction, the most rapid growth is against, or into, the direction of the current. This is likely to be most marked in nearly isotropic substances, where the crystal itself does not contribute greatly to differential growth rates. In highly anisotropic substances, surface energy and supply factors may tend to cancel or reinforce each other's effect. The side of the crystal facing the direction of flow is often referred to as the *stoss* side. Where such a crystal has developed growth zones, these are broadest on the stoss side and hence may indicate in a general way the local direction of flow.

Flow directions are also indicated by growth steps. Where growth is due substantially to the accretion of layers, these develop fastest, and hence tend to form a tier, toward the direction of supply. Such steps are convex away from the flow direction and are approximately concentric. The direction of flow is indicated by the common radius, as shown in Fig. 8-16. It is because of the distorting effects of concentration gradients that artificial crystals are usually rotated as they grow.

DENDRITIC GROWTH FORMS

If, as we have seen, growth of a crystalline body is controlled by diffusion of material or heat through an otherwise homogeneous environment, subtle surface differences may result in locally accelerated growth, yielding protuberances that grow linearly much faster than the adjacent parts. This gives rise to the third basic structural shape developed by free growth—the branched body, or dendrite.

Just as "solid" (i.e., compact) crystals develop in isotropic, spherical form or anisotropic, polyhedral form, dendrites may develop without any directional anisotropy, or they may be of highly regular and clearly anisotropic character.

Two-dimensional dendritic growths of manganese oxides are well known and are characteristically isotropic, or irregularly branching.   Native silver, on the other hand, forms fine three-dimensional dendrites of highly regular nature.   In all true dendrites the stems and primary, secondary, and higher order branchings are in crystallographic continuity and are parallel to prominent directions in the "solid" crystal of which the dendrite constitutes a skeletal version.

Clearly a branched structure of this kind has a very high ratio of surface area to volume and, unlike the simple sphere and polyhedron, is inherently unstable.   Each part of it reflects an event in its history, the whole growing as a continuously connected unit because the nucleation of a new branch of the existing structure is easier than the formation of an independent nucleus.   As emphasized by Smith (1964), conditions that give rise to branching exist only during growth and only at the periphery of the growing mass.   Within it the branches may continue to grow, however—as we have already seen—filling or almost filling the skeletal voids.   If conditions of crystal growth change, this may be accompanied by the development of a spherical or polyhedral outer surface, in accordance with the usual minimum surface-energy requirements.

MASS STRUCTURES

Although this chapter has had as its primary concern the free growth of single crystals, it is obvious that numbers of these may develop in close proximity and that eventually some may impinge.   Such neighboring growth and impingement often leads to the formation of clearly recognizable group relationships, which may be referred to as *mass structures*.

PRINCIPAL STRUCTURES

Some mass structures are quite gross and have developed along easily recognized lines.   The lining of a vein wall by numerous well-formed prismatic crystals is often referred to as *comb structure*: any section through such a vein shows a crudely parallel, toothed, arrangement that might be likened to the teeth of a comb. Many vein fillings commence with the deposition of numerous crystals of a single mineral, and these grow and impinge to form a compact lining on the vein wall. A second mineral may then grow as a compact lining on the first, a third on the second, and so on, until a grossly banded structure is built up.   This is known as *crustification*.   The structure is frequently developed on the two opposing walls more or less equally, yielding a symmetrical arrangement.   In other cases it is asymmetrical.   Sometimes, where the veinlets develop within brecciated fault zones, crustification develops about nuclei provided by breccia fragments, giving rise to rounded forms that in section are seen to have a *cockade structure*.

In other cases the mass structures are quite fine and require the microscope for resolution.   These finer aggregates usually grow on a nucleus provided by a small particle of some other substance, and develop as rounded bodies composed of compact radiating crystals, the whole mass forming a *concretion*.   These may approach sphericity or be columnar (as in the case of stalactites), and they may have smooth or moderately rough surfaces. In some instances the surface is smooth to the point of appearing polished. In such cases a gel origin, with subsequent crystal growth from the periphery inward, has been suggested (cf. Edwards, 1954, p. 20).

Many concretionary structures show concentric zoning (i.e., color and/or compositional zoning normal to the direction of growth of the constituent crystals), presumably due to growth pauses and changes in conditions of growth. Where primary deposition as a gel has been postulated, these zones have been identified as former Liesegang rings produced by diffusion of the appropriate impurity into the gel before the latter crystallized. The best known example of this is *schalenblende*, a layered arrangement of sphalerite and wurtzite, in which the reddish wurtzite layers contain notably more iron than the yellow sphalerite layers. This ore type has been described in some detail by Ehrenberg (1931), who showed for one example that the wurtzite layers have a composition

$$Zn = 63\%$$
$$Fe = 4\%$$
$$S = 33.3\%$$

that is, $ZnS = 94\%$ and $FeS = 6\%$, and the sphalerite layers have a composition

$$Zn = 67.7\%$$
$$Fe = 0.41\%$$
$$S = 32.8\%$$

that is, $ZnS = 99.6\%$ and $FeS = 0.4\%$.

Ehrenberg suggested that the host gel originally crystallized to the wurtzite structure and then attempted to transform to the stable low-temperature polymorph, sphalerite. This was achieved in the layers low in iron but inhibited or prevented in those rich in it, leading to the preservation of a banded high-iron–wurtzite, low-iron–sphalerite, arrangement.

IMPINGEMENT AND FREE GROWTH

While this appears to be an immediate contradiction of terms, it is not entirely so. The case of impingement in a completely solid medium—the antithesis of "free" growth—is considered in the following chapter. In the present instance growth leads to impingement along only *parts* of the crystal surfaces, leading to the cessation of growth in some directions but permitting continued growth in others— i.e., "free" growth of a directionally limited kind. This leads to the development of some interesting impingement structures and to the crowding out of some crystals by others.

It has already been shown that different crystal faces (forms) differ in their surface free energies and that those with the greater energies grow fastest. In completely free growth this leads to the elimination of the most energetic faces. With partial impingement, on the other hand, it may lead to their exaggerated development.

A little thought will suffice to show that although certain faces grow faster

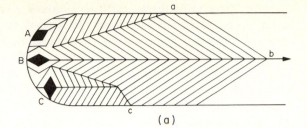

(a)

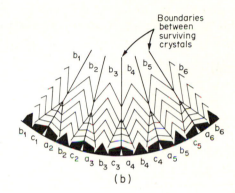

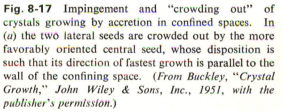

(b)

Fig. 8-17 Impingement and "crowding out" of crystals growing by accretion in confined spaces. In (a) the two lateral seeds are crowded out by the more favorably oriented central seed, whose disposition is such that its direction of fastest growth is parallel to the wall of the confining space. (*From Buckley, "Crystal Growth," John Wiley & Sons, Inc., 1951, with the publisher's permission.*)

than others, the directions of most rapid enlargement of the crystal will not be the normals to the fastest growing faces. They will be the normals to the *edges* formed by the intersection of two rapidly growing faces or the directions of corners (coigns) formed by the intersection of three or more rapidly growing faces. As has been pointed out by Buckley (1951, p. 262), if one considers a section through a cube, the diagonals of a square so derived will be directions of greater growth velocity than the face normals. For a lozenge-shaped crystal section (e.g., of arsenopyrite), the major diagonal will be the direction of greatest growth velocity.

Such considerations determine which of a number of adjacent freely growing crystals will grow fastest, which will survive, and which will be overwhelmed by their neighbors. The principle involved is shown in Fig. 8-17. Figure 8-17a shows, on a relatively large scale, the case of three similar but differently oriented crystals. Crystal *B* is clearly the most favorably oriented with respect to the host fissure and grows almost unimpeded into the parent medium. Crystals *A* and *C* are less favorably oriented and after a brief period of growth are overwhelmed by crystal *B*. A common result of this principle is shown in Fig. 8-17b.

In many veins there are large numbers of crystals more or less similarly oriented, and this leads to semiparallel growth and hence to the development of *columnar structures* of varying degrees of perfection. These are a natural development of heavily nucleated comb structures. Where nucleation has led to a wide variety of orientations the result is, of course, a disordered array of dominated and dominating growth forms. A special case of noninterference yields the radiating crystalline concretion; here the crystal nuclei all develop with a given crystallographic direction directed radially from the particle acting as a base (i.e., central nucleus). Radial growth velocities are essentially equal, and while each crystal impinges laterally with its neighbors none is crowded out. The result is a radial, columnar structure, with a highly even, rounded surface of cotermination.

RECOMMENDED READING

Buckley, H. E.: "Crystal Growth," John Wiley & Sons, Inc., New York, 1951.

Gilman, J. J. (ed.): "The Art and Science of Growing Crystals," John Wiley & Sons, Inc., New York, 1963.

Read, W. T., Jr.: "Dislocations in Crystals," McGraw-Hill Book Company, New York, 1953.

Smith, C. S.: Some elementary principles of polycrystalline microstructure, *Met. Rev.*, vol. 9, no. 33, pp. 1–48, 1964.

9
Growth and Growth Structures in Polycrystalline Aggregates

To this point we have been concerned with the growth of a rigid crystalline particle into a nonrigid environment of gas, vapor, or liquid. While this is clearly important in principle, it is a rather special case in nature. Natural substances, of which ores happen to be a group, are usually *polycrystalline*. That is, they are composed of aggregates of crystalline particles, like or unlike, which usually do not show crystal faces and which adhere to each other along *grain boundaries*. In this case the growth of any particular particle is not "free" but is in the rigid environment of its neighbors. We are therefore concerned with solid-solid interfaces.

Much of the investigation of interfaces has been concerned with metals, and it appears that quite a number of the metallurgical findings can, by analogy, be applied to a wide variety of polycrystalline matter. Before going further, however, it is perhaps well to remember that the drawing of analogies is not without danger when details become involved. While there does indeed appear to be an excellent overall analogy between metal interfaces and those of other substances, there are several important matters that must be kept in mind. Most metals have rather simple, close-packed structures and, as was pointed out in Chap. 3, are characterized by a rather diffuse and directionless kind of bonding. They thus possess a very low anisotropy and a very high atomic mobility. Silicates, however, are commonly highly anisotropic, and strong and highly directional bonding is almost the rule.   Thus, whereas metal grains are able to readily adjust themselves to their own and their neighbors' requirements, silicates are unlikely to be so tractable. Sulfides and related minerals are probably somewhere in between.   Most of them contain some metallic bonding, and most may be regarded as close-packed structures of sulfur atoms with highly mobile interstitial metal lattices.   The analogy between metals and many ore minerals is therefore likely to be close, and this has in fact been found to be so.   We may therefore consider some of the principles of polycrystalline microstructure as determined in metals and then, keeping in mind the probability of some quantitative divergences, apply these to the study of ore minerals.

GRAIN BOUNDARIES

McLean has defined a grain boundary in a metal as "the boundary separating two crystals (or 'grains') that differ either in crystallographic orientation, composition, or dimensions of the crystal lattice, or in two or all of these properties" (1957, p. 1).   Clearly this definition excludes the bounding surfaces of the free-grown crystals already discussed and is concerned only with solid-solid contacts.

Under the microscope a grain boundary looks so simple—a line discontinuity—that it is usually barely considered by mineralogists.   When one examines the matter† a little more carefully, however, it is soon seen that if they are simple, their simplicity is far from obvious.

An early suggestion (Rosenhain and Humfrey, 1913; see McLean, 1957) was that crystalline grains of metal were each surrounded and cemented together by amorphous material of the same composition—the *amorphous cement theory*. In 1924 Jeffries and Archer suggested three alternatives:

1. The grain boundary may simply be a void between the two crystalline grains.
2. The grain boundary is a zone in which some of the atoms are held in both of the adjacent lattices, these therefore being distorted along the contact.
3. There is a zone of amorphous material.

The first of these does not appear to hold, since it implies a much lower cohesion of grain boundaries than that observed.   The third is, of course, the amorphous cement theory.   The second became known as the *transition lattice theory*, in which it was postulated that adjacent grains possessed their own lattice structures right to the boundary, which consisted of one or perhaps two layers of atoms in compromise positions.   The actual boundary was thus regarded as having quite a high degree of order and an energy of distortion or strain corresponding to the degree of misorientation of the two grains concerned.   It was presumed that the transition layer structure would be such as to represent the lowest possible energy under the conditions concerned.   This appears to have been the first occasion on which energy was associated specifically with an interface.

† For an excellent review, see McLean, "Grain Boundaries in Metals," Oxford University Press, Fair Lawn, N.J., 1957.

However, neither the amorphous cement nor the transition layer theory could satisfactorily explain the properties of grain boundaries. The former appeared to satisfy observations of grain-boundary sliding but could not explain the later discovered dependence of grain-boundary energy on orientation difference. The latter readily accounted for the energy differences but did not explain the mechanical behavior. The fact that grain-boundary energies were sensitive to changes in orientation did, however, indicate that the boundary was probably not more than two or three atom diameters wide.

Mott, in 1948, put forward a compromise transition layer theory in which he suggested that a boundary might consist of "islands" of good atomic matching separated by regions in which matching is poor. In this hypothesis, areas of good and bad fit are regarded as sharply defined and not merging into each other. An energy-orientation relationship is thus allowed for, and sliding is accounted for by postulating negligible resistance in the zones of good fit and disordering or melting—leading to local elimination of resistance to shear—in the zones of bad fit.

Ideas on grain boundaries became more sophisticated with the recognition of the nature and widespread incidence of dislocations. Figure 9-1 shows how a wall of edge dislocations causes a sharp change in lattice orientation. Where dislocations gather together in this way the resultant change in orientation is equivalent to a

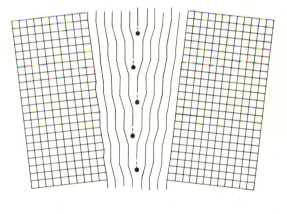

Fig. 9-1 Simple tilt boundary consisting of a "wall" of edge dislocations. The boundary causes a change in orientation and the development of a bend in the crystal. (*From McLean, "Grain Boundaries in Metals," Oxford University Press, Fair Lawn, N.J., 1957, with the publisher's permission.*)

rotation about an axis within the new boundary and parallel to some simple crystallographic direction common to both lattices. The new interface is referred to as a *simple tilt boundary*. Where a screw dislocation is involved the boundary is referred to as a *twist boundary*. In this case the orientation difference between the two lattices is again equivalent to a rotation about a simple crystallographic axis that, however, is normal to the new boundary.

One of the most informative ways of studying possible grain-boundary configurations is through "bubble models." Limitations of these are that they can be studied in only two dimensions and hence can be used only in connection with tilt boundaries, but they are very helpful nonetheless. When two close-packed "rafts" of bubbles (or marbles, metal balls, etc.), representing two adjacent crystal lattices, are joined together, their interface is a two-dimensional grain boundary, and the structure of this can be studied with progressive change in the orientation of the two rafts. In this method it appears that where the angle of misfit is small, i.e., where there is a *low-angle boundary*, it condenses into a stack of edge dislocations. However, with *high-angle boundaries* (greater than about 25°) dislocations do not form, and the lattices are perfect almost to the interface. In such cases McLean (1957, p. 40) suggests that the boundary structure is better described as a thin misfit layer having short intervals of good fit.

From this and other more detailed considerations it is possible to recognize three general types of boundary situation in metals. These result from variations in the lattices concerned and from the degree of mutual tilt or twist they have developed.

1. Where two adjacent grains have lattice planes of essentially identical spacing and where these are appropriately oriented, the matching may be good. The degree of *misfit* is negligible, and the boundary is said to be *coherent*. These may be "ordinary" grain boundaries, but they are most frequently those between twins or a precipitate and its matrix. As will be shown later, these are boundaries of characteristically low energy; atoms are matched, bonds are satisfied, and any distortion is essentially elastic. (Over long distances, coherent boundaries, other than perfectly fitting twin boundaries, may, however, degenerate into dislocation boundaries—for detail, see McLean, 1957, p. 38.)
2. Where grains meet in low-angle boundaries of simple tilt type, these appear to consist of single rows of dislocations. More complicated low-angle boundaries may or may not consist of dislocations, but it appears likely that they do.
3. High-angle boundaries may have a random misfit structure resembling a liquid and appear to be no more than three atoms wide.

Since any disruption or discontinuity in a crystal structure raises its potential energy, the incidence of grain boundaries, like surfaces, leads to an increase in the energy of the system. In the case of surfaces—solid-fluid interfaces—the free energy is that of unsatisfied bonds at and near the surface—the surface free energy.

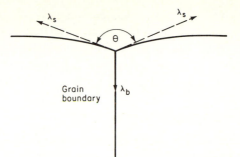

Fig. 9-2 Section normal to a grain boundary, showing the groove formed by thermal etching.

In the case of grain boundaries the energy is that of bonds distorted or left unsatisfied at a solid-solid interface.   This is called the *grain boundary*, or *interfacial, free energy*.

For liquid-vapor interfaces (i.e., "surfaces") surface free energy is numerically equal to the surface tension.   This is because the liquid is highly mobile and atoms can diffuse to or from the surface layers quickly enough to maintain the original density as the surface area increases or decreases.   Clearly metals and other solids do not have this mobility.   This is not to say that solid surfaces and interfaces do not possess a tension; they do, and this may be demonstrated by thermal etching.

In this, a specimen of metal† is heated in a vacuum or some suitable inert atmosphere.   Grooves develop where the grain boundaries meet the surface, giving rise to a "thermal etch."   It appears that the grooves form as a result of the equilibrating of the two surface free energies and the grain-boundary free energy.   Thus from Fig. 9-2:

$$\lambda_b = 2\lambda_s \cos \frac{\Theta}{2} \tag{9-1}$$

where Θ = angle of the groove
λ_b = grain-boundary free energy
λ_s = surface free energy

This, of course, assumes a symmetrical groove, which in turn presupposes that the grain boundary is normal to the surface and that there is no crystallographic orientation effect.   Departures from these requirements cause a scatter of values of Θ, though this is slight in metals where grain boundaries are close to normal to the surface.   The value of Θ may be determined with a goniometer in reflected light or by interferometry.

However, as a result of the lack of mobility of solids, the tensions of solid-solid interfaces are normally not numerically equal to their free energies.   Equality may be approached in metals (and probably in metallic sulfides) at temperatures close to their melting points, because in this case the atomic arrangement along the grain boundary tends to disorder due to thermal agitation.   In spite of all

† Stanton (1966, unpublished) has also achieved thermal etching of galena on heating at 600°C *in vacuo.*

this, however, it has been shown for metals that the discrepancy between interfacial tension and interfacial energy, while existent and important in principle, is usually not great, and in the following the two will be regarded as being equal.

GRAIN SHAPE

FOAM STRUCTURES

The shapes of grains in a polycrystalline aggregate are influenced by two principal factors: the requirements of interfacial energies and of space filling.   In minimizing their free energies, grains will adjust their shapes to minimize the total area of interface, particularly the areas of high-energy (high-angle-misfit) interfaces.   All such adjustment must, however, conform to the geometrical constraints of space filling. For example, a spherical grain shape might be that of minimum interfacial energy, but an aggregate of spheres certainly does not fill space.   All grain boundaries must be interfaces; no voids are permitted.   When an aggregate of like grains— as in a metal or a monomineralic rock—has adjusted itself in this way, its structure may be said to have *matured*, and its configuration is then very similar to that of a foam.   The development of such mature structures in solids is a part of the process known as *annealing*.

The foam analogy has been recognized for a long time and has been illustrated particularly effectively by Smith (1964).   It must be remembered, of course, that crystal interfaces and bubble walls are very different things.   Crystal interfaces are more or less structured solids whose energies are dependent on orientation. Bubble walls, however, consist of thin layers of structureless liquid whose energies are independent of orientation and are proportional only to area.   In spite of these differences, however, a soap froth provides a remarkably good model of grain shape and, as will be shown later, of grain growth.   Figure 9-3 shows a two-dimensional aggregate of bubbles.   Figure 9-4 shows the surface of a piece of aluminum sheet heated just to the point of incipient melting (after Smith, 1964) and an etched section of polycrystalline galena.   The geometrical similarity of the metal and sulfide with the soap froth is clear.

Triple junctions   Inspection of all three substances of Figs. 9-3 and 9-4 shows that, in two dimensions, all bubbles and grains meet three at a time in a point. Careful examination of a soap froth shows that in three dimensions the great

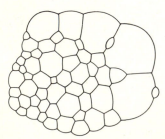

Fig. 9-3   Two-dimensional diagram of an aggregate of bubbles.   (*From Smith, "Metal Interfaces," American Society for Metals, Cleveland, Ohio, 1952.*)

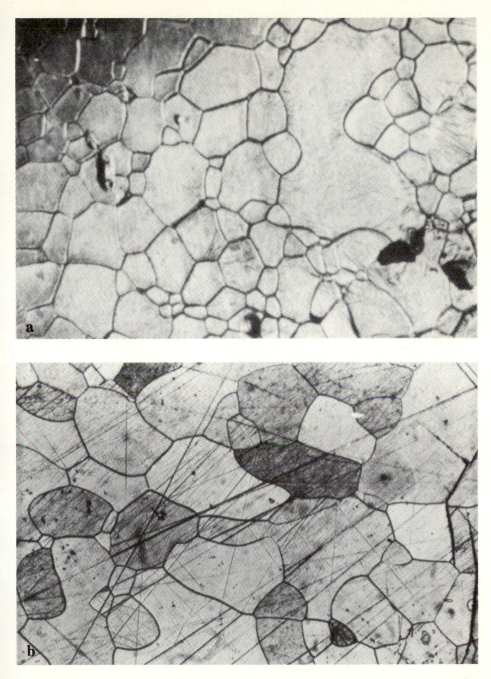

Fig. 9-4 (*a*) Polycrystalline aluminum sheet, and (*b*) polycrystalline galena, both showing foam structures (the etchant has emphasized cryptic scratches in the galena). (*Aluminum photograph from Smith, Met. Rev., 1964.*)

majority of bubbles—and hence of grains—meet in threes along lines.   Occasionally four or more are contiguous in a point, but this is unusual.   A line along which three bubbles or grains meet is called a *triple junction*.   When intersected by a surface (such as the plane of a microscope section) such junctions appear as points, which are referred to as *triple-junction points*.

It is also apparent from Fig. 9-4 that the angles between grain boundaries are approximately equal and hence all close to 120°.   When viewed in three dimensions, e.g., by using once again our transparent soap froth, the equality of the angles is confirmed.   Deviations from equality observed in plane sections of a mature structured solid are, of course, due to the generation of "apparent" angles by the random plane.   In a mature single-phase foam structure all triple-junction angles are in fact 120° when measured normal to the junction concerned.   Such an arrangement results from the requirements of minimum area and the balancing of forces at triple junctions.

A further obvious feature of Fig. 9-4 is that many of the froth or grain boundaries are curved.   This is clearly required for space filling, as illustrated particularly by the two- and three-sided grains.   In fact all grains other than those with six sides—and in some cases with five sides—must develop curved boundaries to fill space.†   The *degree* of curvature is, however, an energy requirement.   In the froth of our present analogy the curvature is proportional to the pressure across the bubble wall; in single-phase solids it is proportional to the free energy of the interface.

If a metastable aggregate is heated to a temperature at which there is substantial atomic mobility, the grain boundaries will move and adjust themselves in such a way as to approach these requirements.   They behave as if conforming with a triangle of forces, and move into equilibrium so that

$$\frac{\lambda_1}{\sin \Theta_1} = \frac{\lambda_2}{\sin \Theta_2} = \frac{\lambda_3}{\sin \Theta_3} \tag{9-2}$$

where λ_1, λ_2, λ_3 are interfacial free energies, and Θ_1, Θ_2, Θ_3 are the corresponding angles between grain boundaries.   Where there is no anisotropic effect, as is the case with the close-packed structures of most metals and the more common oxides and sulfides, single-phase aggregates adjust so that, as we have seen, $\Theta_1 = \Theta_2 = \Theta_3 = 120°$.   Again it must be emphasized that most solids only *approach* this configuration since orientation effects—even if only ever so slight—are always present.   In practice, however, the approximation is remarkably good.

The progressive adjustment of triple junctions with time in a metal has been studied by Aust and Chalmers (1950), who annealed a tricrystal of tin in which the triple junction was initially not in equilibrium.   Their result is shown in Fig. 9-5. It can be seen that the angle changed rapidly at first and then slowly settled down to a constant equilibrium position.   The same phenomenon has been demonstrated

---

† These are arcs of circles.   The geometrical and topological relations are dealt with very clearly by Smith (1950, 1964).

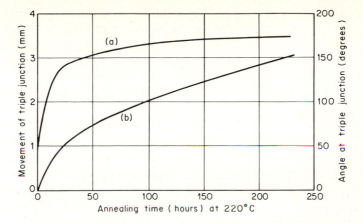

Fig. 9-5 Change with time (*a*) in angle and (*b*) in position of a triple junction in a tin tricrystal heated at 220°C.   (*From Aust and Chalmers, Proc. Roy. Soc. London, Ser. A, 1950.*)

in sulfides by Stanton and Gorman (1968) by the progressive measurement of single triple junctions in galena and also in polycrystalline single-phase aggregates of galena, sphalerite, and chalcopyrite by the application of a statistical procedure developed by Harker and Parker (1945).

The latter procedure provides a means of determining characteristic angles in polycrystalline aggregates by the measurement of angles revealed on random-plane surfaces such as those of microscopical polished sections.   A little considera-tion suffices to show that random sections through any solid angle will yield

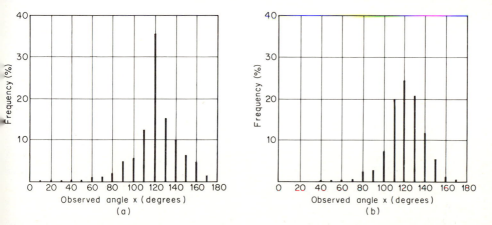

Fig. 9-6 (*a*) Theoretical frequency of an observed angle *x* on a random plane for a true angle *b* of 120°.   (*b*) Measured frequency of observed angle *x* on a random plane for a true angle *b* of 120° (these measurements were made on a polished surface of brass).   (*Both from Smith, Trans. Amer. Inst. Mining Met. Engrs., 1948.*)

apparent angles—as traces on the plane of sectioning—ranging from 0 to 180°. However, it has been shown (see Smith, 1948) that if a large number of such angles (say, 200) are measured, the one most frequently observed is the true one.   Figure 9-6a shows the theoretical frequency of observed angles on a random plane in the case where the true solid angle is 120°.   Figure 9-6b shows an actual measured frequency distribution for such a case.   The decrease in sharpness of B as compared with A may be attributed to the effect of orientation on interfacial energies and hence on triple-junction angles, though incomplete annealing could have the same effect.

Such blunting of frequency distributions provided the basis for Stanton and Gorman's second series of experiments, on the adjustment of triple junctions in polycrystalline aggregates of single sulfides.   The sharp theoretical peak of Fig. 9-6a is based on the assumption of full annealing of essentially isotropic grains. Granting a close approach to the latter condition, the less annealed the structure, the more likely is it that many true angles will not be 120°.   For example, a given triple junction might show (as a result of deformation or other cause) nonequilibrium arrangements such as 120°:110°:130° or 140°:110°:110°.   Such true angles will give a frequency distribution of apparent angles that, although still possibly showing a mode of 120° and an essentially symmetrical distribution, has a flatter peak.   In other words decrease in constancy of the true angle will affect the kurtosis of the distribution, which will become correspondingly platykurtic.

A convenient way of measuring such flattening, where the number of apparent angles measured is kept constant, is the determination of the standard deviation of

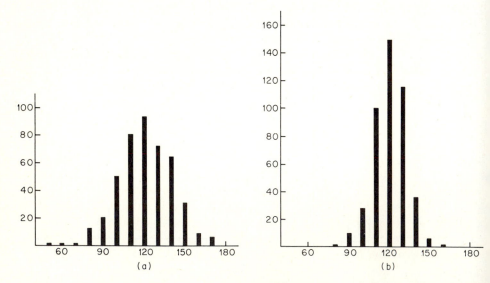

Fig. 9-7   (a) Frequency distribution of 450 angles measured on a polished surface of natural chalcopyrite from Mount Isa, Queensland.   (b) Frequency distribution of 450 angles measured on the same specimen after heating at 300°C for 49 days.   (*From Stanton and Gorman, Econ. Geol., 1968.*)

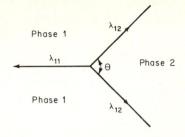

Fig. 9-8 Junction of one grain of phase 2 with two grains of phase 1.   The junction line is normal to the plane of the page, and θ is the true dihedral angle.

the angles as measured.   Where the true angles vary considerably about a mean of, say, 120°, the distribution of apparent angles will be rather flat and the standard deviation large.   Where the true angles all equal or closely approach 120°, the distribution will be sharp and the standard deviation notably smaller.   Thus progressive annealing of a moderately deformed aggregate can be studied by determining the change in the standard deviation of a set number of randomly measured apparent angles.   Figure 9-7 shows the frequency distributions for a specimen of Mount Isa chalcopyrite prior to heat treatment and after 49 days at 300°C.   The similarity of this distribution and that of Fig. 9-6 is clear and demonstrates again the analogy between metals and sulfides, as well as the usefulness of the statistical approach.

DIHEDRAL ANGLES

This is a special case of the triple-junction angle of the foam structure—that in which the three grains meeting at the triple junction *are not all of the same phase.*

Where two substances are involved, any three grains will clearly include two grains of phase 1 and one grain of phase 2.   In such a case the configuration of Fig. 9-4 changes to that of Fig. 9-8, in which we have two equal angles and one angle, Θ, of different value.   Resolving "tensions" parallel to λ_{11} we have

$$\lambda_{11} = 2\lambda_{12} \cos \frac{\Theta}{2} \tag{9-3}$$

as in Eq. (9-1).

This relationship was first recognized by Smith (1948), who termed Θ the *dihedral angle.*   He very quickly saw the metallurgical significance of the principle involved and stated: ". . . the *dihedral* angle of phase *b* in equilibrium against a grain boundary in phase *a* is a definite property of any given system and is the principal factor in determining the structure of annealed polycrystalline multiphase alloys" (1964, p. 25).   There is no doubt that this also holds for many sulfide and related minerals, as will be shown a little further on.

The relation between $\lambda_{11}/\lambda_{12}$ and Θ is simply calculated from

$$\frac{\lambda_{11}}{\lambda_{12}} = 2 \cos \frac{\Theta}{2} \tag{9-4}$$

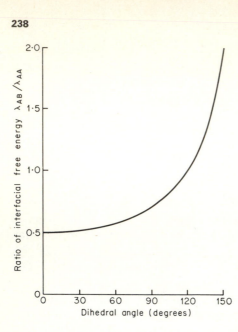

Fig. 9-9 Graph of the relation between the dihedral angle θ and the value of the interfacial free-energy ratio. (*From McLean, "Grain Boundaries in Metals," Oxford University Press, Fair Lawn, N.J., 1957, with the publisher's permission.*)

and is as shown in Fig. 9-9. Under special conditions Θ may be measured directly. In general, however, it is determined from random measurements of apparent angles on random planes, as already outlined for triple-junction angles in foam structures. In all cases, whatever the solid angle, that most frequently observed on the random plane is a close approximation to the true value. Table 9-1 shows values for dihedral angles and interfacial free-energy ratios in some common alloys, as determined by Smith (1950, 1964).

Smith (1948) has also shown how values obtained for a two-phase alloy may be checked by determining dihedral angles at appropriate three-phase triple junctions. Where equilibrium has been attained in a configuration such as that of Fig. 9-10, we have

$$\frac{\lambda_{12}}{\sin \Theta_3} = \frac{\lambda_{13}}{\sin \Theta_2} = \frac{\lambda_{23}}{\sin \Theta_1}$$

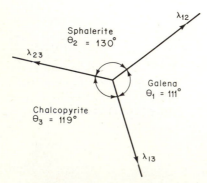

Fig. 9-10 Three-phase triple junction, involving galena, sphalerite, and chalcopyrite as examples. (*After Stanton, Nature, 1964.*)

Table 9-1 Dihedral angles and interfacial free-energy ratios for randomly oriented boundaries in some common alloys. (After Smith, 1964, p. 31)

System	Temp., °C	Interface between Phase A	Phase B	Grain boundary used as a comparison interface C	Dihedral angle Θ, deg	$\dfrac{\gamma A/B}{\gamma C}$
Cu–Zn	700	αfcc†	βbcc	α/α	106	0.83
	700	αfcc	βbcc	β/β	120	1.00
Cu–Al	600	αfcc	βbcc	α/α	90	0.71
	600	βbcc	γcc	γ/γ	100	0.78
Cu–Sn	750	αfcc	βbcc	α/α	98	0.76
	750	αfcc	βbcc	β/β	115	0.93
Cu–Ag	750	α(Cu)fcc	β(Ag)fcc	α/α	80	0.65
	750	α(Cu)fcc	β(Ag)fcc	β/β	95	0.74
Cu–Si	845	αfcc	βbcc	α/α	40	0.53
	845	αfcc	βbcc	β/β	130	1.18
	830	αfcc	χhcp	α/α	105	0.82
	830	αfcc	χhcp	χ/χ	110	0.87
Zn–Cu–Al	375	ηhcp	εhcp	ε/ε	115	0.93
(three-	375	ηhcp	εhcp	η/η	115	0.93
phase	375	εhcp	$\beta_1$fcc	β_1/β_1	110	0.87
alloy)	375	εhcp	$\beta_1$fcc	ε/ε	95	0.74
	375	ηhcp	$\beta_1$fcc	β_1/β_1	120	1.00
	375	ηhcp	$\beta_1$fcc	η/η	110	0.87
Zn–Sn	160	β(Sn)bct	α(Zn)hcp	α/α	95	0.74
	160	β(Sn)bct	α(Zn)hcp	β/β	130	1.18

† fcc = face-centered cubic.
bcc = body-centered cubic.
cc = complex cubic γ brass structure.
hcp = hexagonal close packed.
bct = body-centered tetragonal.
or = orthorhombic.

Suppose that it is wished to check a value already obtained for $\lambda_{11}/\lambda_{13}$ against values for $\lambda_{11}/\lambda_{12}$ and Θ_1, Θ_2, and Θ_3 obtained independently. Now

$$\frac{\lambda_{11}}{\lambda_{13}} = \frac{\lambda_{11}}{\lambda_{12}} \frac{\lambda_{12}}{\lambda_{13}}$$

$$= \frac{\lambda_{11}}{\lambda_{12}} \frac{\sin \Theta_3}{\sin \Theta_2} \qquad (9\text{-}5)$$

and this value can be compared with that found by the direct method. For the $\alpha\beta$ and $\alpha\alpha$ boundaries in brass, Smith (1948, p. 15) found

$$\frac{\lambda_{\alpha\beta}}{\lambda_{\alpha\alpha}} = 0.74$$

and for leaded brass

$$\frac{\lambda_{\alpha Pb}}{\lambda_{\alpha\alpha}} = 0.707$$

From the three-phase configuration $\alpha\beta$ and lead analogous to that of Fig. 9-10 and Eq. (9-5), the ratio of $\lambda_{\alpha\beta}/\lambda_{\alpha Pb}$ was found to be

$$\frac{\sin 110°}{\sin 120°} = 1.085$$

and thus

$$\frac{\lambda_{\alpha\beta}}{\lambda_{\alpha\alpha}} = 0.707 \times 1.085 = 0.766$$

in good agreement with the directly determined value of 0.74.

Measurements by Stanton (1964) on a number of two- and three-phase sulfide ores show that where these have been naturally annealed, given sulfide systems are also characterized by quite highly reproducible dihedral angles and hence the boundaries involved have characteristic interfacial energies. Not all mineral groupings have been reproduced experimentally, but those which have yield similar values to those obtained for natural material. Table 9-2 lists values for natural and synthetic sulfides by Stanton (1964, pp. 67–68).

Stanton has also confirmed values for certain ratios using the three-phase triple-junction method of Smith. For galena, sphalerite, and chalcopyrite the angles are as shown in Fig. 9-10:

$$\frac{\lambda_{sc}}{\lambda_{ss}} = \frac{\sin 111°}{\sin 119°} \frac{\lambda_{sg}}{\lambda_{ss}}$$

$$= 1.068 \times 0.79$$

$$= 0.84$$

which compares with 0.85 determined directly.

Table 9-2 Dihedral angles and interfacial free-energy ratios in some common sulfides. (After Stanton, 1964, pp. 64–65)

| System | | Interface between | | Grain boundary used as a comparison interface C | Dihedral angle Θ, deg | $\dfrac{\gamma A/B}{\gamma C}$ |
		Phase A	Phase B			
PbS–ZnS	(N)†	PbS	ZnS	ZnS	103	0.80
PbS–ZnS	(N)	PbS	ZnS	PbS	134	1.28
PbS–ZnS	(S)	PbS	ZnS	PbS	130	1.18
$CuFeS_2$–ZnS	(N)	$CuFeS_2$	ZnS	ZnS	108	0.85
$CuFeS_2$–ZnS	(S)	$CuFeS_2$	ZnS	ZnS	106	0.83
$CuFeS_2$–PbS	(N)	$CuFeS_2$	PbS	PbS	125	1.10
ZnS–FeS	(N)	ZnS	FeS	FeS	107	0.84
ZnS–FeS	(S)	ZnS	FeS	FeS	108	0.85
ZnS–FeS	(N)	ZnS	FeS	ZnS	107	0.84

† (N) denotes natural material; (S) denotes synthetic material.

Similarly

$$\frac{\lambda_{gc}}{\lambda_{gg}} = \frac{\sin 130°}{\sin 119°} \frac{\lambda_{sg}}{\lambda_{gg}}$$

$$= 0.8758 \times 1.28$$

$$= 1.12$$

which compares with 1.10 found directly.

Two features of dihedral angles in alloys have been commented upon particularly by Smith (1964, pp. 31–32). Firstly, the range of dihedral angle in simple two-phase systems is quite small. Secondly the two-phase interface commonly has a lower energy than that of the grain boundaries of the phases composing it—not higher as might have been expected from crystallographic and chemical differences. Clearly both of these observations apply also to the common sulfides.

GRAIN GROWTH

The total energy of a polycrystalline aggregate is the sum of the energies of the lattices, the grain boundaries, and the surfaces. Ignoring the last, the relationship between lattice and grain-boundary energies may be briefly considered.

When a metal or, under appropriate conditions, any other polycrystalline aggregate, is plastically deformed, the lattice structures become distorted, or strained. Extensive "cold-work" may cause heavy distortion. Much of the energy expended by such cold-working is thus transferred to the crystal lattices, and this then becomes the *stored energy of cold-work*, or *strain energy*. While some energy of this kind is stored by deformation of grain boundaries, most is retained in the distorted lattices. Under these conditions most of the latent driving force for later structural modification lies *within* the grains, and grain-boundary energies are relatively unimportant. This is considered in detail in the next chapter.

In annealed metals, however, the lattices are in equilibrium and their energies are therefore much less. Under these conditions grain-boundary energies become relatively important and assume the dominant role in structural modification. Given temperatures high enough for atomic mobility, such modification follows the usual tendency toward lower energy states. Since the total grain-boundary free energy is proportional to the area of boundary, modification will progress by decreasing this area and hence by increasing grain size.

Increase in grain size involves boundary migration, and the prerequisite for this is boundary curvature. The "driving force" concerned—the grain-boundary free energy—is proportional to $1/R_1 + 1/R_2$, where R_1 and R_2 are the radii of curvature of the grain in two mutually perpendicular directions. The curvature inherent in the foam arrangement is thus all-important. It imparts an instability to the structure that adjusts to decrease energy, and such adjustment takes the form of movement of curved interfaces toward their centers of curvature. By this means small grains, with small radii of curvature, tend to contract, and large grains tend to grow correspondingly at the small grains' expense.

Once again the process in solids may be visualized more readily by use of the froth analogy. Here boundary free energy is represented by pressure difference

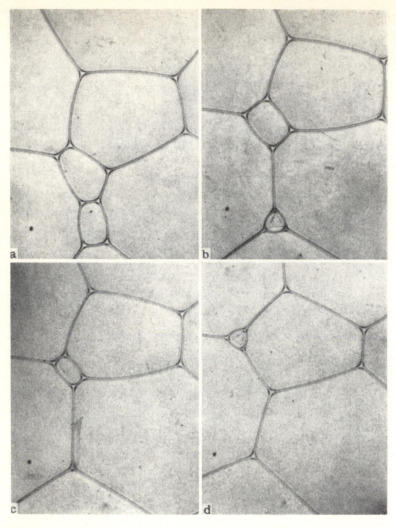

Fig. 9-11 Sequence of boundary positions in disappearance of bubbles during normal growth. (*From Smith, "Metal Interfaces," American Society for Metals, Cleveland, Ohio, 1952.*)

across the bubble wall, and transfer of material between grains is represented by the diffusion of gas from one bubble to another. The process of bubble growth has been illustrated very beautifully by Smith (1964, fig. 8), who points out that this involves both continuous and discontinuous motion of the boundaries:

> The smaller, high-pressure triangular bubbles (tetrahedral in three dimensions) shrink and disappear without disturbing the neighboring interfaces, but simply extending them. One three-sided bubble vanishes and three adjacent ones lose one side each . . . .   However, when a bubble with more than three sides shrinks, it must at some point lose a side, and, for a moment, the four adjacent films must meet in an unstable arrangement. This is followed by rapid readjustment to a new equilibrium with a new face at right angles to the one that has disappeared and shared by two adjacent bubbles (1964, p. 14).

The process is illustrated by Fig. 9-11, due to Smith (1964, plate 3, fig. 8).

Looking at the process in a less dynamic but more precise way, the initial movement is as shown in Fig. 9-12. Starting with the necessary curved faces and a triple junction out of equilibrium (> 120° in this case), the boundaries *AC* and *BC* move down *CD*. In this way the triple-junction area migrates through positions *b*, *c*, . . ., until the solid angle is 120°. The boundaries *AC* and *BC* have then taken on a greater curvature and migrate toward their centers. This then increases the angle at the triple junction, and so on. Through continuous readjustment of this kind the boundary migrates into the grain with the smaller radius of curvature until this is finally consumed.

In general, equilibrium is attained only when all faces have become flat and all angles 120°. However, it will be seen that a stable state is not easily reached or held. If a substantial number of proximate junctions reach equilibrium at the same moment, the structure may stabilize. But should one junction reach an equilibrium position before another, the changes necessary to equilibrate the second junction may well induce instability in the first. The situation is neatly put by McLean:

> Stability of grain structure is therefore an intricate state of affairs in which persistence of equilibrium at any given point depends on there being no change in the neighbourhood, which implies equilibrium in the neighbourhood. This in turn demands equilibrium still further from the given point, and so on, so that in the last resort equilibrium at any one point is dependent on there being equilibrium everywhere. Conversely, upsetting the equilibrium at one point is liable to upset it everywhere (1957, p. 89).

There are several special circumstances by which trends toward equilibrium can be disrupted and by which usually unstable arrangements can be locked. Detailed consideration of these is, however, beyond present concern. The principal factors leading to cessation of boundary migration and the limiting of grain growth are the diminution of interfacial free energy as growth progresses, and the hindering effect of impurities. In many cases the two act together: while the driving force for migration is high, boundaries are able to slip round impurity particles and continue moving, but eventually it diminishes to the point where the boundary can no longer do this and becomes locked against the inclusion in its path.

As a result of the sum of these influences, grain growth proceeds by the absorption of smaller grains by larger ones. This was indicated by the froth model

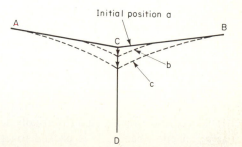

Fig. 9-12 Detail of progress of grain-boundary migration.

and follows from the fact that in general the larger the grain, the greater the number of faces bounding it. Only those grains with six sides (i.e., in two dimensions) can have straight boundaries all meeting adjacent boundaries at 120°. Those with less than six must have convex boundaries, and those with more than six must have concave, if these are to meet the boundaries of adjoining grains at 120°. As boundary migration proceeds toward the centers of curvature, the grains with more than six edges (i.e., in general, the larger) will grow, and those with less than six (the smaller) will disappear.

POLYCRYSTALLINE GROWTH STRUCTURES

The range of microstructures commonly developed in polycrystalline matter has been well illustrated by Smith (it is almost impossible to consider any aspect of microstructure without reference to C. S. Smith), to whom Fig. 9-13 is due, in his "gallery of structures" (1964, p. 47, fig. 1). Most of these structures were first recognized in metals, but many are now known to be developed in ceramics (see Kingery, 1960), and there is little question that many occur in rocks, particularly those termed "ores."

RANDOM SINGLE-PHASE AGGREGATES

Two principal structures fall within this group: growth impingement boundaries (Fig. 9-13, A1 and A2) and the foam structures already discussed (Fig. 9-13, A3).

The place of growth impingement boundaries is somewhat ambiguous. They have already been considered in connection with growth in open space, and it is probable that most of them result from nucleation, growth, and impingement in fluid media—vapor, hydrothermal or other solutions, and igneous melts. Some, however, undoubtedly develop as a result of massive transformations—i.e., replacement uninfluenced by preexisting structures. Examples of this are the replacement of feldspar by unoriented epidote or chlorite granules, of single coarse grains of galena or chalcopyrite by finer mosaics of chalcocite, or of arsenopyrite by sphalerite. In each case the new phase begins as a number of separate nuclei in the old, each such nucleus growing quite independently of the structure of the host until the latter is removed and the new grains find themselves in contact with each other. The result is a truly random impingement structure. Provided the aggregate is not subjected to temperatures permitting atomic mobility, it will not be annealed and the structure will persist in a metastable state.

Foam structures are common in natural materials. In some silicates their development is often inhibited by anisotropy of growth, as in feldspar aggregates, but in others the structure is quite usual. Much metamorphic quartz, for example, shows apparently excellent foam structure, as does the olivine of olivine nodules and the apatite of metamorphosed phosphate concentrations.

Foam structures, or close approximations to them, are also very common in polycrystalline aggregates of the more common sulfides. Polished sections of almost any monomineralic aggregate of galena, sphalerite, and chalcopyrite show the familiar polyhedra and 120° triple-junction angles. Even pyrite, arsenopyrite, cobaltite, and others which normally tend to idiomorphic form show the same arrangement when each is packed together with its own kind. The orientation

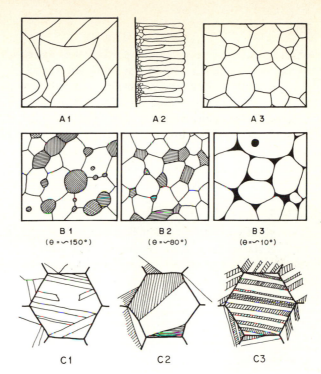

A 1   A 2   A 3

B 1   B 2   B 3
($\theta = \sim 150°$)   ($\theta = \sim 80°$)   ($\theta = \sim 10°$)

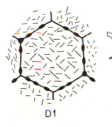

C1   C2   C3

D1   D2   D3

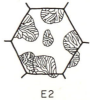

E1   E2

F1   F2

Fig. 9-13 C. S. Smith's "gallery of structures," developed by growth and transformation "in the solid" in polycrystalline aggregates. (*From Smith, Met. Rev., 1964.*)

effects, which are so pronounced when these minerals grow in matrices of other minerals, seem to dwindle to insignificance when like meets like.

Unless the mineral concerned is optically strongly anisotropic, the structure is not apparent in untreated polished section, and it is for this reason that it has been missed by many microscopists. It is quite clear in pyrrhotite or niccolite, for example, when these are viewed between crossed nicols, but it is not apparent in galena, sphalerite, and others like them unless these are suitably etched. Many particles of such minerals have been considered single grains when viewed under the microscope when in fact they were polycrystals. This has undoubtedly led to many erroneous textural interpretations; what have been regarded as boundaries between grains have in fact been boundaries between *aggregates* of grains, and the textural interpretation has suffered accordingly.

It should perhaps be mentioned here that monomineralic sulfides frequently show two-dimensional webs very like a foam structure but depart from this as a result of deformation. When these are carefully measured it is found that many of the triple-junction angles depart from 120°, and it was with material of this kind that Stanton and Gorman studied the adjustment of triple-junction angles. Extensive deformation leads to the development of a schistose structure that is clearly different from the annealed foam configuration. This is discussed in the following chapter. Many aggregates of sulfides and other ore minerals, however, are substantially annealed and may be said to exhibit *foam structure*.

RANDOM DUPLEX AGGREGATES

In these the development of structure follows directly from the dihedral angle. Variation in structure results from variation in dihedral angle from one system to another and from variation in the proportions of the phases concerned (Fig.

Fig. 9-14 Pyrrhotite (light) as the minor phase showing the development of dihedral angles at grain boundaries in sphalerite (shades of gray, variably twinned), the major phase (black patches are pits). Ore from Cobar, N.S.W.

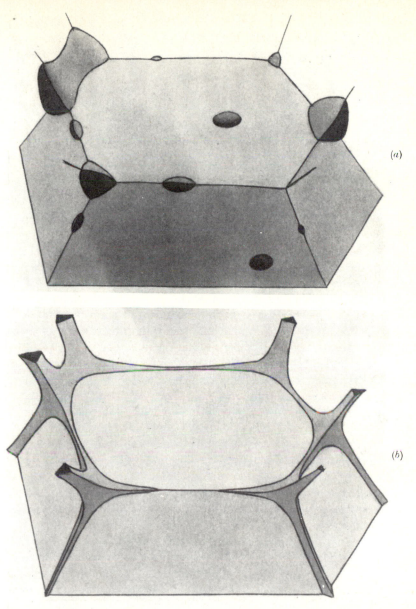

(a)

(b)

Fig. 9-15 Three-dimensional view of a minor phase at (a) triple junctions and (b) along grain edges in a major phase. (*From Smith, Trans. Amer. Inst. Mining Met. Engrs., 1948, as drawn by C. S. Barrett.*)

9-13, B1-B3).   While most structures have been studied in two-phase systems, the same principles hold for polyphase aggregates in general.

For ease of discussion we shall assume that one of two phases is distinctly in excess.   This has no effect in principle, but it facilitates description of results.

Suppose a two-phase polycrystalline aggregate is initially in a disequilibrium state and is then heated so that it begins to anneal.   Grains of major and minor phases will begin to segregate and adjust to each other so as to reduce the energy of

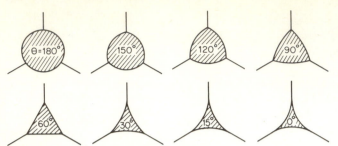

Fig. 9-16 Schematic diagram showing shapes of a minor phase corresponding to different dihedral angles. (*From Smith, Trans. Amer. Inst. Mining Met. Engrs., 1948.*)

the system. It may be shown that it is energetically far more favorable for the minor phase to develop along grain boundaries than within the grains of the major phase; the latter involves the generation of about twice as much interfacial free energy as the former. Formation at triple junctions is even more favorable than along two-grain boundaries. As a result, progressive annealing leads to the segregation of the minor phase at triple junctions and grain boundaries in the major phase. The retention of real "inclusions" within matrix grains is comparatively rare. The general nature of the situation is illustrated in Fig. 9-14, which shows sphalerite and pyrrhotite as major and minor phases, respectively, in ore from Cobar, N.S.W.

As this segregation proceeds, the shape of the particles of the minor phase develops according to the ratio of phase-boundary to grain-boundary free energies [Eq. (9-3)]. Where this ratio is high, Θ is large and the minor phase tends to nearly spherical, or equant, grains; where it is low, Θ becomes small and the minor phase tends to concave, cuspate bodies or even films. This is illustrated in three dimensions in Fig. 9-15. Those few grains retained within a matrix grain are usually spherical or nearly so. (In some cases this is a little surprising, since in noncubic crystals some anisotropy of interfacial energy might be expected. However, this does not seem to manifest itself, at least to a readily detectable extent.)

Looking at this quantitatively, at $\Theta \rightarrow 180°$, $\lambda_{12}/\lambda_{11} \rightarrow \infty$, and the minor phase approaches spherical shape (assuming that there is no orientation effect). With decrease in $\lambda_{12}/\lambda_{11}$ the dihedral angle decreases, and the minor phase begins to move out along the grain boundaries of the major phase, becoming a convex polyhedron. Assuming a normal triple junction, four grain boundaries are involved, and so the minor phase assumes the form of a convex tetrahedron—a convex triangle in two dimensions, as shown in Fig. 9-16. As $\lambda_{12}/\lambda_{11}$, and hence Θ, decrease progressively, the apices of the tetrahedron move out until at $\lambda_{12}/\lambda_{11} = 1/\sqrt{3}$, $\Theta = 60°$. At this stage its faces become plane, and in two dimensions it appears as an equilateral triangle. With $\lambda_{12}/\lambda_{11} < 1/\sqrt{3}$ and hence $\Theta < 60°$, the polyhedron (and the triangle produced by sectioning) become progressively more concave, the apices continuing to move further and further out along the matrix grain boundaries. Eventually $\lambda_{12}/\lambda_{11} = \frac{1}{2}$ so that $\Theta = 0$. Clearly at this point it is energetically just as favorable to have a phase boundary as it is to have a matrix grain boundary, and as a result the minor phase moves right out along the grain boundaries and forms an intergranular film.

The relationship between shape and dihedral angle in two dimensions is illustrated diagrammatically in Fig. 9-16.

GRAIN SHAPE IN SULFIDE AGGREGATES

Extensive studies of the more common sulfides (Stanton, 1964) have shown that many ores have structures quite closely analogous to those of the random duplex aggregate of the metallurgist.   When carefully examined it may be seen that most of the mineral particles in polycrystalline sulfide and oxide ores terminate in tiny cusps, and when the surrounding material is lightly etched it becomes apparent that these cusps meet grain boundaries of the matrix, forming triple junctions. What previously may have appeared to be a massive band of mineral *A* with a cuspate boundary against a massive band of mineral *B* is usually revealed as a pair of polycrystalline layers, the cusps on the boundary grains of each phase matching with grain boundaries in the other, forming a series of phase-boundary–grain-boundary triple junctions.   Most "inclusions" are similarly seen to be intergranular and to develop a series of triple junctions with the three or more neighboring matrix grains.   The resulting shapes of minor-phase particles are thus much the same as in metals.   With high interfacial-free-energy ratios the dihedral angle is large, and the particle approaches either sphericity or—in the case of highly directional crystals—idiomorphism.   At lower free-energy ratios the dihedral angle is correspondingly smaller and the particle develops concavity, sometimes to the point of forming a film.

One of the best examples is that of galena-sphalerite.   Examination of almost any lead-zinc ore shows the two minerals to develop distinctly different shapes relative to each other when each in turn constitutes the minor phase.   Sphalerite as a minor phase in galena occurs as isolated rounded bodies, whereas galena as a minor phase in sphalerite forms wispy concave particles that often link together to form a network through the major phase (Fig. 9-17).

This has frequently been regarded as a replacement phenomenon: the wispy

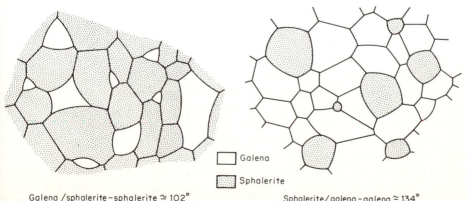

☐ Galena

▒ Sphalerite

Galena /sphalerite–sphalerite ≃ 102°   Sphalerite/galena–galena ≃ 134°

Fig. 9-17 Schematic diagram of grain shape when galena occurs as a minor phase in sphalerite and vice versa.

occurrence of galena as the minor constituent has been interpreted as representing the onset of replacement of the sphalerite matrix by galena along grain boundaries and the rounded form of the minor sphalerite as that of residual material remaining after extensive replacement by the galena. However, in all cases investigated this has been disposed of by etching and measuring the two occurrences of sphalerite. Were the rounded sphalerite particles all that remained after extensive replacement —replacement that began at grain boundaries and progressively consumed and rounded off the original grains—the grain size of the isolated material should be appropriately smaller than that of the immediately adjacent matrix. Etching of the grain boundaries shows that this is not the case.

Etching of the major phase in each case shows the form of the minor phase to reflect the dihedral angle. Cusps in the minor phase almost invariably meet a grain boundary in the major phase, giving a phase-boundary–grain-boundary triple junction. As shown in Table 9-2, the characteristic dihedral angle for galena-sphalerite–galena-galena is about 134°, giving the characteristic rounded form of the sphalerite.

It appears that the high dihedral angle may manifest itself in either near sphericity, as with sphalerite in galena or pyrrhotite in pentlandite, or in a tendency to idiomorphism, as in the case of pyrite, sperrylite, arsenopyrite, and so on. Presumably in the latter case interfacial energies are very substantially lower (though, of course, still high relative to the matrix grain boundary) parallel to the faces concerned.

The common pair sphalerite-chalcopyrite also shows good random duplex structures. However, although chalcopyrite is common as a minor phase in sphalerite, the reverse relationship is comparatively infrequent, and it is not so easy to compare the two as in the case of galena and sphalerite. Chalcopyrite as the minor phase readily develops triangles at triple-junction points and lenses along grain boundaries in the sphalerite matrix. Grain-boundary segregation seems rather more common here than with the galena-sphalerite pairing. Localization along twins is also frequent, and this is dealt with in more detail a little further on. Sphalerite and pyrrhotite often show good random structures, the phase-boundary–grain-boundary energies being much the same ($\lambda_{sp}/\lambda_{pp} = 0.84 = \lambda_{sp}/\lambda_{ss}$) whichever constitutes the minor phase. However, this relationship is sometimes complicated by the anisotropism of the pyrrhotite, which may grow as laths with their long dimensions, and straight sides, parallel to their crystallographic c axis. The cause of such variation in habit of the pyrrhotite is not known but is very possibly related to compositional variation in the pyrrhotite or sphalerite, or both. It may well be that a quite slight impurity or deviation from stoichiometry is sufficient to trigger anisotropic growth. Chalcopyrite as a minor phase in galena forms convex bodies having dihedral angles of about 125°. Pentlandite as a minor phase in pyrrhotite occurs largely as the "network texture" of Edwards (1954). This is the result of a low dihedral angle formed by the pentlandite-pyrrhotite phase boundary against the pyrrhotite grain boundary. The angle approaches zero, and the pentlandite forms an intergranular film about the rounded particles of pyrrhotite.

Although other mineral pairs still await measurement, it is clear that many of them occur together as random duplex aggregates similar in principle to those of metal alloys.

EFFECT OF IMPURITY AND TEMPERATURE
ON SHAPE IN RANDOM AGGREGATES

The effects of slight compositional differences in the liquid phase of metallic two-phase systems may be substantial, and as a result, they have been studied in some detail by metallurgists. Where such liquids have a high dihedral angle against a metal grain boundary, the latter is not penetrated and the properties of the solid-phase polycrystal are unaffected. Where, however, the addition of small quantities of impurity to such liquids reduces Θ to zero, the grain boundaries are penetrated, and the metal becomes brittle.

It is possible that this principle operates in some instances during the deposition of ores from solutions or melts. For example, certain minerals sometimes occur as grain-boundary films in a single phase to which they do not appear to be chemically related. These may have been deposited from solutions that had a low dihedral angle against the grain boundaries in question and that hence were able to migrate round the preexisting minerals. It has also been suggested (Smith, personal communication) that such solutions might facilitate the growth of idioblasts of pyrite, arsenopyrite, and other minerals by providing a liquid medium for crystal growth. If this were the case the crystal would be virtually entirely surrounded by liquid that would dissolve, and carry away, surrounding material, in turn allowing growth of the crystal without restraint.

The effect of impurities on the development of grain shape as developed "in the solid" is hardly known but almost certainly constitutes a very large field for investigation—one perhaps analogous with that of the effect of impurity on modification of habit during free growth, already discussed in Chap. 8. It has already been noted that pyrrhotite often forms random aggregates with sphalerite but that in other, by no means rare, cases it forms laths cutting across the grain boundaries of its host. Since sphalerite is so prone to compositional variation and since pyrrhotite is frequently nonstoichiometric, one's immediate suspicion is that the

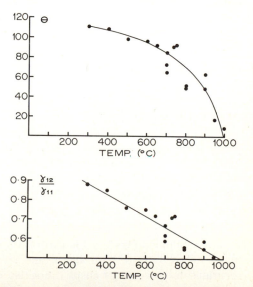

Fig. 9-18 Variation of the dihedral angle Θ and of interfacial free-energy ratios with temperature for galena against two sphalerite grains. (*From Stanton and Gorman, Econ. Geol., 1968.*)

pyrrhotite habit is related in some way to aberrations in the composition of either
—or both—phase.   The effect in this case would be to make the interface with the
basal plane a stable, slow-growing one and that with the prism a high-energy one,
leading to rapid growth and the development of elongated habit.   Similarly
pyrite in one sphalerite matrix may be conspicuously idiomorphic, whereas in
others it is quite rounded.   Such effects are quite gross and easily discernible.
Whether more subtle ones are to be found in the modification of dihedral angles is
not known, but it seems very likely that they occur.

The effect of temperature on dihedral angles in metals does not appear to be
great.   Temperature does affect both phase- and grain-boundary energies, but
apparently both change proportionately with rise in temperature so that there is
little or no change in their ratio and hence in the dihedral angle concerned.
Little is known about the effect of temperature on sulfide interfaces, though Stanton
and Gorman have investigated sphalerite-galena between 300 and 1000°C and have
obtained what appears to be a very clear diminution of Θ with increase in temperature.   This is shown in Fig. 9-18.   Details of the runs are given in Table 9-3.   For
the natural sphalerite-galena aggregate concerned (compact fine-grained sulfide
from Rosebery, Tasmania), Θ decreases slowly between 300 and 600°C and then
appears to fall rapidly to zero at a temperature between 950 and 1000°C.   The
variation in values obtained for different runs at a given temperature are in part
statistical, but there is little doubt (from direct observation) that it is due in large
part to the effect of irregularly disposed small amounts of impurity.   The effects of
both temperature and impurity on sulfide interfacial-free-energy ratios—and hence
the development of shape—looks to be a fertile one for future investigation.

ORIENTED STRUCTURES RESULTING FROM
NUCLEATION DURING GROWTH

Up to this point no consideration has been given to the effect of relative crystallographic orientation on grain-boundary energies.   It has been tacitly assumed that
orientation has no significant effect on energy, growth, and shape and that given
appropriate annealing conditions, a polycrystal would simply adjust to a random
arrangement of lowest energy.   For simple cubic close-packed structures this may
be approximately true, but orientation does have a measurable effect on boundary
energies, and in certain cases this may have important implications in growth and
the development of shape.

The general nature of the relation between relative orientation and interfacial energy is not difficult to visualize.   Juxtaposition of two like crystals in
parallel orientation results in a single crystal—the boundary is completely coherent
—and there is zero interfacial energy.   As the crystals are rotated with respect to
each other, however, the coherence is lost; there develops a significant misfit
across the boundary, and an increment of interfacial free energy develops.   For
two face-centered-cubic crystals, for example, rotation on a {100} plane about the
[100] axis yields increasing misfit until a maximum is reached at 45°.   Further
rotation leads to decreasing misfit until at 90° a complete match is achieved again.
In such a simple operation the theoretical grain-boundary energy rises to a maximum, drops to zero, rises again, and so on, giving a series of energy maxima
separated by low-energy cusps (Shockley and Read, 1949).   This has been con-

Table 9-3 Dihedral angles and interfacial free-energy ratios for galena-sphalerite phase boundaries versus sphalerite grain boundaries at progressively increasing temperatures. (From Stanton and Gorman, 1968, p. 920)

Temperature, °C	Number of measurements	Dihedral angle Θ, deg	$\dfrac{sg}{ss}$
300	300	110	0.88
400	300	108	0.85
500	300	98	0.76
600	300	96	0.75
650	300	92	0.72
700	300	73	0.62
700	300	64	0.59
700	300	83	0.67
735	300	91	0.71
750	300	92	0.72
800	300	49	0.55
800	300	52	0.56
900	300	49	0.55
900	300	63	0.59
950	300	18	0.51
1000	300	9	0.50

firmed by experiment (e.g., McCarthy and Chalmers, 1958). Clearly the situation becomes more complex with rotation on less simple crystallographic planes or when two unlike crystals are involved. In the latter case structures may be similar, with differences only in atomic spacings, or the structures may be quite different from each other. It has been shown theoretically, however (e.g., Fletcher, 1965), that rotation on any given pair of planes yields energy maxima separated by cusped minima (whose periodicity is dictated by the structure of the interface concerned). Investigations have been carried out on such elements as tin, lead, iron, and silicon using tricrystals similar to those used by Aust and Chalmers to study rates of triple-junction equilibrium. In the present case, however, the three crystals are grown from seeds in specified orientation, and the three angles are measured in a plane normal to the triple junction. By regarding one of the grain-boundary energies as unity and substituting in

$$\frac{\lambda_{12}}{\sin \Theta_3} = \frac{\lambda_{13}}{\sin \Theta_2} = \frac{\lambda_{23}}{\sin \Theta_1}$$

the relative grain-boundary energies are obtained. These may then be plotted against the relative rotation of the grains, giving results such as those of Fig. 9-19 (McCarthy and Chalmers, 1958).

Examination of these curves shows that energy rises rapidly with the first few degrees of misorientation and that the slope then decreases substantially until the next cusp is approached. If we take the first 15° of twist as being the really significant interval from the point of view of energy increase, it is apparent that only a small proportion of grains are in low-energy orientation with respect to each other. This situation may be changed very sharply, however, by a substance's capacity

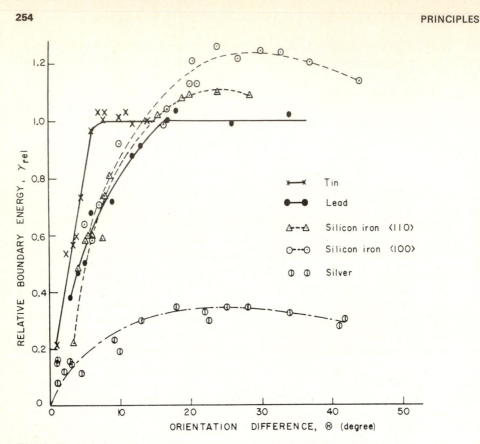

Fig. 9-19 Relative grain-boundary energies as a function of orientation differences in some industrial metals. (*From McCarthy and Chalmers; reproduced by permission of the National Research Council of Canada from the Canadian Journal of Physics, vol. 36, p. 1646, 1958.*)

to develop twins. Since twin planes are planes of close matching, they are boundaries of low energy. Thus substances that form twins have more low-energy orientations than substances that do not. This may have a pronounced effect on growth and subsequent structure.

Suppose one grain of a single-phase polycrystal possesses a normal high-energy boundary with an adjacent grain and that a temperature allowing atomic mobility is approached. One grain is about to grow at the expense of the other, and this, for most orientations, may be expected to take place in accordance with curvature. Suppose that this orientation, however, is close to that for the formation of a twin. Since a very large reduction of energy can be achieved by the conversion of a grain boundary to a twin boundary, the first few atoms involved in growth will assume a twin orientation with respect to the grain that grows. These atoms then act as a nucleus for the remainder of the grain being consumed, and these are eventually converted to a twin of the grain that has grown. Such twins are known as *annealing twins* (Fig. 9-13, C1).

Clearly any grain is likely to make a number of such encounters during growth, and a number of twins may develop.   In general the number of twins should be proportional to the number of encounters made, and hence there should be a general direct relation between grain size and the number of twins per grain.   It should not be thought from this that *all* encounters result in twins, which is far from the case.   Only a few encounters will involve orientations close enough to potential twin orientations for twins to form.   The more encounters that are made, however, the greater the chance of such orientations being involved, and hence the relationship between grain size and twin incidence.   Development of twins is favored by a low twin to grain-boundary–free-energy ratio so that the lower the value for this, the greater the incidence of twins.   Since the twin orientation is a low-energy one, it carries with it a low driving force for growth, and so those substances that develop annealing twins tend to have smaller grain sizes than those that do not. The relationship between twin frequency and grain size in a single-phase metal is shown by the work of Hu and Smith (1956) on α brass (Fig. 9-20).   This appears to be independent of annealing temperature, as might have been expected.

Annealing twins are common in some sulfides, particularly sphalerite.   It is in fact very difficult to find a zinc ore in which the sphalerite does not show at least some twins that appear to result from grain growth.   Such twins have been as-cribed to deformation (e.g., Edwards, 1947) by either tectonic stresses or "mutual interference of individual grains during growth" (1947, fig. 34).   While some twins in sphalerite do indeed appear to be deformational, most are not.   The annealing twins are usually evenly spaced, straight-sided, and coherent—i.e., do not end within a grain.   Most deformation twins are unevenly spaced, many are bent, and many are incoherent, tailing off in irregular fashion *within* the grain concerned.

Stanton and Gorman (1968) have studied the incidence of annealing twins in

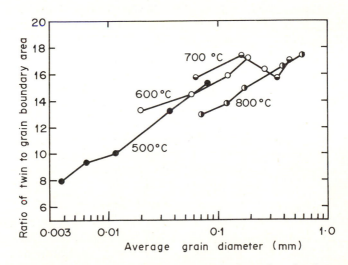

Fig. 9-20 Frequency of twins as a function of grain size in annealed α brass. (*From Hu and Smith, Acta Met., 1956.*)

sphalerite and have shown this to be related to grain size, as in metals. Such twins are always parallel—no more than one set of twins is present in any one grain—and are always coherent. The proportion of twinned grains is usually of the order of 80 per cent.

Chalcopyrite also forms annealing twins, though less extensively than sphalerite. It is rare to find more than 50 per cent of grains twinned, and the number of twins per twinned grain is much lower than for sphalerite.

None of the other common sulfides have been found to develop annealing twins. Galena and pyrrhotite might have been expected to do so but do not. Pyrite does not show twins of any kind in most ores.

Structures analogous to annealing twins may develop in two-phase aggregates, and the compound particles so formed are referred to as *coherent duplex grains*. The mineralogical analogy appears to be the oriented intergrowth of certain metamorphic minerals, such as the micas. In metals the structure develops in stable two-phase alloys where the two phases possess planes of similar configuration and spacing. If equilibrium concentrations are established and the specimen is brought to a temperature at which annealing and grain growth can occur, the latter proceeds by the usual transfer of material from smaller grains to larger ones. In this diffusion process there is always a chance that the concentration of one phase in the other will, in a very small segment, rise just to the point where the first forms a nucleus. Such nucleation may take place coherently on the grain boundary of the other phase, in which case both phases then grow together as two crystallographically distinct, but coherent, units. The two then grow together in the same way as the twinned grains of a single-phase metal (Fig. 9-13, C2 and C3).

This kind of structure is not common among the ore minerals, but it does seem that the process concerned may operate in the development of some sphalerite-chalcopyrite and chalcopyrite-bornite intergrowths. These can be produced synthetically at high temperatures and perhaps significantly are found most commonly in high-temperature veins. Such a relationship is of course quite in accordance with crystallographic requirements. In the case of sphalerite-chalcopyrite there is a near identity in the sulfur configuration on the {111} planes of the two minerals. This is, as already noted, the plane on which each mineral individually forms annealing twins. For chalcopyrite the interatomic spacing between sulfur atoms on this plane is about 3.7 Å, and for sphalerite it is 3.83 Å—a difference of about 4 per cent. Chalcopyrite and bornite also have this plane in common (cf. the "blende" group of Chap. 4), the interatomic spacing for the relevant sulfur atoms being 3.84 Å. It is therefore not surprising that sphalerite and bornite both develop coherent duplex grains with chalcopyrite.

DUPLEX STRUCTURES FORMED BY PRECIPITATION

These constitute the important group of structures generally referred to as "exsolution," or "unmixing," textures. They develop as a result of the cooling of solid solutions and assume a variety of forms, depending on the structures of the phases concerned and the site of nucleation. In most cases the final phases are separated by visible boundaries, though in some the structure is less distinct.

In some of the earlier literature (cf. Edwards, 1947) it was inferred that mineralogical solid solution involved the dissolving of one *mineral* in another, i.e., that there existed some kind of molecular solution. In fact—and as was pointed out in Chap. 3—the process involves incorporation (at elevated temperatures) of impurity *atoms* of variety appropriate for the formation of certain *potential* minor phases. Should physical and/or chemical conditions change, and structural and nucleating factors prove suitable, these may precipitate.

Structures formed by precipitation are common in all classes of crystalline material, including metals, ceramics, and rock-forming minerals. They are particularly common among sulfides and oxides, and are almost ubiquitous among ores formed at high temperatures. Most of them are oriented, and most are nucleated on fairly readily identified features in the host, though this is not always the case. Because of its capacity for incorporating large quantities of impurity, sphalerite is the most conspicuous host among the sulfides. Among the more common species chalcopyrite and pyrrhotite are, after sphalerite, the most frequent hosts. The group of oxides magnetite-hematite-ilmenite-rutile also appear to develop a variety

Table 9-4 Exsolution pairs and groups observed in natural oxide and sulfide-type ores

Component†	Investigator
Oxides	
Cassiterite-tantalite+	Edwards (1940)
Chromite-hematite	Ramdohr (1931)
Chromite-ilmenite	Ramdohr (1940)
Corundum-hematite	Bray (1930)
Hematite-rutile+	Newhouse (1936)
Hematite-ilmenite+	Ramdohr (1926)
Hematite-ilmenite-rutile	Edwards (1938)
Hematite-sitaparite	Mason (1944)
Ilmenite-pyrophanite	Edwards (1938)
Magnetite-franklinite-jacobsite-hausmannite-hetaerolite	Orcel-Pavlovitch (1931)
Magnetite-hematite	Greig (1935)
Magnetite-ilmenite+	Ramdohr (1926)
Magnetite-pyrophanite	Schneiderhöhn-Ramdohr (1931)
Magnetite-rutile(?)	Newhouse (1936)
Magnetite-spinel	Ramdohr (1925, 1940)
Manganosite-zincite	Frondel (1940)
Tantalite-columbite+	Pehrman (1929)
Tantalite-rutile	Edwards (1940)
Tantalite-ilmenite	Edwards (1940)
Sulfides and related minerals	
Alabandite-pyrrhotite	Schneiderhöhn-Ramdohr (1931)
Bismuthinite-argentite	Schwartz (1931)
Bismuthinite-emplectite	Ramdohr (1931)
Bornite-chalcocite+	Schwartz (1928)

† Host named first except where indicated +, in which case either may apparently act as host.

Table 9-4   (continued)

Component	Investigator
Sulfides and related minerals (*continued*)	
Bornite-chalcopyrite[+]	Schwartz (1931)
Bornite-klaprothite	Krieger (1940)
Bornite-tetrahedrite	Edwards (1946)
Chalcocite-covellite	Bateman-Lasky (1932)
Chalcocite-stromeyerite	Schwartz (1935)
Chalcopyrite-cubanite	Schwartz (1927)
Chalcopyrite-chalcopyrrhotite	Borchert (1934)
Chalcopyrite-pyrrhotite[+]	Hewitt-Schwartz (1937)
Chalcopyrite-tetrahedrite	Edwards (1946)
Cubanite-pyrrhotite	Newhouse (1931)
Galena-argentite	Missen-Hoyt (1915)
Galena-galenobismutite	Berry (1940)
Galena-matildite[+]	Ramdohr (1938)
Galena-tetrahedrite	Guild (1917)
Galena-pyrargyrite	Schneiderhöhn (1922)
Galena-proustite	Guild (1917)
Linnaeite-millerite	Schneiderhöhn-Ramdohr (1931)
Pyrrhotite-chalcopyrrhotite	Bochert (1934)
Pyrrhotite-pentlandite	Newhouse (1927)
Sphalerite-chalcopyrite[+]	Buerger (1934)
Sphalerite-pyrrhotite	Stillwell (1926)
Stannite-chalcopyrite	
Stannite-sphalerite[+]	
Stannite-tetrahedrite	Schneiderhöhn-Ramdohr (1931)
Stannite-pyrite	Ahlfeld (1934)
Stannite-cubanite	
Silver-dyscrasite	Carpenter-Fisher (1932)
Tennantite-galena	Ramdohr (1931)
Tennantite-chalcopyrite	Ramdohr (1931)
Bornite-linnaeite	Edwards (1938)

of precipitation structures among themselves.   The principal established and inferred pairings are given in Table 9-4 (after Edwards, 1954) and clearly involve a large proportion of the more common sulfides, oxides, and related minerals.

1. Growth from numerous independent nuclei   In this case the minor phase precipitates as numerous small particles throughout the body of the host grain (Fig. 9-13, D1).   The habit and distribution of such a minor phase may vary considerably, both for a given pair and, particularly, from one pair to another. Perhaps the most common of all sulfide pairings is that of chalcopyrite in sphalerite. Concentration of the minor phase varies substantially from one specimen to another but is rarely more than 10 per cent.   Particles may be distributed more or less evenly throughout the host grain, but in some instances there is a marked impoverishment near the grain boundary.   This has been attributed to a sudden rise in diffusion rates toward the edge of the host crystal, with consequent impoverishment in this region, but the frequent lack of segregation *at* the grain boundary in many such cases seems to preclude this as a general explanation.

The chalcopyrite may occur as small spherical particles in apparently random arrangement, or it may appear as lenses whose long axes and distribution are in parallel arrangement. The former is usually referred to as an *emulsion texture*. The latter often leads to the development of a *seriate texture*, consisting of a number of parallel linear groups of blebs showing somewhat regular size distributions. When etched these seriate patterns are frequently seen to be related to twinning in the sphalerite; the extremities of the chalcopyrite lenses are in fact dihedral angles, each lens is bisected by the host twin plane, and the linear arrangement simply follows the length of the twin. The degree of coherence between the phases here is apparently very high—the precipitate is often twinned and this twin plane always appears to be coplanar with that of the host (Fig. 9-21).

Where the appropriate impurities occur, stannite also forms extensive precipitate structures in sphalerite. Its form and distribution are, however, quite different from that of chalcopyrite. The particles are usually rather larger and are rarely spherical or lensoidal. Rather the stannite forms asymmetrical particles often with a planar "leading edge" behind which the rest of the particle trails out irregularly. Etching of the sphalerite shows that this leading edge is usually not coincident with a sphalerite twin plane, as might have been expected by analogy with chalcopyrite. Since stannite has a sphalerite structure, this is rather surprising. The plane common to both is the {111} sulfur plane of the sphalerite—the latter's twin plane—and close coherence might have been expected here. Apparently, however, there is another plane on which nucleation takes place more readily in this case.

Pyrrhotite is also commonly precipitated from sphalerite, and tetrahedrite appears occasionally. Apparently pyrrhotite is not precipitated as a result of simple supersaturation of iron accompanying a fall in temperature. As noted in Chap. 5, Barton and Toulmin (1966) have shown that pyrrhotite particles occur in sphalerite matrices whose iron contents are far below saturation levels. These investigators point out, however, that by adding sulfur to the system (see Chap. 5), pyrrhotite may be precipitated in spite of low iron concentrations, and this, presumably, is a frequent cause of its occurrence in sphalerite in nature. Such pyrrhotite may form more or less random rounded blebs in the manner of chalcopyrite, though in many cases it shows a strong tendency to form flat platelets parallel with and aligned along sphalerite twins. In some cases the blebs join in a somewhat irregular fashion to form *segregation veinlets*—a tendency shared by chalcopyrite.

In the majority of sphalerite-chalcopyrite intergrowths sphalerite is the host. Occasionally, however, chalcopyrite fills this role. Somewhat surprisingly the relations that develop are not the simple inverse of those developed when chalcopyrite is the minor phase. It is very clear that the concentration of sphalerite blebs in chalcopyrite is usually far lower than vice versa. Also the shapes of the blebs are quite different: instead of a spherical form, the sphalerite usually develops as somewhat cellular star-shaped bodies, the sides of which presumably develop

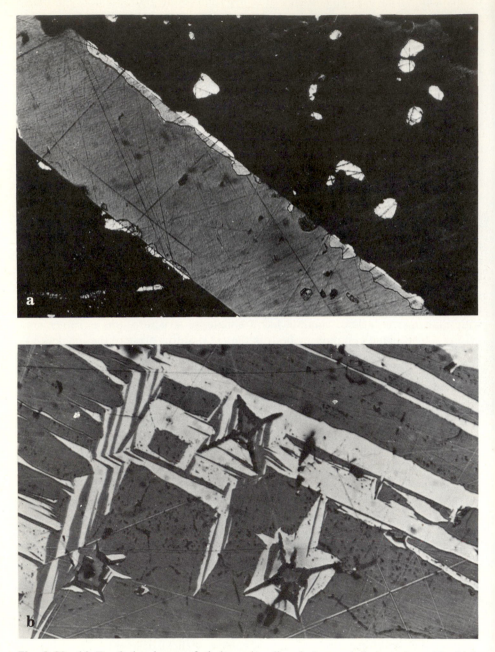

Fig. 9-21 (*a*) Exsolution lenses of chalcopyrite aligned along sphalerite twins. The high degree of coherency between the two minerals is indicated by the relationship of twinning in the chalcopyrite to the localizing twins in the sphalerite host. (*b*) Deformation twins in chalcopyrite apparently caused by contraction about star-shaped sphalerite inclusions.

parallel with the {111} planes of the host.   It is clear that in at least some cases the nucleation and growth of such sphalerite particles takes place at quite high temperatures, as evidenced by the frequent development of deformation twins in the chalcopyrite immediately surrounding the sphalerite (Fig. 9-21).   There seems little doubt that these have been induced by lattice stressing about the sphalerite inclusion as the whole cooled.

Pentlandite frequently develops as exsolution bodies in pyrrhotite as a result of diffusion of nickel from the pyrrhotite.   This has been dealt with in some detail by Hawley and Stanton (1962, pp. 56–57 and figs. 47–53).   In this case the nickel appears to diffuse to, and accumulate in, the basal planes of the pyrrhotite, leading to the development of platelets of pentlandite parallel to the (0001) plane of the host.   When viewed at right angles (or at a high angle approaching this) these platelets are seen to be somewhat cellular and of nearly circular outline.   In the simplest case these occur in linear arrangement following one or more (0001) planes in a given pyrrhotite grain.   In others they may grow from a twin plane into the (0001) planes on either side of this, giving rise to a herringbone pattern.   They may also project into (0001) planes from grain boundaries, particularly where the host pyrrhotite is itself enclosed in a matrix of coarser pentlandite.   Such matrices are often in the form of "networks" cementing together the larger rounded grains of pyrrhotite, and they may be largely the result of exsolution themselves.   In these cases the pyrrhotite platelets may represent the penultimate stage of exsolution, or—as is suggested a little further on—they may be analogous to the bainite plates developed in certain steels.

The crystallographic basis of the pyrrhotite-pentlandite intergrowth is not so neat as that of sphalerite-chalcopyrite but is quite clear nonetheless.   In the hexagonal pyrrhotite the (0001) plane is one of alternating Fe and S planes.   In the cubic pentlandite the {111} planes are Fe, S, S, Fe, the distance between two adjacent S planes being twice that between an S plane and an adjacent Fe plane.   As shown by Gruner (1929) adhesion of the two minerals on S planes is possible but unlikely.

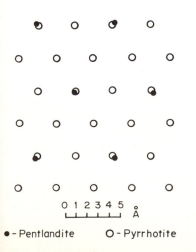

0 1 2 3 4 5 Å

● - Pentlandite   O - Pyrrhotite

Fig. 9-22   Relative arrangements of sulfur atoms in the (0001) and {111} planes of pyrrhotite and pentlandite, respectively.   (*From Gruner, Amer. Mineralogist, 1929.*)

If one of the S planes of the pentlandite is superimposed on the (0001) S plane of the pyrrhotite, a good fit is obtained, but there are many sulfur deficiencies on the pentlandite side. The same is the case with the other, adjacent, S plane of the pentlandite. However, if, as in Fig. 9-22 (after Gruner, 1929, pp. 234–236, fig. 5) the two sets of fits are observed together, it is seen that they are complementary; i.e., the holes left by one are filled by the other so that the two {111} sulfur planes of the pentlandite *together* just match a single (0001) sulfur plane in the pyrrhotite. Interatomic distances, too, are fairly close when the two sulfur planes of the pentlandite are projected onto a single plane. It is doubtful, however, if a second coordination of this kind would be sufficient to develop the necessary coherence. On the other hand the pyrrhotite (0001) Fe plane is a good fit with respect to the {111} Fe plane of the pentlandite, and coherence on this family of planes seems most likely.

2. Branched crystallographic growth Although structures of the type just discussed are commonly related to the crystallography of the host, they are typically discontinuous and are characterized by a multiplicity of nuclei. There are other precipitation structures, however, that appear to grow from comparatively few nuclei and that develop a high degree of continuity. These are the *Widmanstatten* structures of certain natural (meteoritic) and synthetic alloys and the *lattice intergrowths* of a number of sulfide and oxide minerals (Fig. 9-13, D2).

According to Smith (1964) such structures are not merely composites of many separately nucleated plates of the type considered under the preceding heading but are a rather special kind of dendrite. In the first case the abundant, independent nuclei do not meet because each drains the space about it. In the case of the Widmanstatten structure, growth commences at relatively few nuclei and then proceeds, under conditions of slow cooling, along appropriate crystallographic planes in the host. The precipitate extends itself along one plane and then may branch out along crystallographically similar planes (e.g., the octahedral planes of iron, magnetite, or sphalerite) to form a lattice intergrowth composed of one or a very few crystals. The final structure is essentially that of a dendrite that has, however, grown in a solid rather than in the usual fluid medium. Its identity as a single crystal is often shown up clearly by continuity of double refraction where optically anistotropic substances are involved.

One of the better examples among sulfides is that of cubanite in chalcopyrite, where precipitation along the {111} planes of the chalcopyrite leads to the development of a three-dimensional reticulated arrangement. The continuity of the cubanite blades is often emphasized by its double refraction when the structure is viewed between crossed nicols. Occasionally chalcopyrite forms this kind of structure in sphalerite (it develops readily in high-temperature syntheses which are allowed to cool fairly slowly) and in bornite. Chalcocite also occurs as lattice intergrowths in bornite. Among the oxides both hematite and ilmenite are known to grow in this way along the octahedral planes of magnetite, and there are no doubt other examples.

3. Oriented parallel wedge-like plates   Among metals an example of this kind of structure is provided by *bainite*.   This is a "transformation" structure in certain carbon steels and takes the form of aggregates of thin wedge-like plates that grow into the host grain in parallel arrangement from grain boundaries.   The composition of the material so formed has been thought to be a mixture of ferrite (solid solution of carbon in body-centered cubic iron) and finely divided Fe_3C particles or simply ferrite supersaturated with carbon.   Its growth appears to be limited by lattice strain, and its appearance is illustrated in Fig. 9-13, D3.

The only likely analogy among the sulfides appears to be that of pentlandite in pyrrhotite—where, as already described, pentlandite projects into the host pyrrhotite from the pyrrhotite's grain and twin boundaries.   In both cases the pentlandite takes on a finely serrated form, presumably in conformity with the planes of the pyrrhotite that it follows.   This serrated form may reflect the presence of numbers of parallel, fine plates.

REPLACEMENT STRUCTURES

With the emphasis placed on the replacement theory of ore formation during much of the twentieth century, great importance has been attached to structures that might indicate the nature of the process.   However, it appears that at least some of the structures diagnosed as resulting from replacement are not in fact due to this.   The example of galena in sphalerite has already been mentioned; in some cases, at least, the grain-boundary relations between these two are those of simple random duplex aggregates resulting from grain growth or annealing and are not due to replacement.   It seems likely that at least many of the structures ascribed to replacement up to about 1955 are due rather to simple grain growth, precipitation, annealing, and so on.

It is perhaps significant that in spite of the emphasis placed on replacement in primary ores, most good examples of its operation have been found in secondary (weathered) occurrences.   One is led to the suspicion that this may be indicative of the relative importance of the process in the two environments.   Present concern is not with such speculation, however, but with the microstructures that the process may produce.

As with the other growth structures, the process of replacement is one of diffusion.   However, whereas the structures discussed to this point result from physical changes in isochemical situations, replacement results from changes in environments where both physical *and* chemical conditions may alter and which are chemically "open"—i.e., capable of accepting and disposing of materials involved in chemical change.   For example, the formation of an exsolution structure involves no addition or subtraction with respect to a grain's surroundings, but the conversion of FeS to FeS_2 requires gross addition of sulfur or subtraction of iron.   This involves a system that is open on a scale larger than the grain.

Replacement may involve cations, anions, or both.   It may or may not be oriented with respect to the crystal structure of the host.   There appears to be no information on the degree of coherence between the structure of the host and replacing substance (*metasome*), though this is probably poor or absent in many cases. In general the process is "volume-for-volume" rather than "atom-for-atom," and it may be highly selective. The resulting structures may be conveniently grouped as *massive, oriented,* or *eutectoid.*

1. Massive replacement In this case the metasome grows into the host apparently uninfluenced by any directional properties of the latter, as in Fig. 9-13, E1. Replacement usually proceeds from a matrix grain boundary, a cleavage, or a crack and develops as an irregularly rounded embayment or a scalloped front. Some sphalerite embayments in pyrite and arsenopyrite are probably of this type, and there are many examples among secondary ores. Covellite and chalcocite may replace primary copper minerals, galena, sphalerite, and others in this way; secondary pyrite or marcasite replacing pyrrhotite often does so similarly, leading to the development of a rim of FeS_2 about an FeS nucleus, often referred to as a *bird's eye* structure; and oxidation of iron-bearing minerals frequently proceeds in the same fashion.

In some cases the metasome may develop in the core or within certain zones of the host rather than as a grain-boundary rim. Some of the core and zoned arrangements of galena and sphalerite in pyrite of Mount Isa, Queensland, have been ascribed to this, and if the interpretation is correct they are fine examples of zonal replacement. These were described by Grondijs and Schouten (1937) and were regarded by these investigators as due to zoned nonstoichiometry in the pyrite, leading to correspondingly zoned variation in susceptibility to replacement. The structure, however, is open to an alternative explanation: the zones may represent original microconcretions composed principally—though not entirely—of pyrite that has coarsened and developed faces during regional metamorphism. In this case the present galena and sphalerite zones would simply represent original concretionary layers of this composition.

Massive replacement can be highly selective, as the first of the above hypotheses implies. Where chalcopyrite and pyrite occur together, covellite and chalcocite frequently replace the chalcopyrite without any modification of the pyrite, which remains as sharp crystals surrounded by the secondary sulfide. Galena associated with sphalerite may be selectively replaced by secondary copper sulfides, and bornite in chalcopyrite may act similarly. Selective replacement is probably the rule rather than the exception.

2. Oriented replacement In some cases the metasome shows an orientation dependence with respect to the host. This is usually not so distinct as that of precipitates but may be quite well developed nonetheless.

Perhaps the best example is provided by the oxidation of magnetite to martite (the term for hematite formed in this way) along the octahedral planes. The boundaries of the old and new phases are sharp and straight, and in this case it may be difficult to distinguish between precipitated and replacement structures. Secondary copper sulfides sometimes develop at a chalcopyrite grain boundary and then extend along {111} planes in the host to give parallel and reticulated (*lattice replace-*

ment) structures. This may combine with some replacement independent of orientation to yield a structure somewhat reminiscent of a eutectoid. The cleavages of galena, pentlandite, and various other oxide and sulfide minerals may also influence the development of orientation-dependent replacement structures.

3. Eutectoid replacement Solidification of eutectics often yields distinctive *eutectic structures*. These have the form of intimate intergrowths of plates, rods, interlocking club-like bodies, or spiral cones.

Clearly the development of a eutectic intergrowth involves the contemporaneous formation of two solid phases at the expense of a third, liquid, phase. It is thus, in a way, a replacement process. A number of theories have been put forward to explain the intimacy and regularity of the structures, and it is only with the comparatively recent recognition of the importance of energy-orientation relationships that the problem has been resolved.† An early suggestion was that the two solid phases crystallized alternately. In this, one of the phases would nucleate and grow first. This would lead to an enrichment of the surrounding liquid in the second component, which would then proceed to nucleate and grow as a second phase upon the first, and so on. The ensuing process was thus seen as the alternating formation of two sets of thin plates or other forms. Another suggestion was that some alloys formed by the rapid dendritic growth of one phase and the entrapment—and subsequent solidification—of the second phase in the interstices between the branches of the first. Subsequent analysis and experiment has shown that in most cases growth of the two phases is simultaneous and that their structural relations are due to systematic adherence to, or avoidance of, particular crystallographic orientation relationships. The structure begins as a duplex nucleus, and this is followed by the development of branches or other forms such that:

1. Each phase retains continuity with the original nucleus.
2. Spacing between branches conforms to the requirements of diffusion and the release of heat of solidification.
3. Morphology follows the requirements of surface-energy–orientation relationships.

Eutectics involve the growth of solid phases at the expense of a liquid phase. *Eutectoids*,‡ on the other hand, involve growth of new solid phases at the expense of an older *solid* phase (Fig. 9-13, E2). The nature of eutectoid growth in metals has been stated by Smith:

> A necessary requirement for the growth of a well-formed lamellar eutectoid structure is that the interface between the two growing phases and the parent phase be an incoherent one. Like the liquid in the case of a eutectic, it must provide a ready channel for diffusion but introduce no crystallographic restraint. In polycrystalline material, suitable

† An excellent treatment of eutectic growth and structure in metal alloys is given by G. A. Chadwick, Eutectic alloy solidification, *Progr. Mater. Sci.*, vol. 12, 1963.
‡ Often referred to as *pseudoeutectics* in the mineralogical literature.

interfaces always pre-exist as grain boundaries. The transformation commonly begins by a coherent fluctuation in one grain giving a low-energy nucleus which then grows incoherently into its neighbour, not into the grain in which it first formed. Although the two new phases in the eutectoid are often closely related to each other to minimize the energy between them, they of necessity avoid all low-energy crystallographic relationships with the grain into which they are growing, for such interfaces have much lower diffusion rates (1964, p. 42).

Eutectoid intergrowths are well known among sulfide and related minerals, and a list of recorded binary and ternary groupings is given in Table 9-5. Most of the minerals concerned are cubic or derivatives of cubic structures, though other

Table 9-5 Eutectoid pairs and groups observed in natural oxide and sulfide-type ores

Components	Investigator
Bornite (h)†-chalcocite	Lindgren (1930)
Bornite (h)-chalcopyrite	Van der Veen (1925)
Bornite (h)-galena	Lindgren (1930); Schwartz (1930)
Bornite (h)-native Ag	Guild (1917)
Bornite (h)-stromeyerite	Guild (1917)
Galena (h)-argentite	Schwartz (1930)
Galena (h)-cosalite	Anderson (1934)
Galena (h)-covellite	Schwartz (1930)
Galena (h)-meneghinite	Stillwell (1926)
Galena (h)-proustite	Lindgren (1930)
Galena (h)-pyrargyrite	Anderson (1934)
Galena (h)-stromeyerite	Guild (1927)
Galena (h)-tetrahedrite	Lindgren (1930)
Gersdorffite (h)-niccolite-maucherite	Hawley, Stanton, and Smith (1961)
Gersdorffite (h)-niccolite-maucherite-chalcopyrite	Hawley, Stanton, and Smith (1961)
Gersdorffite (h)-niccolite-pyrrhotite	Hawley, Stanton, and Smith (1961)
Gersdorffite (h)-maucherite-pyrrhotite	Hawley, Stanton, and Smith (1961)
Gersdorffite (h)-niccolite-pyrrhotite-chalcopyrite	Hawley, Stanton, and Smith (1961)
Marcasite (h)-native Ag	Neumann (1944)
Niccolite (h)-chalcopyrite	Lausen (1930)
Niccolite-pyrrhotite	Michener (1940)
Niccolite-maucherite	Hawley and Hewitt (1948)
Niccolite-maucherite-chalcopyrite	Hawley and Hewitt (1948)
Owyheeite (h)-pyrargyrite	Anderson (1934)
Sphalerite (h)-chalcopyrite	Lindgren (1930)
Sphalerite (h)-galena	Lindgren (1930)
Sphalerite (h)-jamesonite	Colony (1928)
Tennantite (h)-stromeyerite	Lindgren (1930)
Tennantite (h)-chalcopyrite	Lindgren (1930)
Tetrahedrite (h)-galena	Anderson (1934)
Tetradymite (h)-altaite	Lindgren (1930)

† Denotes apparent host phase.

structures and symmetries are certainly involved. They appear to result from at least two "transformation" processes:

1. "Replacement by substitution," involving the addition of new components from an external source and the conversion of the host phase to two new (eutectoid) phases
2. "Replacement by decomposition," involving breakdown of a preexisting phase, loss of components, and conversion of the older phase to a eutectoid intergrowth of two or more new phases

The first of these appears to be the more common of the two and results from replacement processes taking place during primary ore deposition or during later weathering and secondary ore formation. The second, although less common, has been identified as the process responsible for the development of some of the important arsenide-sulfide eutectoids of the nickel ores of Sudbury, Ontario (see Hawley, Stanton, and Smith, 1961).

The Sudbury arsenide textures are beautiful examples of eutectoids formed by breakdown and are the more interesting because some of them have been reproduced experimentally—by thermal decomposition. The groupings observed in nature are given in Table 9-5. All occur here and there in the arsenic-rich selvages developed sporadically along the margins of some of the major nickel ore bodies. The gersdorffite of these selvages occurs in two forms:

1. Idiomorphic and subidiomorphic crystals and compact anhedra
2. Spongy and cellular masses of irregular outline

The spongy gersdorffite of type 2 occurs only in association with the arsenide-sulfide eutectoids, and conversely the latter are found only in the vicinity of the spongy gersdorffite. One of the simple natural eutectoids at Sudbury is maucherite-pyrrhotite, which partly pseudomorphs subidiomorphic crystals of gersdorffite. An artificial niccolite-maucherite intergrowth can be formed readily at the expense of natural gersdorffite by heating the gersdorffite at 750°C for 24 hours in arsenic vapor. The first reaction appears to have been

$$Ni_{1-y}Fe_yAsS \rightarrow (1 - y)NiAs + yFeS + yAs + (1 - y)S$$

Ferroan gersdorffite $\qquad$ Niccolite $\qquad$ Pyrrhotite

and the second,

$$12NiAsS(+As) \rightarrow NiAs + Ni_{11}As_8 + 3As + 12S$$

CRYSTALLOGRAPHIC TRANSFORMATION STRUCTURES

A number of elements and compounds are polymorphous, and reference has already been made (Chaps. 3 and 4) to high-low temperature polymorphism among a number of sulfides. Inversion from one polymorph to another may be induced by change in temperature or pressure, or both. High-pressure polymorphism is an important consideration in speculations on the mineralogy of the earth's interior, such transformations always involving change to a higher density form with increase in pressure.   High-temperature polymorphism is encountered in lavas and certain high-temperature ores and is of great practical importance in metals and other industrial crystalline materials.   Inversion from one crystal lattice configuration to another involves distortion and shear, and these may produce characteristic structures (Fig. 9-13, F1 and F2).

The most common of these is the inversion twinning induced by the cubic → tetragonal inversion in certain substances of the diamond-structure group, of which several common sulfides are, as we have noted, prominent members.   The high-temperature disordering of the metal components of chalcopyrite and stannite, for example, causes the symmetries of these minerals to change from tetragonal to cubic.   Fall in temperature leads to reordering, and the shearing involved induces the formation of inversion twins.   In general these have a somewhat patchy distribution, producing a mosaic of areas with differently oriented sets of twins (Fig. 9-13, F2).   Such twinning often provides a useful geothermometer.   Approximate inversion temperatures of chalcopyrite and stannite are 550 and 600°C, respectively.

RECOMMENDED READING

Margolin, H. (ed.): "Recrystallization, Grain Growth and Textures," American Society for Metals Seminar, 1966.

Shockley, W. (ed.): "Imperfections in Nearly Perfect Crystals," John Wiley & Sons, New York, and Chapman and Hall, Limited, London, 1952.

Smith, C. S.: Some elementary principles of polycrystalline microstructures, *Met. Rev.*, vol. 9, no. 33, pp. 1–48, 1964.

Stanton, R. L.: Mineral interfaces in stratiform ores, *Trans. Inst. Mining Met.*, vol. 74, pp. 45–79, 1964.

10
Structures due to Deformation and Annealing

EARLY OBSERVATIONS ON DEFORMED ORES

The fact that some ores have been deformed during regional metamorphism and related events has been recognized for a long time. One of the earlier references to it in English is that of Emmons (1909) in connection with some of the Appalachian deposits, particularly those of Maine and New Hampshire. The metamorphosed nature of these deposits seems to have been suspected very much earlier, however, and was at least hinted at by J. D. Whitney in 1854. Emmons noted of them that:

> The centre of the lode is often massive sulphides which appear to be in all respects like any unmetamorphosed sulphide deposit in the younger rocks of the West. In hand specimen it shows no sign of schistosity or lamination, but along the margins of the deposit where portions of the country rock are involved with the sulphides, thin pyrite bands and quartz bands alternate with bands of schist and all are crumpled together (1909, p. 758).

Shortly after this Lindgren and Irving published their conclusion that the deposits of Rammelsberg were "of epigenetic origin, in which the structure of the ore has been profoundly changed by dynamo-metamorphism" (1911, p. 313). Again, however, this simply reaffirmed the earlier suspicions of others. A number of European geologists had, well back in the previous century, regarded the deposits as of sedimentary origin and hence, *ipso facto*, as being metamorphosed.

In 1917 Uglow published a description of "gneissic galena ore" from Slocan, British Columbia, in which he showed quite clearly that the foliated ore of the Slocan district owed its present structure to deformation. While this conclusion was not, of course, new in a general sense, Uglow's paper was important for its reference to the likely effects of deformation in modifying paragenetic features. His statement was highly prophetic, and anticipated by some 30 to 40 years the widespread recognition of this problem:

> The paragenesis of minerals is an important study which aids in the determination of the origin of ores. Some of the chief criteria used in outlining the paragenesis are: (1) The presence of minerals of one kind enclosed in those of another kind; (2) the gradual substitution of one group of minerals by others by means of replacement, and (3) the occurrence of certain minerals in small stringers and veins intersecting other minerals. Provided an ore deposit has not suffered from dynamic metamorphism and is still in its original condition, the time relations of the various minerals may be determined by the use of the above and other criteria. On the other hand, if an ore body is located in an area which has been subjected to great earth stresses, it is natural to expect that its physical character, as well as that of the country rocks, may have been changed. If the dynamic metamorphism were not intense enough to produce flowage conditions for all the constituent minerals of the ore body, the result would be the fracturing of the more resistant group and the flowage of the less resistant ones. It is quite possible that the oldest minerals of the first phase, that is, previous to the metamorphism, might adapt themselves to the new conditions by flowage, while the younger minerals of the first phase might be deformed by fracture. In such a case, the minerals that were subjected to flowage conditions would appear to be the younger in the anamorphosed ore deposit, because they flow about and fill in the cracks in the minerals that were deformed by fracture. Therefore, unless it were known that the ore body were an anamorphosed one, the paragenesis of the minerals as worked out from their mutual physical relations would be exactly the reverse of the correct order. Consequently, secondary deformation of ore bodies of mixed minerals may produce conditions which suggest a mineral paragenesis quite different from the true one. Therefore in dealing with the ore bodies of pre-Cambrian age, in particular, it seems to the writer that the possibility of secondary deformation of the ores is one which should be emphasized (1917, pp. 657 and 662).

The work of Uglow was reaffirmed by Bateman, who, as we noted in Chap. 2, observed the extreme distortion and flowage of the Slocan galena and who clearly recognized that "age relations of different minerals . . . in shattered ores of this type may be no criterion whatever of the original sequence of mineral deposition —a fact which should not be overlooked in building deductions upon mineral sequences" (1925, p. 562). In the same year Waldschmidt attributed the structures of much of the Coeur d'Alene ores to deformation.

The processes visualized by all these investigators, from Emmons to Waldschmidt, are fairly simple. The ores were regarded as having been subjected to compression or shearing, or both, leading to fracture in brittle minerals and plastic flow in the softer, more ductile ones. Among the former, pyrite, pyrrhotite, arsenopyrite, and sphalerite were the principal common minerals. Among the latter, galena was recognized as being the most important, though chalcopyrite appeared to yield plastically to some extent. Because of the contrasting behavior of these "hard" and "soft" minerals, deformation led to the development of distinctive textures. Ores containing only brittle minerals fractured and fragmented, only later to be "healed" by minerals such as quartz. Ores consisting substantially of galena, on the other hand, "flowed" by movement on slip planes and by lattice bending, developing foliation, intrusion veinlets and related structures. Where ores contained both hard and soft members, the former fractured and then became strung out in the latter as it flowed normal to the direction of compression, forming strongly foliated, often "knotty" ore, of clearly schistose or gneissic aspect.

The first reference to recrystallization of natural sulfides as a process distinct from prior deformation seems to be that of Newhouse and Flaherty (1930), who reviewed certain features of replacement and deformation banding in several well-known ore bodies and who noted clear evidence of recrystallization in both galena and chalcopyrite from Rammelsberg and Coeur d'Alene. In both ores initial deformation was revealed by the development of flattened and ribbon-like crystals. Galena was generally highly flattened and chalcopyrite less so but with its major dimension as much as 15 times the minor. Both minerals, however, were found to show the development of many small equidimensional grains within the outlines of the preexisting flattened ores, and it was recognized that these represented the formation of strain-free material at the expense of the earlier deformed matrix.

Since this time there have been a number of references to natural deformation and recrystallization of sulfides, among the more important of which are those of Ramdohr (1950) and Richards (1966) on Broken Hill, Australia, Kanehira (1959) on some of the Japanese copper ores, and Schachner-Korn (1948), Grigor'yev (1958), Siemes (1964), and Stanton (1959, 1964) on more general aspects. The importance of deformation in the development of many structures in ores is now well established.

EARLY EXPERIMENT AND THEORY

These observations led naturally to experimentation and the formulation of theories of deformation to explain the features found in ores.

The first studies of plastic deformation in "mineral" crystals were those of Reusch, in 1867, on sodium chloride. Since that time extensive studies of deformation in metals and ceramics, as well as in the alkali halides, have been carried out, largely in connection with industrial applications. In particular, accumulation of data and the development of theory have accelerated greatly since the 1930s, largely due to the recognition of dislocations.

Apart from Reusch's experiments on rock salt, and some related work reported by Mügge in 1898, the first systematic work on mineral deformation appears to have been that of Adams and Nicolson, who reported an experimental investigation into the flow of marble in the *Philosophical Transactions of the Royal Society, London*, in 1901. This work was essentially "phenomenological"—i.e., concerned with exploratory experimentation—rather than theoretical and, of course, involved crystalline aggregates rather than single crystals.   The same approach was followed by Adams in his paper "An experimental investigation into the action of differential pressure on certain minerals and rocks, employing the process suggested by Professor Kick" (1910), which dealt with deformation in both single- and multiphase polycrystalline aggregates.   Adams noted that under the conditions of his experiments some minerals (diopside, calcite) showed evidence of internal movement, while others deformed essentially by intergranular movement—cataclasis and crumbling.   Under comparatively low pressure he produced indistinct foliation in some of the harder rocks.

Following this there were few developments for almost 20 years, leading Buerger to comment in 1930 that although the metallurgists had made considerable progress in demonstrating that "the plastic properties of metals are dependent upon, and to a remarkable degree, predictable from their respective crystal structures," the same could not be said about the then current state of knowledge of the mechanisms of plastic deformation in crystals other than those of metals.   Buerger had, however, already made several important contributions himself—on translation and strain hardening in crystals and on plastic deformation of ore minerals, all published in 1928.   These studies, together with another published in 1930, showed that ore minerals deformed by slip, twinning, and fracture and that the mode, planes, and directions of yielding were tied to crystallography from one substance to another.   He showed that galena tended to deform by slip on {100} planes along ⟨110⟩ directions, that sphalerite deformed by twinning on {111} planes, and that certain other sulfides yielded characteristically.   Buerger's findings unfortunately were not followed by the burst of activity that has characterized the fields of metals and ceramics since 1930, and the only precise experimental work carried out on ore minerals since appears to be that of Lyall (1966) and Lyall and Paterson (1966) on galena.   These have explored the deformation of galena in much greater detail than Buerger set out to do and showed that galena deformed readily by slip in several directions, by twinning, and by kinking.

DEFORMATION IN SINGLE CRYSTALS

All this leaves no doubt as to the possibility, and indeed the importance, of deformation in ore minerals.   However, as in the previous chapters on growth structures, its detailed consideration is greatly hampered by the fact that comparatively little work has been done on the deformation of this class of materials explicitly.   One is forced again to accept that ores are simply a class of crystalline materials and then to consider them in the light of *likely* analogies with substances—metals and ceramics—whose behavioral characteristics are better known.

Such substances are known to yield by five principal mechanisms: slip (translation gliding), twinning, lattice bending, kinking, and fracture.   They may also deform by creep, a somewhat ill-defined process that operates very slowly and in accordance with an exponential law at stresses lower than those inducing slip. Fracture may take place by cleavage or irregular rupture.   Slip, twinning, bending, and kinking constitute *ductile* failure, whereas cleavage and irregular fracture are manifestations of *brittle* failure.

SLIP

The nature of ductile failure in crystals other than metals was stated clearly by Buerger in 1928. In substances containing more than one kind of atom or ion, the possibility of movement, and its direction, is dictated by the configuration of the lattice and by charge distribution. Where it can be assumed that the lattice planes are "flat," movement is dictated by charge distribution. Whether the bonds are ionic, covalent, or a combination of the two, the particles at the lattice points are all characterized by some degree of positiveness or negativity. Although cohesion is aided by such forces as gravitational and magnetic forces, the electrostatic forces are quantitatively the most important. In ductile failure the bonds suffer no permanent breakage: as one lattice layer moves with respect to another, bonds are lost and regained consecutively by each particle so that the first and final dispositions of particles and of bonds are identical. There are two requirements for movement of this kind:

1. The direction of movement is such that the initial and final disposition of the differently charged ions is the same.
2. No inherently unstable charge configuration is approached during the movement in question.

The latter is illustrated diagrammatically in Fig. 10-1 (after Buerger, 1928). In (*a*) it may be seen that movement of a positive particle *x* through a distance *m* to a position *y* opposite a negative particle yields a final disposition of charges identical with the original, without approaching an unstable configuration in the process. On the other hand, in (*b*) it is clear that while movement of particle *x* a distance *n* to a position *y* opposite a negative particle gives a final disposition of charges identical with the original, an unstable situation would be encountered in doing this. Clearly such movement would involve momentary juxtaposition of like charges: the large total repulsive force involved would cause immediate fracture, and the ductility of movement (*a*) would be lost.

This leads to the conclusion that ductile movement can take place only along rows of consecutively like-charged particles. The direction or directions of possible movement are therefore fixed and characteristic for a given crystalline substance. Each is termed a *slip direction*. Slip can take place in any plane containing a slip direction, and those that do become so activated are termed *slip planes*. While several crystallographic planes may contain the slip direction, the development of any one of them as a slip plane depends upon their relative resistance to slip under the conditions prevailing. Since ease of movement depends generally on the strength of bonding across the plane in question and since this tends to decrease with increase in bond length, planes of greatest spacing are those along which slip most frequently occurs. A plane of slip and a slip direction lying in that plane may be said to constitute a *slip system*.

In crystals of rock-salt structure the slip directions are parallel to the face

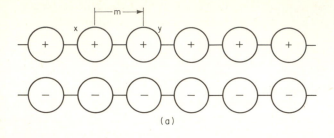

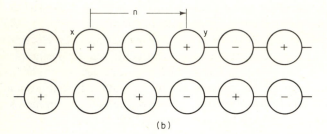

Fig. 10-1 The nature of slip in crystals, showing (*a*) a possible direction and (*b*) an impossible direction. (*Adapted from Buerger, Amer. Mineralogist, 1928.*)

diagonals and hence are six in number.   The plane of widest spacing is the {100}, the next is the {110}, and that of least spacing is the {111}.   The most frequent slip system found in galena is thus the direction $\langle 110 \rangle$ in the cube plane.   Face-centered-cubic structures of diamond structure yield along the {111} (close-packed) plane, probably in the $\langle 11\bar{2} \rangle$ direction.   In conformity with this, hexagonal close-packed structures have their basal (0001) plane as the slip plane and yield in the $\langle 2\bar{1}\bar{1}0 \rangle$ direction.

In metals it appears that the choice of slip plane is often influenced by factors such as temperature, impurities, and the nature of earlier deformation.   The slip direction, however, is quite constant, and in fcc, bcc, tetragonal, hexagonal, and trigonal metal crystals the direction of slip is always that of the most closely spaced row of atoms in the lattice.

Slip in metals appears to be quite abrupt and is accompanied by an audible tick when the deformation is small.   In experiments it is identified by the appearance of *slip lines* that trace the intersections of the slip planes with the surface of the specimen.   The lines are in fact fine ridges formed by the upward movement of one part of the crystal with respect to the other.   Where slip occurs on only one set of planes, the lines usually appear very straight.   Where two or more planes are involved or, particularly, where little patches of slip occur on numerous adjacent planes of the one family, the slip lines may take on a wavy appearance.

Individual slip lines usually show a tendency to cluster into groups, which then constitute *slip bands*.   Distinction between individual lines and such bands is often a matter of magnification, and what appear to be lines under low magnification may be resolved into bands at higher powers.   In certain cases there appears to be a relationship between the number of lines per band and the distance between bands.   In highest purity aluminum, for example, it has been found that single lines form first and that, as deformation proceeds, further lines form until a minimum spacing is achieved that depends on temperature.   Further stress results in adding to the number of lines per band.

The displacement on individual slip lines appears to vary, as might be expected.   Measurements on metals have given distances ranging from about 0.2 to 0.1 mm.   In all such cases the movement amounts to hundreds or thousands of atom diameters.   Whether this all occurs on visible slip lines or whether part is contributed by minute yielding in the zones between them is not known.

For any given crystalline substance there appears to be a clearly defined stress at which yielding proceeds at an appreciable rate.   This is termed the *critical resolved shear stress for slip*.   Below this critical stress any yield is very slow and is referred to as *creep*.   Above it, yielding is quite rapid and is called *slip* in the present sense.   While a variety of stresses are required to cause slip in differently oriented specimens of a given crystal, it has been found among metals and ceramics that, when resolved on the slip plane and in the slip direction, the stress required to just induce slip is always the same for the substance concerned.   This is the critical resolved shear stress and is readily calculated for any given load as follows.   If, as in Fig. 10-2, a force F is applied to a crystal of cross-sectional area A, and the slip plane is inclined at an angle Θ to the compressional axis, we have

$$\text{Area of slip plane} = \frac{A}{\cos \Theta}$$

$$\text{Force/unit area} = \frac{F \cos \Theta}{A}$$

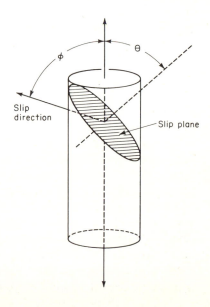

Fig. 10-2 Geometry of resolved shear stress. (*Adapted from Barrett, "Structure of Metals." Copyright McGraw-Hill Book Company, 1943 and 1952, with the publisher's permission.*)

If ϕ is the angle between the slip direction and the axis of compression, the resolved shear stress in the slip direction is

$$\tau = \frac{F}{A} \cos \Theta \cos \phi$$

By appropriate variation of Θ, ϕ, and load (F), the relation between the critical resolved shear stress and orientation may be determined.

Neglecting surface effects, this is liable to modification by impurity and temperature. Impurity *increases* the critical stress. In general the hardening effect is greater for soluble impurities than for insoluble ones—i.e., in alloys in which the impurity forms a solid solution in the main phase, hardness is increased much more than in those in which the impurity forms a distinct and separate phase. As might be expected the critical resolved shear stress for metals decreases with temperature and drops to zero at the melting point. The relationship is not, however, a simple one.

The development of slip lines in sulfides has been shown very well by Buerger (1928) and by Lyall (1966) and Lyall and Paterson (1966). However, virtually all this work has involved galena, and little is known of the behavior of other sulfides, particularly at high hydrostatic pressures.

Before leaving the subject of slip it is worth noting that slip in metals is accompanied by *strain hardening*. Although there is no quantitative information on such an effect in sulfides and related minerals, there is no doubt that strain hardening does occur. The incidence of natural strain hardening, and its experimental elimination, has been investigated by Stanton and Willey (1970). This is discussed further on in this chapter, under "Annealing" (page 285).

TWINNING

The development of twins by crystallographic reorientation during growth has already been discussed in Chaps. 8 and 9. Twins may also be developed by homogeneous shear during deformation, and these are referred to as *deformation twins*, or *secondary twins*.

The mechanism of movement and directional constraints are essentially the same for deformation twinning as for slip. Movement occurs along lattice planes in only those directions that do not involve the development of unstable configurations. The difference between the two phenomena is in the *amount* of movement on a single lattice plane. Whereas in slip such movement may be of the order of thousands of lattice spacings on a single plane, in twinning the displacement is of one lattice spacing only per lattice plane. As a result every atom moves a distance proportional to its distance from the twinning plane in the ideal case. The development of a twin by the apparent rotation resulting from such homogeneous shear is illustrated in Fig. 10-3. In metals such twins appear first as thin lamellae that broaden progressively as deformation proceeds.

Figure 10-4 shows their formation and growth in zinc: *A* represents an early stage of deformation in which the twins are quite thin, and *B*, *C*, and *D* show the progressive broadening of the same twins resulting from continued deformation.

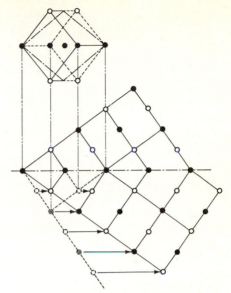

Fig. 10-3 Shear movements for twinning in a body-centered cubic crystal. Twin plane is normal to the page, intersecting it along the heavy dot-dash line in the lower drawing. (*From Barrett, "Structure of Metals."* Copyright McGraw-Hill Book Company, 1943 and 1952, with the publisher's permission.)

Among the common sulfides, sphalerite, chalcopyrite, and pyrrhotite commonly deform by twinning. Secondary twinning is also observed in galena, though this mineral slips so easily that the twin configuration is generally lost almost immediately. Characteristics of deformation twins in these minerals are their rather irregular spacing within individual grains and their incoherent nature. By the latter is meant that the twins in question do not extend right across the grain but taper off within it, requiring misfit, or incoherence, with the enclosing crystal lattice. This internal termination and spindle shape distinguish deformation twins from growth and annealing twins, which maintain sharply parallel sides and extend right across the crystal in question from edge to edge. In addition many deformation twins are involved in lattice bending and are bent themselves. Sphalerite was found by Buerger to twin on the {111} planes—presumably by homogeneous {111} $\langle \bar{1}2\bar{1} \rangle$ shear. Interestingly this type of twinning is not conspicuous among fcc metals. Barrett (1952) suggests that this may be due to the great similarity of forces involved in twinning and slip in this particular case. As the atoms of the moving layer find the same number of nearest neighbors at the same distance at any of the permitted stopping places, it appears that the untwinned state is to only the very smallest degree energetically more favorable than the twinned state. Thus it seems likely that the shifting layers would stop in more or less unpredictable positions and that usually they would be irregularly spaced, leading to faulty stacking rather than a twin.

As in slip, there is a *critical resolved shear stress for twinning*. This follows the same principles and is determined in the same way as for slip. Unlike slip, twinning in metals does not appear to be particularly sensitive to impurity. Apparently the distortions due to foreign atoms are not so effective in inhibiting movements of one lattice spacing as they are those of large numbers of such spacings. This seems intuitively reasonable.

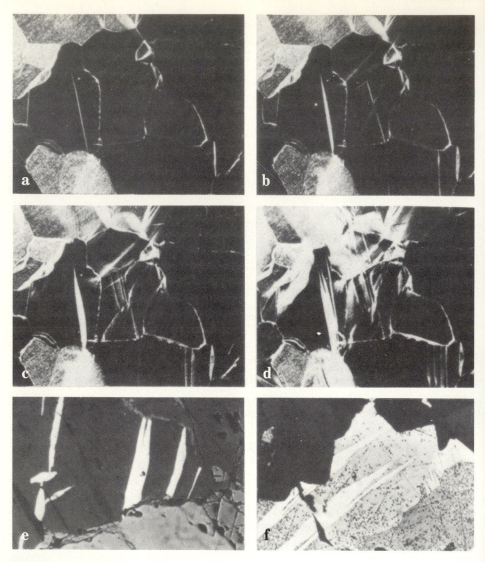

Fig. 10-4 Deformation twinning in metal and metallic sulfide. *A, B, C,* and *D* show progressive development of twins in zinc. (*From Barrett, "Structure of Metals." Copyright McGraw-Hill Book Company, 1943 and 1952, with the publisher's permission.*) *E* and *F* show analogous twins in natural pyrrhotite and chalcopyrite.

Such a principle has been used by Buerger (1928) to account for the different yield mechanism of sphalerite as compared with, for example, galena and chalcopyrite. The latter two minerals are not notably impure in nature, and hence their slip planes and directions might be expected to be essentially perfect from a physical point of view. There is no impediment to lattice movement, and hence these minerals readily deform by slip. Sphalerite, however, is characteristically impure and might be expected to have various structural irregularities due to disparity in size between zinc and the various substituting atoms. Specifically this would lead to "roughness" of slip planes and directions, which would inhibit movement. Thus according to Buerger, sphalerite deforms by the development of a series of twins rather than by movement on a small number of slip planes. However, while this is a very reasonable explanation it now appears that at least much of the twinning in natural sphalerite is of growth-annealing type and that deformation twins may be no more abundant here than in chalcopyrite.

LATTICE BENDING

Deformation is not always as simple and regular as that involved in slip and twinning, particularly in individual grains of polycrystalline aggregates. Here stresses may become quite heterogeneous, leading to the irregular development of a range of crystallographic orientations within individual grains. This rather heterogeneous straining, which for the most part is not defined by sharp surfaces, is a familiar feature of mica, quartz, and other translucent minerals, where it manifests itself in bent cleavages, undulose pleochroism, or particuarly, undulose extinction.

Such irregular lattice bending is well known in metals and commonly is developed conspicuously in sulfide and related minerals. Galena often shows very beautiful bending of cleavages on etched polished surfaces, and indeed this is frequently obvious in the hand specimen in coarse ores, such as those of Broken Hill. Molybdenite also displays bent cleavage, and in sections transverse to the cleavage plane has an appearance remarkably reminiscent of the micas. Highly pleochroic minerals such as covellite, molybdenite, and stibnite reveal bending by pleochroism, and all these, together with pyrrhotite among the common sulfides, display conspicuous undulose extinction where the ore concerned has been stressed.

KINKING

When a specimen is compressed it may yield by kinking rather than by simple slip or twinning. Such kinking occurs where a plane of weakness (i.e., a potential slip or twin plane) is at only a small angle to the axis of compression. The kink develops as a well-defined band, the margins of which are usually nearly, but not quite, parallel and whose length is at a high angle (approaching 90°) to the axis of compression (Fig. 10-2). Kinking is particularly common in galena, stibnite, and molybdenite and is also quite common in pyrrhotite, sphalerite, and chalcopyrite. Kink bands may be easily mistaken for slip bands (to which they are related), but it should be remembered that whereas the planes of movement in slip bands are essentially parallel to the margins of each band, in kinks the slip movements are at quite a high angle to the margins of the kink concerned. Careful observation (particularly of galena, after etching) often reveals the fine high-angle slip lines within the kinks.

In some cases kinks take the form of single bands, but in most cases they are multiple. In galena this can lead to a clearly discernible zigzag pattern of fine slip lines across the kinked surface, each change of direction of the zigzag being delineated by the sharp boundary of the two contiguous kink bands concerned.

FRACTURE

Actual rupture may take place by cleavage, shear fracture, or intergranular fracture.

1. *Cleavage.*   Although many minerals are readily made to cleave in the laboratory, cleavage does not seem to be a common mechanism for yielding under natural conditions.   Those minerals capable of cleaving do so along crystallographic planes of low indices characterized by comparatively wide lattice spacing and weak bonding.   In nonmetallic substances, layer lattices such as graphite and molybdenite always cleave parallel to the layers.   Cleavages do not break radicals in ionic crystals or molecular groupings in covalent crystals.   In purely ionic crystals—which have no radicals—cleavage generally develops in such a way as to expose planes of anions.   Whether a crystal fails first by slip, twinning, or cleavage depends on which of the critical resolved shear stresses is first exceeded.   This is, as already mentioned, affected notably by temperature and composition in metals.   Decrease in temperature increases resistance to slip and hence leads to a greater tendency to brittle failure by cleavage rather than ductile failure by slip.   A similar tendency develops with increase in the amount of impurity.

      Few minerals show clear failure by cleavage.   Pentlandite commonly does, in which case the cleavages are frequently "healed" by associated sulfide, particularly pyrrhotite or chalcopyrite.   Galena may do so, though shear (see below) is much more common.   Occasionally pyrite appears to have cleaved, though this mineral usually fails very irregularly.

2. *Shear.*   A frequent mode of failure is by "shearing off" as a result of continued slip.   For example, galena, stibnite, and a number of other soft minerals deform first by extensive slip, and this then gives way to a gross rupture by shearing.   The resulting break is, of course, far less regular than that produced by cleavage.   In the harder minerals, such as pyrite, arsenopyrite, cobaltite, and others, it appears that shear rupture occurs preceded by little or no slip.   Buerger found no experimental evidence for slip in pyrite, which breaks simply by brittle shear followed by granulation.   Microshearing is extremely common in galena.

3. *Intergranular fracture.*   In this case failure occurs by cracking along grain boundaries, a process that appears to be of little importance in ores.

CREEP

If a substance is subjected to a continuing small stress—one distinctly lower than that required for slip—it may yield by very slow plastic flow.   Such a phenomenon is termed *creep*.   Creep is of considerable technical importance in both metals and ceramics and has been studied in detail in both of them.   Although little is known of the phenomenon in minerals—particularly sulfides—it may well be of importance here and invites investigation.

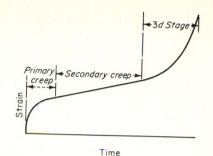

Fig. 10-5 The three stages of creep. (*From Barrett, "Structure of Metals." Copyright McGraw-Hill Book Company, 1943 and 1952, with the publisher's permission.*)

Measurement of creep is normally carried out at elevated temperatures and with constant stress applied over an extended period. Initial application of stress causes elastic yielding that is followed by a period of gradually decreasing strain referred to as *transient*, or *primary*, *creep*. This usually is followed by a longer— perhaps *much* longer—period during which deformation continues at a slower and essentially constant rate. This is referred to as *steady-state*, or *secondary*, *creep*. If the stress is sufficiently high this may be followed by a short period of accelerated creep, known as *tertiary creep*, that terminates in fracture. The general relationships of the three phases of the process are shown in Fig. 10-5. As might be expected the rate of creep is increased by both rise in temperature and increase in load.

Creep occurs in both single crystals and polycrystalline aggregates. Study of zinc crystals has shown that primary creep is accompanied—and presumably at least partly accomplished—by slip and twinning. In polycrystalline aggregates these mechanisms appear to be augmented by grain-boundary flow. This divergence of mechanism may be responsible for the fact that transient creep is "recoverable" in polycrystalline metals but apparently not in single crystals. (By *recovery* is meant that upon removal of the load the newly acquired strain is eliminated, usually in a time about equal to that over which such strain was induced.) Such recoverable creep may be largely that which has taken place by grain-boundary flow rather than by intracrystalline movement. During grain-boundary flow, residual stresses are built up that cause reversed flow when the load is removed, leading to recovery. Similar stresses may build up around slip bands, but presumably these are of a lesser order of magnitude.

Steady-state creep in metals and ceramics also proceeds by slip and grain-boundary flow, but at higher temperatures their relative importance changes. Lattice distortion becomes less important, much more movement is concentrated along grain boundaries, and there is a greatly increased development of subgrains. In some instances creep rate has been found to be heavily dependent on grain size.

At higher temperatures, diffusion may make an important contribution to creep, giving rise to diffusional, or *Nabarro-Herring*, creep. This process is purely theoretical, but seems to be a requirement for the development of certain slow, high-temperature strains in metals and ceramics and for the development of many of the textures of metamorphic rocks. It is presumed that self-diffusion occurs within individual grains, allowing the material to yield to stress by transposition of matter; any residual forces developed within the grains or along grain boundaries are simply eliminated by diffusion. In this way material diffuses from those boundaries suffering a normal compressional force to those areas in tension, causing an elongation or flattening of the grain concerned. A process of this general kind has of course long been thought to be responsible for the development of foliation in many metamorphic silicate rocks.

While it seems almost certain that sulfides and related minerals should deform by creep, there is little or no *definite* evidence for it. No systematic experimental work has been reported so far, though investigations along the lines developed for metals and ceramics should not be too difficult. Largely because of this lack, the evidence of natural material is rather equivocal. Galena shows widespread development of subgrains, which may be a manifestation of creep or partial annealing. One suspects that it may often result from creep but this is no more than an intuitive speculation. Quite a number of ores possess disequilibrium triple-junction and dihedral angles (dealt with in more detail in the following section), and it is possible that these result from extremely slow, low-temperature creep that continues at the present day.

There is little doubt that a better knowledge of creep and creep structures in sulfides would help in deciphering the history of many ores.

DEFORMATION IN POLYCRYSTALLINE AGGREGATES

Although the principles governing the deformation of single crystals must hold in the case of aggregates, the aggregates pose a number of additional complications. The most important factor in this is the occurrence of grain boundaries. Variation in orientation, grain size, and the disposition of phases also contribute substantially to the more complex behavior of an aggregate as compared with a single crystal.

INFLUENCE OF GRAIN BOUNDARIES

Two important matters follow from the existence of a grain boundary: the thin zone of lattice disorder along the boundary itself and the change in lattice orientation going from one grain to the other. Both affect the deformational behavior of the aggregate. Careful measurement of the strain at increasing distances from the grain boundary in coarse-grained tensile specimens by Aston (1927) has shown that there is less deformation near the boundary than in the center of a grain in metals. Apparently the boundary tends to inhibit plastic flow, causing this to take place at the greatest possible distance from it—i.e., toward grain centers. There is also a tendency for the equalizing of strain on either side of the boundary so that if one of two neighboring grains is notably weaker than the other, deformation near the boundary of the stronger grain may exceed that at its center. In inhibiting flow generally, the incidence of boundaries increases the strength of the material concerned.

Strength is also increased by the variation in orientation implicit in the granular structure (the increase being greatest in random aggregates), lessening with increase in degree of preferred orientation. Chalmers (1937) has shown that there is in fact a systematic relationship between resistance to plastic deformation and relative orientation of grains. He took a series of tin bicrystals in which the boundary extended longitudinally through each specimen and in which the relative orientations of the two crystals were systematically varied. Yield point was found to be at a minimum in parallel orientation and at a maximum at maximum tilt. This indicates that when the two slip systems are at high angles it is difficult for yield in one member to be transmitted to the other.

Since a decrease in grain size increases the grain-boundary area for unit volume of material, it might be expected that increasing fineness would be accompanied by an increase in resistance to deformation. This appears to be so in a number of metals, several of which show systematic increases in tensile strength and indentation hardness with decrease in grain size. No information along these lines has been obtained for mineral aggregates.

The most obvious manifestation of deformation, whether this takes place by compression, tension, shearing, or rolling, in polycrystalline aggregates is the flattening, or elongation, of grains. This change in shape—i.e., change in the *configuration of grain boundaries*—may be expressed quantitatively (1) by the ratios of length to breadth in section and (2) by the broadening of frequency distributions of triple-junction-angle values. Many thousands of measurements of grain dimensions in samples of schistose galena from Coeur d'Alene, Idaho, have shown length-breadth ratios from about 1.8 to 3.0, and values similar to these are common in ductile sulfides that have been rendered schistose. The modification of triple-junction angles has already been referred to in Chap. 9. Clearly all angles in a single-phase aggregate are changed from the equilibrium value of 120°, some junctions being modified to two angles less than 120° and one angle greater than 120° and vice versa.

In addition to flattening, deformation may induce preferred orientations analogous to those of "worked" metals and metamorphically deformed silicate and carbonate rocks. As with these other materials, the original lattice orientations of the sulfide grains are changed to new ones in which particular lattice directions are aligned with the principal directions of "flow" in the aggregate. The nature of the final preferred orientation (*deformation texture* of metallurgical usage) is characteristic of the particular sulfide and the relative magnitude of the three principal strains (see Turner and Verhoogen, 1960; Barrett, 1952).

Investigation of preferred orientation in natural sulfide aggregates has been rather limited, though there have been a few notable contributions. Kanehira (1958) has reported that the ore of the Chihara mine in Japan shows distinct preferred orientations in polycrystalline sphalerite, chalcopyrite, pyrite, and hematite and that these orientations are systematically related to preferred orientations in quartz, calcite, muscovite, and chlorite of the enclosing nonsulfide rocks. Pyrrhotite of the Sullivan mine, British Columbia, exhibits a marked preferred orientation, with the basal plane parallel to the bedding of the ore body. At least some of the galena of the deformed veins of the Coeur d'Alene area exhibits a preferred orientation, as does some of the galena from Broken Hill.

INFLUENCE OF IMPURITIES

Impurities may be of two kinds: those existing as discrete minor phases and those in solution. Those existing as minor phases increase or decrease susceptibility to deformation according to their own physical properties and the manner in which they are distributed through the matrix. Hard components, particularly where these aggregate at triple junctions as bodies of high dihedral angle, increase resistance to deformation. Softer components, particularly where these develop low dihedral angles, soften the whole. There is a general tendency for dissolved impurities, on the other hand, to harden metals. The effect of different solutes appears to be related to the difference in the atomic radii of the solute and solvent. Where the solute atom is of similar size to that of the solvent, little lattice distortion is involved, extensive solid solution is possible, and there is little effect on hardness. However, where the solute atom is notably the larger, lattice distortion may be considerable, and hardening occurs. Because such a disparity in size decreases solubility, increased hardness and decreased solubility are related. *Solution hardening* of this kind in metals is detected both in indentation and tensile tests. Little quantitative work has been done on this aspect with sulfides and related minerals, though it seems very likely that analogous effects should occur. Indeed it is almost certain that at least some of the hardness variations in individual minerals detected by indentation tests are due to solution hardening.

GROSS EFFECTS IN TWO- AND POLYPHASE AGGREGATES

In many cases a second, or several, minor phases are present in amounts much too large to be described as "impurities." The effects of these in deformation are much more obvious than the effects of dissolved impurities or of small discrete bodies occurring along grain boundaries and triple junctions in the major phase. It was with deformational features in these distinctly polyphase aggregates that the earlier investigators such as Emmons were chiefly concerned.

The principal effects in aggregation of this kind arise from the coexistence of comparatively hard and comparatively soft minerals. Most commonly galena is the principal "soft" mineral, and this usually constitutes a plastic medium in which the harder constituents move. In nonpyritic lead-zinc ores, sphalerite constitutes hard nuclei about which galena flows. Sphalerite deforms by slip, twinning, kinking, and granulation, but each individual grain usually holds together and develops augen rather than separating out into granules. Galena bends and shears and is drawn out round the sphalerite, the most intense deformation occurring close to the sphalerite. Farther away the galena deforms by twinning, less pronounced bending, and kinking. In ores containing very hard minerals with no pronounced slip or twin planes—pyrite and arsenopyrite are good examples—the hard component deforms virtually entirely by brittle fracture and then separates out in the "flowing" galena, forming strings of sharp granules. The flowage and shearing-out of the galena, the development of augen by the sphalerite, and the formation of these strings of granules all combine to produce a variably schistose structure overall.

ANNEALING

Following deformation, residual stresses remain within the materials in which strain has been induced. *Annealing* is the process by which these stresses and the accumulated strain can, under appropriate conditions, be reduced and eliminated. While it may be thought of in a very general way as a single process, annealing in fact embraces several processes. Each of these makes its own contribution to the elimination of the deformation.

In most cases annealing is accomplished by structural modification. If the material is heated to a temperature permitting atomic mobility, crystal lattices adjust themselves so that in one way or another strained material is replaced by that which is strain-free. This proceeds through a series of stages and leads to the development of a number of characteristic annealing structures. These are obviously closely related to, and indeed are often greatly influenced by, the deformation structures they erase. Thus while they might be regarded as the very antithesis of the structures with which the first part of this chapter is concerned, it is most appropriate to consider annealing phenomena at this present stage.

Again the process must be considered in the light of probable analogies with metals and ceramics. In this connection exhaustive reviews have been written by Beck (1954) and Margolin (1966), both of whom consider the case of metals in great detail. Annealing of these is divisible into three processes: recovery, subgrain growth, and recrystallization.

RECOVERY

This has already been referred to briefly in connection with creep. The term refers to the reversal of flow that may take place following the removal of a deforming stress and that involves the elimination of at least a part of the strain induced in the crystal structure by the deforming process.

In addition to the development of slip lines and other visible manifestations of deformation, straining leads to hardening (*work hardening*), broadening of x-ray lines, increase in electrical resistivity, and accumulation of *stored energy of cold work*. Recovery involves the progressive reduction of these as the crystal structure "recovers" its earlier unstrained state; it may therefore be measured by decrease, with time, in hardness, resistivity, x-ray-line broadening, and stored energy.

The kinetics of such recovery are characterized by rapid initial change, rate of change decreasing with time. Figure 10-6 illustrates the elimination of strain hardening in galena. The general nature of the kinetics, and the analogy between metal (see Beck, 1954) and metallic sulfide, is close. The time-temperature pattern of reduction of x-ray-line broadening in sulfides has not yet been studied, though Paterson (1959), Gross (1965), and various other investigators have established a clear parallelism between the behavior of limestone (calcite) and metals in this respect. Analogous behavior by sulfides is, however, almost certain and is strongly indicated by the comparison shown in Fig. 10-7. Relations among time,

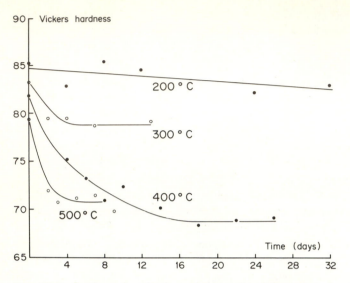

Fig. 10-6 Isothermal softening of naturally deformed and work-hardened galena from Broken Hill, New South Wales. (*From Stanton and Willey, Econ. Geol., 1970.*)

temperature, and reduction of electrical resistivity—and stored energy—in sulfides also seem likely to follow the pattern shown by metals, though the behavior of even the common sulfides still awaits elucidation. This offers an intriguing field for experiment.

The recoverable fraction of work hardening, x-ray-line broadening, etc., in metals increases with decrease in temperature of deformation and with increase in temperature of annealing. That the latter holds also for galena seems to be indicated by Fig. 10-6.

A further process, which may possibly constitute an additional feature of the recovery stage, is the reequilibration of triple-junction angles in slightly strained aggregates—a process already referred to in Chap. 9. Here the straining of the aggregate appears to have been sufficient to modify the grain-boundary angles, but insufficient to require full recrystallization for the lattice readjustment of the grains concerned. As shown in Fig. 9-7, measurements in the strained state give somewhat broadened frequency distributions of angle values. Heating systematically decreases the spread of the distribution to a terminal value. The grain-boundary movement involved does not, however, extend along the full lengths of the grain boundaries concerned. As far as can be seen it occurs only over a minute length, very close to the triple junction.

SUBGRAIN DEVELOPMENT

It is perhaps usual to think of the smallest units of a polycrystalline aggregate as being the individual grains. However, these are often divisible into smaller units called *subgrains*, which are delineated by *subboundaries*.

The first reference to the existence of subgrains was by Rawdon and Berglund in 1927, who detected them in ferrite deformed at high temperature. Northcott (1936) observed intragranular lines in several metals. He referred to these as *veining* and observed that they provided a locus for ready oxidation. Many more observations followed, and Guinier (1950) has treated these comprehensively.

Subboundaries are revealed on polished surfaces by simple oxidation, by etching, or by the segregation of exsolution particles. They define small elements of each grain which are slightly misoriented with respect to each other—generally by a matter of a few minutes of arc. The grain thus consists of a number of minute tilted blocks, which are called *subgrains*. These have been detected in a number of metals, including aluminum, magnesium, zinc, cadmium, nickel, iron, and silver.

Within a single polycrystalline aggregate it is common to find some grains showing subboundaries while others do not. Within an individual grain, subgrain size may be fairly uniform or highly variable. In metals the tilting of the blocks may be quite random, or there may be a degree of preferred orientation.

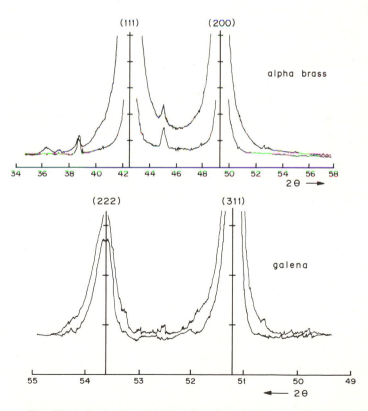

Fig. 10-7 Reduction of x-ray-line broadening, accompanying annealing, in experimentally deformed brass and in naturally deformed galena. (*From Barrett and Massalski, "Structure of Metals," after Warren, 1959; Stanton and Willey, Econ. Geol., 1970.*)

Characteristically subboundaries stop at grain boundaries, indicating that their formation is related to the present grain structure. While subboundaries often appear to localize impurities, their formation does not appear to be connected with them. Subgrains are found in metals of the very highest purity.

In addition to the methods already mentioned, subgrains may be detected, and their nature more precisely determined, by the more delicate effects of x-ray diffraction and optical interference. The existence of mutual tilt can be shown very elegantly by reflecting a finely focused x-ray beam from a polished surface and photographing the image with a Laue camera. For a single structurally perfect grain, a Laue photograph with clear, sharp spots results. When such a grain has suffered bending, but still exists as a single structure, the Laue spots are drawn out into streaks, a phenomenon termed *asterism*. Where, however, the grain is divided into subgrains, each Laue spot is converted into several. (This is well illustrated by Guinier, 1950, pp. 423, 424.) Specifically, if the beam can be made to fall exactly on a subboundary, straddling it, two Laue spots appear. This indicates a tilted relationship, and the degree of tilt can be determined by appropriate measurement of the photograph.

The existence of subgrains is also shown very delicately by double refraction, where the crystal concerned is optically anisotropic. Among minerals, quartz frequently shows subgrains by double refraction of transmitted light, and similarly pyrrhotite does so in reflected light.

It appears that subboundaries probably develop during *both* creep and annealing. Guinier has suggested that those developed during creep are in fact simultaneous annealing effects, and since annealing may accompany deformation during creep, such a suggestion is not without reason.

In the case of creep, if slip or twinning did not occur, early deformation might be expected to take place simply by lattice bending. With continued stress the structure might begin to undergo disruption by the formation of dislocations, which, however, might be expected to develop in a rather ordered way as connected arrays rather than as isolated individuals or clusters. Such a process would be expected to yield walls of dislocations—subboundaries—separating slightly tilted blocks of the original crystal structure—the new subgrains.

The development of such structures by annealing has been suggested by Guinier to take a slightly different path. In this "model," plastic deformation leads to the development of dislocations,[†] scattered more or less randomly through the crystal. The situation in a single bent lattice is shown in Fig. 10-8a. Clearly this state is one of comparatively high energy, and if the deformed material is heated sufficiently to allow atomic mobility, the crystal will, as usual, attempt to reconstruct to a lower energy configuration—i.e., to anneal. It may do this by the movement of the randomly distributed dislocations into arrays. This yields the state of affairs portrayed in Fig. 10-8b. The dislocations are now localized into walls, and the formerly bent crystal is segmented into blocks of unstrained

† Different mechanisms of deformation will yield dislocations of different sign; for detail on results of this, the reader is referred to Guinier (1950, pp. 428–429).

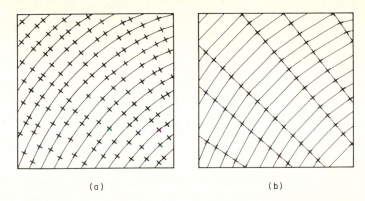

(a)   (b)

Fig. 10-8 Development of subgrains; distribution of dislocations in a bent crystal (*a*) before polygonization and (*b*) after polygonization, according to Orowan and Cahn. (*From Beck, Advan. Phys., 1954.*)

material. The walls of dislocations then constitute subboundaries and the unstrained blocks the subgrains. This particular process of reordering during annealing is known as *polygonization*. The derivation of the term is clear from Fig. 10-8*b*.

Although the formation of subgrains in metals appears to be independent of temperatures, their *size* is temperature- and time-dependent. It has been shown that the subgrain size resulting from creep in pure polycrystalline aluminum increases with increasing temperature of deformation to at least 350°C and with decreasing rate of deformation. In annealing it has been shown that there is an increase in subgrain size with increase in annealing temperature and also with increase in time at constant temperature. Subgrain size also appears to be affected by the presence of solute atoms: the small amount of data available indicates that the presence of solute atoms causes a decrease in size, as might have been expected.

There is now very clear evidence that subgrain development is common in a wide variety of naturally occurring minerals. Much of the quartz of metamorphic —and plutonic igneous—rocks shows abundant substructure, which also appears in many other silicate and related minerals. Among the sulfides, galena and pyrrhotite commonly show subgrains. Many galena grains show subgrain boundaries almost identical in appearance to those in aluminum. The number of subgrains varies from three to four to perhaps a dozen per grain, and they are beautifully delineated by an HCl-thiourea etch. Their boundaries are readily distinguished from grain boundaries after etching: as in metals the grain boundaries are quite wide, whereas the subgrain boundaries are much finer and develop as a web clearly defined by the outline of the parent grain. In addition the two are easily distinguished by subtle variations in etch tarnish: adjacent grains commonly tarnish to slightly different degrees, whereas subgrains within any given single grain do not. In the absence of x-ray measurements there is no precise information on the degree of misorientation in this case, though it is undoubtedly small.

Bending of the lattices of the galena grains is often shown up very clearly by curved cleavages: these always seem to cross subgrain boundaries with no deflection, indicating—on a microscopic scale—a high degree of coherence and small misorientation. Subgrain boundaries always terminate at grain boundaries, and in general the subgrains are essentially equidimensional.

Subgrains in pyrrhotite are best revealed by double refraction. They are less common and less regular than those in galena, but there is no doubt of their identity, and they are by no means rare. What may be subgrains also appear in chalcopyrite and, less commonly, in sphalerite. However, for chalcopyrite these always appear on surfaces of similar orientation, as revealed by etching, and the suspicion arises that they may be some kind of surface artifact developed by flaking, or some similar process, during polishing. These require much more careful investigation. Subgrains have not been found in pyrite, arsenopyrite, magnetite, and others among the harder minerals.

Whether these structures have developed by creep or annealing is by no means clear, and the solution of the problem awaits appropriate experiment. In at least a number of sulfide masses they are common throughout—for the great Broken Hill lead-zinc deposit the author has been unable to find a single polished surface, out of scores examined, that does not show subgrains—so that there is no doubt that they may be developed through some very general and nonlocalized process. The preservation of lattice bending in many cases suggests creep, since the development of subgrains here has not led to the total elimination of strain. Then again, they might represent *partial* annealing.

RECRYSTALLIZATION

This is a term that has been applied rather loosely in the geological literature, usually in connection with grain growth accompanying presumed temperature increases. The grain growth in limestone yielding marble, and the coarsening of quartz and other minerals during metamorphism, is often referred to as *recrystallization*.

The term is also not completely without ambiguity in the metallurgical and ceramic literature, though usage here is a good deal more explicit than in geology.

According to Beck (1954) the classical picture of recrystallization involves the nucleation and growth of new strain-free grains at the expense of a deformed matrix. Here three kinds of physical change have been visualized as taking place simultaneously:

1. Release of stored energy of deformation ("cold work" in metals†) in proportion to the volume of deformed matrix replaced by new strain-free grains.

---

† "Cold work" refers to mechanical modification in the cold state. This produces deformation that is retained after working, to persist indefinitely or to be modified by later treatment. In "hot working," temperatures are such that annealing accompanies deformation so that no or little effects of the latter are retained.

2. Softening, as "work-hardened" material is replaced by the softer annealed material. (As pointed out by Beck, in partially recrystallized metals, hard and soft areas may occur side by side, and this manifests itself in varying microhardness values. The same situation also arises in ore minerals, adding to or subtracting from the effects of anisotropy, varying degrees of solid solution, etc., on hardness readings.)

3. Local reorientation, as high-angle boundaries migrate and deformed grains are replaced by new, undeformed, ones.

More recent work shows that the three do not necessarily occur together. Softening in metals may clearly precede boundary migration, and release of perhaps a substantial portion of the stored energy of cold work may occur before softening and reorientation. Softening is associated in part with polygonization, which frequently takes place prior to any movement of grain boundaries. All the above processes do, however, involve boundary migration of some kind, and it is this that will now be regarded as defining recrystallization *sensu stricto*. In this context the five processes of principal importance are nucleation, primary recrystallization, normal (or gradual) grain growth, secondary recrystallization (or coarsening), and differential recrystallization.

1. Nucleation We return to one of the early considerations of Chap. 8: the formation of a nucleus that is stable and grows. For crystallization from vapors, solutions, and melts it was pointed out that stability of a nucleus was attained only when this reached, and just exceeded, a certain critical size. The same holds for the reordering of a strained crystal.

Several theories for the development of nuclei in solid matrices have been proposed.

a. Nuclei are formed locally by thermal activation, with a probability that increases with increase in local lattice strain.

b. Nuclei are small portions of the lattice that have retained their strain-free state while the major part of the crystal has undergone distortion. Under appropriate annealing conditions these "islands" of unstrained material could be expected to grow and replace their strained surroundings. This is referred to as the *low-energy-block theory*.

c. By some annealing process there may be a gradual, *uniform* elimination of strain over comparatively large volumes, yielding large nucleus blocks. Such a process differs from that of a (the *conventional nucleation theory*). This is referred to as the *high-energy-block theory*.

d. Certain subgrains act as nuclei. In this case subgrains formed during either deformation or annealing form essentially strain-free nuclei. Some may grow at the expense of others until a nucleus is developed to a size sufficient to initiate grain reorientation. The mechanism visualized is indicated in Fig. 10-9. In (a), two grains A and B are each subdivided into subgrains and

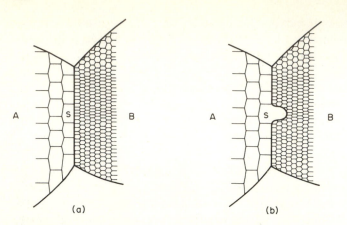

Fig. 10-9 Schematic drawing of two grains with their subgrains (*a*) before and (*b*) after the beginning of grain-boundary migration. Such a grain-boundary migration is nucleated by subgrain *S* of grain *A*, which is notably larger than the subgrains of grain *B* and occurs immediately adjacent to the original grain boundary. (*From Beck, Advan. Phys., 1954.*)

meet along a high-angle boundary. The subgrains of *A* are generally larger than those of *B*. Because of the high angle of the boundary and the disparity in size of the subgrains, one of the *A* subgrains may grow at the expense of one of the *B* subgrains, forming a "bud," as shown in (*b*). As this continues to grow it absorbs the subgrains of both *A* and *B*, and, since it originates in *A*, grain *A* replaces grain *B*.

The nucleating process or processes in sulfides and other ore minerals have not yet been investigated thoroughly. However, the author's observations of natural material, and of specimens involved in experiments concerned with other phenomena, suggest that sphalerite and stibnite, for example, recrystallize from new nuclei (presumably by processes such as in *a*, *b*, and *c*), whereas galena recrystallizes by growth of particular subgrains. The investigation of mechanisms of sulfide recrystallization should not be too difficult and offers an intriguing prospect—and one of considerable relevance to ore-genesis theory.

2. Primary recrystallization Following the development of appropriate nuclei in essentially statistical distribution, the strained matrix is progressively replaced by strain-free grains. As the first formed of these new grains grow, others are nucleated, and for a period nucleation and grain growth proceed concurrently. In metals this leads to a sigmoidal recrystallization pattern such as that shown for aluminum in Fig. 10-10. Initially, during nucleation and very early growth, the development of strain-free material is slow. However, as the grains grow, their growth rates accelerate, and this is added to by the nucleation of more grains.

The curve steepens and continues steeply until nucleation ceases, deformed material is entirely consumed, the new grains impinge, and growth rate flattens off.   This sequence constitutes *primary recrystallization*.   It ends with the complete impingement of the new grains, which then may proceed to grow at each others' expense.

In general the grains developed by primary recrystallization are much smaller than the original ones.   The initial result of recrystallization is therefore a *decrease* in grain size.   The absolute value of the grain size, however, is dependent on both the degree of deformation and the temperature of annealing.   Rates of nucleation and growth both increase with temperature, with that of growth the faster of the two.   Thus increase in annealing temperature leads to a larger grain size at the end point of primary recrystallization.   The growth rate also increases with increase in deformation, but apparently nucleation accelerates even more rapidly.   Thus in metals, grain size tends to *decrease* with increase in deformation.   Evidence of primary recrystallization appears to be preserved in some ores.   It is often shown very beautifully by stibnite and also by pyrrhotite and chalcopyrite (cf. Newhouse and Flaherty, 1930).   The new grains are conspicuously smaller than the earlier ones, which, of course, is quite the reverse of the coarsening often referred to as "recrystallization" in marble and other metamorphic substances—an effect more accurately referred to as *normal grain growth*.

As in the case of recovery, recrystallization is accompanied by softening. Softening associated with recrystallization is rapid and takes the material—metal or sulfide—to its minimum hardness.   If a strained specimen is heated at a series of successively higher temperatures—in each case for a set period of time—softening can be shown to take place in two stages, the first corresponding to recovery, the second to recrystallization.   Figure 10-11 shows this phenomenon in galena

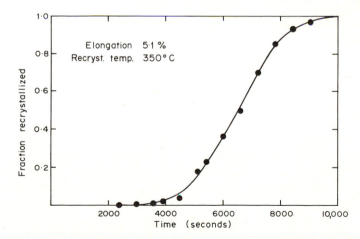

Fig. 10-10   Isothermal recrystallization curve at 350°C for pure aluminum extended 5.1 per cent.   (*From Beck, Advan. Phys., 1954.*) Qualitative observation indicates that sulfides (e.g., galena and stibnite) conform to a similar recrystallization pattern.

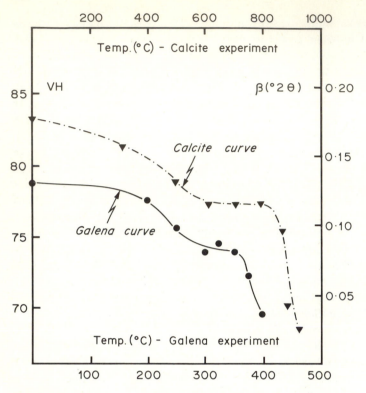

Fig. 10-11 Reduction of hardness (Vickers hardness, VH) in galena (*from Stanton and Willey, Econ. Geol., 1970*) and of x-ray-line breadth (β) in limestone (*from Gross, Phil. Mag., 1965*) accompanying recovery and recrystallization. Note the two-stage nature of both curves.

(Stanton and Willey, 1970) and compares it with the pattern of reduction of x-ray-line broadening in experimentally deformed polycrystalline calcium carbonate (compressed by 5.6 per cent) as found by Gross (1965, p. 805). The two phases of annealing, separated by a pause, are clear and remarkably similar in the two materials.

As might be expected, recrystallization temperatures vary from one substance to another, and this is certainly the case with the common sulfides. Under evacuated-tube conditions, some strained galena may be made to completely recrystallize at 250°C in about 20 days, whereas coexisting sphalerite would require a temperature of 500 to 600°C to recrystallize in this time. This pair is referred to again a little later in this chapter in connection with differential annealing. No experimental data are available for stibnite, but observation of natural material suggests very low recrystallization temperatures, probably below 150°C. Davies (1964) found that chalcocite recrystallized very rapidly—to completion in 2½ hours—at 465 ± 5°C under less than 1 atm vapor pressure. He ascribed the rapidity of the process at this temperature to triggering by the hexagonal-cubic inversion. No doubt chalcocite would recrystallize at considerably lower temperatures, given longer time.   Gill (1968) has shown that at a pressure of 550 bars chalcopyrite recrystallizes rapidly above about 565°C, and pyrrhotite does so between 525 and 660°C.

It must, of course, be kept in mind that temperature is not the only variable affecting recrystallization.   Confining pressure, amounts of distribution of impurities, and especially, degree of prior deformation can also influence the onset and rapidity of recrystallization.   Under a given set of conditions, that portion of a grain or of a single-phase aggregate that has been deformed most severely will recrystallize most readily.   This is often illustrated by galena in deformed ores: frequently those galena grains surrounding hard inclusions appear quite unstrained, whereas those at some distance from the inclusions—and hence those that would be expected to have suffered lesser deformation—are quite notably strained.   In such cases the galena immediately adjacent to the hard inclusions was indeed the more heavily distorted, but this has led to its readier recrystallization so that the original picture of relative deformation becomes inverted.   That this has in fact been the case is readily demonstrated experimentally simply by heating natural material and observing the more highly deformed grains recrystallizing first.

We have already noted that, just as with metals and metamorphic silicates and carbonates, preferred orientations may develop in sulfides during deformation. A further development of preferred orientations can occur during recrystallization, yielding "textures" (i.e., preferred orientations) that in some cases are similar to, and in others quite different from, the preexisting deformational preferred orientations.   This behavior is well known in metals (see Barrett, 1952; Barrett and Massalski, 1966) and has also been experimentally induced in quartz (Hobbs, 1968). Apparently these annealing preferred orientations develop by either (1) initial

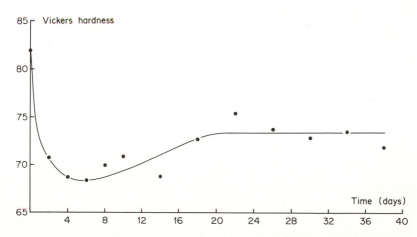

Fig. 10-12   Apparent hardening following initial softening of galena.   It is thought that such hardening is only apparent and is due to the development of a preferred orientation in a polycrystalline aggregate of a mineral possessing hardness anisotropy—in this case galena.   (*From Stanton and Willey, Econ. Geol., 1970.*)

nucleation in preferred orientation or (2) preferred, i.e., more rapid, growth of grains in particular orientations. Stibnite, which as we have already seen begins recrystallization by the nucleation of new grains, develops at least some preferred orientation at the nucleation stage. This is clearly revealed by double refraction, when the mineral is viewed under the microscope between crossed nicols. Galena, on the other hand, may develop a preferred orientation by process (2)—differential grain growth. Any preferred orientation in galena cannot, of course, be detected by double refraction, but it can be determined quite readily by measuring the traces of cleavages, and the attitudes of cleavage surfaces, in the triangular pluck marks of polished sections. Hardness variation during heating may also give qualitative evidence of preferred orientation in galena. In spite of its cubic symmetry this mineral has quite a marked hardness anisotropy, with $\{100\} > \{110\} > \{111\}$. Figure 10-12 shows hardness changes with time at 400°C in a specimen of naturally deformed galena from Broken Hill. Rapid recovery and recrystallization leads to an initial sharp fall in hardness to an early minimum. This is followed by a rise and a subsequent leveling of the hardness curve. It seems likely that the minimum represents the hardness of a near-random aggregate formed by primary recrystallization and that the subsequent rise represents the development—by differential growth—of a preferred orientation whose hardness *in the plane of the polished surface* is greater than that of the random aggregate *in that same plane*.

3. "Normal" (or "gradual") grain growth The development of these larger grain sizes results from the next stage of recrystallization—*normal grain growth*, or *gradual grain growth*, which follows the principles already discussed in Chap. 9. As the point of the completion of primary recrystallization is reached (this need not, of course, occur simultaneously throughout the strained matrix—recrystallization is usually somewhat "patchy"), the new grains meet in growth impingement boundaries, which represent an unstable configuration. Onset of normal grain growth involves the conversion of these *irregular* impingement boundaries to *smoothly curving* ones and the modification of triple-junction angles to values as close as possible to 120°. A new foam structure thus begins to emerge and this is inevitably accompanied by grain growth—that is, by *normal*, or *gradual*, *grain growth*.

Grains grow in such a way as to minimize the total energies of their boundaries—i.e., the total area of interface and, particularly, the area of higher energy interfaces. Larger grains, i.e., those of lesser curvature, grow at the expense of smaller ones of greater curvature. Assuming an essentially random orientation of the nuclei, the grains are absorbed and grow at similar rates, eventually yielding material of more or less even grain size bounded by slightly curved surfaces meeting in 120° triple junctions.

In metals progressive growth of this kind does not appear to disrupt the pattern of grain-size distribution developed in the early stages. The proportion of larger grains increases and that of smaller grains decreases, but within the distribution the loss of grains from one size interval to the next, larger, one is balanced by a gain from the one below. This is shown clearly in Fig. 10-13. To the limited extent to which they have been studied, sulfides behave similarly. Figure 10-13 shows

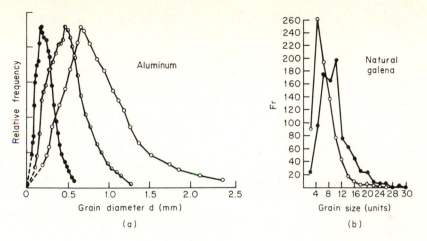

Fig. 10-13 Change in frequency distribution of grain size with grain growth, accompanying heating in (*a*) pure aluminum heated at 500°C for 25 minutes and (*b*) galena from Mount Isa, Queensland, heated at 600°C for 100 days. (*From Beck, Advan. Phys., 1954; Stanton and Gorman, Econ. Geol., 1968.*)

the behavior of aluminum and natural galena from Mount Isa, Queensland (Beck, 1954, and Stanton and Gorman, 1968, p. 918). The overall shape of the distribution remains quite constant, in spite of the increase in absolute grain size. [This particular galena specimen showed a very even (natural) grain size, which accounts for the slow growth rate at a temperature as high as 600°C.]

For metals, growth rates increase with increase in temperature of annealing. Figure 10-14*a* shows isothermal grain growth in high-purity aluminum and Fig. 10-14*b* that in natural galena. Although grain growth is clearly very much faster in

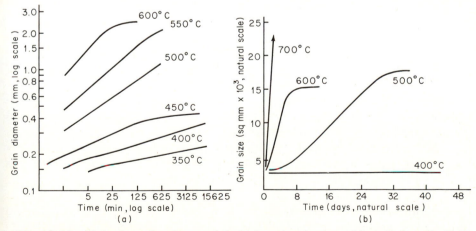

Fig. 10-14 (*a*) Isothermal grain growth in high-purity aluminum (*from Beck et al., 1948*) and (*b*) isothermal grain growth in natural galena from Coeur d'Alene, Idaho (*from Stanton and Willey, unpublished*).

the metal than in the sulfide, the general similarity is clear. Grain growth continues in this way until its driving force is equaled by some inhibiting factor. The latter may be simple locking of triple junctions or the incidence of impurity specks just sufficient to hold up the boundaries, preventing further movement and hence growth.

In most cases normal grain growth in sulfides leads to the development of more or less equidimensional grains and a fairly even grain size. However, in some matrices recrystallization and grain growth combine to deviate from this simple pattern and may yield notably elongated grains and quite uneven grain-size distributions.

Development of nonequidimensional grains has been noted to result from recrystallization and grain growth in *kinked matrices*. Recrystallization of kinked quartz (Hobbs, 1968) and kinked galena (Stanton, 1970) leads to the initial development of flat grains, the larger dimensions of which are essentially parallel to the original kink bands. This leads in turn to the development of a distinctly schistose appearance and, at least in some cases, to preferred crystallographic orientations in the newly developed polycrystalline aggregate. Galena that has recrystallized in this way under experimental conditions appears at first sight to have developed its present structure by deformation, but this is not so; the grains have *grown* (by recrystallization and normal grain growth) as flat bodies, and the shape has only a secondary relationship to deformation. As might have been expected several features of the grain structure of such aggregates are quite incompatible with an immediate deformational origin. The grains, though flat, show little internal evidence of distortion or strain hardening. The only indication of slight strain and/or annealing is the development in some cases of subgrains. Grain boundaries are usually smooth and triple-junction angles approximate to 120°. As an example of this, it has recently been found that the foliated structure of some of the "schistose" galena from Coeur d'Alene, Idaho, is almost certainly the result of recrystallization rather than of compression or shearing as formerly believed, and indeed the structure has been reproduced virtually perfectly by heating a single naturally kinked galena crystal from Broken Hill, N.S.W., at 400°C for 23 days (see Stanton, 1970). In this case the single grain recrystallized to about 2000 smaller grains having a length-breadth ratio of approximately 2:1—almost identical with that of the "schistose" galena aggregates from Coeur d'Alene. Experimental investigation of Coeur d'Alene ore by the author and Helen Willey has shown that the foliated structure is probably quite stable at low temperatures (ca. 200 to 300°C) and hence should persist indefinitely under such conditions. At higher temperatures ($T > 350$ to 400°C), however, the grain configuration becomes unstable, the slightly larger grains grow laterally at their neighbor's expense, and the aggregate assumes an equidimensional foam structure.

4. Secondary recrystallization In some instances growth does not yield essentially even grain sizes but is characterized by the development of conspicuously larger grains that appear among, and grow at the expense of, the "normal" matrix grains. The development of such larger bodies is known as *coarsening, discontinu-*

ous grain growth, or *secondary recrystallization*.   This is common among metals and ceramics, is the process involved in the formation of porphyroblasts in metamorphic silicate and carbonate rocks, and also appears to be responsible for a number of structures commonly found in ores.

It has already been noted that growth of the "gradual" kind progresses with little or no effect on the nature of grain-size frequency distributions.   This is not the case with discontinuous grain growth.   Here a relatively few grains grow much faster than those surrounding them, leading to a disruption of the earlier size distribution pattern.   The latter splits into two portions, reflecting the *duplex* nature of the new structure.   The phenomenon has been studied in detail in metals and appears to arise from two distinct causes: the dependence of boundary mobility on orientation and the inhibition of boundary movement by impurities.

Those boundaries are most likely to move—and move fastest—that are of highest potential energy, and—keeping in mind the effects of curvature—the boundaries of highest energy tend to be those of greatest misfit, i.e., those of comparatively high-angle misorientation.   Where a high degree of preferred orientation occurs, the orientations of most of the grains of the aggregate are, by definition, very nearly the same.   The driving force for boundary migration among them is therefore low.   If, however, there occur here and there in such a matrix stray grains of orientation different from the rest, these will possess high-angle boundaries and hence a strong driving force for growth.   They will therefore grow while the others do not, leading to differential coarsening and a characteristic duplex structure.   Such secondary recrystallization is referred to as *orientation-dependent*.

Through an entirely different mechanism a similar result may be induced by the occurrence of a minor phase.   Where small particles of the latter are dispersed through the matrix, they impede the movement of any boundaries that encounter them.   Clearly if the driving force for movement is relatively small, and if a number of such impurity particles are encountered at about the same time, the grain boundary concerned will cease to migrate.   In some alloys and ceramics (and ores) a minor phase occurs as numerous small particles along almost—but not *quite*—all the matrix grain boundaries.   As a result most of these are locked and cannot move, but where, here and there, a few grains are not hemmed in in this way, the boundaries are able to migrate and the grains to coarsen.   This may lead to the development of a conspicuously duplex structure.   Coarsening of this kind is referred to as *inhibition-dependent*.   Whether or not growth occurs is dictated by the interplay of orientation *and* inhibition, as the reader will probably have sensed. With a high boundary energy a boundary may be able to sweep past several such particles, whereas with a lower driving force for grain-boundary movement, a single minute particle may completely lock the system.   The process is influenced by both orientation *and* inhibitor, the two varying in importance.

Coarsening of both types appears to be common in both synthetic and natural materials.   It is well known in metals, and numerous examples are found among ceramics.   Sintering of such minerals as alumina, spinel, and barium titanate may produce quite spectacular secondary recrystallization.   It is notable that there is commonly a clear tendency for the development of plane sides—crystal faces—on the coarse particles as they grow.   In this case, as already pointed out in Chap. 8, growth is influenced by the anisotropism of the crystal.   The crystal faces constitute low-energy boundaries in single-phase aggregates—they have essentially zero curvature—and it might have been expected that they would move very slowly as a result, inhibiting rather than contributing to coarsening.   It has been suggested (Kingery, 1960) that the reason for the apparent reversal may be found in the occurrence of *very small* quantities of impurity in the ceramics concerned.   If such an impurity formed an intermediate boundary phase of zero dihedral angle between the surfaces of the large and small grains, the energy differences might lead to its acting as a transfer medium.   This explanation is not entirely convincing, but the fact remains that the development of many of these coarser grains *is* accompanied by the development of crystal faces, and there must be some reason for it.

Such ceramic structures appear to have immediate analogies among metamorphic rocks.   Porphyroblasts of garnet, feldspar, quartz, chiastolite-kyanite, and others are *ipso facto* products of coarsening, though in general these occur in multiphase environments and are not directly comparable with the systems being considered at present.   Garnet in quartz is usually much coarser than its matrix, but garnet in garnet is usually of fairly uniform size.   Cases of genuine coarsening are fairly frequent in quartz, however.   In quartzose patches of quartz-rich metamorphic rocks, it is common to find a fine matrix studded with notably coarser grains, almost certainly the result of discontinuous grain growth.

Both orientation- and inhibition-dependent coarsening occur in natural sulfide aggregates dominated by a single phase.   Pyrrhotite commonly develops a high degree of preferred orientation [that of the Chihara mine, Japan (Kanehira, 1959), and of the Sullivan mine, British Columbia, are good examples] that slows grain growth in the major part of the material concerned.   However, here and there in such aggregates grains deviate from the general orientation, and these absorb the surrounding matrix to form coarse patches—orientation-dependent coarsening.   These stand out not only because of their size but also because of their sharply different interference tints when viewed between crossed nicols. Chalcopyrite also appears to coarsen by this mechanism, showing the same relationship between size and orientation—in this case best revealed by etching. Among the other common sulfides, galena, sphalerite, and pyrite do not seem to show orientation-dependent growth, possibly because of their cubic symmetries.

The copper-lead-zinc ore of Rammelsberg provides excellent examples of inhibition-dependent coarsening.   Much of the sphalerite here is accompanied by fine particles of chalcopyrite, which are abundantly distributed along sphalerite grain boundaries and triple junctions.   Although fine itself, the sphalerite is much coarser than the associated chalcopyrite and for the most part is quite even.   Here and there, however, chalcopyrite is absent and the sphalerite has coarsened, giving very clear duplex—i.e., porphyroblastic—structure.

5. Differential recrystallization   It is well known that from one metal to another there may be considerable variation in the conditions required for recrystallization.   For high-purity materials, the principal factor is temperature.

Fig. 10-15 Differential recrystallization in galena-sphalerite ore from Broken Hill, New South Wales: (*a*) original, in which both minerals are deformed, and (*b*) heated at 325°C for 45 days, in which galena is recrystallized but sphalerite retains its deformation. (*From Stanton and Willey, Econ. Geol., 1970.*)

For example, deformed lead and zinc recrystallize at room temperature, but copper, iron, titanium, and other commercial metals require higher temperatures, usually some hundreds of degrees Celsius.   Suppose, therefore, that we take two or more mutually insoluble metals and render them into a polycrystalline aggregate.   The whole is then deformed.   For a particular two of the deformed components, one, *A*, might have a recrystallization temperature (i.e., a temperature at which recrystallization begins and proceeds rapidly) of say, 200°C, whereas the other, *B*, might require a temperature of 500°C.   We then take the deformed aggregate and heat it to some intermediate temperature—say, 350°C.   Component *A* recrystallizes, component *B* does not.   We have induced *differential recrystallization:* metal *A* now shows little or no sign of having been deformed, whereas metal *B* has its deformational history still clearly imprinted upon it.   Were *A* and *B* two mineral components of a rock, most textural interpretations would maintain that *A* had escaped a deformational episode that *B* has passed through, and hence that *A* was younger than *B* and had been introduced after the metamorphism concerned. The behavior of metals indicates how fallacious this line of reasoning may be.

In view of the many analogies we have already noted between metals and natural metallic sulfides, it is not surprising that the sulfides also undergo differential recrystallization.   There are quite a number of ores (e.g., Buchans in Newfoundland, Balmat in New York State, Rammelsberg in Germany) in which chalcopyrite occurs as an essentially fine, unstrained mosaic, whereas the accompanying sphalerite is conspicuously twinned, kinked, and bent.   The same holds for numerous galena-sphalerite aggregates—the galena is comparatively fine and unstrained, whereas the sphalerite is coarser and quite obviously deformed.   While there may be age differences as suggested above, there is a strong suspicion that present differences are due quite simply to differential annealing.

That this may well be the case has been demonstrated by Stanton and Willey (1970), who took a deformed galena-sphalerite aggregate (shear zone ore from Broken Hill) and heated this at 325°C, i.e., just above the recrystallization temperature of the galena, for 45 days.   Figure 10-15a shows the original, with galena and sphalerite both deformed.   Figure 10-15b shows the same material after heating, with galena recrystallized but the sphalerite still quite clearly bent, kinked, and twinned.   The significance of this in the determination of paragenetic sequences is very clear.   This principle of differential recrystallization clearly has profound implications not only in ore studies but in metamorphic petrology in general.

RECOMMENDED READING

Barrett, C. S., and T. B. Massalski: "Structure of Metals: Crystallographic Methods, Principles and Data," 3d ed., McGraw-Hill Book Company, New York, 1966.

Beck, P. A.: Annealing of cold-worked metals, *Advan. Phys.*, vol. 3, pp. 245–324, 1954.

Margolin, H. (ed.): "Recrystallization, Grain Growth and Textures," American Society of Metals Seminar, 1966.

Stanton, R. L., and H. Gorman: A phenomenological study of grain boundary migration in some common sulphides, *Econ. Geol.*, vol. 63, pp. 907–923, 1968.

Stanton. R. L., and H. G. Willey: Natural work-hardening in galena, and its experimental reduction, *Econ. Geol.*, vol. 65, pp. 182–194, 1970.

part two

Important Associations

11
Ores in Igneous Rocks 1: Ores of Mafic and Ultramafic Association

In this chapter we shall consider two important ore types that seem to be characteristically associated with igneous rocks of high mafic mineral content. The first of these two groups of ores is characteristically rich in chromite, in turn often accompanied by substantial quantities of nickel in associated silicates, and sometimes by platinoid metals. The second group is one of iron-nickel-copper sulfide ores, which also often contain notable quantities of the platinoids.

Inevitably both groups have their variants. The chromium ores tend to fall into two categories, or subgroups—those associated with the more mafic "strata" of large layered igneous complexes and those associated with "alpine-type" ultramafic intrusions. The sulfide ores show less well-defined petrological associations than do the chromium ores but exhibit notable variants in nickel-copper ratios and in their platinoid contents.

The related silicate assemblages embrace almost the complete range of mafic and ultramafic rocks. Peridotites, dunites, pyroxenites, and to a lesser extent gabbroic and anorthositic types, are the normal hosts to the chromium-nickel-platinoid ores. The sulfide ores, on the other hand, show a more variable association and a rather stronger affinity with rocks of gabbroic composition—though more highly mafic and ultramafic types are commonly present.

THE ULTRAMAFIC-MAFIC–CHROMIUM-NICKEL-PLATINOID ASSOCIATION

The ultramafic rocks of present concern occur in two main forms and environments:

1. *Layered intrusions.* These are regarded as plutonic and are one of the most clearly defined types of igneous intrusion known. Most (though not all) are of Precambrian or Lower Paleozoic age. Most of them are large, and as a class they exhibit forms ranging from saucer shape to that of a steep-sided cone. They are characteristically layered, compositions of layers ranging from peridotite at the base to granites or granophyres at the top. Where a chilled border phase is present this usually has the composition of a gabbro. The layering, which may possess a high degree of perfection and continuity, and which may show band widths ranging from inches to hundreds of feet, is thought to result chiefly from differentiation *in situ*. In some instances some layers may result from later concordant intrusion.

2. *Alpine-type intrusions.* These are also generally regarded as plutonic, though their provenance is by no means clear. They are much more prominent in geologically and physiographically young terrains than they are in older, deeply eroded ones, and in the former they are generally conspicuously associated with basaltic and andesitic volcanic rocks and their derivatives. This suggests that intrusions of this type may in fact be comparatively shallow subvolcanic products of some kind rather than the result of deep plutonic activity. Most of the world's alpine-type occurrences are Paleozoic or younger, many of them Mesozoic or Tertiary.

   Individual intrusions are small (usually of the order of a few tens of square miles or less in area) and most are of conspicuously elongated lens shape. Large numbers of these small bodies occur as components of narrow belts—serpentine belts—the elongation of each individual lens running more or less parallel to the trend of the belt at the point concerned. Such belts are associated with zones of regional thrust faulting, related in turn to the trends of earlier geosynclines. The rocks are usually layered, but the layers rarely show the perfection, and never the continuity, of those of the layered intrusions. Layers range in width from inches to a foot or so, and are short lenses rather than extensive sheets. Compositions range from dunite to gabbro; more siliceous types characteristically do not occur. The origin of the layering is not known, but is probably largely the result of flow.

In addition to these differences of morphology and environment—and perhaps age—the two types of occurrence show quite gross compositional differences. The "mean" composition of any large individual layered intrusion approximates to that of a gabbro, whereas that of an alpine-type intrusion is distinctly more basic —that of a peridotite or picrite. Peridotites and related rocks, although present in both types of intrusion, bulk much larger in the alpine type than the layered type. In sympathy, MgO is much higher, and Al_2O_3 much lower, in the alpine-type intrusions. These differences will be referred to again a little later.

Chromium occurs abundantly in both environments, virtually entirely as chromite. Nickel occurs in both, but its concentration in ores is at least an order of magnitude lower than that of the chromium of chromite deposits. Its usual abundance in alpine-type intrusions is about 0.2 to 0.3 per cent and rather less than this in layered intrusions. For the most part it occurs as "silicate nickel" in olivine $[(Mg,Fe,Ni)_2SiO_4]$, though minor amounts are known to occur as sulfide or arsenide. The platinoid metals also occur in both, usually native in the alpine-type intrusions and most frequently as sulfides, arsenides, and antimonides in the layered intrusions.

The *chromite* of layered and alpine-type intrusions constitutes the world's sole source of chromium metal, and so deposits of this group are of substantial economic importance. We shall consider these first.

DISTRIBUTION OF CHROMITE CONCENTRATIONS

Layered intrusions are themselves of very sporadic and localized occurrence, and in addition only a few of them contain sufficient concentrations of chromite to be considered sources of chromium ore. About six important layered intrusions are known in the Precambrian and Paleozoic terrains of the world, and there are one or two—including the famous Skaergaard intrusion of Greenland—in Tertiary rocks. Of these, only three—the Bushveld complex of the Transvaal, the Great Dyke of Southern Rhodesia, and the Stillwater complex of Montana—are known to contain really substantial concentrations of commercial chromite (Fig. 11-1). We may therefore say of the distribution of layered-type chromite concentrations that they are few and isolated, though as sources of chromium they are very important.

Chromite in *alpine-type intrusions*, however, is widespread and virtually ubiquitous (Fig. 11-1). There is hardly an orogenic belt, particularly in post-Precambrian terrains, devoid of "serpentine," and there is probably no known serpentine belt that does not have, somewhere along its length, chromite concentrations of sufficient size and richness to constitute ore. In addition, even those ultramafic lenses with no actual ore deposits contain conspicuous chromite, including local concentrations that though rich are of insufficient size to be economic. The largest known deposits are those of the ultramafic belts of the Ural Mountains of Russia, the Philippines, and Turkey. Substantial deposits are well known in ultramafic belts in Cuba, New Caledonia, India, Pakistan, Eastern Europe (Yugoslavia and Greece), and Brazil. Minor deposits are known in the ultramafic belts of California-Oregon, the Maritime Provinces of the United States and Canada, and the Paleozoic belts of eastern Australia.

FORM OF CHROMITE CONCENTRATIONS

The sharpest and most obvious differences between the two types of occurrence ie in their respective forms.

Those of *layered intrusions* are the epitome of regularity; they form layers and groups of layers of great vertical regularity and wide lateral extent. In the vertical sense the layers behave as strata, "interbedded" with bands of other members of

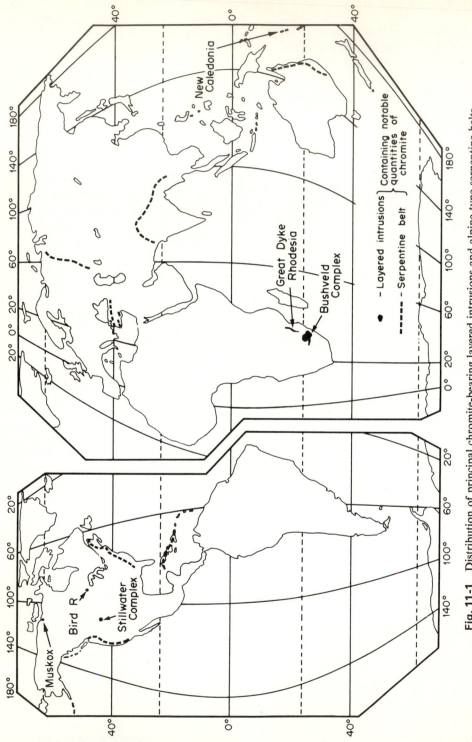

Fig. 11-1 Distribution of principal chromite-bearing layered intrusions and alpine-type serpentine belts.

the mafic-ultramafic zones of the intrusions concerned. Dunites, peridotites, pyroxenites, and harzburgites commonly form the associated strata, the structure of such sequences often being referred to as *pseudostratification*. The thickness of chromite bands ranges from a few inches to 50 feet or so. In the more important areas minable band widths are usually in the range 5 to 20 feet. In a lateral sense individual bands often extend for thousands of feet, and in some cases for miles. The intervening igneous stratigraphy is equally persistent, and as a result stratigraphic columns can be drawn up for particular areas of a given intrusion and stratigraphical correlations made between different areas of the same mass. Figure 11-2, a comparison of stratigraphical columns through the main chromite subzone of the Bushveld complex, illustrates this. Distances between the section localities—180 miles and 60 miles—are very substantial and indicate just how persistent the stratigraphy can be. Like those of "ordinary sediments," individual units of such igneous sediments thicken and thin, giving rise to the expansion and contraction of columns shown in Fig. 11-2.

In detail individual bands may show a variety of features, all apparently analogous to those found in ordinary sedimentary rocks:

1. *Gradational boundaries.* Chromitite (essentially pure chromite rock) may, for example, grade through olivine chromitite to an adjacent band of dunite or peridotite.
2. *Lenses and wedges.* Most chromitite beds are in fact very extensive, flat lenses and taper out at their extremities. In some cases irregularities—particularly gentle domes and basins—have developed on the floor of the intrusion, leading to local lensing and wedging out of chromitite and associated bands. "Sedimentary overlapping" is a common result.
3. *Intraformational contortion.* This is found occasionally and is analogous to that of sedimentary rocks, i.e., contorted chromite layers occur between other chromite layers showing no contortion (see Cameron, 1964, p. 152).
4. *Splits, scour, and fill.* Individual bands of chromitite may split, with the development of thin interfingering bands of silicate. In some cases what appear to be small scour-and-fill structures are developed.
5. *Grouping of strata.* Groups of chromitite bands often occur close together in the manner of groups of coal seams within coal measures.

In contrast to this generally high degree of order in layered intrusions, the chromite concentrations of *alpine-type* occurrences display what must be almost the highest possible degree of *disorder*.

In the first place the chromite shows an extremely wide range of grain size and concentration. It occurs as sparsely disseminated fine grains, through all gradations to dense, coarse-grained concentrations of virtually 100 per cent chromite. Much of it, particularly the lower concentrations, is often distributed in discontinuous bands and stumpy lenses, in conformity with the banding and general structure of the enclosing ultramafic rock. The margins of such bands may be diffuse, but more commonly they are quite sharp, the only diffuse portions

COMPARISON OF GEOLOGICAL COLUMNS
THROUGH MAIN CHROMITE SUB - ZONE

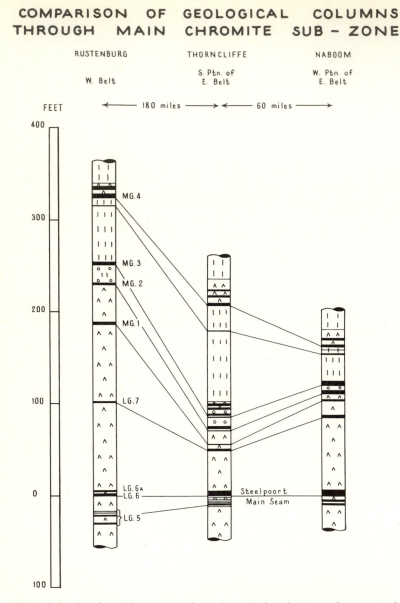

Fig. 11-2 Stratigraphic sequences in, and correlations between, three areas of exposure of the main chromite subzone of the Bushveld complex. [*From Cousins, in Haughton (ed.), "The Geology of Some Ore Deposits in Southern Africa," vol. II, The Geological Society of South Africa, 1964.*]

being the lateral extremities of the lenses, which are normally frayed and gradational. When the chromite is more highly concentrated, shape seems to become less regular. Some of the high-grade material is banded, but in alpine-type occurrences this is the exception rather than the rule. Many of the ore bodies are in the form of pipes, tapering or sharply cut off at the ends.   The long dimensions of such pipes are often parallel to a lineation in the containing rocks; in other cases they appear to be localized by the intersection of faults.   Many ore bodies are much less regular in shape and of much less obvious structural affiliation.   Frequently they occur as pods (aptly referred to as *podiform* by Thayer, 1964), often with their long dimensions parallel to the layering in the enclosing ultramafic silicates but sometimes oblique to it.   Quite a large proportion occur in highly irregular masses referred to—again very aptly—by Sampson (1942) as *sackform masses*.   These often exhibit what appear to be completely random, in some instances quite grotesque, shapes that seem unrelated to any feature of the lithology or structure of the enclosing rock.   Often such masses vary considerably in richness from massive to disseminated material, though there is rarely, if ever, any indication of a reason for such variation.

It is not uncommon for most or all these variants to have developed in a single province.   In New Caledonia, for example, there are abundant layered disseminations following the banding of the enclosing rocks.   There are also some podiform concentrations following the peridotite layering and pipe-like deposits apparently localized by a series of fault intersections.   In other areas of the island there are concentrations that can only be described as sackform and are apparently devoid of structural or close lithological control.   The same kind of variation can be found in virtually all the important areas of alpine-type chromite occurrence.

SETTING OF CHROMITE CONCENTRATIONS

Considerations of distribution and form have inevitably involved minor aspects of setting—sufficient to indicate quite clearly that in their setting, too, the two types of chromite occurrence differ markedly.   This seems to hold from the broadest features of environment right down to the details of lithological association.

The great *layered intrusions* and their constituent layered chromitites all appear to have been injected into comparatively stable tectonic environments.   While all have suffered some folding, faulting, or tilting, none has suffered extensive deformation and metamorphism, indicating that they have been emplaced comparatively late in the tectonic history of the area concerned.   As already noted the bulk composition of each intrusion as a whole seems to be gabbroic; i.e., although the chromitites are now associated with ultramafic layers, the containing masses as a whole are mafic, not ultramafic.   Finally, while the intrusions may have been affected by late-stage faulting, their original emplacement does not seem to have been grossly influenced by major faults.

The *alpine-type intrusions*, on the other hand, have clearly been injected into highly unstable environments, in most cases into geosynclinal zones of volcanism, folding, and faulting.   The time of initial injection, too, is usually close to the most mobile period in the development of the geosyncline concerned.   The bulk composition of the intrusions is peridotitic, and while the chromite may be associated with a variety of rock compositions, the containing masses as a whole are quite definitely ultramafic.   In addition a high degree of serpentinization is virtually ubiquitous in contrast to the layered intrusions, in which serpentinization of magnesian silicates is usually rather minor.   Virtually all intrusions of alpine type are clearly related to early faulting of major scale, and most of them have been modified by faulting and squeezed and moved about by plastic flow—probably semicontinuously ever since their first injection.

It is now appropriate to turn to examples of our different types of occurrence —examples that we can use from now on for comparison when considering features of setting, composition, and origin.

The two most famous *layered intrusions* notable for their chromite content are the Bushveld and Stillwater complexes.   The "Great Dyke" of Rhodesia (one of the world's great sources of chromite) and the Bird River sill of Manitoba are also important, but we shall confine our attention to the first two.

The *Bushveld complex* has been referred to as the world's most spectacular igneous assemblage (Knopf, 1941), and in addition it contains what must be among the world's most spectacular layered ore deposits.   The complex forms a great lobate mass occupying some 26,000 square miles of the Central Transvaal.   Its principal dimensions are 288 miles east-west and 153 miles north-south—and its maximum thickness is about 25,000 feet.   In outcrop it appears as a connected group of lobes, and in section it is believed to be of layered saucer (lopolithic) form (Fig. 11-3).   According to Willemse (1964) the geological history of the area concerned is

1. *The deposition of the Transvaal System,* beginning with the Black Reef Series (quartzites and shales), followed by the Dolomite Series (dolomitic limestone and chert), and ending with the Pretoria Series (shales, quartzite, and andesitic lava), the whole aggregating some 20,000 feet.
2. *A sill phase of diabase sheets* injected into the sedimentary rocks of the Pretoria Series.
3. *An epicrustal phase* involving extrusion and injection of felsites, granophyres, and related rock types.
4. *The Main Plutonic Phase* in which the granodiorites, diorites, gabbros, and ultramafic rocks of the Bushveld complex were injected and differentiated.
5. *A later plutonic phase* during which the Bushveld granite was intruded.

The age of the Bushveld granite has been determined as 1950 $\pm$ 50 million years.

The Main Plutonic Phase has been arbitrarily divided into five principal zones, the younger four of which are each separated by the *main chromitite band,* the *Merensky Reef,* and the *main magnetite band,* respectively.   The principal features of the zones representing the Main Plutonic Phase are lised in Table 11-1, and a simplified picture of their arrangement is shown in Fig. 11-3.   According to Willemse (1964, p. 102) an idealized version of the sequence, proceeding from the *Chill Zone* at the base, is harzburgite or dunite; pyroxenite; chromite-rich pyroxenite; feldspar pyroxenite; norite; gabbro; ferriferous or magnetite-rich gabbro;

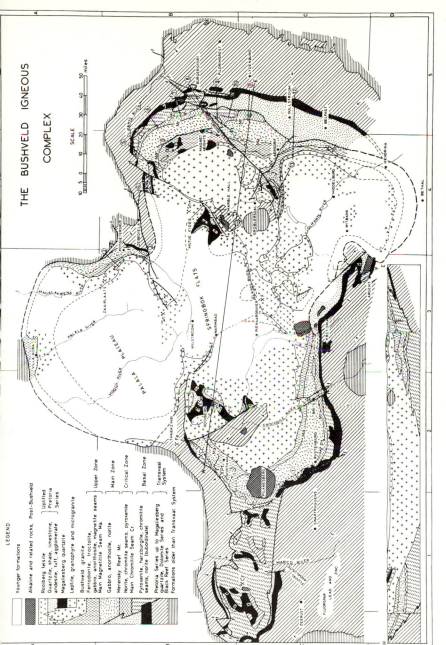

Fig. 11-3 Geological map of, and section through, the Bushveld complex. [*From Willemse, in Wilson (ed.), Econ. Geol. Monograph no. 4, 1969.*]

Table 11-1   Zones of the Main Plutonic Phase of the Bushveld complex.   (From Willemse, 1964, p. 101)

Rock types	Maximum thickness, feet	
Granodiorite	6300	Upper Zone
Olivine diorite		
Gabbro (anorthosite)		
Magnetite bands at different horizons		
Main magnetite band		
Gabbro (anorthosite)	17,000	Main Zone
Thin magnetite band near top		
Norite (subordinate)		
Merensky Reef		
Norite (anorthosite)	3500	Critical Zone
Pyroxenite (subordinate)		
Chromitite bands at different horizons		
Main chromitite band		
Pyroxenite	5000 (excluding the	Basal Zone
Norite (subordinate)	ultramafic rocks of	
Peridotite	the far western	
Chromitite bands mostly near top of zone	Transvaal)	
Norite	Erratic	Chill Zone

olivine diorite; and finally granodiorite.   The *Critical Zone* is, of course, the interval of principal economic interest and is defined as the layered succession between the main chromitite band at the base and the Merensky Reef at the top.   According to Willemse (1964, p. 105) the Critical Zone contains various noritic, anorthositic, and pyroxenitic rocks, clearly banded and often alternating.   A more detailed division of the Critical Zone, as it occurs in the central areas of the complex, by Cameron (1964, pp. 137 and 139) is:

Unit	Description
MR	Pegmatitic diallage norite or feldspathic pyroxenite.
X	Interlayered medium-grained to coarse-grained, massive to regularly or irregularly banded pyroxene anorthosite, noritic anorthosite, and norite; pyroxenite and mafic norite with thin chromite seams near base.
W	Feldspathic pyroxenite with thin chromitite seams at and near base.
R	Upper 158 feet anorthosite, lower 52 feet pyroxenite, thin layers of anorthosite, and mafic norite.
O	Feldspathic pyroxenite.
M	Mostly anorthosite and anorthositic norite with subordinate norite and mafic norite, mostly in lowest portion. Thin layers of chromitite and pyroxenite near top.

L	Mafic norite grading downward into massive pyroxenite. Two persistent groups of thin chromitite seams in middle and upper parts in north; at least one in south. From Doornbosch southward has thick chromitite seam near base.
K	Interlayered pyroxene anorthosite and anorthositic norite in various proportions.
J	Mafic norite and norite on Jagdlust and Winterveld (343) with a thin chromitite near the base and another nearer the middle.
H	Variable. Anorthosite or interlayered anorthosite and norite, with or without thin chromitites.
G	Mafic norite and feldspathic pyroxenite.
F	In northern part of sector consists of norite separating two units of anorthosite with associated thin or impure chromitites. Southward consists of anorthosite or interlayered anorthosite and norite with one or more impure chromitites.
E	Pyroxenite and mafic norite interlayered with chromite pyroxenite and chromitites.
D	Leader and Steelpoort chromitite seams separated by $2\frac{1}{2}$ to 3 feet of pyroxenite.
C	Mostly mafic norite and pyroxenite interlayered at intervals with chromitites and chromite pyroxenite. Includes a zone of troctolite, and a zone of interlayered peridotite and pyroxenite.

This is illustrated in Fig. 11-4, which shows quite clearly the general nature of the lithological setting of the principal chromite seams.

The *Stillwater complex* is very similar petrographically to the Bushveld complex. It is a steeply dipping layered sheet of overall gabbroic composition intruded into Precambrian metamorphic rocks of the eastern fringe of the Rocky Mountains in Montana. Although now dipping nearly vertically, the complex is thought to have been injected, and to have undergone differentiation, as an essentially horizontal sheet. Its present shape and surroundings are shown in Fig. 11-5. The maximum strike length now exposed is about 30 miles, both ends being sharply terminated by faults. *As it now appears* it is therefore much less extensive than the Bushveld complex. The maximum exposed "stratigraphical" thickness is now about 18,000 feet, and it is estimated that some 5000 to 15,000 feet have been lost from the top by erosion, so in *thickness* it was probably quite comparable with the Bushveld.

The intrusion has a well-developed "stratigraphy" and has been divided (Jackson, 1961, p. 2) into three principal lithological zones:

1. *Basal Zone.* This is composed of all rocks underlying the stratigraphically lowest harzburgite. The base of the zone is chilled and is a fine-grained ophitic gabbro; above this chilled margin, the zone is composed of pyroxene gabbros, norites, and feldspathic bronzitites; maximum observed thickness approximately 700 feet.
2. *Ultramafic Zone.* This is composed of all rocks between the base of the stratigraphically lowest harzburgite in the complex and the base of the stratigraphically lowest norite in the Banded Zone. The Ultramafic Zone is

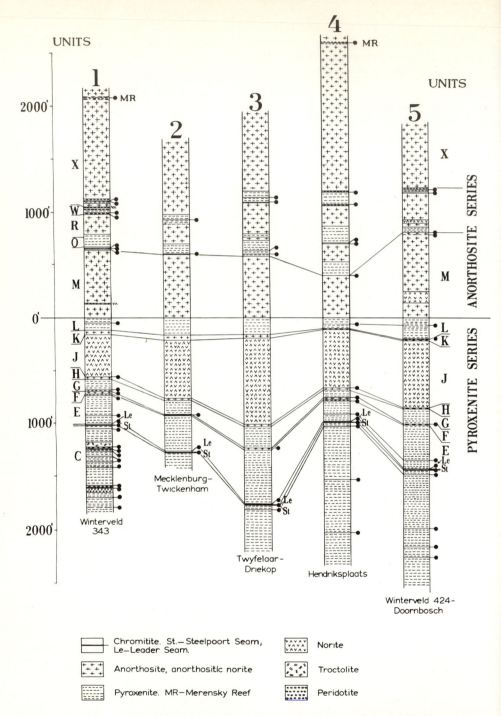

Fig. 11-4 Five stratigraphic columns through the Critical Zone of the Bushveld complex, showing the lithological associations of the chromitite layers. [*From Cameron, in Haughton (ed.), "The Geology of Some Ore Deposits in Southern Africa," vol. II, The Geological Society of South Africa, 1964.*]

composed of dunites, chromitites, harzburgites, and bronzitites; average thickness approximately 3500 feet; it may be divided into two subzones:

a. The Peridotite Member.   This comprises the lower two-thirds of the Ultramafic Zone and is composed of alternating, conformable layers of dunite, chromitite, harzburgite, and bronzitite.

b. The Bronzitite Member.   This comprises the upper one-third of the Ultramafic Zone and is composed of a single thick layer of bronzitite.

3. *Banded Zone.*   This is composed of all the rocks lying stratigraphically above the Ultramafic Zone; it is composed of alternating layers of norite, gabbro, and anorthosite; the maximum exposed thickness is approximately 14,000 feet.

Most of the chromite occurs in the Peridotite Member of the Ultramafic Zone.

There are no *alpine-type* deposits that can be said to stand out as examples of their type, as with the Bushveld and Stillwater occurrences among the layered ores. The alpine-type ores have a setting that is absolutely constant on a large scale—i.e., they always occur in highly deformed and serpentinized peridotitic rocks of alpine type—but in detail their environment is as variable as the rocks that enclose them. According to T. P. Thayer:

The distribution† of podiform chromite deposits is one of the most puzzling geologic problems of the alpine peridotites and peridotite-gabbro complexes.   Size of intrusive mass and chromite content seem to be entirely unrelated.   Although all podiform deposits are associated with dunite, some of the most dunitic massifs, such as Dun Mountain, New Zealand, contain very little chromite.   Systematic distribution of podiform deposits has been recognized only in some alpine complexes that comprise gabbro as well as peridotite.   The major chromite deposits in the Zambales Complex in northern Luzon occur in dunitic border zones of peridotite near contacts with gabbro.   In the Camagüey district of central Cuba, 94 per cent of the known chromite was found in peridotite within half a mile of the gabbro contact; this information guided gravity prospecting in the district by the U.S. Geological Survey.   Smith found that the chromite deposits in the Bay of Islands complex are in the transition zone between peridotite and gabbro.   In some such complexes, however, the chromite deposits appear to be related more closely to other contacts or are scattered at random in the peridotite.   The relation to gabbro contacts is a very effective guide for general prospecting where geologic mapping shows that it is applicable, but it cannot be used to locate individual ore bodies.

Geologic mapping in the United States at scales ranging from 1:10,000 or larger to 1:50,000 or smaller has revealed no systematic areal distribution of podiform deposits in masses that consist entirely of peridotite.   In some districts the deposits are restricted to major zones or masses of dunite but in others the principal deposits are not in the main dunite bodies, although they are surrounded by dunite halos.   In general, chromite deposits are likely to occur in groups in the more dunitic parts of peridotite masses, or strung out as horses along shear zones.   Deposits of massive and disseminated ore may occur together or separately, in any part of a peridotite mass.   In districts where the composition of the chromite itself varies widely, as in Oregon and Cuba, the extreme

† Thayer uses the word *distribution* here in the sense of occurrence as related to the features of the enclosing rock, i.e., *setting* as we are using the term in this book.

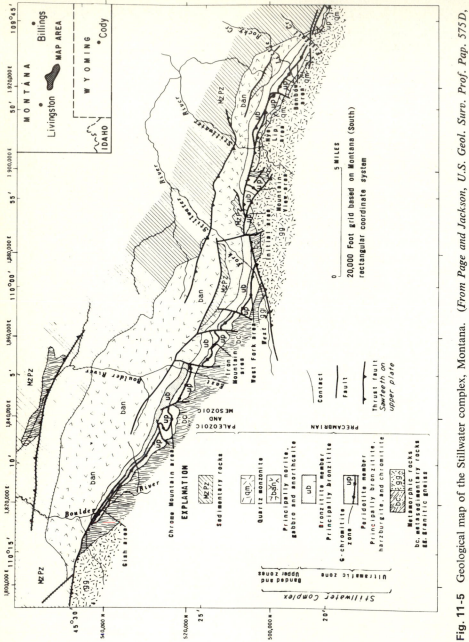

Fig. 11-5 Geological map of the Stillwater complex, Montana. *(From Page and Jackson, U.S. Geol. Surv. Prof. Pap. 575D, 1967.)*

variants may occur in neighboring deposits.   Examples of large solitary deposits might be cited, and it is common experience to find small pods widely scattered through harzburgite.

All podiform chromite deposits are associated with dunite as gangue, as a halo, or as country rock.   Halos range from a few centimeters to many meters thick, and may be sharply defined or gradational (1964, pp. 1501–1502).

CONSTITUTION OF CHROMITE CONCENTRATIONS

The constitution of chromite, and its compositional relations with the other spinels, has been considered in Chap. 4.   A chromium *ore* is, of course, composed of chromite together with one or more silicates, and this has to be kept in mind when considering constitutional features.   Once again the layered and alpine types of occurrence are sufficiently different to warrant separate consideration.

The chromite of *layered intrusions*, while usually favoring the olivine-rich layers, may occur in significant quantity in almost any mineralogical association, provided the rock concerned is basic in terms of its SiO_2 content, that is, up to about 54 per cent, as in the case of anorthositic layers.   The Bushveld chromite, for example, occurs in substantial proportions in all rock types except the diorites and granodiorites.   This is clearly illustrated in Fig. 11-6, which is a section through the F, G, and H units at Jagdlust, in the northeast section of the complex (after Cameron, 1962, p. 98).   Clearly the range of lithological association of the chromite is wide—from a mafic norite to anorthosite in this example.   For layered intrusions in general, the range is dunite-peridotite-pyroxenite-gabbro-anorthosite.

Although *alpine-type intrusions* also show a substantial range of lithology, ultramafic types inevitably bulk much larger in them than in the layered masses. It is also noticeable that the alpine-type chromite concentrations show a very much closer association with olivine-rich rock types than do the layered ones.   When examined microscopically it is rare to find alpine-type chromite with anything other than olivine or serpentinized olivine as its immediate associates.   While there are instances where pyroxene is associated (e.g., the Marias Kiki ore body in New Caledonia), this is the exception rather than the rule.

The two types of occurrence also show gross differences in the composition of the chromite itself.   These cannot be demonstrated in any exact way, since the mineral may vary substantially within each type of occurrence and indeed within individual deposits.   However, the following generalizations may be made:

1. In their MgO/FeO and $Cr_2O_3/Fe_2O_3/Al_2O_3$ relations, chromites from the two types of occurrence overlap extensively.   However, there is a *strong tendency* for the various ratios to differ, as follows.
2. RO: the alpine-type chromites usually have significantly higher MgO/FeO ratios than do the layered ones.   For those of *alpine type* the ratio falls approximately in the range MgO/FeO 1:1 to 7:3, and for *layered* chromites MgO/FeO is approximately 3:5 to 1:1.
3. R_2O_3: In *alpine-type* ores, Fe_2O_3 is characteristically low (usually less than 8 weight per cent); Al_2O_3 and Cr_2O_3 show reciprocal relations—according to Thayer (1964) Cr_2O_3 ranges from about 65 to 16 weight per cent and Al_2O_3

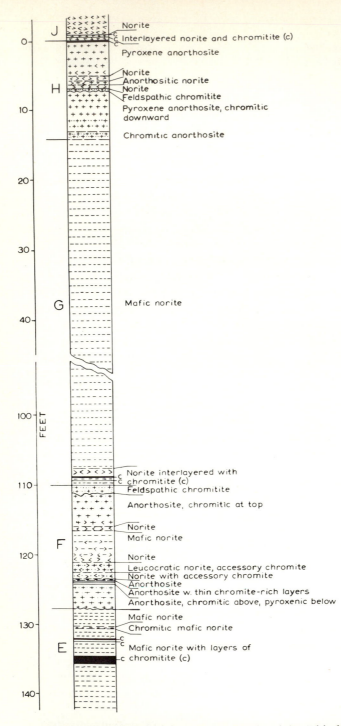

Fig. 11-6 Section through the *F, G,* and *H* units of the Critical Zone (cf. Fig. 11-4), showing the principal lithologies and their sequences. (*From Cameron, in Mineral Soc. Amer. Spec. Pap. no. 1, 1963.*)

320

from about 6 to 52 weight per cent.   In *layered* chromites, Fe_2O_3 is usually notably higher—within the approximate range 10 to 24 per cent. Al_2O_3 and Cr_2O_3 range between wide limits, Al_2O_3/Cr_2O_3 tending to be higher than in the alpine-type chromites.

4. Cr/Fe ratios: It follows from 2 and 3 that these are usually much higher in *alpine-type* chromites (1.5 to 4.5; mean about 2.5) than in *layered* chromites (0.75 to 1.75; mean about 1.0).

In both types of occurrence the composition of the chromite may vary considerably within a single intrusive mass, and even within single ore bodies.   Limited analytical figures suggest that chromite compositions in *layered intrusions* show substantial similarity *within* layers but frequent substantial variation *between* them. *Alpine-type deposits*, with their great variation in shape and internal structure, are naturally more difficult to study.   In general, individual deposits of this type show little internal variation in the composition of their chromite, and indeed in any given area groups of such deposits usually exhibit quite similar chromite compositions.   In some cases, however—notably in the Caribbean, as observed by Thayer (1946, 1964)—there is considerable variation.   Thayer notes that in some districts, such as those of

> . . . Grant County, Oregon, and Sagua de Tanamo, Cuba, the chromite mineral in separate but otherwise apparently identical deposits may differ by 20 percentage points in Cr_2O_3.   Within such districts or regions there is likely to be a uniform pattern of compositional variation.   In the Caribbean region, including Cuba, Guatemala, and Venezuela, the molecular percentage of MgO in total RO varies only between 65 and 75, although the Cr_2O_3 in total R_2O_3 ranges between 25 and 80 mol percent . . . (1964, p. 1501).

All this refers to *mean* compositions, i.e., the compositions of individual bands and pods as indicated by analyses of bulk samples.   On the finer scale of the chromite grain itself, compositions are often far from constant.   Occasionally individual grains appear homogeneous when examined microscopically, but most commonly they are zoned.   In alpine-type chromites, for example, it is very common to find the centers of grains translucent in rich cherry red and of high MgO and Cr_2O_3 content, while the outer zones are opaque with distinctly higher FeO and Fe_2O_3 content.   Occasionally the increase in iron is sufficient to give the outer zone a perceptibly higher reflectivity, in which case the material probably approaches magnetite in composition.   Naturally the inner zones of these chromites have high Cr/Fe ratios—usually greater than $3:1$—whereas the outer zones may have Cr/Fe equal to or less than $1:1$.

Various mechanisms for the development of these iron-rich outer zones have been suggested:

1. Exsolution of iron from the chromite, with concentration near the chromite grain boundaries.
2. Deposition of iron oxide (thrown down during serpentinization) round chromite

grain boundaries; such iron oxide might form marginal granules or a cement or might enter the crystal structure of the chromite itself.

3. Al_2O_3 and MgO might have been subtracted from the outer zones of the chromite during serpentinization, causing a relative enrichment in iron.

4. Primary growth zoning: increase in the proportion of iron may be a simple result of fractional crystallization, analogous to the increase in Fe/Mg ratios found in zoned olivines.

Any of these processes may have operated in different individual deposits; however, the last—growth zoning—seems likely to be the most widespread cause of iron enrichment.

The textures of chromites in different deposits are commonly quite distinctive. In *layered ores* there is a strong tendency toward euhedralism, particularly where the "settled" particles are surrounded by relatively large amounts of interstitial material. Of the Stillwater rocks, Jackson notes:

> In many rocks of the Ultramafic Zone, the individual grains of chromite, olivine, and bronzite are completely bounded by crystal faces, but every gradation between euhedral and anhedral grains exists. Euhedral grains are associated with rocks that include relatively large amounts of interstitial material. With reduction in the amount of interstitial material due to enlargement of the settled constituents, these crystals develop generally polygonal interference boundaries. The degree of automorphism varies from layer to layer, but it is relatively constant in any given layer and can be predicted by knowledge of the amount of interstitial material within that layer (1961, p. 13).

Thus where the chromite and other early formed minerals have crystallized under conditions of "free growth" in the body of the melt, they have, as might have been predicted from the considerations of Chap. 8, developed faces; where they have grown in close proximity to each other and impinged, then cooled slowly to equilibrate boundaries, they have formed "polygonal interference boundaries"— the foam structures of Chap. 9. In the first case, where there is substantial interstitial material, chromite crystals often just touch, giving rise to *network* and *chain* arrangements.

The chromite of *alpine-type ores* usually develops in a variety of ways, mostly quite different from the forms found in the layered types. The development of crystal faces is unusual. Where surrounded by silicate the chromite is almost always rounded, whether it be in the form of fine disseminations or coarse, relatively rich ore. Where it occurs as massive single-phase aggregates, it again, of course, develops foam structures. The rounded forms range from fine (less than 1.0-mm grain size as in much fine *disseminated ore*) to quite large rounded and ellipsoidal bodies 3.0 cm or more across. Chromite of this kind with a grain size around 3.0 to 5.0 mm is termed *pique ore*. That coarser than about 1.5 cm is often referred to as *leopard*, or *grape*, ore. In most cases all these rounded grains are single crystals. Occasionally they are composed of several concentric shells of chromite and serpentine, in which case they are termed *orbicular*.

One feature of the alpine-type chromites to which Thayer (1964) draws particular attention is the development in them of "pull-apart" structures. These appear as transverse silicate-filled fractures in the chromite, and there seems no doubt that they develop as a result of elongation of the chromite grains and their matrix and are a manifestation of the deformation that is a ubiquitous feature of alpine-type occurrences.

These differences in form of the chromite in layered intrusions as compared with that of alpine-type masses are distinctive and widespread and appear to have clear significance concerning origin.

ORIGIN OF CHROMITE CONCENTRATIONS

This appears to be fairly clear—at least from the point of view of general principles —for the layered deposits but notably unclear for those of alpine type.

In the case of *layered intrusions* the mechanism appears to be one of fractional crystallization leading to the settling of successive crops of crystals on the progressively rising floor of the intrusive sheet concerned. Jackson, following work of G. M. Brown (1956) on the layered ultrabasic rocks of the island of Rhum, Inner Hebrides, suggests that the layering in the Ultramafic Zone of the Stillwater complex (and, by analogy, presumably the layering of intrusive sheets in general) may have developed in a manner analogous to that responsible for the development of layering in evaporites:

> The Stillwater ultramafics and evaporites appear to have the following common features: (1) both have a bedded distribution acquired by crystallization and settling of primary precipitates from a saturated solution; (2) both are constructed of compositional layers, commonly monomineralic or bimineralic, which are derived by fractional crystallization . . .; (3) in both, the compositional layers are commonly repeated in cyclic fashion . . .; and (4) primary settled grains within the layers are single crystals, although more material of the same composition may be added as crystallographically oriented overgrowths after deposition . . . (1961, pp. 99–100).

Other suggestions for the development of stratification in layered intrusions (see Turner and Verhoogen, 1960, pp. 304–305) include: (1) development of large convection cells, with formation of crystals in the convecting magma as it moves across the top and down the sides of the chamber and sedimentation of the crystals as the magma moves across the floor; (2) multiple injection of magmas drawn periodically from a deeper magma as this differentiates; (3) changes in mineral equilibrium boundaries accompanying changes in H_2O, O_2, and other partial pressures, leading to the formation and precipitation of different minerals in alternate or cyclic fashion.

For the very beautifully layered sheet intrusions, having large areal dimensions compared with their thickness, the third possibility—with which Jackson's hypothesis agrees—seems likely to be the major mechanism. The development of convection cells and/or the later injection of minor intrusive sheets may well account for minor local aberrations in the layering.

The formation of the *alpine-type* intrusions, their structures, and their various types of chromite concentration seems a vastly more complex problem—or set of problems.   It also looks, despite many suggestions to the contrary, to be quite a different one.   The sharp differences from the layered intrusions in overall bulk composition, abundances of ultramafic types, olivine and pyroxene compositions, and chromite compositions and textures, all argue against a common origin.   In the light of the considerations of Chaps. 8 and 9 it is interesting to note that in the environment existing in cooling layered intrusions it was crystal faces that yielded the energetically most economical solid-liquid interfaces during crystal growth, whereas in the environments of formation of the alpine-type rocks it was the spherical shape that gave the lowest-energy interfaces.   This might well form the basis for some interesting and significant future experiments.   Until recently there has been a very general belief that the alpine-type rocks were of very fundamental nature—probably mantle material pinched into, and squeezed up, very deep faults. Lately there has been the alternative suggestion, supported by a number of field and compositional features, that they are in fact comparatively shallow differ- entiates of volcanic magma chambers, later serpentinized and squeezed into the great faults that had earlier localized the volcanism concerned.   Whichever—if either—of these possibilities is the case, the derivation of the two classes of rocks, and their chromite, do seem to be different.   In addition, whereas the layered intrusions occur virtually in their pristine state apart from simple faulting and tilting, the alpine-type intrusions are metamorphosed.   The short-range lensoidal banding of the ultramafic host rocks, the extreme variation in the form and structural setting of the ore bodies, and the obvious pull-apart and other deformation structures in the chromite ores themselves all point to substantial physical modification after original formation.

NICKEL

As a primary constituent nickel is a very minor component of both layered and alpine-type intrusions.   Some alpine-type masses are important sources of nickel, but its concentration in this case is due to weathering and hence is secondary.

Most *layered intrusions* contain small concentrations of nickel, usually as sulfides and arsenides and usually fairly well confined to a few of the more mafic bands of the sequence.   The sulfide assemblages are commonly pyrrhotite- pentlandite-chalcopyrite, generally with trace quantities of other nickel and copper minerals such as gersdorffite, cubanite, bornite, and valleriite.   Pyrite is sometimes an accessory.

The Skaergaard intrusion contains no more than minute traces of sulfides, apparently all cupriferous.   Wager, Vincent, and Smales (1957) mention bornite, digenite, chalcopyrite, and covellite; virtually no nickel occurs in the Skaergaard sulfides ". . . because, by the time an immiscible iron sulphide liquid separated, nickel had been reduced to very low amounts in the magma as a result of abundant entry into early olivine and pyroxene" (1957, p. 856).   The principal sulfides of Stillwater are chalcopyrite, pyrrhotite, and pyrite, which are fairly conspicuous near the floor of the complex and at one or two of the higher horizons.   Again nickel is not conspicuous.

What appear to be the two best examples of nickel concentration in layered complexes are in southern Africa—the Bushveld complex and the great lopolithic intrusion of Insizwa Range, in Cape Province.

In the Bushveld complex the nickel occurs as a constituent of the Merensky Reef, already noted as the layer forming the upper limit of the Critical Zone of the intrusion. This layer contains substantial amounts of the platinoids and other precious metals, and as accessories, sulfides occur to the extent of 0.5 to 2.0 per cent. The principal species are pentlandite, chalcopyrite, pyrrhotite, and valleriite in order of abundance.

In the Insizwa Range small quantities of pyrrhotite, pentlandite, chalcopyrite, cubanite, and an assemblage of minor associates (Scholtz, 1936) occur in the basal portions of the lopolith.

By far the major part of the nickel of *alpine-type* intrusions is "silicate nickel" occurring as a minor constituent of olivine and serpentine derived from the latter. As was mentioned in Chap. 3 the nickel is incorporated in the olivine principally at the expense of Fe^{++}. Its abundance is usually of the order of 0.3 weight per cent of the whole rock and clearly far too low for the latter to be regarded as an ore at the present time. Substantial concentration may be achieved during weathering and soil formation, however, and residual deposits containing 1.5 to 2.5 per cent nickel are often extensive.

A minute amount of nickel may also occur as disseminated sulfide in alpine-type ultramafic rocks—usually pentlandite or heazlewoodite.

PLATINOID AND OTHER PRECIOUS METALS

The collective term *platinoids* includes platinum itself, palladium, rhodium, ruthenium, osmium, and iridium. Mafic and ultramafic rocks of the present associations yield a substantial proportion of the world's supply of these metals and, with them, some gold and silver. In some ways the mineral assemblages of the deposits concerned are rather similar to those of the next association we are to consider—the mafic-ultramafic–iron-nickel-copper sulfide-platinoid association—but there are also distinctive differences. One is that the platinoids of the present association are generally closely tied to the incidence of chromite, which is not the case with the other association.

The outstanding example of platinoid occurrence in chromite-bearing *layered intrusions* is the Merensky Reef, already referred to. Although termed a "reef" it is in fact one of the layers of the Bushveld complex—a pegmatitic pyroxenite occurring at the top of the Critical Zone. Normally this pyroxenite is bounded at its top and bottom contacts by a concentration of chromite crystals so that the ore-bearing layer is usually demarcated by two well-defined thin chromitite seams. The reef has been traced over a total of 132 miles, though it has been mined along only 11 of these to the present time.

According to Cousins (1964, p. 230), concentration of the precious metals achieves maxima within or adjacent to the top and bottom chromitite bands. The known platinum-group minerals and their associates are ferroplatinum (platinum metal alloyed with varying amounts of iron), nickeliferous braggite [(Ni,Pt,Pd)S$_2$], cooperite (PtAsS), laurite [(Ru,Os)S$_2$], stibiopalladinite (Pd$_3$Sb), sperrylite, gold (alloyed with zinc), pentlandite, chalcopyrite, pyrrhotite, valleriite, and chromite.

While small variations occur along the layer, the overall proportions of the platinoids and gold approximate to Pt, 60 per cent; Pd, 27 per cent; Ru, 5 per cent; Rh, 2.7 per cent; Ir, 0.7 per cent; Os, 0.6 per cent; Au, 4 per cent. The chromite defining the layer is reported by Cousins (1964, p. 231) as being a high-iron variety: Cr$_2$O$_3$, 40.5 per cent; total iron as FeO, 32.6 per cent; Al$_2$O$_3$, 15.2 per cent; MgO, 9.7 per cent; TiO$_2$, 2.0 per cent; Cr/Fe, 1:1. The chromium, nickel, copper, and platinoid metal abundances vary sympathetically. Cousins suggests that this, together with the great evenness of values over large areas, suggests precipitation from a liquid and concomitant crystal settling. However, he notes:

> Numerous problems are, however, unsolved. There are the questions as to why the sulphides and platinum should be concentrated in one thin heave of magma; why the reef should have crystallized as a pegmatite; why chromite should be concentrated at *both* contacts; why the sulphides should consist predominantly of nickel and copper . . . all that can be said at present is that the evidence points to magmatic differentiation under a suitable, but as yet unproven, set of conditions during emplacement (1964, p. 236).

Similar concentrations of the platinoids have not been found in *alpine-type* intrusions, though platinoids are certainly well known to occur in ultramafic belts. The principal metals are platinum, osmium, and iridium, and rather than occurring as arsenide-sulfide concentrations as in the layered intrusions, they are found native, in disseminated distribution. Economic concentrations are produced later by weathering and subsequent placer accumulation.

THE MAFIC-ULTRAMAFIC–IRON-NICKEL-COPPER SULFIDE-PLATINOID ASSOCIATION

Although some deposits of this association show minor similarities to the small sulfide-platinoid occurrences in the layered intrusions of the association we have just considered, there are a number of differences. The principal divergences seem to be:

1. The ores of the present association are dominantly *sulfide*—oxide and arsenide are usually very minor components.
2. The present ores occur neither in large conspicuously layered intrusions nor in alpine-type masses. Their principal habitat is unbanded, essentially irregular gabbro—particularly noritic gabbro—intrusions. A substantial proportion is associated with ultramafic rocks, but these are neither bands of layered complexes nor intrusions of alpine type. Some indeed may be basic to ultrabasic lavas.

3. While variable amounts of magnetite, titanomagnetite, and ilmenite are common minor associates, chromite is conspicuous by its absence or near absence.

While it must be acknowledged that this group of deposits shows some variation in form and environment and clear, if minor, differences in constitution, there is no question that there is a very strong thread of similarity running through it. Such deposits are familiarly known as *sulfide nickel ores*, and they have now been found on most of the continents.

DISTRIBUTION

Until recently it appeared that virtually all the world's "sulfide nickel" occurred in the Northern Hemisphere; ores of this kind seemed to be absent from the Southern Hemisphere until the discovery of the Kambalda deposits, in Western Australia, in 1966. These are now known to be substantial, and their discovery has greatly altered the earlier picture of the distribution of the world's sulfide nickel.

By far the largest part of the world's known resources, however, occurs in Canada, whose production of sulfide nickel easily exceeds that of the rest of the world together. A clear—though conservative—picture of Canada's dominance is given by Table 11-2, which includes production from both sulfide and silicate (laterite) ores. Since Canada's production is entirely from sulfide, while that of most of the other countries is principally from silicate, Canada's preeminence as a possessor of sulfide ores is very marked indeed. Not only has Canada *many* ore bodies of the present type, but it possesses some very *large* ones; those of Sudbury, Ontario, for example, easily overshadow all other known deposits, of either kind, in the rest of the world. In addition to the deposits of Sudbury, substantial ore bodies are mined at Thompson and Lynn Lake in Manitoba. What may amount to large resources occur in the Ungava district, in northern Quebec. Smaller deposits are mined at Marbridge and Lorraine in southern Quebec, Texmont, and Gordon Lake in Ontario, and at Hope, British Columbia (Fig. 11-7).

Table 11-2   Estimated nickel production.   (Adapted from Boldt and Queneau, 1967; U.S. Bureau of Mines Minerals Year Book, 1965)

	Yearly average, 1961–65		1965	
	Short tons ×10³	%	Short tons ×10³	%
Canada	236	56.87	269	56.99
U.S.S.R.	88	21.20	95	20.13
New Caledonia	46	11.08	57	12.08
Cuba	18	4.34	19	4.03
U.S.A.	12	2.89	14	2.97
South Africa	3	0.72	3	0.64
Eastern Europe and Brazil	12	2.89	15	3.18
Estimated world total	415		472	

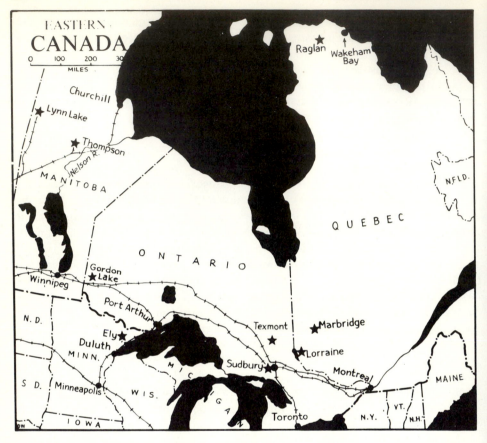

Fig. 11-7 Distribution of the more important sulfide nickel deposits (shown by stars) in eastern Canada and U.S.A. (*From Boldt and Queneau, "The Winning of Nickel," The International Nickel Company of Canada Limited and Methuen & Co., Ltd., London, 1967, with the publishers' permission.*)

Several important deposits are known in northern Russia; these are in the Pechenga and Monchegorsk districts of the westerly part of the Kola Peninsula and the Noril'sk district of northwestern Siberia (Fig. 11-8). The Australian deposits, which are the focus of extremely active exploration and hence which may be added to substantially in the future, all occur in the vicinity of Kalgoorlie (Fig. 11-12).

Most of the ore bodies occur in Precambrian rocks. A few are found in Paleozoic intrusions in Precambrian terrains and, here and there, in post-Paleozoic intrusions in post-Precambrian terrains.

FORM

As a class these deposits display a wide variety of form both in outline and in the distribution of ore types within the individual ore bodies themselves. Most commonly they are in the form of distorted thin sheets localized along the contact of

the associated intrusion with the intruded rock.   Distortion may have resulted
from original irregularity of the contact, from later folding of the contact or, to a
lesser extent, from faulting.   Some of the deposits are in the form of dikes, or of
sulfide-cemented breccia zones, forming "offsets" from the main intrusive body.
Here the deposits again usually take the form of thin sheets.   In other cases the
ore bodies occur as irregular, rounded, or lobate bodies within the intrusive adjacent to its margin, in the intruded rock, or straddling the contact between the two
(see Boldt and Queneau, 1967, pp. 32–38).

   Very commonly the ore of individual deposits varies in type, and the masses
of the different ore types develop a variety of shapes and spatial relations with each
other.   The lower-grade ores consist of rounded—sometimes quite globular—
particles of sulfide disseminated among, and molded about, the silicates of the
associated igneous rock.   This is termed *disseminated ore*.   By increase in the
proportion of sulfides, the globules inevitably begin to touch, producing a network
of sulfide through the silicate host.   Further increase in sulfide progressively
closes up the network until the material is virtually 100 per cent sulfide, at which
point it is termed *massive ore*.   Although the two are often gradational, the ore of
an individual deposit is usually divided, for mining purposes, simply and arbitrarily into two categories—*massive* and *disseminated*.   The proportions of the two
broad ore types vary enormously from one deposit to another, in keeping with their
great variety of spatial relations and boundary shapes.   The only generalization

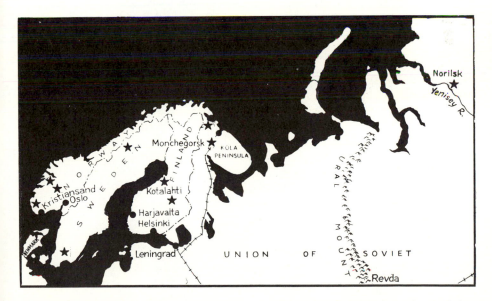

Fig. 11-8   Distribution of the more important sulfide nickel deposits (shown by stars) in
Russia and Scandinavia.   In addition several other smaller deposits occur in Norway and
Sweden.   (*Adapted from Boldt and Queneau, "The Winning of Nickel," The International
Nickel Company of Canada Limited and Methuen & Co., Ltd., London, 1967, with the pub-
lishers' permission.*)

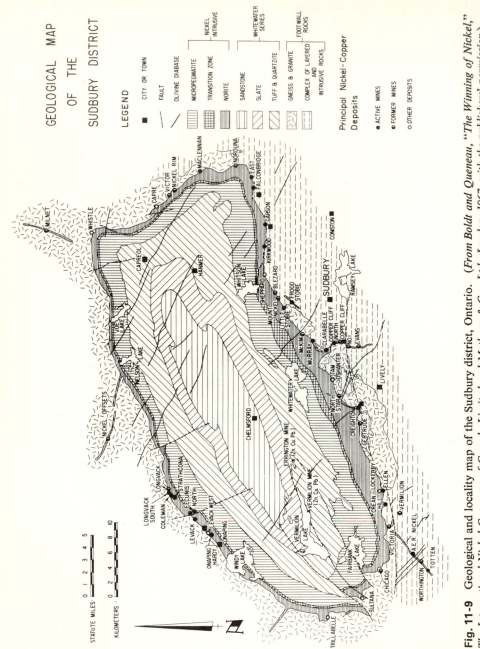

Fig. 11-9 Geological and locality map of the Sudbury district, Ontario. *(From Boldt and Queneau, "The Winning of Nickel," The International Nickel Company of Canada Limited and Methuen & Co., Ltd., London, 1967, with the publishers' permission.)*

that can be made is that the two types are usually in contact.   The disseminated ores often appear to represent an early stage of the process that formed the massive ores.

SETTING

As in their form, these deposits show considerable variety in the detail of setting, but there is a very strong thread of general similarity running through the group as a whole.   Table 11-3 gives the nature of the associated intrusive rocks for a number of documented occurrences (it should be kept in mind that there may well be instances where igneous rocks now regarded as intrusive will in the future be found to be extrusive).   It will be seen that noritic gabbros are prominent among the intrusive rocks, though other types—notably ultramafic rocks and their serpentinized equivalents—are often involved.   Among the adjacent rocks mafic pillow lavas and "greenstones" (metamorphosed mafic lavas and associated pyroclastic rocks) are prominent, though again, other types, including a variety of sedimentary rocks, may be involved.

Four major occurrences demonstrate the range of environment in which sulfide nickel deposits are found:

1. Sudbury, Ontario   As already noted this is easily the largest current source of sulfide nickel.   It embraces about 40 known deposits, some of which are very large indeed.   The principal features of the geological setting are shown in Fig. 11-9.   The sulfide ores occur in and adjacent to the bottom contact of an elongated, basin-shaped intrusion of rock loosely termed norite.   This intrusion outcrops as a northeast-trending elliptical ring, 1 to 4 miles in outcrop width and with major and minor axes measuring about 37 and 17 miles, respectively.   Within the ring itself there are three principal "facies" of igneous rock, arranged more or less concentrically.   These are an inner (upper) zone of micropegmatite, a zone of noritic gabbro, and along the base, a discontinuous basic "noritic sublayer" (Souch, Podolsky, and others, 1969).   The sublayer is characterized by conspicuous inclusions of peridotite, pyroxenite, and gabbro as well as xenoliths of the adjacent rocks.

These three layers are referred to simply as *micropegmatite, norite,* and the *sublayer.*   Boundary relations of the norite with the overlying micropegmatite are gradational, and for this reason a further layer—the *transition layer*—is specifically shown on geological maps.   The rocks outside the basin are gneisses and granites and, along much of the southern margin, steeply dipping metamorphosed volcanic and sedimentary types.   Those inside the basin are principally pyroclastic rocks and their resedimented derivatives, collectively termed the *Whitewater Series.*   Here and there all the rocks immediately surrounding the basin have suffered conspicuous brecciation, in which case they are referred to as *Sudbury breccia.*   Perhaps surprisingly, neither the nickel intrusion nor the Whitewater rocks have been involved in this brecciation.   All three—the intrusive, the surrounding rocks, and the Whitewater Series—have, however, been offset together by a number of faults (Fig. 11-9), which are particularly conspicuous along the northern and western edges. Radioactive dating (Rb/Sr and lead) indicates an age of 1700 to 2000 million years for the various layers of the intrusion (Souch, Podolsky, and others, 1969).

The sulfides occur at least in small quantity virtually continuously round the outer rim of the intrusion, locally achieving concentrations (Fig. 11-9) sufficiently high to be termed ore bodies. However, while occurrence on this scale is apparently very uniform, on a smaller scale there is extensive variation. Much of the ore lies within the sublayer, but large quantities occur at the contact of sublayer and intruded rocks, within zones of Sudbury breccia and in the underlying gneiss, and in quartz diorite dikes emanating from the sublayer along faults and breccia zones. The sulfides therefore occur variably as disseminated, network, and massive ore within the intrusive, as massive and vein ore along the contact, and as the cementing material of a variety of associated breccias.

2. Moak Lake-Setting Lake, Manitoba (the Thompson mine) Substantial and extensive deposits of iron-nickel-copper sulfides occur in an area about 250 miles southwest of Churchill on Hudson Bay (Fig. 11-7). The geological

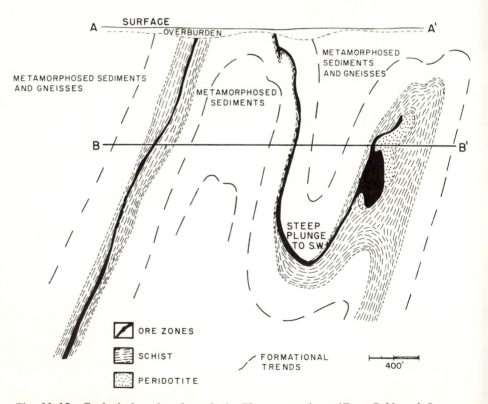

Fig. 11-10 Geological section through the Thompson mine. (*From Boldt and Queneau, "The Winning of Nickel," The International Nickel Company of Canada Limited and Methuen & Co., Ltd., London, 1967, with the publishers' permission.*)

setting of the ores is rather different from that at Sudbury, and in particular, the sulfides appear to be associated with *ultramafic* material rather than with mafic and more felsic rocks as at Sudbury.

According to Zurbrigg (1963) the locus of the ore at the Thompson mine (the largest deposit of a number known in the area) is a schist zone that is conformable with the overlying and underlying formations.   Most of the Moak Lake mineralized area is underlain by gray and pink felsic gneiss, with which are associated minor concordant bands of hornblende-plagioclase gneiss and amphibolite. Peridotite lenses are a conspicuous feature of the district—virtually all of them spatially associated with linear zones of metasediments that in turn grade into, or form sharp contacts with, enclosing gneiss.   The peridotite is composed of pseudomorphs of serpentine after olivine and orthopyroxene and appears to represent a genuine ultramafic rock.   The metasediments comprise quartzite, subgraywacke, limestone (including skarn), iron formation, biotite schist, amphibolite, and minor amounts of metamorphosed basic lava ("greenstone").

In some areas the metasedimentary bands are strongly folded.   Zurbrigg (1963, p. 2) states that the particular structure controlling the distribution of ore at Thompson is an anticlinal fold striking northeast and plunging steeply to the south.   The schist associated with the ore is continuous over the full length of the structure, but the peridotite, which occurs as a stumpy lens, is confined to one end of the ore body (Fig. 11-10).   The Thompson ore zone itself has an overall length of $3\frac{1}{2}$ miles.   Within it, the sulfides occur as disseminations and stringers in the serpentinized peridotite and as massive bodies and stringers in the schist and adjacent metasedimentary rocks and gneisses.

For the Moak Lake-Setting Lake area generally, Boldt and Queneau note that nickel sulfide occurrence

> . . . is associated in a broad way with bodies of peridotite.   Some occurs in and on the margins of the peridotite itself.   The bulk of the known ore, however, is in the enclosing sedimentary rocks.   These enclosing rocks have been folded, converted by metamorphism into schists and gneisses, and injected by pegmatite.   The principal ore-bearing formation is a group of beds of biotite schist and impure quartzite, sandwiched between metamorphosed iron formation and recrystallized impure limestone, called skarn.   The ore minerals replace, and fill fractures in, the biotite schist and to a minor degree the enclosing beds of quartzite.   Thus the ore is for the most part concordant with the folded formations (1967, p. 51).

3. Lynn Lake, Manitoba (Sheritt Gordon mines)   Like the Sudbury and Moak Lake-Setting Lake areas, that surrounding Lynn Lake contains a number of separated nickel sulfide bodies.   The area concerned (Fig. 11-7) is about 170 miles WNW of Moak Lake and is close to the Manitoba-Saskatchewan border.

We have seen that the ore bodies of Sudbury are associated with an elliptical, basin-shaped intrusion and those of Moak Lake with a series of essentially concordant peridotite pods.   In contrast the Lynn Lake deposits are found within two of a series of intrusive plugs (Fig. 11-11).†   These range in composition from

---

† A sill form has, however, also been suggested for these intrusions; see Naldrett and Gasparrini (1971).

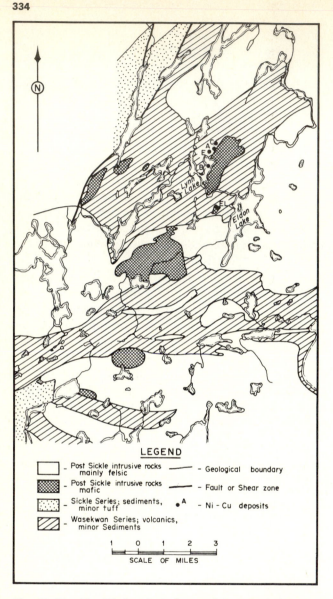

Fig. 11-11 Simplified geological map of the Lynn Lake area, Manitoba. (*Adapted from Boldt and Queneau, "The Winning of Nickel," The International Nickel Company of Canada Limited and Methuen & Co., Ltd., London, 1967, with the publishers' permission.*)

peridotite through norite to diorite and have been intruded into a variety of pre-existing rock units. As shown in Fig. 11-11, these rock units consist of granites, granodiorites, sedimentary and pyroclastic rocks, and lava flows.

The principal ore bodies occur in "amphibolite," a metamorphosed gabbroic rock.   This and the contained sulfides have suffered heavy post-ore faulting so that the deposits now consist of a group of large irregular fault blocks, with smaller fault blocks in between.   The sulfides are found in massive and disseminated form and as masses of stringers forming stockworks within the enclosing intrusion.

4. Kambalda, Western Australia   Kambalda lies about 34 miles south of Kalgoorlie, in the southern part of Western Australia.   Although the Kambalda area remains Australia's richest known nickel locality—six separate high-grade ore bodies have already been found and there is clear promise of more—other deposits

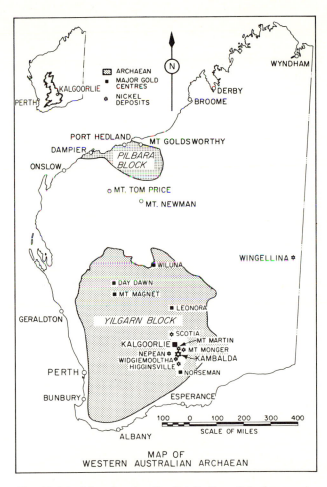

Fig. 11-12   Map showing the distribution of Archean rocks in Western Australia and the occurrence of nickel sulfide deposits in the Kalgoorlie-Kambalda area.   Further deposits in this area have been found since this map was drawn.   (*From Woodall and Travis, Commonwealth Mining Met. Congr. Trans., 1969.*)

have since been found to the south and west of Kambalda itself and well to the north of Kalgoorlie (Fig. 11-12). It may therefore be that in time the area of nickel mineralization will be better referred to as the *Kalgoorlie area*—the district originally made famous by its gold occurrences—rather than Kambalda.

As in the Keewatin Province of Canada, most gold and base metal sulfide deposits of the Precambrian of Western Australia are associated with "greenstone belts" (consisting of a variety of volcanic and sedimentary rocks) that occur as large isolated masses within major granitic-metamorphic terrains. Some of the lavas and pyroclastic rocks of these belts are very highly mafic indeed and appear to be intruded here and there by ultramafic sills, now substantially serpentinized. The Kambalda deposits occur in a serpentinized ultramafic rock mass, though whether this is intrusive (a sill or sills), or a group of lavas, or both, is still not really clear. According to Woodall and Travis (1969) the mass is a sill, injected between two basalt flows. It is approximately conformable with the enclosing basalts and is a layered mass consisting of varying amounts of serpentine, amphibole, talc, chlorite, carbonate, and biotite, together with accessory magnetite and very minor chromite. The associated basalts, which show much excellent relict pillow structure, are now fine-grained amphibole-saussuritized feldspar rocks with variable contents of chlorite, biotite, and carbonate. Lenses of shale and banded chert-iron formation occur conspicuously on both the upper and lower contacts of the ultramafic body, and discontinuous chert bands up to 50 feet thick are present *within* the ultramafic itself. (The remarkable parallelism of these included sedimentary bands with the trends of the enclosing ultramafic belts, and among themselves, and the semicontinuity of many of them along strike seems to indicate strongly that much of the "ultramafic" was originally extrusive rather than intrusive; this would imply that the rocks were picritic—and even more olivine-rich—basalts similar to some of the very high-magnesian basalts found in the Tertiary to Modern volcanoes of the Solomon Islands, Hawaii, and elsewhere at the present day.) Some of the sedimentary bands are very rich in sulfides and contain up to 50 per cent pyrrhotite, with trace to minor quantities of pyrite, sphalerite, chalcopyrite, and galena. The age of the rocks has been determined, by Rb/Sr measurement (Turek, 1966), as about 2700 million years.

The outcrop pattern shown in Fig. 11-13 reflects a domal structure, the ultramafic layer dipping outward at a general angle of about 40°. The structure of the lower contact—that most intensively studied so far—is, however, not nearly so simple as the overall structure would suggest. Indeed it is quite complex in parts, as shown in Fig. 11-14. Woodall and Travis (1969) consider that the shape of the contact cannot be explained by drag folding or other tectonic distortion and suggest that some of the structures may be due to erosion, i.e., scouring, of the basalt by the ultramafic as the latter was injected. An alternative explanation, if the host rock is a flow rather than a sill, is that the "structures" are original surface irregularities in the underlying volcanic flow.

The nickel and associated sulfides occur in the lower part of the ultramafic sheet, and most of the ore grade material is at or close to the basal contact. This is referred to as *contact mineralization*. Some of this sulfide can actually be seen

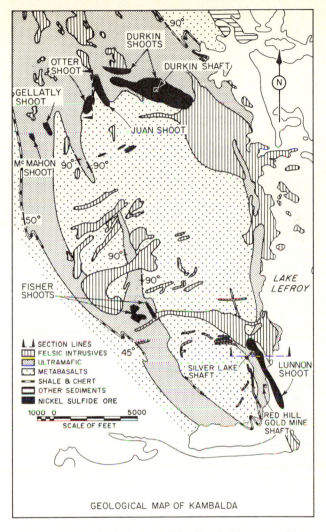

Fig. 11-13 Geological map of the Kambalda area. (*From Woodall and Travis, Commonwealth Mining Met. Congr. Trans., 1969.*)

fingering down between the pillow structures of the underlying basalts. A lesser but still substantial amount of sulfide occurs well within the ultramafic and hence stratigraphically above and separate from the contact deposits. These higher masses are referred to as *hanging-wall mineralization*. As in the case of Sudbury, some nickeliferous sulfide occurs in nearby shear zones, usually where the latter intersect contact ore bodies. By obvious analogy such occurrences are termed *offset mineralization*. The sulfides are present both in massive and disseminated form, and where both types occur in a single ore body, the disseminated material

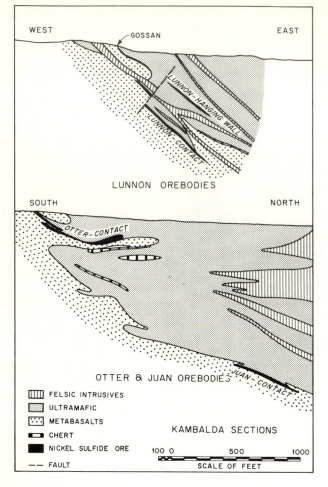

Fig. 11-14 Sections through the northern and southern parts of the Kambalda dome structure shown in Fig. 11-13. (*From Woodall and Travis, Commonwealth Mining Met. Congr. Trans., 1969.*)

overlies the massive. The relationship of the two types, however, is apparently not gradational: the massive sulfides, whether part of the hanging wall or a contact ore body, have sharp upper and lower contacts.

CONSTITUTION

It is here that the ore bodies of this association show their greatest similarity. The principal components are invariably pyrrhotite, pentlandite, and chalcopyrite, commonly in that order of abundance. Pyrite and magnetite are frequently—though not invariably—present as substantial accessories, as is cubanite. (Pyrite is a major constituent at Kambalda, as is cubanite in some parts of Sudbury.) A variety of nickel and iron arsenides and sulfarsenides often occur as minor components. The platinoid metals, particularly platinum and palladium, may occur as important traces, usually as arsenides and bismuthides. Gold and silver occur native and as tellurides and selenides.

The proportion of iron among the sulfide metals is variable, but iron sulfide —chiefly pyrrhotite—is characteristically the dominant opaque mineral. This is almost invariably followed by pentlandite and then by the copper minerals, chiefly chalcopyrite, often with cubanite. Although nickel is almost always more abundant than copper, Ni/Cu ratios are very variable. As shown in Table 11-3, Ni/Cu ratios of between about 2:1 and 3:1 are the most common. However, in certain cases copper exceeds nickel in abundance—as at Noril'sk—while in other cases, such as Thompson and Kambalda, copper is very low and Ni/Cu ratios are as high as 15:1. Similarly the platinoids may be notable or insignificant in amount, and gold and silver, while virtually ubiquitous, may vary substantially.

Several of the occurrences whose settings have been used to illustrate environmental variation also provide quite a representative range of constitution.

1. Sudbury This has a most extensive mineralogy, the primary ore-mineral components of which have been grouped by Hawley and Stanton (1962) as major, minor, and "precious metal" constituents:

a. *Major constituents.* Pyrrhotite, pentlandite, chalcopyrite, cubanite.
b. *Minor constituents.* Magnetite, ilmenite and magnetite-ilmenite intergrowths, pyrite, gersdorffite, niccolite, maucherite, heazlewoodite, bornite, valleriite,

Table 11-3   Nickel and copper contents and principal host rocks of some important sulfide nickel occurrences

Deposit	*Ni%*	*Cu%*	*Ni/Cu*	*Principal host*
Sudbury, Ontario	1.62	1.29	1.3:1	Mafic-noritic
Gordon Lake, Ontario	1.4	0.6	2.3:1	Ultramafic
Marbridge, Quebec	2.3			Ultramafic
Lorraine, Quebec	0.6	1.6	0.4:1	Mafic
Ungava, Quebec	1.8	0.7	2.6:1	Ultramafic
Lynn Lake, Manitoba				
"A plug"	1.2	0.6	2:1	Mafic-noritic
"EL plug"	4.5	1.5	3:1	Mafic-noritic
Moak Lake-Setting Lake (Thompson mine), Manitoba	1.9	0.13	14.6:1	Ultramafic
Bird River, Manitoba	1.0	0.3	1:3	Mafic to ultramafic
Hope, British Columbia	0.8	0.3	2.7:1	Ultramafic
Yakobi, Alaska	0.3	0.2	1.5:1	Mafic-noritic
Kambalda (Lunnon), W. Aust.	5.9	0.5	11.8:1	Ultramafic to mafic
Noril'sk, Siberia	0.4	1.0	0.4:1	Mafic
Kotalahti, Finland	0.8	0.3	3.8:1	?Ultramafic
Petsamo, N.W. Russia	3.8	1.8	2.1:1	Ultramafic

sphalerite, stannite, galena, parkerite, tetradymite, native bismuth, bismuthi-nite.

c. *Precious metal constituents.*   Native gold, native silver, sperrylite, michenerite ($PdBi_2$?), froodite (α $PbBi_2$), hessite, schapbachite.

Apart from the silicates, the four major sulfides form the main bulk of the ore, chalcopyrite and cubanite tending to vary inversely.   In some parts, e.g., the lower levels of the Frood mine, cubanite may exceed chalcopyrite in amount. In general the pentlandite occurs as networks and irregular masses about and between grains of the dominant pyrrhotite.   A minor amount of the pentlandite is found as exsolution bodies within pyrrhotite, as described in Chap. 9.   Cubanite occurs partly as random duplex and polyphase aggregates with the other major sulfides, partly as oriented blade-like intergrowths in chalcopyrite.

Of the minor minerals the arsenides—particularly gersdorffite—often occur as arsenic-rich selvages at and near the edges of some of the ore bodies.   Niccolite and maucherite are often associated with the gersdorffite, frequently as beautiful eutectoid intergrowths with each other and with chalcopyrite.   Pyrrhotite is also an occasional constituent of such intergrowths, which appear to have developed by the chemical breakdown and replacement of the gersdorffite (Hawley, Stanton, and Smith, 1960).   Platinoid minerals, particularly sperrylite and the palladium bis-muthides, also seem to be preferentially developed in this type of ore.

While some of the Sudbury ore shows features attributable to solidification and annealing, much of it shows deformation structures and has clearly been modified to at least some extent by later stress.

2. Thompson   The mineralogy here is far less complex than at Sudbury, and the spectacular arsenide-platinoid and copper-rich ores of the latter are lacking. According to Zurbrigg (1963) the Thompson ore is composed of pyrrhotite, pent-landite, and pyrite, with minor chalcopyrite and marcasite and trace quantities of nickel arsenides (niccolite and gersdorffite).   Marcasite-carbonate veinlets cut the ore, the marcasite probably representing primary pyrrhotite dissolved and re-deposited in veins under a later low-temperature regime.   Of the primary ore, Zurbrigg notes:

> The sulphides occur as fine-grained masses, lenses and veinlets in the schist, as coarse-grained masses and stringers where the host rock is less schistose, and as stringers, veinlets and disseminations in peridotite.   Both the coarse-grained and the fine-grained sulphide masses are termed breccia sulphide, in recognition of the numerous rock in-clusions or remnants contained in them.   The remnants range in size from mere specks to several inches in diameter, and, particularly in the fine-grained breccia sulphide, they have extremely irregular contacts which suggest progressive replacement of rock by sulphides (1963, pp. 235–236).

The general macroscopic appearance of the ore is such as to suggest that the present sulfide-silicate-boundary relationships might have developed by late-stage deformation rather than by replacement.   Much more work, particularly micro-scopical, remains to be done on this ore.

3. Kambalda Like the Thompson ore, that of Kambalda has a high Ni/Cu ratio and consists principally of pyrrhotite, pentlandite, and pyrite, with minor chalcopyrite. Only the Lunnon contact (Figs. 11-13 and 11-14) has been studied in detail mineralogically.

The Lunnon deposit is conspicuously zoned parallel to the base of the ultramafic layer and is unusual in its high pyrite content. This is shown diagrammatically in Fig. 11-15. Woodall and Travis note that the sulfide zone here consists of a disseminated sulfide-silicate layer averaging about 7 feet in thickness lying above a layer of massive sulfide averaging about 4 feet in thickness. Locally the disseminated ore increases up to 18 feet and the massive material to about 10 feet.

The *massive ore* is 90 to 95 per cent sulfide and consists of a pyritic zone underlain by a rich layer of banded pyrrhotite-pentlandite ore. The latter consists of pentlandite-rich lamellae, 1 to 3 mm thick and parallel to the general attitude of the massive layer, set in a pyrrhotite-rich matrix. The mean pyrrhotite-pentlandite ratio in this zone is about 2.7:1. Pyrite commonly occurs as schlieren up to 5 cm long, roughly parallel to the pentlandite bands. Chalcopyrite is often associated with this pyrite, with which it sometimes forms eutectoid intergrowths. Minor

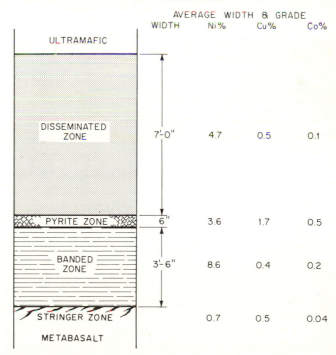

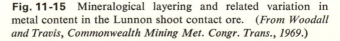

Fig. 11-15 Mineralogical layering and related variation in metal content in the Lunnon shoot contact ore. (*From Woodall and Travis, Commonwealth Mining Met. Congr. Trans., 1969.*)

constituents of the banded ore are magnetite (with some magnetite-chromite inter-growths), altaite, galena, and sphalerite.   The pyrrhotite-pentlandite zone becomes notably pyrite rich near its top and grades into the massive pyrite zone, which consists of 60 to 80 per cent of coarse (3 to 7 mm) pyrite set in a matrix of pentlandite, fine-grained pyrite, and chalcopyrite.

The overlying *disseminated ore* usually consists of about equal amounts of sulfide and silicate and variable amounts of carbonate.   Two broad types occur:

1. Fine-grained pyrrhotite-pentlandite-serpentine-talc-chlorite assemblages, with minor carbonate.
2. Fine-grained pyrrhotite-pentlandite-talc-chlorite-dolomite-ankerite assemblages containing prominent rounded magnesite bodies up to 1.0 cm across.   The latter contain abundant rounded, fine sulfide inclusions ranging in diameter from 0.1 mm down to submicron specks.

The disseminated ore shows no obvious spatial variation in mineralogy other than its increase in sulfide-silicate ratios toward its contact with the underlying massive ore.   Very minor quantities of magnetite, magnetite-chromite, and ilmenite are trace accessories.

The hanging-wall mineralization (i.e., that within the body of the ultramafic layer rather than at the basal contact) is similar mineralogically to the contact ore and varies from wide zones of low-grade disseminated sulfide, through richer disseminated ore, to high-grade massive material.   Offset mineralization is composed of the same group of major minerals as the other forms of occurrence but includes in addition a number of trace species: millerite, molybdenite, sphalerite, gersdorffite, galena, melonite, and gold.   Of millerite, Woodall and Travis note particularly that it is a common accessory and, occasionally, is the major nickel sulfide of offset mineralization in the footwall basalts.

ORIGIN

The derivation and mode of formation of these ores seems, in some respects, to have been considered largely on the basis of the features of one area of occurrence—Sudbury.   The origin of the Sudbury ores has received great attention for a long time, but it is only comparatively recently that the ore type as a category has attracted systematic investigation.   Inevitably the following account reflects this.

Several theories have been put forward to account for the whole or parts of the Sudbury occurrence:

1. Sulfide-silicate immiscibility, and consequent magmatic differentiation *in situ.*
2. Magmatic differentiation—again due to sulfide-silicate immiscibility—*at depth*, with subsequent injection of sulfides as magmas in their own right.
3. Hydrothermal activity, with the deposition of the sulfides principally by replacement of rocks adjacent to the norite contact.
4. "Sulfurization" (i.e., the late addition of sulfur) to a cooling melt, with consequent abstraction of iron, nickel, copper, etc., as the sulfides.

Table 11-4 Schematic presentation of crystallization history of the silicate, silicate-sulfide, and sulfide rocks of the Sudbury intrusion, as deduced by Hawley (1962)

Liquid fractions		Crystallized products		Examples
				In part at:
Silicate-rich fraction, minor sulphides	(Aplite?) Micropegmatite Norite-quartz gabbro and altered equivalents		*Disseminated Sulphides*	Creighton
	Quartz diorite	At base and in offsets with residual sulphides in Q.D. and in Q.D. Breccia, irregular segregations and stringers of sulphides where fractured	*Disseminated Ore* in Q.D. *Disseminated Ore* in Q.D. Breccias *"Granite" Breccia Ore*	Frood-Stobie Frood-Stobie Hardy, Levack
Sulphide-rich fraction, minor silicates	Sulphides crystallized *in situ*	Non- or slightly differentiated. Common pyrrhotite-pentlandite-chalcopyrite ore with or without silicate blebs	*Massive Ore* *Breccia Ore* *Chalcopyrite Stringer Ore*	Creighton Frood
		Differentiated by crystallization with silicate blebs, grading from low copper to high copper	*Silicate-Sulphide Ore* *Massive and Breccia Ore* *Cubanite Zone* *Arsenide Zone* *Siliceous Mineral Zone* with high precious metals Bi, Te, Pb, (Sn)	Frood-Stobie
	Injected sulphides	Along faults, breccias, and contact zones (largely undifferentiated)	*Massive Ore* *Breccia Ore* *Disseminated Ore* in walls *Minor Stringer Ore*	Falconbridge, Murray, Garson, parts of Hardy, Levack?Creighton, and others

Sudbury *Irruptive*

5. Meteorite impact.   The whole Sudbury intrusion, the Sudbury breccias, and
   the Whitewater Series are the result of meteorite impact, the sulfide ores
   representing original meteoritic material localized along the base of the crater.
   Alternatively, the impact of a meteorite may have triggered terrestrial mag-
   matism which then yielded the sulfide concentrations.

The hypothesis of *differentiation in situ* is based upon the experimental work
of Vogt (cf. Chap. 2) and his successors on the immiscibility of sulfides in silicates
and upon the restriction of the sulfide concentrations to the sublayer and its derivatives, such as the quartz diorite dikes of the Frood, and to adjacent structures.
The sulfides are seen as having come out of solution in the sublayer rocks during
cooling of the melt *after* the latter intruded into its present position; the sulfides
migrated to, and built up along, contact interfaces and accumulated within contact
depressions.   Here and there they were reintruded to form dikes, breccia fillings,
and veins in adjacent structures.   As the sulfide segregated from the main mass of
noritic liquid, silicates separated in minor quantities from the developing sulfide
masses also.   A schematic presentation of the crystallization history of the Sudbury
intrusion (termed "irruptive") and its ores as deduced by Hawley (1962) is given in
Table 11-4.   The development of the two-liquid system—"sulfide-rich" and "sili-
cate-oxide" liquids—is portrayed diagrammatically in Fig. 11-16.
   Not only does this theory have an appealing simplicity, but it also accounts
very nicely for the fact that the sulfides are sparse in the micropegmatite, increase
in abundance slightly in the transition zone and norite, achieve the status of disseminated *ore* in some of the higher parts of the sublayer, and then develop as
massive ore here and there along the base of the intrusion.   This seems to fit very
nicely a picture of gravity separation during and following the separation of the
sulfide-rich liquid.
   There are, however, several features that may mitigate against this "simple"
differentiation theory.   The ratio of sulfide to silicate in some areas is high—too
high for complete solution at the likely temperatures of intrusion involved—and
some of the sulfide must almost certainly have been injected with the silicates as
suspended droplets, i.e., as an already separate phase.   In some cases the sulfides
apparently localized at the base of the intrusive are found between the contact
rock and a *chilled margin* of norite.   Thus some deposits that appear at first sight
to be simple segregations represent reinjection or quite later, separate, injection.
Finally although there is a general similarity from one deposit to another at
Sudbury, there are certainly some compositional differences, which may be hard
to explain on the basis of segregation *in situ* from a single sublayer.
   The theory maintaining *segregation of most of the sulfide from silicate at
depth, with the injection of the two as already separated phases*, is hard to fault.   If
the two were largely injected together in a turbulent fashion, much sulfide would
probably have been caught up in suspension in the silicate magma.   This would
then have begun to segregate, giving the nice pattern of separation now seen in so
many of the deposits.   Separate, slightly later, injection of some of the sulfide
would account for the ore now found along chilled norite contacts and much of the

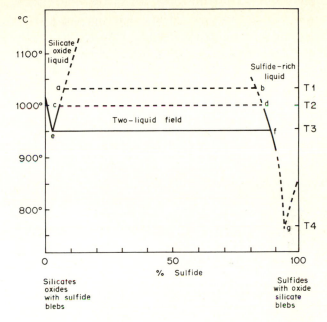

Fig. 11-16 Development of a "two-liquid" system in the cooling of a sulfide-rich intrusion such as that of Sudbury, as visualized by Hawley. The diagrammatic section illustrates cooling, separation of immiscible silicate-rich and sulfide-rich liquids, and crystallization, treated as a binary system. Point *e* lies close to left ordinate. Crystallization at *e* or *g* would extend over a range of temperatures. Points *d* and *c* represent approximate composition of "immiscible-silicate-sulfide" ores. (*Redrawn from Hawley, Can. Mineralogist, 1962.*)

breccia and vein ore in the underlying rocks. To the question, "Why, if these later injections could come to rest along the lower contact of the sublayer and in the underlying rocks, did they not form dikes in the norite, micropegmatite, and Whitewater Series?" the answer may be given that the intrusion as a whole was still largely liquid at the time of the final injections of the sulfides.

Although fashionable in a minor way in the late 1950s, the theory of *hydrothermal replacement* has little to support it as a major process at Sudbury. There seems good evidence to suggest that some of the minor veins, particularly at Falconbridge, are hydrothermal, but this is no reason why the idea should be extrapolated to embrace the whole Sudbury mineralization. In most of the sulfide ores there is no suggestion of corrosion of associated silicates, and indeed where silicates are entirely enclosed within sulfide they often occur as beautifully faceted crystals, particularly the feldspars. The case for localization of the sulfides by replacement on any significant scale is therefore weak. This is not to say that there has been no hydrothermal activity; it might be expected as a very minor process in the final consolidation of a basic magma, and it seems almost certain that it has been responsible for the formation of some hydrous silicates, carbonate, quartz, and small quantities of sulfide.

The theory of *sulfurization* arose during the early 1960s (e.g., Kullerud, 1963, p. 187) and has been defined by Kullerud as ". . . the reaction between sulfur from an external source and cations such as iron, nickel and copper in solid solution in common rock-forming minerals or in igneous magma" (1965, p. 178).

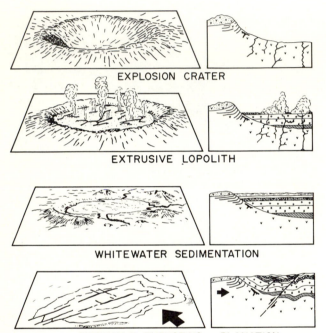

Fig. 11-17 Possible sequence of events involved in the meteorite-impact theory, as visualized by Dietz (1964). *Explosion crater:* A crater is formed by an asteroid striking the earth. The crater walls are upturned, a nest of breccia occupies the crater bottom, and liquefied meteorite substance is injected into the walls. *Extrusive lopolith:* Magma rises into the crater and is ponded into a saucer-shaped pool which differentiates. Chilling against the atmosphere, a welded tuff is laid down as a surficial crust. *Whitewater sedimentation:* After cooling and congealing of the extrusive lopolith, a saline lake or arm of the sea occupies the remnant crater. The Whitewater sediments are laid down, capping and burying the lopolith. *Grenville thrusting and planation:* The Grenville orogeny generates thrusting from the southeast which converts the circular filled-crater scar into an oval form. Erosion truncates the structure to the form seen today. (*Redrawn from Dietz, J. Geol., 1964.*)

This idea was based on experiments by Kullerud and Yoder (1963, pp. 215–218; 1964, pp. 218–222) that showed that reactions between iron-bearing silicates and sulfur result in the formation of pyrrhotite and/or pyrite. In one of their experiments, where the silicate was fayalite containing minor nickel, pentlandite was also a product of the reaction. Applying this to Sudbury, one would assume that the intrusion was first emplaced and differentiated into micropegmatite-transition-zone–norite-sublayer units, the latter having unusually high concentrations of nickel and copper (and perhaps iron) in solution in the silicate melt. Following this, as a late stage of the intrusive activity but prior to complete consolidation of the melt, sulfur vapor was introduced, combining with the nickel, copper, and much of the iron. The resulting sulfides, together with their minor associates, then separated out as droplets that accumulated here and there along the basal contact as large segregations. Such a simple sequence would not, of course, account for the breccia sulfides or for those injected along chilled contacts. In such cases some late squeezing and reinjection—or injection of sulfides segregated quite separately at greater depths—would have had to occur.

The *meteorite-impact theory* is a relatively recent one and has been put forward in its most comprehensive form by R. S. Dietz (1964). Dietz suggests that the Sudbury basin as a whole is an asteroid-impact structure, or "astrobleme." He postulates that such a structure might have been formed by an asteroid about 4 km in diameter traveling at 15 km/sec, the impact of which would have generated some 3×10^{29} ergs. The details of Dietz' argument—a very novel one—are too complex to be considered here, but the elements of his proposed history are depicted diagrammatically in Fig. 11-17.

Dietz' speculations appear to be supported by recent petrographic observations by French (1966), who has studied microstructures in rocks toward the base of the Whitewater Series, finding features similar to those known to be developed in rocks subjected to shock pressures generated by meteorite impact or artificial explosions. He suggests that his observations "establish a strong similarity between Sudbury and other structures for which an impact origin is either proven or strongly indicated" and that this might imply that "meteorite impacts may either produce or trigger igneous processes involving large volumes of magma and economic ore deposits" (1966, p. 2).

These five lines of approach cover the full range of ideas so far applied to the Sudbury problem. However, while Sudbury still towers over all other known occurrences as far as sheer tonnage is concerned, in principle it has now become just one of a number of examples of a clearly recognizable "type" of ore occurrence. We have therefore reached the point where any given theory can be tested quite widely and where valuable deductions can be made on the basis of broad comparisons.

Recognition of this fact places us in a very much better position to consider origin. Taking them in a slightly different order from that in which they have just been presented, we may examine our five principal theories against a background of a number of occurrences.

While the theory of *meteorite impact* may have evidence to support it at Sudbury, there is no indication that any such phenomenon has occurred in the vicinity of most other deposits of the present kind. Thus although it must be conceded that an impact may have triggered appropriate magmatic activity at Sudbury, the wider evidence of occurrence indicates that meteorite impact is not of primary significance in the generation of iron-nickel-copper sulfide ores as a class. This is supported by constitutional features; while what little is known of lead and sulfur isotope values in these deposits do not contradict a meteoritic origin, the relatively great abundance of copper most certainly appears to do so—a point clearly recognized by Dietz himself. Whereas the general level of Ni/Cu ratios in nickel-iron meteorites is of the order of 100:1, that in the ores is very variable and ranges up to 15:1 (Table 11-3).

The idea of *sulfurization* has an attractive simplicity, but it too encounters difficulties, particularly on constitutional grounds. Desborough (1966) has drawn attention to two principal points:

1. *Low zinc sulfide content.* Zinc is present at about the same level as copper and nickel in mafic rocks, and where the latter contain minor disseminated sulfides, sphalerite is normally present in much the same amount as chalcopyrite and pentlandite. However, the iron-nickel-copper sulfide ores possess only the most minute traces of zinc. This appears to be evidence against both sulfurization *and* magmatic segregation *in situ*.
2. *Low magnetite content.* Sulfurization of olivine at high temperatures and pressures inevitably leads to the generation of large quantities of magnetite, according to the reaction

$$4FeMgSiO_4 + S \rightarrow FeS + Fe_3O_4 + 4MgSiO_3$$

An orthopyroxene of the form $(Fe,Mg)SiO_3$ would presumably react in an analogous way, yielding FeS, Fe_2O_3, and enstatite (as above) or an orthopyroxene heavily depleted in iron. However, in either case the sulfurization process should lead to the formation of magnetite in about the same amount as pyrrhotite. While many iron-nickel-copper sulfide deposits carry conspicuous magnetite, this never seems to occur in quantities anywhere near approaching those of pyrrhotite.

The first point—low zinc content—probably does not present too serious a difficulty where ultramafic rocks, as distinct from mafic ones, are involved, since zinc is very low in ultramafic intrusions. Copper is also low in ultramafic intrusions so that its relatively high abundance in ores associated with ultramafic rocks might be an embarrassment (though Table 11-3 does suggest that copper is distinctly higher in the ores associated with mafic rocks than in those associated with ultramafic rocks). The most critical difficulty lies in the relatively low magnetite content, and it is difficult to see how this could be accounted for in terms of the sulfurization hypothesis.

There are undoubtedly some cases where at least some of the ore of particular deposits as these now occur was deposited—or redeposited—by *hydrothermal activity*. The development of cavity fillings and late-stage hydrous minerals in some cases indicates at least a small amount of solution activity in the deposits concerned. The presence of uncorroded crystals suspended in sulfide has already been mentioned as evidence against extensive hydrothermal replacement. Perhaps the strongest evidence against hydrothermal activity seems to be provided by occurrence, when a large number of deposits are considered together. The ores are always almost entirely *within* the mafic or ultramafic igneous rocks. Why, if they have been deposited from migrating hydrothermal solutions, have they not come to rest in a wide variety of hosts? The answer seems to be that the sulfides have a very close affiliation with their hosts and had insufficient mobility—certainly far less than a dilute aqueous solution would have given them—to move far from their source.

There then remains the possibility of *magmatic segregation, in situ* or *at depth*. While some deposits, such as those of Kambalda, might represent separation *in situ* of sulfide from genuine high-temperature silicate-sulfide solutions, it seems certain that this is not always the case. Occurrence of sulfides as sheets between chilled margins of the intrusive and the intruded rock as in some Sudbury localities, and very high ratios of sulfide to silicate as at the Thompson mine, are two very substantial pieces of evidence against formation entirely by segregation *in situ*. These difficulties do not arise, however, if segregation has taken place at least partly in some deeper, larger body of magma. In this case:

1. Sulfide-bearing silicate liquid *and* discrete sulfide liquid are injected together; segregation occurs and the sulfide concentrates at the base of the intrusion or along suitable interfaces.
2. Silicate and sulfide liquids are injected separately from time to time, both, however, following essentially the same conduits and coming to rest in the same locale.

CONCLUDING STATEMENT

Although one set of deposits—those of Sudbury—appears to be related to a layered gabbroic intrusion, it is very doubtful whether ores of the present group have any general close tie with layered intrusions. Most of the deposits are associated with highly mafic to ultramafic rocks and there do not seem to be any felsic associates, such as are conspicuous in the Bushveld and Stillwater masses. Derivation from a large intrusion of Bushveld-Stillwater type would seem to require the occurrence of abundant anorthosite and leucogabbro bodies in the same locality, but these are usually inconspicuous or absent. Even in the case of Sudbury the "sublayer" contains abundant peridotite inclusions, as well as pyroxenite and gabbro fragments, indicating some highly mafic to ultramafic affiliation. As a class, too, deposits of the present group are usually associated with plugs and thin sheets, the latter rarely showing obvious signs of differentiation.

One notable feature of the occurrence of these ores is their remarkably frequent association with volcanic rocks, particularly pillow basalts. The environments usually look to have been highly unstable until only shortly before, and even during, the emplacement of the nickel-bearing rocks. One suspects that the latter may represent a late-stage tapping of some very deep source that had already supplied the volcanic rocks of the surrounding terrain. In this connection perhaps it should be kept in mind that *some* of the nickel-bearing sheets may themselves be lavas rather than intrusions: sulfides may have been extruded, as components of thick flows, rather than intruded, and their localization in such cases has been due simply to irregularities on the surface over which the compound silicate-sulfide lavas were poured.

If we regard the association of these ores as an ultramafic rather than a mafic one, the usual high nickel-to-copper ratio is not surprising. This impression is strengthened by the fact—already noted—that it is the ores associated with a higher proportion of gabbroic rocks that show the higher copper. There is some suggestion that the platinoids are preferentially associated with those ores showing higher copper-to-nickel ratios. Sudbury and Noril'sk, both of noritic association and with relatively high Cu/Ni, are notable sources of the platinoids. Thompson and Kambalda—both of ultramafic affiliation—have conspicuously low Cu/Ni ratios, and both are also conspicuously low or lacking in platinoids.

Perhaps the most remarkable constitutional feature of these iron-nickel-copper sulfide ores is their very low chromite content. This is an extraordinary—and, one suspects, a most highly significant—characteristic. We have already seen that many of the great layered mafic intrusions, and almost all alpine-type ultramafic intrusions, contain substantial chromite. Many students of the mafic-ultramafic iron-nickel sulfide ore type would consider the latter to be derived either from deeply differentiating layered intrusions or from ultramafic sources such as those that have provided alpine-type peridotites. However, either derivation implies an obvious chromite component—a component in which our ores are conspicuously lacking. One's suspicion is that this is a critical feature, the elucidation of which may well take us much closer to understanding the history of iron-nickel-sulfide ores.

RECOMMENDED READING

Boldt, J. R., and P. Queneau: "The Winning of Nickel; Its Geology, Mining and Extractive Metallurgy," The International Nickel Company of Canada Limited and Methuen & Co., Ltd., London, 1967.

Haughton, S. H. (ed.): "The Geology of Some Ore Deposits in Southern Africa," vol. II, The Geological Society of South Africa, Johannesburg, 1964.

Hawley, J. E.: The Sudbury ores: their mineralogy and origin, *Can. Mineralogist*, vol. 7, pt. 1, 1962.

Naldrett, A. J., and E. L. Gasparrini: Archean nickel sulfide deposits in Canada: their classification, geological setting and genesis with some suggestions as to exploration, *Geol. Soc. Australia Spec. Pub.* no. 3, 1971.

Thayer, T. P.: Some critical differences between alpine-type and stratiform peridotite-gabbro complexes, *Intern. Geol. Congr. 21st, Copenhagen, 1960, Rept. Session Norden*, pt. 13, pp. 247–259, 1960.

Thayer, T. P.: Principal features and origin of podiform chromite deposits, and some observations on the Guleman-Soridag District, Turkey, *Econ. Geol.*, vol. 59, pp. 1497–1524, 1964.

Wilson, H. D. B. (ed.): Magmatic ore deposits, *Econ. Geol. Monograph* no. 4, 1969.   (This very valuable monograph was published well after Chap. 11 was written and it has therefore not been possible to refer to several very valuable papers it contains.   The reader is strongly advised to refer to this excellent publication.)

12
Ores in Igneous Rocks 2:
Ores of Felsic Association

This chapter is concerned with three ore associations of igneous affiliation. The first may be termed the *carbonatite association*, which includes a complex association of rare earth, thorium-uranium, and base metal compounds, and a host of other substances, with carbonatite masses of a wide variety of occurrence and composition. The second is termed the *anorthosite–iron-titanium oxide association* and includes a number of layered and discordant bodies of oxide associated with anorthosite-norite occurrences of both layered and massif type. The third—the *quartz monzonite-granodiorite–copper-molybdenum sulfide association*—is the class of ore deposits now becoming well known as the "porphyry coppers."

While it is convenient to refer to this group as being of *felsic association*, it must be kept in mind that igneous rocks cannot be realistically divided into sharply defined categories, and so at least limited overlap between igneous rock types of the present chapter with certain mafic types, and particularly with some of those of Chap. 11, is inevitable.

THE CARBONATITE ASSOCIATION

It may be fairly said that the carbonatites are one of the most fascinating, yet one of the most perplexing, of all rock types. Their outstanding characteristic is, as their name suggests, that they all contain significant quantities of carbonate.   Here, however, their constitutional uniformity ends.   The carbonate itself varies widely in composition and abundance, and a formidable range of associated accessory minerals occur in a wide variety of assemblages.   Pecora has referred to the carbonatites as ". . . essentially carbonate-silicate rock with a great variety of other minerals" (1956, p. 1537).   The breadth of this definition indicates clearly the complexity of the group and the difficulty of its classification.   A rather more specific definition than Pecora's is given by Verwoerd: "*Carbonatite* may be defined as a granular rock consisting of primary calcite, dolomite, ankerite or other rock-forming carbonates as principal constituents, with subordinate apatite, magnetite, silicates and accessories, and exhibiting the primary features of intrusive rocks" (1966, p. 121).   That is, in their occurrence they *appear to be* genuine igneous rocks, apparently crystallized from a carbonate-rich melt and intruded into the rocks in which they are now found.   In addition they always appear in close association with igneous silicate rocks of low-silica–high-alkali type, and with these, the carbonatites show a strong tendency to occur along zones of rifting in old, stable shield areas of continents.

The carbonatites first attracted serious petrological interest about 1920, and their study is thus quite a young one.   Indeed most work on them has been done since the end of the Second World War, and the rate of discovery and investigation appears to be steadily accelerating.   Pecora reported in 1956 that some 30 occurrences were known; Verwoerd estimated that about 120 were known in 1966.   The reason for this great acceleration of discovery is primarily economic.   In addition to their abundant carbonate minerals, the carbonatites often contain, among their minor components and "accessories," abundant phosphate, concentrations of iron and manganese oxides, and small but highly significant quantities of compounds of niobium, uranium, thorium, rare earth elements, copper, and others.   What at first appeared to be petrological oddities are now becoming important economic targets, and there is no doubt that ores of the "carbonatite association" have been established as an important type.

DISTRIBUTION

Carbonatites are widely distributed in both space and time.   They have now been found on all the continents except Antarctica and range in age from Precambrian to Modern.

It seems likely that the first carbonatites to be recognized as such were those of Kaiserstuhl, in Germany, descriptions of which began to appear in the literature about 1846.   In 1884 Bose noted rocks now known to be carbonatites in west central India, and then in 1895 Högbom discovered the now famous carbonatite dikes of Alnö Island, near Sundsvall in Sweden.   Some 25 years later W. C. Brögger (1921) described the carbonatites and associated rocks of the Fen area, in Telemark in Norway, and suggested that they had formed from a magma consisting essentially of carbonates.   Since that time numerous occurrences have been found in the Americas, Africa, Russia, India, and now in Australia.   By far the most important area of occurrence appears to be central and southern Africa, and it is

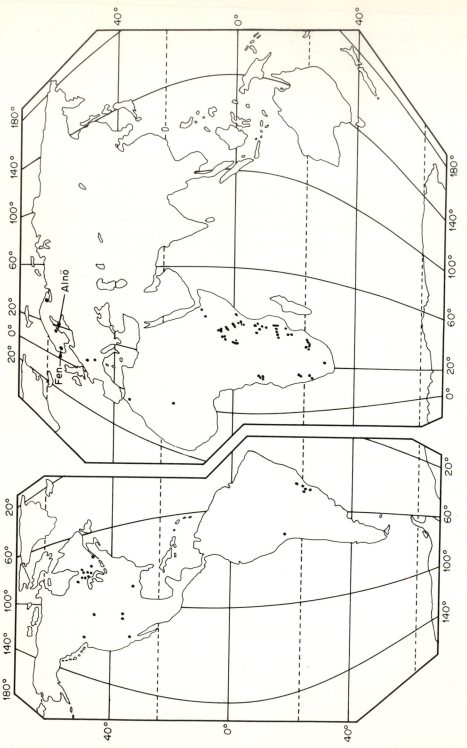

Fig. 12-1 Distribution of the more important carbonatite localities of the world.

here—in the volcano of Oldoinyo Lengai in Tanganyika—that carbonatites have actually been observed forming as lava flows (Dawson, 1966).

The more important carbonatite localities of the world are shown in Fig. 12-1.

Although occurrences range in age from Precambrian to Modern, their distribution in time—as in space—is by no means regular.   Several of the well-known intrusions are Precambrian: Mountain Pass, California, and Loolekop, Spirekop, and Magnet Heights in the Transvaal, are the principal examples.   A number of others occur in Precambrian areas but as components of intrusions very much younger than the host terrain.   Similarly there are well-known occurrences of Paleozoic age, such as the classical ones of Fen and Alnö, but these again are relatively few.   The great age of carbonatite development seems to run from the beginning of the Mesozoic era to modern times.   These include the historic Kaiserstuhl occurrences (Tertiary), the great Tanzania-Uganda-Kenya-Nyasaland province of east Africa (Mesozoic-Tertiary-Modern), The Sao Paulo-Minas Gerais-Santa Caterina belt of Brazil (probably early Jurassic), and the Monteregian province of eastern Canada (Mesozoic).

We may therefore say that carbonatites are widespread in space and time but that they have achieved their most conspicuous concentration in east Africa during the last 150 million years.

FORM AND SETTING

The form and setting—or at least the immediate, local, setting—of carbonatites are inextricably bound together and it is therefore most appropriate to consider one with the other.

Carbonatites are found in the form of intrusions and lavas.   They may also occur as quantitatively significant components of pyroclastic formations.   They have as ubiquitous associates alkaline igneous rocks, usually highly undersaturated with respect to silica and often quite highly mafic.   They are also usually associated with intense "alkali metasomatism"—i.e., alkali metal enrichment apparently induced by solution activity—which has led to the formation of a group of rocks termed *fenites*.   The hypothetical process of alkali metasomatism is termed *fenitization*.

Where the carbonatite occurs as an intrusion it may take the form of a plug, a dike, or a sill.   Dikes are quite common and there are a few important occurrences apparently in the form of sills.   However, by far the most common—and indeed almost the characteristic—mode of occurrence is as plugs and curved bodies forming part of rounded, intrusive masses of conspicuously alkaline igneous rocks.   These alkaline intrusions usually take the form of complexes, containing quite a variety of alkaline rock types, and are mostly conspicuously oval to near circular in plan.   The area of such complexes at the surface varies from about 1 to over 20 square miles.   Within these complexes the carbonatites normally occur as irregularly rounded masses.   Continuous or semicontinuous concentric rings and dikes within the complex are also common (dikes may extend out into the

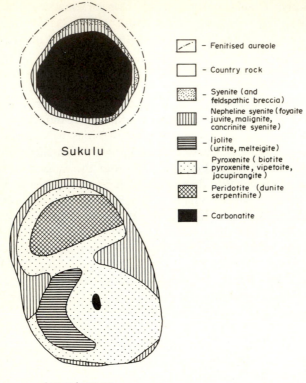

Sukulu

Jacupiranga

- Fenitised aureole

- Country rock

- Syenite (and feldspathic breccia)

- Nepheline syenite (foyaite juvite, malignite, cancrinite syenite)

- Ijolite (urtite, melteigite)

- Pyroxenite (biotite pyroxenite, vipetoite, jacupirangite)

- Peridotite (dunite serpentinite)

- Carbonatite

Fig. 12-2 Highly simplified geological maps of the Sukulu (Uganda) and Jacupiranga (Brazil) intrusions, illustrating the great potential variation in the proportion of carbonatite in alkaline complexes. (*From Verwoerd, Ann. Univ. Stellenbosch, 1966.*)

surrounding country rocks, but this is the exception rather than the rule).   Perhaps surprisingly the size of a carbonatite does not seem to be related to the size of the alkaline complex enclosing it.   Large complexes commonly possess quite small carbonatite masses and vice versa.   Figure 12-2, showing the Jacupiranga and Sukulu (Uganda) intrusions, illustrates this very clearly.   Many alkaline plugs do not, of course, contain carbonatites at all, and very occasionally, e.g., Nooitgedacht, carbonatite occurs virtually devoid of associated alkaline silicate rocks.   The larger carbonatites are of the order of 2 to 5 square miles in area, and depths of up to at least some thousands of feet are indicated.   The Palabora (Loolekop) carbonatite (Fig. 12-3), which has been extensively drilled for economic exploitation, covers an area of only 2250 × 1200 feet but has been proven for a depth of at least 3000 feet.   This is clearly a deep vertical pipe, and it seems likely that most of the plug carbonatites are of this form and of corresponding depth.

Occasionally major carbonatites are found as sills, and among these the Kaluwe intrusion of Zambia is a good example (Fig. 12-4). Alnö Island itself, and also the Glenover occurrences, are excellent examples of the dike form (Fig. 12-5). The semiconcentric dikes have been suggested by von Eckermann (1948, p. 85; 1966, p. 10) to be cone sheets. Transgressive dikes are usually radial and presumably fill tension fissures.

Although less common than the intrusive types, carbonatite lavas and pyroclastic rocks are well known. Their lesser abundance may, of course, be no more than an erosion effect; they may have formed in large quantities but, being surface features, have been eliminated by erosion far more quickly than their intrusive counterparts. As already noted, the formation of carbonatite lavas and pyroclastic rocks has actually been observed. The form of the lavas appears to be essentially similar to that of normal silicate lavas. According to J. B. Dawson,

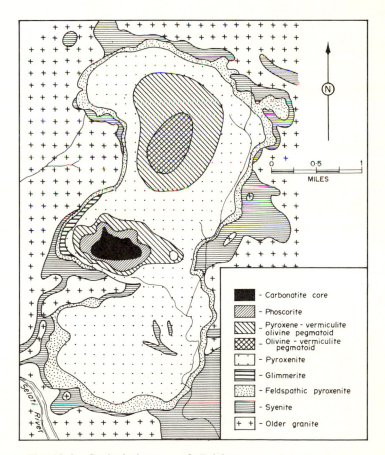

Fig. 12-3 Geological map of Palabora complex, northeastern Transvaal. (*Adapted from Russell et al., Trans. Geol. Soc. South Africa, 1954.*)

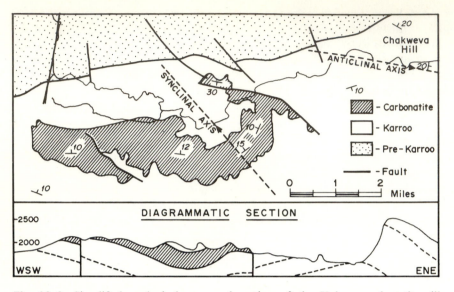

Fig. 12-4 Simplified geological map and section of the Kaluwe carbonatite sill, Zambia. [*Redrawn from Bailey, in Tuttle and Gittins (eds.), "Carbonatites," John Wiley & Sons, New York, 1966, with the publisher's permission.*]

the modern carbonatite volcano of Oldoinyo Lengai has produced common lava structures:

> Lavas consisting mainly of soda and carbon dioxide were extruded on to the floor of the crater in January, June, September and October 1960, and August 1961, . . . They were extruded both as highly mobile *pahoehoe* flows and as viscous, blocky *aa* flows, . . . In every way they behaved like silicate lavas, the *pahoehoe* lavas showing ropy structures, small squeeze-ups and minor driblet cones. . . . Cracks on the *pahoehoe* flows were coated with sublimated nahcolite ($NaHCO_3$). A noticeable feature of the flows is that although they were black in colour when first extruded, they began to turn white after a very short time (24–36 hours). After a matter of 6–7 days the lava becomes light grey-white in colour; in this state it is deliquescent (1966, pp. 161–162).

Little need be said of the pyroclastic rocks other than that they are common —carbonatitic volcanic activity is characterized by an abundance of volatiles and is often highly explosive—and range from coarse agglomerates to fine tuffs.

In addition to being invariably volcanic, the association of the carbonatites is also invariably an undersaturated one: the silicate rocks of the plugs and related bodies are always deficient in silica with respect to Al_2O_3, total iron, the alkaline earths, and the alkalis. Collectively these rocks show a wide variation in ratios of felsic to mafic components, yielding an extensive range from very light to very dark varieties. The most abundant light minerals are potassium feldspar and nepheline, and the principal dark ones are pyroxene (usually diopsidic to aegeritic), olivine, and biotite. Since the range of mixtures is so wide and because it is gradational, there has been a tremendous proliferation of names, each coined to denote a particular variant. The more felsic members are, of course, syenites, nepheline syenites, tinguaites, and related types. With decrease in the proportion of the light components these change to more mafic types and eventually to rocks such as mica pyroxenites and mica peridotites. The latter, however, differ from normal picrites, pyroxenites, and peridotites in that they contain much greater amounts of the alkali metals—they are "alkaline ultramafic" rocks. The whole range of silicate rocks may also contain carbonate, and by increase in this carbonate component they achieve the identity of silicate-carbonatites and carbonatites of various types. This is discussed in more detail in the next section.

The third feature of the setting of carbonatites that, though very common, has not quite the universality of the volcanic-subvolcanic environment and the undersaturated alkaline association, is the presence of *fenite*. As already noted, the fenites, which vary enormously in composition and structure, are the product of the process of *fenitization*, defined in turn by Verwoerd as "a metasomatic process responsible for the alteration of older rocks occurring in contact with alkaline intrusions and carbonatite, with or without *substantial* material exchanges" (1966, p. 122). The original coinage and definition of the terms *fenite*, *fenitize*, and *fenitization* were the work of Brögger and derived from his investigation (1921) of the Fen occurrence. Put simply, fenitization is a type of *wall-rock alteration* characteristically induced about alkaline complexes—so characteristic that its recognition always constitutes prerecognition of the presence of an adjacent alkaline intrusion. Where material changes have occurred these involve, from one

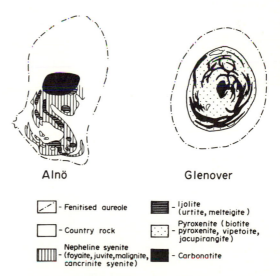

Alnö   Glenover

- Fenitised aureole

- Country rock

- Nepheline syenite
  (foyaite, juvite, malignite, cancrinite syenite)

- Ijolite
  (urtite, melteigite)

- Pyroxenite (biotite pyroxenite, vipetoite, jacupirangite)

- Carbonatite

Fig. 12-5 Highly simplified geological map of the Alnö (Sweden) and Glenover (Transvaal) complexes, illustrating the semiconcentric dike form. Legend as for Fig. 12-2. (*From Verwoerd, Ann. Univ. Stellenbosch, 1966.*)

locality to another, a variety of additions of such as CO_2, F, H_2O, Fe, Ca, Mg, Na, K, Ti, Ba, and P. Silica is usually subtracted—as is Na in some cases—from the rocks undergoing fenitization. Relict structures are often preserved in the final fenite. Like the alkaline intrusions that have induced their formation, the fenites vary enormously in composition and structure, and this has resulted in another complicated nomenclature. Many fenites are very similar to certain igneous rocks, and have been named accordingly, e.g., *syenite fenite*. As suggested by Verwoerd (1966, p. 122) it is probably better to identify fenites as we do metamorphic rocks in general—by using the principal minerals as modifiers. Thus just as a certain metamorphic rock might be a garnet-orthoclase-quartz-sillimanite gneiss, so we may have a rock that we call not a "syenite fenite" but an "orthoclase-albite-aegerine fenite."

The fourth feature of the setting of carbonatites is, in contrast to the three already discussed, a very large-scale one. In many cases—perhaps most—the alkaline complexes with their constituent carbonatites occur along major rift zones cutting old, stable shield areas. The close spatial relationship of the east African carbonatites to parts of the east African rift system appears to be a fine example of this association.

CONSTITUTION

The principal components of carbonatites are calcite and dolomite. Normally there are at least some silicates present, and as we have already noted, increase in these yields silicate carbonatites and, eventually, alkaline igneous rocks. In addition to minor silicate, carbonatites usually carry small to trace quantities of rare metal carbonates, a variety of oxides, together with barite and a variety of other unusual minerals. Sulfides are often present in trace quantities, and in one instance—Palabora—these occur as quite an extensive assemblage and include economic quantities of copper.

As a general rule, carbonatites *sensu stricto* are made up of 80 per cent or more of carbonate, including both common and rare varieties. Among the common carbonates, calcite is the most important, and rocks composed principally of this mineral are termed *sövites*. Dolomite is the next most common, and rocks in which this dominates are termed *beforsites*. (It should be emphasized that the use of the names sövite and beforsite alone is a gross simplification of a highly complex nomenclature; simplification of carbonatite nomenclature generally is, however, widely favored.) Iron- and manganese-bearing carbonates, including siderite, ankerite, and rhodochrositic types, are also fairly common but are much less abundant than calcite and dolomite.

Pecora reported in 1956 that over 50 different minerals had been reported from carbonatites. The total by 1966 (Heinrich, 1966) was over 160, the principal ones of which are given in Table 12-1 (adapted from Heinrich, 1966, pp. 158–160).

Apart from the Ca-, Mg-, Fe-, and Mn-bearing carbonates, the following minerals are particularly common, and some may occur in quantities up to 20 per cent in various carbonatite occurrences: apatite, barite, fluorite, strontianite, pyroxene, olivine (including serpentine), nepheline, K feldspar, albite, phlogopite,

Table 12-1   Minerals known to occur in carbonatites.   Some very rare primary species have been omitted.   (Adapted from Heinrich, 1966, pp. 158–160)

Mineral	Abundance†	Mineral	Abundance†
Native elements		**Oxides, hydroxides** (*continued*)	
Graphite	R	Perovskite	M, C
Gold	VR	Pyrochlore	C
Silver	VR	Pandaite	M
Carbide		Betafite	M
Moissanite	VR	Columbite	M
Sulfides, etc., and sulfosalts		Mossite	VR
Chalcocite	VR	Fersmite	M
Digenite	VR	Fergusonite	VR
Bornite	VR	**Halides**	
Galena	M	Fluorite	C, E
Sphalerite	M	Sellaite	R
Chalcopyrite	M	Cryolite	VR
Pyrrhotite	M	**Carbonates**	
Valleriite	VR	Calcite	E
Millerite	VR	Siderite	M, E
Pentlandite	VR	Rhodochrosite	R(?)
Cubanite	VR	Magnesite	VR
Covellite	VR	Aragonite	VR
Linnaeite	VR	Witherite	VR
Stibnite	VR	Strontianite	M
Pyrite	C	Dolomite	E
Bravoite	VR	Ankerite	E
Marcasite	R	Röntgenite	VR
Arsenopyrite	VR	Bastnaesite	M, E
Molybdenite	R	Thermonatrite	E
Tetrahedrite	VR	Unnamed Ca–Na–K carbonate	E
Oxides, hydroxides		**Sulfates**	
Periclase	VR	Barite	C, E
Corundum	VR	Celestite	R
Hematite	C, E	Anhydrite	R
Ilmenite	M, C	**Phosphates**	
Rutile	M, C	Monazite	M
Cassiterite	VR	Florencite	R
Anatase	R	Cerian goyazite	VR
Brookite	R	Apatite	C, E
Cerianite	VR	**Borate**	
Baddeleyite	M	Ludwigite	?
Thorianite	VR	**Silicates**	
Brucite	VR	Olivine	M
Goethite	R	Monticellite	R
Spinel	VR	Topaz	VR
Magnesioferrite	VR	Andradite	VR
Magnetite	C, E		

† E = may be essential; C = common; M = moderately common; R = rare; VR = very rare; and ? = abundance in doubt.

Table 12-1   (continued)

Mineral	Abundance†	Mineral	Abundance†
Silicates (*continued*)		Silicates (*continued*)	
Sanidine	VR	Actinolite	R
Anorthoclase	VR	Soda tremolite	R
Albite	M	Soda actinolite	R
Scapolite	R	Hastingsite	R
Natrolite	R	Barkevikite	VR
Thomsonite	VR	Arfvedsonite	VR
Heulandite	VR	Riebeckite	M
Chabazite	VR	Glaucophane	R
Melanite	VR	Chrysotile	M
Grossularite	VR	Antigorite	M
Epidote	VR	Melilite	R
Allanite	R	Talc	R
Cerite	VR	Muscovite	R
Zircon	M	Biotite	R
Thorite	R	Phlogopite	M
Thorogummite	R	Manganophyllite	VR
Sphene	M	Stilpnomelane	VR
Wollastonite	M	Vermiculite	M
Beryl	VR	Chlorite	M
Vesuvianite	M	Clinochlore	VR
Prehnite	VR	Dickite	VR
Diopside	M	Nepheline	M
Augite	M	Analcite	VR
Aegirine-augite	C	Nosean	VR
Aegirine	C	Quartz	C
Hornblende	VR	Orthoclase	M
Tremolite	R	Microcline	M

Table 12-2   Maximum modal mineral percentages in carbonatites and silicocarbonatites of Alnö and Fen.   (From Pecora, 1956, p. 1543, after Von Eckermann, 1948; Brögger, 1921)

	Alnö	Fen
Calcite	97	93
Dolomite	81	90
Alkali feldspar	72	65
Pyroxene	31	50
Biotite (and phlogopite)	47	27
Olivine (incl. serpentine)	15.8	?
Apatite	16.4	16.4
Magnetite	9.5	8
Melanite	19	5.8
Perovskite	4.2	?
Pyrochlore	2.1	1.4

riebeckite, wollastonite, bastnasite, pyrochlore, and iron-titanium oxides. Pyrite and quartz are quite common. The *maximum* modal percentages of the more abundant minerals in the classical Fen and Alnö occurrences determined by Brögger (1921, pp. 401–403) and von Eckermann (1940, pp. 64–140) are given in Table 12-2. Table 12-3 (after von Eckermann, 1948, pp. 111–137) gives the modal composition of selected sövitic and beforsitic dikes at Alnö. This gives an idea of the amount of variation likely to occur in one occurrence. Clearly differences *between* occurrences are likely to be even greater.

Variation in mineralogy is necessarily reflected in chemical constitution. Table 12-4 gives the ranges of the major components in the carbonatites at Fen, Alnö, and East Africa (as at 1956). Table 12-5 gives averages in major and minor elements for a large number of sövitic and beforsitic carbonatites collectively and for four sodium carbonatites from Oldoinyo Lengai. Other elements, some of which, such as Ba and Sr, vary from trace to major amounts from one occurrence to another, are given in Table 12-6.

Of the chalcophile metals other than copper, minute quantities occur in galena, tetrahedrite, molybdenite, bornite, chalcocite, and valleriite. These sulfides are normally accompanied by small quantities of pyrite, pyrrhotite, and of course, chalcopyrite.

The principal mineral products of carbonatites are probably apatite and

Table 12-3 Modal compositions of some sövitic and beforsitic dikes, Alnö. (After Von Eckermann, 1948)

	1	2	3	4	5	6	7	8	9	10
Calcite	23.5	45.2	60.0	67.2	93.3					
Dolomite						80.0	71.7	65.5	57.1	55.8
Olivine	9.8			11.7				5.3		
Melilite			8.5						14.0	
Pyroxene		9.1								
Riebeckite								10.6		
Biotite	17.6	24.0	20.1	9.5			15.0	10.3	20.3	33.2
Garnet	13.0									
Orthoclase	23.5					6.2				
Nepheline	2.8									
Quartz						0.9	1.8			
"Opaques"					1.1	2.8		2.7		3.5
(chiefly magnetite and other than pyrite)		1.7	2.5							
Apatite	2.5	20.0	6.1	11.2	1.7	6.3	6.5	4.5		6.5
Pyrite	1.8		1.0	0.4	1.0	1.9	2.5	0.7		1.0
Perovskite			0.8			1.0				
Corundum	2.8									
Anatase									0.7	
Fluorite					0.3			0.4		
Barite							0.2		5.4	
CO$_2$ (vacuoles)						0.9	2.3		2.5	

Table 12-4   Ranges of composition of carbonatites of Fen, Alnö, and East Africa.   The Alnö figures also include silicocarbonatites and carbonatized silicate rocks.   (After Pecora, 1956, p. 1544)

	Alnö (49 analyses)	East Africa (21 analyses)	Fen (10 analyses)
SiO_2	0.1–49.6	0.3–17.2	0.73–35.86
TiO_2	Tr–4.5	0.2–0.4	0.12–2.51
Al_2O_3	0.2–18.2	0.18–3.7	0.60–6.57
Fe_2O_3	0.1–8.0	0.12–9.4	0.53–7.26
FeO	0.1–10.5	0.23–8.9	0.67–9.14
MgO	0.1–15.3	0.8–1.88	1.62–10.59
CaO	4.0–55.4	24.1–54.5	25.35–50.47
Na_2O	0.03–2.3	0.01–0.9	0.01–2.21
K_2O	0.03–12.8	0.02–1.2	0.17–3.62
P_2O_5	0.6–4.8	0.13–5.5	0.95–6.92
CO_2	3.1–43.1	24.2–43.7	6.99–42.88
H_2O+	0.1–2.3	0.2–1.5	0.04–0.37
S	0.01–20.3	0.16–0.44	0.07–0.67
F	0.06–2.4	0.02–0.3	0.08–0.56

Table 12-5   Average abundances of major and minor elements in carbonatites other than natrocarbonatites (column 1) and natrocarbonatites (column 2).   (Largely after Heinrich, 1966, pp. 222, 223)

	1	Number of analyses	2	Number of analyses
SiO_2	10.29	140	0.58	4
TiO_2	0.73	131	0.09	2
Al_2O_3	3.29	136	0.09	2
FeO	3.60	131	1.02	4
Fe_2O_3	3.46	149	0.30	2
MnO	0.68	111	0.14	2
MgO	5.79	129	1.17	4
CaO	36.10	116	15.54	4
BaO	0.40	117	1.02	4
SrO	0.46	101	1.05	4
Na_2O	0.42	105	29.56	4
K_2O	1.36	97	7.14	4
F	0.81	82	2.26	4
Cl			2.90	4
P_2O_5	2.09	109	0.95	2
S	0.56	84	0.10	2
SO_3			2.46	4
CO_2	28.52	135	31.72	4
H_2O (tot.)	1.44	94	5.15	4

Table 12-6   Comparison of minor and trace element abundances (in ppm) in carbonatites with those in igneous rocks and limestones.   (After Pecora, 1956; Gold, 1963; and Heinrich, 1966)

	All igneous rocks	Gabbros	Ultramafic to felsic alkaline rocks	Limestones	Carbonatites
RE_2O_3 (tot.)	50	5	100–500		$200 - n \times 100{,}000$
BaO	300	200	1000–10,000		5000–100,000
SrO	300	200	1000–10,000		5000–20,000
Nb_2O_5	5	20	10–50		10–5000
Ta_2O_5	2	1	1–10		0.1
ZrO_2	300	100	400–2000		10
TiO_2	8000	15,000	5000–20,000		1000–30,000
Sc	13			1	10
Co	18			0.1	17
Ni	100			20.0	8
Cu	70			4	2.5
Ga	26			4	1
Zr	170			19	1120
Mo	1.7			0.4	42
Sn	32			1	4
Cr	117			11	48
La	40			1	516
Ce	40			11.5	1505
P_2O_5	3000	3000	2000–20,000		$1000 - n \times 10{,}000$
F	500	300	100–2000		200–24,000
S	500	2000	1000–3000		10,000–100,000

limestone—the former for phosphatic fertilizer and the latter for cement. Fluorite, barite, and strontianite are additional "nonmetallic" products of various carbonatites. The principal metallic elements are niobium, copper, thorium, uranium, zirconium, and rare earths. Some of these are obtained from weathering products, in association with residual iron and/or manganese. (Deans, 1966, table 1, gives an excellent list of industrial products obtained from some of the African carbonatites up to 1966.) The following are the principal metals obtained from carbonatites:

1. *Niobium.*   The principal niobium-bearing mineral in carbonatites is pyrochlore. The niobium occurs as a substitute for titanium in the pyrochlore structure, and thus, not surprisingly, it also occurs to a minor extent in rutile, brookite, and anatase. According to Deans (1966, p. 395) pyrochlore appears to belong to restricted phases of carbonatite formation; it seems to be absent from complexes where carbonatite is subordinate to silicate rocks (syenite, pyroxenite, dunite, etc.) and is also very low in late-stage rare earth-strontium-barium carbonatites and the modern sodium carbonatites of Oldoinyo Lengai. The major carbonatitic niobium occurrences of Africa are in

relatively large plugs (and the Kaluwe sil!) in which related silicate rocks are subordinate or absent.   Known primary "ore-grade" occurrences are

a. Panda Hill, Tanganyika: pyrochlore-bearing sövite carrying 0.3 per cent Nb_2O_5; pyrochlore-bearing biotitic contact rock carrying 0.70 per cent Nb_2O_5.
b. Chilwa Island, Tanganyika: pyrochlore and niobium rutile-bearing sövite carrying about 1 per cent Nb_2O_5.
c. Oldoinyo Dili, Tanganyika: Pyrochlore sövite carrying about 0.35 per cent Nb_2O_5.
d. Kaluwe, Zambia: pyrochlore sövite; large "low-grade" reserves.
e. Kaiserstuhl, Germany: pyrochlore-perovskite in sövite(?) carrying > 0.2 per cent Nb_2O_5.
f. Fen, Norway: pyrochlore-bearing sövite carrying 0.2–0.5 per cent Nb_2O_5.
g. Oka, Quebec: niobian perovskite and niocalite.

2. *Rare earths.*   Although alkaline igneous rocks are well known to be enriched in the rare earths, they are quite overshadowed in this respect by the carbonatites, which, as a class, contain greater abundances of rare earths than any other known rocks.

   Pecora has pointed out (1956, p. 1546) that the chemical affinity of rare earths for phosphorus and fluorine in an environment rich in Ca and CO_2 greatly influences their final distribution, and accordingly rare earths are the principal metals of a number of carbonates, fluophosphates, and phosphates of carbonatites.   Where a carbonatite contains little or no material of this kind, the rare earths ordinarily substitute for calcium in pyrochlore, perovskite, apatite, and to a lesser extent, calcite.

   The Sulphide Queen ore body, at Mountain Pass, California, exhibits a range of rare earth metal content from less than 1 to 38 per cent.   The rare earth elements here are chiefly of the cerium subgroup, occurring principally in bastnaesite containing La_2O_3, 29.6 per cent; CeO_2, 50.3 per cent; Nd_2O_3, 14.3 per cent; Pr_6O_{11}, 4.4 per cent; Sm_2O_3, 1.3 per cent; and Y_2O_3 not detected.   The host carbonatite consists of calcite 40 to 75 per cent, barite 15 to 50 per cent, bastnaesite 5 to 15 per cent, and a large number of minor species.   According to Olsen and others (1954) the bastnaesite content may reach as high as 60 per cent.   In the Bearpaw Mountains, groups of small carbonatites contain up to 3.5 per cent rare earths.

   In Africa several carbonatites contain significant quantities of bastnaesite or monazite or both and may carry up to 20 per cent combined rare earths. At Wigu Hill in Tanganyika (Harris, 1961; McKie, 1962), rare earth-rich zones up to 4 feet wide and up to 500 feet long and containing bastnaesite, monazite, and cerian goyazite $[Sr(Al,Ce)_3(PO_4)_2(OH)_5H_2O]$ occur in dolomitic carbonatite dikes.   These may contain $La_2O_3 > 10$ per cent.   Also at Nkombwa Hill, Zambia, monazite is abundant in magnesian-rich (ankerite and ferroan magnesite) carbonatites, and at Kangankunde Hill, Malawi, ankeritic carbonates carry a large assemblage of minor minerals, among them monazite. Deans (1966, p. 403) notes that at Kangankunde, ore widths of 25 to 100 feet in places carry averages of 5 to 10 per cent monazite, 10 to 30 per cent strontianite, and 2 to 5 per cent barite.

3. *Uranium and thorium.* Many carbonatites are distinctly radioactive, and radiometric surveys often outline their boundaries—and those of their parent complexes—extraordinarily well. The principal radioactive minerals are thorium-bearing pyrochlore and uranothorianite. Uraninite has not been detected. The quantities of the two metals are usually very small, and the principal value of the U-Th minerals seems to be as indicators of carbonatites rather than as sources of radioactive metals.

4. *Copper and other sulfide metals.* A number of carbonatites—notably those of Loolekop in the Palabora complex and the Sulphide Queen in the Mountain Pass occurrence—contain minor but conspicuous sulfides. Of these, pyrrhotite and pyrite are the most widespread and usually the most abundant, but sphalerite, galena, chalcopyrite, and other copper sulfides, nickel sulfides, and others are by no means uncommon as traces. Most of the occurrences, however, are sporadic and low grade and of no current economic interest.

The Loolekop (Palabora) carbonatite is a notable exception and is now a major source of copper. Though of low grade (average 0.69 per cent to a depth of 3000 feet) it is of large tonnage and hence, from the mining point of view, analogous to the low-grade "porphyry coppers"—a type of occurrence that we shall be considering in a later section and which constitutes one of the world's most important classes of copper ore.

It is interesting to note that the Palabora (also spelled Phalaborwa) ore body was found virtually by accident as a result of drilling designed to test the mass as a possible source of uranium. The carbonatite core of the complex had given promising results from a radiometric survey, and it was thought that the uranothorianite that it contained might be present in economic quantities. This was found not to be so; the drilling revealed large reserves of magnetite-rich carbonatite containing only uneconomic grades of uranothorianite, but somewhat surprisingly it disclosed the presence of very persistent small amounts of copper sulfide.

It is now known that the main copper ore body is identified with the central vertical carbonatite plug. This is approximately 300 × 1800 feet in plan and averages about 1 per cent copper, in chalcopyrite and subordinate cubanite. There is also an outer zone of banded carbonatite and phoscorite (the latter is described by Deans as a ". . . heterogeneous, often brecciated, assemblage of apatite, titanomagnetite and serpentinized olivine, with subordinate vermiculite and calcite," 1966, p. 390), carrying about 0.5 per cent copper. The ore body as a whole is now known to extend to at least 3000 feet and to have reserves of 315 million tons of ore averaging 0.69 per cent Cu available for opencut mining to 1200 feet. The ore also contains minor gold and silver, and the low-Ti magnetite—about 25 per cent of the ore—is a valuable source of iron.   Other primary sulfides present are pyrrhotite, pyrite, marcasite, valleriite, sphalerite, and the nickeloan sulfides pentlandite, millerite, bravoite, violarite (secondary?), and linnaeite.

ORIGIN

It has already been said that the carbonatites are one of the most perplexing of all rock groups.   Thus, like most enigmas, they have aroused great curiosity and have provoked the development of a wide variety of hypotheses.   A full account of these theories, and the arguments for and against, are quite beyond the scope of this book, but it is worthwhile for the economic geologist to grasp at least the main threads of the debate.   We shall therefore look briefly at what appear to be the four principal possibilities:

1. Carbonatites represent solidified igneous carbonate magma.
2. Carbonatites result from hydrothermal activity and replacement.
3. Carbonatites are xenoliths of sedimentary limestone caught up in a rising silicate melt.
4. Carbonatites result from "gas transfer"—i.e., result from the injection into a silicate melt of large quantities of extraneous CO_2 and a variety of P- and rare element-bearing gases.

Solidification from magma   The plug, dike, and sill form of carbonatite bodies, together with their usual rather coarse, granitoid textures, have suggested that most of them have been intruded as igneous melts.   This is supported by the observation of pyroclastic counterparts and, in a most conclusive way, by the actual observation of the extrusion of present-day sodium carbonatite lavas.

Originally the evidence against a carbonate magma was experimental: Smyth and Adams (1923) found the melting point of calcite to be 1389°C at 1000 atm pressure—prohibitively high.   In support of this it was later found that even in atmospheres of H_2O *or* CO_2 at 5000 bars, no sign of melting of calcite could be seen at temperatures up to 800°C.   All this made the magmatic hypothesis apparently untenable (observation of the Oldoinyo Lengai effusions did not come until 1960).

This situation changed dramatically with the observation by M. S. Paterson (1958) that rapid and complete melting of calcite occurred at 1000°C, and incipient melting at 900°C, at a *combined* H_2O–CO_2 pressure of only 50 bars.   This has been followed by intensive experimentation, particularly by Wyllie and Tuttle (1960a, 1960b), who have shown that melts can persist down to temperatures close to 600°C. Although the experimental data do not as yet prove or disprove any particular petrological process, it do confirm (Wyllie, 1966, p. 351) that many of the mineralogical features observed in carbonatites *can be explained* in terms of carbonatite magmas.   Virtually all the work so far has been concerned with the carbonate-silicate component, though the behavior of phosphate and fluorine has been investigated in a preliminary way (the system $CaO–CaF_2–P_2O_5–CO_2–H_2O$).   In this connection Wyllie notes:

> The experimental data establish the facts that a carbonatite magma containing initially more than a few percent of P_2O_5 would begin to precipitate apatite before calcite, and that calcite and apatite could be co-precipitated through a wide temperature interval. The synthetic carbonatite magmas are very fluid, and crystal settling occurs with both calcite and apatite.   Given a carbonatite magma precipitating these minerals, there is a good chance that partial segregation of the minerals could occur as a result of crystal settling during quiet periods.   With the onset of explosive activity, producing movement within the magma column, any accumulations of apatite-rich mixtures would become streaked out forming bands parallel to the flow structures in the crystallizing magma. There is little difficulty in accounting for the observed flow banding and segregation of apatite on the basis of a magmatic carbonatite intrusion (1966, p. 327).

The field evidence provided by older carbonatites, observations of present-day carbonatite lavas, and experimental results thus make the case for magmatic activity—at least as one important process of carbonatite formation—impregnable.

Hydrothermal addition and replacement   The suggestion that carbonatites might be hydrothermal phenomena arose as a result of the apparently excessively high temperatures required for the generation of carbonatite magmas, as indicated by the experiments of Smyth and Adams in 1923.   As we have seen, this difficulty has now been removed and the necessity to invoke hydrothermal activity to explain carbonatites proper is no longer present.

However, the very high volatile content of carbonatite magmas would suggest the likelihood of pneumatolytic and hydrothermal effects in the surrounding rocks, and indeed these do seem to be almost ubiquitous.   As we have already noted, this activity generally involves addition of H_2O, CO_2, F, P_2O_5, Fe, Ti, Ca, Ba, and K and some elimination of Na and Si; the result is a fenite.

Xenolithic derivation   In this postulated process, the carbonate simply represents older, deeply buried sedimentary carbonate that has been cut, and rafted up, by the rising igneous magma.   Two lines of evidence indicate that the carbonatites are *not* xenoliths:

1. *Trace element abundances.*   Table 12-6 gives the average abundances of a number of trace elements in igneous rocks, carbonatites, and sedimentary limestones.   Clearly carbonatites as a class bear no resemblance to limestone in this respect.
2. *Strontium isotope* ($^{87}Sr/^{86}Sr$) *abundances.*   As with the sulfur and lead isotopes discussed in Chap. 7, the ratios of $^{87}Sr/^{86}Sr$ can be used as traces in determining geological processes. ^{87}Sr is radiogenic and derived from ^{87}Rb, whereas ^{86}Sr is nonradiogenic so that the ratio of the two can be used to determine the original Rb/Sr ratio of any given rock.   Using this approach,

Powell, Hurley, and Fairbairn (1966) determined $^{87}Sr/^{86}Sr$ in undoubted xenolithic limestone, normal sedimentary carbonate rocks, carbonatites, and in alkaline rocks with which these carbonatites occurred. Their principal results and conclusions were:

a. Diffusion of strontium in limestone xenoliths: even where the latter are small and completely restructured by the heat of the host melt, diffusion of strontium is negligible.

b. $^{87}Sr/^{86}Sr$ in carbonatites is quite different from that of limestones in the rocks intruded by the carbonatites concerned.

c. $^{87}Sr/^{86}Sr$ in carbonatites as a class appears to be significantly different from $^{87}Sr/^{86}Sr$ in sedimentary limestones as a class.

d. In three carbonatite-alkaline complexes studied (Magnet Cove, Arkansas; Iron Hill, Colorado; Spitzkop, Transvaal), $^{87}Sr/^{86}Sr$ in the carbonatites was identical with those in the associated alkaline silicate rocks.

e. The range of $^{87}Sr/^{86}Sr$ in 21 carbonatites studied was similar to that observed in basalts as a class.

f. $^{87}Sr/^{86}Sr$ in carbonatites as a class is distinctly lower than that of upper crustal materials in general.

From this it appears that:

1. Carbonatites are not xenolithic.
2. Carbonatites are (genetically) intimately related (i.e., comagmatic with) the alkaline silicate rocks of the complexes in which the carbonatites occur.
3. Carbonatites and their associated alkaline silicates are not derived from the upper crust but from a region—in the lower crust or upper mantle—similar to that from which basalts come.

Gas transfer   Many carbonatites, through their association with large quantities of pyroclastic material and their occurrence in or close to breccia pipes, appear to have had close ties with volcanic activity of a very highly explosive kind.   In addition their most prominent minerals are compounds of "volatile" substances: CO_2 in the abundant carbonate, F in apatite, S in barite and sulfides, and so on. All this suggests that carbonatite formation is associated with the movement of large quantities of gas (or potentially gaseous matter)—a deduction supported by observations on modern volcanoes of alkaline affinity.   It has been suggested that carbonatites might have developed by the injection of large quantities of CO_2, H_2O, S, F, Cl, B, and heavy metal compounds into a basic melt.   If these substances were *retained* by the melt until it had completely crystallized, a carbonatite with a rare element assemblage would result.   Fenites would then be a related manifestation of the gas transfer process.

Quite apart from the mechanics of the transfer process there is the question of the source of the CO_2.   There is no clear evidence as to where this and the associated minor elements come from.   It has been variously suggested that the CO_2

is of genuinely juvenile origin and/or the result of assimilation of limestone—though as von Eckermann colorfully puts it, petrologists working on carbonatites have ". . . walked like cats round hot porridge and evaded expressing any view on the origin of the carbon" (1961, p. 35).   However, as we have already seen there is weighty geochemical evidence against significant assimilation of limestone.   On the other hand, CO_2 is well known to be an abundant product of volcanism—including basaltic volcanism—all over the world, and it is now largely accepted that the CO_2 of carbonatites could quite easily have been supplied by the relevant magma chamber.

CONCLUDING STATEMENT

In spite of their constitutional variability there is no doubt that the carbonatites constitute a distinctive rock group and, in consequence, a distinctive ore type or *ore environment*.   On origin, there seems little question that they are igneous and of volcanic and subvolcanic nature.   The reason for the very high carbonate is not clear: keeping in mind that the carbonatitic masses usually represent late phases of the development of the alkaline complexes, it seems most likely that they are late fractions in which much volcanic CO_2 has remained trapped or that they are late fractions into which such CO_2 has been injected.   Whatever the case, there is no doubt—from the evidence of constitution, eruptive phenomena, and the occurrence of fenites—that potentially volatile substances have been abundant and highly active.   Such activity carries with it the likelihood of the development of *concentrations* of mineral matter, and thus the carbonatites as a class are promising targets for mineral exploration.

THE ANORTHOSITE–IRON-TITANIUM OXIDE ASSOCIATION

We have already seen, in Chap. 11, that some ore associations overlap slightly: the highly chromiferous mafic and ultramafic rocks commonly contain very minor quantities of Fe–Ni–Cu sulfides, and conversely some of the great Fe–Ni–Cu sulfide deposits contain a little chromite.   In a similar way the present association has affiliations with two of the associations already considered.   Some anorthositic ores consist largely of magnetite and apatite and are thus similar constitutionally to some of the magnetite-apatite concentrations of carbonatite complexes.   Others occur as anorthosite–iron-titanium oxide bands in large layered intrusions and are thus related to the chromitites often developed in these masses.   Indeed ultramafic chromitite bands and anorthosite-magnetite-ilmenite bands are sometimes found within a single layered intrusion.   The Bushveld complex is an outstanding example.

There appear to be two distinct types of anorthosite occurrence (Turner and Verhoogen, 1960):

1. Those occurring as layers within stratified basic sheets, as in the Bushveld, Stillwater, and Duluth complexes.   The feldspar here is a highly calcic plagioclase, usually bytownite.

2. Those occurring as large "plutonic" intrusions—batholiths or stocks—in metamorphosed terrains. The feldspar here is less calcic than that of the layered rocks and is usually in the andesine-labradorite range.

The layered anorthosites, which have already been referred to in connection with stratiform chromite, appear to have developed by fractional crystallization *in situ* of basaltic magma. For the most part their genesis is known and understood. The "plutonic" anorthosites, on the other hand, are not at all well understood and are recognized as one of the major problems of silicate petrology. Anorthositic masses of this type possess the following features:

1. They appear to be confined to Precambrian terrains.
2. The larger occurrences are in the form of domed intrusions of batholithic size (several have exposed areas of over 1000 sq km).
3. They are usually coarse grained and commonly show conspicuous micro-deformation.
4. The principal silicate constituent is plagioclase ranging from about An_{35} to An_{60}. Hypersthene, augite, and olivine are minor associates. According to Turner and Verhoogen (1960, p. 322), variation in felsic-mafic ratios may be expressed by the following nomenclature: where the mafic mineral component is less than 10 per cent, the rock is an *anorthosite proper*; between 10 and 22.5 per cent, *gabbroic anorthosite*; between 22.5 and 35 per cent, *anorthositic gabbro*. With further increase in the mafic component, the rock becomes a gabbro, usually of noritic type.
5. Most large anorthositic intrusions display a range in composition from anorthosite proper to noritic gabbro. With their conspicuously anhydrous nature, they are often regarded as being of charnockitic affinity.

There have been a variety of suggestions—both magmatic and metamorphic —as to the derivation of these anorthosites. Recently Philpotts (1966) has presented evidence from some of the extensive Canadian occurrences and suggests their derivation from calc-alkali *acid dioritic* magma (compare with the *basaltic* parent of the large layered complexes) injected during a period of orogenesis. Plagioclase-rich crystal accumulations formed by fractional crystallization and gravitational differentiation during the earliest stages of solidification, and these were then progressively purged of their iron-rich interstitial residual liquid by filter pressing. Such a process could, of course, occur at any time in the earth's history and does not explain the restriction of plutonic anorthosites to the Precambrian. Philpotts suggests that a magma of the type postulated would be most likely to yield anorthositic rocks where the parental melt was intruded:

> . . . during orogeny into high-grade metamorphic rocks (anhydrous) under great load pressure an anorthosite-mangerite suite of rocks would develop. However, it appears that the high pressures necessary for the formation of these rocks are likely to occur only at the base of mountain belts. This may be the reason therefore for the restriction of

these rocks to Precambrian terrains, for it is only in such areas that sufficient erosion will have occurred to expose the roots of orogenic belts (1966, pp. 60–61).

To date no systematic comparative study has been made of the iron-titanium oxide concentrations associated with the two major types of anorthosite occurrence. We shall therefore consider the two together, bearing in mind, however, that future study may show the two to be distinctive ore types.

DISTRIBUTION

The distribution of the principal known iron-titanium deposits of anorthositic affiliation are shown in Fig. 12-6. Some of the principal anorthositic masses not yet known to contain economic concentrations are also shown, since these are naturally important from the mineral exploration point of view. The most important ore bodies are those of southern Quebec (Allard Lake), the Adirondack area of northeastern U.S.A., and Scandinavia (Egersund-Sogndal district of Norway and the Kiruna district of Sweden). Other occurrences are found in southern Ontario, Minnesota (Duluth gabbro), the Transvaal (Bushveld complex), Mexico, Bengal, and elsewhere. In space, therefore, deposits of the present affiliation are widespread, but sharply localized within and adjacent to the associated anorthositic complexes. In time, they are found principally, though not quite exclusively, in the Precambrian. Those associated with the *plutonic* anorthosites appear to be entirely Precambrian.

FORM AND SETTING

Form is variable. In *layered complexes* the deposits often occur as extensive thin sheets parallel to the banding of the enclosing silicate rocks, as in the case of the Bushveld intrusion. Form in this case is identical with that of the layered chromite described in Chap. 11. In other instances the ore bodies take the form of a series of lenses rather than semicontinuous sheets. Such lenses are parallel but occur in stepped or *en echelon* arrangement. This type of occurrence is exemplified by the deposits forming part of the Duluth gabbroic complex. The latter includes two main zones of magnetite-ilmenite concentrations: a lower (North Range) zone at about 10,000 feet above the base of the complex and about 30 miles in length and an upper (South Range) zone about 5 miles in length. Each zone consists of several "belts," each belt being made up of several series of lenses lying end-to-end but separated along strike by several hundred feet of gabbro. According to Lister (1966, p. 292), individual lenses range up to about 15 feet in thickness and are up to several hundred feet in length and width; i.e., they form thin lenses rather than sheets. The zones, belts, and individual lenses run parallel to the overall layering of the enclosing gabbro.

Although by far the major part of the iron-titanium oxide concentrations of layered complexes is in the form of these sheets or thin lenses concordant with the igneous layering, a minor amount occurs as discordant, dike-like bodies. Subsidiary occurrences of this kind occur, for example, in the Bushveld complex.

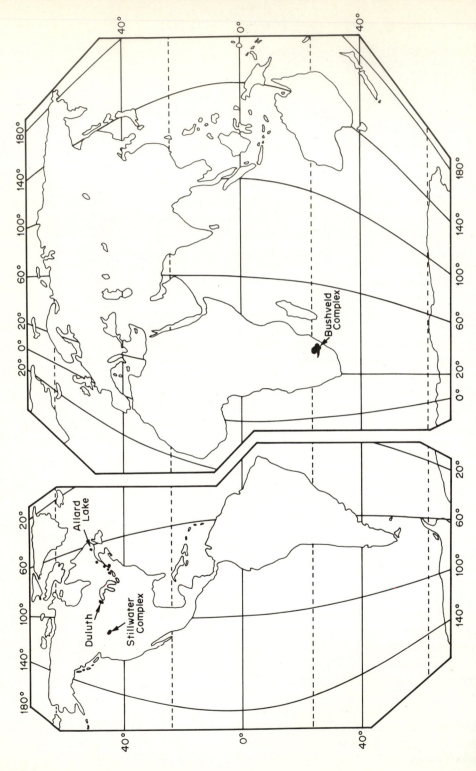

Fig. 12-6 Distribution of anorthositic, and anorthosite-bearing, masses containing the more important anorthositic iron-titanium oxide deposits of

The deposits associated with the *plutonic anorthosites* are typified by those of Allard Lake and some of the Adirondack localities. Individual deposits are essentially discordant and within a given area may vary considerably in shape and attitude. Some concentrations have the form of dikes, while others occur simply as tabular or lenticular masses showing no particular dispositional relationship with the anorthositic host. Deposits of this general kind may be very large, and indeed the Lac Tio deposit at Allard Lake is one of the largest iron-titanium deposits known. This is a tabular, almost flat-lying mass some 3600 feet long, 3400 feet wide, and over 300 feet thick, containing more than 125 million tons of ore averaging 36 per cent Fe and 32 per cent TiO_2 (Hammond, 1952; Lister, 1966).

Of *setting*, little need be said concerning the ores of the *layered complexes*. In general this is analogous to that of the chromite bands of such complexes, except that the chromite tends to occur in ultramafic layers, whereas the iron-titanium oxides tend to occur in the anorthosite-norite layers. Figure 11-3 and Table 11-1, showing the "stratigraphy" of the Bushveld complex, illustrate the closeness in principle of the two ore types. The *Main Magnetite Band* (containing TiO_2 up to 25 per cent) lies some 20,000 feet above the *Main Chromite Band* and defines the base of the *Upper Zone*. The outstanding feature of the latter is the widespread occurrence in it of magnetite-rich layers which range in thickness up to about 8 feet. Magnetite is certainly not confined to the Upper Zone, however; bands of iron-titanium oxide ore in the Main and Critical Zones have been exploited for many years.

The boundaries between oxide and silicate masses in layered complexes range from extremely sharp to gradational. Bateman (1951, p. 419) has observed a major oxide band of the Bushveld complex whose lower boundary is of "knife-edge sharpness" but whose upper boundary is quite gradational. Conversely, Lister has noted of some of the Duluth lenses that:

> Contacts between magnetite-ilmenite rich lenses and their host rock are gradational across several inches. The upper contact of most lenses is sharper than the lower contact and the upper part of most lenses contains a greater amount of oxide minerals than the lower part. The lenses contain from 20 to 70 percent oxide minerals and their host rock contains about 5 to 10 percent oxide minerals (1966, p. 292).

The regional setting of the deposits of *plutonic anorthosite* affiliation is, of course, one of Precambrian gneisses. Within this they occur within or very close to the associated anorthosite-anorthositic gabbro intrusions. Although a few deposits are wholly or partly outside the anorthosite, almost all of them are well within it—often, as is the case of Lac Tio, several miles inside the present exposed margin of the intrusion concerned.

Commonly the oxide concentrations contain inclusions of the anorthositic rock that encloses them. Such inclusions are in the form of irregular blocks and slices of varying size and shape. At Lac Tio, blocks range up to 15 feet in largest dimension and appear to have highly variable orientations. Contacts between oxide masses and the enclosing silicates are, like those of layered complexes,

Table 12-7   Principal minerals of some anorthositic iron-titanium oxide ores

Deposit	Magnetite	Hematite	Maghemite	Ilmenite	Rutile	Ulvöspinel	Sulfides	Apatite
Bushveld, Transvaal	Minor	n.d.†	Major	Major	n.d.	n.d.	V. minor	n.d.
Allard Lake (Lac Tio), Quebec	Minor	Major	n.d.	Major	n.d.	n.d.	n.d.	n.d.
St. Charles, Quebec	Major	n.d.	n.d.	Major	n.d.	n.d.	n.d.	Major
St. Urbain, Quebec	n.d.	Major	n.d.	Major	n.d.	n.d.	n.d.	n.d.
Ivry, Quebec	n.d.	Major	n.d.	Major	n.d.	n.d.	n.d.	n.d.
Degrosbois, Quebec	Major	n.d.	n.d.	Major	n.d.	n.d.	Minor	Minor
Eagle Lake, Ontario	Major	Minor	n.d.	Major	n.d.	n.d.	n.d.	Minor
Pusey, Ontario	Major	n.d.	n.d.	Major	Minor	n.d.	n.d.	Minor
Little Pic River, Ontario	Major	n.d.	n.d.	Major	n.d.	n.d.	n.d.	Minor
Seine Bay Range, Ontario	Major	n.d.	n.d.	Major	n.d.	n.d.	n.d.	Minor
Duluth, Minnesota	Major	n.d.	n.d.	Major	n.d.	Minor	n.d.	n.d.
Fiskå, W. Norway	Major	Minor	n.d.	Major	n.d.	n.d.	n.d.	n.d.
Verkshaugen, W. Norway	Major	Major	n.d.	Major	n.d.	n.d.	n.d.	n.d.
Øyen, Tafjord, W. Norway	Major	n.d.	n.d.	Major	n.d.	n.d.	Minor	n.d.
Øvre Røddal, Tafjord, W. Norway	Major	n.d.	n.d.	Major	n.d.	Minor	n.d.	Minor
Pluma Hidalgo, Mexico	Major	n.d.	n.d.	Major	Major	n.d.	n.d.	Major

† n.d. = not detected.

variable.   Some—particularly in the case of dike-like masses—are sharp.   Others, as at Lac Tio, are substantially gradational; here, according to Lister:

> The upper part of the sheet is a series of alternating layers of almost pure hemo-ilmenite and of anorthosite with disseminated hemo-ilmenite.   The layered portion of the sheet has a maximum thickness approaching 200 feet.   Most of the ilmenite layers are a few inches thick but some are several feet thick.   The ilmenite layers contain numerous small angular to sub-rounded anorthosite inclusions.   Between the ilmenite layers are layers of anorthosite containing 5 to 50 percent disseminated ilmenite (1966, p. 284).

CONSTITUTION

This is far from constant; the host rock itself may vary from almost monomineralic anorthosite through to anorthositic gabbro and norite, and the ores show substantial mineralogical and chemical variation both within and between individual ore bodies.   The overall "ore" assemblage found in the iron-titanium oxide anorthosite association includes magnetite, hematite, maghemite, ilmenite, ulvöspinel (Fe_2TiO_4), rutile, a variety of spinels poor in iron and titanium, minor sulfide, and apatite.   Table 12-7 gives the principal minerals present in the ores of a number of the more important occurrences.   Assemblages vary somewhat from one individual deposit to another, but clearly magnetite, ilmenite, and hematite are overwhelmingly the most important.   Apatite is sometimes abundant, particularly in the discordant concentrations, and may constitute up to about 35 per cent of an ore.

The ores also—necessarily—show considerable variation in their chemical constitution, though there are a number of broad consistencies, some of which appear to give significant information on origin.

The major metals are, of course, iron and titanium (Table 12-8).   In a broad way there appears to be an increase in Fe/Ti ratio going from deposits associated with the more mafic members of the family, e.g., norites and anorthositic gabbros, to those associated with the highly felsic anorthosites (and in some cases syenites). In some discordant *plutonic* deposits, such as Ellen Lake (Lister, 1966), this trend also develops upward from the base of the oxide mass, i.e., Fe/Ti increases progressively from base to top.   In concordant deposits in *layered complexes* the reverse seems to hold, however.   This may result from changing oxygen partial pressures and is referred to again a little further on.

The principal minor constituents of the oxides are vanadium, chromium, aluminum, manganese, and magnesium. V^{3+}, Cr^{3+}, and Al^{3+} occur virtually entirely as substitutions for Fe^{3+} in magnetite and are therefore heavily enriched in the magnetite of coexisting magnetite-ilmenite assemblages. Mn^{++} and Mg^{++}, on the other hand, are incorporated in the ores by substituting for Fe^{++} in the ilmenite, leading to a pronounced preferential enrichment in this phase.   Lister (1966, p. 303) notes that in a number of ores where magnetite and ilmenite have developed as a coexisting pair:

1. Magnetite contains 1.5 to 8 times more V, 1 to 9 times more Cr, and "more Al" than the ilmenite.

Table 12-8   Analyses of iron-titanium oxide-rich rocks (including ores) from Tafjord, Norway, and the Bushveld Complex, Transvaal

	(1)	(2)	(3)	(4)	(5)	(6)	(7)
SiO_2	22.45	35.16	25.27	2.77	1.46	1.63	1.50
TiO_2	7.97	5.13	6.16	20.41	18.82	13.69	18.76
Al_2O_3	11.80	9.18	5.39	1.40	2.83	0.74	3.12
Fe_2O_3	24.62	9.00	3.10	58.46	55.58	68.67	60.68
FeO	24.06	21.53	40.88	12.07	17.53	9.48	11.50
MnO	0.29	0.73	1.08	0.43	0.33	0.30	0.20
MgO	3.22	4.24	7.12	0.70	0.93	1.04	1.49
CaO	2.46	6.97	6.29	0.56	0.50	0.45	0.16
Na_2O	0.80	2.24	0.55				
K_2O	0.85	1.67	0.50				
P_2O_5	0.05	3.24	4.19	Trace	Trace	Trace	Trace
H_2O+	1.38	0.57	0.22	2.61	1.46	2.11	1.01
Total	100.03	100.03	100.83	99.63	99.49	98.16	98.50

Note: Analyses 1 to 3 are from Gjelsvik, 1957, p. 490, and analyses 4 to 7 are from Schwellnus and Willemse, 1943, p. 30, and are information from Geological Memoirs reproduced under the Copyright Authority of the Government Printer of the Republic of South Africa No. 4252 of 13:1:1970.

(1). Average ore, Øyen, Tafjord.
(2). Ferrogabbro, Øvre Røddal, Tafjord.
(3). Oxide rich ferrogabbro, Øvre Røddal, Tafjord.
(4). Titaniferous magnetite ore, Upper Band, Bushveld complex.
(5). Titaniferous magnetite ore, Middle Band, Bushveld complex.
(6). Titaniferous magnetite ore, Lower Band, Bushveld complex.
(7). Titaniferous magnetite ore, Critical Zone, Bushveld complex.

2. Ilmenite contains 1 to 8.5 times more Mn and 1 to 6 times more Mg than the magnetite.

Figure 12-7 (Lister, 1966, p. 304) shows how the minor element abundances vary with variation in Fe/Ti ratio—in turn a reflection of progressive changes in magnetite-ilmenite ratios. As would, of course, be expected, hematite-ilmenite ores do not show these partitioning effects and the trends arising from them.

Microstructures of the ores are fairly consistent. Most of the assemblages are dominated by magnetite-ilmenite, and for the most part these occur together as random aggregates. If the present grain-boundary shapes were the result of crystallization from a liquid followed by slow cooling—and frequent coarse grain size indicates that in most cases cooling was indeed slow—the grain boundaries should be straight and smoothly curving, and the triple junctions should exhibit characteristic equilibrium angles. Very limited investigation indicates that this is often the case, though much more work is needed on this aspect of the micro-structures. In some cases, such as the Magnet Heights deposit of the Bushveld, associated plagioclase shows conspicuous layering, with the large dimensions of the *subidiomorphic* plagioclases lying parallel to each other and to the upper and lower margins of the ore body.

Microstructures of ores containing oxides other than, or in addition to, magnetite and ilmenite are governed largely by the solid solutions indicated by the joins in Fig. 12-8. These in turn are governed by the proportions of available iron and titanium and the prevailing oxygen partial pressure. Two of the solid-solution series are of paramount importance: magnetite-ulvöspinel and hematite-ilmenite, both of which are complete.

Small amounts of titanium substituting in magnetite yield titanomagnetite, a

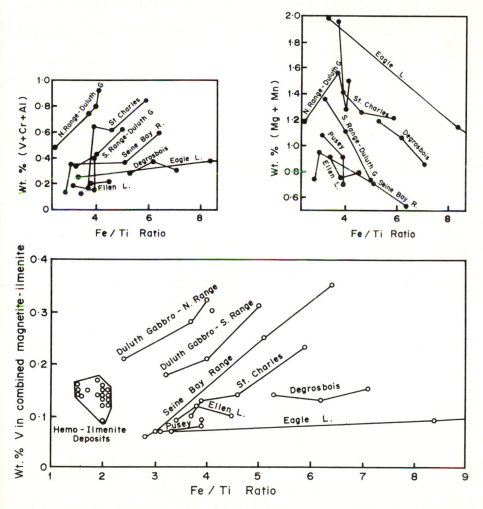

Fig. 12-7 Relations between Fe/Ti ratios and weight per cent minor R^{3+} elements (that is, V + Cr + Al), minor R^{++} elements (that is, Mg + Mn), and vanadium, in combined magnetite-ilmenite of samples from a number of layered and plutonic deposits. (*From Lister, Econ. Geol., 1966.*)

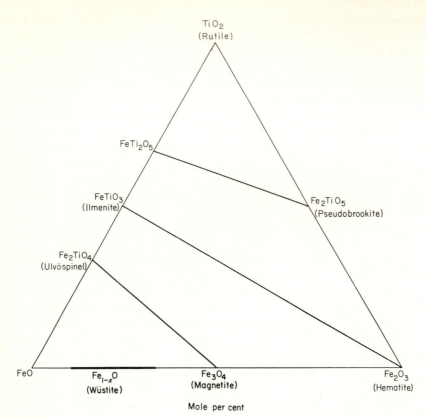

Fig. 12-8 Phases in the system FeO–Fe$_2$O$_3$–TiO$_2$, showing the three major solid-solution series (mole per cent). (*From Buddington and Lindsley, J. Petrol., 1964.*)

product characterized by slight optical anisotropism. On cooling, a separate phase, ulvöspinel, may develop and segregate into the octahedral planes of the magnetite, yielding a lattice intergrowth. Since Ti is soluble in the Fe$_2$O$_3$ structure (cf. Chap. 4), slowly cooled solid solutions may—according to the proportions of available Fe and Ti—yield ilmenite that contains hematite exsolution plates parallel to the basal planes or vice versa. Ilmenite is sometimes found as apparent "exsolution intergrowths" in magnetite, which in view of the distinctiveness of the magnetite-ulvöspinel and ilmenite-hematite solid-solution series, is at first sight surprising. Experimental results indicate, however, that this ilmenite is the result of later oxidation of an earlier formed intergrowth of *ulvöspinel* in magnetite. Occasionally small quantities of magnetite occur as microintergrowths in ilmenite-bearing hematite and hematite-bearing ilmenite, but it seems likely that this is due to later small-scale reduction. On occasions, too, the pair magnetite-rutile appears instead of the usual Fe$_2$O$_3$–FeTiO$_3$ solid solution or its exsolved equivalent. This may represent a low-temperature equilibrium.

Valuable information on temperatures and oxygen partial pressures of formation can be obtained by careful determination of magnetite and ilmenite compositions where these have formed in equilibrium. "Magnetite" is, of course, a spinel solid solution (magnetite-ulvöspinel) and "ilmenite" a rhombohedral Fe–Ti oxide solid solution (ilmenohematite or hemoilmenite). Buddington and Lindsley (1964) have shown that where spinel and rhombohedral phases coexist within the composition ranges $Mt_{90}Usp_{10}$ to $Mt_{20}Usp_{80}$ and $Hem_{15}Ilm_{85}$ to Hem_3Ilm_{97}, the compositions are unique functions of temperature and oxygen fugacity. These investigators have applied their experimental results to carefully analyzed coexisting pairs from many localities and have estimated temperatures ranging from about 600°C to about 800°C, and $\log_{10} fO_2$ from about -13.6 to 18.5 are indicated (1964, p. 335).

ORIGIN

There seems little doubt that the formation of these ores is tied in some fundamental way with the formation of the associated anorthositic rocks. The principal hypotheses put forward for the concentration of the oxides are

Magmatic   1. Fractionational crystallization and gravitative differentiation
2. Residual liquid segregation (including pegmatitic and certain eutectic liquids), forming concordant or discordant bodies
3. Late-stage magmatic-hydrothermal activity leading to deposition by replacement

Metamorphic   Expulsion of Fe and Ti from ferromagnesian crystal structures during granulite metamorphism, with subsequent ionic migration to concentrating rock structures

There seems no doubt that the great *layered complexes* and their contained oxides are magmatic. This is also almost certainly the case with at least many of the *plutonic anorthosite* masses, though it is in connection with these that metamorphic origins have been suggested.

1. Fractional crystallization   Here the iron-titanium oxides are presumed to have crystallized in abundance early, or at particular stages, in the cooling history of the parent igneous mass. Just as, for example, early formed olivines appear to have sunk to the base of many differentiated intrusions, forming olivine-rich layers, so it is suggested that many oxide-rich layers may have developed by settling. This theory is widely accepted for stratiform chromite and, with its neat analogies to silicate petrogenesis, has an appealing simplicity. At first sight it appears to be a good explanation for the iron-titanium oxide-rich layers of complexes such as that of Duluth.

However, as we noted in the section on "Constitution," textural relations developed between silicates and oxides often indicate that the oxides crystallized *with* or *later than* the associated silicates (cf. subidiomorphic plagioclase within oxide), clearly precluding concentration by sinking of early formed oxide crystals.

In those cases where the Fe–Ti oxides occur as well-formed crystals (as, for example, in the Stillwater chromite layers), gravitative differentiation may well have been the concentrating process. However, in the many cases where the oxide forms an allotriomorphic matrix for idiomorphic single crystals of silicate, concentration by some other process is clearly indicated.

2. Residual liquid segregation Although it has often been suggested that the occurrence of oxide as particles interstitial to silicate indicates that the oxide crystallized from interstitial residual liquids, this is not necessarily true. As we have seen in Chap. 9, such an arrangement can be merely the expression of solid-solid interfacial energies and resultant dihedral angles. It is when the oxide forms embayments or veins in *single crystals* of silicate that later solidification of the oxide may be indicated. It should always be kept in mind (and this is referred to again in Chap. 18) that transgressive textures of this kind could be due to distinctly postformational plastic deformation, but evidence of deformation in the silicates should enable the observer to quickly recognize this. In some cases where deformation has not occurred, embayments are presumably due to reaction with a late oxide liquid, and the oxide is hence genuinely late.

Bateman (1951) has been a strong proponent of the residual iron-titanium oxide liquid hypothesis and refers to the primary concentrating process as *late gravitative liquid accumulation* (1951, p. 406). He postulates that in certain cases differentiation has led to extreme iron-titanium enrichment in the later fractions:

> With progressive crystallization and enrichment of the residual liquid, its density would come to exceed that of the silicate crystals and its composition might reach the point of being chiefly oxides of iron and titanium. . . . At this stage three possibilities may occur: Final freezing may occur to yield a basic igneous rock with interstitial oxides, as mentioned above; the residual liquid may be filter pressed out of the crystal mesh and be injected elsewhere; or the enriched residual liquid may drain downward through the crystal interstices and collect below to form a gravitative liquid accumulation (1951, pp. 406–407).

Figure 12-9 (Bateman, 1951, fig. 1) illustrates these postulated processes. Bateman suggests that the *Palisades sill* of New Jersey provides a good example of interstitial iron enrichment. This doleritic intrusion has been shown by Walker (1940) to be well differentiated, its upper, later crystallized part showing marked oxide enrichment. In the coarsest upper portions this oxide, though interstitial, constitutes a major component. While it is far from being an ore, Bateman points out that had the sill been much thicker, and had it cooled more slowly, a strongly iron-enriched residual liquid might well have developed. He considers that the Fe/Ti-oxide layers of the *Bushveld* represent the undisturbed accumulation of late-stage oxide liquid into layers and that the various discordant concentrations in the plutonic anorthosites of *Quebec*, the *Adirondacks*, and *southern Sweden* exemplify the results of filter pressing and residual liquid injection.

This principle—the concentration of many iron-titanium oxide ores by the development of appropriate residual liquids rather than by crystal settling—has

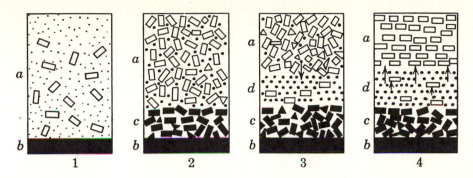

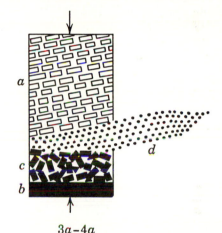

$3a-4a$

Fig. 12-9 Idealized diagrammatic representation of late gravitative liquid accumulation. *1*, Early stage of crystallization of basic magma *a* after formation of chill zone *b*; *2*, layer of sunken early formed ferromagnesian crystals *c*, resting on chill zone *b*, with mesh of later silicate crystals above, whose interstices are occupied by residual magma enriched in ore oxides; *3*, mobile, oxide-rich, residual liquid draining down to layer *d* and floating up later silicate crystals; *4*, formation of a concordant oxide ore body in which a few late silicate crystals are trapped; *3a–4a*, mobile, enriched gravitative accumulation *d* squeezed out or decanted to form late magmatic injections. (*From Bateman, Econ. Geol., 1951.*)

been supported by a number of investigators (see particularly Lister, 1966) and now seems well established. In addition it has now been suggested that some of these residual liquids may actually be eutectics. Recently Philpotts (1967), who has investigated extensively some of the Quebec anorthosites, has drawn attention to the substantial amounts of apatite found in a number of anorthositic oxide concentrations. He notes that rocks consisting mainly of iron-titanium oxides and apatite often occur as small dike-like bodies in anorthosites, that these dikes have quite a consistent composition of two-thirds by volume of oxides and one-third apatite, and that such dikes always seem to have associated with them other dikes

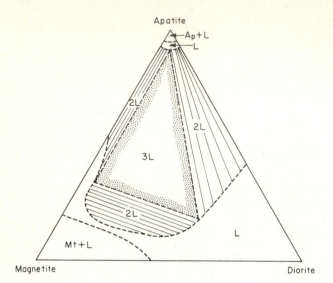

Fig. 12-10 Schematic isothermal section (at 1420°C) through the system magnetite-apatite-diorite, showing the possible nature of the extent of the three- and two-liquid fields. (*From Philpotts, Econ. Geol., 1967.*)

rich in ferromagnesian minerals and apatite.   On the basis of a series of experiments on the magnetite-fluorapatite system, he has established that a eutectic of composition approximately two-thirds oxide–one-third apatite can indeed be formed readily and furthermore that such eutectic mixtures of magnetite and apatite form immiscible liquids with silicate melts having the composition of dioritic dike rocks usually associated with the oxide-apatite concentrations.   Philpotts' results, which must be emphasized as being exploratory rather than a serious attempt at exactitude, are summarized in his schematic phase diagram (1967, p. 311) shown in Fig. 12-10.

3. Hydrothermal replacement   Although, as we have already noted, there are commonly clear indications of corrosion and replacement of feldspar by oxides in these deposits, there is little to suggest that such replacement is of the *hydrothermal* or *pneumatolytic* kind.   In particular there is usually little or no alteration of the associated feldspars and ferromagnesian minerals such as would be expected to result from the passage of large quantities of aqueous solution.   Thus while hydrothermal activity may have played a very minor part in the development of some deposits of the present kind, it does not seem to have made any quantitatively significant contributions.

4. Metamorphic migration   It has been suggested by Hans Ramberg (1948) that many charnockitic suites have acquired their present mineralogical character during metamorphism at pressure-temperature conditions corresponding to granu-

lite facies.   While anorthositic massifs are often found in complexes of charnockitic affinity, they are not an essential component of such suites; however, when
plutonic anorthosites do appear, they are, according to Ramberg (1948, p. 554),
almost invariably among rocks of charnockitic affinities.   If one accepts Ramberg's
contention that the charnockitic rocks (and hence the plutonic anorthosites and their
associated Fe–Ti oxide concentrations) are metamorphic:

> . . . the question arises whether the formation of titanic iron ore in such provinces is
> pre-metamorphic, syn-metamorphic, or post-metamorphic.
>   If the ore be pre- or post-metamorphic it might have formed by magmatic differ-
> entiation or by hydrothermal solutions and replacement.   Because recent ilmenite-rich
> sediments occur at several places it might be tentatively suggested that the ore in char-
> nockitic rocks is highly metamorphosed sedimentary titanic ore.
>   A pre- or post-metamorphic formation of the ore implies that the titanic iron ore in
> hypersthene-bearing rocks and anorthosites in granulite facies has no genetic relationship
> to the recrystallization processes which give rise to the typical character of the silicate
> rocks.   In this case, therefore, one should expect to find ilmenite ore equally often in
> lower facies, viz. in amphibolite facies, in epidote-amphibolite facies, or in green-schist
> facies.   This is not so; segregations of ilmenite ore occur almost solely in granulite
> facies, and generally in connection with basic rocks and anorthosites.   It therefore may
> be assumed that segregation of the ore has some genetic relationship to the recrystalliza-
> tion processes which form the several hypersthene-bearing rocks in granulite facies (1948,
> pp. 556–557).

Ramberg points out that where metamorphism is absent or of low grade,
substantial quantities of Fe and Ti are accommodated in minerals such as sphene,
biotite, hornblende, and titanaugite.   However, if they become involved in high-
grade metamorphism (e.g., granulite facies), these minerals break down, the
reactions involved leading to the development of new crystal structures in which
iron and titanium atoms are less acceptable and hence to the liberation of ions of
these two metals.   Some reactions (Ramberg, p. 558) involved are

1. $\underset{\text{Hornblende}}{NaCa_2(Mg,Fe)_4Al_3Si_6O_{22}(OH)_2} + \underset{\text{Garnet}}{(Mg,Fe)_3Al_2Si_3O_{12}} + 5SiO_2 \rightleftharpoons$

$$\underset{\text{Plagioclase}}{NaCa_2Al_5Si_7O_{24}} + 7\underset{\text{Hypersthene}}{(Mg,Fe)SiO_3} + H_2O$$

2. $\underset{\text{Hornblende}}{NaCa_2(Mg,Fe)_4Al_3Si_6O_{22}(OH)_2} + 4SiO_2 \rightleftharpoons$

$$\underset{\text{Plagioclase}}{NaCaAl_3Si_5O_{16}} + 3\underset{\text{Hypersthene}}{(Mg,Fe)SiO_3} + \underset{\text{Diopside}}{Ca(Mg,Fe)Si_2O_6} + H_2O$$

3. $\underset{\text{Hornblende}}{NaCa_2(Mg,Fe)_4Al_3Si_6O_{22}(OH)_2} + \underset{\text{Biotite}}{K(Mg,Fe)_3AlSi_3O_{10}(OH)_2} + 7SiO_2 \rightleftharpoons$

$$\underset{\text{Plagioclase}}{NaCaAl_3Si_5O_{16}} + \underset{\text{Diopside}}{Ca(Mg,Fe)Si_2O_6} + \underset{\text{Orthoclase}}{KAlSi_3O_8} + 6\underset{\text{Hypersthene}}{(Mg,Fe)SiO_3}$$

In most cases Ti and Fe are set free on the right-hand side of these equations.
Having been so liberated, Ramberg suggests, the two metals diffuse as atoms, ions,
or molecules through the transforming rock mass, eventually to concentrate as gross
oxide masses in appropriate structural or chemical environments.

CONCLUDING STATEMENT

There seems little doubt that the iron-titanium concentrations of the layered complexes have formed by crystal and/or heavy liquid settling, with minor later injection in some cases. Minor deformation is probably fairly widespread. The *modus operandi* in the plutonic anorthositic complexes is, however, not clear at all, though the formation of an immiscible oxide-apatite eutectic may explain the segregation of the ores in some cases. The ores of the plutonic anorthosites also appear to have suffered variable and sometimes quite substantial deformation. While there is clearly much to be done in comparing the compositions of the ores of the two types of occurrence, there is probably also a great deal to be learned from their microstructures. Depending on the relative importance of "free" crystal growth in the melt, later growth in the solid, and deformation, a wide variety of grain-boundary configurations and intracrystalline structures may have developed. Keeping in mind the implications of the features discussed in Chaps. 8, 9, and 10, careful study of microstructures may well give much additional information on formational processes and on postformational events.

THE QUARTZ MONZONITE-GRANODIORITE– COPPER-MOLYBDENUM SULFIDE ASSOCIATION

Apart from their almost ubiquitous affiliation with igneous rocks of conspicuously porphyritic texture, the outstanding features of deposits of this association are their very large size and their very low grade. Some of them contain hundreds of millions of tons of ore, but compared with many of the world's copper deposits their grade is extremely low; overall copper content for most "porphyry coppers" is between 0.5 and 1.0 per cent, and some of them can be profitably worked only because of enrichment induced by weathering processes. *In toto* they make a very large contribution to the world's copper supply, and they yield the major part of all copper produced in the United States. As their name suggests, they are principally sources of copper, though as a class they now yield significant quantities of molybdenum. Gold and silver, and occasionally other base metals, are minor products.

With improved methods of mining and concentration, ores of low grade are becoming increasingly important as sources of metal, and this has led to the successful exploitation of porphyry copper deposits that only a few years ago would have been too low in grade for economic working. Indeed mines in porphyry coppers are now in the forefront in the development of low-cost mining methods.

Although until recently their geographical distribution appeared to be highly restricted, modern exploration methods are now showing porphyry copper deposits to be quite widespread, and their importance as suppliers of copper on a world scale is becoming very great indeed.

DISTRIBUTION

Originally almost all known porphyry copper occurrences were confined to the southwestern corner of the United States. More intensive search has, however, now indicated that their distribution is not nearly so limited as this and indeed

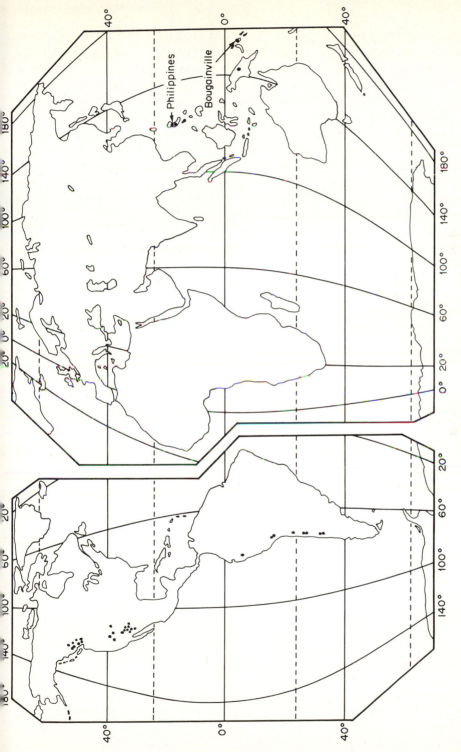

Fig. 12-11 Distribution of principal known porphyry copper deposits.

that they may well be quite widely distributed in space. The principal known occurrences are indicated in Fig. 12-11. Clearly there is a marked concentration in and about Arizona, though western North America generally is well endowed with them. The occurrences in the Philippines and Solomon Islands appear rather isolated, but the number of discoveries in the Southwest Pacific area generally are likely to increase. New discoveries in New Guinea, for example, seem almost certain. Analogy suggests that with the further application of modern geochemical and geophysical methods, further finds will be made in other young orogenic areas.

While the distribution of known occurrences is spreading in a geographical sense, this is not the case with their distribution in time. There is a marked restriction—and this applies to new discoveries as well as old—to the Mesozoic and Tertiary eras, particularly to the Cretaceous and early to middle Tertiary periods.

FORM AND SETTING

Form and setting are most satisfactorily considered together. We shall first consider the broad pattern of occurrence of the deposits and then the more detailed aspects of setting and form.

The most conspicuous feature of the regional setting of the porphyry coppers is that they occur within, or very close to, orogenic zones. Most of the American deposits occur within the Wasatch-Jerome belt (Fig. 12-12). Those that do not, e.g., Ely and Santa Rita, are, however, quite close to the orogenic borders on a regional scale. The distribution of the Canadian deposits is analogous to these, the major occurrences in this case being more or less confined to the Mesozoic "Interior" orogenic belt of British Columbia (Brown, 1969). The deposits of the Philippines, New Guinea, and the Solomon Islands are in orogenic zones that are still highly active and that date back only to Mesozoic or Tertiary time.

Within the orogenic belts the deposits appear to show a very marked affinity for *eugeosynclinal* zones—those parts of the geosynclinal prism in which volcanic rocks are prominent. Almost all the American deposits are conspicuously associated with volcanic rocks, and the Canadian deposits show a similar eugeosynclinal affinity. The Philippine and Solomon occurrences both occur in environments composed almost entirely of pyroclastic rocks and lavas—and are within a few miles of currently active volcanism that probably has a semicontinuous relationship with the Tertiary activity.

Almost as conspicuous as the volcanism itself is the fact that the volcanic rocks generally appear to be of "calc-alkali" type—the andesites, latites, dacites, and rhyolites that seem to be such a characteristic feature of the later stage of the evolution of orogenic belts all over the world. Some basalt is also commonly found in the region of the deposit, though this is usually of minor bulk relative to the more siliceous types. The rocks associated with the Toledo and Pangunu deposits are principally andesites and basaltic andesites.

There is some indication that there is a further regional association—major faulting. Schmitt (1966, p. 17) and others suggest that the American deposits at least are localized about the intersection of the host orogen with a major zone of

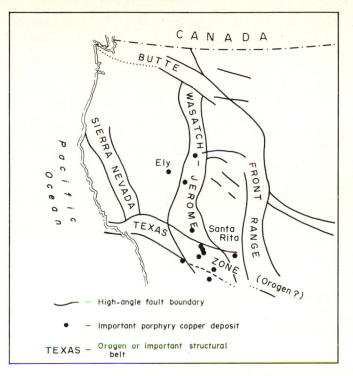

Fig. 12-12 Relation of porphyry copper occurrence to the Wasatch-Jerome belt, southwestern United States. The heavy lines indicate high-angle-fault boundaries. [*Adapted from Schmitt in Titley and Hicks (eds.), "Geology of the Porphyry Copper Deposits: Southwestern North America," University of Arizona Press, Tucson, 1966, with the publisher's permission.*]

transcurrent breakage—the Texas fault zone (Fig. 12-12). Schmitt puts forward the suggestion (1966, p. 30) that this fault zone, being a large and hence probably a deep structure, has extended into and tapped the mantle, from which the copper may then have been abstracted. Chapman (1968) tentatively suggests that some of the British Columbian deposits may have a similar kind of structural affiliation and derivation. Whether major deposits elsewhere can be related to this kind of intersection is not yet known, though it would not be surprising if it were found that they could.

The volcanic rocks of the regional association may be older and/or younger than the porphyritic intrusions and the ores, and some may be essentially contemporaneous. Various nonvolcanic sedimentary rocks such as limestones and shales are, of course, commonly associated with them. The intrusive rocks and their associated ores are thus found in a variety of environments on a local scale: volcanic, volcanic-sedimentary, sedimentary, and here and there, in the relevant metamorphic equivalents. It may be said that on a local scale the deposits are associated with almost any of the rock groups of the geosyncline, though there is a conspicuous tendency for volcanic rocks to be prominent.

We now come to the ore-bearing intrusions themselves. By definition the ores occur largely in these, though in most cases some sulfides are found in the intruded formations. The intrusions are usually of comparatively small outcrop area and fall into the category of *stocks* or *bosses*. Their composition ranges from granite to diorite, though quartz monzonite and granodiorite are by far the most common hosts. In a few cases diabase dikes contain significant amounts of ore mineral. In many localities the stocks are composite, and/or there are several discrete intrusions of somewhat different composition. Such compositional variations commonly embrace a range such as quartz diorite-monzonite-quartz monzonite-granodiorite, and there are often several textural variants of one or more of these compositional types.

A conspicuous feature of all such intrusions or intrusional groupings is that *at least one member is porphyritic.* In any individual locality there may be granitoid rocks and/or phanerocrystalline rocks with coarse phenocryst development, each of which appears to be of the required *composition* for ore occurrence. Sulfides rarely appear, however, unless at least one member of the suite consists of fine phenocrysts set in an aphanitic or semiaphanitic groundmass; i.e., unless at least some intrusive material conforms with what is commonly termed *porphyry*. Stringham notes that if distinctly older Precambrian granites be disregarded, examination of the 24 principal American occurrences shows that

> . . . in five large districts—i.e. Bisbee, Christmas, Mission, Morenci and San Manuel— porphyry only is present. In the remaining 19 districts large intrusive bodies of rock having porphyritic texture coexist with intrusives exhibiting granitoid texture. In none of the 24 districts are granitoid intrusives present without an accompanying major porphyry phase (1966, p. 37).

Stringham also notes (p. 37) that he has found it impossible to obtain microscopic criteria for distinguishing between aphanitic ore-bearing intrusive porphyries and lavas of similar composition and that he has found no clear microscopic features to indicate whether or not an intrusive porphyry is likely to carry ore. From a broad study of the American occurrences he suggests the following as qualitative requirements for porphyry copper development:

1. Intrusives should not be more basic than andesite-diorite, although there seems to be preference for quartz monzonite-quartz latite associations.
2. Intrusive porphyry is absolutely a necessary feature existing with or without associated granitoid types.
3. Passive intrusive emplacement is the most desired structural condition.
4. Where granitoid and porphyritic-textured intrusives coexist, the porphyry should be late in the development of the complex.
5. Sharp boundaries between granitoid and porphyritic types are more favorable, although some gradation is not entirely unfavorable.
6. Wall or country rock may be of all lithologic types, thicknesses, and ages except perhaps for the Pleistocene.

7. For disseminated deposits, intrusive or highly siliceous metamorphic and sedimentary rocks are the most favorable host rocks (1966, p. 39).

The ore minerals occur in the upper portions of the host intrusive and to at least a minor degree in the immediately adjacent intruded rocks. As regards *form*, in both cases the sulfides occur as disseminations through the body of the rock concerned and as constituents of small irregular veinlets following a variety of cooling or deformational fractures. In most instances the major part of the ore occurs within the intrusive rock, but there are some cases—such as the Bingham and Santa Rita deposits—where a very large proportion of the ore body as a whole

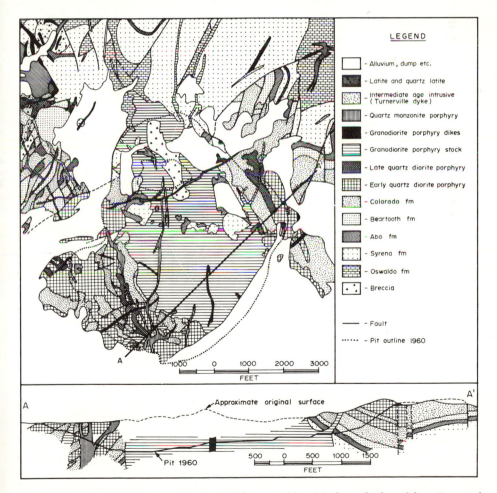

LEGEND

- Alluvium, dump etc.
- Latite and quartz latite
- Intermediate age intrusive (Turnerville dyke)
- Quartz monzonite porphyry
- Granodiorite porphyry dikes
- Granodiorite porphyry stock
- Late quartz diorite porphyry
- Early quartz diorite porphyry
- Colorado fm
- Beartooth fm
- Abo fm
- Syrena fm
- Oswaldo fm
- Breccia

- Fault
- Pit outline 1960

1000   0   1000   2000   3000
FEET

A   Approximate original surface   A'

Pit 1960

500   0   500   1000   1500
FEET

Fig. 12-13 Geological map of the Santa Rita area, New Mexico. [*Adapted from Rose and Baltosser, in Titley and Hicks (eds.), "Geology of the Porphyry Copper Deposits: Southwestern North America," University of Arizona Press, Tucson, 1966, with the publisher's permission.*]

occurs in the surrounding sedimentary and volcanic formations.   Figure 12-13 (after Rose and Baltosser, 1966, p. 212, fig. 3) shows how the Santa Rita ore body (as indicated by the outline of the pit) occurs mainly in a central granodiorite porphyry stock but also, to a very significant extent, in the adjacent sedimentary units.   The Bingham ore body occurs partly in quartzites and limestone; according to Peters, James, and Field (1966, p. 168), at the pit surface about 32 per cent of the ore body is in granite, 27 per cent is in granite porphyry, 16 per cent is in other intrusive rocks, and 25 per cent is in sediments.

In most cases the intrusive host constitutes the central core of the ore body, which extends—for different individual deposits—for a variety of distances out into the intruded rocks.   There is thus a strong tendency for deposits to be near circular or rounded in form and to show a concentric zonal pattern of host rock, grade, and sulfide mineral composition.   We may thus say of *form* that porphyry copper deposits are usually subcircular masses of rock containing disseminated and fine vein sulfides and are centered in small stocks and bosses containing conspicuously porphyritic members.

CONSTITUTION

There are two principal constitutional aspects: the ore minerals themselves and the rock alteration that invariably accompanies them.   The primary nature of the host rocks has already been considered.

The *ore-mineral assemblages* are simple.   Pyrite is almost always (though not quite invariably) the most abundant sulfide.   Pyrrhotite is rarely if ever present in notable quantities.   Chalcopyrite usually follows pyrite in abundance and hence is the most important primary copper mineral.   Bornite is minor, though almost ubiquitous, and may occur as exsolution intergrowths with the chalcopyrite. Chalcocite and to a lesser extent covellite may occur as very minor primary constituents in some deposits, though they usually occur only as weathering products. Molybdenite is a common minor constituent and as already noted often occurs in sufficient quantity to merit commercial extraction.   Rather than being simply "porphyry copper deposits," a number of occurrences might now be more accurately termed "porphyry copper-molybdenum deposits."   As already noted minor quantities of gold and silver are usually present and are often commercial.

In some deposits—notably Bingham—sphalerite and galena may occur in substantial quantities, and with them there is usually a suite of trace constituents such as tetrahedrite-tennantite, enargite-famatinite, arsenopyrite, and boulangerite. Commonly there is more or less concentric zoning of ore types.   Again Bingham provides a good example (Fig. 12-14).   Here there is a central zone of disseminated pyrite, chalcopyrite, and molybdenite, with minor bornite, chalcocite, and covellite. The molybdenite here shows a tendency to occur in veinlets rather than as disseminations.   Around this but still within the intrusion is a zone with a greatly increased pyrite content—a "pyritic halo."   About this in turn, and developed principally in the sedimentary envelope, is a zone of lead-zinc veinlet mineralization.   Farther out again—at distances of 1000 to 10,000 feet from the central copper-molybdenum

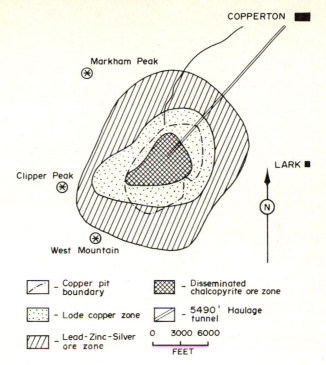

COPPERTON ◼

Markham Peak
✖

Clipper Peak
✖

LARK ◼

N

West Mountain
✖

▱ – Copper pit boundary   ▥ – Disseminated chalcopyrite ore zone

▦ – Lode copper zone   ◰ – 5490' Haulage tunnel

▥ – Lead-Zinc-Silver ore zone

0   3000 6000
⊢——⊣——⊣
FEET

Fig. 12-14 Metal zoning at Bingham, Utah. [*From Peters, James, and Field, in Titley and Hicks (eds.), "Geology of the Porphyry Copper Deposits: Southwestern North America," University of Arizona Press, Tucson, 1966, with the publisher's permission.*]

ore body—is a diffuse zone of lead-zinc ore bodies that have formed by fissure filling and, apparently, by replacement.

Extensive *hydrothermal alteration* seems to be an outstanding and characteristic feature of porphyry copper ore bodies. Such alteration affects both sulfide-bearing (host) rocks and those immediately surrounding them. The principal products of this alteration include clay minerals (of both kaolinite and montmorillonite groups), muscovite-sericite, biotite, quartz, and K feldspar. Chlorite, carbonate, and epidote minerals may also be important. So far as the intrusive rocks are concerned, high-temperature relatively anhydrous igneous assemblages are converted to low-temperature hydrous assemblages. Several broad types of alteration have been recognized for a long time, and Creasey (1966, p. 58) suggests a division into three principal types:

1. *Propylitic alteration.* Quartz and muscovite are widespread, and with this pair four principal assemblages occur: (1) chlorite-calcite-kaolinite, (2) chlorite-calcite-talc, (3) chlorite-epidote-calcite, and (4) chlorite-epidote.

Table 12-9 Summary of the hydrothermal alteration of the principal porphyry copper deposits in Arizona. (After Creasey, 1966, p. 63)

Deposit	Host rocks	Alteration type	Critical assemblages	Primary sulfides, oxides, and others
Ajo	Cornelia Quartz Monzonite, volcanics	1. Potassic 2. Propylitic?	Muscovite-biotite-K feldspar chlorite-zoisite	1. Magnetite 2. Pyrite, chalcopyrite, bornite, tetrahedrite, sphalerite, specularite, molybdenite
Bisbee (excluding replacement ore)	Granite porphyries	1. Argillic 2. Propylitic	1. Muscovite-kaolinite 2. Chlorite-epidote-calcite	Pyrite, chalcopyrite, bornite
Globe-Miami district	Lost Gulch Quartz Monzonite, Pinal Schist	1. Argillic 2. Propylitic 3. Quartz-sericite	1. Muscovite-montmorillonite 2. Chlorite-epidote-calcite 3. Quartz-sericite	Pyrite, chalcopyrite, molybdenite, sphalerite, galena
Morenci	Granite porphyry, diorite porphyry	1. Argillic 2. Propylitic 3. Quartz-sericite	1. Muscovite-kaolinite-montmorillonite	Pyrite, chalcopyrite, molybdenite, sphalerite
Ray	Granite Mountain Porphyry, Pinal Schist, diabase	1. Potassic 2. Argillic 3. Propylitic	1. Muscovite-biotite-K feldspar 2. Muscovite-kaolinite 3. Chlorite-epidote-calcite	Pyrite, chalcopyrite, molybdenite
Silver Bell	Alaskite, dacite, quartz monzonite porphyry	1. Potassic 2. Argillic 3. Propylitic 4. Quartz-sericite	1. Muscovite-K feldspar 2. Muscovite-kaolinite 3. Chlorite-calcite-montmorillonite 4. Quartz-sericite	Pyrite, chalcopyrite, bornite, tetrahedrite, molybdenite

2. *Argillic alteration.* This is indicated by the presence of clay minerals and evidence of strong leaching of calcium. This loss of calcium is reflected chiefly by the absence of carbonate and epidote minerals. Quartz is a ubiquitous mineral in altered rocks of this kind and is usually accompanied by either of two principal assemblages: (1) muscovite-kaolinite-montmorillonite or (2) muscovite-chlorite-montmorillonite. K feldspar is not stable here.
3. *Potassic alteration.* This is indicated by the assemblage muscovite-biotite-K feldspar or any two of these three phases. Creasey points out that chalcopyrite seems to be the only widespread primary copper mineral in zones of potassic alteration.

In addition there is a fourth type of assemblage—quartz-sericite-pyrite—that does not fit into any of the above.

Table 12-9 (adapted from Creasey, 1966, p. 63, table 1) summarizes host-rock-alteration–sulfide assemblages for several of the more important American occurrences. Table 12-10 (after Bauer et al., 1966, p. 240, table 3) illustrates the chemical changes induced by alteration of the monzonitic stocks of the Robinson district of Nevada. (The classification of alteration here is a local one and deviates somewhat from that of Creasey but does not necessarily contradict it.) In most cases alteration develops a zonal pattern of intensity and type, and in

Table 12-10 Analyses of monzonitic rocks and their altered counterparts, Robertson Mining District, Nevada. (Adapted from Bauer et al., 1966, p. 240)

	Sericitic alteration assemblage (3 analyses)	Biotite-argillic alteration assemblage (8 analyses)	Unaltered monzonite and monzonite porphyry (6 analyses)
SiO_2	68.87	66.02	58.84
Al_2O_3	14.33	14.02	17.35
Fe_2O_3	4.19	1.39	3.17
FeO	1.49	1.92	2.86
MgO	1.01	1.28	1.98
CaO	0.15	1.14	6.15
Na_2O	0.15	1.20	3.18
K_2O	4.74	7.15	4.22
MnO	Trace	Trace	0.16
H_2O+	2.73	1.36	0.79
TiO_2	0.50	0.50	0.71
P_2O_5	0.15	0.26	0.35
CO_2	0.01	0.19	0.14
F	0.35	Trace	0.10
Cl	0.01	Trace	0.03
S	4.79	2.29	0.13
Cu	0.73	1.09	Trace

many instances the abundance of sulfide, and the proportions of the different sulfides, are directly related to such variations.†

ORIGIN

The porphyry coppers are one of the few categories of ore deposits on whose origin there is virtually complete unanimity.   There seems no doubt that they are of igneous origin, that they are part of, or are derived from, the principal host stock of the locality concerned, and that in any particular area they represent a closely related suite of orthomagmatic, contact-metamorphic, and metasomatic ores.

The source of the porphyritic rocks and their intrusive associates is not known, though very limited $^{87}Sr/^{86}Sr$ data suggest that at least some may come from the lower crust or mantle. $^{32}S/^{34}S$ ratios in the sulfides of the Bingham stock are very uniform and have a mean value close to meteoritic, which does not contradict the evidence of the $^{87}Sr/^{86}Sr$ ratios given above.   This isotope information is, however, far too sparse to be unequivocal.

Two features of the intrusions and their setting stand out and may well be significant.   These are the distinctly porphyritic nature of the intrusions and the presence of abundant volcanic rocks in the intruded or overlying formations.   The porphyritic character indicates two phases of crystallization: a comparatively deep phase, yielding coarse crystals, followed by rapid injection to higher levels that with the concomitant sharp falls in pressure and temperature, led to a second comparatively shallow phase of crystallization, this involving rapid congealing of the remaining liquid and the development of the characteristic fine groundmass. The association with volcanic sequences containing much material comparable in chemical composition with that of the intrusions suggests that the intrusions were related to (though not necessarily feeders of) at least part of the volcanic activity concerned.   A volcanic affiliation of this kind would also imply a rapid rise of the magmas, with correspondingly rapid diminution of confining pressure.   Since the solubility of the sulfides in these magmas is likely to be directly related to confining pressure, a sudden fall in the latter would cause the separation of large numbers of small sulfide globules throughout the portion of the melt concerned.   Two principal possibilities then appear:

1. The whole mass of melt congeals rapidly and virtually as one; the resulting rock consists of early formed silicate phenocrysts set in a fine groundmass, and the sulfides occur as small disseminated particles throughout this groundmass. (Freezing of the intrusion would have to be rapid otherwise the molten sulfides would begin to collect to form larger, massive segregations.)
2. Solidification of the mass as a whole commences with the rapid solidification of

† For further reading on alteration-mineralization zoning in and about porphyry copper deposits, the interested reader is particularly referred to the paper "Lateral and vertical alteration-mineralization zoning in porphyry ore deposits," by J. D. Lowell and J. M. Guilbert, in *Econ. Geol.*, vol. 65, pp. 373–408, 1970.   This publication appeared after the present chapter had been written.

an outer, upper crust, followed by slower congelation of the inner, lower portions. Here we might expect the chilled margin to be essentially the same as the material of 1 and to possess similarly disseminated sulfide. On further cooling, the crust begins to fracture, yielding numerous small, often irregular, cracks; i.e., it becomes "crackled." If the vapor pressure of the underlying molten material exceeds that of the confining pressure, volatiles (sulfur gases, halides, sulfates, heavy metal compounds, CO_2, H_2O, etc.) separate from the melt and form bubbles in it. These stream to the upper levels of the magma chamber, into the innumerable fractures of the "crackled" crust, and in many cases, through to the country rocks adjacent to the intrusive contact. In this case:

a. The upper portions of the intrusion beneath the chilled crust may be be enriched in disseminated sulfide by *replacement*.
b. The chilled crust, already containing early formed particles of sulfide in its groundmass, receives additional sulfides deposited by *fissure filling* in the myriads of small cooling cracks and by lesser *replacement*.
c. The surrounding rocks—limestones, detrital sediments, earlier volcanics —receive sulfide as small replacement particles and pore-space fillings. Depending on the distribution of rock types and structure, this *contact* or *pyrometasomatic* or *pneumatolytic* mineralization develops as partial or complete halos about the source stock or follows structural breaks and the intersections of these with favorable host rocks. The mineralization as a whole is, of course, made up of the sulfides and related metallic minerals, together with materials such as quartz, barite, and carbonate.
d. The gases and liquids (largely H_2O and CO_2) involved in the lower-temperature phases of the mineralizing processes attack the silicates of both intruded and intrusive rocks, leading to the ubiquitous hydrothermal alteration so characteristic of the porphyry coppers.

CONCLUDING STATEMENT

Most of the porphyry coppers appear to involve quite shallow, and probably rapidly injected, intrusions. Their rounded outlines and their very clear spatial tie with volcanic rocks also suggests that at least many of them may have been subvolcanic. Such a shallow, subvolcanic nature is beyond any doubt in the case of the Pangunu deposit of the Solomon Islands, and a little license allows one to see the likelihood of similar situations in many of the slightly older porphyry copper environments. The occurrence of the youngest known deposits—those of the Solomons, New Guinea, and the Philippines—in modern island arcs and the close tie between the North American deposits and eugeosynclinal zones strongly suggests a fundamental connection between many porphyry coppers and volcanic island arc structures. (The possible tie between porphyry copper incidence and regional faulting in some older areas certainly does not contradict this. Where the porphyritic rocks intrude rocks notably older than themselves, this may reflect sporadic but long continued activity of the great fault zones that originally coincided with very old volcanic arc structures.)   The volcanism of arcs is well known to evolve from mafic and ultramafic type in the early stages of activity to andesitic, dacitic, and rhyolitic type in the later stages of arc development.   It would therefore appear that further deposits of this type are most likely to be found in the more mature portions of modern volcanic arcs—the larger, older islands and the extensions of the arcs into what are now continental margins—and in older (e.g., Mesozoic and Paleozoic) arc structures that have somehow escaped erosion or whose underlying faults have remained active over very long periods of time.   Among the modern arcs, islands such as Sumatra, New Guinea, Japan, and Cuba stand out.   The Kamchatka Peninsula and southern Alaska are good examples of areas in which the mature portions of modern arcs have become "continental."   These, together with areas of older arcs exhibiting the above characteristics, look to be promising areas of potential porphyry copper occurrence.

RECOMMENDED READING

A. Carbonatite association

Heinrich, E. W.: "The Geology of Carbonatites," Rand McNally & Company, Chicago, 1966.

Tuttle, O. F., and J. Gittins: "Carbonatites," Interscience Publishers, New York, 1966.   (See especially T. Deans, Economic mineralogy of African carbonatites, pp. 385–413.)

Verwoerd, W. J.: South African carbonatites and their probable mode of origin, *Ann. Univ. Stellenbosch*, vol. 41, ser. A, no. 2, 1966.

B. Anorthosite association

Buddington, A. F., and D. H. Lindsley: Iron-titanium oxide minerals and synthetic equivalents, *J. Petrol.*, vol. 5, pp. 310–357, 1964.

Lister, G. F.: The composition and origin of selected iron-titanium deposits, *Econ. Geol.*, vol. 61, pp. 275–310, 1966.

Philpotts, A. R.: Origin of certain iron-titanium oxide and apatite rocks, *Econ. Geol.*, vol. 62, pp. 303–330, 1967.

C. Porphyry copper association

Lowell, J. D., and J. M. Guilbert: Lateral and vertical alteration-mineralization zoning in porphyry ore deposits, *Econ. Geol.*, vol. 65, pp. 373–408, 1970.

Titley, S. R., and C. L. Hicks (eds.), "Geology of the Porphyry Copper Deposits: Southwestern North America," University of Arizona Press, Tucson, 1966.

13
Iron Concentrations
of Sedimentary Affiliation

Almost all sedimentary rocks contain readily detectable quantities of iron. Some contain less than 1.0 per cent, others as much as 65.0 per cent, and a few even more. Between these extremes there exists a complete range of iron-bearing sedimentary rocks that, when iron assumes notable proportions, are usually qualified as *ferruginous*—ferruginous shales, ferruginous cherts, and so on. These may merge into others, perhaps of only slightly higher iron content, that are classified as "iron ore." Indeed many common sedimentary rocks—shales, sandstones, limestones—are known to grade, by the simple addition of iron minerals, through "ferruginous" analogs to low- and then high-grade iron ore and finally to essentially pure iron mineral.

Such complete gradation from "ordinary" sedimentary rocks to "iron ores" naturally suggests that the iron itself is sedimentary, and indeed such a view has been held by most geologists for a very long time. As early as 1833, Edward Hitchcock (whose perspicacity has already been referred to in Chaps. 2 and 10), in his studies of some of the iron ores of the New England region, maintained that these were originally deposited as sediments: "At all these localities the ore is found in distinct beds in the strata; and sometimes it has a slaty structure, having the appearance of a contemporaneous origin with the rock" (1833, p. 356). In 1836 the German D. C. G. Ehrenberg found that bacteria played an important part in the formation of limonitic iron ores that had long been known to be actively accumulating in bogs and marshes in Europe. At much the same time a number of the oolitic iron ores of England were being actively investigated and scientific publication on the Northampton Sand Ironstone, for example, began about 1860. The European and English deposits studied at this time were mostly contained in Mesozoic to Recent strata; all were of conspicuously sedimentary affiliation and all seem to have been regarded quite simply as iron-rich sedimentary rocks.

This general view has prevailed and has been strengthened in the ensuing 100 years. There now seems no question that at least most of the major primary iron concentrations in sedimentary rocks are sediments themselves. While much is still to be learned concerning source and processes of derivation, transport, and deposition of the iron—problems with which much of this chapter is concerned— the general nature of deposition is clear: the iron has been segregated by sedimentary processes, and these processes are dominantly chemical. We are thus largely concerned with low-temperature solution chemistry in natural environments, that is, with the principles that concerned us in Chap. 6.

PRINCIPAL MINERALS OF IRON-RICH SEDIMENTARY ROCKS

Although hematite and magnetite are quantitatively—and economically—the most important iron minerals of sedimentary iron concentrations, they are by no means the only ones. As a class, such iron concentrations contain a wide variety of iron-bearing species—oxides, hydroxides, carbonates, silicates, and sulfides—and many individual deposits have been found to possess a substantial range of these compounds. In most cases the mineral assemblages of a given deposit appear to reflect original conditions of deposition. In other cases, however, diagenetic, and particularly metamorphic, processes have clearly been important in the development of the assemblages as they now occur.

OXIDES

The principal *oxides* are *hematite*, *magnetite*, and *goethite* (including *limonite*). *Hematite* occurs in a wide variety of forms—as layers interbedded with fine silica (chert), as ooliths, as replacements of fossils, and sometimes simply as an earthy matrix to other minerals. The first type of occurrence is general in Precambrian deposits, whereas the oolitic and replacement forms are most abundant in post-Precambrian ores. *Magnetite* is abundant and virtually ubiquitous in Precambrian occurrences; it is not nearly so abundant, though still quite common, in younger ores. *Goethite* of sedimentary ores almost characteristically occurs as ooliths in which it forms concentric layers in association with layers of the silicate, chamosite. Such ooliths are bound by a matrix of clay, calcite, siderite, and/or earthy limonite. Ores of this kind are almost entirely confined to post-Precambrian strata and are characterized by abundant excess water and a variety of impurities including phosphate.

CARBONATES

Next in importance to the oxides is *carbonate*, principally *siderite* ($FeCO_3$). This is usually impure and in most instances is $(Fe,Mg,Ca,Mn)CO_3$, with Fe highly dominant. In the case of the Lake Superior ores, James (1954) estimates $FeCO_3$ to be about 70 per cent of the carbonate molecule. It must be kept in mind, of course, that a certain amount of carbonate in carbonate ores—and particularly those of post-Precambrian age—is calcite, which is a common minor constituent of the matrices of oolitic ores.

SILICATES

Next in importance to the carbonates—though of rather lesser overall importance —are the *silicates*. Some of these are doubtless primary, in the sense that they formed at the time of sedimentation, while others have almost certainly formed during diagenesis and/or metamorphism. According to James (1966) the only important iron silicates definitely of primary sedimentary origin are chamosite, greenalite, and glauconite. Thuringite, minnesotaite, and stilpnomelane appear to be of diagenetic–low-grade metamorphic origin. Iron-bearing minerals such as grunerite and almandine are, of course, clearly products of regional metamorphism.

Chamosite is the most important silicate of post-Precambrian sedimentary iron concentrations and in some cases is the major iron-bearing constituent of the ore concerned. It is a dark green mineral of rather uncertain crystal structure, regarded variously as comparable with chlorite, kaolinite, and antigorite structures. Microscopically it is pleochroic yellow to green. It occurs almost invariably as a constituent of ooliths, though it may occur as scattered flakes in the matrices of oolitic sediments and in chamositic mudstones.

Greenalite is another important iron-bearing silicate. Notable occurrences are in the Mesabi and Gunflint districts of the Lake Superior iron ore province. The "classical" occurrence of greenalite may be said to be that of the Biwabik iron formation of Minnesota (it was from material from here that the mineral was first described). In contrast to chamosite, which seems to be typically post-Precambrian, greenalite is almost entirely restricted to Precambrian iron concentrations; in the Biwabik it may constitute up to about 70 per cent of a sample, where it occurs typically as small rounded granules. These are dark green and variably anisotropic. Early work by Gruner (1933) suggested greenalite to be essentially an iron serpentine, with a crystal structure similar to antigorite and a formula approaching $9FeO \cdot Fe_2O_3 \cdot 8SiO_2 \cdot 8H_2O$. More recent investigations indicate a grouping with amesite, chamosite, and cronstedite as a member of the septechlorite group (Fig. 13-1). (These are closely related chemically to the chlorites and structurally to the serpentine minerals.)

Glauconite is also important. It is a mica mineral that occurs generally, though not exclusively, in marine sediments and sedimentary rocks, particularly "greensands." Its formula may be written $(K,Na,Ca)_{1.2-2.0}(Fe^{3+},Al,Fe^{++},Mg)_4$ $(Si_{7-7.6}Al_{1-0.4}O_{20})(OH)_4 \cdot nH_2O$, and its relations with some of the clay minerals and other micas are illustrated in Fig. 13-2 (after Burst, 1958). In its optical

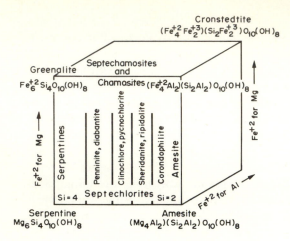

Fig. 13-1 Compositional relationships among chlorites, septechlorites, and serpentines. (*From Deer, Howie, and Zussman, "Rock Forming Minerals," vol. 3, Longmans, Green & Co. Ltd., London, 1962, with the publisher's permission.*)

properties and crystal structure glauconite is somewhat similar to biotite.   Its color is related to its Fe^{++}, increase in this leading to a progressive deepening of its green tint.   An interesting observation connected with this is that glauconites from older sediments tend to contain less ferrous iron and hence are paler green than those of younger terrains: the reason for this is not known.   Whereas chamosite and

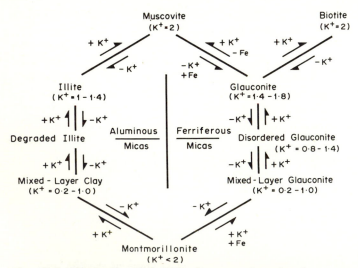

Fig. 13-2 Diagenetic relationships between mica minerals, as suggested by Burst. (*From Burst, Amer. Mineralogist, 1958.*)

greenalite are essentially ferrous minerals, glauconite is dominantly ferric, with a ferric-ferrous ratio of about $7:1$ according to James (1966). The mineral thus does not form in strongly reducing environments, though judging by its associates— sedimentary ferrous sulfide and organic remains—it forms in what must be at least *moderately* reducing environments. As a constituent of modern sediments it has been found in sandy sediments ranging from shallow near-shore environments to depths of over 14,000 feet. Its principal habitat in modern sediments appears to be the depth interval 60 to 2400 feet.

Thuringite (and its close relative bavalite) are iron-rich chlorites of frequent occurrence in sedimentary iron deposits. As such they are close relatives of chamosite; a schematic diagram due to Hey (1954), showing relationships between the "oxidized chlorites," is given in Fig. 13-3. The formula for thuringite as determined by Engelhardt (Engelhardt, 1942; see James, 1966) is $(Si_{4.8}Al_{3.2})$ $(Mg_{1.4}Fe_{7.4}^{++}Fe_{1.5}^{3+}Al_{1.7})(OH)_{16.0}O_{20.0}$ and for bavalite is $(Si_{4.5}Al_{3.5})(Mg_{0.7}$- $Fe_{9.5}^{++}Fe_{0.1}^{3+}Al_{1.6})[(OH)_{13.9}O_{0.9}]O_{20.0}$. Bavalite thus contains a lower propor-

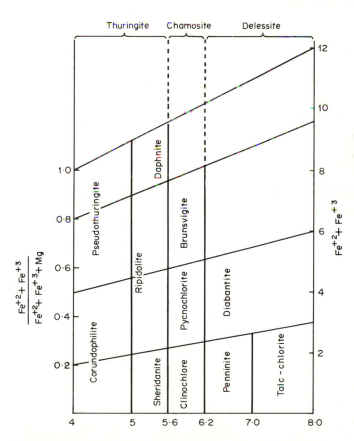

Fig. 13-3 Nomenclature of orthochlorites and oxidized chlorites. (*From Deer, Howie, and Zussman, "Rock Forming Minerals," vol. 3, Longmans, Green & Co. Ltd., 1962, with the publisher's permission.*)

tion of Mg and a much higher ferrous-ferric iron ratio than thuringite, though the two minerals are otherwise of very similar chemical composition.

Minnesotaite was identified by Gruner (1944) as an iron-rich talc having the formula $(OH)_{11}(Fe^{++},Mg)(Si,Al,Fe^{++})O_{37}$. A modified formula is $(Fe,Mg)_3Si_4$·$O_{10}(OH)_2$, which compares with "normal" talc, $Mg_3Si_4O_{10}(OH)_2$. Minnesotaite is gray-green in hand specimen and yellow-green to colorless under the microscope. The more deeply colored varieties are pleochroic. It occurs characteristically as microscopic needles and plates and usually resembles sericite and talc, except that it often occurs as minute rosettes in addition to the sheaf structures shown by the other two. Although regarded by Gruner as a primary mineral, it has been suggested by other investigators to be a very low-grade metamorphic product.

The two principal associates of minnesotaite are greenalite, already considered, and *stilpnomelane*. The latter appears to have a layered structure closely related to that of talc, and the formula (Gruner, 1944, p. 298) is $(OH)_4(K,Na,Ca)_{0-1}$·$(Fe,Mg,Al)_{7-8}Si_8O_{23}$·$(H_2O)_{2-4}$. Stilpnomelane was earlier thought to be typically a mineral of iron-rich, low-grade, regionally metamorphosed sedimentary rocks and associated veins, but it is now known to be a widespread and often abundant constituent of chlorite- and glaucophane-bearing schists. The reason for the earlier nonrecognition is its general similarity to biotite, for which it was frequently mistaken. Minerals of the stilpnomelane group vary from dark green to reddish brown and black, which, with their pleochroism and micaceous appearance, accounts for the confusion. Stilpnomelane is not nearly so elastic as biotite, however, and for this reason has been referred to as "brittle mica."

SULFIDES

Several iron sulfides have been identified in sedimentary iron concentrations. Pyrite is overwhelmingly the most important, but *marcasite, pyrrhotite,* "*hydrotroilite*" ($FeS \cdot nH_2O$ or $FeS \cdot nH_2S$), "*melnikovite*" ("gel pyrite"), and *greigite* (Fe_3S_4) have also been recorded.

Pyrite concentration ranges from sparse disseminations to massive beds. The former are, of course, common and consist of pyrite, as cubes or spheres, or both, scattered through dark, often heavily carbonaceous black shale. Although pyrrhotite is sometimes abundant in iron sulfide-rich beds, it is apparently not a primary mineral and is found only where the iron-rich beds have suffered some metamorphism. One also gains the impression that it occurs only in conspicuously volcanic sediments or sedimentary associations, though this is not easy to substantiate. Marcasite is not common and seems to be restricted to certain types of iron concentration, particularly the blackband ironstones. The other iron sulfides are not now important constituents of ores, and indeed most of them have been found in, and studied from, disseminations in modern marine sediments.

SOURCE, TRANSPORT, AND DEPOSITION
OF IRON IN NATURE

The incidence and movement of iron at and near the earth's present surface is now fairly well known and understood. However, this knowledge is of somewhat doubtful help in solving problems of sedimentary iron concentration; there is, unfortunately, no iron concentration approaching the size and richness of most "ores" known to be accumulating in sediments at the present day so that the relevance of present-day processes is by no means certain. It is therefore not surprising that the elucidation of earlier processes of iron sedimentation has its problems. Some of these are most appropriately considered in connection with the development of specific ore types, but before looking at them it is helpful to consider the more general facts and processes of source, transport, and deposition.

SOURCE

There are two principal sources of iron at the earth's surface: continental erosion and volcanic activity. To these may be added a third—the floors of oceans and lakes.

1. Continental erosion The greater part of the iron of the continental masses is contained in silicates—olivines, pyroxenes, amphiboles, micas, and others. A lesser, though very substantial, part is in the form of oxides. Most of this oxide iron of igneous rocks occurs in magnetite, with smaller amounts in ilmenite and titanomagnetite. Sedimentary rocks contain much hematite and goethite, with lesser magnetite and iron-titanium oxides. The incidence of oxide iron in metamorphic rocks is much the same as in sediments. Total iron expressed as FeO rarely exceeds 14 per cent of "normal" igneous rocks, and about 12 per cent is a general maximum. Basalts and their intrusive equivalents commonly contain 11 to 13 per cent iron as FeO and andesites and their intrusive equivalents about 6 to 9 per cent. Sedimentary and metamorphic rocks have highly variable iron contents. Some "averages" listed by Poldervaart (1955) are given in Table 13-1.

2. Volcanism Volcanic activity yields iron in the rocks—both lavas and pyroclastics—themselves and as exhalations. Where these are produced on the land surface the iron of the rocks is, of course, trapped in the silicate, oxide, and associated minerals and that of the exhalations is either deposited on the walls of fissures or lost to the atmosphere. Where volcanism occurs beneath the sea, however, some of the iron of the lavas and tuffs is leached by the water and that of the exhalations is trapped and held—at least momentarily—in the body of the ocean. Much of the exhaled iron is probably given off as ferrous chloride, though there is a lack of precise information on this point. That the amount may be considerable on a local scale has been established by observations of Zelenov (1958) about the Ebeko volcano in the Kurile Islands and by observations of others in the Greek islands (notably Santorin) and the volcanic islands of the Pacific. In a number of instances the sea (particularly in bays) has been observed to be discolored red as a result of oxidation of volcanic iron exhaled on the sea floor or from the submarine slopes of volcanoes. In this connection Bonatti and Joensuu (1966) have found substantial quantities of goethite near submerged volcanic cones on the floor of the east Pacific Ocean, and Bostrom and Peterson (1966) have found much iron in sediments sampled along several traverses over the East Pacific Rise. All these deposits

Table 13-1   Iron contents of some igneous, sedimentary, and meta-morphic rocks.   (Percentages, water free, from Poldervaart, 1955)

FeO	Fe₂O₃	Rock type and reference
I. Igneous rocks		
1.0	1.4	102 rhyolites; Daly, p. 9, 1933
1.8	1.6	546 granites; Daly, p. 9, 1933
2.6	1.3	137 granodiorites; Nockolds, p. 1014, 1954
2.7	1.7	40 granodiorites; Daly, p. 15, 1933
3.1	3.4	87 andesites; Daly, p. 16, 1933
4.5	3.2	70 diorites; Daly, p. 16, 1933
5.5	3.5	49 andesites; Nockolds, p. 1019, 1954
7.0	2.7	50 diorites; Nockolds, p. 1019, 1954
9.1	2.9	137 "normal" tholeiites; Nockolds, p. 1021, 1954
9.8	1.1	"Average tholeiite"; Poldervaart and Green, from Poldervaart, 1955
6.5	5.4	198 basalts; Daly, p. 17, 1933
10.0	3.7	43 plateau basalts; Daly, p. 17, 1933
8.8	4.5	182 ultramafic igneous rocks; Nockolds, p. 1032, 1954
9.9	2.5	23 peridotites; Nockolds, p. 1032, 1954
II. Sedimentary rocks		
0.3	0.2	"Average orthoquartzite"; Pettijohn, p. 241, 1949
1.1	2.3	"Average quartzite, Finland"; Sederholm, p. 4, 1929
0.3	1.1	"Average sandstone"; Clarke, p. 30, 1924
	2.4	"Average arkose"; Pettijohn, p. 259, 1949
	3.7	"Average Mississippi silt"; Clarke, p. 509, 1924
4.3	1.0	"Average greywacke"; Pettijohn, p. 250, 1949
3.1	4.3	51 Paleozoic shales; Clarke, p. 552, 1924
2.6	4.3	"Average shale"; Clarke, p. 30, 1924
1.9	4.3	27 Mesozoic and Cenozoic shales; Clarke, p. 552, 1924
	0.5	"Average limestone"; Clarke, p. 30, 1924
III. Metamorphic rocks		
3.8	2.9	22 slates; Clarke, p. 631, 1924
5.4	3.3	61 slates; Poldervaart, p. 136, 1955
4.8	3.0	50 phyllites; Poldervaart, p. 136, 1955
4.6	2.1	103 mica schists; Poldervaart, p. 136, 1955
2.0	1.6	250 quartzofeldspathic gneisses; Poldervaart, p. 136, 1955
7.8	3.6	200 amphibolites; Poldervaart, p. 136, 1955

appear to result from submarine volcanic–hydrothermal activity.   This is referred to again in Chaps. 14 and 15, but for present purposes there seems no doubt that volcanic activity constitutes a substantial source of "free" iron—i.e., iron in simple combinations available for transport and deposition elsewhere.

3. Sea-floor detritus   That the source of iron of marine sedimentary iron concentrations might lie in iron-bearing detritus accumulating on the floor of the ocean itself has been suggested by several investigators, notably Borchert (1960).

Hough (1958) and Govett (1966) have suggested that some lacustrine iron concentrations may be similarly derived from iron-bearing detritus on lake floors. Borchert has pointed out that most sedimentary iron concentrations occur in sequences that also contain—albeit frequently at some distance from the iron deposits themselves—carbonaceous units. These represent the accumulation and anaerobic decay of organic matter in *deeper* parts of the off-shore floor and indicate that during sedimentation in these areas, reducing conditions, and perhaps notable enrichment of H_2S, prevailed. At the other end of the scale the *shallow* waters of the near-shore zones would have been well oxygenated:

> Between the two kinds of water—the oxygenated water of the shallow seas and the H_2S-rich bottom water above the sapropel—there is a *transition zone* which is *of the utmost importance in connexion with the dissolution and mobilization of iron from ordinary sediments.*
>
>   Between the bottom zone where the organic matter suffers no oxidation but merely some bacterial decomposition, and the shallow water zone where there is ample oxygen to cause virtually complete oxidation of all organic substances, there must exist a *carbonic acid zone rich in* CO_2 which is conducive to the partial oxidation of organic matter so as to produce abundant CO_2 while maintaining a more-or-less strongly reducing milieu. Within this zone iron may be dissolved in such quantities as to yield concentrations of 50 mg iron bicarbonate per cu. dm (1960, p. 266).

In this way, according to Borchert, iron is removed from iron-bearing minerals in erosional detritus, then to be moved with the currents until it encounters chemical conditions causing precipitation. Derivation from this type of source is illustrated in Fig. 13-4 (after Borchert, 1960, fig. 5).

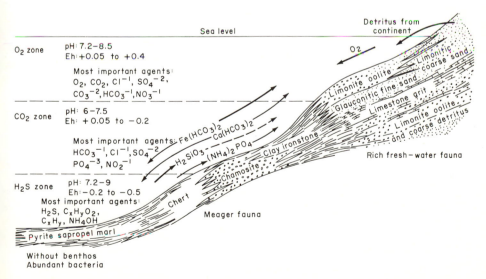

Fig. 13-4 Schematic section showing relations between ironstone facies and physicochemical conditions, and the chemistry of transport of iron derived from sea-floor detritus, as visualized by Borchert. (*From Borchert, Trans. Inst. Mining Met., London, 1960.*)

TRANSPORT

The transport of submarine volcanic iron presents no problem. Exhaled material may remain in the ferrous form (chloride or bicarbonate) and be carried away in solution, or it may be oxidized and removed, in finely particulate form, as a suspension. In a general way the same holds for iron derived from the leaching of submarine lavas and tuffs and from accumulated erosional detritus.

In contrast to this the derivation of iron from land surfaces poses one of the most difficult problems in the whole of iron ore-genesis theory.

1. Transport in solution On the surface, transport necessarily awaits the weathering of the source rocks. There is no difficulty in this. That iron may be abstracted from a variety of minerals and rendered into simple compounds there is no doubt; it is a matter of everyday observation and may proceed with great rapidity in appropriate climates. The sulfides of iron, particularly marcasite and pyrrhotite, are highly susceptible to weathering oxidation and go to ferrous and ferric sulfates and sulfuric acid. Silicates containing ferrous iron are readily weathered to yield ferrous ions that may be removed as the bicarbonate. The oxides of iron, such as magnetite and ilmenite, are expectably fairly resistant, but even these are broken down in time under appropriate conditions, yielding their ferrous ions. What *does* pose the difficulty is the *movement* of the iron following its initial release by weathering. As we have already seen from Chap. 6, in oxidizing environments and at neutral and higher pH—the dominant state of affairs in surface waters—iron forms ferric hydroxide, a compound of negligible solubility. To form high iron concentrations in the sediments they are producing, rivers must carry much iron but little detritus—that is, to be most effective they should carry iron in solution. If, under natural conditions, iron cannot be kept in solution, how does transport take place?

As yet there seems to be no clearly satisfactory answer to this question. A very *general* answer, however, may be that significant quantities of iron may possibly be kept in solution by appropriate small deviations from the "normal" surface environment and in subsurface waters. For example, Castano and Garrels have shown (1950, in connection with the Clinton ironstones) that under experimental conditions, significant quantities of ferrous iron can be carried in solution in aerated river waters of pH 7 or lower. They point out that where such waters enter a marine environment with solid calcium carbonate in equilibrium with the sea water, the iron is virtually completely precipitated as ferric oxide, both in the water itself and as a replacement of the carbonate. It also appears that considerable quantities of iron may be held in solution in natural waters in organic complexes and as ferric colloids peptized by organic acids—a possibility first actively investigated by Gruner almost 50 years ago. In particular it has been found that there may be strong complexing of ferrous iron in concentrated solutions of tannic acid. This may be a significant phenomenon in forested regions.

The transport of iron derived by carbonic acid leaching of sea- (or lake-) floor detritus follows naturally from the movement of bottom currents.

2. Transport in suspension   In addition to movement in solution, iron can, of course, be transported mechanically as fine oxide or hydroxide.   While, with higher Eh and pH, the oxide forms and the iron is thrown down, there is no reason why such precipitates should not be carried along, as very fine suspensions, by the slowest moving rivers.   It may also be moved as a constituent of fine clay particles —as an essential constituent of the clay, i.e., as a substituting impurity in the clay lattice or, most importantly, as an iron oxide coating on the surface of the mineral platelets.   This might involve no more impurity than is observed in certain kinds of sedimentary iron concentration.

There seems little doubt that several of these solution and suspension mechanisms operate to a greater or lesser extent in any individual case and that all contribute in the movement of erosional iron over the world as a whole.

3. Movement of iron contributed in volcanic solutions   As has already been pointed out, the transport of iron derived from volcanic orifices is in many respects quite a different matter.   Where it is contributed to the sea floor, this may occur in an oxidizing or a reducing environment and, where the volcanic products are plentiful enough to seriously modify the natural oceanic milieu, in one in which the pH may be very low indeed.   Where the volcanic additions are substantial, pH and Eh are probably low and the iron drifts away in solution, with the ferrous ions in equilibrium with SO_4^{--} or Cl^-.   As such solutions become dissipated into the body of the ocean (with its much higher Eh and pH), the iron would be oxidized and begin to settle through the water column.   If the volcanic orifice is on the land surface, initial transport is in what may be the highly acid waters of the hot spring. Alan, Zies, and Fenner (1919), for example, observed enormous quantities of acid (HCl, H_2SO_4, HF, etc.) given off in the Valley of Ten Thousand Smokes in Alaska, and highly acid conditions have been observed in many volcanic waters throughout the world.   In these situations large quantities of iron may be carried in solution in equilibrium with sulfate and/or chloride ions.   Such solutions persist until mixture with normal river waters or the body of the sea causes neutralization, oxidation, and precipitation of a ferric compound.   The case of the contribution of iron to the near-shore waters of the Okhotsk Sea (Zelenov, 1958) by springs associated with the Ebeko volcano has already been mentioned.

DEPOSITION

The mechanism of deposition of iron in sediments is naturally dependent on the form in which the iron is contributed, on the conditions prevailing on the sea or lake floor, and on the conditions immediately below the mud-water interface of the body of water concerned.†   Whether the iron is contributed as ferrous complexes or ferrous ions, free oxide/hydroxide particles, ferric coatings on clay micelles,

---

† It must be emphasized that it may be very difficult to tell whether a given mineral is truly sedimentary in the sense that it was as it is now precisely at the moment it arrived at the bottom or whether it is diagenetic.   Confusion of the two may well lead to erroneous conclusions concerning modes of primary deposition.

or as an integral part of iron-bearing mineral detritus, the iron compound formed just before, at, or just after arrival at the bottom depends on the chemical and physical conditions at and just above the mud-water interface. We are therefore immediately concerned with the considerations of Chap. 6—the Eh-pH conditions of the environment concerned and the amounts of the constituents with which iron may combine. Since sedimentary iron ores are composed of oxides, carbonates, silicates, and sulfides, the aqueous systems of immediate concern are those containing one or more of dissolved oxygen, carbon dioxide, silica, and sulfur.

1. Processes As already noted in Chap. 6, plotting in two dimensions of a system possessing a large number of variables requires that all but two be arbitrarily fixed. Such arbitrary values may be changed, of course, where one is prepared to draw up a number of diagrams representing some particular range of conditions. Figure 6-4 (after Garrels, 1960, p. 157, fig. 6-21) shows the stability fields of some important iron minerals as functions of Eh and pH in an aqueous system at 25°C, 1 atm total pressure, total dissolved sulfur $= 10^{-6}\, m$, total dissolved carbonate $= 10^0\, m$. For these values (and it must be emphasized again that any change in such values would modify the stability fields to at least some extent) the diagram gives a picture of the mineral groups that are capable of coexisting under given conditions of Eh and pH and hence that might be expected to develop together within the same sedimentary environments.

The effect of changes in our variables is well illustrated by the influence of higher concentrations of silica on the high pH-low Eh portion of the present system. The modification may be seen easily by comparing Fig. 6-4 with Fig. 13-5. Where silica is absent this area of the system is characterized by the stable existence of magnetite, and depending on Eh, pH, and the values assigned to the other variables, magnetite may coexist with hematite, siderite, or pyrite, or with siderite and hematite, or with siderite and pyrite. When silica is added, however, the iron combines with this, and in the present case, ferrous metasilicate ($FeSiO_3$) forms instead of magnetite. Thus, as is noted by Garrels, if we are concerned only with iron oxides and iron silicate in the presence of amorphous silica, magnetite is eliminated as a solid phase; where natural bottom waters are saturated with amorphous silica, and where there is sufficient silica to satisfy all the available iron, iron silicate will form in preference to magnetite.

While theoretical and experimental work of this kind clearly lacks the complexities inevitable in nature, it does at least indicate that different environments of iron sedimentation should be characterized by different assemblages of iron minerals. That different environments do indeed possess different but characteristic assemblages is now well established, and with the appropriate application of solution chemistry it is not too difficult to reconstruct some of the depositional processes concerned. This is referred to again shortly.

In addition to these physical (i.e., inorganic) controls, biological influences are also important in the formation of some iron ores and indeed have almost certainly operated to at least some extent in the development of most of them. One of the earlier investigators of biological iron deposition has already been men-

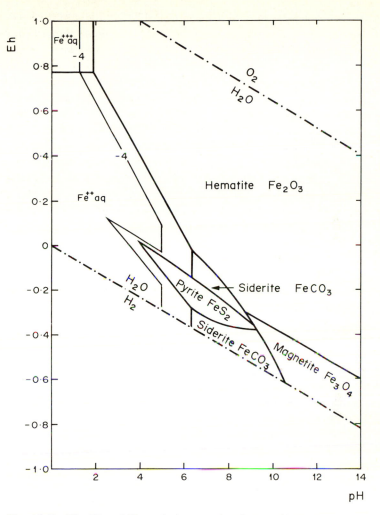

Fig. 13-5 Eh-pH stability relations among iron oxides, pyrite, and siderite at 25°C, 1 atm total pressure, total dissolved carbonate = 10^0 m, total dissolved sulfur = 10^{-6} m. Compare with Fig. 6-4. [*Redrawn from Garrels, "Mineral Equilibria," after Fig. 6.21 (p. 157), Harper & Brothers, New York, 1960.*]

tioned—D. C. G. Ehrenberg, who in 1836 pointed out that organisms, particularly bacteria—played an important part in the precipitation of limonitic ores in bogs and swamps. Probably the most extensive treatment of this field in English is the detailed paper of E. C. Harder (1919), which reviews much of the earlier work and in which Harder reports much of his own experiment.

There are three possible ways in which organisms—principally bacteria— can precipitate iron compounds:

1. The organism modifies the physical-chemical nature of its surroundings as a result of its own body functions, thus creating conditions suitable for the precipitation of an iron compound The best examples of this are the "sulfate-reducing" bacteria. These live on organic detritus on the sea or lake floor and are able to reduce certain oxygen-bearing compounds, notably sulfates. The bacteria also produce H_2S by the destruction of proteins in the organic material on which they live. This production of a reducing environment, rich in S^{--}, leads to the precipitation of FeS and other sulfides.

2. The organism accretes an iron compound as its skeleton or as an encrustation about its body. The accretion of siliceous skeletons by diatoms and radiolarians, and the formation of calcium carbonate and calcium phosphate hard parts by a variety of organisms, is analogous. The use of iron compounds for the construction of hard parts is not common, though it is just possible that some minute spherical pyrite particles of marine black muds are formed in this way.

3. The organism acts as a catalyst—i.e., it causes or accelerates formation and precipitation of a compound that is stable in the environment concerned. A well-known example of this is the formation of copious, often flocculent, precipitates of ferric hydroxide about some natural springs. In this case the organism uses the oxidation of ferrous to ferric iron as a source of energy (an electron), thereby accelerating the process and causing the copious precipitation often observed.

The importance of organic activity in iron precipitation is not known in any precise way; one can say little more than that it *does* cause some precipitation and that it is important in *some* instances. There is no doubt that it is very important indeed in the precipitation of limonite in bogs and marshes and FeS and FeS_2 in both salt- and fresh-water anaerobic environments. How significant it has been in the development of iron concentrations in other environments is not known, however. There may well be a large field for investigation here.

2. Environments Some of the principal environments of iron accumulation and their processes of fixation are

a. *Unrestricted marine environments with little or no bacterial activity.* Assuming free circulation, the sea water at the mud-water interface has the normal Eh and pH of the open ocean, i.e., an Eh of $+0.1$ to $+0.3$ volt and a pH of 7 to 8. In this case ferric compounds are stable and the characteristic minerals are goethite, hematite, and various indefinite ferric compounds.

b. *Near-shore marine environments with facies-controlled bacterial activity.* In this case the development of a facies pattern leads in turn to the development of sedimentary zones in which bacteria are active. Such activity may lead to a reduction of Eh, but this is unaccompanied by any significant change in pH. Referring to Figs. 6-4 and 13-5, it may be seen that, for the normal sea water pH of 7 to 8, progressive reduction of Eh in this way would lead to the formation of hematite in the zone of highest Eh, then siderite, and then pyrite at an Eh of around -0.2 volt. That is, the development of the "normal" facies of sedimentation would be accompanied by the development of parallel *facies of iron formation*—a concept elegantly developed by H. L. James (1954).

c. *Restricted shallow-water marine environments with bacterial activity.* These include estuaries, large enclosed bays, such as those of the Dutch coast, and barred lagoons. Organic matter is abundant in such places, leading to intense bacteriological activity and concomitant reduction of Eh. Ferric compounds are reduced, and if the sulfide ion is present, sulfide forms readily. Such environments, with their frequently intense bacterial activity, might be particularly important in the development of large accumulations of the silicates chamosite and glauconite. Both of these are, as we have already seen, highly aluminous, and one would immediately presume that they formed by the reaction of clay, or other aluminous particles, with iron. This may occur if alumina and iron are delivered together, as ferric oxide-coated clay particles, to the correct environment of deposition—i.e., restricted shallow-water marine environments in which bacteria are active. Chamosite of the Northampton Sand ironstones is associated with siderite, indicating an environment defined by a pH of 7 to 8 and an Eh of $+0.05$ to -0.10 volt. A small amount of pyrite is associated as well, suggesting that the environment had an Eh just below zero for the most part. Glauconite forms from iron-rich muds in the same kind of environment as chamosite, but at a lower Eh. According to Carroll (1958) its stability field is defined by a pH of 7 to 8 and an Eh near -0.20 volt.

d. *Lakes and bogs of tundra type fed by subsurface bicarbonate solutions and streams and lakes receiving acid volcanic effluent containing iron in solution.* In general the iron is precipitated by oxidation, though carbonate may be precipitated by loss of CO_2 in anaerobic environments.

PRINCIPAL KINDS OF IRON-RICH SEDIMENTARY ROCKS

There is a wide variety of iron-rich sediments and sedimentary rocks. The greatest modern concentrations of iron-rich material are probably beach, and raised beach, deposits of magnetite and associated ilmenite. Detrital concentrations of this sort are not, however, extensively preserved in older rocks, whose iron minerals are, as we have already noted, almost entirely of the chemically precipitated kind.

In modern marine sediments, such "chemical" iron generally increases going from coastal regions to the deeps, where most of it is bound in red muds or clays. The total iron content of such clays is remarkably uniform at about 8.5 per cent, though some more calcareous deep-sea clays have been found to contain up to about 16 per cent Fe_2O_3 (Revelle, 1944). Such levels of iron are, of course, far below those exhibited by most of the iron concentrations in older rocks. Present-day sediments of the Black Sea—which, because of the highly reduced and sulfurous bottom waters might have been expected to show particularly high (sulfide) iron contents—are even lower in iron than the deep sediments of the open sea. Total iron in the Black Sea muds ranges from about 3 to 5 per cent, and of this between one-quarter to one-half is contained in detrital material. However, as already noted, higher concentrations of chemically precipitated iron have recently been found adjacent to submarine and seaboard volcanoes and over broad areas of submarine hot-spring activity.

While these recent discoveries are most encouraging and hold out hope that mechanisms of sedimentary iron concentration may soon be studied at firsthand, there is still no firm evidence of the present-day formation of iron accumulations of the richness and magnitude of many found in older rocks. The only major class of iron accumulation that has so far been found to have modern analogs (indeed most deposits of this class *are* modern) is that of the *bog iron ores*, examples of which are well known in the glacial bogs and lakes of Europe, Russia, and Canada. Of the major types, however, these are by far the least significant as sources of iron.

Although the lines of division are not always clear and although there are the inevitable instances of apparently hybrid types, iron concentrations in sedimentary rocks fall generally into three great classes:

1. *Iron formations*, which are by far the largest and most widespread and which supply by far the major part of the world's iron.
2. *Ironstones*, which are of lesser, though still very substantial importance, and which have yielded the bulk of the indigenous iron of England and Europe.
3. *Bog iron ores*, which are of much lesser importance than the other two and whose exploitation has been largely restricted to Europe and Scandinavia.

Each of these is characterized, and distinguished from the others in a general way, by differences in geological setting and age, form, mineralogy, and chemistry, and each appears to set different problems of derivation and mode of deposition.

BOG IRON DEPOSITS

We may proceed by examining the least important—and, as it happens, the least problematical—of our three classes first, so clearing the way for weightier considerations to come.

Distribution, form, and setting   The principal habitats of this group of sedimentary iron concentrations are the swamps, lakes, and sluggish streams of recently glaciated areas. They are abundant and are virtually restricted to recently glaciated tundra areas of the Northern Hemisphere. Many such areas have not yet undergone full erosional regrading since glaciation, and as a result there is widespread development of internal drainage systems. It is principally in the lakes and swamps constituting the centers of such systems that bog ores develop. Minor variants of the type are known to form in volcanic streams and lakes and are therefore found in volcanic provinces such as Japan and the Kurile Islands; others, formed in association with coal measures, are now found as minor features in Carboniferous and Permian coal-bearing sequences in various parts of the world.

Of the form and setting of these deposits there is little to be said. They are simply lake or swamp sediments particularly rich in iron. Their form is that of a fine lacustrine sediment developed over the whole or part of—particularly the marginal areas—the lakes and swamps concerned. Investigations in Scandinavia indicate that the incidence of bog ores in any given region has little or no relation to the nature of the bedrock of the lakes or, in most cases, to the mouths of contributary streams.

Harder (1919) divides bog ores into two principal types:

1. *Lake ores.* These are well developed in northern Ontario and Quebec in Canada, in the eastern United States, and in Scandinavia. They appear to have formed in turbulent or agitated water round the margins of lakes.
2. *Marsh, or peat, ores.* These are common in swamps and peat bogs in cool areas. They are found in shallow depressions, either at the surface or, very frequently, beneath a few feet of porous surface soil. The iron minerals are frequently mixed with swamp humus or peat.

The two principal variants referred to above are

1. *Volcanic lake ores.* These are formed in lakes (sometimes in old calderas) fed by acid, iron-bearing waters from thermal springs and are particularly well developed in the middle part of Honshu in Japan.
2. *Blackband ironstones.* These characteristically occur in association with coal measures and hence are most extensively developed in sedimentary sequences of Carboniferous and Permian age. Some are, however, younger. They are composed principally of carbonate, which forms nodules, discontinuous lenses, or thin beds a foot or so thick associated with the coal beds.

Constitution Goethite is the only major identifiable iron mineral of bog deposits *sensu stricto.* In blackband ironstones the iron occurs as siderite, and in the Japanese volcanic lake ores the mineral is again goethite.

The *lake ores* of glaciated regions consist of oolitic or pisolitic grains of goethite about 0.1 to 0.4 inch across that are often cemented together to form disks ranging from an inch or so to about a foot in diameter. Grains and disks, together with small quantities of clay and other fine detritus, are in turn cemented again to form solid lenses and bands of ore. Ljunggren (1955), in a differential thermal analysis and x-ray study of some Swedish lake ores, found a substantial variation in degrees of crystallinity of such material: the hard ores gave moderately sharp thermal peaks and good x-ray diffraction patterns, whereas the soft, earthy limonitic ores gave different thermal peaks and quite diffuse x-ray diffraction patterns. Apparently the hard ore is quite well-crystallized goethite, whereas the soft material is a more primitive, and substantially amorphous, mixture of oxide and hydroxide.

Lake ores contain variable, and sometimes substantial, quantities of manganese, and MnO_2 contents of over 40 per cent have been measured in some occurrences. Ores with high manganese also carry substantial trace metal assemblages, including radioactive elements. Presumably these have been incorporated by adsorption on the manganese oxide. A very minor amount of the iron of these ores occurs as carbonates and vivianite (a blue ferrous phosphate $[Fe_3^{++}(PO_4)_2 \cdot 8H_2O]$).

The *marsh*, or *peat bog*, ores are composed mainly of earthy, partly pisolitic, yellow "limonite" and are often very impure, with a high percentage of detrital material and plant remains. Both of these are encrusted and cemented together with limonite and other, minor, iron minerals. Ores of this type also differ from the lake ores in that they often contain quite substantial quantities of carbonate and phosphate. Table 13-2 (after Clarke, 1924, p. 538) gives analyses of four bog ores and illustrates the very high carbonate and phosphate contents sometimes attained.

Blackband ironstones are, as already noted, essentially carbonate ores. Siderite is the essential constituent and occurs both in massive fine-grained form and as a cement for quartz and other clastic particles and organic matter. These fresh-water carbonates seem to have a notably higher iron content than their marine counterparts in ironstones and iron formations. MnO, CaO, and MgO are all comparatively low. Analyses of blackband material are given in Table 13-3. Phosphate is variable but generally low; sulfur may be substantial in some cases and represents minor marcasite; and carbon, expectably, is often quite high.

Table 13-2   Analyses of bog iron deposits from Ederveen, Germany.   (Quoted by Clarke, 1924, p. 538)

	1	*2*	*3*	*4*
Fe_2O_3	10.58	2.49	8.0	36.49
$FeCO_3$	20.77	37.70	30.6	6.12
$MnCO_3$	4.04	0.67		2.91
$CaCO_3$	·2.27	4.46	4.0	4.10
$MgCO_3$	0.17	0.10		0.21
$Fe_3(PO_4)_2$	4.30		2.9	5.47
$Fe^{3+}PO_4$		1.75		1.76
$CaSO_4$	0.07			
Al_2O_3	0.93	0.21		0.60
KCl	0.03	Trace		Trace
NaCl	0.23	Trace		Trace
SiO_2	49.30	50.84	49.1	25.60
Org	1.57	0.03	1.8	1.20
H_2O^+	2.06	1.12	}3.3	4.00
H_2O^-	3.68	0.95		12.10
Total	100.00	100.32	99.70	100.56

Table 13-3   Analyses of "blackband ironstones"

	(1)	(2)	(3)	(4)	(5)
SiO_2	6.46	13.35	10.04	5.10	8.67
Al_2O_3	2.64	5.79	5.57	2.35	4.47
FeO	42.08	41.03	37.99	50.92	43.11
Fe_2O_3	6.85	0.41	1.49	Nil	0.42
MgO	1.76	3.36	3.37	0.30	2.09
CaO	3.87	3.00	4.59	0.70	5.15
MnO	2.32	0.55	1.51	0.58	2.07
Na_2O					
K_2O		0.86	0.55		
TiO_2	0.21				
P_2O_5	0.65	0.70	0.80	0.33	1.18
CO_2	32.70	28.49	29.92	31.80	32.74
S				0.18	0.18
SO_3	0.20		Trace		
FeS_2	0.11		0.06		
C		0.07	1.42	5.00	
H_2O^+	Trace	1.36	1.47	} 2.70	} 0.67
H_2O^-	0.15	0.57	0.74		
Total	100.00	99.54	99.52	99.96	100.75

Explanation
(1). Clay ore.   Ashburnham, Sussex; from bed in Cretaceous Wadhurst Clay (Lamplugh, in Lamplugh, Wedd, and Pringle, p. 227, 1920).
(2). Clay ore.   Rosser Vein Mine, Dowlais, South Wales; from Carboniferous coal measures (Strahan, in Strahan and others, p. 114, 1920).
(3). Blackband ore.   Brown Rake, Butterby, Yorkshire; from Carboniferous coal measures (Gibson, in Strahan and others, p. 52, 1920).
(4). Blackband ore.   Prestwich Mine, Natal; from coal beds of Carboniferous-Triassic Karov System (Wagner, pp. 127–128, 1928).
(5). Blackband ore.   Holmes County, Ohio; from Pennsylvanian coal measures (Stout, pp. 168–169, 1944).

Information on the *volcanic lake ores* of Japan and the Kurile Islands is not readily available; however, Zelenov notes of limonitic material derived from thermal springs on one of the Kurile islands:

It should be emphasized that such limonites are extremely pure, with hardly any additions.   Chemical analysis of a fresh limonite specimen from the Limonitovyy Protok, "Bogdan Khmel'nitshiy" caldera, is as follows (percentage of dry sample): SiO_2—0.36; Al_2O_3—0.55; Fe_2O_3—72.92; FeO—4.06; CaO—0.13; MgO—0.14; Na_2O—0.09; K_2O—0.15; P_2O_5—0.93; SO_3—5.14; H_2O^+—13.29; CO_2—0.19 and C_{org}—1.50 (1959, p. 48).

Where such material is carried to and deposited in lakes, there is an inevitable varying admixture with clastic and organic matter; Zelenov quotes an average, for one of the Japanese volcanic bog ores, of 49.5 per cent iron.

Origin   Since bog ores are commonly seen in the process of formation, they are one of the few ore types whose origin is beyond controversy.   Tundra environments, with a cold humid climate, high water table, and frequent development of substantially closed drainage systems, are by far the most favorable for bog iron accumulation.   Solution, transport, and precipitation of the iron follows the standard pattern of reduction-solution-transport-oxidation-precipitation, but depending on the precise nature of the local physiography and bedrock, these may lead to the development of slightly different kinds of concentration.

Under marsh conditions, the accumulation of decaying plants induces reducing conditions in the surface waters, and leads to the formation of abundant

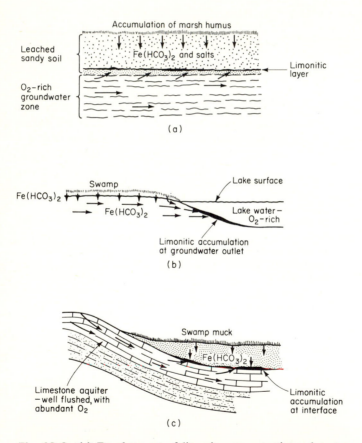

Fig. 13-6   (*a*) Development of limonite concentrations close to the water table in a marsh environment; (*b*) deposition of iron oxides from underground waters debouching on lake floor upon interaction with oxygenated near-shore lake water; (*c*) deposition of iron oxides due to interaction of marsh soil waters with underlying well-oxygenated water of limestone aquifer.   (*Adapted from Borchert, Trans. Inst. Mining Met., London, 1960.*)

CO_2. Any iron in the surface waters or in the vadose zone is thus reduced and combines to form ferrous bicarbonate [$Fe(HCO_3)_2$]. If, however, the ground water beneath the shallow vadose zone moves fairly freely and contains oxygen, inter-action of this with the descending vadose ferrous bicarbonate solutions leads to the local oxidation of the latter and the formation of a layer of limonite—marsh, or peat bog, ore—along the water table. This is illustrated in Fig. 13-6a.

If such marshy conditions are of regional extent and incorporate lakes, a similar process may lead to the development of lake ore. In this case the ferrous bicarbonate solutions move underground, for considerable distances, eventually to discharge through bottom sediments into the body of a lake. Where such lake's waters are well oxygenated, the discharging ground waters are oxidized and the iron precipitated on the lake floor as limonite. Since such discharge tends in most instances to occur near the margins of the lakes concerned, bog ores of this kind are commonly best developed near the edges of the lakes and may be absent from the center. This is illustrated in Fig. 13-6b.

A third sequence may develop in the presence of limestone. Through the development of extensive solution cavities, limestone may act as a conduit for comparatively fast-moving, well-oxygenated water. Where the intake beds of a limestone aquifer form high ground, substantial hydrostatic pressures—and hence an artesian system—can develop. This may lead to the situation illustrated in Fig. 13-6c, in which well-oxygenated artesian water is forced upward, so inter-acting with the reduced, ferrous bicarbonate-charged ground waters descending from the marsh milieu above. The bicarbonate solutions are oxidized along the zone of interaction, and subsurface limonite deposits form.

The origin of blackband ores is not so clear. Some of it is probably of dia-genetic origin. Some may be of primary nature, formed by loss of CO_2 from ferrous bicarbonate and protected from oxidation by the decay of the associated swamp flora. Their loci of deposition may be brackish—presumably semistag-nant—waters and marine swamps.

The mode of formation of volcanic lake ores seems fairly clear. Iron, probably chiefly as the ferrous chloride, is contributed by volcanic springs. On emergence, and also probably in the resultant streams and in the lakes themselves, the chloride is oxidized and the resultant hydroxide precipitated.

IRONSTONES

These constitute the lesser of the two more important types of sedimentary iron accumulation. Although, as discussed a little later in this chapter, some investi-gators have suggested that ironstones and iron formations may not be distinct entities, there are many *general* differences, and a general distinction between them seems justified. Whereas iron formations characteristically contain much "chert" and hence have high silica-iron ratios, the ironstones contain little if any chert and have much lower silica-iron ratios. In addition, virtually all iron formations are of Precambrian age, whereas the ironstones are almost always post-Precambrian. A number of other differences are considered a little further on.

Distribution, form, and setting Ironstones are of quite widespread occurrence and some have been important sources of iron.   They are sometimes divided into two broad types, differences between which are probably due to the differences in derivation.   By far the more important of the two is the *minette*, or *Lorraine*, type, which appears to have been deposited in quite normal near-shore marine sedimentary sequences.   The other, referred to as the *Lahn-Dill* type, is less common and less extensive and occurs in conspicuously volcanic sediments.   The

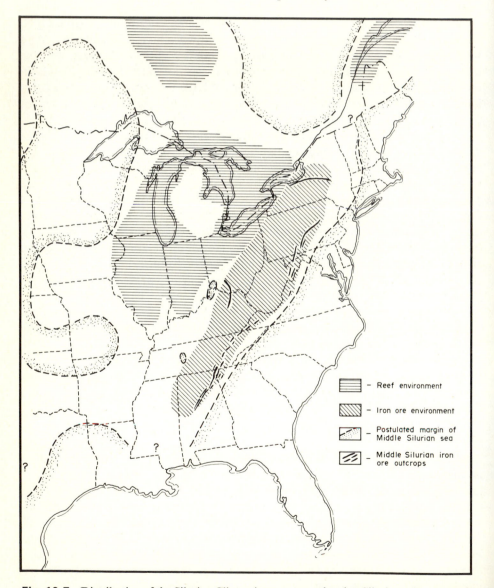

Fig. 13-7 Distribution of the Silurian Clinton iron ores as related to Silurian paleogeography and marine environments. (*Redrawn from Ellison, Econ. Geol., 1955.*)

"type occurrence" of the minette ores might be said to be those of the Jurassic of Alsace-Lorraine. After these perhaps the best known are the Silurian "Clinton" ironstone beds of the United States, which occur as sporadic iron-rich beds extending from New York State in the north to Alabama in the south (Fig. 13-7). At Wabana in Newfoundland and in related sediments in Nova Scotia, Ordovician sandstones and shales contain important quantities of oolitic chamosite and hematite of ironstone type. Oolitic limonite ores of a similar kind occur in the Permian sedimentary rocks of the Desert Basin in northern Australia. In the Lower Jurassic beds of Frodingham and Cleveland in England (Fig. 13-8) there are important bedded concentrations of oolitic limonite, siderite, and chamosite, and the Middle Jurassic of England and Europe contain two very well-known and important occurrences—the oolitic limonite-chamosite-siderite ores of the "Northampton Sand Ironstones" of England and the generally similar "minette" beds — already referred to—of France (Lorraine), Germany, Belgium, and Luxembourg. Ores of this type are in fact known in many countries and in sedimentary beds ranging in age from Cambrian to Pliocene.

Deposits of this kind may be extensive, i.e., continuous over distances of miles, or they may consist of restricted lenses appearing here and there individually or as groups within a restricted sequence of sedimentary rocks. Their thickness ranges from a few inches to a few tens of feet. The iron-rich beds of the Clinton ores and those of the Australian Desert Basin, for example, range up to about 30 feet. Some Ordovician oolitic limonite ores in Wisconsin achieve thicknesses of about 50 feet, and this seems to be close to the maximum for beds showing true ironstone characteristics. Most commonly the ores are in the form of groups of beds confined to a relatively small stratigraphic interval in the sequence concerned. The Wabana deposits, for example, are in the form of 12 principal beds, which can be divided into five zones; individual beds range from about 5 to 30 feet in thickness. The minette deposits of western Europe are in the form of some 10 beds that occur in a number of Jurassic sedimentary basins distributed along a zone near a margin of the Tethys Sea. In this case individual beds vary up to about 28 feet in thickness. The related English Northampton Sand deposits are up to 25 feet thick.

Most ironstone deposits occur in near-shore facies of shallow-water marine sequences. The Clinton ores are associated, and interbedded, with limestones, siliceous limestones, calcareous siltstones and shales, sandstones, and conglomerates. Their relation with the Silurian shoreline and reef zones of the eastern United States is shown in Fig. 13-7. The Frodingham, Cleveland, and Northampton ironstones are associated with marine clays, sandstones, limestones, and marls, and in the case of the Northampton ores, two units of estuarine sedimentary rocks. In this case the iron-rich unit is overlain by up to 70 feet of massive yellow to orange-brown sandstones (the *Northampton Sand*) which in turn are succeeded by the *Lower Estuarine Series*. The latter sediments are composed of sands, silts, and clays, some of which are carbonaceous. According to Hollingworth and Taylor (1951) vertical rootlet markings are a common feature, and there seems no doubt of a brackish estuarine environment. This section of the English Jurassic was deposited near an oscillating shoreline, and the iron beds now constitute a shallow-

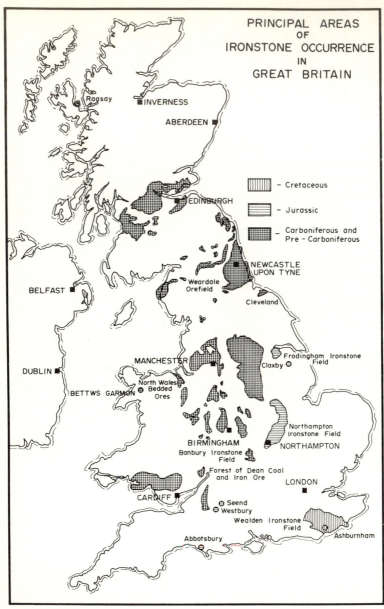

Fig. 13-8 Distribution of iron ore deposits (ironstones) of Great Britain.

water, near-shore marine component. The minette deposits of Europe have a generally similar association and were apparently deposited in the shallow waters of a series of Jurassic basins, as already noted.

The only well-documented examples of ironstones deposited in fresh water are those of the northern Turgai and northern Aral'sk districts of Kazakhstan in

Russia, described by A. L. Yanitsky (Davidson, 1961). These are oolitic ores associated with Middle Oligocene sediments of old river valleys and associated lakes. Related sediments are quartz gravel, shingle, and sands—all of which contain ooliths—and siltstones and clays. The iron concentrations are found in fluviatile, deltaic-lacustrine, and lacustrine facies and were apparently formed in low-lying river valleys cut into the uplifted surface of marine Paleogene sediments of the areas concerned. Locally they contain shells of the fresh-water mollusk *Pisidium*, and there are also abundant fossil tree trunks and branches, spores, and pollen grains. The largest deposit—that of Lisakovsk—extends along a former river valley for a length of 100 km with a breadth of from 2 to 8 km. Apparently the oolitic ironstones here have been deposited in marshes and the swampy margins of lakes as well as in the river channels. Since the latter inevitably meandered, ore beds of lake type are now found cut by those of stream type and vice versa.

Constitution The ironstones exhibit quite a wide variety of mineral assemblages, presumably reflecting a variety of depositional conditions and diagenetic processes. Oxides, carbonates, and silicates are all represented, and sulfide is known here and there as a minor constituent of reduced facies.

Among the oxides, hematite and limonite (goethite) are important. While magnetite occurs quite widely in deposits of this type it generally does so only in small amounts and overall is far less important then the other two oxides. Siderite and related iron-rich, mixed carbonates are important in some deposits and are well known in the Mesozoic deposits of England and western Europe. Silicates are abundant. According to James (1966) chamositic rocks are one of the most characteristic ore types of the post-Precambrian ironstones. Glauconite is widespread but rarely appears as an important component of an individual deposit. Apart from one or two exceptions, such as the Upper Cretaceous "greensands" of New Jersey, U.S.A., and the Upper Cretaceous and Tertiary of New Zealand, glauconite occurs as little more than disseminations.

The oxide ores consist of a major fraction of limonite and/or hematite, most of which is typically in oolitic, pelletal, or sometimes, pisolitic form. Ooliths have ordinarily developed round cores of clastic material and are cemented together by a matrix of chamosite, calcite, or powdery oxide. Chamosite is commonly associated with primary oxide in the ooliths themselves, though a similar association is often achieved by the secondary alteration of some chamosite to limonite. Some ores of this type are dominantly limonite, others are dominantly hematite, the principal chemical difference between the two being little or no more than the lower water content of the hematitic types. Ironstones of both kinds are rather heterogeneous, and it seems certain that the original chemical precipitates have been disturbed and redistributed by water movement in many cases. Magnetite ironstones are known but are rare and of small extent.

Analyses of limonitic and hematitic ore types are given in Table 13-4. When a number of such analyses are examined, there is a suggestion of a general crude inverse relation between SiO_2 and CaO, reflecting the dominance of either clastic

Table 13-4   Analyses of limonite (goethite) and hematite ironstones

	(1)	(2)	(3)	(4)	(5)	(6)	(7)	(8)
SiO_2	8.60	17.75	6.55	10.94	4.66	13.60	2.59	46.79
Al_2O_3	3.72	3.38	6.91	3.72	3.05	5.17	0.75	0.25
FeO	9.57	7.37	0.18	7.67	21.17	0.23	0.88	0.40
Fe_2O_3	35.82	47.36	62.88	42.88	52.08	55.56	52.56	50.80
MgO	2.10	1.72	1.10	5.90	1.14	1.49	0.21	0.02
CaO	15.90	5.25	6.06	10.80	2.88	10.26	22.40	0.80
MnO	0.39	0.45	0.38	0.29	0.78	0.14	0.11	0.01
Na_2O				0.05		0.07	0.05	0.05
K_2O				0.02		1.22	0.03	0.04
TiO_2				0.14	0.28	0.22	0.10	0.05
P_2O_5	1.67	2.20	1.83	0.71	2.11	1.80	0.08	0.12
CO_2	15.10	6.08	3.33	14.98	10.78	7.36	18.50	0.06
S	0.18	0.04				0.03	0.31	0.03
SO_3							0.01	0.17
C							0.08	0.09
H_2O^+	}6.84	}8.23	10.24	1.76	1.72	1.88	1.12	0.27
H_2O^-			1.44	0.12	0.27	0.80	0.21	0.08
Total	99.89	99.83	100.90	99.98	100.92	99.83	99.99	100.03

Explanation

(1) and (2). Oolitic limonite ore of minette type; from Middle Jurassic Landres-Amermont basin, Lorraine (Coche et al., 1954, quoted by James, p. W18, 1966).

(3). Oolitic limonite ore of minette type; from Middle Jurassic ironstone beds of Luxembourg (Lucius, 1945, quoted by James, p. W18, 1966).

(4). Oolitic hematite ironstone of minette type, Kirkland, New York; in calcitic ore from the Kirkland bed of the Silurian Clinton Group (James, p. W19, 1966).

(5). Oolitic hematite of minette type; from Ordovician Wabana deposits, Newfoundland (Hayes, pp. 52–53, 1915).

(6). Oolitic hematite, associated with chamosite, dolomite, calcite, and quartz in hematite ironstone of minette type. Gila County, Arizona, from the Devonian Martin Formation (James, p. W19, 1966).

(7). Hematite-calcite ironstone of Lahn-Dill type (volcanic association). Lahn-Dill region, Germany; from Devonian iron-rich beds of the Koenigszug mine (Harder, p. 60, 1954).

(8). Hematite-quartz ironstone of Lahn-Dill type, Lahn-Dill region, Germany; from Devonian iron-rich beds of the Constanze mine (Harder, p. 61, 1954).

quartz or calcite (the latter clastic or interstitial).   In minette ores, P_2O_5 is typically high and is usually located in pelletal collophane. Al_2O_3 is also usually quite high (2 to 10 per cent) and occurs in detritus and, particularly, in the chamosite.   In hematitic ores of the "Lahn-Dill" type, P_2O_5, Al_2O_3, and MgO are significantly lower than in ores of minette type.   This feature may be due to the differences in derivation.   Analyses of two magnetite-rich ironstones are given in Table 13-8.

Silicate ores are an important component of the ironstones as a class and, as we have already noted, chamositic types are virtually characteristic.   Such chamosite occurs as ooliths, which may contain layers of hematite or limonitic material as primary components or—with magnetite—as secondary alterations.   Analyses of some chamositic ironstones are given in Table 13-5. Although they are not so heterogeneous as the oxide types, it is clear that the chamositic rocks are by no means homogeneous either mineralogically or chemically. There are notable increases in SiO_2 and Al_2O_3 and a sharp decrease in Fe_2O_3, reflecting the presence of ferrous silicate in the place of ferric oxide. Substantial amounts of SiO_2 and carbonate may be present as detrital particles, and there are often extensive diagenetic modifications, such as the development of siderite and the conversion of chamosite to magnetite already referred to. P_2O_5 is again quite high.

Analyses of carbonate (sideritic) ironstones are given in Table 13-6. In conformity with James' facies concept, ferrous iron greatly dominates here. However, once again there is substantial heterogeneity in chemical and mineralogical

Table 13-5   Analyses of chamositic ironstone

	(1)	(2)	(3)	(4)	(5)	(6)
SiO_2	26.20	41.80	16.22	15.60	21.78	49.64
Al_2O_3	15.71	16.18	7.65	6.96	10.67	8.82
FeO	28.45	24.28	35.38	18.25	22.70	16.70
Fe_2O_3	3.70	4.28	2.99	2.24	6.20	8.71
MgO	1.36	0.43	1.84	8.83	3.61	3.19
CaO	6.50	1.00	4.01	17.64	12.25	2.10
MnO	1.67	1.86	3.12	0.56	0.08	0.23
Na_2O				0.01	0.08	
K_2O				0.02	0.09	
TiO_2	0.10	0.05	0.61	0.18	0.53	
P_2O_5	3.93	1.12	4.91	0.80	1.62	0.91
CO_2	3.00	Nil	16.64	24.99	8.45	5.15
S	1.39†	0.56†			0.22	0.72
SO_3					1.47	
C					0.50	
H_2O+	} 7.90	} 8.44	2.61	3.38	8.36	5.01
H_2O-			0.78	0.14	1.01	‡
Total	99.91	100.00	96.76	99.60	99.62	101.18

† FeS_2.
‡ Sample dried prior to analysis.

Explanation

(1). Oolitic chamositic ironstone. Llangoed, Anglesey, Wales; from Ordovician iron-rich beds (Strahan et al., p. 14, 1920).

(2). Oolitic chamositic ironstone. Bonw, Mynydd-y-Garn, Anglesey, Wales; from Ordovician iron-rich beds (reference as for 1).

(3). Oolitic chamositic ironstone. Wabana, Newfoundland; from zone 4, Scotia bed, Ordovician age (Hayes, p. 58, 1915).

(4). Oolitic chamositic ironstone. Westmoreland, New York, U.S.A.; from top 6 inches of Westmoreland bed of Silurian Clinton Group (James, p. W22, 1966).

(5). Chamositic ironstone. North Göttingen, Germany; from Lower Jurassic (Harder, 1951, from James, p. W22, 1966).

(6). Sandy chamositic ironstone. Landres-Amermont Basin, Lorraine, France; Middle Jurassic minette beds (Coche et al., 1954).

Table 13-6   Analyses of carbonate ironstones

	(1)	(2)	(3)	(4)	(5)	(6)
SiO_2	4.88	9.20	8.03	8.51	7.56	18.60
Al_2O_3	3.38	8.95	8.86	6.12	4.10	5.31
FeO	49.32	39.53	27.58	36.91	43.86	23.06
Fe_2O_3	1.20	2.64	23.76	1.77	1.83	21.47
MgO	1.51	2.06	3.13	3.75	3.92	3.26
CaO	3.46	6.73	4.65	5.54	2.90	4.47
MnO	0.37	0.05	0.17	0.42	0.13	0.76
Na_2O				0.05	0.11	
K_2O				0.03	1.13	
TiO_2				0.36	0.18	
P_2O_5	0.62	2.70	1.84	1.30	0.11	0.60
CO_2	32.70	23.01	17.35	20.70	32.84	16.30
S	0.12	0.04	0.01	0.05		0.21
C				0.27		
H_2O+	} 2.33	} 4.32	} 3.95	4.05	0.81	} 5.77†
H_2O-				10.00	0.28	
Total	99.89	99.23	99.33	99.83	99.76	99.81

† Sample dried prior to analysis.

Explanation

(1). Northampton Sand Ironstone; sideritic mudstone with minor chamosite (Stewarts & Lloyds Ltd., in Taylor, p. 61, 1949).

(2). Northampton Sand Ironstone; chamositic carbonate rock (Stewarts & Lloyds Ltd., in Taylor, p. 60, 1949).

(3). Northampton Sand Ironstone; sideritic rock containing ooliths of limonite and lesser chamosite (Stewarts & Lloyds Ltd., in Taylor, p. 60, 1949).

(4). Cleveland Ironstone; siderite-chamosite rock from "main seam" (siderite and chamosite estimated as 34 per cent each) (Hallimond, p. 51, 1925).

(5). Siderite bed in the Silurian Brassfield Dolomite, Kentucky, U.S.A. (James, p. W25, 1966).

(6). Minette ore from the Jurassic ironstones of Luxembourg (James, p. W25, 1966).

constitution.   This is illustrated by the Northampton Sand Ironstone, of which (1), (2), and (3) of Table 13-6 are typical analyses.   As with the other ironstones, Al_2O_3 is quite high and represents detritus and the chamosite fraction. P_2O_5 is usually also high and reflects the incidence of collophane.   CaO and MgO are moderately abundant, with CaO/MgO ratios usually, though not invariably, greater than 1.0.   As noted by James (1966, p. W24) there is a substantial overall spread of chemical composition due to gradations into chamositic and limonitic facies and into normal mudstone, sandstone, limestone, and dolomite.

Quantitatively, sulfide ironstones are not important.   Some are associated with base metal sulfide ore deposits, and in such cases they constitute the pyritic or pyrrhotitic portions of such occurrences.   Where, however, iron sulfide is associated with other ironstone lithologies, it is generally as pyritic black shale or as minor pyritic variants of carbonate or silicate ores.   The outstanding exception to this among post-Precambrian ironstones is the pyritic member of the Wabana deposits in Newfoundland.   There, pyrite is oolitic, and pyrite-rich layers alternate with fissile carbonaceous shales.   The latter exhibit an abundant graptolite fauna, with which brachiopods, orthoceratites, and other fossils are associated.   The pyrite spherules themselves consist of layers of pyrite, commonly alternating with concentric layers of phosphatic material.   Many other ironstone occurrences contain minor pyrite (very occasionally, pyrrhotite); this is normally associated with, or is a very minor part of, carbonate or silicate ore types.   Analyses of highly pyritic material are few since it does not usually constitute "ore."

Origin   The derivation and mode of deposition of the ironstones is neither so simple nor so clear as that of the bog ores, but as we shall see a little later in this chapter, they are a good deal less controversial than the banded iron formations.

1. *Derivation.*   It is likely that, as a class, ironstones derive their iron from all three sources discussed earlier: continental erosion, submarine volcanic hot springs carrying iron salts, and upwelling currents containing iron salts abstracted from deeper portions of the ocean floor.   There is good reason to suspect that these are not mutually exclusive—that some ironstones may have one source, others another, and that in many cases there has been a contribution from more than one source.

      Derivation by continental erosion poses certain requirements of terrain and climate.   To form high sedimentary concentrations of iron it is necessary for the rivers concerned to deliver to the sea large quantities of iron but comparatively little detritus.   This matter does not, of course, arise where the iron is delivered subaqueously by regional subterranean aquifers.   As already noted the iron of rivers may be carried in suspension as an insoluble compound in finely particulate form or in solution.   That of ground waters is in solution.   Whether the iron is carried in suspension or solution, or both, a high relative iron content in the final sediments is most likely to be achieved where the source terrain is senile and stable—i.e., produces little detritus—and the climate is warm and humid—i.e., promotes deep weathering and solution activity.   On arrival at the sea any ferrous iron is oxidized, and thus all iron is sedimented; separation from normal clastic material could take place by the sifting action of currents or by the entrapment of coarser material in lagoons where rivers debouched into these rather than into the open sea.

      Provided submarine volcanism can supply the quantities of iron now observed in ironstone deposits, derivation from such a source poses no conceptual difficulties.   Since iron forms an insoluble compound of one kind or another over almost the full Eh-pH range of marine environments, it will precipitate whatever the chemistry of the local environment happens to be.   Derivation of iron from adjacent deeper parts of the ocean floor, as postulated for marine sedimentary iron ores in general by Borchert, could probably

supply the quantities of iron found in most ironstone deposits.   Borchert's
ideas have already been illustrated in Fig. 13-4.

2. *Deposition.*   In many cases it is far from simple to distinguish results of sedimentation—in the strict sense of mechanical deposition—from those of diagenesis.   Undoubtedly the mineral assemblages of many ironstones result
from both processes.

A very fine example of the development of a facies pattern in modern
iron-bearing (oxide-silicate) marine sediments has been described by Porrenga
(1967).   The sediments concerned are those of the Niger delta and show a
clearly zoned distribution of pelletal goethite, chamosite, and glauconite.
A diagrammatic section through the delta is given in Fig. 13-9.   Although,
as is also usual among the actual ores themselves, the recent Niger delta
sediments do not show a complete array of facies, they do demonstrate
conditions under which typical ironstone mineralogies may develop.   The
environment is an open marine one.   Brown pellets and organic fillings consisting of goethite and occasional small amounts of chamosite predominate
in a narrow zone of near-shore waters less than about 10 meters deep.   This
is succeeded by a broad band of chamositic sediments that persists to about
50 meters.   Between 50 and about 125 meters there is no conspicuous
authigenic iron mineral, but below this, glauconite appears and persists to
some 250 meters.   Although the sediments involved in Porrenga's work were
by no means "ores," they do show very nicely that a facies pattern may develop
essentially at the time of sedimentation.   Assuming that the iron has been

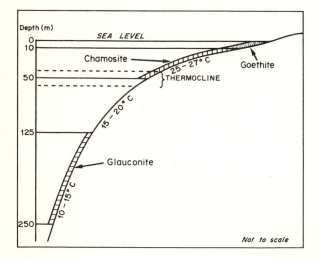

Fig. 13-9 Schematic diagram (vertical scale greatly
exaggerated) showing a section through the Niger delta,
with the depth distribution of goethite, chamosite, and
glauconite and bottom water temperatures as measured in
January-February 1959. (*From Porrenga, Marine Geol.,
1967.*)

contributed by continental erosion via the Niger (clearly volcanic solutions are not involved in this instance), there are two paths by which such could have arrived at the site of deposition:

  a. Directly from the continent, either in solution or as fine particles of a simple iron compound, e.g., $Fe(OH)_3$. In these circumstances the iron mineral that forms at or just below the mud-water interface is the stable reaction product for the environment (i.e., sedimentary facies) concerned.

  b. Indirectly, via the sea-floor decomposition of continental detritus delivered by the Niger and the formation of sea-floor solutions containing dissolved iron.

The order of events of (*a*) should lead directly to the development of a facies pattern such as that observed in modern sediments by Porrenga and deduced in older ones by James. The order of events of (*b*) should lead to a generally similar pattern through a sequence of events such as that deduced by Borchert.

BANDED IRON FORMATIONS

These constitute by far the most important of all classes of iron concentrations. In fact there is no doubt that deposits of this group exceed in importance all others combined—igneous segregations, other classes of sedimentary deposit, and any that may have resulted from metamorphic processes. Banded iron formations are known in every Precambrian continental region and occur in enormous abundance in almost every case. In addition to their high frequency of occurrence, they are, individually, often huge. However, in spite of this common occurrence and widespread nature, and the fact that mining has provided excellent exposures and much mineralogical and chemical data, the origin and mode of deposition of these deposits are still quite uncertain. It is probably no exaggeration to say that they are little better understood now than they were 50 years ago, although they have been investigated—and written upon—intensively during that period.

Distribution, form, and setting The greatest known Precambrian banded iron formation province is undoubtedly that of the Labrador-Quebec-Ontario-Minnesota belt (Fig. 13-10)—a province that has become a "classical" one through the long history of research and controversy centered about the deposits of its southwestern extremity in the Lake Superior region (Fig. 13-11). The great occurrences of Brazil, India, and southern Africa are also notable and have been subjects of substantial research in iron formation theory. More recently, a very extensive banded iron formation province in Western Australia (Fig. 13-12) has begun to receive industrial and scientific attention and seems likely to figure prominently in future investigation. Other, lesser, iron formations are known in virtually every other Precambrian area in the world. Because they are so widespread, it is not surprising that several "local" names for them have arisen. They are often referred to as *taconite* in North America, *itabirite* in Brazil, *hematite-quartzite* in India, *banded jaspilite* in Australia, and *quartz-banded ore* in Scandinavia. In southern Africa they may be referred to as *banded ironstone*, but such occurrences should not, through this term, be confused with "ironstones" as described in this chapter.

One of the outstanding characteristics of this class of deposit is that they are almost entirely confined to the Precambrian. In fact, apart from one or two minor and controversial deposits, they are *all* of Precambrian age, and there is a remarkable clustering—involving deposits from all continents—about an age of 2200 million years.

Whereas bog deposits are rarely more than a few miles in extent, and ironstones a few tens of miles, the major iron formations appear to have had depositional continuity over hundreds of miles. The Mesabi Range occurrence (Fig. 13-11) is over 100 miles long, and iron formation of the Labrador Trough of Quebec

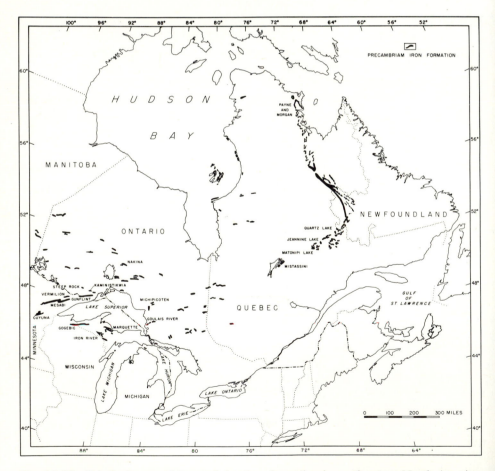

Fig. 13-10 Principal areas of Precambrian iron formation in northeastern North America. (*From Lepp and Goldich, Econ. Geol., 1964, adapted from Geological Survey of Canada Map 1045A-M4.*)

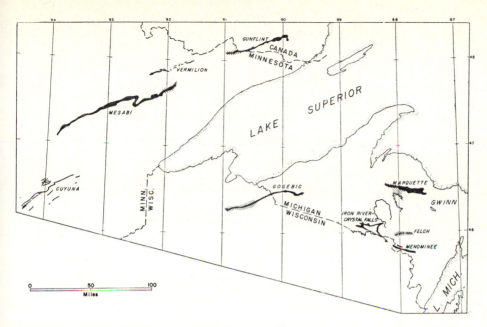

Fig. 13-11   Principal areas of Precambrian iron formation in the Lake Superior region. (*From James, Econ. Geol., 1954.*)

and Newfoundland extends semicontinuously for well over 500 miles (Fig. 13-10). Thicknesses range up to almost 2000 feet—at least an order of magnitude greater than the thicknesses of bog and ironstone deposits.   In general, banded iron formations seem to have one large dimension and two small ones—the length by far the largest, the width larger than, but of the same order of magnitude as, the thickness.   Such a configuration suggests a relationship with troughs, or perhaps with comparatively narrow zones running parallel to former shorelines.

Iron formations typically show well-developed banding—i.e., bedding.   In most cases this is delineated by the alternation of iron-rich beds with beds of "chert"—fine, dense, cryptocrystalline silica.   Sulfide iron formation, a relatively unimportant type, is the exception; in this case chert is scarce or absent, and the bedding is typically that of fine, pyritic carbonaceous shale.   In addition to the major chert-iron mineral-chert stratification of most iron formations, there are usually developed various finer laminations—fine bedding due to the greater or lesser abundance of oxide, carbonate, silicate, or silica in the iron-rich beds and to minor inhomogeneities in the chert.   In some cases the iron minerals and chert occur as very flat lenses rather than continuous beds, in which case the stratification takes on a pronounced "wavy" appearance.

If there is a consistency in the geological setting of iron formations, such consistency is by no means obvious.   In some cases sedimentation appears to have been marine, in others perhaps estuarine or fresh water.   At present it is not known whether any one of these environments is vital for iron formation development, or

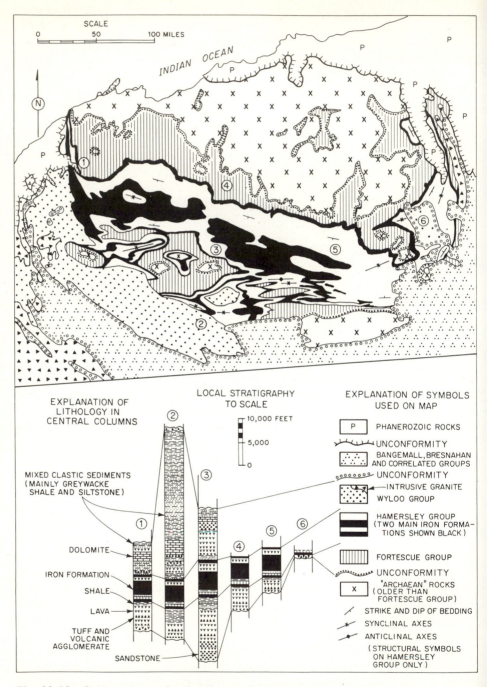

Fig. 13-12 Geological map showing the distribution of iron formations and associated rocks in the Hamersley iron province of Western Australia. (*From Trendall, Geol. Soc. Amer. Bull., 1968.*)

even if any one of them is of overwhelming importance.   Similarly some iron formations occur in conspicuously volcanic sequences and are clearly associated with
some volcanic unit or units.   In other instances there seems to be no sign—at
least in the present state of knowledge—of volcanic affiliation.

Constitution   In one deposit or another most of the major minerals of iron are
developed abundantly in iron formations.   Hematite and magnetite constitute by
far the major part of these deposits as a group, and *in toto* the two are probably
more or less equally abundant.   Carbonate is of common occurrence and may be
present in enormous quantities as the major constituent, as at the Helen mine in
Ontario.   The silicates are probably of much less overall abundance in this class of
deposit than in the ironstones, though very occasionally they are major constituents; samples of the Biwabik iron formation have been found to contain up to 70
per cent greenalite.   Sulfide iron formations are less important, though there are
some instances of substantial occurrences.   Again the Helen mine of Ontario
distinguishes itself, with thicknesses of pyritic ore ranging up to 50 feet.

It is self-evident that all banded iron formation is characterized by abundant
iron mineral.   Less evident, but of about the same abundance and undoubtedly of
similar importance genetically, is the "chert" already referred to.   This material is
an important constituent of all iron formations.   While such chert is conspicuously
less rich in iron than the interbedded iron-rich strata, it all contains *some* iron, and
in many instances it is most accurately described as "ferruginous chert."

Although all the different mineralogical types of iron-bearing sediment are
developed in the ironstones, it is in the banded iron formations that the different
mineralogies and chemistries are developed to greatest degree.   These have been
described and their relationships synthesized with great clarity by James (1954,
1966).   Several accounts of occurrences in Canada, supporting and adding to the
observations of James, are due to A. M. Goodwin (1956, 1962, 1964).

As was pointed out previously, the field occurrences of the different mineralogical types of iron formation, coupled with our knowledge of the chemical stabilities of the minerals concerned, suggest that the nature of any given deposit—or part
of a deposit—is related to the chemical characteristics of the environment in which
it accumulated.   Thus whatever the origin of the iron and whatever its precise
mode of deposition, it appears to have accumulated in conformity with a chemical
facies pattern.   Such a "facies concept" provides a simple and convenient division
of the different mineralogical types—a division that, though constructed on a
hypothesis, is really quite independent of the latter, based as it is on the observable
facts of mineralogy and chemical composition.   We may therefore now look in
some detail at the constitutional characteristics of these "facies of iron formation,"
leaning heavily on the works of James and of Goodwin.

General.   As in the case of the ironstones, the different facies of iron formations are rarely, if ever, found as neatly arranged, clear-cut contiguous zones.
Many of the rocks were deposited in fluctuating environments, with resultant
interlaying and modification of contrasting materials, and processes of both contemporaneous clastic sedimentation and closely following diagenesis have usually

added at least some complications to the mineralogies that might have been expected from quiet and uninterrupted chemical precipitation.

In spite of this the oxide (hematite), carbonate, and pyritic facies are generally quite sharply delineated from each other. As shown originally by Krumbien and Garrels (1952) there is little or no overlap in the theoretical stability fields of these minerals in low-temperature–low-pressure aqueous environments (see, again, Figs. 6-4 and 13-5), and apparently reflecting this, primary hematite, carbonate, and sulfide appear to be mutually exclusive within a given layer of iron formation. Correspondingly, adjacent beds seem to contain minerals with adjacent stability fields; magnetite and carbonate beds may succeed each other, and so may carbonate and sulfide, but sulfide and oxide beds are rarely if ever in juxtaposition. Oxide, carbonate, and sulfide thus separate out rather nicely in terms of our "facies" division.

The behavior of the silicates is not nearly so clear-cut, and the silicate facies overlaps substantially with those of oxides and carbonate. Eh—the major environmental control of iron mineralogy—is clearly the factor involved. Whereas the formation of one or other of the oxides, carbonate, or sulfide is affected by only small differences in Eh, the silicates as a group precipitate over quite a wide range of Eh—at least from $+0.05$ to -0.2 volt. As a result primary iron silicate of one species or another may be found interstitial to any of the other major iron minerals or as layers within beds of oxide, carbonate, or sulfide. Its most frequent habitats are the oxide and carbonate facies, suggesting that most silicate material is deposited close to zero Eh.

Figure 13-13 (after James, 1954, p. 242) shows, in highly diagrammatic form, the general nature of iron formation facies relations. Silicates may be visualized as being deposited over the whole range of facies, with their major development somewhere near the zone of oxide-carbonate interdigitation. The development of all facies in one sedimentary unit, as shown in Fig. 13-13, rarely if ever occurs, but the occurrence of two of the zones is common and demonstrates the general validity of the facies principle.

Oxide facies. There appear to be two principal variants of the oxide facies: banded hematite rock and banded magnetite rock. Each consists of alternating bands of chert and the relevant iron oxide, and the two may be regarded as "subfacies" of the oxide facies of iron formation.

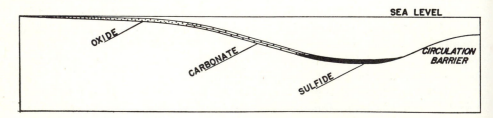

Fig. 13-13 Diagrammatic view of the relation between mineralogical type and shoreline and water depth in the development of facies of iron formation as postulated by James. (*From James, Econ. Geol., 1954.*)

Table 13-7   **Analyses of hematite iron formations**

	(1)	(2)	(3)	(4)	(5)	(6)
SiO$_2$	40.1	46.94	28.30	51.56	44.80	39.72
Al$_2$O$_3$	0.8	Nil	3.86	Trace	0.37	0.45
FeO	1.6	1.41	2.57	0.90	0.71	1.03
Fe$_2$O$_3$	50.1	51.00	63.39	47.48	53.08	58.58
MgO	2.0	Nil	0.56	Nil	0.07	0.07
CaO	1.4	Trace	0.86	Nil	0.28	0.03
MnO	0.2	0.02	Nil		Trace	Nil
Na$_2$O		Nil				
K$_2$O		Nil			0.12	0.07
TiO$_2$		Nil			0.07	Trace
P$_2$O$_5$	0.07	0.39	0.09			0.02
CO$_2$	2.6					0.40
S	0.01					
SO$_3$						0.03
H$_2$O+		} 0.68	} 0.72		0.04	0.28
H$_2$O−					0.03	0.02
Total	98.88	100.44	100.35	99.94	99.57	100.70

Explanation
(1). Menominee district, Michigan; from Precambrian hematite iron formation (James, p. 260, 1954, and p. W20, 1966).
(2). Middleback Ranges, South Australia; from Precambrian "hematite quartzite" of Camel Hill (Edwards, p. 468, 1953).
(3). Zululand, South Africa; from Precambrian banded iron formation of the Swaziland System, Umhlatuzi Valley (Wagner, p. 71, 1928).
(4). Johannesburg, South Africa; from Precambrian banded iron formation of the Witwatersrand System (Wagner, p. 73, 1928).
(5). Ukraine, Russia; from Precambrian banded hematite iron formation of the Krivoi Rog Series (Semenenko et al., 1956, from James, p. W20, 1966).
(6). As for (5) (Semenenko et al., 1956, from James, p. W20, 1966).

The hematite variety consists of thin alternating layers of chert and hematite, the latter of which may occur as finely crystalline laminae or as ooliths. Ooliths are very common indeed, and their structure has been found to survive even quite strong metamorphism. Bedding is always well developed, though layers may pinch and swell over quite short distances, giving the rock the wavy appearance already referred to. Analyses of samples from a number of hematite iron formation occurrences are given in Table 13-7.

Magnetite iron formation consists of alternating thin layers of chert and magnetite—or, more correctly, layers of quartz-rich and magnetite-rich material. Rocks of the magnetite subfacies are apparently much more complex than those of the hematite subfacies and generally contain substantial impurities of sideritic carbonate and of iron silicate. Representative analyses are given in Table 13-8. Keeping the above gradational tendencies in mind, it is apparent that the magnetite-banded rocks have rather more MgO, CaO, and CO$_2$ than the hematite-banded ones and, of course, a much higher FeO/Fe$_2$O$_3$ ratio.

Table 13-8   Analyses of magnetite iron formation and ironstone

	(1)	(2)	(3)	(4)	(5)	(6)	(7)
SiO_2	51.52	45.66	39.50	34.44	35.86	6.2	7.03
Al_2O_3	0.08	0.28	0.44	0.85	1.57	5.9	7.13
FeO	10.24	21.28	18.51	22.06	20.26	34.8	25.92
Fe_2O_3	35.37	19.16	29.21	30.54	38.56	18.2	32.56
MgO	0.20	2.73	2.00	2.30	1.74	2.4	1.82
CaO	0.02	1.04	2.71	1.72	0.51	4.6	2.84
MnO			0.12	0.21	0.16	0.3	0.22
Na_2O							
K_2O				0.13	0.02		
TiO_2			0.04	0.02	0.04		
P_2O_5	0.06	0.09		0.07	0.14	1.8	1.57
CO_2	1.06	7.54	6.22	7.36	0.60	19.0	9.38
S			0.13	0.01		0.1	0.02
SO_3					0.17	0.04	
C		0.12	0.02	0.04			0.56
H_2O+	1.48†	1.54†		0.44	0.60	4.6	4.88
H_2O-				0.17	0.06	1.3	6.20
Total	100.03	99.44	98.90	100.36	100.29	99.24	100.13

† Samples dried at 100°C before analysis.
Explanation
(1). Mesabi district, Minnesota; banded magnetite-chert from the Precambrian Biwabik Iron Formation (Lower Cherty division) (analysis of 30 feet of diamond drill core, from Gruner, p. 58, 1946).
(2). As for (1) (analysis represents 20 feet of diamond drill core).
(3). Lake Albanel district, Quebec; banded magnetite-chert from the Precambrian Temiscamie Iron Formation (analysis of 58 feet of diamond drill core, from Quirke, p. 312, 1961).
(4). Gogebic district, Michigan; banded magnetite-chert from the Precambrian Ironwood Iron Formation (Huber, p. 100, 1959).
(5). Ukraine, Russia; banded magnetite-quartz with minor iron silicates and carbonate, from the Krivoi Rog Series (Semenenko et al., 1956, quoted by James, p. W21, 1966).
(6). Easton Neston, England; magnetite-siderite-chamosite rock from Middle Jurassic Northampton Sand Ironstone (Taylor, p. 60, 1949).
(7). West Rosedale, Cleveland, England; magnetite ironstone from the Jurassic "Dogger Seam" (average of numerous analyses, given by Lamplugh et al., p. 60, 1920).

A notable difference between the oxide facies of iron formations and ironstones is the apparent total lack of goethite in the former. In sympathy with this the water content of the iron formations is much less than that of the ironstones. Al_2O_3 and CaO are also far lower in the iron formations, and there is a corresponding drop in CO_2.

Carbonate facies. Iron formation of this type occurs in very large amounts in some localities and is a major lithology in the Lake Superior regions of the United States and Canada. Earlier investigations suggested that most if not all of the hematite-banded ore of these areas was produced by the weathering of sideritic carbonate, and so material of this kind was regarded as being by far the dominant

primary type.   More recent studies, however, have shown that much of the hematite is primary, thereby reducing the importance of carbonate.   Nonetheless the
latter still bulks large in some localities, and although it is much less important
overall than the oxides, it is still a substantial source of sedimentary iron.

Carbonate iron formation *sensu stricto* consists of interbedded chert and carbonate and hence in this form is a simple carbonate counterpart of hematite-banded
iron formation.   The carbonate bands are fine-grained and are light to dark gray
or brown in color.   Interbedded chert is fine and dense and often appears darker
than the carbonate bands due to its semitransparency.   Since carbonate is deposited
in an Eh range between the ranges required for oxide and sulfide precipitation and
heavily overlapping with that for iron silicate formation, it is not surprising that
carbonate iron formation may grade into or be interbedded with any of the other
types.   Close carbonate-sulfide relations are common, and interbedding of fine
carbonate and pyrite layers have been observed frequently in the appropriate
sedimentary environment.   At the Helen mine, Michipicoten (Ontario), virtually
all the minable iron is contained in the *Helen siderite-pyrite member*—a great
lenticular body up to 350 feet thick lying beneath an iron-rich banded chert and
containing large carbonate and sulfide bands that interdigitate at their junction
(Fig. 13-14).

Although the change carbonate → sulfide is common, that of carbonate →
hematite is, expectably, rare.   Where iron formation of carbonate facies gives way
to mineralogies produced by higher Eh, it usually does so by grading from the

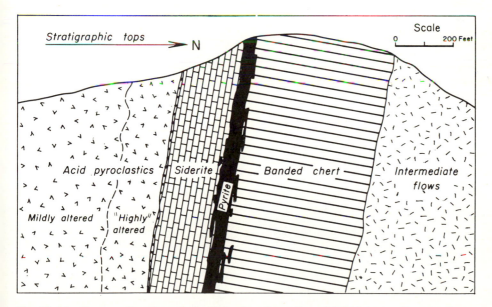

Fig. 13-14   Cross section through the Helen iron formation of Michipicoten, Ontario, showing the notable development of carbonate and sulfide types of iron formation (and the presence
of volcanic wall rocks).   (*From Goodwin, Econ. Geol., 1964.*)

normal carbonate-chert to complex silicate-magnetite-carbonate-chert members. Rocks of the latter kind make up much of the iron formations of the Mesabi, Gogebic, and Marquette Ranges of northern Minnesota and Wisconsin. Such magnetite-silicate rocks are apparently products of environments intermediate between carbonate and oxide (hematite subfacies) zones.

Table 13-9 gives analyses of a number of carbonate iron formations from different parts of the world.

Sulfide facies. The close connection between this and the carbonate facies is apparent from the above discussion of the carbonate rocks. The two ore types have close chemical affiliations and, not surprisingly, are common associates.

Iron sulfide is widespread in small amounts in virtually all sedimentary rock types, pyrite being far the most commonly occurring mineral. In spite of this

Table 13-9   Analyses of carbonate iron formation

	(1)	(2)	(3)	(4)	(5)	(6)
SiO_2	40.09	30.62	26.97	46.46	24.25	13.91
Al_2O_3	2.23	0.41	1.30	0.24	1.71	3.63
FeO	25.84	34.50	39.77	26.28	35.22	28.62
Fe_2O_3	9.96	0.46	2.31	0.64	0.71	3.28
MgO	2.79	4.41	1.99	3.10	3.16	5.73
CaO	0.50	1.45	0.66	1.87	1.78	4.34
MnO	1.25	0.93	0.29	0.21	2.11	1.73
Na_2O	0.03		}0.09		0.04	0.94
K_2O	0.40				0.20	0.45
TiO_2	0.23	0.02		Trace		0.09
P_2O_5	0.13	0.10	0.03	0.13	0.91	0.23
CO_2	14.38	27.03	26.20	19.96	27.60	27.06
S	0.01	0.07		0.11		
FeS						8.43
C	0.05	0.06			1.96	
H_2O+	1.80	0.02	0.51	1.15		1.30
H_2O-	0.56	0.03	0.10	0.07	0.21	0.10
Total	100.25	100.11	100.22	100.22	99.86	99.84

Explanation

(1). Iron County, Wisconsin; chert-carbonate from the Precambrian Ironwood Iron Formation—drill hole no. 121, 301–313 feet (Huber, p. 90, 1959).

(2). Gogebic County, Michigan; chert-carbonate from the Ironwood Iron Formation—drill hole no. 1201, 786–798 feet (Huber, p. 90, 1959).

(3). Marquette district, Michigan; carbonate iron formation from the Precambrian Negaunee Iron Formation (Van Hise and Bayley, p. 337, 1897).

(4). Gunflint district, Ontario; carbonate iron formation from the Precambrian Gunflint Iron Formation (James, p. W26, 1966).

(5). Iron River district, Michigan; chert-carbonate from the Precambrian Riverton Iron Formation—3 feet of drill core (James, p. 257, 1951).

(6). Michipicoten district, Ontario; lower portion of carbonate body, Helen mine (Collins, Quirke, and Thomson, p. 69, 1926).

widespread nature, however, pyritic beds are quantitatively the least important of the four principal iron formation types. The two most notable occurrences of sulfide iron formation are probably those of the Helen mine and of the Wauseca Pyritic Member of the Dunn Creek slate, underlying the sideritic Riverton Iron Formation in the Iron River district of Michigan.

A conspicuous and important difference between sulfide iron formations and all others is that the sulfide is associated with black, carbonaceous organic shale—not the chert that is such a characteristic feature of the other three facies. Such carbonaceous shale is usually finely banded and fissile. Chert layers may occur here and there, but these are usually small and of irregular occurrence and are comparatively rare.

The pyrite itself is usually very fine and is often barely visible. However, it may constitute up to 40 per cent or more of a specimen, in which case sulfide-rich layers stand out clearly in polished sections. Crystal outlines are often well developed, suggesting that *in its present form* the pyrite is substantially diagenetic. Any pyrrhotite usually occurs as small plates having their major dimensions parallel to the bedding of the containing shale. Increasing metamorphism of sulfide iron formations naturally leads to coarsening of grain, and even at very low metamorphic grade, pyrite usually develops as conspicuous subidiomorphic cubic (and occasionally octahedral) crystals.

In addition to the Helen and Wauseca occurrences, what appear to be excellent examples of sulfide iron formations occur close to, or as part of, some of the world's largest Precambrian base metal sulfide ore bodies. Mount Isa, Queensland, is a notable example, and there are many others, including the Roan copper deposits of Zambia. Sulfide iron concentrations of this kind are, however, left to be considered in Chap. 15.

Chemical analyses of some sulfide iron formations are given in Table 13-10.

Silicate facies. Consideration of this facies has been left until last in view of its pervasive occurrence. As already noted, it occurs in association with almost all the other iron formation lithologies—a habit resulting from the wide Eh range of iron silicate stability and, perhaps, to a greater general complexity of control than with the oxides, carbonates, and sulfides.

James (1954, p. 267) maintains that at least two major types of silicate iron formation can be distinguished in the Lake Superior region:

1. Type 1, important in the Mesabi and Gunflint Ranges, is a dark-green, fine-grained, granular rock containing ellipsoidal granules of iron silicate. Much of this is greenalite, though minnesotaite and stilpnomelane are prominent—presumably as products of very low-grade metamorphism. The granules are about 0.5 mm in diameter and are crudely spherical, ellipsoidal, or irregular in shape. They consist of two or more of magnetite, carbon, carbonate, chert, and iron silicate and are bound together by chert or other silicate material.
2. Type 2 is nongranular and typically finely bedded, usually consisting of thin laminae of iron silicate with magnetite and carbonate as impurities in some

Table 13-10 Analyses of sulfide iron
formation (i.e., pyritic shales and slates
associated with banded iron formation
occurrences)

	(1)	(2)
SiO_2	36.67	51.81
Al_2O_3	6.90	6.78
FeO	2.35	19.83
Fe_2O_3		3.44
MgO	0.65	3.46
CaO	0.13	0.37
MnO	0.002	0.49
Na_2O	0.26	
K_2O	1.81	
TiO_2	0.39	0.93
P_2O_5	0.20	0.16
CO_2		4.30
S		1.94
FeS_2	38.70	
SO_3	2.60	
C	7.60	2.54
H_2O+	1.25	4.32
H_2O-	0.55	0.33
Total	100.062	100.70

Explanation
(1). Iron River district, Michigan; pyritic
graphitic slate associated with Precambrian iron formation (James, p. 255,
1951).
(2). Iron County, Wisconsin; pyritic slate
from the Ironwood Iron Formation—
drill hole no. 121, 715–736 feet (Huber, p.
101, 1959).

of the layers.   Pyrite is a common accessory and may develop as pyritic
layers or as flat lenses.

As might be expected from its great variability in mineralogy, the chemical
composition of silicate iron formation varies widely.   Not only are there variations
in the proportions of iron silicate and chert, there are also substantial yet fluctuating
amounts of carbonate and magnetite.   Table 13-11 lists analyses of some silicate
iron formations.

Origin   It has already been said that in spite of their ubiquity and abundance in
the Precambrian rocks of the world, and in spite of the wealth of exposure and
constitutional information provided by mining, the derivation and precise mode
of deposition of iron formations remains unclear.   It is in fact one of the most
tantalizing problems in ore-genesis theory—a problem that defies solution while

continuing to mock the geologist with its apparent simplicity. James has nicely understated the position in saying that the "subject of origin [is] a fertile one for speculation" (1966, p. W47).

Such speculation began—on an intensive scale—during the second half of the nineteenth century, and so it has continued for almost 100 years. As a result there has been the inevitable recycling of ideas just as in the broader field of ore genesis generally. Before looking at these ideas, however, it is worth pausing briefly to set down the principal differences between iron formations and the other kinds of sedimentary iron concentration—particularly the ironstones of the widespread minette type. Some of these apparent differences are not too easily sustained and have in fact been challenged on points of detail, though one must

Table 13-11   Analyses of silicate iron formation

	(1)	(2)	(3)	(4)	(5)	(6)
SiO_2	51.18	48.11	50.96	29.35	30.26	52.18
Al_2O_3	11.95	3.27	1.09	0.70	2.26	3.08
FeO	12.15	16.69	30.37	39.51	37.00	16.68
Fe_2O_3	8.09	13.62	5.01	4.41	16.47	18.30
MgO	2.42	2.91	5.26	3.81	2.09	4.36
CaO	1.12	0.80	0.04	2.10	Trace	2.90
MnO	2.71	3.27		1.02	0.78	
Na_2O	2.12	0.24			0.16	
K_2O	1.86	2.32			0.14	0.58
TiO_2	0.51	0.52		0.01	0.03	
P_2O_5	0.54	0.44		0.14	0.01	0.28
CO_2	3.70	5.62		16.37	Trace	†
S			Trace	0.06		0.32‡
C			0.21	0.08		
H_2O^+	1.19	1.74	6.41	2.79	8.73	}0.94
H_2O^-	0.07	0.44	0.75	0.06	2.15	
Total	99.61	99.99	100.10	100.41	100.08	99.62

† Lost on ignition.
‡ Sample dried prior to analysis.
Explanation
(1). Iron River district, Michigan; from iron formation of the Precambrian Stambaugh Formation (James, p. 271, 1954).
(2). As for (1).
(3). Mesabi district, Minnesota; greenalite rock from the Biwabik Iron Formation (James, p. W23, 1966).
(4). Gogebic district, Michigan; from silicate-carbonate iron formation—drill hole no. 1201, 651–665 feet (Huber, p. 91, 1959).
(5). Roper River, Northern Territory of Australia; oolitic greenalite-magnetite rock (Cochrane and Edwards, p. 17, 1960).
(6). Verkhtsevskovo region, Russia; stilpnomelane-actinolite-magnetite-quartz rock from the Krivoi Rog Series (Semenenko et al., 1956, quoted by James, p. W23, 1966).

keep in mind that iron formations and ironstones have fairly wide ranges of chemistry, mineralogy, and geological setting, and so some overlap—and hence breakdown of distinctions on matters of detail—is more or less inevitable.

Nonetheless there seems no question that a number of broad differences do exist—the extraordinarily low Al_2O_3 and P_2O_5 contents of the iron formations as compared with the ironstones, the remarkable interbedding of iron mineral and chert in the iron formations as compared with the relative lack of bedding and absence of chert in the ironstones, the predominantly Precambrian age of the iron formations as compared with the predominantly post-Precambrian age of the ironstones—and that these are of great significance genetically. The following, patterned on an excellent summary by James (1966, pp. W46–47), are the main points of difference and similarity.

1. *Distribution and setting*
    a. *Geographical distribution.* No distinct differences, other than those imposed by age; both are found on all continents.
    b. *Age. Iron formations:* Cambrian to early Precambrian, with their major incidence distinctly Precambrian; principal formations in Australia, India, North America, Russia, and southern Africa are of remarkably similar age—about 2200 million years. *Ironstones:* Middle Precambrian to Pliocene, with their major incidence distinctly post-Precambrian; the major occurrences are Paleozoic and Mesozoic.
    c. *Geological association.* While there may be differences of association, these are not distinct or easy to sustain. Superficially the associated rocks are similar—shales, sandstones, carbonate rocks, and graywackes. However, *some iron formations* are conspicuously associated with volcanic rocks—tuffs and flows—and in some cases where volcanic rocks were not immediately recognized, careful microscopic work is revealing the presence of much very fine volcanic tuff interbedded with the iron minerals. There is thus a conspicuous *tendency* for *iron formations* to have a volcanic association, which the *ironstones* (apart from the less common Lahn-Dill type) do not. In addition, whereas the carbonate rocks associated with *ironstones* are limestones and/or dolomites, those associated with *iron formations* are almost always dolomite.
2. *Form*
    a. *Shape and extent.* Both types occur as flat lenses conformable with the enclosing sediments, and both tend to be elongate, i.e., to have one of their larger dimensions much greater than the other. *Iron formations:* These have their major axis up to more than 100 miles in length. *Ironstones:* These are less extensive, with their major axis of the order of a few tens of miles.
    b. *Thickness. Iron formations:* The major iron formations have thicknesses from about 150 feet to about 2000 feet and most are in the form of one, or a very few, major layers. *Ironstones:* These are much thinner, major

occurrences ranging from about 5 feet to about 50 feet, and many are in the form of a number of comparatively thin layers confined to a particular member of a sedimentary sequence.

c. *Physical features.   Iron formations:* These are finely and regularly banded, consisting of alternating iron-rich (i.e., high-iron–low-silica) and silica-rich (i.e., high-silica–low-iron) bands.   Oolitic structure is rare except in hematite iron formations.   *Ironstones:* The banding is absent or poor; oolitic and pisolitic structures are typical in oxide and silicate ironstones and may occur in sulfide ironstones.

3. *Constitution*

a. *Chemical.   Iron formations:* These consist essentially of free silica, iron, and the radical appropriate to the facies, that is, CO_2^{--}, SiO_3^{--}, etc. Al_2O_3, P_2O_5, Na_2O, and K_2O are typically very low, as are trace elements.   MgO/CaO is generally greater than 1.   *Ironstones:* SiO_2/Fe is much lower than for iron formation; Al_2O_3 ranges up to about 17 per cent in chamositic rocks and is commonly in the 5 to 10 per cent range. P_2O_5 is notably higher than in iron formations; MgO/CaO is generally less than 1.

b. *Mineralogical.   Iron formations:* The order of abundance of iron compounds is oxide, carbonate, silicate; hematite and magnetite are about equal in amount overall and goethite is absent.   Siderite is the dominant carbonate, dolomite is common but subordinate, and calcite is rare or absent.   Greenalite is the only primary silicate; chamosite, glauconite, and collophane are absent.   Chert is typical of, and always present in, all but sulfide facies; in amount such chert is approximately equal to, though slightly less than, the associated iron mineral.   *Ironstone:* The order of abundance of iron compounds is probably oxide, silicate, carbonate; goethite is the dominant oxide, hematite is usually present but subordinate, and magnetite is very minor or absent.   Chamosite is the principal primary silicate, and glauconite is common but subordinate. In addition to sideritic carbonate, calcite and dolomite are common and may be abundant.   Collophane is usually present and is sometimes abundant.   Interbedded chert is absent.

Rather than considering separately source, transport, and deposition as we have done for bog ores and ironstones, it is perhaps easier, in the case of banded iron formations, to examine each genetic theory as a whole.   Naturally these theories have a close connection with those proposed for ironstones—continental weathering, erosion, and marine deposition; seaboard and submarine volcanism and marine deposition; sea-floor leaching and redeposition, and so on—but the differences between the two classes of occurrence necessarily require some variations on these themes.

1. *Continental erosion: marine deposition.*   In this case, as with the ironstones, iron is thought to have been released by weathering of land surfaces and

transported to the sea by slow-moving rivers and in underground waters. On arrival at the sea it is contributed to near-shore zones, particularly basins, troughs, and partly barred areas of the ocean, where it is precipitated as oxide, carbonate, silicate, or sulfide in accordance with the Eh-pH characteristics of the prevailing sedimentary facies pattern. The silica of the chert bands is regarded similarly as an erosion product, carried in solution by surface and underground water.

Three principal ideas have been put forward to explain the interbedding of iron mineral and chert:

a. *Regular fluctuation in supply due to seasonal climatic changes.* Here it is suggested (e.g., Sakamoto, 1950) that the transport of iron and silica, particularly in underground water, followed a seasonal cycle. Assuming sharply defined and pronounced wet and dry seasons: (1) in the wet seasons ground waters tend to become acid—iron may therefore have been dissolved and carried as the bicarbonate in subsurface solutions, eventually to be discharged into lagoons or the sea and (2) in the dry seasons the zone just above the water table becomes neutral to alkaline —iron is no longer dissolved but silica is, and this is carried by the subsurface waters to the sea. In this way iron and silica could have been delivered alternately to the site of deposition, leading to a fine and regular banding.

b. *Regular fluctuation in deposition due to seasonal change in biological activity.* It has been suggested by Krauskopf (1956), Huber (1959), and James (1966) that heavy deposition of silica may result from the explosive growth of silica-accreting organisms in response to seasonal (or longer cycle climatic change) increase in temperature and/or nutrient content of surface waters at the site of deposition. In this case the rate of deposition of iron mineral would remain essentially constant, whereas the chert layers—which always contain minor iron—would represent sudden, brief showers of organically precipitated silica contributed from the surface layers and superimposed periodically on the steadily accumulating iron. The development of alternating iron mineral–ferruginous chert bands would follow.

c. *Regular fluctuations in deposition due to pulsatory tectonic movements.* It has been pointed out of the Griquatown Ironstones by Cullen (1963) that there appear to be two sets of chert bands: a fine set, which may be seasonal, and a much broader set—incorporating the finer bands— which Cullen suspects may be related to tectonic movements. The precise nature and effects of these movements are not clear, though they are regarded as "essentially epeirogenic" and presumably affect the incidence of silica through their influence on erosion.

Whichever—if any—of these postulates is correct, the continental erosion theory encounters one outstanding difficulty. This is the transport of the enormous quantities of iron and silica involved *without* the concurrent movement of other materials.   While the amount of detritus, for which Al_2O_3 is a good index, is quite low in ironstones, it is almost vanishingly small in the banded iron formations.   It is thought by some, notably Borchert (1960), that such near absence of detritus unquestionably precludes an erosional source, no matter how mature the source landscape or how sluggish the streams of the time.   However, a number of investigators have suggested that, while it would indeed be very difficult to transport large quantities of iron—much of it necessarily in suspension—without substantial amounts of accompanying fine detritus under modern conditions, conditions prevailing in the Precambrian might have been significantly different.   In particular it has been proposed that the earth's atmosphere during the periods of iron formation deposition had a higher CO_2 content than during later periods. As Rubey (1951) and James (1966) have pointed out, if the partial pressure of CO_2 in the Precambrian atmosphere were 0.03 atm—two orders of magnitude greater than the present 0.0003 atm—the equilibrium pH of surface waters could have been as low as 6.1 as compared with the present 8.2.   It could have been even lower if, as is quite possible, the total volume of water on the earth's surface were less than now.   Under such conditions surface waters would have had a much greater capacity for leaching iron and for transporting it in solution than they have at present.   Thus large quantities of iron might have been moved in surface solution, just as in acid ground waters of the present day, without accompanying detritus.

In a similar vein it has been suggested that the Precambrian atmosphere may have contained little or no oxygen.   This immediately removes the chief precipitant of iron and hence the principal inhibitor of its movement.   Cloud (1965), for example, has put forward the idea that oxygen was generally absent but was produced locally by biological photosynthesis.   In this way it is possible to account for large-scale surface movement of iron in solution and, at the same time, for *local* heavy precipitation of its oxides to form iron-rich sediments.

An ingenious suggestion for the separation of iron and alumina during erosion, without postulating a Precambrian atmosphere significantly different from the present one, has been put forward by Govett (1966).   Very briefly, Govett suggests that, given a deeply weathered senile land surface and rivers so sluggish that they could carry nothing coarser than finely particulate iron oxide, the only aluminous matter that could be carried in suspension would be fine clay.   However, Govett points out, there is evidence to suggest that kaolin can form only through an organic intermediary, and since there was probably little or no terrestrial plant life in Precambrian time, there may well have been no clay.   The problem thus appears solved at a blow!   However —and so neat is Govett's idea that one cannot help but be a little wistful that it were so—some banded iron formations contain or are bordered by shales containing quite high alumina.   The sulfide iron formations are in fact pyritic shales.

2. *Continental erosion: lacustrine deposition.*   Although the evidence of associated lithologies indicates that most banded iron formations were deposited under marine conditions, there are cases where this is not certain.   In addition it is possible to extend the definition of "lake" to include saline bodies of water having a close, though restricted, connection with the sea and showing certain lacustrine characteristics.   For example, Govett, in the investigation referred to above, noted that:

> . . . a lacustrine environment, as distinct from a marine environment, will be taken to mean a restricted basin that lies essentially within a single climatic regime wherein a seasonal stratification could develop in the water, and not necessarily imply fresh water. Whereas the hypothesis to be proposed is developed upon the basis of a fresh-water, warm, monomictic lake cycle, the same conclusions could be derived equally well from consideration of a saline body of water.   In the latter case the seasonal variation in surface density could arise from accession of fresh water during a rainy season that would overlie denser, saline water.   Evaporation during a dry season would increase the salinity, hence the density of the surface waters until they become similar in density to the deeper water, and circulation would become possible.   Whatever the precise mechanism, the essential prerequisite is only that there shall be at least one period during the year when the waters are stratified and an oxygen deficiency can occur in the deep water and at least one period during the year when circulation is possible and oxidizing conditions can develop in the deeper water (1966, p. 1199).

The lacustrine hypothesis encounters the same difficulties as the marine hypothesis as far as the lack of detrital impurity is concerned: the problem of the extraordinarily low content of aluminous matter remains.   The main feature of the lacustrine hypothesis is that it provides an alternative explanation (see Hough, 1958) for the iron mineral-chert banding.   This explanation rests on the development of seasonal temperature-density stratification in some lakes and the regular destruction of this stratification by seasonal overturn—the series of events referred to by Govett above.

The timing of stratification and overturn in dimictic lakes is shown diagrammatically in Fig. 13-15.   (*Dimictic* lakes are deep lakes occurring in the temperate zones and characterized by twice yearly overturn—in spring and autumn.   *Monomictic* lakes are deep lakes occurring in the warm-temperate and subtropical zones that undergo overturn once a year—in winter.)

The upper (warmer and hence less dense) stratum is termed the *epilimnion* and the lower stratum (colder and denser) the *hypolimnion*.   Where lakes are rich in nutrients they are able to support a high level of biological activity, and this together with the rain of dead (and hence oxygen-consuming) organisms from the near-surface waters often leads to reducing conditions in the hypolimnion.   Thus where iron and silica are contributed to the lake in summer, the iron tends to go into solution on reaching the hypolimnion, whereas the silica is deposited more or less straightaway.   At the onset of autumn overturn the iron is oxidized and sedimented, the seasonal cycle leading in this way to a silica-iron mineral-silica alternation.   Superimposed on this

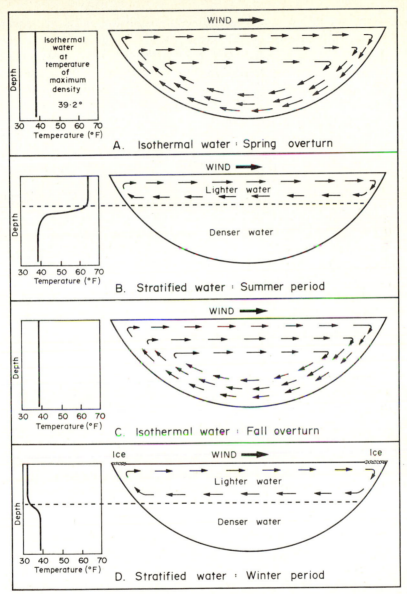

Fig. 13-15 Seasonal timing of stratification and overturn in a dimictic lake. (*Adapted from Hough, J. Sediment. Petrol., 1958.*)

fluctuation in the depositional environment may also be, of course, the effects of the seasonal cycle of ground-water activity suggested by Sakamoto. During the wet summer, ground waters and lakes tend to be acidified and carry and hold iron in solution. During the dry winter they tend to alkalinity, carrying and holding silica but precipitating iron. Thus the process postulated by Sakamoto would be in phase with, and would emphasize, the effects of overturn.

3. *Volcanic exhalation: marine and/or lacustrine deposition.* In this case, as with the ironstones and volcanic lake ores, iron (and, in the present case, silica) is contributed virtually directly to the site of deposition via volcanic hot springs. Leaching of iron from hot submarine lavas by sea water has also been proposed as an ancillary process. The principal early proponents of the volcanic hypothesis were Van Hise and Leith (1911), but the idea has had many adherents and Goodwin (1956, 1962, 1964) has recently put forward much evidence in favor of a volcanic source for some of the Canadian Lake Superior occurrences.

Transport and the lack of detrital impurities are not problems here. Banding is attributed to pulsing of volcanic activity and to differential flocculation and settling of the two principal components. Each pair of iron-chert bands would represent a single contributory episode, iron and silica being emitted together during each burst of exhalative activity. The iron is quickly precipitated and, with the comparatively high density of its compounds, settles rapidly to form bands of composition appropriate to the facies of sedimentation concerned. The silica flocculates less rapidly and, with its lesser density, settles after the iron, leading to an iron mineral-chert alternation.

This idea has an appealing simplicity and its attractiveness is increased by the fact that iron deposition is known to result from such a process at the present day. There are, however, several objections:

a. For some large iron formations evidence of associated volcanism, if there, is not apparent.
b. If the Precambrian volcanic hot-spring waters contained amounts of iron and silica comparable with those of modern thermal springs, impossibly large quantities of solution would be required for the contribution of the amounts of iron and silica now found in the iron formations.
c. Marine and seaboard volcanism are widespread at the present time, but no deposits comparable with the Precambrian ores have been found in modern sediments.

In reply to these points it may be said that more careful investigation is beginning to show clear evidence of pyroclastic material (and hence concurrent volcanic activity) in material associated with some iron formations and, until recently, loosely labeled "shale"; that the amount of analytical data on the iron and silica content of modern thermal springs is still sparse; and that since the major part of the sea floor is still unexplored, there may well be modern accumulations as yet unknown. On the last point, indeed, oceanographic samplings are beginning to reveal the presence of rich iron oxide deposits in areas of sea-floor volcanic activity. However, much further investigation is required on these three aspects.

4. *Sea-floor leaching and redeposition.*   As already noted, Borchert (1960) maintains that the near absence of detritus from banded iron formations unquestionably precludes their derivation by erosion.   As an alternative, he
   suggests that they are derived from the sea floor itself by the process outlined
   earlier.
        This, like the volcanic hypothesis, has an appealing neatness.   How-
   ever, while it might be a good explanation for some *ironstones*, it does not
   really seem adequate to account for the *banded iron formations*.   It does
   not preclude the deposition of detritus, and most importantly, it provides
   no mechanism for the development of the iron mineral-chert banding that is
   so characteristic a feature of the iron formations.

CONCLUDING STATEMENT

We have now come to the point where we may attempt to review and sum up present
knowledge of iron concentrations of sedimentary affiliation.   This is not an easy
task—what at the outset showed promise of being quite simple may now be seen
to be very complex indeed—but it is valuable to marshal our facts and look at
some of the possibilities that they suggest.
     Clearly the only sedimentary iron concentrations whose origin and mode of
deposition can be said to be really understood are the marsh and lake ores—the
bog iron ores of present usage.   There seems no doubt that these are derived by
subsurface weathering and leaching, transported as bicarbonate and humates in
ground water of low pH and Eh, and deposited in lakes and marshes by loss of
CO_2 and, in most cases, oxidation.   Organic (bacterial) activity is undoubtedly
important in this environment.
     The *ironstones,* and particularly the *banded iron formations*, remain enig-
matical.   The only aspect of the formation of these in which any understanding
can be said to have been achieved is that of their deposition.   In spite of some
disagreement concerned principally with the relative importance of diagenetic
change, it appears that the distribution and mineralogical constitution of the ores
is governed by sedimentary facies patterns through the variation in chemical
environment—particularly Eh—going from one facies to the next.   Diagenesis
may, however, modify this to some extent.
     From whence the iron came, and how it was transported from its source to
its site of deposition, remains, however, unclear.   This is not to say that we do not
know where iron *may* come from or how it *may* be transported.   The unsettling
fact is that although several sources and transporting processes can be proposed
and although these can be observed yielding and moving iron at the present day, no
modern (i.e., currently accumulating) deposits *of the size and purity of the older ores*
have yet been found.   Porrenga (1967), for example, has shown a very nice facies
pattern of iron oxide–iron silicate deposition in modern sediments derived from
continental erosion—but the iron minerals there are no more than disseminations in an enormous mass of normal detritus.   Similarly a number of volcanic-
exhalative iron concentrations are known, and while some of these are quite rich

they are nowhere near as extensive as the older ironstone and iron formation occurrences.

The ironstones look to be the lesser of the two problems. Certainly they look more like "normal" chemical sedimentary rocks than do the iron formations. With their quite high Al_2O_3, clastic quartz, Ca–Mg carbonate, and P_2O_5 contents, they could be fairly easily visualized as forming where iron compounds were contributed in abundance to situations in which detrital, and other chemical, sedimentation proceeded normally but slowly just as in many appropriate localities today. Some of the smaller ironstone deposits occur in conspicuously volcanic sequences, and there is little doubt that they represent heavy deposition of exhalative iron superimposed on minor volcanic detrital (subaqueous pyroclastic) accumulation. The major ironstone occurrences—such as the Jurassic of Europe and England and the Clinton beds of America—do not have a volcanic association, but as is suggested by James, they may well be large analogs of the bog concentrations. As we have seen, this important group of ironstones occurs in estuarine areas, near-shore zones characterized by the presence of small islands, channels and closed depressions, or in restricted basins. It seems well within the bounds of reason to suggest that most of their iron was contributed by ground waters debouching directly onto the near-shore sea floor. The iron in this case would have been dissolved from soils and rocks by carbonated waters and transported underground as bicarbonate and humates. There is thus no problem of source and transport—particularly if substantial artesian aquifers are involved—and the product is a mixture of minor "normal" sediment and iron compounds appropriate to the facies concerned.

The banded iron formations appear to be a much tougher problem. There seem to be two outstanding—and perhaps significant—features of these ores:

1. The remarkably low alumina and minor element content of the oxide, carbonate, and silicate facies, in contrast to the fairly "normal" alumina content of most sulfide facies

2. The abundance of interbedded silica in the oxide, carbonate, and silicate facies, in contrast to its virtual absence—i.e., as well-bedded chert—in most occurrences of the sulfide facies

One cannot help being sorely tempted to agree with Borchert that the lack of Al_2O_3 in ores of the three principal facies indicates that the principal components of these ores were not derived by continental erosion. No matter how mature a landscape and how sluggish its river systems, it seems quite impossible to visualize a surface erosional process that did not shift aluminous matter. That aluminous material *was* in fact comminuted and moved at much the same time as the iron was accumulating is shown by the occurrence of shale lenses within the iron formations. Huber's analyses (1959, p. 102) of two "argillites," containing 10.18 per cent Al_2O_3 in one case and 14.82 per cent in the other, seem to negate both the general proposition that aluminous matter might not have been carried in significant quantities by the Precambrian rivers and Govett's suggestion that clays as a class might not have developed in the Precambrian. Such doubts about the nontransportation of Al_2O_3 are reinforced by the frequent occurrences of conglomerates and other coarse detritals adjacent to, and even within, the iron formations. One cannot help suspecting that the low content of Al_2O_3 is due to either of two possibilities:

1. The iron and its associates are not products of surface erosion.
2. The Al_2O_3 and other minor impurities *are* products of surface erosion and slow deposition but have been swamped by the (comparatively) very rapid deposition of iron and silica from some other source.

The latter possibility is favored here—albeit very tentatively. There seems little doubt, particularly from the work of Goodwin, that some of the great iron formations are volcanic-exhalative. Recent work by La Berge (1966) has shown that other iron formations, thought to be nonvolcanic through an apparent lack of associated volcanism, are in fact riddled with previously unrecognized, fine, volcanic detritus. In addition those ironstones known to be of volcanic derivation —the Lahn-Dill type—are suspiciously similar to iron formations in their conspicuously low content of Al_2O_3 and P_2O_5. If the iron formations were derived from exhalations, low Al_2O_3 might be accounted for by its near absence in the exhalations or by the swamping of slowly accumulating erosional alumina by rapidly accumulating exhalative iron and silica. The comparatively high Al_2O_3 content of *one* of the banded iron formation facies—the sulfide facies—has an intriguing look of significance here, and we shall take this up in a moment.

The very large amount of silica, the nature of this, and the extraordinarily regular banding it develops with the iron compounds looks to hold a positive bonanza of clues, and it is perhaps the most exasperating part of this whole tantalizing problem that the recognition of these clues continues to elude us. We may not, however, be too far from the beginning of a solution; one cannot help suspecting that Krauskopf—and, following him, Huber and James—may have had the first vital flash of insight that will lead to the revelation of the clues and the solution of the problem: is it that the bulk of this silica is not inorganic at all but due to the periodic explosive proliferation, and then death and accumulation, of silica-accreting microorganisms? Krauskopf's findings on the solubility of silica, combined with our knowledge of the ability of diatoms and other modern marine organisms to virtually strip the sea of its silica, suggest very strongly that organic activity may be the beginning of the answer. It may well be that, periodically, suitable temperatures, optimum concentrations of nutrients, and buildup of volcanic(?) silica concentrations in bays and other confined coastal waters led to a rapid and extensive blooming of silica-accreting organisms. If their proliferation and accumulation were rapid, silica sedimentation may have temporarily overwhelmed that of iron, leading to the development of high-silica–low-iron "chert" beds.

It is here that the sulfide facies looks so interesting, and for two reasons— its comparatively high Al_2O_3 content and its lack of chert. The Al_2O_3 content— analyzed values range up to almost 9 per cent—shows that a significant amount of aluminous matter *was* being deposited during iron formation time. There is no reason to suspect that Al_2O_3 deposition was slower in the oxide, carbonate, and silicate facies than in the sulfide facies, indicating that in the first three its *concentration* was depressed by the much more rapid accumulation of something else—the iron and silica.   If indeed there has been such a swamping of detrital sedimentation by chemical sedimentation *nearer the shore*, the iron and silica surely cannot have been erosional products.   The sedimentation pattern linked with continental erosion is just the opposite of this—a fact specifically utilized by Cullen (1963) in his explanation of the Griquatown Ironstones.   This line of thought points to extensive volcanic exhalations, or perhaps extensive artesian debouchment, as the most likely source of the chemical products.

Whereas the high content of Al_2O_3 in the sulfide facies may provide an insight into the *rate* of iron and silica sedimentation and hence, perhaps, *source*, the near absence of chert from rocks of this facies may give a clue to the *nature* of the silica-depositing process.

The development of facies of iron formations is dictated substantially by Eh.   While, as we have seen, there may be slight seasonal changes in the pH of contributions made to them, large bodies of water remain close to neutral and, certainly in the sea, pH plays no significant role in the development of chemical facies patterns.   The solubility of silica, however, is not affected by Eh but by temperature and, to a minor extent, by pH.   Thus if silica settled by simple flocculation or, as is perhaps more likely, by the sinking of silica-accreting microorganisms from near-surface layers, there seems no reason why it should not accumulate in the sulfide facies just as it does in the others.   In fact one might expect accumulation to be favored in the sulfide facies, since the greater coldness of the deeper waters concerned would retard the dissolution of silica and hence lead to its apparent more rapid accumulation here.   Certainly derivation from organic remains implies descent from near-surface waters (the locale of high rates of photosynthesis), hence the traversing of a deeper water column and therefore greater opportunity for dissolution in transit, but it is hard to see why this would cause a conspicuous cutoff in chert deposition going from carbonate to sulfide facies.

One is therefore led to the suspicion that precipitation of silica was in some way related to bottom chemistry: that it did not initially precipitate in the body of the overlying water simply to rain down onto whatever bottom situations it happened to encounter, but that precipitation was in fact heavily influenced by conditions in the water just above the mud-water interface.   Any such influence on inorganic precipitation of silica seems unlikely, since we have already seen that its solubility is little affected by pH and Eh and that the lower temperatures of the sulfide zones would favor its precipitation in just the facies from which it is now conspicuously absent.   If on the other hand—as Krauskopf suggests—precipitation of the silica has been biological, we may have an answer.   Perhaps the silica has been accreted by organisms operating at and close to the floor, sensitive to Eh, and unable to live in environments in which free oxygen is absent and/or the sulfide ion is abundant.   The abundance of chert in some nonpyritiferous black shales suggests that it may be the sulfide ion that is important and hence that the relation to Eh is indirect.

Thus the present evidence suggests that the iron and silica may have been provided by sea-floor volcanic exhalation or ground-water debouchment on a grand scale. The iron very probably precipitated quantitatively by inorganic means as dictated by Eh, pH, and the incidence of CO_3^{--} and S^{--} ions and as continuously as it was supplied. Although it also probably accumulated continuously, the silica may have precipitated in neither a steady fashion nor by inorganic means. Its heavy deposition may well have come only with the periodic confluence of suitable temperatures, nutrient concentrations, salinities, and so forth—conditions leading to the sudden abundant but brief blooming of silica-accreting benthonic organisms. These might then have caused the rapid deposition of silica wherever bottom conditions permitted their growth—i.e., in all environments not "poisoned" by the sulfide ion—and thus, temporarily, the swamping by silica of the otherwise steadily accumulating iron. If (1) the iron were provided by terrestrial degassing (i.e., volcanic exhalation) and if such degassing has decreased with time and (2) the suggested silica-accreting benthonic organisms were at the zenith of their evolutionary pattern in the middle Precambrian, we may have a reason for the seeming preferred association of banded iron formations for strata of Precambrian age.

The development of ideas on these remarkable deposits has indeed followed a twisting and tortuous path—one that may be compounded of several paths— and it seems clear that we have come to the end of none of them yet. There is probably no better way of ending this chapter than by quoting some concluding words of H. L. James: "These suggestions for [the] origin of ironstone and iron-formations should be taken for what they are—that is, plausible speculations . . . most surely they are not final answers" (1966, p. W51).

RECOMMENDED READING

James, H. L.: Sedimentary facies of iron formation, *Econ. Geol.*, vol. 49, pp. 235–293, 1954.
——: Chemistry of the iron-rich sedimentary rocks, *U.S. Geol. Surv. Prof. Pap.*, 440, Chap. W, 1966.
Trendal, A. F.: Three great basins of Precambrian banded iron formation deposition: a systematic comparison, *Geol. Soc. Amer. Bull.*, vol. 79, pp. 1527–1544, 1968.
Winchell, N. H., and H. V. Winchell: The iron ores of Minnesota, Bull. no. 6, Geological and Natural History Survey of Minnesota, Harrison and Smith, State Printers, Minneapolis, 1891.

14
Manganese Concentrations of Sedimentary Affiliation

Just as almost all sedimentary rocks contain substantial quantities of iron, so do they contain readily detectable—though generally much lesser—amounts of manganese. Through their close association as transition elements in the periodic table the two show much the same behavior in the cycle of weathering and sedimentation, and for this reason they tend to accumulate in the same or similar sedimentary environments. Hence they often form mixed, or closely adjacent, concentrations.

One very conspicuous difference between the two elements in common sedimentary rocks is their abundance. Manganese contents are generally an order of magnitude less than iron, and manganese/iron ratios of between 1:40 and 1:60 are fairly general in "ordinary" sediments and sedimentary rocks. Total iron and manganese contents, and Fe/Mn ratios, of some common igneous and sedimentary rock types and of some modern sediments are given in Table 14-1. However, as with iron, manganese may reach substantial concentrations in certain sediments, and it is well established that in some localities there are complete gradations from "normal" sedimentary rocks, to those containing notable manganese, and finally to those containing sufficient manganese to be regarded as ores. There is thus ample evidence that many concentrations of manganese compounds in sedimentary rocks are, like those of iron, lithified chemical sediments—a fact

Table 14-1   Iron and manganese in some igneous and sedimentary rocks and in some modern sediments.   (See Strakhov, 1966, for more detailed information on modern sediments)

Material	Fe^{3+}	Fe^{++}	Total Fe	Mn	Fe/Mn	Authority
Rhyolite	1.09	0.70	1.79	0.08	22.38	Daly, 1933
Andesite	2.64	2.17	4.81	0.15	32.07	Daly, 1933
Andesite	2.72	3.85	6.57	0.15	43.80	Nockolds, 1954
Basalt	4.20	4.55	8.75	0.23	38.04	Daly, 1933
Granite	1.24	1.26	2.50	0.08	31.25	Daly, 1933
Granodiorite	1.01	1.82	2.83	0.08	35.38	Nockolds, 1954
Diorite	2.49	3.15	5.64	0.08	70.50	Daly, 1933
Ultramafic rocks	3.50	6.15	9.65	0.15	64.33	Nockolds, 1954
"Average magmatic rock"	2.39	2.66	5.05	0.19	26.58	Clarke and Washington, 1924
Shale	3.26	1.82	5.08			Clarke, 1924
Sandstone	0.86	0.21	1.07			Clarke, 1924
Graywacke	0.78	3.01	3.79	0.08	48.38	Pettijohn, 1949
Subgraywacke	0.70	1.89	2.59	0.15	17.27	Pettijohn, 1949
Limestone	0.39		0.39	0.08	4.88	Clarke, 1924
Mississippi silt	2.88		2.88	0.08	36.00	Clarke, 1924
Terrigenous mud	4.28	1.75	6.03	0.08	75.38	Clarke, 1924
Marine "blue mud"	6.45		6.45	0.23	28.04	Sujkowski, 1952 (in Poldervaart, 1955)
Radiolarian ooze	7.38		7.38	0.93	7.94	Poldervaart, 1955
Globigerina ooze	3.81		3.81	0.46	8.28	Poldervaart, 1955

that has been established for well over 100 years.   We are thus concerned once again with chemical processes of sedimentation and diagenesis—reactions at low temperatures and pressures in aqueous media—and hence with the principles that concerned us in Chap. 6.

CRYSTAL CHEMISTRY OF MANGANESE

Like those of iron, sedimentary concentrations of manganese as a class contain oxides, hydroxides, carbonates, silicates, and sulfides.   Individual deposits may contain one or more of these groups.   The oxides (including the hydroxides) are by far the most abundant overall, carbonates follow, and these in turn are succeeded by the silicates and finally by the sulfides, which are quite rare.

Manganese is the major metallic component of a very large number of minerals.   For example, in 1943 Fleischer and Richmond listed 34 confirmed varieties of oxide alone, and there have been a number of additions to this total since.   As carbonate it appears as a component, in varying amounts, in mixed-crystal series of the rhombohedral carbonates, and mention has already been made of its presence as a minor constituent of some of the Lake Superior iron formation carbonates. It occurs in a wide variety of silicates, from "manganese" silicates such as *braunite*

($3Mn_2O_3 \cdot MnSiO_3$) and *rhodonite* ($MnSiO_3$) through to minerals such as *mangan-hedenbergite* [$(Fe,Ca,Mn)Si_2O_6$] and other ferromagnesian species containing manganese as a minor component. It has three sulfides and substitutes in substantial amounts in those of several other metals—notably ZnS.

This ability of manganese to find a place in a multiplicity of compounds and crystal structures results from its highly variable crystal chemistry—the result in turn of the diversity of valence states it is able to adopt. The ionic radii (standard six-fold coordination) for the three valence states of manganese, and of the cations with which it is most frequently associated in crystal structures, are

Mn^{++}	0.91 Å	Mn^{3+}	0.70 Å	Mn^{4+}	0.52 Å
Fe^{++}	0.83	Fe^{3+}	0.67	V^{4+}	0.60
Mg^{++}	0.78	Ti^{3+}	0.69	W^{4+}	0.67
Ca^{++}	1.06	Cr^{3+}	0.64	Mo^{4+}	0.67
Zn^{++}	0.78	V^{3+}	0.65		
		Al^{3+}	0.58		

The manganous form occurs in sulfides, silicates, carbonates, and oxides. The monosulfide, MnS, is known to have at least three crystal modifications. *Alabandite*, the rare natural sulfide, has an NaCl (i.e., PbS) structure, as noted in Chap. 3. According to Goldschmidt another form of MnS, of semimetallic character, appears to develop and to concentrate in the troilite phase of meteorites; it is not known as a terrestrial form, however. A further form—or pair of forms —has been prepared synthetically; these are modifications of sphalerite and wurtzite structure, respectively. Goldschmidt has pointed out that the interatomic distance Mn—S in both of the latter is 2.44 Å, intermediate between Zn—S, which is 2.33 Å for both forms, and Cd—S, which is 2.5 Å for both forms, so that this type of MnS readily develops mixed crystals with both ZnS and CdS in the corresponding structures. The disulfide, *hauerite*, has already been referred to as a member of the pyrite group in Chap. 3.

Manganese sulfides are, however, quite rare, and the total amount of manganese occurring in them in nature must be infinitesimal. By far the major part of the divalent manganese of the earth's crust occurs in silicates. The size of the Mn^{++} ion, 0.91 Å, is close enough to those of Mg^{++}, 0.78 Å, Ca^{++}, 1.06 Å, and Fe^{++}, 0.83 Å, to allow quite substantial substitution and isomorphous entry into silicates such as the pyroxenes, amphiboles, and garnets. For the same reason there is considerable isomorphous entry of divalent manganese into the rhombohedral carbonates, though with its intermediate ionic radius it is difficult to say whether it substitutes for Ca or the Fe–Mg component in the dolomite arrangement. Manganous oxide and hydroxide are rare in nature—any oxidation tends to take the manganese quickly to the tri- or tetravalent state—and the total amount of manganese in this form is, as with the sulfides, vanishingly small.

Manganese in the trivalent state is much more common and expectably is far more abundant in oxide form than the manganous ion. It is well known in *manganite* (Mn_2O_3) and in the iron manganese oxide *bixbyite* [$(Fe,Mn)_2O_3$], and there are several mixed oxides and hydroxides containing it. The Mn^{3+} ion is the largest trivalent ion among aluminum and several important members of the iron family:

Al^{3+}	Cr^{3+}	V^{3+}	Fe^{3+}	Mn^{3+}
0.58	0.64	0.65	0.67	0.70 Å

and for this reason—as pointed out by Goldschmidt (1954, p. 625)—it is just too large to form an oxide of the corundum structure. Thus instead of coordinating the six oxygens in the hexagonal structure of Al_2O_3, it does so in the cubic structure of scandium trioxide (Sc^{3+}, 0.83 Å; Sc_2O_3, $a_0 = 9.79$ Å, C-type structure of the rare earth oxides).

Because of its unsuitable size Mn^{3+} does not replace four-coordinated Al^{3+} in aluminosilicates, but as with Fe^{3+}, it does in some cases replace six-coordinated Al^{3+} (one of the features of the crystal chemistry of aluminum is that its ionic radius is such as to be very close to the limits of the stability ranges for both four- and six-fold coordination with oxygen and it thus assumes whichever of these coordinations is appropriate). Manganese substituting in this way character-istically imparts a pink color to the mineral concerned.

As might have been expected mangano-manganic compounds also occur in nature, and a well-known mineral of this type is the double oxide *hausmannite* ($MnO \cdot Mn_2O_3$). It was noted in Chap. 3 that this (the manganese equivalent of magnetite) has a tetragonal-deformed spinel structure, the deformation being due to the large size of the Mn^{3+} ion. Braunite is another common mangano-man-ganic compound.

Development of the tetravalent state leads to the further contraction of the ion, to 0.52 Å. This is much smaller than the ions of other common tetravalent elements, and so there is little ionic substitution by, or for, Mn^{4+}. This ion is therefore virtually confined to minerals that are simple oxides of manganese. The principal mineral is probably the best known of all natural manganese compounds—*pyrolusite* (MnO_2).

In summing up the disposition of Mn^{++}, Mn^{3+}, and Mn^{4+} in the earth's crust, Goldschmidt (1954, p. 626) notes that although there are many exceptions, there is a general tendency for the different manganese ions to be associated with natural environments as follows:

1. Mn^{++} is the predominant manganese ion in the primary manganese compounds of magmatic rocks.
2. Mn^{4+} predominates among the manganese compounds of the sedimentary cycle in contact with the atmosphere and well-aerated surface waters.
3. Mn^{3+}, the intermediate stage of oxidation, tends to develop where Mn^{++} compounds are being oxidized or Mn^{4+} compounds are being reduced. It therefore develops in part as an intermediate compound in the oxidation of primary Mn^{++} compounds by underground (including hydrothermal) solu-tions and in part as a product of the partial reduction of sedimentary Mn^{4+} compounds during deep metamorphism.

PRINCIPAL MINERALS OF MANGANESE-RICH SEDIMENTARY ROCKS

The principal minerals of manganese, i.e., those important either for their abundance in sedimentary concentrations or for their significance in indicating the nature of sedimentary ore-forming processes are†

Alabandite (MnS) and *hauerite* (MnS$_2$). Rare; never abundant; formed only under anaerobic conditions in the presence of high S^{--} concentrations.

Rhodochrosite (MnCO$_3$). Uncommon, but manganese does commonly occur as a constituent of mixed Ca, Mg, Fe^{++}, Mn^{++} rhombohedral carbonates, in which the Mn^{++} may be the dominant metallic ion; formed at moderately low Eh in the carbonate facies of sedimentary accumulations and in vein and replacement deposits and their metamorphosed equivalents.

Rhodonite (MnSiO$_3$). Fairly common as a product of the metamorphism of sedimentary manganese ores, and hence often associated with manganese-rich garnet of the spessartite type [Mg$_3$Al$_2$(SiO$_4$)$_3$]; also occurs in vein and replacement deposits.

Manganosite (MnO). Uncommon; not known to occur in abundance; probably almost always formed by magmatic (including hydrothermal and hence exhalative) activity.

Hausmannite (MnO·Mn$_2$O$_3$). Common and may occur in great abundance as the principal mineral of an ore deposit; formed by subsurface solution activity, chemical sedimentation, and surface weathering of manganese compounds.

Pyrochroite [Mn(OH)$_2$]. Probably uncommon and not known to be abundant; thought to be generally a product of subsurface solution activity, but may form by weathering and sedimentation of earlier primary material.

Manganite (Mn$_2$O$_3$·H$_2$O). Probably rare; never found in abundance; usually a weathering product and hence may be sedimented; found as a metamorphic mineral at Franklin Furnace, New Jersey. The mineral *groutite* has a similar composition and appears to be a weathering product also.

Pyrolusite (MnO$_2$). One of the most common of the manganese ore minerals and is often abundant; frequently found as a weathering product, but numerous occurrences have been interpreted as resulting from subsurface solution activity. The mineral *ramsdellite* has a similar composition and like pyrolusite is most commonly a weathering product.

Wad (variable composition conforming generally with MnO·MnO$_2$·H$_2$O). The term wad is applied to a dark-brown or black, earthy, impure aggregate of manganese oxide; composition is variable; crystalline structure generally amorphous; the manganese is apparently present in more than one oxidation state, and water is present in variable amounts. Hewitt and Fleischer (1960) suggest that it is probably the first compound formed (1) in the surface oxidation of manganous minerals and (2) by the precipitation of manganese oxide from surface waters. It has considerable powers of cation adsorption.

---

† Most of these minerals may also form by processes other than those of sedimentation, and this is mentioned where appropriate.

Psilomelane [Ba(Mn^{++},Mn^{4+})$_9$O$_{18}$·2H$_2$O].   Common, with cryptomelane, as a weathering product of primary manganese compounds (however, other manganese compounds have often been mistaken for psilomelane, which is therefore not as common as thought earlier); also occurs as a subsurface mineral and is the most common oxide in a group of exhalative veins in Tertiary volcanic rocks in the southwestern United States.

Cryptomelane [K(Mn^{++},Mn^{4+})$_8$O$_{16}$].   Essentially a potassium-bearing mineral of psilomelane type that has probably frequently been identified as psilomelane; very common; often abundant and frequently a secondary weathering product; well known as an exhalative (hot-spring) product and is reported by Hewitt and Fleischer (1960, p. 10), for example, as being the principal mineral of the tufa apron of a hot spring in Saline Valley, California.

Bixbyite [(Mn^{3+},Fe^{3+})$_2$O$_3$].   Also known as sitaparite in Africa, Europe, and India; common in some areas; appears to be largely of subsurface formation.

Braunite [3(Mn^{3+},Fe^{3+})$_2$O$_3$·Mn^{++}SiO$_3$].   Common and sometimes abundant; apparently formed in a number of ways; occurs in exhalative-sedimentary ores as a product of metamorphism of manganiferous sedimentary rocks; Hewitt and Fleischer (1960, p. 11) consider it may have formed by the action of hot ground waters on stratified manganese carbonate in a number of localities in southwestern United States; the mineral is also known to develop abundantly by weathering processes.

Jacobsite [(Mn^{++},Fe^{++})O·(Mn^{3+},Fe^{3+})$_2$O$_3$].   Already referred to as a member of the spinel group in Chap. 3; fairly common but not abundant; probably a product of metamorphism of sedimentary ores in many instances.

Franklinite [(Zn^{++},Fe^{++},Mn^{++})O·(Fe^{3+},Mn^{3+})$_2$O$_3$].   Already referred to as a member of the spinel group in Chap. 3; known only at Franklin and Sterling Hill, New Jersey, where is is thought by some investigators to be a product of regional metamorphism of a Precambrian sedimentary deposit.

In addition to the above there are a large number of more complex minerals containing substantial manganese.   Most of these are double oxides of manganese and copper, zinc, lead, iron, and titanium.   Some contain substantial quantities of other metals.   The mineral *coronadite* [Pb(Mn^{++},Mn^{3+})$_8$O$_{16}$], for example, contains some 20 weight per cent lead, and in some localities the ore as mined yields several per cent lead in addition to manganese.   In addition to these "foreign" metals present in the combined state, the manganese oxides often contain substantial impurities of elements such as lithium, cobalt, nickel, copper, zinc, lead, thallium, vanadium, tungsten, molybdenum, arsenic, and antimony.   These appear to have been acquired by adsorption, presumably followed by the formation of the oxides and hydroxides of the relevant metals.

A final word concerning manganese mineralogy: the oxides—and associated hydroxides—often occur as extremely fine-grained aggregates of several different species, in which state they are notoriously difficult to determine.   Even where grains can be readily distinguished under the microscope, the optical properties of some of the oxides are so similar that they cannot be differentiated one from the other. Under these conditions x-ray diffraction is required for positive identification.

SOURCE, TRANSPORT, AND DEPOSITION
OF MANGANESE IN NATURE

As a result of their chemical similarity manganese and iron behave in almost parallel fashion in the formation and dissolution of rocks of most kinds. As far as the cycle of sedimentation is concerned, they have similar sources, they are transported in analogous compounds, and they are precipitated mainly as the oxides and carbonates. The words *almost parallel* have, however, been chosen carefully; while manganese and iron not infrequently occur together or adjacent in sedimentary deposits, there are many notable cases where one occurs to the almost complete exclusion of the other. It is clear that in such instances some vital difference—even if only of degree—has manifested itself at some critical stage of the weathering-erosion-transport-deposition process and that this has led to almost complete segregation of the two elements. We shall therefore consider first the source, transport, and deposition of manganese and then look for possible ways in which sedimentary processes might separate it from its sister element iron.

SOURCE

Like iron, manganese destined for sedimentary deposition may be derived from continental erosion, from volcanic activity, or from deeper parts of the floor of the body of water in which it is finally to be deposited.

Manganese occurs in small quantity in virtually all igneous rocks—to the extent of some 0.1 to 0.3 per cent expressed as MnO. It is most abundant in basic rocks such as basalts, decreasing with increase in SiO_2 content of the containing rock to rather less than 0.1 per cent in rhyolites and related types. Manganese varies enormously in sediments, from the high concentrations of sedimentary ores down to the very low contents of some sandstones and derived quartzites. Larger amounts of manganese occur in shales and siltstones and in the cherts and jaspers of volcanic marine sequences. Metamorphic rocks exhibit essentially the same abundances, and the Mn/Fe ratio of crustal rocks averages about 1:50.

Manganese compounds are well known as exhalative products in volcanic regions. It has been observed as a sublimate round Vesuvius and Etna, in the Valley of Ten Thousand Smokes of Alaska, and about volcanoes in Japan, the islands of the Southwest Pacific, and those of many other areas. It is also well known, and often abundant, in the aprons of hot springs. Table 14-2 (after Hewitt and Fleischer, 1960, p. 40) gives partial analyses of manganese-bearing hot-spring waters, and of manganese-bearing travertine and tufa formed about such hot springs. In Japan, in the modern volcanic islands of the Southwest Pacific, and in other areas of late Tertiary to Modern volcanism, concentrations of manganese oxides are common in beds of agglomerate and tuff and in small pyroclastic lenses between lava flows. The remarkably high content of manganese in deep-sea sediments of the Pacific Ocean is partly derived from volcanic ash emitted from explosive andesitic volcanoes of that region. The average manganese content of these sediments, expressed as MnO_2 and on a $CaCO_3$-free basis, is about 5.5 per cent. This is about 10 times the figure for the equatorial Atlantic and is probably due to the unusually intense volcanic activity around the western margin of the

Table 14-2 Manganese, iron, SiO_2, and calcium in some hot-spring waters and sinter aprons. Contents given as ppm in waters, weight per cent in sinters. (After Hewitt and Fleischer, 1960, pp. 40–41)

	Mn	Fe	SiO_2	Ca
Ouray, Colorado:				
Active hot-spring water	0.92	0.42	49	376
Sinter	50.0	0.1		
Poncha, Colorado:				
Active hot-spring water	0.10	0.46	84	17
Sinter	30.0	0.5		
Abraham, Utah:				
Active hot-spring water	0.75	0.0	75	352
Sinter	33.24	12.55	2.42	3.93

Pacific. Ash and sublimate have apparently drifted with the wind, finally to settle in the ocean water and to build up high concentrations in deeper areas where other sedimentation has been slow.

This brings us to the third source of manganese—Mn-bearing detritus on the floors of the oceans and lakes themselves. Manganiferous volcanic ash is perhaps the most important class of detritus here, but the more basic igneous (particularly volcanic) terrains generally, and those of earlier Mn-rich sedimentary rock, contribute manganiferous detritus that may become source material for the extraction and chemical redeposition of manganese in the developing sedimentary pile. The manganese is extracted in sea-floor areas of lower Eh-pH, the processes of leaching and removal being essentially analogous to those proposed for iron by Borchert (1960) and illustrated in Fig. 13-4.

TRANSPORT

A large number of manganese compounds are soluble in neutral and acid solutions; oxides, manganous bicarbonate, chlorides, sulfates, and others are all soluble at appropriately low pH values. In alkaline solutions, however, dissolved manganese is readily precipitated as hydroxide, carbonate, silicate, or sulfide, depending on the exact value of Eh and pH and on what anions happen to be present.

Because of its close similarity to iron, manganese is transported by mechanisms generally analogous to those already discussed for iron—with the provisos that (1) manganese oxidizes more slowly and less completely than iron for any given set of conditions and (2) manganese oxides and hydroxides are more soluble than their iron counterparts at any given Eh and pH: manganese will tend to go into solution more readily, and remain in solution longer, than iron. Apart from these differences—which are clearly of degree rather than of kind—manganese follows the same migratory paths as iron. It is carried as the bicarbonate in anaerobic ground waters and probably, to some extent, in surface waters of slightly acid pH. Examples of the latter are streams draining areas of heavy tropical jungle, where water has initially had to percolate through deep mats of decaying humus. In such areas, particularly where the streams traverse basic volcanic terrains, it is common to find that most of the cobbles of river beds are coated with manganese oxide. Apparently the manganese has been carried some distance in solution as bicarbonate or humate and then has been adsorbed on cobble surfaces, there to be oxidized and form a surface film of oxide. Thus in addition to being transported in solution and as fine suspended particles, the manganese may also be moved downstream as coatings on detritus and as an adsorbed component of clay particles.

The transport of manganese in acid volcanic and other hot-spring waters is, of course, no problem.

In summary it may therefore be said that the transport of manganese is similar to that of iron in principle, but because of slight differences in stabilities the movement of manganese is easier, and less problematical, than that of iron.

DEPOSITION

Again as in the case of iron, manganese delivered to a site of sedimentation may accumulate as oxide, hydroxide, carbonate, silicate, or sulfide, depending on the physical-chemical characteristics of the sedimentary and diagenetic environments. That is, there does appear to be some development of "facies of manganese formation" just as there are facies of iron formation. However, the arrangement is less well marked in the case of manganese. There does not, for example, appear to be a recognizable large-scale segregation of the different oxidation states in manganese oxide beds as in the case of hematite and magnetite banded iron formations (though certain oxides, which are not capable of stable coexistence, are segregated in nature and do not occur together as primary associates). The manganese sulfides do not develop to an extent approaching anywhere near the concentrations achieved by pyrite in sulfide iron formations, and this facies could almost be said to be absent from sedimentary manganese concentrations. Carbonates may be well developed, and the best sedimentary zoning found in manganese-bearing beds is that developed by associated oxide and carbonate. In most instances the oxide and carbonate pass from one to the other through a zone of mixed and interdigitating oxide and carbonate. Silicates may be well developed and may occur in either of, or straddle, the oxide and carbonate zones.

SEPARATION OF MANGANESE FROM IRON IN THE SEDIMENTARY CYCLE

We have now seen that the chemical similarities of manganese and iron are such as to suggest—or, indeed, to require—that the two should move and be precipitated either with or close to each other. In a number of instances this does appear to have been the case. In the Cuyuna area of the Lake Superior iron province, manganese occurs abundantly with some of the iron formations and constitutes over 20 per cent of some of the ores of this district. Many other such composite occurrences are known. In other areas, however, iron ores contain no more than a trace of manganese, and many manganese ores contain almost no iron. Clearly in these cases there has been a natural segregation of manganese from iron, and whatever process has induced such segregation has operated with remarkable efficiency.

The nature of this process is unknown, and the establishing of its identity is perhaps the most difficult of the problems associated with manganese sedimentation.

Figure 14-1 shows the stabilities of anhydrous manganese and iron compounds at specified concentrations of total carbonate, total sulfur, and total silica at 25°C and 1 atm pressure. Keeping in mind the limitations of this kind of diagram in portraying systems as they operate in the more complex situations in nature, we may examine the likely comparative behavior of the two metals in weathering and sedimentation.

When the two equilibrium diagrams are examined together it can be seen that the wide Fe_2O_3 stability field covers all the manganese mineral fields down to Mn_3O_4. Thus wherever iron and manganese occur together, hematite (or limonite) should be stable in the presence of any of the manganese oxides down to mangano-site. Therefore—if the diagrams are a reasonable indication of stabilities in the natural environment—it appears virtually impossible to precipitate manganese without precipitating iron and vice versa, if the two are present in solution together. The fact that each *has* precipitated without the other in many instances in nature is thus a very difficult problem.

There appear to be three principal possibilities:

1. Segregation at source One possibility is that manganese, but not iron, has been extracted from the source material. Clearly one possible explanation is that the two were not in fact in solution together—that manganese was much more readily dissolved from the source rock or was present to the exclusion of iron in hot-spring waters so that no iron was being transported to the site of deposition anyway. This idea has been put forward by several authorities, including Gold-schmidt (1954, p. 633). The latter has pointed out that since the ionic radius of Mn^{++} is much larger than that of any other divalent element of the iron family, its ionic potential is relatively low and hence it is more susceptible to leaching by weakly acid solutions than compounds of the other elements of the family. It has also been pointed out that whereas iron in igneous rocks occurs in difficulty soluble oxides as well as in ferromagnesian minerals, manganese occurs almost entirely in the less stable ferromagnesians—a state of affairs leading to the readier liberation and removal of manganese than of iron. The fact that the Mn/Fe ratio in soil waters is commonly higher than in igneous rocks has been taken as evidence of such readier dissolution of manganese. However, such differential leaching seems unlikely to account for more than a few per cent difference in the overall extracta-bility of the two elements, and the incidence of the relatively higher Mn/Fe in soil waters is very likely due to early precipitation of iron rather than to preferential dissolution of manganese.

Krauskopf has carried out some simple empirical (but nonetheless highly informative) experiments to test the proposition that manganese dissolves more readily than iron from igneous rocks. The material used was from Paricutin volcano—a basaltic andesite averaging 56.4 per cent SiO_2, 5.23 per cent iron, and 0.09 per cent manganese and with Fe/Mn = 58. This was subjected to leaching by H_2CO_3, H_2SO_4, HCl, and sea water at temperatures from room temperature to

Table 14-3   Extraction of manganese, iron, and silica from Paricutin basaltic andesite.   (From Krauskopf, 1957, p. 70)

	pH	Temp, °C	Time, days	Amount of lava, g	Vol. of solution, ml	Mn extracted mg	Mn extracted % of total	Fe extracted mg	Fe extracted % of total	Ratio Fe/Mn	SiO_2, mg
H_2CO_3 alone	5.5	23	1	20	40	0.016	0.09	0.41	0.04	26	
H_2SO_4 alone	0	23	1	20	40	0.44	2.4	28	2.7	63	
HCl alone	3.0–5.3	210	8	10	50	0.5	5.5	24	4.6	48	9.3
HCl alone	1.0–2.8	210	4	5	50	0.27	6.0	25	9.6	93	20
Sea water alone	7.2	210	5	10	50	0.2	2.2	5	1.0	25	9.2
Sea water + HCl	0.9–5.0	210	8	7	50	1.2	19	65	17.5	54	16
Sea water + HCl	1.0–4.4	210	4	5	50	0.38	8.5	30	11.5	79	15
Sea water + HCl	1.0–1.2	100	2	30	750	5.9	22	280	18	49	870
HCl + steam + sea salt		300	0.04	64		4.0	7.0	243	7.3	61	
Molten lava + dist'd H_2O	6.7				1000	<0.01		0.04		>4	1
Molten lava + sea water	8.4				1000	<0.01		0.04		>4	2
Molten lava + sea water + HCl	1.3				1000	0.01		0.6		60	2

220°C for periods of from 1 to 8 days.   Among the experiments were ones in which molten basalt was poured into sea water and into sea water with added HCl and in which vapor from boiling HCl was passed through fragmented basalt. Results (after Krauskopf, 1957, p. 70) are given in Table 14-3.   These show that *both* iron and manganese were readily abstracted from the basalt.   While in one or two cases a slightly higher proportion of manganese than of iron was dissolved, the ratios of the two elements in the various solutions was not much different from that of the parent rock.   Extreme values for Fe/Mn in solution were 93 and 25, which compares with 58 in the original rock.   Thus experiments seem to strongly support the theoretical prediction that separation at source—by differential solution—is very unlikely indeed.

A somewhat different "at source" separation has been suggested by Hewitt, who has emphasized the importance of volcanic solutions as sources of sedimentary manganese ore.   Hewitt proposes (1966 and earlier publications) that as igneous hydrothermal solutions rise from their deep source there occurs a progressive stripping of their iron, leading to a dominance of manganese—and perhaps an almost complete absence of iron—by the time such solutions reach the surface as volcanic hot springs.

Hewitt supports his contention with two lines of evidence: analyses of the hot springs themselves (cf. Table 14-2) and the incidence of iron and manganese in ore deposits thought to have been formed at different depths within the crust. He points out that of almost 60 analyzed hot-spring waters from the western United States that contained both iron and manganese, some two-thirds showed more manganese than iron.   Similarly a review of the iron and manganese content of 860 hot-spring waters in Japan by Nishimura (1952) showed a majority containing more manganese than iron, and in fact large quantities of manganese, almost free of iron, were being mined from aprons adjacent to active hot springs on Hokkaido.   Turning to iron-manganese abundances in hydrothermal ores deposited beneath the surface, Hewitt notes that there appears to be a progressive change from high iron–low manganese in the deep deposits to low iron–high manganese in the shallower ones.   He considers that in the deepest and hottest zone—that of pyrometasomatic and hypothermal ore formation—little manganese is deposited (rarely more than 0.1 per cent of the ore concerned) and pure manganese minerals, chiefly silicates, are rare.   However, much iron may be deposited in ores of this group.   In the higher, cooler zone of mesothermal ores, iron-free manganese minerals are still unusual, though manganosiderite, in which Fe/Mn is usually about 3:1, may be conspicuous.   It is only in the shallow, comparatively cool zone of epithermal ores that iron becomes subordinate and large quantities of almost iron-free manganese minerals, particularly rhodochrosite and rhodonite, appear.

Thus while Krauskopf's analysis and experiment suggest that segregation is unlikely to take place by dissolution at an *erosional source*, Hewitt's observations suggest that it does occur, by prior precipitation of iron, in subsurface hydrothermal conduits, leading to heavy enrichment of manganese in a *volcanic hydrothermal source* of sedimentary manganese.   The chemical reactions responsible for such volcanic-hydrothermal separation, however, are unknown.

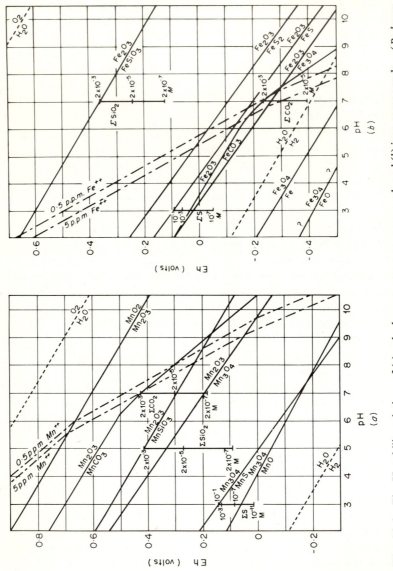

Fig. 14-1 Eh-pH stability relations of (*a*) anhydrous manganese compounds and (*b*) iron compounds. *(Redrawn from Krauskopf, Geochim. Cosmochim. Acta, 1957.)*

2. Segregation by differential precipitation: elimination of iron by prior oxidation

An alternative explanation is that since iron compounds are usually more readily oxidized than their manganese counterparts, natural solutions containing both iron and manganese might, under appropriately balanced conditions, lose iron by rapid oxidation and precipitation, leaving manganese alone in solution to be eventually deposited, free of iron, in some other area of sedimentation. Differential precipitation of this kind might be induced by purely inorganic means or, perhaps, by biological agencies.

The possibility of purely inorganic separation may be considered theoretically by examination, once again, of Fig. 14-1. It is apparent here that for a given pH, iron compounds precipitate at lower Eh than their manganese counterparts or, alternatively, that for a given Eh the relevant iron compound precipitates at a lower pH than does that of manganese. As pointed out by Krauskopf (1957, p. 71) the fact that, for example, $Fe(OH)_3$ is much less soluble than MnO_2 means that the gradual addition of base to a solution containing ions of both metals leads to the virtual quantitative precipitation of iron before the precipitation of manganese begins. [$Mn(OH)_3$ appears to be metastable under conditions such as these; the manganese is precipitated directly as MnO_2 or a hydrate of this.] Thus if a solution containing, say, 10^{-4} mole/liter of both metals is kept at an Eh of 0.2 volt and is made increasingly alkaline, iron begins precipitating at pH 6.1. Manganese, on the other hand, does not begin to precipitate until the pH is much higher—about 9.5—by which time the iron concentration has dropped to 10^{-15} M. Similarly Krauskopf has calculated that if the pH is held at 6.0 and the Eh raised, iron begins to precipitate at a potential difference of $+0.2$ volt, but manganese does not precipitate until this reaches $+0.6$ volt. At this point the iron concentration has again dropped to 10^{-15} M. Thus we may say that for any value of Eh there is a pH range of three units in which iron will be virtually quantitatively precipitated while more than 1.0 ppm of manganese remains in solution in equilibrium with MnO_2 or Mn_2O_3. Thus a limited increase in pH in some natural situation may lead to the selective elimination of iron from iron-manganese solutions, the manganese then continuing on until an even more alkaline environment is encountered. It may then precipitate as oxide, carbonate, or silicate (or, in rare cases, the sulfide), depending on the Eh and the concentration of the anions concerned.

Similarly, biological agencies may cause or accentuate segregation of the two elements. In Chap. 13 it was noted that certain bacteria utilized the energy of reactions involving iron. Similarly it is known that other species are capable of utilizing either iron or manganese reactions but favor those of manganese. It is well known that manganese may be precipitated preferentially in water pipes even where the water carries as much or more iron than manganese. This led Zapffe (1931) to investigate the possible role of bacteria in the precipitation of manganese, and he noted two species (*Crenothrix polyspora* and *Leptothrix ochracea*) that were able to selectively precipitate manganese oxides (1931, p. 828). The manganese oxide coatings of pebbles and boulders in streams in tropical volcanic areas already referred to are almost certainly partly due to oxidation of stream-water manganese by algae. Larger plants may also fix manganese oxides; Ljunggren has observed the precipitation of MnO_2 around the rhizoids of the small moss *Marsupella aquatica* that grows on boulders and pebbles in some of the watercourses of western Sweden (1955, p. 34). That biological agencies may very effectively segregate manganese from iron there is therefore no doubt; whether, however, it can do this on a large scale and over a long period of time in one place is uncertain.

Although the foregoing indicates that manganese might conceivably be separated by inorganic and/or biological depositional processes, there is one piece of evidence weighing heavily against extensive or frequent separation of this kind. This is the absence of notable quantities of iron from the sediments in which the manganese ores have been deposited. We have seen that the iron/manganese ratio in most crustal rocks is between about 20:1 and 80:1, with an average of about 50:1. Thus whatever the source, depositional segregation of the two elements should lead to the development of iron concentrations, or a notably high background level of iron, associated with the manganese. Since iron would be expected to precipitate prior to the manganese, its concentrations or heightened background levels would be expected on the source side of the manganese deposits, probably not too far away and certainly within the same sedimentary unit. There is no question that in some instances—such as Cuyuna—notable quantities of iron *are* associated with the manganese, but in many cases they are not, which suggests strongly that in at least many cases the segregation of manganese from iron is unlikely to have been induced by depositional processes.

3. Segregation by diagenetic processes: selective movement and concentration of manganese by pore-space waters of low Eh

Ideas based on diagenetic segregation have received great impetus from the work of the noted Russian investigator N. M. Strakhov, who has looked to areas of modern sedimentation and manganese concentration to find ways in which iron and manganese may be separated in nature.

In Strakhov's view, few if any manganese and manganese-iron concentrations in sedimentary rocks are strictly speaking sedimentary; rather they are "sedimentary-diagenetic." The history of the ores necessarily begins with sedimentation. In lakes, gulfs, and nearly landlocked seas, iron and manganese are initially concentrated to a minor degree in the deeper, distal parts, due simply to the fineness of their oxide suspensions. In the open oceans initial minor concentrations of the two likewise develop in areas distant from shore, in this case due to the near absence of detrital sedimentation from these parts. Once settled, the oxide particles —particularly in lakes and small landlocked areas of the sea—are incorporated in the accumulating sediments. With a moderately high organic content, oxygen is eliminated and the Eh may be lowered to the point where first the Mn^{4+} and Mn^{3+} and then the Fe^{3+} are reduced. Both metals then go into solution and move upward and laterally, as indicated in Fig. 14-2, movement continuing until such time as the Eh is raised and the two are halted by oxidation. Since iron will always be the last to be reduced, and hence rendered mobile, and the first to be oxidized, and hence rendered immobile, manganese tends to creep ahead all the time, leading to a progressive segregation of the two elements. While such a process is

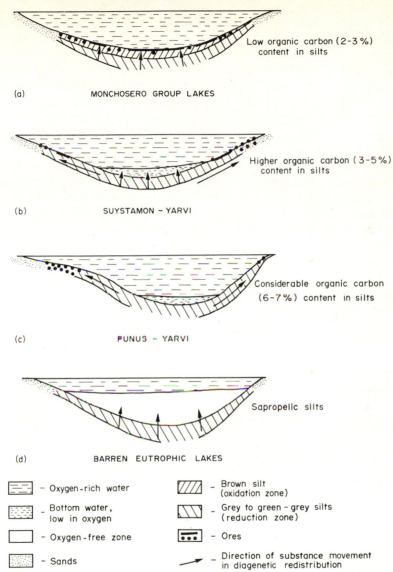

Fig. 14-2 Diagenetic movement and concentration of manganese in lakes; due to the slightly different stabilities of iron and manganese, the manganese tends to migrate a little farther than the iron, leading to *partial* diagenetic segregation of the two elements. (*Adapted from Strakhov, Int. Geol. Rev., 1966.*)

unlikely to completely separate large masses of manganese from large masses of iron, it does lead to partial segregation and to quite substantial increases in the Mn/Fe ratio. In the open oceans this greater geochemical mobility of manganese leads to the development of "manganese nodules" in which Fe/Mn is commonly of the order of 1:1 to 1:2—a very substantial change from our "crustal average" of 50:1.

KINDS OF SEDIMENTARY MANGANESE DEPOSITS

As in the case of iron, manganese has been found to accumulate in both fresh-water and marine milieus. Once again it is the marine occurrences that seem to be by far the most important but that continue to be the least understood. The fresh-water types—the lake and marsh ores of manganese—are recognizably forming at the present day, and largely due to the efforts of the Russian school, a great deal is known about their processes of formation. The marine ores of manganese seem to be more closely related to estuarine and island-dotted shelf environments than do those of iron, though as we have seen some of the important *ironstone* occurrences appear to have been closely tied to environments of this kind. Considerable investigation has been carried out in modern shelf, landlocked sea, and related areas and some of the processes of manganese deposition are being recognized, but much clearly remains to be done. While it is not too difficult to explain *slight* concentrations of manganese and the *tendency* for segregation of manganese from iron, the reasons for the development of large masses of almost pure sedimentary manganese and, in many cases, the almost complete elimination of iron must be acknowledged as remaining an enigma.

Because of its close chemical similarity to iron, analogies can be seen between most sedimentary concentrations of manganese and any one of our three great groups of sedimentary iron ores. Manganese is well known to concentrate as oxide and carbonate in lakes and marshes of tundra regions just as in the case of iron, and indeed some of these deposits are compound concentrations of the two elements. Similarly there are marine deposits of manganese-rich sediments that can be seen—in their form, composition, geological environment, and probably, modes of formation—to be analogs of ironstones or banded iron formations, and indeed, once again, such deposits are sometimes conspicuously compound concentrations of the two elements. Although we are now going to group manganese deposits in a rather different way from the grouping used for iron in Chap. 13, the existence and the validity of these analogies should be kept very much in mind; while it is convenient to group manganese deposits into a larger number of types than were assigned to iron, it will be shown that in most cases such types are analogous to some part of one or another of our three classes of sedimentary iron concentrations. The reason for this more detailed division in the case of manganese lies in the fact that, since manganese is more valuable than iron per unit of weight, much smaller and poorer concentrations enter the category of commercial "ore." Interest in such smaller, but more numerous, deposits opens the way for recognition of "subtypes," and it is largely this that has led to the present grouping of sedimentary manganese ores.

MARSH AND LAKE DEPOSITS

Although these are known to be represented among Tertiary and older sedimentary sequences, most occurrences are Modern and indeed many are known to be in the process of formation at the present day.   As already noted they are unquestionably analogs of marsh and lake iron ores and in many cases are compound; iron figures prominently in virtually all investigations of current fresh-water manganese accumulations.

DISTRIBUTION, FORM, AND SETTING

As with iron, the principal deposits of this class are found in the glaciated tundra areas of the Northern Hemisphere; their main areas of occurrence are therefore Russia, Scandinavia-Europe, and Canada.

Deposits of this type in Canada are numerous but small and of no great importance commercially.   Hanson (1956, pp. 10–11) refers to two principal localities.   In the area of Kaslo, British Columbia, a bog manganese bed covers an area of about 8 acres and lies on glacial drift; it varies in thickness up to about 3 feet, except locally near mineral springs where small mounds occur.   The largest known bog deposit in Canada appears to be that of Dawson Settlement, New Brunswick, where a bed up to 15 feet thick occurs over about 10 acres.   The ore is a mixture of manganese oxides and peat and apparently formed by precipitation from springs debouching at the surface.   The springs contributed iron as well as manganese, the iron tending to precipitate first, leading to an increase in the ratio of Mn/Fe with increased distance from the orifice.   The Finnish deposits are good examples of occurrence in Scandinavia; Vaasjoki observes that "in many lakes, especially those of Central Finland, the presence of iron ores is a common phenomenon; and often these ores contain variable amounts of manganese . . ." (1956, p. 53). These deposits form sheets up to about 20 cm thick on the lake floors and are apparently best developed where the overlying water depth is between 1 and 3 meters.   Similar postglacial marsh and lake deposits are common in Norway and Sweden, and large iron-manganese deposits of this kind were discoverd in the late 1940s in the district of Lake Tisjon in Dalecarlia, Sweden.   The abundant lake ores of Russia have been the subject of much of the best modern work on deposits of this type, and these particularly are referred to in the following pages.

CONSTITUTION

There is relatively little mineralogical and chemical information on marsh and lake ores, due doubtless to their commercial insignificance.   It appears that the variety of manganese minerals deposited in lakes is quite small.   Much of the manganese seems to be in the form of largely amorphous hydrous oxides.   Some early Pliocene deposits of this type in Nevada and Baja California have been found to be still in a substantially uncrystallized state and to have a specific gravity of between 1.3 and 1.6.   In most occurrences such oxides form crusts, cakes, and small oolitic and nodular masses.   With onset of crystallization, which probably follows burial, mineral species such as cryptomelane, psilomelane, and pyrolusite form. Full chemical analyses of bog ores are not readily available, if they exist at all, and there is little information even on their manganese and iron.   Mean Mn is generally in the range of 5 to 20 weight per cent and mean Fe in the range 25 to 40 weight per cent.

ORIGIN

The materials of these ores are derived precisely as are those of bog iron deposits —by solution and movement of Mn^{++} and Fe^{++} as bicarbonates and soluble humates through marshy soils of low pH and Eh. When contributed to oxygenated lakes, the relevant oxides are formed and these then drift and settle as fine precipitates on the lake floor. Where decaying organic matter accumulates in the deeper portions of the lake, reducing conditions develop and manganese and iron may go back into solution. Rendered mobile in this way, they move until oxidizing conditions are encountered again and they are reprecipitated. Movement of this kind naturally takes place from the deeper toward the shallower portions of the lake floor, and with its greater mobility manganese tends to move just a little more readily than the iron, leading to partial segregation of the two (Fig. 14-2). In this way iron-bearing manganese ores, or manganiferous iron ores, may develop in the medium-depth and littoral portions of the lake.

Manganese may also concentrate—by itself but usually with iron—as a subsurface layer along the water table in marshy areas and as layers and pockets where bicarbonate solutions of marsh milieus interact with aerated artesian or subartesian water, particularly in limestone regions. The mechanisms of deposition are analogous to those already described for bog iron ores.

MODERN MARINE DEPOSITS

Manganese is well known to be concentrating on the sea floor at the present time. Within the broad category of marine deposits, there appear to be three broad environmental types, each of which, however, may grade into the others:

1. Accumulation in seas fed only by terrigenous material, i.e., soluble and detrital derivatives of continental, as distinct from primary volcanic, land masses
2. Accumulations in large marine basins receiving both terrigenous and volcanic material, i.e., soluble and detrital derivatives of both continental and active volcanic land masses
3. Accumulations in both confined seas and open oceans in areas of sea-floor hydrothermal activity

At the limits of its range, type 2 clearly merges into types 1 and 3.

DISTRIBUTION, FORM, AND SETTING

Occurrences of the first environmental type are known to be currently developing in a number of arctic and subarctic areas, notably the Arctic Ocean itself and the Baltic, White, Barents, and Kara Seas (Fig. 14-3). There is no active volcanism within or adjacent to any of these bodies of water, and hence all are characterized as receivers of terrigenous material only. The distribution of manganese in the sediments of the Arctic Ocean is shown in Fig. 14-4. Its general level of abundance is clearly very low.

All other oceans have at least some volcanic activity associated with them,

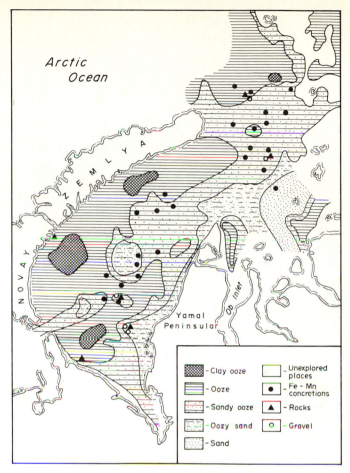

Fig. 14-3 Manganese occurrences and concentration zonation in a confined arctic sea—the Kara Sea. (*Adapted from Strakhov, Int. Geol. Rev., 1966, after Gorshkova, 1957.*)

and in this the Pacific is outstanding. Whereas the Atlantic-Indian Ocean province has had 97 known volcanoes active since A.D. 1500, the Pacific is estimated to have had some 350. Coupled with this the Pacific has a very low ratio of watershed-ocean area so that the terrigenous component of sedimentation is small compared with that of other oceans. Thus in both absolute and relative terms the Pacific is the greatest receiver of volcanic materials, and this appears to be reflected in the high manganese content of its sediments. This is clearly an environment of type 2, and the distribution of manganese in it is shown in Fig. 14-5.

The Pacific is also known to possess manganese accumulations of our third type, i.e., deposits formed in association with sea-floor hydrothermal activity. Oceanographic dredgings carried out over the East Pacific Rise by members of the

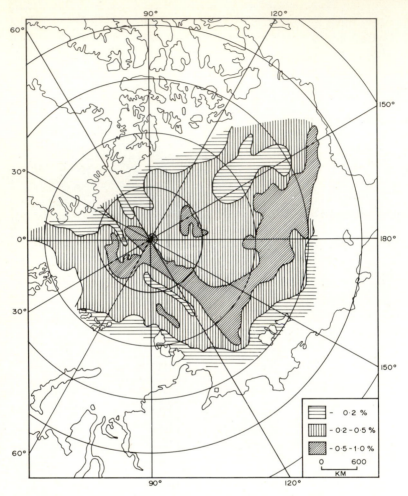

Fig. 14-4 Manganese occurrence and concentration zonation in the Arctic Ocean. (*Adapted from Strakhov, Int. Geol. Rev., 1966, after Belov and Lapina, 1961.*)

Scripps Institute (Bostrom and Peterson, 1966, p. 1258) showed the presence of abundant manganese clearly related to a zone of high sea-floor thermal gradient. The positions of the traverses, and the relation of iron-manganese abundance to sea-floor heat flow and to the topography of the Rise, are shown in Fig. 14-6.

In their form the three environmental types are fairly similar: the manganese occurs—virtually entirely as oxide-hydroxide—as coatings and impregnations of bottom fragments and outcrops, as crusts and slabs, and as sheets of concretions and nodules. Nodules are by far the most important and were first found to be widely distributed in the major oceans by the *Challenger* Expedition in 1873. Subsequent oceanographic expeditions have found them in virtually every dredging of the Pacific floor, and it has been estimated that tens of millions of square miles of the Pacific Ocean—not to mention the other oceans, which also possess nodules —are covered with them.   Nodules generally vary in size between 1.0 and 30.0 cm in largest dimension, with a mean of the order of 3.0 cm.   As reported by Mero in 1962 (p. 749) the largest nodule found to that time weighed 850 kg.   Shapes are variable; most are single spherical bodies but many are compounded of numerous spheres, shape probably depending largely on those of the nucleus or nuclei.

The Arctic Ocean manganese is associated with clayey and sandy oozes. The color of the sediments is yellow to brown, indicating a low organic content and a lack of intensive reduction.   In some of the confined seas of the arctic region, organic matter is more plentiful, the sediments are darker, and there is evidence of the diagenetic movement of iron and manganese already referred to in connection with lake ores.   The Pacific manganese is associated with a range of sediments. Figure 14-5 clearly incorporates manganese accumulation of both of our types 2 and 3; the area over the East Pacific Rise is, as we have noted, at least partly of a direct hydrothermal nature, and there are doubtless others.   Of the three

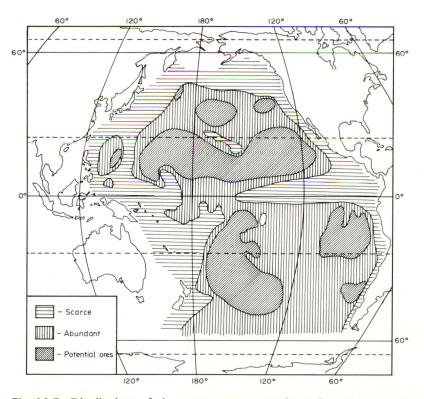

Fig. 14-5 Distribution of iron-manganese concretions (i.e., "manganese nodules") and a qualitative estimate of their abundance on the floor of the Pacific Ocean. (*Adapted from Strakhov, Int. Geol. Rev., 1966, after Skornyakova and Andruschenko, 1946.*)

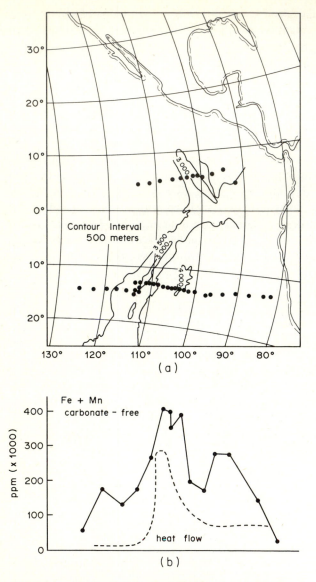

Fig. 14-6 (*a*) Map of the East Pacific, showing traverses and positions of bottom samplings and readings, and (*b*) diagram showing covariation of heat flow and (Fe + Mn), carbonate free, for the southern traverse. (*Redrawn from Bostrom and Petersen, Econ. Geol., 1966.*)

levels of manganese abundance indicated in Fig. 14-5, the first is essentially periph-eral and is characterized by terrigenous, terrigenous-diatomaceous, and carbonate sediments and is low in total manganese, which occurs as films on detritus rather than as concretions; the second is characterized by red clays and radiolarian and, occasionally, globigerina oozes and contains higher total manganese that occurs largely as concretions; the third is characterized by deep-water red clays and radio-larian oozes and a high level of manganese that is present almost entirely as con-cretions.

CONSTITUTION

Although a very small amount of carbonate has been detected in some of the deep sediments of the Baltic, virtually all modern marine manganese is in the form of oxide-hydroxide.   The oxide nodules and slabs are brown to black, with an earthy appearance, and are usually somewhat friable.   The general range of density is 2 to 3, indicating quite high porosity.   Internally the structure of the nodules is concretionary or colloform, and banding of different widths and delicacy stands out by variation in reflectivity, porosity, or impurity.   Variation in reflectivity results, of course, from variation in the manganese minerals concerned or variation of mixtures of them.   In all cases iron oxides are present, and in some they are abun-dant, such impurity again varying from band to band.   A generalized view of the abundances of the principal constituents of marine nodules is given in Table 14-4.

In addition to the manganese and manganese-iron compounds, there are normally impurities such as clay, barite and very fine quartz, feldspar, apatite, rutile, and so on.   A general picture of the overall chemical composition of Pa-cific and Atlantic nodules, for which sampling is most reliable, is given in Table 14-5. Like the lake and marsh ores, marine manganese also characteristically carries substantial traces, particularly of cobalt, nickel, and copper.   In general it appears that these, like the manganese itself, are less abundant in the shallow-water terrig-enous deposits than they are in pelagic ones.   Abundances of traces are also given in Table 14-5.   According to Strakhov (1966, p. 1188) there appear to be some genuine differences in the amounts and nature of traces from ocean to ocean, no

Table 14-4   Maximum, minimum, and average weight percentages of the major components of ocean-floor manganese nodules.   (From Mero, 1965, p. 179)

Material	Weight percentages†		
	Maximum	Minimum	Average
MnO_2	63.2	11.4	31.7
Fe_2O_3	42.0	6.5	24.3
SiO_2	29.1	6.0	19.2
Al_2O_3	14.2	0.6	3.8
$CaCO_3$	7.0	2.2	4.1
$CaSO_4$	1.3	0.3	0.8
$Ca_3(PO_4)_2$	1.4	Traces	0.3
$MgCO_3$	5.1	0.1	2.7
H_2O	24.8	8.7	13.0
Insoluble in HCl	38.9	16.1	26.8

† On a total weight of air-dried-sample basis.

Table 14-5 Maximum, minimum, and average weight percentages of 27 elements in manganese nodules from the Pacific and Atlantic Oceans. (From Mero, 1965, p. 180)

| | Weight percentages (dry-weight basis)† | | | | | |
| | Pacific Ocean, 54 samples | | | Atlantic Ocean, 4 samples | | |
Element	Maximum	Minimum	Average	Maximum	Minimum	Average
B	0.06	0.007	0.029	0.05	0.009	0.03
Na	4.7	1.5	2.6	3.5	1.4	2.3
Mg	2.4	1.0	1.7	2.4	1.4	1.7
Al	6.9	0.8	2.9	5.8	1.4	3.1
Si	20.1	1.3	9.4	19.6	2.8	11.0
K	3.1	0.3	0.8	0.8	0.6	0.7
Ca	4.4	0.8	1.9	3.4	1.5	2.7
Sc	0.003	0.001	0.001	0.003	0.002	0.002
Ti	1.7	0.11	0.67	1.3	0.3	0.8
V	0.11	0.021	0.054	0.11	0.02	0.07
Cr	0.007	0.001	0.001	0.003	0.001	0.002
Mn	41.1	8.2	24.2	21.5	12.0	16.3
Fe	26.6	2.4	14.0	25.9	9.1	17.5
Co	2.3	0.014	0.35	0.68	0.06	0.31
Ni	2.0	0.16	0.99	0.54	0.31	0.42
Cu	1.6	0.028	0.53	0.41	0.05	0.20
Zn	0.08	0.04	0.047			
Ga	0.003	0.0002	0.001			
Sr	0.16	0.024	0.081	0.14	0.04	0.09
Y	0.045	0.016	0.033	0.024	0.008	0.018
Zr	0.12	0.009	0.063	0.064	0.044	0.054
Mo	0.15	0.01	0.052	0.056	0.013	0.035
Ag	0.0006		0.0003‡			
Ba	0.64	0.08	0.18	0.36	0.10	0.17
La	0.024	0.009	0.016			
Yb	0.0066	0.0013	0.0031	0.007	0.002	0.004
Pb	0.36	0.02	0.09	0.14	0.08	0.10
L.O.I.¶	39.0	15.5	25.8	30.0	17.5	23.8

† As determined by x-ray emission spectrography.
‡ Average of 5 samples in which Ag was detected.
¶ L.O.I. = Loss on ignition at 1100°F for 1 hour. The L.O.I. figures are based on a total weight of air-dried sample basis.

doubt representing derivational differences between the environmental types already referred to.

ORIGIN

The derivation of the manganese of modern marine deposits seems perfectly clear. In environments of the first type (purely terrigenous) the manganese is contributed by continental erosion; in those of the second type it comes partly from the continents and partly from volcanic detritus and solutions; and in the third type it is contributed virtually exclusively from sea-floor volcanic hot springs.

That derived from the continents is delivered in very low concentrations, and when, as in the case of the Arctic Ocean, it is accompanied by abundant terrigenous detritus, its level in the sediments is initially very low.   If—again as in the case of the Arctic Ocean—there is little organic accumulation in the muds, the manganese and iron remain oxidized and hence immobile.   Because of this lack of movement there is no opportunity for enrichment or segregation of the two elements.   Accumulation of organic matter in the bottoms of some of the marginal arctic seas does, however, lead to some migration, concentration, and segregation in these. Reflecting this, the highest manganese abundances found in Arctic Ocean muds is about 1.0 per cent, in contrast to the Kara Sea sediments, which contain up to 9.0 per cent manganese.

Manganese concentration on the continental shelves and slopes of other oceans follows essentially the same principles as those operating on the floors of lakes and of the Kara and other marginal arctic seas; enrichment takes place by diagenetic movement induced by lowered Eh-pH in areas of organic accumulation. The reasons for concentration in the deeper regions, however, are rather different. Taking the Pacific again as our example, we have already noted of this ocean that its overall rate of sedimentation is very low, that it receives a relatively large volcanic component, and that it can be divided into three broad provinces (Fig. 14-5). According to Strakhov (1966, p. 1191) the deep, pelagic areas are in preferential receipt of the very finest fractions of terrigenous material—material of which manganese and iron colloids and suspensoids form a large component.   The deeper reaches therefore receive a very large share of total manganese contributed to the oceans.   However, since the deep sea floor receives negligible organic matter, concentration of the manganese and iron, once these reach the bottom, is not induced by organic reduction as in the littoral zones and shallow seas.   Rather does movement occur by very slow diagenetic redistribution, the slightly higher mobility of the manganese leading to its enrichment over iron in the nodules.   The concentration of the two elements, and their associated traces, is then greatly emphasized by the extremely low rate of detrital sedimentation in the pelagic areas.   Such scarcity of detritus also assists in the building up of high concentrations in areas of deep-sea hydrothermal activity, as in the case of the East Pacific Rise (Fig. 14-6).

MANGANESE DEPOSITS OF THE ORTHOQUARTZITE-GLAUCONITE-CLAY ASSOCIATION

We now turn from modern deposits, whose formational processes may be observed in progress, to those of ancient sediments—older, consolidated deposits whose derivation and concentration are matters of conjecture.   In some cases such older deposits are, in spite of some uncertainties, fairly well understood.   Others remain highly problematical, and among these are the concentrations of the orthoquartzite-glauconite-clay association.

Deposits of this group constitute the largest individual "ore" concentrations known, and in bulk they constitute some 70 per cent of the world's total known

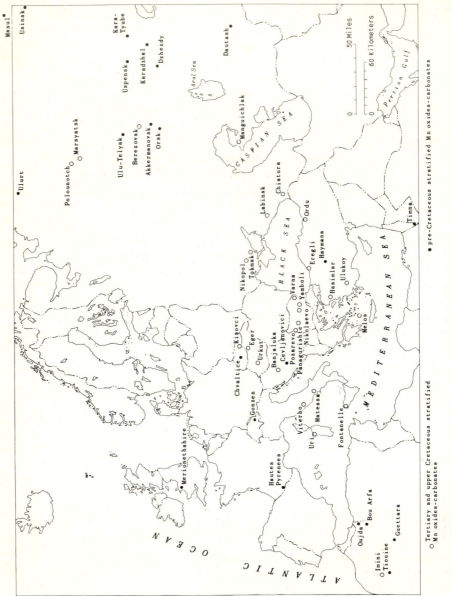

Fig. 14-7 Distribution of principal manganese deposits in Russia, the Middle East and eastern Europe, and North Africa. *(From Hewitt, Econ. Geol., 1966.)*

reserves of manganese.   Varentsov (1964, p. 57) estimates their manganese content
to be approximately 794 million metric tons.

DISTRIBUTION, FORM, AND SETTING

The principal area of occurrence—and overwhelmingly the most important one—
is southern Russia, along the northern hinterland of the Black Sea.   The principal
localities here are Nikopol', Bol'she-Tokmaksk, Labinsk, and Chiatura.   The
Nikopol' and Chiatura deposits are the largest of all known manganese ore bodies.
Deposits of similar kind, though of lesser importance, occur elsewhere in Russia,
particularly at Mangyshlak, Polunochnoe, and Marsyatskov (see Fig. 14-7).   All
are of Tertiary age.   Others of similar nature and association are known in the
Timna region of southern Israel, Turkey, Bulgaria, and elsewhere along the northern side of the Mediterranean, again in Tertiary strata.   A very large deposit,
probably of the present association, is also known in Lower Cretaceous beds of
Groote Eylandt, northern Australia.

   The form of the deposits is quite simple: where they are not complicated by
folding they have a thin lens shape conformable with the bedding of the enclosing
strata.   Where folding has occurred, the shape is, of course, modified, though relations with the enclosing beds remain essentially the same.   An apparent discordancy may arise where a deposit (e.g., Nikopol') consists of a number of small
lens-shaped beds; if these have formed in some definite relation to a shoreline that
is, for example, receding, the deposit *as a whole* will appear to be shallowly discordant.   In such cases the disposition of the individual manganiferous beds is
concordant—it is only the outline of the extremities of all the small lenses that is
discordant.

   In most cases deposits are made up of a number of contiguous beds within
a small interval of the sequence of sedimentary rocks concerned.   For example,
in the case of the Chiatura deposit the sequence begins with about 100 feet of
coarse sandstone and conglomerate; this is then overlain by the "ore member," a
unit varying from 4 to 25 feet in thickness and containing at any one place three to
18 distinct beds of manganese ore.   Such beds range in thickness from about 1
inch to about 3 feet and alternate with thin layers of sandstone, clay, and glauconite.

   The setting of the deposits is fairly simple and characteristic.   All appear to
occur on a stable basement—either a rigid Precambrian or Paleozoic platform or a
stable intermontane region of a geosynclinal area.   Of the two, the rigid platform
setting is the more common.   The ore body may be in direct contact with the
basement or may be separated from it by considerable thicknesses of sediment.

   Paleogeographically the deposits are shallow-water marine to estuarine.   Of
the Nikopol' occurrence, Varentsov notes: "The sediments of the zone of deposits
formed in an island-dotted sea.   Discoveries in the ore-bearing sediments of numerous remains of crustaceans, squamate fishes, sharks' teeth, pelecypod flaps and
sea urchins indicate that the part of the basin dotted with islands was in fairly
shallow water" (1965, p. 28).

   The sedimentary rocks associated with deposits of the Nikopol' type are sandstones, siltstones, clays, glauconitic beds, marls, limestones, conglomerates, carbonaceous shales, and in some cases, coaly beds. Figure 14-8 illustrates the lithological and structural setting of the Nikopol' deposits. The disposition of the old land areas, and the nature and disposition of the sediment types, indicates that the manganese-bearing beds accumulated near shore in shallow island-dotted areas and in littoral parts of marine basins and lagoons. This seems to be typical for ores of this group.

CONSTITUTION

Although oxide ores are by far the most abundant, carbonate and mixed oxide-carbonate ores are often present and are important in some instances. In general the occurrence and disposition of ore type appears to reflect facies of chemical sedimentation, just as in the case of the iron-rich sediments, and indeed facies of

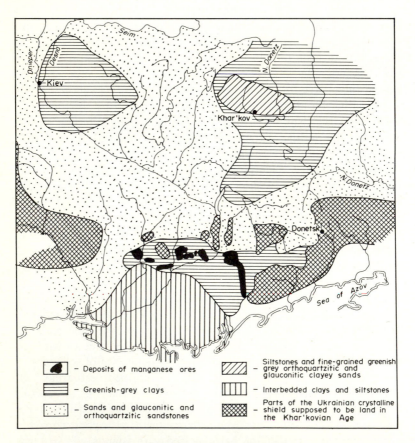

Fig. 14-8 Lithological and part paleogeographical map illustrating the setting of the Nikopol' manganese deposits in Tertiary rocks just north of the Black Sea. (*Adapted from Varentsov, Econ. Geol. U.S.S.R., 1965.*)

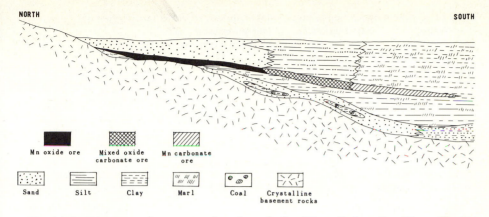

NORTH   SOUTH

| Mn oxide ore | Mixed oxide carbonate ore | Mn carbonate ore |
| Sand | Silt | Clay | Marl | Coal | Crystalline basement rocks |

Fig. 14-9 Diagrammatic cross section through the Nikopol' deposit. (*From Hewitt, Econ. Geol., 1966, after Varentsov, 1964.*)

iron and of manganese are interleaved in some deposits. The general nature of a typical manganese facies pattern—that of Nikopol'—is shown in Fig. 14-9. The ore consists of irregular concretions, nodules, and rounded earthy masses of manganese oxide and/or carbonate enclosed in a silty or clayey matrix, and according to Varentsov (1964, p. 7) it falls into three zones. The first, or oxide zone, appears to have been deposited nearest the strand of the original marine basin and is composed of pyrolusite and psilomelane with associated marls, clays, and sands. Proceeding further from the old shoreline and diverging from the basement, carbonate appears and a mixed oxide-carbonate zone develops. This is composed mainly of psilomelane, manganite, manganocalcite, and rhodochrosite. Further again from the old strand the oxide eventually disappears and the ore minerals are entirely carbonate—calcian rhodochrosite and manganocalcite, with substantial glauconitic and siliceous impurity. The carbonate of this third zone then thins out and gives way to deeper water green-blue clays containing occasional small nodules and concretions of manganocalcite. Similarly the Chiatura deposit consists of a number of manganiferous beds arranged in an oxide–oxide-carbonate–carbonate facies pattern, the ore minerals in this case being associated principally with glauconitic clay. The ore beds give way to sandstones (often containing sponge spicules and sharks' teeth), glauconitic sands and clays, and other shallow-water rock types.

The chemical constitution of ores of the orthoquartzite-glauconite-clay association as illustrated by those of Nikopol' and Chiatura is indicated in Table 14-6. Examples of oxide and carbonate ores are given for each occurrence. It is clear that the manganese content of these ores is far higher and the iron content far lower than in the Modern concentrations. Varentsov (1964, p. 14) estimates that in ores of Nikopol' type, the oxide ores generally show $Mn/Fe = 8$ to 10, commonly ranging up to 15 to 20 or, as in Table 14-6, even higher. This is in sharp contrast to lake ores ($Mn/Fe = 0.1$ to 2.5 approximately) and the oceanic manganese

Table 14-6   Chemical composition (partial analyses) of principal ore types of the Nikopol' and Chiatura deposits.   (After Varentsov, 1964)

	(1)	(2)	(3)	(4)	(5)	(6)
SiO_2	19.68	17.41	8.20	8.42	6.41	7.31
Al_2O_3	7.95	1.29	4.62	0.80	4.54	1.47
TiO_2		0.31				
FeO		Fe = 1.07		Fe = 0.1	1.5	Fe = 1.2
Fe_2O_3	0.64		1.92		0.72	
MgO	0.30	0.37	1.73	2.00	1.52	
CaO	0.90	0.81	2.26	0.32	13.46	22.65
MnO	0.05	0.85	30.55	29.30	32.59	30.72
Mn_2O_3	68.58	74.15	35.10	37.60		
Na_2O					0.74	
K_2O						
P_2O_5			0.46		0.79	
SO_3	0.15		0.12		0.12	
CO_2		1.54		0.15	35.72	33.91
H_2O	0.82		1.94	8.41	0.44	1.71
Ig. loss	0.18		11.88			
Total	99.25	98.33	98.78	87.10	98.20	99.27
Mn/Fe	83	44	47	473	19	19

Explanation
(1). Pyrolusite-psilomelane ore, Nikopol'.
(2). Pyrolusite-psilomelane ore, Chiatura.
(3). Manganite ore, Nikopol'.
(4). Manganite ore, Chiatura.
(5). Carbonate ore, Nikopol'.
(6). Carbonate ore, Chiatura.

nodules (Mn/Fe ranging from less than 1.0 to about 3.0).   In short the concentration of manganese and the separation of manganese from iron are very much greater in the older ores than in the modern ones.

ORIGIN

The lack of abundant associated volcanic rocks with the major examples of this type of occurrence seems to preclude a volcanic derivation, at least in the sense that the manganese was derived from volcanic activity going on at the same time as the deposition of the manganese-bearing sediments.   In this connection, however, Hewitt (1966, p. 445) has pointed to the occurrence of small amounts of andesitic tuff in sandstone overlying the Chiatura manganese member and of tuffaceous material forming the matrices of some of the manganese oxide nodules. However, the Russian investigators have not stressed this, and the prevailing opinion among them seems quite clearly to be that the southern Ukrainian ores were derived by weathering and erosion from the Precambrian crystalline shield of that region.   The rocks concerned are Proterozoic (1400 to 1750 $\times$ 10^6 years old) and include the Krivoi Rog formation, which is well known to contain abundant banded iron formations and ferruginous quartzites and schists. Beneath this formation is a thick succession of spilitic rocks and of sediments of substantially volcanic derivation. All these units could have yielded substantial manganese by erosion. Detailed paleobotanical work has shown that manganese deposition coincided with a marked change of climate in the region—from humid subtropical to a drier cold-temperate one—and it may be that this resulted in a changed pattern in the weathering-out and transport of manganese. Certainly the coincidence of manganese ore formation with a climatic change seems to suggest an erosional source.

The mechanism of deposition of these beds so remarkably rich in manganese is by no means clear. The abundance of glauconite supports the paleobotanical evidence of a cool climate, if the ideas of Porrenga (1967) mentioned in Chap. 13 are correct. The low detrital content of the beds themselves suggests that erosion was not vigorous at the time. It is noteworthy that although the amount of iron in the sediments is low, the *patterns* of iron and manganese distribution, where investigated, are similar. This, together with the remarkably localized high concentration of manganese, may suggest the operation of two "influences":

1. A broad factor or set of factors—the development, over quite a large area, of Eh-pH conditions suitable for the sedimentation of manganese and iron compounds.
2. A local factor or set of factors—restricted areas in which the deposition of manganese and iron was favored to the point where the sedimented manganese began to catalyze—and greatly accelerate—further manganese oxide and carbonate formation. While such catalysis might have been substantially inorganic, the localized intense activity of bacteria could well have made an important contribution. Species such as the *Crenothrix polyspora* and *Leptothrix ochracea* noted by Zappfe, given the onset of a suitable climate, the availability of abundant decaying shallow water fauna, and suitable Eh, pH, and salinity, might have developed explosively, carrying out rapid and selective precipitation of manganese compounds.

The zoning of the ores would, as with iron, reflect a progressive small change in Eh going toward deeper water—i.e., a facies arrangement. The pisolitic and nodular form of the ores is presumably partly a result of depositional processes and partly one of diagenetic, and later, migration and accretion.

MANGANESE DEPOSITS OF THE LIMESTONE-DOLOMITE ASSOCIATION

Deposits of the orthoquartzite-glauconite-clay association are by definition of conspicuously detrital association. The group of manganese ores that we are to consider now has a more chemical—or biochemical—affinity; the ores are conspicuously related to marine carbonate sedimentation and are found in a variety of limestone-dolomite settings.   Varentsov (1964) divides deposits of this type into two principal categories:

1. Manganiferous limestone-dolomite formations developed on stable basement platforms ("Moroccan type").
2. Manganiferous limestone-dolomite formations of geosynclinal zones: an "Appalachian type" formed in miogeosynclines and a "Usinsk type" formed in eugeosynclines.

While these types exhibit certain differences—chiefly in their geological setting—such differences are largely of degree rather than of kind.   We may therefore consider them together, simply noting divergences where these appear.

DISTRIBUTION, FORM, AND SETTING

Like the orthoquartzite-glauconite-clay group, deposits of the limestone-dolomite association are not really widespread; although they occur quite widely as isolated small concentrations, their really significant occurrences are sharply localized. Indeed they are almost entirely confined to North Africa, in particular Morocco. The deposits here range in age from Carboniferous to Upper Cretaceous.   Deposits of this kind are known also in the Appalachian zone of North America, in eastern Australia, on some of the islands of the Southwest Pacific, and in China.   Generally speaking these are small and of geological rather than industrial interest.   The Usinsk deposit of the Lower Cambrian of southwestern Siberia is an exception; the manganiferous zone here is over 1000 feet thick and some 15,000 feet long and is hence an important source of manganese.

In their form the deposits vary from small lenses to quite extensive and continuous beds of manganese-rich sediment.   Usually an individual deposit takes the form of a series of lenses of varying size occurring close together in a "manganese member,"i.e., a manganese-bearing unit of the sedimentary sequence concerned. In some cases the manganiferous lenses may occur within an old reef structure, in which case the shapes of the individual ore bodies are often quite irregular.

The main features of the geological settings of the principal types are

1. *Stable platform* ("*Moroccan*") *type*.   A general view of this type of occurrence is given in Fig. 14-10, a schematic east-west section through the Imini deposit of southern Morocco.   According to Varentsov (1964) several features are characteristic.   Such deposits occur, usually discordantly, on the eroded surfaces of stable platforms and adjacent to old ridges and uplifted blocks that formed land areas at the time of manganiferous and associated sedimentation.   The stratigraphic succession is usually divisible into three distinct sections: (1) a lower, usually red, terrigenous member; (2) the "ore member," usually composed of manganese oxide ores with dolomites, limestones, subor-

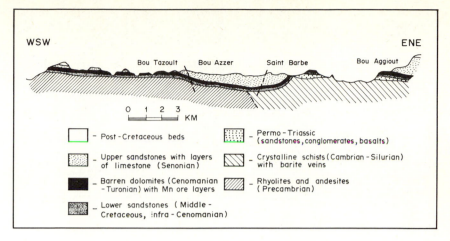

Fig. 14-10 Schematic section through the Imini deposit in southern Morocco, illustrating the general nature of the geological setting of Varentsov's "stable platform," or "Moroccan," type of manganese concentration. (*Redrawn from Varentsov, "Sedimentary Manganese Ores," Elsevier Publishing Company, Amsterdam, 1964, from Bouladon and Jourovsky, 1952, with the publisher's permission.*)

dinate red clays, and sometimes gypsum; (3) an upper terrigenous member, which is also usually red. Accumulations of this type are thus:

. . . littoral developments in platform basins. The red, terrigenous sediments that form the top and bottom of the formation play an important part in their makeup, and beds of manganese oxide ores are intimately associated with successions of dolomites which commonly include intercalations of gypsum and anhydrite. These formations are characterized by a definite trend of lateral variation: towards the shore their place is taken by red, coarsely clastic rocks, and towards the centre of the basin they pass into rather homogeneous limestone-dolomite sequences (Varentsov, 1964, p. 46).

2. *Geosynclinal type.* Varentsov considers that these may be divided into miogeosynclinal and eugeosynclinal groupings. The former involves no more than a slight lifting of the "background" amount of manganese distributed through a limestone unit and does not merit further consideration here. The eugeosynclinal group includes, and is typified by, the Usinsk deposit, which is illustrated diagrammatically in Fig. 14-11. According to Varentsov the sediments associated with this kind of manganiferous limestone-dolomite formation are characteristically complex and discontinuous. In the case of the Usinsk deposit, the carbonate succession lies concordantly on a series of volcanic rocks—chiefly andesitic—and derived sediments. The lower part of the carbonate sequence consists of dolomites, cherty dolomites, and carbonate sedimentary breccia. This is overlain by the "ore member," which consists of manganiferous, bituminous limestones intercalated with black, also bituminous, pyritic slates. This unit becomes increasingly dark and shaly toward the center of the original basin and much lighter and

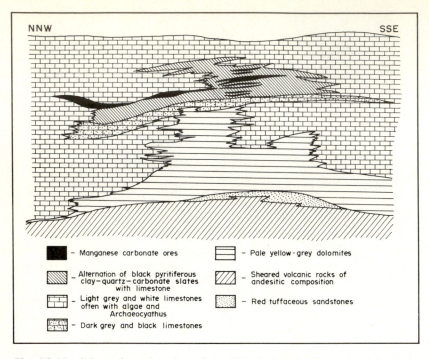

NNW   SSE

- Manganese carbonate ores   - Pale yellow-grey dolomites

- Alternation of black pyritiferous   - Sheared volcanic rocks of
  clay—quartz—carbonate slates   andesitic composition
  with limestone

- Light grey and white limestones   - Red tuffaceous sandstones
  often with algae and
  Archaeocyathus

- Dark grey and black limestones

Fig. 14-11 Schematic section through the Usinsk deposit, illustrating the general nature of the geological setting of Varentsov's "eugeosynclinal" type of manganese concentration. (*Redrawn from Varentsov, "Sedimentary Manganese Ores," Elsevier Publishing Company, Amsterdam, 1964, with the publisher's permission.*)

obviously algal toward the old shore. The whole is overlain by the third member—a light-colored algal limestone into which the ore member grades conformably.

The fact that this succession lies on and eventually gives way laterally to volcanic units, coupled with the variability and discontinuity of the associated sediments, indicates quite clearly that deposition of the manganiferous sediments took place in a eugeosynclinal environment.

CONSTITUTION

Deposits of the Moroccan type are composed almost entirely of oxides. While a little carbonate may occur, this is greatly subordinate. In general the ores have very high manganese contents and are low in iron, aluminum, and phosphorus ($Fe_2O_3 < 2$ per cent; $Al_2O_3 < 2.5$ per cent; $P_2O_5 < 0.1$ per cent). Thus again the ores have no more than a very small detrital component, and there has been a substantial natural segregation of manganese from iron. Heavy metal impurities, on the other hand, are characteristic and often quite large; barium expressed as BaO ranges up to about 7.0 per cent, lead as PbO to about 6.5 per cent, copper as CuO to about 0.4 per cent, and zinc as ZnO to about 0.1 per cent.

There is little to be said concerning Varentsov's miogeosynclinal type.   His example, the Shady Dolomite Member of the Lower Cambrian succession at Bumpass Cove, Tennessee, contains little more than heightened background contents of manganese: 45 analyses of diamond drill core gave a range of approximately 0.1 to 0.9 per cent, with a mean of 0.38 per cent MnO.

The Usinsk ores consist of manganese-bearing carbonates together with a variety of products of a reduced environment.   The principal ore minerals are calcian and ferroan rhodochrosites, which occur as ooliths and very fine laminae intercalated with manganoan stilpnomelane.   Carbon is ubiquitous and often in the form of anthracite-graphite.   Pyrite and pyrrhotite may exceed 8.0 per cent of the rock.   Algal remains and sponge spicules are common, the latter often replaced by manganiferous carbonate.   Chemically the Usinsk ores are rather different from those of Moroccan type.   Manganese is still high and Al_2O_3 low, but iron oxide may exceed 15.0 per cent and P_2O_5 ranges up to about 0.5 per cent. Barium, lead, copper, and zinc do not appear in notable amounts.

ORIGIN

An absence of associated volcanic rocks seems to be one of the distinctive features of the Moroccan platform type of manganiferous limestone-dolomite occurrence. The two other principal features are the clear association with terrigenous red beds and, in some cases, with beds of gypsum.   It therefore appears that the manganese here is a product of weathering and erosion in an arid climate.   In fairness it should, however, be noted that some geologists suspect that the ores may have some volcanic affiliation, though the evidence for this is apparently not strong. Again the depositional process is not clear.   The low to negligible detrital component, coupled with the very high Mn/Fe ratio, suggests a highly selective biological-biochemical mechanism operating in some clear, undisturbed shallow-water environment.

The very low-grade and dispersed manganese of the miogeosynclinal type of occurrence also seems most reasonably ascribed to a weathering-erosional source. Deposition in this case is probably one of simple incorporation in the sedimentary carbonates during or shortly after deposition.

The eugeosynclinal type is, however, not nearly so easily accounted for, and there are clearly two possibilities as far as source is concerned: contemporaneous volcanism or weathering and erosion.   The presence of volcanic rocks under, and laterally adjacent to, the Usinsk carbonate formations makes a direct volcanic source seem likely.   However, the Russian geologists, who have obviously been in the best position to study these ores, favor derivation by later erosion of these volcanic rocks and their associates.   The problem is still unresolved.   The relatively high iron content of these ores (Mn/Fe ranges from 2:1 to 10:1, approximately) coupled with their occurrence as carbonate and silicate associated with black bituminous limestones and pyritic carbonaceous shales suggests fairly direct debouchment of manganese-iron-bearing solutions into a reducing bottom environment.

MANGANESE DEPOSITS OF VOLCANIC AFFILIATION

As a petrological—as distinct from economic—group, this is by far the most important class of manganese concentration. Indeed there are some who suspect that virtually all manganese deposits in premodern rocks have a volcanic affiliation; Hewitt, for example, in emphasizing the difficulties involved in the segregation of manganese from iron by surface processes has, as we have already noted, suggested that the very high concentrations of manganese—and high Mn/Fe ratios—in many deposits may be most satisfactorily explained by derivation from hot springs contributing manganese almost exclusively. He points out that many occurrences —such as those of Morocco, Chiatura, and Nikopol'—have minor volcanic matter associated with them and that this may have more genetic significance than has been generally thought. While acknowledging the possible correctness of Hewitt's ideas, this section is concerned only with those deposits of clear volcanic association and those for which a close and direct volcanic affiliation seems beyond doubt.

DISTRIBUTION, FORM, AND SETTING

Of all manganese concentrations these are by far the most widespread. They are abundant on all continents and are an obvious feature of innumerable andesitic volcanic islands. They appear in volcanic-sedimentary successions of a wide range of ages, from Precambrian to Recent. At least some of the manganiferous sediments of the Lake Superior iron province appear to be of this type, and those associated with the Cuyuna Range iron ores are, for example, known to contain up to 20 per cent manganese. Such concentrations are widespread in the lava and pyroclastic sequences of Paleozoic geosynclines in western North America, the Urals, eastern Australia, and elsewhere. They appear again in great abundance in the Tertiary volcanic areas of Japan, Indonesia, and the West Indies, and they are common in Tertiary, Pleistocene, and Recent volcanic sequences in all the larger islands of the Southwest Pacific. Not only are deposits of this group widespread, but when they appear they do so in abundance. In many cases they occur literally in dozens, scattered here and there in part or the whole of the volcanic-sedimentary sequence concerned. In this they are in sharp contrast to most other types of manganese concentrations, which, as we have seen, are fairly narrowly restricted in space and time and are usually represented by relatively few deposits.

However, in spite of their ubiquity and abundance, deposits of this group are rarely of economic size. While the total amount of manganese represented by them must be enormous, it is dispersed among the multitude of small occurrences and vast volumes of material that can only be called *manganiferous sediment*. The manganese concentrations are in all cases simply lenses of sedimentary rock of particularly high manganese content forming part of a sequence characterized by an anomalously high manganese content throughout. The ore is usually bedded, and the high-grade manganese dies out—gradually or abruptly—in an essentially concordant fashion. The lenses may occur singly or as groups and often show a marked preference for a limited part or parts of the manganiferous unit concerned.

The most detailed treatment of volcanic-sedimentary manganese deposits as a class is that of N. S. Shatskiy (1964, p. 1030), who divides them into two main groups: (1) "volcanogenic formations of the greenstone type" and (2) "volcanogenic formations of porphyritic type." Of these two the "greenstone" group is by far the most important. The volcanic rock types concerned here are basalts, spilites, keratophyres, and quite commonly, andesites and andesitic tuffs. Acidic types are not common and the association is essentially a basic to a basic intermediate one.

Within the greenstone group Shatskiy recognizes three lithological-environmental groupings: (1) the *jasperoid formation*, (2) the *siliceous-shaly formation*, and (3) the *remote siliceous-shaly formation*. Of these the jasperoid formation is by far the most important. Indeed we may compound Shatskiy's nomenclature and say that the whole volcanic class of manganese deposits is dominated by those of greenstone-jasperoid association. In environments of this kind the jaspers themselves are conspicuously manganiferous almost throughout, and where concentration rises, pods and layers of "ore" result. The association is really just a part of the wider and very well known one of the "Steinmann Trinity"—pillow lavas (often spilitic), radiolarian cherts (including jaspers and white and greenish cherts), and serpentine belts. Large quantities of andesitic volcanic matter—usually disguised as graywacke and associated shale—are a common part of the sequence, and there may also be stratiform pyritic bodies containing sulfides of copper, zinc, lead, and other minor metals.

The other volcanic associations are much less important. One important manganese province that Shatskiy (1964, p. 1041) suggests as partly of porphyritic-siliceous-shaly association is that of Central Kazakhstan in Russia. Here a series of deposits occurs along a single, clearly delineated stratigraphic horizon at the base of the Carboniferous succession. The associated rocks are siliceous, tuffaceous shales and quartz porphyritic lavas.

CONSTITUTION

Oxide ores are greatly dominant in the volcanic group of deposits as a whole, though there are occasional occurrences of carbonate and silicate. The ores of the jasperoid association are *ipso facto* members of a highly oxidized assemblage; the jaspers themselves may be regarded as a variety of highly siliceous iron formation of hematite facies, and so manganese deposits of this environment are clearly related to the most highly oxidized facies of iron formations.

The deposit that seems to be the best described example of mineralogical zoning in a manganese concentration also happens to be one of confused association. This is the Karadzhal deposit of Central Kazakhstan. The manganese here is associated with abundant limestones, scattered red beds, igneous rocks of basalt-andesite-trachyte type (principally as porphyritic dike rocks), minor jasper, and abundant iron formation. Some of the containing sediments contain abundant glass shards, indicating contemporaneous volcanic activity. Shatskiy (1964, p. 1041) regards Karadzhal as being part of a "volcanogenic formation of porphyritic type," but this seems by no means certain. The zoning is clearly a facies arrangement and is interleaved with a facies pattern of iron formation; proceeding from the original shore the sediment change accompanying increasing water depth appears to have been silts, shales → sandy limestones → limestones, with silica and hematite impurity → jasper → magnetite, iron silicate, siderite → hematite → manganese oxides → manganese and iron carbonates. (In the light of chemical equilibrium relationships in low-temperature aqueous systems and the normal disposition of facies in iron formations, the interpolation of a substantial hematite zone between those of magnetite-chlorite-siderite and manganese oxide-carbonate seems anomalous. The reason for its appearance here, however, is not known.) The disposition of the zones is indicated in Fig. 14-12 (after Maksimov, 1960, fig. 3).

The principal iron and manganese minerals at Karadzhal are hematite, magnetite, siderite with associated calcite, thuringite, braunite, hausmannite, and jacobsite. Minor sulfides, associated with small carbonaceous intercalations and the carbonate ores, include pyrite, arsenopyrite, chalcopyrite, galena, sphalerite, and tetrahedrite. Quartz and chalcedony are virtually ubiquitous, finely admixed with the ore minerals and as the major constituents of jasper bands and siliceous intercalations. Barite and calcite occur as lenses and minor secondary veinlets.

The abundant manganese deposits of jasper association on the Japanese islands are good "young" examples of the greenstone-jasper association, though they are perhaps most accurately regarded as manganiferous iron ores. They are associated with basaltic lava—often pillowed—and minor basaltic tuff, together with jasper. The latter consists of fine opal and chalcedony and radiolarian skeletons, all dusted with fine hematite. With increase in manganese oxides and hematite, this grades into ore. Minor amounts of chlorites and carbonate and trace amounts of pyrite, chalcopyrite, galena, and sphalerite are common. Chemically the ores commonly range from 25 to 35 per cent Fe and 5 to 10 per cent Mn; calcium is present up to about 10 per cent and phosphorus to about 1.0 per cent. Ore from the Kokuriki (A) and Nikura (B) deposits of Hokkaido have chemical compositions as follows (after Suzuki and Ohmachi, 1956; K. Mon, analyst):

	A	B
SiO_2	15.90	11.80
Al_2O_3	5.44	2.64
FeO	0.12	0.22
Fe_2O_3	53.64	53.26
MgO	1.66	1.08
CaO	3.68	6.24
MnO	13.71	18.30
S	0.09	0.03
P	0.67	0.77
H_2O+	3.41	3.44
H_2O-	1.28	1.25
	99.60	99.03

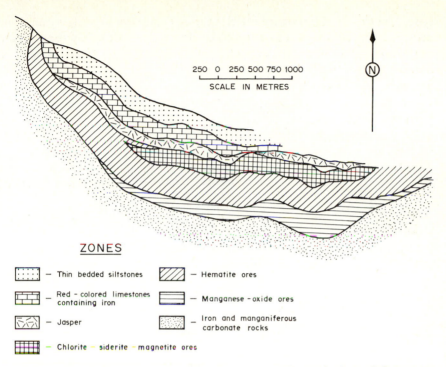

250 0 250 500 750 1000

SCALE IN METRES

ZONES

— Thin bedded siltstones		— Hematite ores
— Red - colored limestones containing iron		— Manganese - oxide ores
— Jasper		— Iron and manganiferous carbonate rocks
— Chlorite – siderite – magnetite ores		

Fig. 14-12 Schematic diagram of facies pattern in the ore horizon of the western Karadzhal manganese deposits. (*Redrawn from Maksimov, Int. Geol. Rev., 1960, after Dyugayev.*)

ORIGIN

Manganese concentrations of volcanic environments form one of the few groups of ore deposits whose origin is no longer really controversial. There seems no question that they are dominantly "exhalative-sedimentary"; that is, the manganese has been contributed by volcanic hot springs discharging directly on to the sea floor or about the coasts of areas of seaboard volcanism. A minor amount is probably derived from sea-water leaching of submarine lavas and tuffs and through the weathering of these rocks where they are exposed on the land surface. Hot springs, however, are almost certainly the major source.

Concentration is, once again, probably the result of several factors. It is notable that with the volcanic ores—in contrast to those of the orthoquartzite-glauconite-clay and limestone-dolomite associations—iron is almost always present in substantial quantity, either admixed with the manganese or as immediately adjacent concentrations (cf. Karadzhal). The lack of segregation here may well result from the contribution of the two elements together—via hot-spring solutions—and rapid coprecipitation. If the oxides were rapidly precipitated near the source orifices, there would be no opportunity for separation during surface transport and little chance for sedimentary or diagenetic processes—either inorganic or bacteriological—to induce segregation. In those cases where ores of volcanic association show a clear concentration of manganese with respect to iron, Hewitt's suggestion (1966, p. 258)—that the iron may have been selectively subtracted from the hot-spring waters at some depth, leaving the solutions relatively rich in manganese by the time they debouched at the surface—may hold.

RECOMMENDED READING

Hewitt, D. F.: Stratified deposits of the oxides and carbonates of manganese, *Econ. Geol.*, vol. 61, pp. 431–461, 1966.

Krauskopf, K. B.: Separation of manganese from iron in sedimentary processes. *Geochim. Cosmochim. Acta*, vol. 12, pp. 61–84, 1957.

Mero, J. C.: "The Mineral Resources of the Sea," Elsevier Publishing Company, Amsterdam, 1965.

Shatskiy, N. S.: On manganiferous formations and the metallogeny of manganese, Paper I. Volcanogenic-sedimentary manganiferous formations, *Intern. Geol. Rev.*, vol. 6, pp. 1030–1056, 1966.

Varentsov, I. M.: "Sedimentary Manganese Ores," Elsevier Publishing Company, Amsterdam, 1964.

15
Stratiform Sulfides of Marine and Marine-Volcanic Association

Although much still has to be learned concerning the source, transport, and accumulation of many iron and manganese deposits of sedimentary affiliation, there is little doubt that they are themselves sedimentary rocks.   While there may also be doubt concerning the relative roles of sedimentation and diagenesis in determining the final details of composition and structure of these deposits, opinion again seems virtually unanimous that most of their compositional and other characteristics developed during sedimentation and that any modification of these primary features was completed prior to lithification.

Unfortunately there is no such unanimity concerning the deposits we are now to consider.   As with the concentrations of iron and manganese, all occur in sedimentary rocks and almost all are clearly stratiform and conformable with the sequences that enclose them.   However, there is almost no aspect of their origin that is not controversial.   Indeed it is deposits of this group that have been chiefly concerned in the more recent disputes outlined in Chap. 2.   In the opinion of some, the sulfides are late introductions and hence partial pseudomorphs of the beds in which they occur; in the opinion of others they are, quite simply, sediments themselves.   Since deposits of this group are one of the principal sources of nonferrous metals available to man, the resolution of their origin and habit of occurrence are

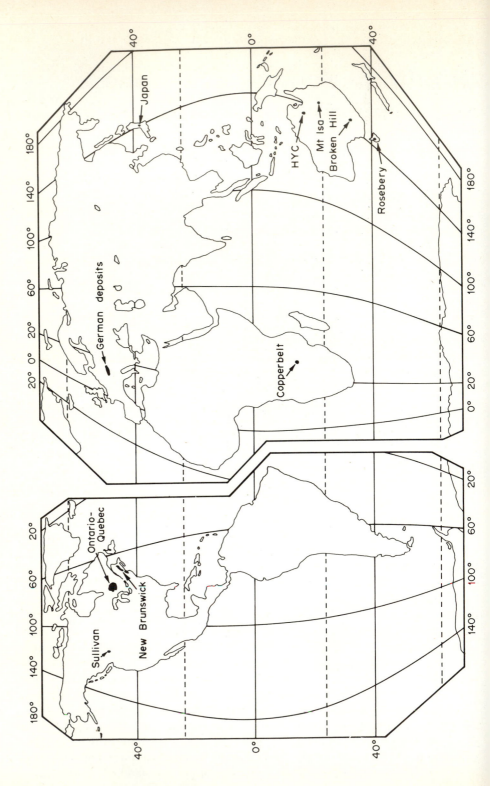

clearly matters of utmost importance.   It is for this reason that they have become one of the central problems of mid-twentieth century ore-genesis theory.

Before examining them in detail it is necessary to qualify just a little the group of ores we are concerned with.   We have defined them as *of marine and marine-volcanic association*, and indeed most of them occur in marine sequences of conspicuously volcanic affiliation.   The class does, however, embrace what is perhaps best described as a spectrum of occurrence.   It includes a number of deposits enclosed in sedimentary rocks that are clearly marine but that are not known to contain volcanic rocks—or at least no rocks *so far recognized* as volcanic.   A few stratiform sulfide deposits are conspicuously associated with volcanic rocks that, however, show no clear evidence of having been deposited on the floor of the sea. Between these "end members" there is a whole range of deposits—by far the major proportion of the class—enclosed in rocks showing, with greater or lesser emphasis, the marks of combined marine-volcanic environments.

DISTRIBUTION

Sulfide concentrations of this type have a very wide distribution both in space and time.   Indeed it is probably true to say that they are the most widely distributed of all ores.   They are most abundant in the Precambrian and are a conspicuous feature of virtually every shield area of the world.   They are also very important in a number of Paleozoic eugeosynclinal belts and are found—particularly in the West and Southwest Pacific areas—in Mesozoic, Tertiary, and sub-Recent terrains.

The positions of some of the more important occurrences are shown in Fig. 15-1.   Unfortunately there is little information available for the Asian continent, but the examples indicated show how widespread the class is.   The following list, while by no means exhaustive, includes the principal occurrences and districts.

1. *Precambrian.   Canada:* Kidd Creek, Timmins, and numerous others in the Superior province of Quebec, Ontario, and Manitoba; Sullivan, British Columbia.   *United States:* Balmat-Edwards and pyritic deposits of the Gouverneur district, New York; *Australia:* Broken Hill, Mount Isa, MacArthur River. *Africa:* the Roan and other ore bodies of Rhodesia and Katanga.
2. *Paleozoic.   Canada:* Bathurst deposits, New Brunswick, and related deposits in Newfoundland and Nova Scotia.   *United States:* continuation of the Canadian Appalachian belt in Maine, New Hampshire, Vermont, New York; deposits in Oregon and Washington.   *Australia:* Rosebery, Captain's Flat, Cobar, and other deposits of the eastern Australian Paleozoic terrain. *Ireland:* Avoca and Tynagh.   *Europe:* numerous deposits in the Norwegian Caledonides; Rammelsberg and the Mansfeld-Kupferschiefer in Germany; Rio Tinto in Spain.   *Russia:* the Urals province.   *Iran:* the Zardu and other deposits of the Bafq district.
3. *Mesozoic and Tertiary.*   Deposits on islands of the West and Southwest Pacific, notably Japan, Fiji, and New Guinea.

FORM

Deposits of this group show a wide variety of forms, apparently depending on the disposition of the particular sedimentary unit in which they occur and, particularly, on the nature of the deformation these sediments may have undergone.

Some occurrences are clearly limited by features of the facies pattern developed in the containing sediments, and their shape is thus limited in the same way as are particular facies of iron and manganese formations (cf. Fig. 14-12). This seems to be the case particularly with the different ore bodies of Mount Isa and of the Roan copper province. Facies constraint of this type usually leads to the development of a thin lens shape elongated parallel to the original facies boundaries— i.e., the depositional strike of the containing sediments. Where the deposit is not influenced in this way, shape is normally that of a thin lens, the two larger dimensions being of the same order of size and lying parallel to the associated stratification. The difference in size between the smaller and the two larger dimensions is most commonly of about an order of magnitude; i.e., if the body has a thickness of some tens of feet its larger dimensions are usually of the order of hundreds of feet; with a thickness of 100 feet the larger dimensions are usually of the order of thousands of feet. In some cases parts of the edge of a lens may be frayed due to interdigitation of sulfide-rich and sulfide-poor sedimentary rock. The deposits of Tynagh in Ireland show this feature well. In other cases, where sulfide lenses are closely tied to a particular facies that has migrated laterally up the sequence concerned, discontinuous or semicontinuous *en echelon* arrangements may be developed, as was noted in the case of some manganese concentrations. An individual "deposit" may be made up of one, two, three, or not infrequently, many lenses occurring close together. Depending on distances between them and the distinctiveness of their features, such lenses may or may not be regarded as separate entities for geological and mining purposes.

A simple lens shape is preserved only where the enclosing rocks have undergone little or no deformation. It follows from their conformability that where the enclosing rocks are folded, the outlines of the sulfide concentrations exhibit fold form too.

The MacArthur River deposit of northern Australia occurs in Precambrian sedimentary rocks that have been barely touched by deformation, and in consequence it lies in an almost horizontal position. The young Japanese occurrences likewise conform to the virtually undisturbed initial dips of the enclosing pyroclastic formations. The Sullivan ore body of British Columbia occurs entirely within a limb of a broad fold in the enclosing Precambrian Aldridge sedimentary rocks and as a result retains its lens form, but with a dip of some 30°. Although the rocks containing the Mount Isa ore bodies have suffered negligible metamorphism, they have been quite intensely folded and as a result the sulfide masses— which, like the Sullivan, occur on a single limb of a large fold—are tilted, in this case to about 80°. The Appalachian deposits of New Brunswick have a variety of shapes, in conformity with the structures developed in the containing rocks. Some, such as the Heath Steele deposits, are virtually simple lenses. One—the Anaconda Caribou—has the simple, broadly curved shape of a gentle fold. Two of the larger masses—the BM & S No. 6 and BM & S No. 12—are involved in tight drag folds.

The copper-iron sulfide concentrations of the Roan province illustrate very well the effects of varying intensity of structure on shape. At the southeastern end of the area (Fig. 15-2) the prevailing synclinal structure is fairly open, and consequently the shape of the "ore horizon" is simple (Fig. 15-3a). At the western end, however, the synclinal structure is tighter, and there has been intense deformation about protuberances in the granite that had earlier formed the basement of Roan sedimentation. In such places (Fig. 15-3b) the ore-bearing sediments are heavily deformed and conspicuously drag-folded, and the outlines of the ore bodies, in conforming with the enclosing structures, are correspondingly complex.

A deposit famous for its complexity as well as its size is that of Broken Hill in New South Wales. The containing sedimentary rocks here have been intensely deformed and metamorphosed to sillimanite grade. Not only is the folding in the environs of the sulfide masses isoclinal, but embroidered on the steep major folds are innumerable minor structural complications. The shape of the deposit reflects this complexity. The "ore body" is in fact made up of six separate major ore zones and a number of smaller ones, and a simplified presentation of its structure is shown in Fig. 15-4.

A conspicuous feature of these stratiform sulfide concentrations as a class is their progressive development of minor discordancy as their structural complexity increases. Those occurring in unfolded or only very slightly folded strata show a

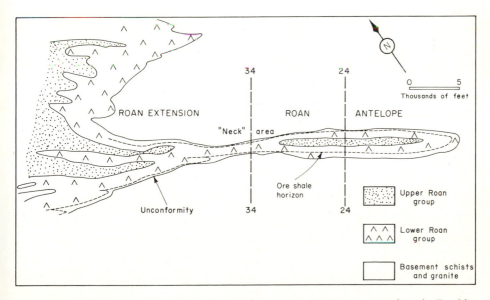

Fig. 15-2 Highly simplified geological map of the Roan Antelope copper deposit, Zambia, showing the positions of the two sections of Fig. 15-3. (*Adapted and much simplified from Davis, Econ. Geol., 1954.*)

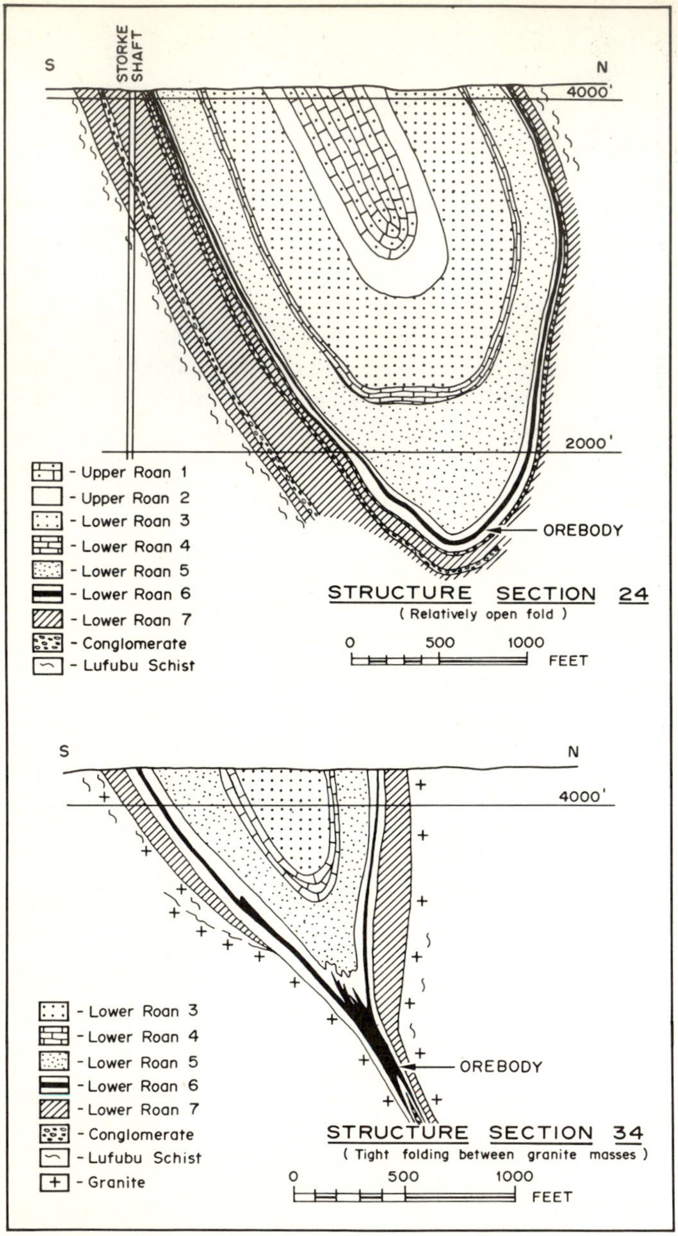

Fig. 15-3 Structure sections 24 and 34 (cf. Fig. 15-2 for positions of these sections in plan) through the Roan Antelope deposit, showing the effect of tightness of fold structure on the shape of the ore body. RL and RU in the legend refer to sedimentary units of the Lower and Upper Roan groups, respectively. [*From Mendelsohn (ed.), "The Geology of the Northern Rhodesian Copper Belt," Roan Antelope Copper Mines Limited and MacDonald & Co. (Publishers) Ltd., London, 1961, with the publisher's permission.*]

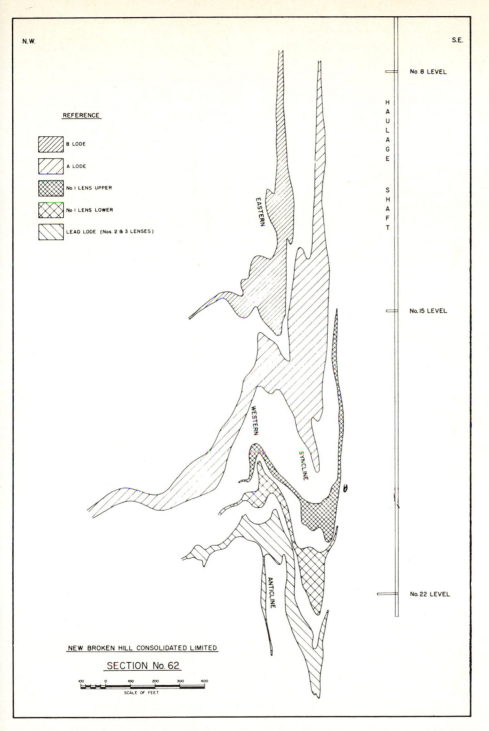

Fig. 15-4 Simplified section through the Broken Hill "ore body" at New Broken Hill Consolidated Limited mine section No. 62, showing the multiple lensoidal nature and tightness of folding of the deposits. (*From Carruthers and Pratten, Econ. Geol., 1961.*)

high degree of concordancy. The ore itself is usually very fine-grained and stratified in apparently perfect parallelism with the associated sedimentary silicates and
carbonates. Such stratification is often very well developed, individual beds
ranging from a foot or so in thickness down to the most delicate fine laminations,
commonly of the order of only a few hundredths of an inch thick. This stratification may reveal itself through variations in the relative amounts of sulfide and nonsulfide or through variation in the amounts of the different sulfides themselves.
Individual laminae of an inch or so in thickness may be continuously traceable for
hundreds, or even one or two thousand feet. In these cases discordancies are
absent or only on the smallest scale—perhaps a few cross-cutting veinlets an inch or
so long and a fraction of an inch thick.

As soon as folding and deformation of the containing rocks become notable,
discordancies—though always minor—become more obvious and frequent. The
Rammelsberg ore bodies, in the Middle Devonian Goslar Slates of Germany, are
concordant with the slates overall but show a number of discordant veinlets and
protuberances along their margins. It was these features particularly that led
Lindgren and Irving, in 1911, to consider the Rammelsberg sulfides to be late introductions rather than sediments. At Mount Isa the lead-zinc ores, though beautifully concordant for most of their extent, are in some places involved in zones of
intense contortion, and here numerous small cross-cutting veinlets—particularly
of galena—appear.

A large proportion of the Paleozoic ore bodies of the present type are found in
well-cleaved volcanic-sedimentary rocks metamorphosed to greenschist grade.
Good examples are found among the deposits of eastern Australia and the northern
Appalachians. In such cases an original stratification is partly preserved, but
some of the sulfide clearly conforms to the cleavage planes rather than to the bedding, and small veins and coarse aggregations of sulfide are fairly frequent. In
such occurrences the sulfide grain size is usually perceptibly coarser than that of the
simpler stratiform ores and is comparable with the prevailing grain size of the
accompanying low-grade metamorphic silicates.

In cases such as Broken Hill, where contortion of the containing sediments is
intense, deviation from conformability is common and sometimes substantial on a
local scale. Here there is virtually no trace of stratification within the ore itself,
and where metamorphism has been of high grade, the sulfides are coarse and
comparable in grain size with the accompanying metamorphic rock minerals.
Within the ore body itself, sulfide and nonsulfide are commonly patchy and appear
to have been drawn out and disordered. About the margins of the main sulfide
mass, local protuberances and cross-cutting veins are common and, as at Broken
Hill, sulfides often appear to "break through" the crests of tight minor folds,
forming disordered masses cutting across the layering of the metasediments lying
above. In cases such as this the overall conformability of the sulfides may be
recognized only after extensive mapping of the ore body and the enclosing rocks or,
as, for example, at Broken Hill and Noranda, where regional mapping reveals the
occurrence of a number of deposits along some preferred stratigraphic horizon.

In such instances deposits of our stratiform class begin to take on characteristics of those we will be referring to as *stratabound* in Chap. 16.

SETTING

The broad setting of these deposits is, by definition, marine and volcanic, though we have already widened our category a little by noting that *either* volcanism *or* evidence of marine sedimentation may, in some cases, be slight or not clearly apparent. In the spectrum of ore occurrence thus created, we find four main, but merging, categories.

1. OCCURRENCES IN WHICH SULFIDE LOCALIZATION APPEARS TO BE SPATIALLY RELATED TO BOTH SEDIMENTARY AND VOLCANIC FEATURES

In this case the sulfide concentrations usually occur within a "normal" sedimentary unit, such as a shale, not far from the boundary of this with tuff or lava. Sometimes they are found just within the volcanic unit. Where a number of deposits occur in a single district, they show a clear affiliation with one or two specific horizons and, usually, for rather specific lithological environments within and adjacent to these horizons. The associated volcanic rocks are usually pyroclastic rather than lavas and range from basaltic to rhyolitic in composition. Andesitic and dacitic tuffs are probably the most frequent volcanic associates.

In their sedimentary affiliations these deposits are often very interesting indeed. Although intense metamorphism and structural deformation may have all but obliterated the original features of the containing rocks in some instances, there are many cases where the nature and environment of deposition of the associated sedimentary rocks can be deduced with fair accuracy. For undeformed deposits such as MacArthur River the setting is preserved and exhibited very clearly indeed.

Wherever sedimentary associations are clear, two things stand out: sedimentation (1) has taken place in fairly shallow water near a shoreline of some kind and (2) has been accompanied by substantial biological activity. Evidence of nearshore deposition lies in the presence of associated coarse sediments, in the short range and often rapid variation in sediment type, frequently with the development of pronounced facies patterns, and in some cases, in the presence of carbonate reef structures. Evidence of biological activity appears in the frequent extensive occurrence of carbon in the finer sediments, in the presence of fossils and sometimes of bituminous matter, and in the development of the above reef structures, algal in the Precambrian occurrences and commonly conspicuously coralline in younger ores.

Naturally no single ore province is likely to show all these volcanic and sedimentary features. Where ores are contained entirely in subaqueous volcanic rocks, the marine aspect of the environment is indicated only by the occurrence of pillows

Table 15-1   Stratigraphic succession embracing the HYC deposit, MacArthur River, northern Australia.   (From Cotton, 1965, p. 197)

Group	Formation	Member	Submember
Roper Group	Crawford Mainoru Limmen Sandstone		
		Unconformity	
MacArthur Group	Lynott	Upper Lynott Lower Lynott	
		Minor unconformity except in HYC trough	
		Reward Dolomite Deep Creek Dolomite	
	Amelia Limestone	Barney	⎰HYC Pyritic Shales ⎱Green Vitric Tuffs ⎰Laminated Dolomite
		(Lower members not listed)	
Tawallah Group			

and occasional minor interbedded sediments in the lava units and by the development of sedimentary structures in pyroclastic units.   Where, however, the ores occur in "normal" sediments *adjacent* to volcanic rocks, the interplay of volcanic and sedimentary processes is often displayed quite vividly.   Perhaps the two finest examples of ores clearly associated with volcanism on one hand and with sedimentary facies patterns on the other are MacArthur River and Mount Isa.

MacArthur River (HYC).   This deposit is a comparatively recent discovery and has been described in only a preliminary way by Cotton (1965).   The age of the ore and of the containing sediments have been determined isotopically as approximately 1600×10^6 years.   The sedimentary sequence consists of limestones, dolomites, shales, and a "green tuff" member.   The succession (after Cotton, 1965) is given in Table 15-1.   The principal unit—the Amelia Limestone —is about 600 feet thick and includes numerous algal biostromes.   As a whole it occupies a broad, shallow-dipping basin, the focus of which is named the Bulburra Depression.   The tuff unit achieves its greatest thickness—about 400 feet—at the center of this depression and thins to zero at its margins.   It is very fine-grained and is composed chiefly of feldspar detritus carrying well-preserved relict volcanic shards.   It is restricted in extent to within a 3-mile radius of the sulfide-bearing beds and grades upward into the overlying *HYC Pyritic Shale Submember*—the ore member.   This is composed chiefly of black carbonaceous-dolomitic-pyritic shales. These are fine and finely bedded but contain some intercalated dolomitic sedimentary breccia and occasional tuff.   According to Cotton:

> The plan outline of the deposit is approximately elongate semi-elliptical. . . .   The orebodies may be visualised as a number of tabular bodies of this shape, superimposed

one above the other, and separated by comparatively barren horizons. There is progressive diminution in area from the lower to the upper orebodies.

The transition upward from the Green Tuff Submember is brought about by an increase in the proportion of interbedded black shale. The barren shales then predominate over minor dolomitic and pyritic shale beds.

Above these beds there is an abrupt change to thin-bedded strongly pyritic black shale which contains thin, conformable, galena-sphalerite laminae. This lowest orebody carries the highest grade mineralization of the deposit (1965, p. 199);

and

There is every evidence that the Bulburra Depression was shallow and subject to hypersalinity during the deposition of the sequence immediately preceding the Barney Member. Numerous algal stromatolite biostromes occur through this sequence. . . . Lenticular algal reefs occur on one horizon within the ore beds. These are diagnostic of a shallow marine or inter-tidal environment. . . . Salt crystal moulds are abundant in certain siltstone beds amongst the lower Amelia dolomites. A marker bed at the base of the Laminated Dolomite Submember carries mud crack impressions. These two features indicate complete marine withdrawal over a wide area very close in space and time to the depositional site of the ore beds. The setting immediately prior to the deposition of the ore beds would thus appear to be a very shallow, hyper-saline barred basin, subject to complete exposure, most likely by evaporation, followed by the occasional influx of sea water which dissolved the salt and deposited thin sandy beds (1965, p. 199).

Mount Isa. While the original setting is not quite so well preserved at Mount Isa as at MacArthur River, it is fairly readily identified nonetheless. A diagrammatic cross section of the immediate mine environs is given in Fig. 15-5. The sulfides and enclosing sediments are about the same age as those of MacArthur River. The sedimentary sequence is given in Table 15-2 and includes abundant volcanic material above, below, and within the ore beds, together with

Table 15-2 Stratigraphic succession embracing the Mount Isa deposits, northwestern Queensland. (From Bennett, 1965, p. 235)

Age	Formation	Thickness, ft
	Western Volcanics	?
	Unconformity?	
	Magazine Shale	700
	Kennedy Siltstone	1020
	Spear Siltstone	560
Mount Isa Group	Urquhart Shale	3000
	Native Bee Siltstone	2600
	Breakaway Shale	3400
	Moondarra Siltstone	4000+
	Judenan Beds	
	Myally Beds	
	Eastern Creek Volcanics	15,000+
	Mount Guide Quartzite	8000
	Argylla Formation	—

The *Age* column is labelled "Lower Proterozoic" spanning all formations.

shales, siltstones, and abundant dolomitic carbonate beds. In the mine area there are two principal rock types:

a. *Carbonaceous black shales.* The principal nonsulfide components of these are fine dolomitic carbonate and extremely fine potassium feldspar, the latter showing abundant relict glass shards. The rock is variably darkened with carbon and normally contains much pyrite. Pyrrhotite is a subordinate iron sulfide, and the whole has the appearance of a large, extremely well-bedded, sulfide iron formation (cf. Chap. 13). Most of this mass contains at least minor quantities of galena and sphalerite, and where these two achieve marked abundance in the beds, the "pyritic iron formation" changes— gradationally and completely conformably—into "lead-zinc ore." These are the Mount Isa lead-zinc ore bodies, which occur as a number of beauti- fully bedded lenses in the pyritic Urquhart Shale Member of the Mount Isa Group.

b. *Silica-dolomite.* This is a somewhat disordered-looking rock consisting of coarsely crystalline dolomite, coarse patches and veins of quartz, and minor quantities of shale. The whole has the appearance of a reef and reef breccia mass. This, too, contains sulfide almost throughout, in this case pyrite, pyrrhotite, and chalcopyrite as the major components. While far less ordered than the lead-zinc-bearing shales, higher concentrations of the economic mineral—chalcopyrite—also develop as lenses of overall conformable dis- position. The ore in this case is quite different from the lead-zinc; it occurs as irregular veins and patches and hence is highly discordant on a small scale.

Although the evidence is not clear, it is suspected that the copper-bearing silica-dolomite is an algal reef and reef breccia. The lead-zinc-bearing shales inter- digitate with the fingers of this "reef" and are thought to be finely bedded, highly organic products of off-reef sedimentation. Their principal detrital component is fine tuff. The interdigitating relations between the two ore types and their host rocks is quite clear from Fig. 15-5. The segregation of copper from lead-zinc is a remarkable feature of the sulfide body as a whole and is considered in more detail in the section on "Constitution."

2. OCCURRENCES IN WHICH SULFIDE LOCALIZATION APPEARS TO BE SPATIALLY RELATED TO VOLCANISM BUT NOT TO ANY SPECIFIC FEATURES OF SEDIMENTATION

There is usually evidence suggesting that the volcanic rocks concerned have come to rest subaqueously, though this is not always absolutely certain, and even where it is certain, there may be doubt as to whether the environment was marine or lacustrine.

The associated volcanic rocks are, again, usually pyroclastic and of andesitic to rhyolitic composition. Perhaps the best known exceptions to this are the pyritic copper lenses of Cyprus, which are enclosed by lavas of basalt to picritic basalt

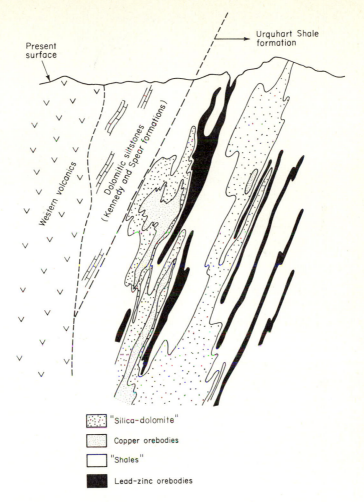

Fig. 15-5 Simplified section through the Mount Isa "ore body," showing the multiple lensoidal nature of both types of ore body and the incidence of the principal enclosing rock types. (*Adapted from Bennett, Commonwealth Mining Met. Congr. Trans., 1965.*)

composition.   These lavas commonly show pillow structure, indicating that they and their pyroclastic associates were deposited under water.   In most provinces, however, the ores are conspicuously associated with andesitic to rhyolitic rocks of the "calc-alkali" lineage and with their pyroclastic derivatives rather than with the lavas themselves.   Those of the Precambrian Superior province of Canada— Noranda-Kirkland Lake, Timmins, Matagami, Chibougamou (see Goodwin, 1961, for a regional account)—occur in basalt-andesite-dacite-rhyolite sequences and show distinct preferences for the more siliceous, pyroclastic members.   At the other end of the time scale the *kuroko* deposits of Japan, and similar deposits of Fiji and the Solomon Islands, are associated with Mesozoic and Tertiary basalt-andesite-dacite volcanism, most frequently with the pyroclastic members.

It is a feature of deposits of this setting that within a given region or province they tend to occur on one, or a very few, "volcanic horizon." This is especially true for the Noranda-Kirkland Lake area and for some of the Japanese deposits. The subaqueous nature of the volcanic rocks is usually indicated by bedding, which may be very delicate in the finer pyroclastic units. However, whether the environments have been marine or lacustrine is, as we have already noted, often far from clear. Evidence is often indirect and unreliable, particularly in older terrains. The extensiveness of a horizon may indicate a large—and hence marine —area of contemporaneous deposition or a very restricted, and hence probably lacustrine or lagoonal, one.

3. OCCURRENCES IN WHICH SULFIDE LOCALIZATION APPEARS TO BE SPATIALLY RELATED TO SPECIFIC ZONES OR ENVIRONMENTS OF SEDIMENTATION BUT THAT SHOW NO APPARENT ASSOCIATION WITH VOLCANISM

This is a small group but one which includes some very important deposits, notably those of Mansfeld, Germany, and of the Roan province in Africa.

The *Mansfeld* ores are better referred to as those of the *Kupferschiefer* or *Kupferschiefer-Marl Slate*. The deposits concerned are in the form of an extensive cupriferous shale—the Kupferschiefer—that was first worked near Mansfeld, at the eastern end of the Harz Mountains, in the thirteenth century. Since that time it has yielded some 2,200,000 tons of copper, together with some lead and zinc. The present margins of the Kupferschiefer are indicated in Fig. 15-6 (adapted from Dunham, 1964). According to Dunham:

> The formation is a bituminous-calcareous or dolomitic shale lying near the base of the Zechstein (Middle Permian) of Northern Europe. The Kupferschiefer or its equivalents can be traced from northern England eastward through the Low Countries, across Germany into Poland. . . . The southern and to some extent the western margins of the Zechstein basin or lagoon in which it was deposited are in part well defined; the northern margin is concealed, probably below the Baltic. In England the Marl Slate (Lower Permian Marl, Hilton Plant Bed, Kirksanton Grey Beds) rest variously on a thin basal conglomerate, on aeolian dune sands (the Yellow Sands) or on red sandstone (Penrith Sandstone). In Germany a basal Zechstein conglomerate may underlie it, but over considerable areas it rests on a red sandstone, shale, and conglomerate formation, the Rotliegende. This may be bleached for a short distance below the shale contact (Weissliegende). Ancient dunes, elongated ESE have been recognized in the Weissliegende Sands. Everywhere, dolomitic limestone, the Magnesian Limestone or Zechsteinkalk overlies the Marl Slate-Kupferschiefer and this in turn succeeded by the great evaporite series now proved in N. E. England, Denmark, Hannover, Werra, around the Harz Mountains and at Strassfurt. The deposition of the Kupferschiefer-Marl Slate was, in short, merely a brief prelude to a prolonged episode of cyclic lagoonal evaporation (1964, pp. 5 and 6).

The sediments in which the copper mineralization occurs were thus deposited in a very specific—and clearly recognizable—environment. Although the immediate succession was widespread (Dunham, 1964, p. 8, estimates that some 8000

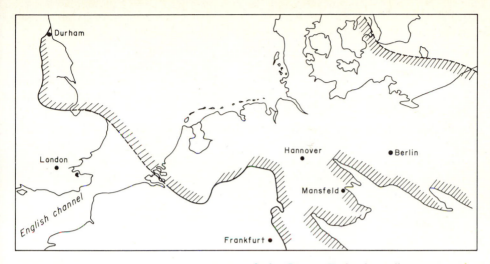

Fig. 15-6 Approximate present margins of the Lower Zechstein sedimentary rocks, showing the position of the principal Kupferschiefer occurrence, at Mansfeld, Germany. (*Adapted and much simplified from Dunham, Econ. Geol., 1964.*)

square miles of black shale greatly enriched in Cu, Pb, and Zn are involved), the areas of bottom stagnation, though numerous, were apparently localized. Several investigators have pointed out that there was no bottom fauna other than anaerobic bacteria in the localities concerned; what fossils occur are of fish, *pelagic* shellfish, and terrestrial plants. The sediments thus accumulated in a series of putrid, shallow, near-shore basins.

The sedimentary succession associated with the deposits of the *Northern Rhodesian Copperbelt* has many features in common with that just described for the Kupferschiefer—a similarity noted very explicitly by W. G. Garlick (in Mendelsohn, 1961, pp. 147–148). The broad structural framework of the region and the position within this of the Copperbelt are shown in Fig. 15-7. A closer view, showing the positions of the main sulfide occurrences, is given in Fig. 15-8. The rocks of the area are of Precambrian age, and the regional geological column is given in Table 15-3.

The stratigraphy of the Lower Roan rocks specifically is (Mendelsohn, 1961, p. 117)

Hangingwall Formation	Argillite and quartzite
Ore Formation	⎧ Hangingwall argillite ⎨ Ore shale—argillite ⎩ Impure dolomite
Footwall Formation	⎧ Footwall conglomerate ⎪ Quartzite and argillite ⎨ Lower conglomerate ⎩ Quartzite (in part aeolian)

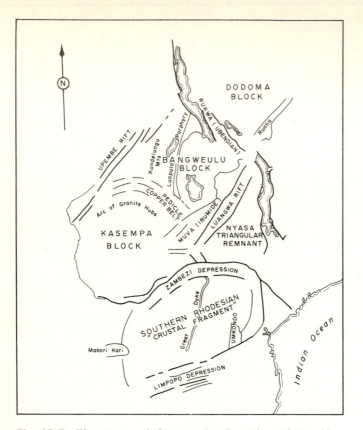

Fig. 15-7 The structural framework of portion of Zambia (Northern Rhodesia). [*From Brock, in Mendelsohn (ed.), "The Geology of the Northern Rhodesian Copper Belt," Roan Antelope Copper Mines Limited and MacDonald & Co. (Publishers) Ltd., London, 1961, with the publisher's permission.*]

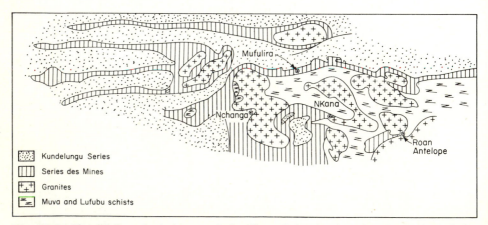

Kundelungu Series

Series des Mines

Granites

Muva and Lufubu schists

Fig. 15-8 Simplified geological map of the Copperbelt, showing the distribution of the principal rock groups and the positions of the major ore bodies. (*Adapted from Gray, Econ. Geol., 1930.*)

Table 15-3 Principal rock types and their age relations in the Katanga System. (After Mendelsohn, 1961)

Gabbro—Intrusive, commonly into Upper Roan dolomites

Series	Group	Formation	Rock Types (minor or rare rock types in parentheses)
	Upper		Shale, quartzite
	Middle		Shale
Kundelungu			Tillite
	Lower	Kakontwe (limestone) Tillite	Shale Dolomite and shale Tillite
	Mwashia		Carbonaceous shale, argillite (dolomite and quartzite)
	Upper Roan		Dolomite and argillite (quartzite, breccia)
			- Argillite and quartzite
Mine		Hangingwall	Quartzite Argillite and feldspathic quartzite (dolomite)
	Lower Roan	Ore	Argillite, impure dolomite, micaceous quartzite (graywacke, arkose)
		Footwall	(Footwall conglomerate) Argillaceous quartzite Feldspathic quartzite Aeolian quartzite Conglomerates

Unconformity
Granite—Intrusive into Lufubu System

The Ore Formation is, as its name suggests, the principal locus for sulfides in the Roan sequence, containing some two-thirds of the known ore of the Copperbelt. It is a well-defined lithological and stratigraphical unit, generally some 50 to 60 feet thick but ranging from a few feet to about 200 feet.   Variation in present thickness is in some places depositional and in others due to later folding and attenuation.   The sulfides occur principally in the middle member—the Ore Shale— and hence are referred to as the *ore shale deposits*.   Lithologically the unit varies from a shale to a fine silt and from slightly to conspicuously carbon-bearing. According to Mendelsohn, free carbon ranges up to about 1.5 per cent.   The rock is finely bedded and consists of mica, quartz, feldspar, carbonate, and sulfide, with minor quantities of chlorite, tourmaline, apatite, the above-mentioned carbon, and trace accessories. While most of the sulfide of the ore formation is found in this shale, some notable quantities occur here and there in the associated dolomites and arenites of this unit.

A minor second group of important deposits occurs stratigraphically distinctly—usually about 80 feet—below the Ore Formation and is invariably within a quartzite of the Footwall Formation. These deposits are referred to as the *footwall (quartzite) deposits*. The rocks concerned are sulfide-bearing micaceous to micaceous-feldspathic medium-grained quartzites.

A third group of ore bodies occurs in the Mufulira area (Fig. 15-8); these ore bodies are contained in quartzites and hence are referred to as *quartzite deposits*. According to Mendelsohn (1961, p. 122) the Ore Formation cannot be recognized in the Mufulira area, but the general position of the ore bodies in the Lower Roan Group here is similar. The ore-bearing unit at Mufulira, which rests on a formation containing a thick sequence of aeolian quartzites, consists of a series of sulfide-bearing quartzites interbedded with normally barren dolomite, shale, and argillite. There is a well-defined and repeated sequence at Mufulira:

> Argillite
> Dolomite
> Graywacke
> Sericitic quartzite
> Gritty quartzite
> Argillite

Some of the quartzites are variably gritty, pebbly, or feldspathic; some are dark-gray to black subgraywackes, and these contain up to 5 per cent of carbonaceous material. The major part of the sulfides occurs in gray sericitic to feldspathic quartzites.

As is clear from the stratigraphic columns already given, the basal members of the Roan Group of sediments were laid down unconformably on the granite-schist complex of the basement. Apparently the schist component had been much more susceptible to erosion than the granite so that the pre-Roan landscape consisted of granite-schist hills separated by low-lying, flat areas of schist. As a result the Footwall conglomerates and quartzites are of very variable thickness and thin out substantially—sometimes completely—over the former granite hills. It follows that the sedimentation was of near-shore shallow-water type, as evidenced by the lithologies. In addition to the deposition of sandy and pebbly beds, carbonate sedimentation was frequent and extensive and largely of algal reef type. Garlick (1964) and Malan (1964) have shown conclusively that "collenia-type" stromatolites are abundant at Mufulira and in some of the related Katangan deposits, and Garlick has suggested, on good evidence, that much of the dolomite in these areas and in other parts of the Copperbelt is slightly metamorphosed algal bioherms. This has been beautifully confirmed by detailed work by Paltridge (1968) on fine structure of an algal biostrome fringe at Mufulira.

This demonstrated that:

a. The coarse-grained sediments beneath the biostrome were cross-bedded and scoured-and-filled, indicating an energetic near-shore environment prior to algal growth.
b. Thin shale zones with mud cracks also occur beneath the biostrome succession, indicating deposition in shallow water subject to intermittent exposure in a littoral environment.
c. The mineralogical and chemical composition of the biostrome changes from the center to the fringe and the rate of algal growth at the fringes was governed by the rate of deposition (and K feldspar content) of the contemporaneous sandstones.

It thus appears that the sediments—and the sedimentary sequences—of the Copperbelt, all of which are apparently remarkably similar from one mineralized locality to another, were laid down in a well-flushed, well-aerated neritic to sub-littoral environment in which photosynthesis, and consequent shallow-water biological activity, proceeded swiftly.

The disposition of sulfides, and their composition, seems to be firmly related to the biostromes, where the latter occur. The biostromes tend to form finger-shaped masses, between which are "interreef" argilites. Table 15-4 and Fig. 15-9

Table 15-4 Chemical compositions of stromatolites and associated sedimentary rocks at Mufulira. (From Malan, 1964, p. 412)

	(1)	(2)	(3)	(4)	(5)	(6)	(7)
Cu	0.25	0.06	0.08	0.15	0.26	0.21	0.10
SiO_2	58.9	78.8	20.9	20.8	52.0	60.4	25.10
Al_2O_3	15.7	5.3	9.1	11.4	13.6	15.0	6.4
MgO	6.4	1.7	13.2	10.1	10.0	5.1	14.9
CaO	6.2	6.4	22.2	21.5	7.9	5.8	20.2
MnO	0.18	0.13	0.21	0.81	0.18	0.18	0.50
FeO	2.9	1.2	2.4	2.5	3.2	3.7	2.2
K_2O	2.89	0.63	1.40	1.35	4.04	2.74	1.73
Na_2O	0.30	0.15	0.20	0.19	0.22	0.19	0.32
P_2O_5	0.15	0.08	0.17	0.10	0.20	0.20	0.08
CO_2	6.4	5.8	33.1	32.7	9.8	5.7	29.6
S	0.08	0.2	0.08	0.09	0.39	0.4	0.08
Total	100.42	100.45	103.04	101.69	101.79	99.62	101.21

Explanation
(1). Interreef argillite.
(2). Coarse feldspathic quartzite lens associated with algal reefs.
(3). Dolomitic margin of biohermal stromatolite.
(4). Argillaceous core of biohermal stromatolite.
(5). Argillaceous fillings between biohermal stromatolites.
(6). Dark argillaceous collenia.
(7). Silty Inter B/C dolomite with algal bedding.

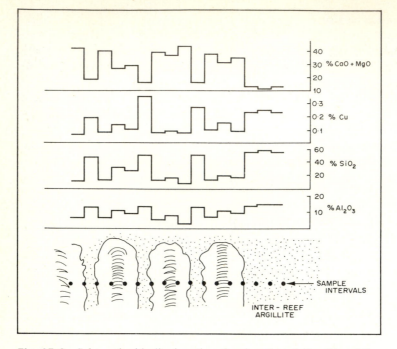

Fig. 15-9   Schematic, idealized section showing the relationship of the distribution of (CaO + MgO), SiO_2, Al_2O_3, and Cu with the incidence of stromatolite bioherms and associated shales.   (*Redrawn from Malan, Econ. Geol., 1964.*)

(both after Malan, 1964) show the composition of the reef and interreef rocks and the relation of Cu and S to the two lithologies.   Sulfide occurs in the shaly interreef rocks rather than in the reefs.   Figure 15-10—a beautifully clear diagrammatic representation by Garlick (1964)—shows the relation of sulfide distribution and mineralogy to one such reef—a "barren dolomite" on a granite-schist rise near

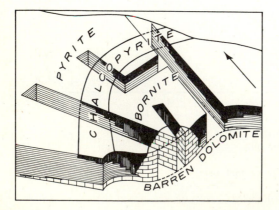

Fig. 15-10   Schematic diagram showing in highly simplified form the probable prefolding relationships between dolomite and sulfide mineral zoning. The dolomite is thought to represent algal reef carbonate originally built up about a "basement high." (*Redrawn from Garlick, Econ. Geol., 1964.*)

the Irwin Shaft of the Roan Extension mine.   It can be seen that the sulfides occur in the shales surrounding the dolomite reef but not in the reef itself, that pyrite is the lowest sulfide stratigraphically and is followed by chalcopyrite then bornite (then chalcocite in some areas), and that this stratigraphic succession leads to the horizontal zoning bornite $\rightarrow$ chalcopyrite $\rightarrow$ pyrite, proceeding from the reef fringe toward progressively deeper water shales.

Thus, like the Kupferschiefer-Marl Slate deposits, those of the Copperbelt show no obvious relation to volcanism but a very pronounced tie with specific facies of sedimentation.   However, the environments of sedimentation of the ore-bearing rocks in the two provinces look very different indeed—at first sight, at least.

4. OCCURRENCES IN WHICH SULFIDE LOCALIZATION APPEARS TO BE SPATIALLY RELATED TO NEITHER VOLCANISM NOR TO SPECIFIC ENVIRONMENTS OF SEDIMENTATION

Although not a large one, this group includes two of the world's largest lead-zinc ore bodies—those of Sullivan and Broken Hill.   The great Sullivan ore body of British Columbia is the outstanding example among those deposits whose surroundings are unmetamorphosed and hence whose primary features are preserved for examination.   The similarly great ore body of Broken Hill occurs in sedimentary rocks that have been metamorphosed almost beyond recognition.

The *Sullivan* deposit—a huge sulfide lens (Fig. 15-11) showing beautiful stratification and conformability with the enclosing sediments through almost the whole of its mass—occurs in the Precambrian Lower Purcell sediments of southeastern British Columbia.   These rocks were laid down in the "Beltian Trough" —a simple elongated geosyncline in which some 45,000 feet of dominantly shallow-water sediments accumulated.   The lowest unit is the Fort Steele Formation, much of which is composed of orthoquartzites showing clear evidence of deposition in shallow, turbulent water.   The upper third of the unit is composed of finer beds —argillites and calcareous argillites—apparently deposited under quieter conditions.   This is conformably overlain by the Aldridge Formation—a unit of some 15,000 feet thickness that contains the Sullivan ore bodies.   The Lower Aldridge (about 400 feet) is composed of fine siltstones and argillites and shows cross-bedding, scour channels, and occasional ripple marks and graded beds.   It is notable for its rusty appearance in outcrop, due to widespread dissemination of pyrrhotite, and lesser pyrite and iron silicate and carbonate.   According to A. C. Freeze: "The widespread distribution of the iron-bearing minerals and their tendency to be associated with primary sedimentary features strongly supports the suggestion that probably most of this iron was deposited contemporaneously with the common detrital constituents" (1966, p. 266).

Near the top of the Lower Aldridge, substantial intraformational conglomerates and boulder beds appear, and these, with the scour channels that are prominent in the underlying materials, indicate extensive slumping.   It is near the transition of the Lower and Middle Aldridge that the ore zone occurs.   The Middle Aldridge (about 9000 feet thick) is more ordered, gives far less indication of turbulence, and

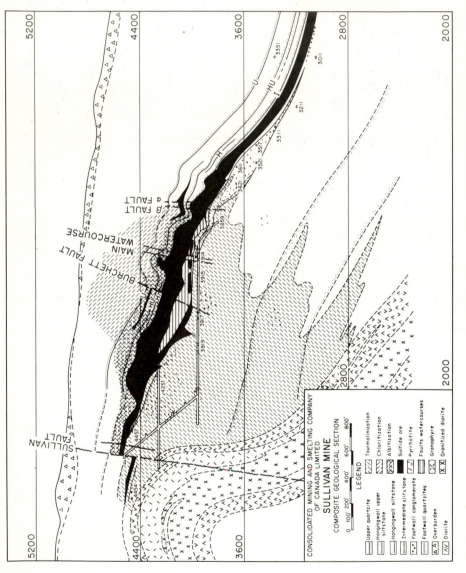

Fig. 15-11 Geological section through the ore body of Sullivan mine, British Columbia. *(From Freeze, Can. Inst. Mining Met. Spec. Vol. no. 8, 1966.)*

contains less disseminated sulfide.   According to Freeze:

> The beds that comprise the ore zone may be up to 300 feet thick and. . . appear to
> have been deposited during a transition from a long period of sedimentation in a shallow
> water environment to an even longer period of sedimentation in a relatively deep water
> environment in which submarine landslips and turbidity currents were prevalent. . . .
>
> The composition of the ore zone beds as deduced from nearby unmineralized or
> weakly mineralized sections is somewhat similar to the rocks in the footwall and hanging-
> wall zones.   That is, they are greyish to greenish argillites, siltstones, quartzites, etc.
> On the whole, the ore zone contains appreciably more argillite than either the footwall
> or hangingwall sections and the proportion of thick massive graded beds separating the
> thinbedded argillites, etc., is greater and better developed.   The characteristics of these
> beds are maintained over a surprisingly wide area in contrast to the variability found in
> the footwall and hangingwall beds (1966, p. 268).

The footwall of the ore body is composed of a conspicuous "Footwall Conglomerate" and lesser orthoquartzite.   The hanging-wall rocks are essentially similar to
those of the ore zones except that those of the latter show rather better sorting.
Carbonaceous material is not conspicuous in the ore zone, and there seems to be no
sign of reef or other carbonate sedimentation.

Thus while much of the sedimentation was probably not too far from a shoreline, there is no evidence yet of any definable facies pattern or of organic activity
related to it.   Neither does there seem to be any clear evidence of volcanism.
No flows or pyroclastic beds have yet been positively identified in the Aldridge
Formation.   It has been suggested that the diorite-granophyre mass just below
the upper tip of the sulfide lens (Fig. 15-11) may in fact be an old volcanic complex,
though this is speculative.

The sedimentary setting of the *Broken Hill* deposit remains an enigma in
spite of Herculean efforts by Broken Hill geologists.   The fact that the sulfide-rich
lenses are generally concordant with the enclosing metasedimentary units has been
recognized since the early 1900s and was referred to very explicitly by E. S. Moore
(1916).   However, the possible implications of this concordancy were not really
recognized until the time of King, O'Driscoll, and Thomson (1953), as has already
been referred to in Chap. 2.   Since then there has been an increasing awareness of
the lithological, as distinct from the structural, environment of the sulfides, and
much further investigation of this lithological aspect has now been carried out,
particularly by Thomas (1960) and Carruthers and Pratten (1961), of the Consolidated Zinc Corporation.

On a broad scale the general association of sulfide concentrations with three
particular rock types stands out.   These are (1) basic *amphibolites*, (2) a conspicuous type of garnet gneiss known as the *Potosi Gneiss*, and (3) a well-bedded magnetite-rich rock referred to locally as *banded iron formation*.   The amphibolites occur
as stratigraphic units and appear to be metamorphosed basic lavas or tuffs (their
iron content is notably higher than commonly found in basalts; a number of analyses show total iron as FeO close to 20 per cent, suggesting that some of them may
have been tuffs to which iron was added during sedimentation).   The Potosi
Gneiss is also distributed stratigraphically and constitutes a range of garnet-
plagioclase-potassium feldspar-biotite-quartz gneisses.   Apatite and zircon are

ubiquitous accessories, and sillimanite, magnetite, ilmenite, pyrite, pyrrhotite, and galena occur sporadically.   Their chemical compositions suggest that they may have originated as andesitic to dacitic tuffs.   The banded iron formations are magnetite-quartz-garnet-apatite rocks, with minor and sporadic ilmenite, gahnite, and sulfides.   With variation in their major constituents, the rocks develop variants such as quartz-magnetite, garnet-quartz, and quartz-apatite rocks.   With other variations in chemistry, the rocks may grade into garnet quartzites, garnet "sandstones" (metamorphic rocks consisting almost entirely of garnet and with a "sandy" appearance and friable texture), garnet-hedenbergite quartzites, manganese silicate rocks, garnet-feldspar quartzites, and in some areas, Potosi Gneiss and amphibolite.   The magnetite-quartz-garnet-apatite members and their immediate variants are usually well banded, the banding having every appearance of being original bedding.   With their almost ubiquitous high Al_2O_3 and P_2O_5 content these rocks are clearly quite unlike banded iron formations as described in Chap. 13.   They are, however, iron-rich sedimentary rocks of some kind and may have affiliations with the ironstones of present usage.   Of the three rock types—amphibolite, Potosi Gneiss, and banded iron formation—the last two are most conspicuously associated with sulfide, though on a semiregional scale the general spatial tie of ore and stratiform amphibolite is very clear indeed.

Thomas (1960) made a particularly close study of the relation between the Broken Hill sulfides and their lithological environment.   He considers that the relation is a close one—that the ores are an original and intrinsic part of the geological environment—and that the development of the different rock types, and the occurrence of sulfide-rich beds within them, exhibits a cyclic pattern.   This "environmental approach" has continued to be developed by Zinc Corporation geologists, notably by King, Carruthers and Pratten (1961), and Carruthers (1965).

A conspicuous feature of the Broken Hill metasedimentary rocks generally is their marked short-range variability.   Most of the types within the broad units occur as lenses and appear and disappear over quite short distances.   Such variability and discontinuity indicates turbulent conditions and a mobile environment during at least much of the period of sedimentation.   This, together with the chemical composition of many of the rocks, suggests the possibility of a nearshore volcanic environment.   It must, however, be clearly kept in mind that, with the metamorphic camouflage and the still sketchy state of our knowledge of Broken Hill, this must be recognized as no more than an interpretation.   At the present time we do not *know* that the sedimentary rocks enclosing the Broken Hill sulfides were deposited in a specific facies or that they had a volcanic connection.   In fact we do not *know* that the banding within individual units is *sedimentary*.

5. GENERAL REMARKS CONCERNING SETTING

From the foregoing examples it can be seen that the setting of stratiform sulfides of marine and marine-volcanic association varies substantially.   Some show clear spatial relations with specific, well-developed facies of sedimentation and with volcanic activity; others are spatially related to one or the other of these phenomena, while others show no clear relation to specific conditions of sedimentation or to volcanism.   Principal features of setting appear to be:

a. The majority of known deposits of this group are associated with *both* specific sedimentary facies and volcanic rocks.   A significant number occur with one or the other, and where volcanic material is not apparent there is room for questioning the completeness of our knowledge of some of the associated fine sedimentary rocks.

b. Although associations with specific sedimentary facies is often unequivocal, such associations are far from constant.   Association may be conspicuously with reduced sediments, as with the Kupferschiefer-Marl Slate and many other deposits in carbon-bearing sediments.   On the other hand, the major part of the associated sediments may be of a clear-water, well-flushed type, as with the biostromal Copperbelt sediments.   Other deposits have a hybrid association; e.g., the Mount Isa copper sulfides occur in what may be algal biostromal carbonate masses, the adjacent lead-zinc sulfides being in heavily carbonaceous reduced shales.

c. Whether or not facies affiliations are clear cut, there is strong evidence that in all cases enclosing sediments have been laid down not far from a shoreline. The evidence may be biological (fossils, abundant carbon, reef carbonate), chemical (associated gypsum, red beds), or mechanical (variability and coarseness of sediments; scour and fill, cross-bedding, and other evidence of turbulence).   The full range of deposits from Mount Isa, MacArthur River, the Kupferschiefer, the Copperbelt to Broken Hill and Sullivan are enclosed in sediments that, in one way or another, show evidence of near-shore character.

d. The associated volcanic rocks may be lavas or pyroclastic types but are most frequently tuffs or tuffaceous sediments.   They fall within the compositional range basalt-andesite-dacite-rhyolite but are most frequently of andesite-dacite type.

CONSTITUTION

As a group these ores cover a substantial range of compositions.   They include a variety of sulfide and sulfide-type minerals, oxides, carbonates, silicates, and others and exhibit considerable variations in sulfide metal abundances.   Nonetheless they show a number of compositional consistencies, some of which are very definite and distinctive indeed.   In this section we shall consider the principal features of element, mineral, and isotope abundances and also, very briefly, the form and associations of the more important mineral constituents.

1. ELEMENTAL CONSTITUTION

Here we are concerned mainly with the abundance of the sulfide metals, though the incidence of others such as calcium, magnesium, manganese, and silicon is also of concern and is known to be important in a number of cases.

The principal metals of the sulfide component of this class of ores are, in

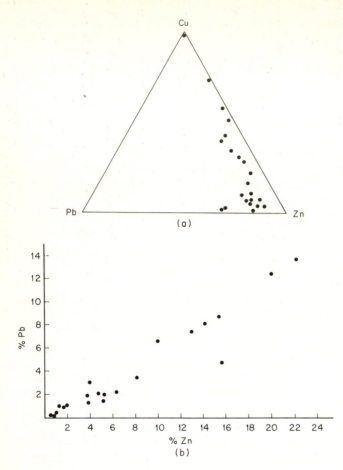

Fig. 15-12 (*a*) Atoms per cent copper-lead-zinc relations and (*b*) lead-zinc ratios (weight per cent basis) in a number of large stratiform ore deposits.

order of importance, iron, zinc, lead, and copper. The relative importance of lead and copper may be disputed, and indeed these two seem to be under some regional influence. For example, in the Precambrian of eastern Canada and eastern Australia, iron and zinc sulfides are prominent in both areas, but copper is conspicuous and lead is low in Canada and lead is conspicuous but copper low in Australia.

Although it is overwhelmingly dominant for these ores as a class, iron ranges from being superabundant among the sulfide metals in some deposits—such as Mount Isa, Sullivan, and New Brunswick—to something less than 5 per cent as at Broken Hill. However, there is a very clear pattern in the incidence of the four metals. Some deposits contain as their sulfide only iron sulfide, and iron is the only metal of the four that occurs alone in this way. Some deposits contain iron and copper—never, however, copper alone.   Some contain iron-copper-zinc—but never zinc without iron and copper.   Others contain iron-copper-zinc-lead—but never lead without the other three, even though the copper may be present in almost vanishingly small amount.   This pattern

> Iron
> Iron-copper
> Iron-copper-zinc
> Iron-copper-zinc-lead

was first noted by R. L. Stanton in 1954 in a study of a group of small Paleozoic deposits in New South Wales, and it has since become obvious that it is a world-wide phenomenon.

Ignoring iron, this leads to the arrangement shown in Fig. 15-12a, which depicts Cu–Zn–Pb ratios (atoms per cent) in a number of well-known ore bodies of the present class in Europe, North America, Africa, Australia, and the Middle East.   Some contain only copper.   Zinc then appears and increases, and then lead, the lead increasing substantially in a number of ores when the Zn/Cu ratio exceeds about 60:40.   Lead does not necessarily appear, however, and some ores (e.g., Balmat-Edwards in New York State) may have zinc as almost their sole non-ferrous component.   A very conspicuous feature of the diagram is the absence of points from the Cu–Pb edge; ores dominated by copper and lead sulfides together are not known among this class of deposit.   The distribution of points in Fig. 15-12a reflects in part a general similarity in lead-zinc ratios in many of the occurrences, and this is borne out by a plot of weight per cent Pb versus Zn, as shown in Fig. 15-12b.   Figure 15-12b also gives some idea of the range of *absolute abundances* of the two metals in this class of ore.

In spite of gradational relationships suggested by Fig. 15-12, there is a strong tendency for the segregation of copper from lead-zinc in these ores.   This is just a little surprising in view of the many chemical similarities (including similarities of crystal chemistry) between copper and zinc and the substantial differences between lead and zinc.   There are several features of this segregation:

a. Deposits of the two types, i.e., of copper and of lead-zinc, commonly occur scattered separately throughout a single district.   Keeping in mind that these stratiform ores generally show clear preferences for specific horizons, it is common to find that the copper deposits are stratigraphically *below* the lead-zinc.

b. Where copper and lead-zinc ores occur in juxtaposition in the one sulfide mass, it is common for the copper ore to occur at the base of the lens and for lead-zinc to develop at a higher stratigraphic level—and in some cases, for the lead-zinc to be overlain in turn by barite concentrations.   These are remarkably widespread and constant features, as is indicated by the examples given in Table 15-5.

c. Copper and lead-zinc may occur as discrete, but immediately adjacent, masses, as at Mount Isa.   In such cases there may be no clear stratigraphic order, but

Table 15-5   Stratigraphic zoning in some stratiform ore deposits.   (Adapted from a compilatic by Brathwaite, 1969)

Deposit	Rosebery, Tasmania	Buchans, Newfoundland	Rammelsberg, Germany	Meggen, Germany
Age	Cambrian	Ordovician	Devonian	Devonian
Zoning from base upward (1 → 3) Up ↑	3. Barite–(Zn–Pb) 2. Zn–Fe–Pb–Cu– (barite) 1. Fe–Cu	3. Barite–(Zn–Pb) 2. Zn–Pb–Fe–Cu– barite 1. Fe–Cu–Zn–Pb– (barite)	3. Barite–(Zn–Pb) 2a. Zn–Pb–(barite) 2b. Zn–Fe–Pb 1. Fe–Cu	3. Barite–(Pb) 2. Zn–Fe–Pb 1. Absent

Korbalikha, Russia	Kuroko Ores, Japan	Heath Steele-B1, New Brunswick	Captain's Flat New South Wales	Noranda & Mataga District, Quebec
Devonian	Miocene	Ordovician	Silurian	Precambrian
3. Barite–Zn–Pb 2a. Zn–Pb–Fe–Cu 2b. Fe–Zn–Pb–Cu 1. Cu–Fe	3. Barite 2. Zn–Pb–barite 1. Fe–Cu	3. Absent 2. Zn–Pb 1. Cu–Fe–Zn–Pb	3. Absent 2. Zn–Pb–(Fe–Cu) 1. Fe–Cu–(Zn–Pb)	3. Absent 2. Zn–Fe 1. Fe–Cu

there is a clear tendency for the two to be separated according to facies—copper in the more carbonate-rich biostromal(?) sediments and lead-zinc in the finely bedded off-reef shales.

d. The copper ores commonly contain 2.5 to 3.5 per cent copper, with less than 1.0 per cent lead and zinc.   The lead-zinc ores contain lead and zinc as indicated in Fig. 15-12b and copper characteristically in about the 0.5 to 1.5 per cent range.   Segregation seems particularly pronounced where the two ore types are adjacent, as at Mount Isa; here lead and zinc constitute only traces in the copper ore, and copper is 0.04 to 0.08 per cent in the lead-zinc ore.

Zoning of Pb/Zn ratios is common in the lead-zinc ores.   At Mount Isa there is a strong tendency for Pb/Zn to be highest near the copper ore bodies, zinc increasing upward and northward.   At Sullivan, Pb/Zn is highest close to the iron-rich core, zinc increasing outward (Fig. 15-13).   At Broken Hill, Pb/Zn ratios generally decrease upward stratigraphically and toward the southern end of the deposit.   Cu/Fe ratios decrease systematically proceeding outward from the old reef structure in some of the Zambian deposits (Fig. 15-10).   Traces such as As, Sb, Cd, and Ag also show quite marked zoning in many ore bodies.

Very little investigation has been done so far on relationships between sulfide and nonsulfide elements in stratiform ores.   This is a field that seems to offer great scope for research and that appears likely to yield much information relevant to genesis.   Examples of close relationships that may exist between abundances of the two classes of materials are those between copper incidence and $Al_2O_3 + SiO_2$ at Mufulira (Fig. 15-9) and between total sulfide sulfur and Al_2O_3 for the two ore types at Mount Isa (Fig. 15-14).   Analyses of sulfide-rich material from a number

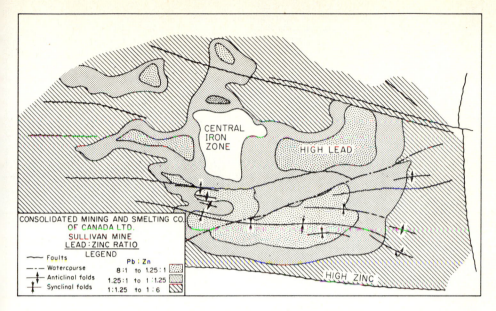

Fig. 15-13 Metal zoning in the Sullivan ore body, British Columbia. (*From Freeze, Can. Inst. Mining Met. Spec. Vol. no. 8, 1966.*)

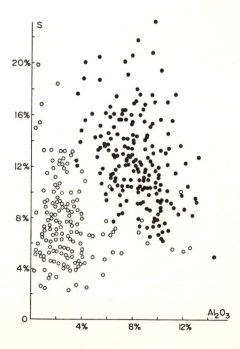

Fig. 15-14 Sulfide sulfur–Al_2O_3 relations in the shales (lead-zinc ore: filled circles) and silica-dolomite (copper ore: open circles) at Mount Isa. (*After Stanton, Trans. Inst. Mining Met., London, 1962.*)

Table 15-6　Compositions of some stratiform and associated deposits

	(1)	(2)	(3)	(4)	(5)	(6)	(7)	(8)
SiO_2	55.46	66.39	24.48	27.0	32.6	46.64	20.02	19.90
Al_2O_3	17.36	5.94	4.24	7.04	3.2	2.08	9.85	4.13
TiO_2	0.63	0.32	0.36	0.34		0.05	0.11	0.03
FeO	3.3		7.20	10.29	13.9	5.75	22.76	21.82
Fe_2O_3	1.4	1.06	3.0					
MgO	4.54	4.20	4.92	1.03	7.3	0.29	1.02	3.84
CaO	1.53	6.02	8.80	0.74	13.1	1.31	1.42	2.36
MnO	0.12	0.18	0.17	0.02		0.32	0.18	0.10
Na_2O	0.34	0.06	0.29	0.96		0.30	0.19	0.05
K_2O	9.88	1.94	3.01	3.24		0.44	2.05	0.10
P_2O_5	0.26	0.09	0.19	0.28		0.61	0.14	0.05
H_2O+	}0.39	0.39		0.25			0.07	0.02
H_2O-		0.02		0.36				
Cu	3.15	2.77	0.01	0.57	3.3	0.10	0.06	0.64
Zn	<0.01		9.50	16.57		20.03	9.92	10.40
Pb			4.25	9.73		8.37	2.55	5.37
S	1.05	1.01	14.36	20.32	7.7	14.10	28.60	27.40
Halogens	0.09					0.11		
CO_2	1.45	9.04	13.66	0.59	19.8	0.06		2.92
Total	100.95	99.43	98.44	99.33	100.9	100.56	98.94	99.13

Explanation
(1). Roan Antelope (average of 4 analyses from Mendelsohn, p. 491, 1961; analyst(s) unknown).
(2). Mufulira (from Mendelsohn, p. 491, 1961; analyst unknown).
(3). Mount Isa lead-zinc ore, No. 7 ore body; analyzed by Mount Isa Mines Laboratory.
(4). MacArthur River lead-zinc ore in bituminous tuffaceous shale (from Croxford, 1968; analyzed by Mount Isa Mines Laboratory).
(5). Mount Isa copper ore, 650–750 ore body (average of 210 analyses, all by Mount Isa Mines Laboratory).
(6). Broken Hill, "B" Lode (analyst, G. Kalocsai).
(7). Rosebery.
(8). Captain's Flat (analysts for 7 and 8, I. Khan and G. Kalocsai).

of ores are given in Table 15-6.　Naturally there are substantial "within-deposit" variations in the compositions of all these deposits, and the analyses are intended only as an indication of compositions.

2. MINERALOGICAL CONSTITUTION

The mineralogy of these deposits is characteristically simple.　For the group as a whole the common sulfides pyrite, pyrrhotite, sphalerite, galena, and chalcopyrite are overwhelmingly dominant.　Occasionally bornite and/or chalcocite are important as primary components, though this is distinctly uncommon.　Arseno-pyrite, tetrahedrite, native bismuth, and bismuthinite are common traces, and other minerals—particularly cobaltian sulfarsenides—are found in minute quantities in some of the ores.　Magnetite is a common associate, generally as magnetite-rich sediment *adjacent to*, rather than within, the sulfide concentrations. Carbonate is ubiquitous and barite and fluorite common. The silicates present are those appropriate to the chemistry and metamorphic grade of the enclosing sedimentary units.

As indicated by their chemistry, the ores fall into two principal types: (1) lead-zinc, with very minor copper, and (2) copper.

Of the two the lead-zinc ores look the simpler. Their principal features are

a. The chief base metal sulfides are sphalerite and galena, the former usually dominant.
b. Pyrite is clearly the typical iron sulfide; pyrrhotite is subordinate to absent (Mount Isa and Sullivan are among those that contain significant pyrrhotite, but even here it is quite subordinate to pyrite in the lead-zinc ores).
c. Except where pyrrhotite is present, the sphalerite is light in color and contains low iron and manganese.
d. The associated nonsulfides are low in iron, e.g., are white mica and/or pale chlorite, rather than green chlorites, at lower grades of metamorphism.
e. The ores are well bedded—often beautifully so. Stratification is developed by variation in sulfide-nonsulfide and sulfide-sulfide proportions from band to band. Bands vary in thickness from a foot or so down to hundredths of an inch. In the finer sequences the bedding may be very delicate indeed.

The principal features of the copper ores are

a. The overwhelmingly most important base metal sulfide is chalcopyrite; in many deposits other base metal sulfides are virtually absent from the copper ores. In some cases, e.g., the Copperbelt, bornite and chalcocite are important primary sulfides, but this is very much an exception.
b. Pyrrhotite is clearly the typical iron sulfide, though pyrite is usually present. In some cases, e.g., the Copperbelt, pyrite is virtually the only iron sulfide; this is quite atypical, however.
c. Any traces of sphalerite in the pyrrhotite ores are dark red in transmitted light and carry high iron.
d. The associated nonsulfides are usually conspicuously more iron-rich than those of the lead-zinc ores; iron oxides—particularly magnetite but sometimes hematite—may be present along with iron-rich dark chlorites and iron-bearing carbonate.
e. The ores are usually *not* well bedded. Again the Copperbelt ores are an exception. Usually the chalcopyrite-pyrrhotite ores, while stratiform in their outlines, are in the form of disordered, irregular veins and patches, associated with similarly disordered veins and patches of carbonate, quartz, and lesser chlorite.

In addition to these two principal ore types, a copper-zinc type is also fairly common (particularly in the Superior province of the Canadian Precambrian and in the Norwegian Caledonides). The sulfide mineralogy is usually sphalerite-chalcopyrite (either may dominate), with pyrite as the iron sulfide. Pyrrhotite also occurs in these ores but is usually minor in amount.

Whereas mineralogical zoning—i.e., statistical trends in metal and hence mineral ratios—is common in the lead-zinc ores, it is uncommon, or at least little known apart from the Roan, in the copper ores. This may be a reflection of economic considerations. Since we are interested in two main components in lead-zinc ores, we are naturally concerned with variations in their proportions; since we have only one important economic component in the copper ores, our one concern here is *grade*—i.e., variation in absolute quantity of the copper alone. Undoubtedly there is frequent systematic zoning of Cu/Fe and (Pb + Zn)/Fe ratios, but because of the uneconomic nature of the iron, this has rarely been studied. Zoning of copper minerals and of copper minerals versus iron is, however, well documented on the Copperbelt (cf. Fig. 15-10) and is recognized in a general, qualitative way in many stratiform pyritic lead-zinc ores.

The microstructures of these ores appear to conform with the principles discussed in Chaps. 8 and 9—particularly the latter. The chief influence on grain-boundary shape has been interfacial free energy, though there are also some nonequilibrium structures that have apparently developed both early and late in the history of the ores.

What appear to be early nonequilibrium structures in the lead-zinc ores are microspherical bodies—particularly of pyrite—that are found in the ores themselves and in the associated barren pyritic zones. These are generally referred to as *framboids* or *framboidal pyrite* and have an apparent cellular structure that gives them a somewhat raspberry-like appearance. These are very common in the Rammelsberg ore and are known, with varying abundance, in many other deposits of this class. They have a distinctly organic appearance and are thought by some to represent original bacterial tests now replaced by pyrite. In addition to these framboids, there are common occurrences of colloform aggregates of pyrite, galena, and sphalerite. Pyrite is the dominant member of these aggregates and contains zones of galena and/or sphalerite, the galena in particular often forming distinctly spongy intergrowths with the pyrite.

The bulk of the sulfide is, however, arranged as single-phase or duplex or polyphase random aggregates. Single-phase aggregates (e.g., initially monomineralic bands) show normal foam structures. Pyrite in matrices of other sulfides (or nonsulfides) is usually subidiomorphic, with variably high dihedral angles against adjacent grain boundaries. As noted in Chap. 9, pyrrhotite may behave in a random fashion or it may develop a lath-shaped habit—variability due, presumably, to effects on interfaces of nonstoichiometry in the pyrrhotite and its hosts and of trace impurities. Sphalerite, galena, and the very minor chalcopyrite of the lead-zinc ores behave randomly. Sphalerite, with its high dihedral angle (ca. 130°), tends to form rounded bodies when isolated in galena, as, much less frequently, does the chalcopyrite (ca. 125°). Galena as a minor phase in sphalerite (dihedral angle ca. 103°) forms concave cuspate bodies as described in Chap. 9.

Ores of this group characteristically contain virtually no structures that could be ascribed to exsolution.   Occasionally numbers of rounded blebs of chalcopyrite are found in single grains of sphalerite, though these may well represent no more than small impurity grains trapped during secondary grain growth ("coarsening") of sphalerite.

Most of the lead-zinc ores show minor, almost cryptic, deformation, which is often unapparent in routine microscopic investigation.   It is, however, virtually ubiquitous and is apparently due to late-stage movement.   It may be revealed by the width of spread of frequency distributions of angles "as measured" at single-phase triple junctions, by displacement of sphalerite cusps from their associated galena grain boundaries, by departure of grain-boundary shapes from equilibrium smooth curves, minor kinks, deformation twins, and so on.   Ores of this kind typically do not show accretionary structures.

The vein-type copper ores often look obviously deformed, in contrast to their lead-zinc associates.   This deformation may not in fact be as intense as it first appears, though careful etching, observation, and measurement usually show at least some effects of movement.   The chalcopyrite and pyrrhotite usually exhibit at least minor deformation twinning and undulose extinction, and non-equilibrium single-phase triple junctions are usually revealed by measurement. The deformation and subsequent synthetic annealing of chalcopyrite from Mount Isa has already been discussed in Chap. 9.   Although these ores have a distinct vein form—in contrast to the bedded nature of the lead-zinc ores—they are similar to the latter in showing virtually no evidence of exsolution or of accretion.   There is no indication that they have grown, in their present form at least, by simple accretion on the walls of the fractures in which they occur.

3. ISOTOPIC CONSTITUTION

Much concerning the isotopic constitution of these ores has already been given in Chap. 7.

The principal features of the isotopic constitution of their *sulfide sulfur* are

a. The mean value of $\delta^{34}S$ for a number of samples from any single deposit has always been found to be heavier than the meteoritic value.   In some cases the degree of "heaviness" is slight—e.g., mean $\delta^{34}S = +1.0$ at Broken Hill; in others it is substantial—e.g., $\delta^{34}S = +26.8$ for one sample of sulfides at Zardu, Iran.   Most deposits fall in the range $3.0 < \delta^{34}S < 15.0$ (Fig. 7-3).
b. For the simpler, bedded lead-zinc ores, the spread of $\delta^{34}S$ values for any single deposit is usually fairly narrow (Fig. 7-3).   The copper ores, however, seem to show a rather greater range.
c. There is a tendency for a systematic partitioning of ^{32}S and ^{34}S as between the major sulfides.   This has been demonstrated most clearly for galena-sphalerite.   Where coexisting pairs have been measured carefully, the sulfur of the galena has almost invariably a higher $^{32}S/^{34}S$ ratio than has the sulfur of the sphalerite.

d. In the few cases so far measured and documented, $^{32}S/^{34}S$ tends to change systematically going upward stratigraphically in an individual deposit.

The principal features of the isotopic constitution of their *lead* are

a. For any individual deposit the values for $^{206}Pb/^{204}Pb$, $^{207}Pb/^{204}Pb$, and $^{208}Pb/^{204}Pb$ are extremely uniform throughout the ore body as a whole. In many cases this within-deposit uniformity is so great that even the most precise mass spectrometry is unable to distinguish between samples.

b. Where a number of deposits occur on a single sedimentary horizon within a mineralized district, there is a high between-deposit uniformity. Again the similarity of ratios may be so great as to completely preclude distinction between the different ore bodies on the basis of lead isotopes.

c. As a group the deposits have $^{207}Pb/^{204}Pb:^{206}Pb/^{204}Pb$ and $^{208}Pb/^{204}Pb:^{206}Pb/^{204}Pb$ relations approximating to a single "growth curve," indicating their possible common origin in a deep source of fairly (though certainly not absolutely) uniform U–Th–Pb ratios.

d. In the one case (Mount Isa) where the lead from the lead-zinc ore and the trace lead from the immediately adjacent copper ore have been analyzed isotopically (Richards, 1967), the two leads are indistinguishable.

ORIGIN

Two principal theories are held for the origin of the sulfides of these ores:

1. They have been derived from intrusive granitic rocks, transported to their present positions in solutions, and precipitated by substitution, or *replacement*, of *preexisting* rock materials. In this view they are *subsurface hydrothermal replacement deposits*.
2. They have been derived from solutions—the sea, rivers, or sea-floor springs—precipitated subaqueously as chemical sediments, and accumulated as an *addition* to rock materials forming *contemporaneously*. In this view they are *subaqueous sedimentary deposits*.

A third theory has developed by combination of these two: iron sulfides have first accumulated by sedimentation, providing sulfide for later localization of copper, zinc, and lead by replacement.

There are a number of variations on these principal themes, the more important being:

1. *Subsurface replacement*
    a. Metals and sulfur are derived from plutonic intrusions and deposited by selective replacement of silicate-carbonate sedimentary rocks.
    b. Metals and sulfur are derived by lateral secretion (including expulsion of pore waters during compaction) of traces from underlying or adjacent

sediments and deposited by selective replacement of silicate-carbonate sedimentary rocks.

  c. Nonferrous metals are derived from plutonic intrusions and/or by lateral secretion from sediments and deposited by selective substitution for the iron of preexisting sedimentary iron sulfide.

  d. In certain volcanic provinces, iron and the nonferrous metals occur in silicate structures and as pore-space compounds in bedded pyroclastic rocks; these are then impregnated by hot sulfur-bearing volcanic gas, rendering the preexisting metal compounds into metallic sulfides.

2. *Subaqueous sedimentation*

  a. Metals and sulfur are derived from the body of the sea; sulfide S^{--} is produced by bacterial sulfate reduction of sea water SO_4^{--} in stagnant areas of the sea floor; and metallic sulfides deposit and accumulate in suitable euxinic environments.

  b. Metals are derived from local continental drainage, S^{--} by bacterial reduction of sea water SO_4^{--}, as in (*a*); the metals accumulate as sulfides in near-shore euxinic environments, as in the highly organic reduced shales developed near organic reefs.

  c. Metals and at least part of the sulfur are derived from sea-floor hot springs (generally thought to be substantially volcanic, though deep geothermal brines may be significant—see Chap. 6); where such contributions are made to euxinic environments, S^{--} is generated bacterially and metallic sulfides are precipitated and accumulate as sediments in volcanic-organic environments.

  d. Metals and sulfur combine as sulfides within submarine volcanic conduits; sulfides are then emitted into the bottom waters as fine suspensoids from hot springs, and/or as tuffs, breccias, and agglomerates—and even perhaps as flows—from sea-floor volcanic fissures and cones.

These clearly constitute a wide range of alternatives, and the choice of the "correct one" has, as we have already noted, been one of the greatest problems of ore petrology. Once again the problem may not be resolvable into a single, simple answer. Our present class of deposits may in fact be a multiple one, and several of the theories may be correct—each for a subclass of our broad class of stratiform sulfide ores.

In considering possible origins we shall therefore try to avoid any suggestion of exclusiveness; i.e., we shall keep in mind that in fact a variety of sources and mechanisms may be involved. We shall also keep in mind that "origin" is an omnibus term of confused meaning and therefore look separately at source, transport, and deposition—the "components" of origin.

1. SOURCE

Virtually all crustal and mantle materials contain the relevant metals and sulfur, and all are thus potential sources. Since the earth's surface was probably molten at some stage, the source of all materials may be said to be *ultimately* igneous. We are, however, more concerned with "immediate" source, and for this, certain compositional features appear to provide the best of a rather poor set of clues.

Many features of composition are unreliable due to obvious segregation during the ore-forming processes. The separation of copper from lead-zinc is an obvious illustration. The tendency for constancy of lead/zinc ratios coupled with their clustering about the meteoritic-average crustal value (Fig. 15-12) might be taken as a hint of a mantle, or highly homogenized deep crustal, source, but this ignores the possibility of deviations from original ratios—a possibility so clearly indicated by the obvious changes in Cu/Zn and Cu/Pb ratios—and assumes that all available lead and zinc has come to rest in the deposit concerned.

There seem to be two more promising indicators at the moment: lead isotopes and S/Se ratios.

a. *Lead isotopes.* We have already noted that within-deposit ratios are always very uniform and that the ratios for a range of deposits fall remarkably close to simple U–Th–Pb decay-growth curves. Both of these suggest that the leads have been derived (with remarkably little contamination) from some *comparatively* homogeneous source or from *comparatively* similar sources. Individual granitic bodies, or local segments of sedimentary prisms, seem far too variable and inhomogeneous to satisfy this requirement. Alternatives seem to be the basaltic layer of the deep crust, the upper mantle, or very large volumes of homogenized metasediment deep in the crust. The author is inclined to favor the basaltic layer or upper mantle.

b. S/Se *ratios.* It has been suggested by J. S. Brown (1965) that some instances of Pb isotope uniformity may be ascribed to derivation from sea water. However, for many of the deposits with which we are concerned, selenium contents are higher by *one or two orders of magnitude* than those of sulfides *known* to have formed by sea-water sulfate reduction. This seems to indicate that the source is neither the body of the sea nor continental effluent (assuming that such effluent has not been derived from unusually highly seleniferous areas such as the present Colorado Plateau—cf. Chap. 7). The S/Se ratios are rather reminiscent, however, of volcanic sulfur and sulfur compounds, suggesting a possible igneous source.

That the sulfide elements have not been derived from the body of the sea is also indicated by a piece of field evidence already remarked upon: the deposits are never as extensive as the facies that encloses them. Indeed they are generally highly localized.

This evidence leads the author to the *tentative* conclusion that at least many of these deposits have a volcanic source in the basaltic layer or upper mantle. Both of the latter have minor constitutional heterogeneities, as any student of basalt compositions knows, but both are *comparatively* uniform. The U–Th–Pb ratios, like those of any others of their constituent elements, for example, Si/Al or Mg/Ca, would have undoubtedly varied—but not greatly. Other minor heterogeneities might be expected to have developed by minor contamination and/or loss in the volcanic conduit.   Keeping these points in mind there seems little doubt that a volcanic source is compatible with the lead isotope abundances.   It is also compatible with the incidence of selenium.

There is, of course, the obvious difficulty of the apparent lack of associated volcanism in some cases.   Sullivan and the Roan are excellent examples.   All that can be said at this stage is that careful search for volcanic detritus is merited. The intimate association of fine tuff with the Mount Isa sulfides was not recognized until 1960, and it was only in 1966 that La Berge drew attention to the shards interbedded with the Transvaal and Western Australian iron formations.

2. TRANSPORT

Again it is isotopic abundances that seem to offer the most reliable clues—in this case those of lead and sulfur.

In Chap. 7 it was noted that in contrast to the rather uniform within-deposit Pb and S isotope ratios of stratiform sulfides, those of many veins showed very clear variability in this respect.   As was pointed out by Stanton and Russell (1959) in connection with lead, there seems no doubt that the components of veins have migrated, often over substantial periods of time, through various kinds of crustal materials.   They are thus exposed to extensive contamination by traces of lead and sulfur in the rocks they traverse and, in the case of sulfur, to a variety of opportunities for fractionation.   On simple theoretical grounds one might therefore expect—within individual vein deposits and between a number of similar vein deposits in a single locality—that lead and sulfur isotope abundances might vary substantially, which indeed we now know them to do.   A little thought suffices to show that the only deposits that might be expected not to show such variations— or to show them to only small degree—are those whose components have not had to migrate slowly through crustal materials.   The rapid and probably closely channeled movement developed in volcanic conduits seems that preeminently most likely to avoid contamination and to give minimum opportunity for sulfur fractionation.   Once again a volcanic affiliation is indicated as a strong possibility. Transfer in this case would be first in the melt; some components might then move to high temperature, rapidly rising gas and some, finally, to near-surface hot-spring solutions.

3. DEPOSITION

These deposits are conspicuously not due to the filling of gross openings; apart from a few copper ores they do not occur as veins, and we have already observed that even these do not show evidence of accretion.   If they have been deposited beneath the earth's surface, the sulfides have formed by replacement and, perhaps to a minute extent, by pore-space filling.   If they have been deposited at the surface (i.e., subaqueously), they have come to rest as chemical sediments, detrital sediments (sea-floor pyroclastic or erosional detritus), or sulfide lava.

Clearly there is a wide range of stages at which sediments (and it must be remembered that the containing materials of the ores are always sedimentary rocks) might be replaced: during diagenesis, during the later stages of lithification, after lithification but prior to metamorphism, after metamorphism, and so on. Evidence for replacement in the earlier stages is quite common and reliable. Soft parts of fishes, plants, and delicate shells and corals are well known to be replaced by sulfides during and shortly after the accumulation of some euxinic sediments. Portions of some pyroclastic cones in Japan (see Kinkel, 1966) are known to be undergoing replacement by late-stage pneumatolytic pyrite at the present time. It is worth noting, however, that most *demonstrable* replacement does seem to be of this "early-stage" kind. There are indeed many cases where replacement has been *postulated* to have affected well-lithified or metamorphosed rocks, but concrete evidence is usually lacking. The only really unambiguous evidence—mineral pseudomorphism—is almost always conspicuous by its absence. For example, the sulfides at Broken Hill are thought by some to have been deposited by late-stage replacement of the enclosing high-grade metamorphic rocks. However, although the latter contain abundant idiomorphic to subidiomorphic garnet, sillimanite, pyroxene, and amphibole, 30 years of intensive microscopic work has failed to reveal a single example of the assumption by sulfide of the crystal outlines of any of these silicates. Where we *do* know that replacement has taken place in ores such as these—that is, during weathering and the formation of secondary sulfides—pseudomorphism is widespread and obvious. Crystal outlines, cleavages, and grain structures in the primary sulfides are often preserved beautifully in the secondary sulfides.

We may therefore say that replacement is well established as a process taking place during and, particularly, shortly after sedimentation of both "normal" and volcanic kind. It cannot, however, be said to be established as a process of any consequence in the deposition of sulfides in deeply buried, highly lithified sedimentary rocks.

We have already noted that if the sulfides were deposited contemporaneously with the associated nonsulfide sedimentary materials—i.e., as sediments themselves —they might have done so as chemical precipitates, pyroclastic or erosional particles, or lavas.

Investigation of sulfide formation—and heavy metal concentration in general —in modern sediments has long been hampered by the great difficulties involved in oceanographic work. It has been known for a long time that heavy metals are notably adsorbed by sea-floor manganese compounds (referred to in Chap. 14), but neither the relative nor the absolute abundances of the metals here match those of stratiform base metal ores. Laboratory studies have shown that sulfate-reducing bacteria are capable of producing enormous amounts of sulfide very quickly, given appropriate conditions and an abundant supply of SO_4^{--}. They have also shown (see Temple and Le Roux, 1964) that such bacteria can survive high concentrations of the heavy metals if they have a sufficient supply of SO_4^{--}; by rapid generation of S^{--} they can render the metals into insoluble form—the sulfides— so rendering them nontoxic. There is also no reason in principle why sulfides should not be precipitated quite abiologically by, for example, sea-floor volcanic H_2S. However, oceanographic work up to about 1965 had shown no more than very minor development of sulfide in modern sediments, and such sulfide always seemed to be pyrite or some form of troilite.

Recent work in the Red Sea (see, for example, Miller et al., 1966; Degens and Ross, 1969) has, however, dispelled all doubt of the possibility of marine chemical precipitation on a large and concentrated scale.   As mentioned in Chap. 6, three hot ($T \simeq 56°C$) brine pools, situated in depressions as shown in Fig. 6-8, are apparently being fed from springs in the sea floor.   Each is now an area of active chemical sedimentation, and extensive deposits are being formed.   The principal elements are Si, Al, Fe, and Ca, but substantial quantities of zinc (up to about 17 per cent so far detected in individual samples), copper (up to 3.5 per cent), and lead (up to 0.75 per cent) are being deposited.   This is indeed a sedimentary base metal deposit *in vivo*.   Sulfide is quite abundant, the principal species identified so far being pyrite, sphalerite, and chalcopyrite.   The abundance relationships of the base metals, and also the patterns of sulfur and lead isotope abundances, bear a truly remarkable similarity to those found in the older stratiform ores.   Cu–Zn–Pb relations in samples of the Red Sea metal-rich sediments are shown in Fig. 15-15. The similarity between this and Fig. 15-12a is striking.   Isotopic studies (Cooper and Richards, 1969) of the leads of the sediments of the brine pools and of the "normal" Red Sea sediments show that the leads of the brine pool material are (1) isotopically uniform and very similar in their other features to the leads of many

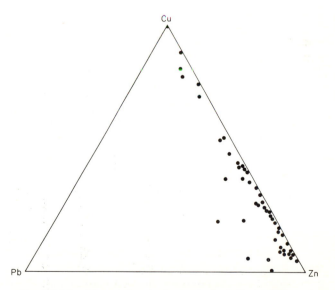

Fig. 15-15 Atoms per cent copper-lead-zinc relations in some metal-rich sediments of the Red Sea hot brine pools. Note the striking similarity between this and the pattern of Fig. 15-12a. [Calculations based on analyses by *Bischoff, in Degens and Ross (eds.), "Hot Brines and Recent Heavy Metal Deposits in the Red Sea," Springer-Verlag New York Inc., New York, 1969, with the publisher's permission.*]

stratiform sulfide ore bodies and (2) isotopically quite different from the "pelagic" leads of the "normal" sediments, which show a conspicuous lack of uniformity. Similar isotopic studies of the sulfide sulfur of the brine sediments and of the adjacent "normal" sediments (Kaplan, Sweeney, and Nissenbaum, 1969) show that in the metal-rich sediments of the Atlantis Deep, $\delta^{34}S$ values are positive and possess about the same degree of uniformity as do most stratiform ores (cf. Fig. 7-3), in contrast to the "normal" sediment sulfides that exhibit highly negative and variable $\delta^{34}S$. (It must be mentioned, however, that one core in the smaller Discovery Deep was found by Kaplan et al. to have sulfide sulfur of quite uniform, but highly negative, $\delta^{34}S$.)

Modern examples of large-scale pyroclastic and lava-type sulfide concentrations are not known [Skinner and Peck (1969) and others have, however, described small iron-nickel-copper sulfide segregations in some Hawaiian lavas]. Some of the sulfide concentrations of northern Quebec and Ontario look distinctly like sulfide tuffs, breccias, and agglomerates, and in at least one case (Vaux mine, Quebec), it has been suggested that the sulfides have accumulated as sinter, quenched lava, and pyroclastic material about a recognizable Precambrian volcanic cone. Some of the pyritic masses of Cyprus (Bear, 1963; Hutchinson, 1965) are in the form of pillows, and this has been taken by some geologists to indicate deposition as genuine submarine sulfide lava flows.

Neglecting for the moment the possibility of sulfide lavas, there is a simple method for determining—at least in a gross way—whether replacement or sedimentary processes have been responsible for the localization of stratiform sulfides in older terrains. The reasoning is as follows:

a. If the sulfides have been deposited by selective replacement, they have formed by *substitutional processes;* i.e., they have formed at the expense of some particular constituent or constituents of the host rock.

b. If the sulfides have been deposited as sediments, they have formed as *additions to* the accumulating sedimentary pile; i.e., they are simple additions to, not substitutions for, associated sedimentary materials. In this case progressive increase in the amount of sulfide should not lead to any systematic change in the proportions of the constituents of the host rock.

Thus suppose that the nonsulfide associates of a certain occurrence of stratiform sulfides are carbonate, quartz, and feldspar (for which CO_2, SiO_2, and Al_2O_3 may be used as indices), that the proportions of these in "barren" sections of the strata vary over the field shown by filled circles in Fig. 15-16a, and that the mean proportions are essentially central to this field. Suppose now that sulfides were introduced into the beds by selective replacement of carbonate. As sulfide sulfur increased, the proportion of carbonate would decrease, causing the field of Fig. 15-16a to shift progressively toward the SiO_2–Al_2O_3 edge of the diagram, as shown. If, on the other hand, the sulfides were introduced during sedimentation, they would—as additions—cause no change in CO_2–Al_2O_3–SiO_2 proportions, and hence progressive increase in sulfide sulfur would leave the field essentially stationary, as in Fig. 15-16b.

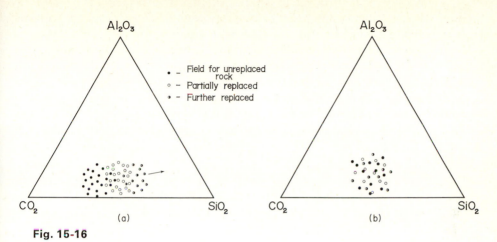

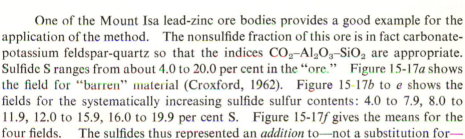

Fig. 15-16

One of the Mount Isa lead-zinc ore bodies provides a good example for the application of the method. The nonsulfide fraction of this ore is in fact carbonate-potassium feldspar-quartz so that the indices CO_2–Al_2O_3–SiO_2 are appropriate. Sulfide S ranges from about 4.0 to 20.0 per cent in the "ore." Figure 15-17a shows the field for "barren" material (Croxford, 1962). Figure 15-17b to e shows the fields for the systematically increasing sulfide sulfur contents: 4.0 to 7.9, 8.0 to 11.9, 12.0 to 15.9, 16.0 to 19.9 per cent S. Figure 15-17f gives the means for the four fields. The sulfides thus represented an *addition* to—not a substitution for—the nonsulfides and are hence sediments.

It should be noted that there are two matters to be watched in the application of this method.

a. *Pore-space filling.* Disseminated sulfides might conceivably have been deposited as fillings of pores in not fully compacted rocks. While this should be readily apparent from microaccretionary structures, it is clear that such a process also involves addition and would yield the same arithmetical results as sedimentation. It is necessary, however, to keep *quantities* in mind here. Many of the bands of stratiform ores are 75 to 100 per cent sulfide—far in excess of the 50 per cent or so permissible on the basis of porosity. Certainly some of the "low-sulfide" bands might, on a purely numerical basis, be ascribed to pore-space filling, but as we have just noted, the microaccretionary structures that would be expected from such a process are rarely if ever observed in these ores. Any banded depositional structures that *do* appear are of a *concretionary* nature—i.e., are related to central nuclei rather than to peripheral grain boundaries, as would be the case had banding resulted from the filling of pores.

b. *Sedimentary facies changes.* The associated sedimentary rocks may show *primary* compositional variations to which variations in the incidence of sulfide are related. Such variations are most likely to reflect original facies

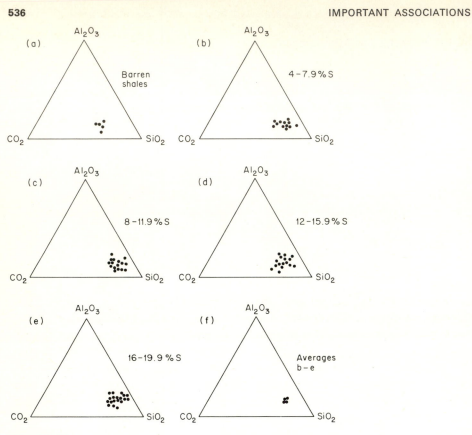

Fig. 15-17

changes and may give spurious indications of substitution, i.e., replacement.
For example, as carbonate tends to accumulate in mildly aerated environments and sulfides in anaerobic ones, increase in sedimentary sulfide may
often be accompanied by a systematic decrease in carbonate.   In such cases
calculations of the present kind would give results suggestive of replacement.
As an example, a number of separated ore bodies at Mount Isa exhibit
between them a systematic drop in the proportion of CO_2 with increase in
sulfide S, as shown in Fig. 15-18.   This suggests replacement, in apparent
contradiction to our earlier finding.   The problem may be resolved in most
cases by simple calculation.

The density of any substance is given by $D = M/V$; hence

$$V = \frac{M}{D}$$

Now for the host (apparently carbonate in the present case),

$$V_1 = \frac{M_1}{D_1}$$

and for the metasome (our sulfides),

$$V_2 = \frac{M_2}{D_2}$$

but since replacement is a volume-for-volume process,

$$V_1 = V_2$$

$$M_1 = M_2 \frac{D_1}{D_2}$$

Substituting a value of 2.90 for D_1 (density of Mount Isa dolomitic carbonate) and 5.29 for D_2 (a mean value for the density of the mixed Fe–Pb–Zn sulfides), we have

$$M_1 = \frac{2.90}{5.29} M_2$$

$$= 0.548 \, M_2$$

and from this the number of weight units of carbonate that would be lost when replaced by a given number of weight units of sulfide can be computed and changes in host-rock mineral ratios, with increase in sulfide, obtained. The four ore bodies

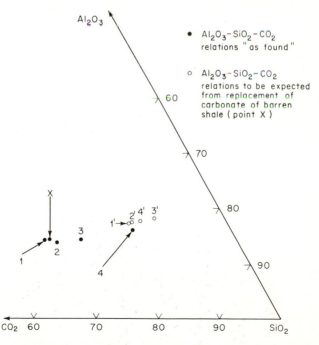

Fig. 15-18 Comparison of actual CO_2–Al_2O_3–SiO_2 relations in the principal lead-zinc ore bodies compared with those theoretically expected as a result of replacement of "average Urquhart shale" by the amounts of sulfide concerned in each case.

represented by points 1, 2, 3, and 4 in Fig. 15-18 contain mean sulfide S of 10.6, 10.8, 13.1, and 12.4 per cent, respectively. Replacement of "barren" shale (marked ×) would have given the points 1', 2', 3', and 4'. Apparently the relationship between 1, 2, 3, and 4 is not a replacement one and, as the field and mine geology also indicate, reflect a simple primary facies change.

A calculation of this kind should always be done as a check where numerical results suggest the possibility of replacement.

In addition there are several possible geochemical indicators of depositional processes.

a. Sulfur isotopes.   The existence of systematic relationships between $^{32}S/^{34}S$ and stratigraphy strongly suggests sedimentation as the depositional process—for the sulfide S, at least.   The heaviness of the sulfur and its tendency in some cases to *lighten* going upward stratigraphically is, however, not easy to reconcile with bacterial sulfate reduction of sea water SO_4^{--}.   Since the $^{32}S/^{34}S$ of such SO_4^{--} appears to have been lower during some geological periods than it is now, it is reasonable to expect that the $^{32}S/^{34}S$ of at least some of these deposits would be comparable with that of modern marine sulfide.   This it conspicuously is not.   Further, if systematic stratigraphic trends were due to progressive ^{32}S removal in a restricted basin, the stratigraphically higher material should have *lower* $^{32}S/^{34}S$—but as we know the opposite is the case at Heath Steele.   Sulfur isotope abundances therefore suggest that in many cases the depositional process was sedimentary but, on present evidence, that it was not a simple marine one and was at least substantially abiological.   This is supported by the recent studies on the Red Sea accumulations.

b. Sulfur/selenium ratios.   These appear (keeping in mind the ambiguities of S/Se ratios already referred to) to corroborate the latter deductions based on $^{32}S/^{34}S$; if the ores were deposited as sediments their materials were not simple sea-water derivatives, and the process of deposition is thus unlikely to have been reduction of sea water SO_4^{--}.

CONCLUDING STATEMENT

It was said at the beginning of this chapter that rather than constituting a simple, consistent, clearly defined category, the stratiform sulfides of marine and marine volcanic association probably included a whole spectrum of types.   There seems no doubt now that this is so.   Some occur in marine sediments showing no hint of volcanism, others are closely tied to volcanic sequences and show negligible relation to specific sedimentary zones and processes.   Others—the majority—seem to show clear associations with both volcanism and specific environments of sedimentation.   Even those that are clearly related to features of sedimentation show wide variation in their affiliations: some are associated with products of well-flushed environments—algal reefs and sandy sediments—while others seem tied to environments of a conspicuously stagnant and putrid kind.

The pronounced stratigraphical affiliations, the tendency for segregation of copper from lead-zinc according to stratigraphy and, in some cases, to facies patterns, the stratigraphic arrangement of metal ratios and the frequent tie between within-deposit stratigraphy and sulfur isotope ratios all suggest that the major part of the sulfides has been deposited during sedimentation.   This is supported by the lack of substitutional relationships between sulfide and host rock in a number of examples.   This is, however, probably where depositional similarities end.   Deposits whose localization and mineralogical zoning are clearly spatially related to sedimentary facies may well be sedimentary precipitates *in toto* and localized at least partly by biological agencies.   Some of those spatially unrelated to specific facies may have resulted from simple abiological precipitation about sea-floor hot springs or from submarine pyroclastic deposition.   The frequent lowness of within-deposit variation in $^{32}S/^{34}S$ ratios, coupled with the fact that the ores usually *expand* the sedimentary sequence in which they occur, suggests very rapid deposition of sulfide; i.e., any richness of sulfide appears due not to diminution of other sedimentation but to the latter's being overwhelmed by very rapid sulfide sedimentation.   A minority of deposits may well have been extruded as sulfide-rich flows under the pressure of a very large water column in deep areas of the sea.   Others may have been deposited initially—by any of the above processes—largely as iron sulfide, the iron of which was later partly replaced by migrating base metals. Such base metals might well have come, by expression of pore waters during compaction, from underlying tuffs or other metal-rich sediments or from late-stage volcanic gases or solutions.   While replacement of this kind would follow burial of the iron sulfide, it would probably take place during the diagenetic and early compactional stages.   Finally, in volcanic areas, volcanic gases escaping to the surface may induce base metal replacement of preexisting stratiform volcanic iron sulfides or the replacement of early pyroclastic silicates by a variety of sulfides.

Deposition may thus be by sedimentation or replacement, or both.   Where the latter operates it does, however, seem likely that it is a shallow phenomenon closely related to sedimentation in time.   As evidence accumulates it seems less and less likely that any of our important stratiform concentrations have been deposited by late-stage plutonic replacement of older sedimentary and metasedimentary rocks.

On *transport* there is not a great deal that can be said.   If the source has been distinct and distant from the site of deposition, the extreme within-deposit uniformity of lead isotope ratios indicates avoidance of migrational contamination and hence rapid and highly confined transport.

On *source* the evidence is rather negative.   The high selenium contents and the nature of the $^{32}S/^{34}S$ abundances suggest that it is neither local continental erosion nor the sea itself.   The strong tendency for the leads of deposits widely spread in space and time to adhere to a U–Th–Pb decay-growth curve suggests derivation from a source a good deal less heterogeneous than the upper crust and the silica-rich "plutonic" rocks it contains.   Such a source may have been the upper mantle, the basaltic lower crust, or some highly homogenized segment of the deep crust.

In the present state of the evidence the author inclines to the view that most of these deposits are tied to one phase or another of volcanism. At the height of a volcanic episode, and close to the center of the activity, sulfide concentrations might be expected to form in all the ways that the silicate rocks do: by immiscible liquid segregation and subsurface injection, by shallow subsurface replacement and cavity filling, as pyroclastic rocks, and as chemical sediments round submarine fumaroles and hot springs. As volcanism waned,† and at greater distances from its centers, hot-spring deposits would predominate, and indeed these might well form almost without any other volcanic products in the beds, suggesting an "avolcanic" environment. It is, of course, the chemical precipitates with which we are chiefly concerned at this moment. They would be expected to form wherever marine and andesitic volcanic environments coincide—island arcs and, in some cases, submarine rift valleys.‡ In the arc environment we have a whole range of situations: the arc in its youth and prime, when volcanic phenomena are of overwhelming importance and a variety of ore types develop; the arc at maturity, with less violent volcanism and increase in importance of chemically sedimented sulfides and with associated minor pyroclastic rocks; and finally, the "island arc" as part of the continental margin, volcanic activity almost passes from recognition and consists of only weak and desultory hot-spring activity, and sulfide deposits—often of widespread low grade—form in shoreline and shallow interisland environments. By this time extrusive and pyroclastic activity have ceased, and the ore-bearing beds contain little to suggest a volcanic affiliation.

All this is, however, no more than a reasonable hypothesis. Clearly there is almost no aspect of the "origin" of these deposits that can be said to have been solved. Indeed we have barely shaken ourselves free of the all-embracing—and hence highly inhibiting—plutonic replacement theory and are hardly past the threshold of a new attack on the problem. Such a stage in the investigation of so important a group of deposits is, however, an intriguing and exciting one.

RECOMMENDED READING

Anderson, C. A.: Massive sulfides and volcanism, *Econ. Geol.*, vol. 64, pp. 129–146, 1969.
Dunham, K. C.: Neptunist concepts in ore genesis, *Econ. Geol.*, vol. 59, pp. 1–21, 1964.
Mendelsohn, F. (ed.): "The Geology of the Northern Rhodesian Copper Belt," Roan Antelope Copper Mines Limited and MacDonald & Co. (Publishers) Ltd., London, 1961.
Stanton, R. L.: Elemental constitution of the Black Star orebodies, Mount Isa, and its interpretation, *Trans. Inst. Mining Met.*, vol. 72, pp. 69–124, 1962.
Tatsumi, T. (ed.): "Volcanism and Ore Genesis," p. 448, University of Tokyo Press, Tokyo, 1970.

† The relationships between the timing of volcanic sulfide ore deposition and the *waning* stages of a volcanic cycle are well dealt with by Kinkel (1966) and Horikoshi (1969).
‡ It should be kept in mind that at least some of the "troughs" associated with island arcs are faulted, i.e., "rift," valleys.

16
Some Stratabound Ores
of Sedimentary Affiliation

All the ores of sedimentary association that we have considered so far are essentially *stratiform*; that is, they occur as layers that are disposed concordantly with respect to the stratification of the enclosing sedimentary materials. In general, too, they are sharply confined stratigraphically; within a given district they occur on one or a very few preferred sedimentary horizons. In some cases the iron, but more particularly manganese and sulfide, have suffered some modification of their conformability as a result of diagenetic changes and tectonic deformation, but these effects are usually local and unimportant in principle.

The ore types we are now to consider also habitually favor specific horizons within local sedimentary sequences and show marked preferences for certain rock types. For the most part, however, they do not occur as concordant layers; they are found, within specific sedimentary units, as cross-cutting veins, pore-space and breccia fillings, cave and solution-cavity linings, and minor layers. They are thus confined to particular stratigraphic horizons on a *large scale* but are usually discordant on a *small scale*. They are therefore referred to as *stratabound* deposits. Clearly *stratiform* ores are also *stratabound*, but *stratabound* ores are not necessarily *stratiform*. There are many differences between these two morphological categories, and we shall look at the more important of them in this chapter.

There is a wide variety of stratabound ores, and their number increases

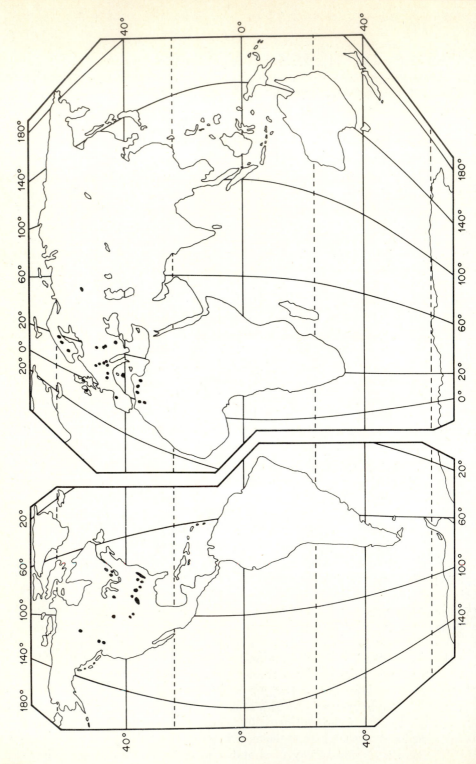

Fig. 16.1. Distribution of major limestones, isolated zone, reef occurrences, and seas of occurrence, throughout the world

steadily as greater emphasis is placed on the regional aspects of ore occurrence in mineral exploration. Three groups stand out, however: the *limestone–lead-zinc*, the *sandstone–uranium-vanadium-copper*, and the *conglomerate-orthoquartzite–gold-uranium-pyrite* associations.

THE LIMESTONE–LEAD-ZINC ASSOCIATION

This constitutes one of the world's great sources of lead and zinc. It is the principal source of these metals in the United States and Europe and yields large quantities of lead and zinc both in Canada and North Africa.

DISTRIBUTION

The more important occurrences are those of North America, Europe, Russia, England, and North Africa, and these are indicated in Fig. 16-1. Deposits of this type are known elsewhere, e.g., Australia, but in the present state of exploration these appear to be small and of low lead and zinc content.

The principal earlier known areas of occurrence were Europe (particularly the region of the Alps) and a region of the United States roughly within the upper and middle areas of the Mississippi Valley. In consequence this kind of deposit became known as "Alpine"- or "Mississippi Valley"-type mineralization. Lesser used synonyms are, respectively, "Silesian type" and "Tri-State type" (an important European area of occurrence is in Silesia, and in the Mississippi Valley much of the earlier mining was close to the tri-state border of Missouri-Kansas-Oklahoma). Other important North American occurrences are in the southern Appalachians and, particularly, in the Pine Point (Great Slave Lake) area of northern Canada. The principal European occurrences are those of Germany, northern Italy, Yugoslavia, Austria, and southern Poland. Several important occurrences are exploited in Algeria and Tunisia.

In addition to exhibiting quite a wide geographical distribution, this type of mineralization shows a fairly wide spread in time. The Precambrian, however, is not prominent: a few deposits occurring in Precambrian rocks, e.g., small showings near MacArthur River and Dugald River in north Australia, are known, but deposits of this age are neither abundant nor large. Substantial deposits first appear in Lower Paleozoic sediments, and important occurrences occur in strata from that point in time right through at least to the end of the Mesozoic era. Some of the more important occurrences and ages of their containing units are

Cambrian: Southeast Missouri, U.S.A.; Norway-Sweden border districts; Sardinia.
Ordovician: Eastern Tennessee, U.S.A.; Siberian platform deposits, U.S.S.R.
Devonian: Pine Point, N.W.T., Canada; Silesia (the Silesian-Cracovian deposits extend from Devonian to Jurassic).
Carboniferous: English Pennines; Ireland; Kazakhstan, U.S.S.R.; Tri-State field, U.S.A.
Permian: Trento Valley, Italy.

Triassic: Eastern Alps (including Bleiberg); Silesia.
Jurassic: Silesian-Cracovian province.
Cretaceous: Northern Algeria and Tunisia.

FORM AND SETTING

These are inextricably tied, the setting influencing form to a marked degree.

By far the majority of these ores occurs within sedimentary carbonate units. A small proportion occur in associated sandy and shaly rocks. Usually the carbonate rock concerned clearly constitutes part of a reef, i.e., it is biohermal. In some cases it is fore-reef breccia or back-reef calcarenite. The rock may be genuine limestone, but magnesian limestones and dolomite are the most frequent hosts. Frequently the three are present. The deposits thus usually occur along belts conforming to ancient elongated reef complexes or as local groups conforming to more localized irregular reefs and banks. The distribution of such reefs and associated carbonate accumulations is necessarily related to ancient shorelines, bottom topographies, and of course, climate. Thus given a climate suitable for reef formation, the distribution of the reefs—the potential host rocks—has been controlled by paleogeography. Hence the *primary regional* control of limestone-lead-zinc occurrence is now paleogeography.

Upon this is superimposed—again in a regional sense—a variable structural factor. A large proportion of known deposits are found not far from the intersection of major, regional faults with the reef complex concerned. As at Pine Point (Fig. 16-2) these faults appear to be old features developed in the basement rocks along which movement has continued to occur during and after the deposition of the reef and its associated sediments. A fault zone trending NE–SW in the Precambrian rocks, and going under the Paleozoic carbonate rocks at Pine Point, may be seen at the upper right (northeast) of the map. It must be emphasized that major structures of this kind are not always apparent; all that can be said is that they appear with rather conspicuous frequency.

We may now decrease our scale from a regional one to that of a district. On this scale it is usually clear that the deposits—and there may be many of them in a district—tend to occur in one or a very few units in the local reef-carbonate succession. Within these units they seem to occur in a variety of situations—sedimentary features, zones of coincidence of sedimentary and tectonic structures, and in later solution—karst—openings. Some of the sedimentary features in which the ores may be localized are shown in Fig. 16-3 (modified after Callahan, 1967, fig. 1). Callahan has listed the principal settings:

   1. Above unconformities in sedimentary environments such as reefs and facies changes (A-1), compaction or drape structures (A-2), in stratigraphic "pinch-outs" (A-3), talus or landslide breccias (A-4). A-3 and A-4 are related to basement topography.
   2. Below unconformities in solution collapse breccias (B-1) related to a karst topography predating the development of the unconformity concerned, or (B-2) formed by thinning of underlying beds of subsurface drainage.
   3. At a facies change in a formation or between basins of deposition (Callahan, 1967, p. 15).

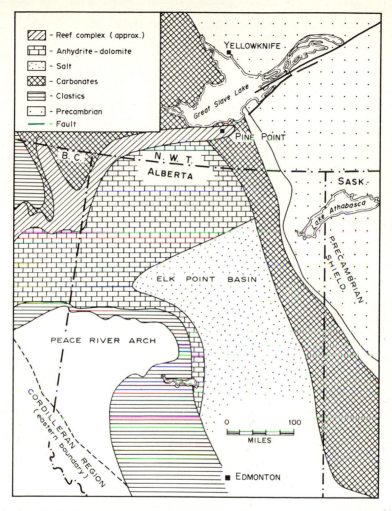

Fig. 16-2 Regional geological setting of the Pine Point limestone–lead-zinc deposits. (*Redrawn from Campbell, Econ. Geol. Monograph no. 3, 1967.*)

A good example of localization according to facies is provided by the "Ladinian plateau reef" type of occurrence, as shown in Fig. 16-4 (after Maucher and Schneider, 1967). The sulfides here are in back-reef sediments, but in other cases they occur within the reef itself, in the fore reef and fore-reef breccias, and so on.

Simple examples of deposition related to a coincidence of tectonic and sedimentary structure are shown in Fig. 16-5. There are numerous variations on this theme. Localization of this kind often occurs at several distinct stratigraphic levels within a given sedimentary unit.

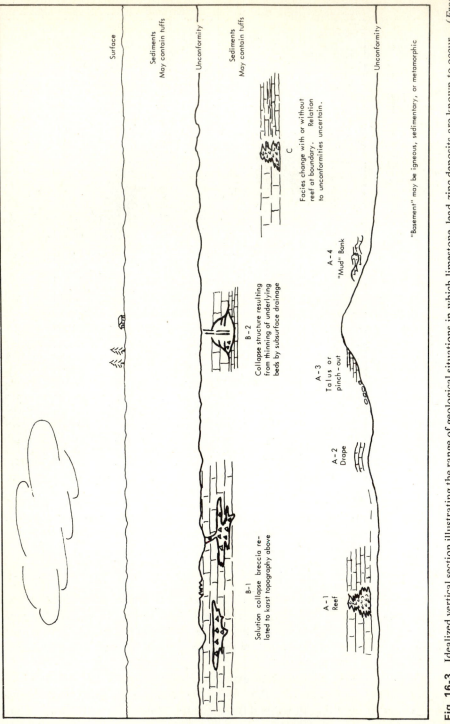

Fig. 16-3 Idealized vertical section illustrating the range of geological situations in which limestone–lead–zinc deposits are known to occur. (*From Callahan, Econ. Geol. Monograph no. 3, 1967.*)

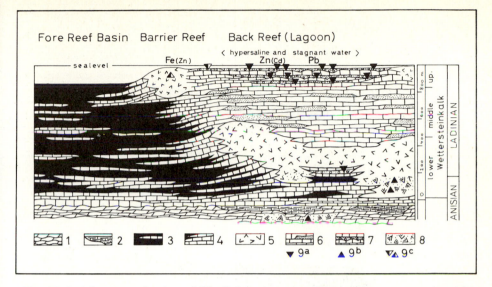

Fig. 16-4 Diagrammatic view of "Ladinian plateau reef" type of occurrence, in which the sulfides are localized principally in the back-reef facies, and to a lesser extent in limestone of the reef wall. 1 = uppermost "Alpine Muschelkalk": wavy-clumpy, thinly bedded, bituminous limestone with chert nodules; 2 = andesitic green tuffs (ash and crystal tuffs, few lapilli with thin layers of marl and limestone); 3 = "Partnach-shales": marls, shales with lenticles of layered limestone (like 4) (Ladinian basin facies); 4 = "Partnach-limestone": bituminous, marly, layered limestones (Ladinian basin facies); 5–8 = different types of "Wetterstein-limestone" (Ladinian reef facies): 5 = massive limestone and dolomite, partly cavernous with relict patterns of bioherms (often coral colonies); 6 = well-layered gray limestone (mainly calcarenite), with debris and colonies of algae (Dasycladaceae), single algal patch reefs; 7 = predominantly thinly layered limestone, with intercalations of the "special facies" in distinct sequences (back-reef units, tuffaceous marls, slump structures, "ore sediments," etc.); 8 = late, diagenetic alteration of the cavernous reef body by recrystallization of dolomite, quartz, and different Fe dolomites; 9a = Pb–Zn(Fe) sulfide ores with sedimentary fabrics; 9b = Pb–Zn sulfide ores primarily enriched in metasomatic replacement bodies, locally associated with small amounts of Cu–Sb–As minerals; 9c = predominantly Fe dolomite and Fe sulfide ores, with small amounts of ZnS(PbS). (*From Maucher and Schneider, Econ. Geol. Monograph no. 3, 1967.*)

Proceeding to a finer scale again—to that one observes on the mine face—the form of the sulfide masses in detail is highly variable.   In some cases the ore is genuinely stratiform on a small scale.   The sulfides appear to occur as beds that may be finely laminated and extend over inches or several tens of feet.   More frequently the sulfides form small cross-cutting veins, linings of cavities, and the cementing material of collapse and other breccias.   In many cases the cavities are still open and surfaces are encrusted with crystals, often of large size and great perfection.   In some cases—notably in some of the southern Appalachian deposits—folding, or faulting, and sometimes intrusion (by igneous melts), of the containing rocks and their ores has led to extensive deformation and metamorphism.

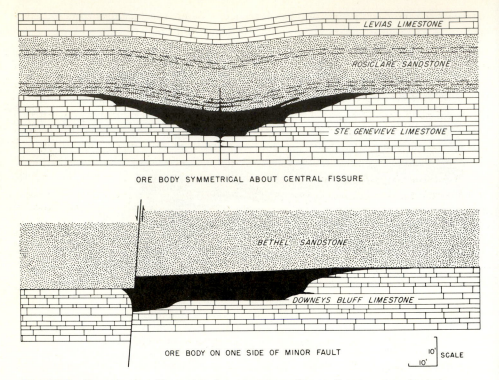

Fig. 16-5 Cross section illustrating the occurrence of some limestone-type lead-zinc ores in which particular tectonic and sedimentary features coincide; examples are from the Cave in Rock district, southern Illinois. (*From Grogan and Bradbury, Econ. Geol. Monograph no. 3, 1967.*)

CONSTITUTION

The characteristic minerals of these deposits are galena, sphalerite, barite, and fluorite. As such they represent remarkable segregation within the earth's crust of lead, zinc, barium, and fluorine. In some deposits lead is present almost to the exclusion of zinc and vice versa, and barite and fluorite may be present in larger or smaller amounts. In general, however, all four are present in one proportion or another. Pyrite and particularly marcasite are common sulfide accessories, and chalcopyrite is often present usually as a minor to trace component. Other non-sulfides are calcite, aragonite, and dolomite and sometimes siderite, ankerite, and colloform silica.

Of the sulfides the sphalerite is characteristically pale in color, with very low iron and manganese contents. Like the sphalerite of the *stratiform* sulfide deposits, it does, however, carry high cadmium and greenockite is an occasional accessory mineral. Galena is characteristically very low in silver in contrast to the galena of the stratiform deposits, which is almost invariably notably argentiferous.

The morphology and textural relations of the sulfides and associated minerals vary with the small-scale form of the occurrence.   Where the deposits are genuinely stratiform or the ore occurs as irregular masses within the host rock, textures are those of polycrystalline aggregates.   The precise nature of the grain boundaries here is not known, but occasional photographs and sketches in the literature suggest that the vein sulfides at least are usually of nonequilibrium, impingement type.   Where the sulfides occur in veins these may be massive or crustified.   Limestone–lead-zinc ores provide some of the classical examples of crustification, which may be symmetrical or asymmetrical and may involve large numbers of bands. One of the most famous structures is that of schalenblende—already referred to in Chap. 8—in which zinc sulfide occurs as alternating bands of sphalerite and wurtzite.   Colloform textures are common in both veins and breccia fillings.   These may involve schalenblende, layered intergrowths of pyrite and marcasite, and the other sulfides and nonsulfides.   Finally the growth of crystals as linings on the walls of solution cavities provides optimum conditions for free growth, and as a result large, well-formed crystals, and some very beautiful impingement structures, are to be found in ores of this type.

Isotope abundances in these ores have received a great deal of attention.

1. *Sulfur.*   It will be remembered that this occurs not only as the sulfide but also as the sulfate—in barite.   We shall concern ourselves chiefly with the sulfide sulfur.   There appear to be five principal features:

   a. For any individual deposit or district, mean $\delta^{34}S$ is almost always greater (heavier) than meteoritic (a feature of the stratiform sulfides discussed in Chap. 15, it may be remembered).   Joplin, Missouri, may be an exception; mean $\delta^{34}S$ here is about -4, though since only a few analyses have been done, this figure is not very reliable.

   b. The spread of $\delta^{34}S$ values within individual deposits is variable but may be quite small and similar to those of many stratiform deposits.

   c. Such within-deposit variation is often largely related to crustification and to growth zoning in large crystals. $\delta^{34}S$ values tend to be uniform within crustified bands and zones but to vary across them (cf. Ault and Kulp, 1960, p. 93; Folinsbee, Krouse, and Sasaki, 1966).

   d Between-deposit variation in $\delta^{34}S$ is substantial.

   e. Although little work has been done on this aspect, there appears to be some relation between $\delta^{34}S$ and sedimentation; Brown (1967) has suggested that at Bonneterre $\delta^{34}S$ tends to increase upward stratigraphically, and Maucher and Schneider (1967) have pointed out that in the Gorno area, Italy, $\delta^{34}S$ is slightly higher in the reef facies than in the basin facies.

Whatever the isotopic composition of the source of the sulfide sulfur and whatever mixing occurred during transport, there has clearly been some *depositional fractionation* during the formation of these deposits.

2. *Lead.*   Like their sulfide sulfur the lead of these deposits is variable:

    a. In some deposits any within-deposit variation is very slight indeed and the leads appear *ordinary*.   Occurrences of this kind are those of the Alps, Silesia and the Benelux countries of Europe, the British Isles, North Africa, and Canada (Pine Point).

    b. Other occurrences show substantial within-deposit variation and the leads are *anomalous*.   In containing large amounts of radiogenic lead the ores appear to be younger than the enclosing rocks and hence have been referred to as J-type anomalous in terms of the alternative usage mentioned in Chap. 7.

    c. In some deposits containing *anomalous* lead, the proportion of radiogenic lead appears to vary systematically; J. S. Brown (1967) has pointed out that in the Bonneterre deposits of southeast Missouri the lead is distinctly more radiogenic at the base of the deposit, decreasing upward, and that for large mineralized areas the lead is generally more radiogenic about the center, decreasing toward the perimeter of the field concerned.   In addition the earlier deposited lead tends to be less radiogenic than that formed later.

These three features indicate that the lead of some limestone–lead-zinc deposits is derived from some comparatively fundamental and uniform source with little or no radiogenic contamination, while that of others has suffered significant radiogenic contamination.   In some cases the isotopic zoning suggests that such contamination took place close to or within the present areas of mineralization.

3. *Lead-sulfur.*   It has been shown by Eckelmann and his coworkers (1961) that the isotopic constitutions of lead and sulfur are related in some of the south-

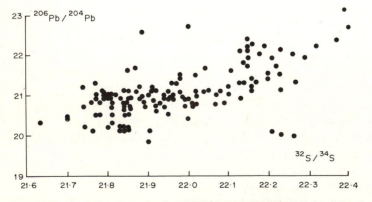

Fig. 16-6   Relation between $^{206}Pb/^{204}Pb$ and $^{32}S/^{34}S$ in a variety of galena samples from deposits in southeast Missouri. (*Modified from Ault and Kulp, Econ. Geol., 1960.*)

east Missouri deposits; the more radiogenic the lead of a galena, the lighter its sulfur tends to be.   This is illustrated in Fig. 16-6 (Ault and Kulp, 1960).

ORIGIN

The following features look as if they may well be of genetic significance and are worth recapitulating:

1. In any one area the sulfides tend to favor particular facies and horizons and to be absent from other, often large, masses of associated carbonate.
2. Large-scale fault structures may or may not be regional associates of ore occurrence.
3. Liquid inclusions indicate that temperatures of formation were generally in the 100 to 150°C range, rarely rose to 200°C, and were commonly less than 100°C and that the liquids from which precipitation took place were highly saline (usually 15 to 25 per cent salt), principally Na–Ca–Cl.
4. Some fluid inclusions possess a high heavy metal content.
5. The principal sulfide metals are lead and zinc (with iron); copper is almost absent, and there are no corresponding "limestone copper" deposits.
6. The host rocks contain organic matter, and indeed some of the fluid inclusions contain or are composed of petroleum.
7. $^{32}S/^{34}S$ ratios are not those of igneous sulfur ($\delta^{34}S \rightarrow 0$) or of sedimentary sulfur ($\delta^{34}S$ tending to negative values).   They are heavy and variable.
8. Lead isotope abundances suggest a simple history in some cases and more complex histories—involving migrational contamination—in others.
9. The sulfide grain boundaries are of impingement and accretionary type; they are not "equilibrium structures," indicating that temperatures required for annealing ($\simeq 200°C$ for galena) have not been attained.

The principal possibilities of origin appear to be

1. Related to sedimentation
   a. Direct precipitation from sea water
   b. Direct precipitation from submarine exhalations
   c. Detrital sedimentation
   d. Movement of sedimentary materials in pore-space liquids during compaction of sediments, with subsequent redeposition
   e. Any of the above with diagenetic modification, including development of veins, replacement masses, grain growth generally.
2. Related to "foreign" solution activity
   a. Igneous solutions.
   b. Other metal-bearing solutions of deep origin (both 2a and 2b much later than the formation of the host rocks, in contrast to 1b and 1d, which are close to the time of formation of the host rocks).

While those parts of these ores that are stratiform may well be sedimentary in their present form, those that occur in veins and cavities are manifestly introduced, at least on a small scale. There is thus no question that much of the sulfide of this class of deposit is migrant, even if only on a small scale.

The very low temperatures indicated by both fluid inclusions and textures indicate that deposition was not a deep, plutonic igneous process. The sulfide sulfur, also, is not that ordinarily found in "igneous-hydrothermal" veins. According to Roedder (1967) salinities indicated by the liquid inclusions are generally considerably higher than those observed in igneous hydrothermal solutions (usually < 5 weight per cent salt), and the presence of oil in some inclusions is hardly suggestive of a purely igneous origin.

While some of the stratiform masses may represent simple marine chemical sedimentation, there is strong evidence—quite apart from the vein form of much of the ore—against the *general* validity of such a hypothesis in the case of limestone–lead-zinc deposits. The temperatures indicated by the fluid inclusions are in this case somewhat too high, as are the salinities. Again the sulfur isotope ratios look too heavy to have been induced by simple sea-floor bacterial sulfate reduction.

A very old suggestion is that some of the deposits have been precipitated from ground (meteoric) waters in much the same way as are bog iron and manganese oxide deposits in limestone areas. While such an explanation may hold in some isolated cases, it does not appear to fit the facts of field occurrence, mineralogy, and chemistry in most cases.

A very appealing explanation—and one that looks as if it may be close to the truth for many occurrences—is that involving transport of the metals by brines expressed from large thicknesses of sediment during compaction (cf. the deep geothermal brines of Chap. 6). This was suggested by W. H. Newhouse in 1932 and has recently received strong support from Roedder (1967) and Jackson and Beales (1967). The latter authors have argued—most eloquently—from the point of view of oil geology.

It is well known that the source material of oil accumulates in off-shore shaly and sandy sediments and that upon compaction of these, oil, gas, and water migrate updip to a variety of traps, important among which are permeable reefs. The latter themselves often contain notable quantities of organic matter—a fact of importance in the present argument. The initial fixation and subsequent movement of the metals is seen as substantially analogous to that of oil. It is thought that the base metal ions are initially fixed in off-shore sediments by adsorption onto clays, organic matter, etc., and also by precipitation directly as sulfides in black shales. Upon burial and compaction the metals are desorbed and dissolved by the pore-space brines and begin to migrate updip and/or along suitable structures. Brines of the oil-field type are known to develop high salinities similar to those now found in the fluid inclusions of the limestone–lead-zinc ores. Thus large volumes of hot, connate brine containing heavy metal ions and abundant sulfate could move up and into reef structures following zones of original permeability, solution cavities, breccia pipes, and so on. At least two possibilities for the precipitation of the metals as sulfides occur:

1. *Biological.* Sulfate-reducing bacteria, living on the organic matter in the limestone, reduce SO_4^{--}, producing S^{--} that combines with the metals. Alternatively the sulfur might come from the organic matter itself.
2. *Abiological.* Methane, common in organic sediments and in the fluid inclusions of the ores, may take part in reactions of the type:

$$CH_4 + ZnCl_2 + SO_4^{--} + Mg^{++} + 3CaCO_3 \rightarrow$$
$$ZnS + CaMg(CO_3)_2 + 2Ca^{++} + 2Cl^- + 2HCO_3^- + H_2O$$

This equation, due to P. B. Barton (1967), accounts for the production of metallic sulfide (zinc is used as the example here—other metals would behave analogously) and also for the generally observed dolomitization of the carbonate.

Such a sequence of events would account for the field occurrence, mineralogy, chemistry, and isotope abundances of many deposits. The abiological reaction would most conveniently explain the heavy sulfur and lack of any substantial within-deposit variation in $\delta^{34}S$. Volcanic source sediments—and a migrational path composed largely of related pyroclastic sediment—would probably provide the best fit for those deposits containing *ordinary* lead. Basement faults—as at Pine Point—active during and after reef formation, would undoubtedly have guided solutions and localized deposition in some cases.

One obvious thing that none of the theories seems to explain is the extraordinary lowness—or absence—of copper. Be this as it may—and it must be highly significant—it is well to keep in mind that a variety of events is probably responsible for the deposits as a group, that several processes have acted here and there, and that to no one of them are these ores to be exclusively attributed.

SANDSTONE–URANIUM-VANADIUM-COPPER ASSOCIATION

Deposits of this general association occur in fluviatile sedimentary rocks in many parts of the world. The greatest development by far, however, occurs in the western and southwestern United States, and much of this present section is based on observations made in that region. Deposits of this group constitute one of the world's principal sources of uranium and by far the most important source in the United States itself. They are also quite important sources of copper and vanadium.

The three metals are rarely, if ever, of equal importance in a given deposit. Some contain uranium almost exclusively, with small amounts of vanadium. Others contain uranium having a large vanadium component. Some are composed almost entirely of copper, and yet others contain Cu–U in varying proportions. A few contain all three. The only combination that does not seem to have been observed is that of copper-vanadium, with uranium absent. This spectrum of ore types is illustrated diagrammatically in Fig. 16-7. In the western United States there appears to be some indication of a broad spatial zoning of types dominant in U, U–V, and Cu.

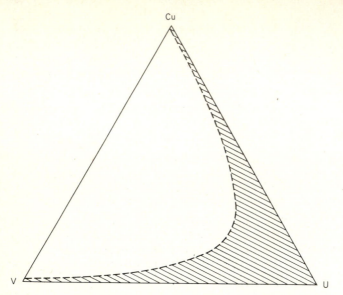

Fig. 16-7 Qualitative portrayal of the general abundance relationships among uranium, vanadium, and copper in sandstone-
type deposits.   The shaded area represents the general nature of
the metal abundance relationships.   Copper and vanadium do
not normally occur together in significant quantities.

DISTRIBUTION, FORM, AND SETTING

The uranium-rich members of this type are virtually confined to the western United
States, where they occur in the Colorado Plateau area of Colorado, Arizona, Utah,
and New Mexico and in Wyoming and South Dakota.   Less well-documented
occurrences which, however, may be of somewhat similar type are known in northern and western Australia, and in arenaceous sedimentary rocks of the Mount
Painter area of South Australia (some of these may, however, have a closer relationship to the conglomerate-orthoquartzite–gold-uranium-pyrite association of
the following section).   The more important vanadium-rich ores of the "Colorado
Plateau" province have been found in Wyoming, South Dakota, and the central
Colorado Plateau area, and the uranium and uranium-copper ores achieve greatest
prominence in the southwestern Colorado Plateau area.   The greatest development of the copper-rich ores is in a broad strip running southeast through
Utah and New Mexico (Fig. 16-8).   Sandstone copper deposits are quite widely
distributed throughout the world, though they are of minor importance overall.

For present purposes we shall regard the Colorado Plateau as the "type
area" and concentrate attention on this.

The age of the deposits ranges from Carboniferous to Tertiary.   The
uranium-vanadium ores seem to achieve their greatest prominence in the Triassic
(Shinarump Conglomerate) and the Jurassic (Morrison Formation).   The copper
deposits, which may also contain some silver, are developed chiefly in the interval
Permian-Triassic, though there are certainly some ores in Jurassic rocks.   It is

notable that within any particular area, deposits are usually restricted to one preferred stratigraphic zone a few feet or a few tens of feet in thickness but are widespread—often over several hundreds of square miles—within such a zone.

Within these zones individual deposits occur as lensoidal to irregular masses,
often greatly elongated so that there tends to be one major dimension and two
minor ones.   Within the deposit the ore minerals occur as fillings of pore spaces in
sandstones and conglomerates, as veinlets, and as replacements of detrital fragments and fossil plant material.   There is often incontrovertible evidence of replacement in the preservation of cell structures of wood.

The characteristic setting of the deposits is well summarized by McKelvey,
Everhart, and Garrels:

> Most of the deposits are in tabular masses, elongated in the direction of the long axes of
> the sandstone or conglomerate lenses in which they occur, or in a direction parallel to
> the orientation of logs and other marks of current lineation.   The host rocks are thus
> interpreted to be fossil stream channel deposits.   Most of the deposits occur in or near
> the thicker parts of the lenses, where mudstone partings or fine debris are present, and

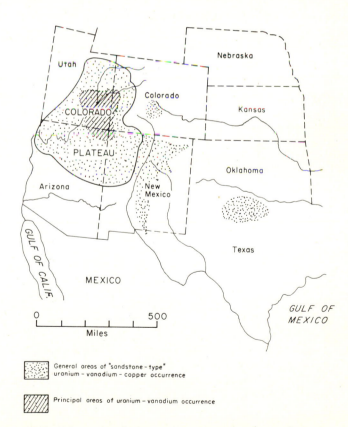

Fig. 16-8 Approximate area of occurrence of sandstone-
uranium-vanadium-copper ores in the western United States.

where logs and other types of carbonaceous matter are abundant. The deposits commonly cut across the bedding, particularly where they form concretion-like structures known as "rolls" (1955, pp. 493–494).

The usual host rocks of these "sandstone" deposits are coarse types such as conglomerates, sandstones, and siltstones. Within these, deposits are related to a variety of sedimentary features: contacts of coarse sediment with mudstone, thickness of coarse unit, "pinch-outs," stream channels and various details of their shape, and fossil logs and other plant remains. Some small—but very valuable— deposits have consisted almost entirely of a replacement of a single large fossil log!

A diagrammatic cross section of deposits in a "paleochannel" (after Mitcham and Evensen, 1955, p. 141) is given in Fig. 16-9. Series of lenses are often related to braided streams, to successions of meander bends, to specific portions of outwash fans, and so on. Commonly the deposits are near mudstone-coarse sediment contacts and are bottomed by a relatively impermeable stratum. In some cases the locus of *uranium* ores of sandstone type is not controlled by a sedimentary feature but is in the form of a "roll"—a body of ore curved in section and of great length compared with its width and depth—usually contained within a single sedimentary unit. These rolls appear to have formed in relation to some feature of a contemporary water table or stream (Fig. 16-10, after Adler, 1964, p. 51).

Such relations between ore and sedimentary features are by far the most obvious ones. However, it has been pointed out that in addition to these various sedimentary "controls," the distribution of ore when viewed *regionally* sometimes gives the impression of a tie with some fairly large-scale structural feature. In various districts there may be a tendency for a concentration of deposits not far from (i.e., within a few miles of) major fault zones, salt anticlines, laccolithic domes, major fold axes, and so on.

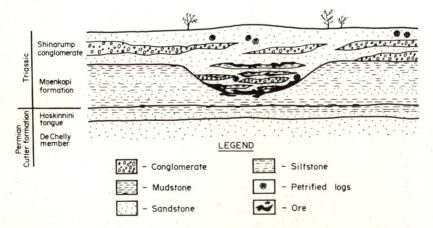

Fig. 16-9 Schematic cross section of a uranium concentration associated with plant material (logs) in a "paleochannel." (*Redrawn from Mitcham and Evensen, Econ. Geol., 1955.*)

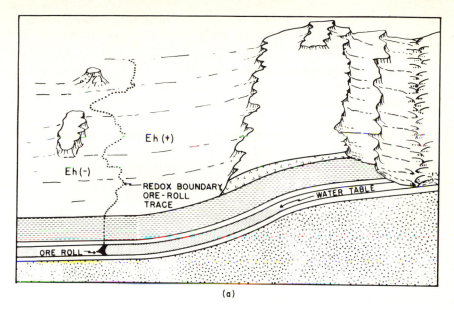

(a)

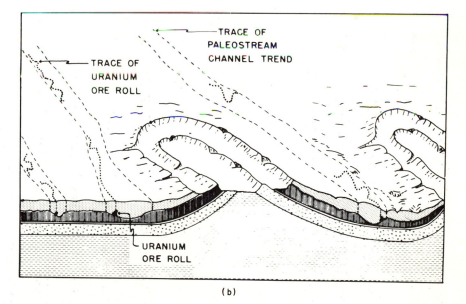

(b)

Fig. 16-10 (*a*) Uranium-ore roll formed parallel to basin rim in paleostream sediments under geochemical conditions conducive to hydrogen-sulfide accumulation, and (*b*) uranium-ore roll trends in paleostream channels more or less paralleling local structures. (*From Adler, Econ. Geol., 1964.*)

In summary we can therefore say that deposits of this group:

1. Occur in continental and marginal-marine clastic sediments deposited under fluviatile-deltaic conditions
2. Usually display a configuration related to lenses of coarse sediment and to the contacts of these with finer sediment
3. In addition to their clear *individual* relation to local-scale sedimentary features, may show broad and diffuse *group* relations with regional-scale tectonic features
4. Are in virtually all cases associated with organic debris

CONSTITUTION

It has already been noted that the ores constitute a gradational series (Fig. 16-7) of constitutional "types"—vanadium-minor uranium to uranium-minor vanadium; uranium-vanadium-copper; uranium-copper; and copper-uranium to copper. The combination that has *not* been observed is high vanadium-high copper; ores high in vanadium are generally low in copper and vice versa. Some of the copper ores also contain notable silver.

The principal ore minerals are

Uranium: Uraninite, coffinite $(USiO_4)_{1-x}(OH)_{4x}$, uraniferous asphaltite (asphaltic organic matter containing uranium)
Vanadium: Roscoelite (vanadium mica), montroseite [$VO(OH)$], mixed-layer vanadium clay minerals
Copper: Chalcocite, covellite, chalcopyrite, bornite

In addition to these minerals, most deposits contain pyrite and/or marcasite. Many occurrences also contain a number of "trace" sulfides that are found in amounts and combinations varying from one deposit to another. Galena is a common one, particularly in the uranium-rich ores. Others are sphalerite, greenockite, gersdorffite, clausthalite, smaltite, molybdenite, and eucairite. Arsenic and selenium are also known. The principal "nonmetallic" associates of the ore minerals and associated sulfides are gypsum, calcite, dolomite, fluorite, barite, and kaolinite.

The quantities of uranium, vanadium, and copper vary enormously within and between deposits. In many cases individual ore bodies fluctuate so much in grade that single analyses do not mean very much. Discussing the Colorado Plateau uranium-vanadium ores, Heinrich notes:

> The grade of the deposits is likewise highly variable, and ranges from 0.1 to 1% U_3O_8, and locally much higher. The grade of many mined [ores] falls in the range of 0.2 to 0.4% U_3O_8. During the mining period before and after World War I, much relatively high-grade material was selectively mined or hand-sorted. In the Maggie C deposit, Long Park, Colorado, the largest of three mineralized fossil logs contained 360 lb of ore reported to contain 20% U_3O_8, as well as other 8 to 10% U_3O_8 ore, making the log worth $5,000 at $5/lb of U_3O_8 (1920). At the Cracker Jack claim, Calamity Gulch,

Table 16-1 Partial analyses of asphaltites from sandstone-type uranium-vanadium deposits in Mesozoic rocks of Utah. (From Hess, 1933, p. 458)

	(1)	(2)
Volatile matter	49.57	67.11
Insoluble matter	46.32	26.54
U	1.13	2.88
V	0.23	1.17
S	4.98	1.37
As	Trace	Trace
Se	None found	None found

Explanation
(1). Specimen from Cowboy claim, Shinarump conglomerate.
(2). Specimen from lower part of the Jurassic sandstone on the west side of Temple Mountain.

Colorado, a log 80 ft long and about 2 ft across, a smaller 18-in. log 30 ft distant, and the intervening sandstone yielded 180 tons averaging 5.65% U_3O_8, 8% V_2O_5, and containing 2.6 g Ra. The gross value was $350,000. Two logs on the Dolores claims were worth $232,900 in U, Ra, and V, all in 1920 prices . . . (1958, p. 383).

Two partial analyses of uranium-vanadium ores are given in Table 16-1 (W. T. Schaller, analyst, in Hess, 1933). The variability of metal content is due, clearly, to the fact that the ore minerals occur as patches ("bunches"), veinlets, and irregular streaks in the containing sedimentary lenses. Likewise copper is very variable; disseminated material might average between a fraction of 1 per cent and, say, 3.0 per cent. However, in rich patches, particularly where chalcocite has formed by local heavy replacement of woody material, copper content may be in the vicinity of 60 per cent. In such cases the silver content of the chalcocite is often high.

Isotope abundances—particularly in the uranium-rich members of the association—have received considerable attention.

1. *Sulfur.* Most of our knowledge on sulfur isotope ratios is due to M. L. Jensen (1958, p. 598; 1967, p. 143), who has investigated a number of both uranium-rich and copper-rich occurrences. $^{32}S/^{34}S$ ratios in almost all deposits are high (negative $\delta^{34}S$) and show substantial within-deposit variability. The latter is commonly of the order of 20 per cent and may be greater. Sulfur isotope abundance behavior in these deposits is thus in sharp contrast to that found in limestone–lead-zinc ores and the stratiform ores of Chap. 15. Between-deposit variation seems to be about the same as for the other two associations, though it may be a little greater.

2. *Lead.* Not surprisingly this has a high radiogenic component and is anomalous in both of the senses of Chap. 7. Isotope ratios and "ages" determined by Miller and Kulp (1958, pp. 938 and 941) on the basis of $^{206}Pb/^{238}U$, $^{207}Pb/^{235}U$, and $^{207}Pb/^{206}Pb$ are highly variable.

ORIGIN

J. W. Finch (1933) summarized very well what appear to be the significant features of the copper-rich ores of this type, and for the most part his observations apply just as well to the uranium-vanadium types.   Of the beds he notes:

> (1) The formations are a great group generally known as the Red Beds.   (2) Those containing copper are clastic sediments.   (3) They accumulated in a climate that was arid to semi-arid, depending upon relief.   (4) They are mostly continental deposits formed by temporary streams and other subaerial agencies.   (5) They characteristically contain vegetal remains, silicified in some localities, in others lignitized.   (6) At many horizons saurian tracks are found and several localities are famous for vertebrate remains.   (7) Lime cementation is characteristic.   (8) Gypsum is common.   It is bedded in the Permian of Texas and New Mexico, also in the Triassic of Wyoming, southeast Nevada, and northern Arizona, but is disseminated or forms later seams in other places.   (9) Many beds have been derived from pre-Cambrian rocks.   (10) The time of sedimentation was remote from igneous epochs of the Rocky Mountain region.   (11) Most of the beds had been formed before the Jurassic orogeny of the Pacific Coast region.   Local beds of volcanic ash are found in the Morrison formation of Utah at the top of the Red Beds (1933, p. 482);

and of the deposits themselves:

> (1) Although they are found in the Red Beds, they seldom occupy strata of red color. Some writers have thought that these beds were bleached by copper-depositing reactions, others that it was done by other agencies, or that the beds were of gray or pale color when deposited.   (2) They differ from hydrothermal deposits in the absence of any related quartz or other gangue.   Calcite and gypsum are found in the beds where copper is absent and where it is present.   (3) Most of the deposits are too remote from Jurassic igneous centers to have been affected by them.   (4) They show no general physiographic, structural or geographic relation to Tertiary igneous regions.   (5) Tertiary hypogene copper deposits of southwest New Mexico and northwest Utah have no physiographic or structural connection with the sedimentary copper deposits.   (6) Near-by deposits definitely attributable to Tertiary magmatic solutions lack copper, therefore the Red Beds copper segregations were not derived from such solutions.   (7) They contain copper sulphides, but not with hydrothermal mineral associations.   (8) They are not restricted to special epochs as are hydrothermal ores, but were formed through several long geologic periods.   (9) The sedimentary copper is not attributable to supergene sulphide mineralization from solutions originating in Tertiary copper districts, because of the wide geographic distribution of the sedimentary segregations and their general remoteness from these districts.   (10) They are distributed over such great areas and with such uniformity of character that all of them must have had a common mode of origin. (11) They were apparently dependent upon local sources of copper that were also local sources of the sediments.   (12) The copper sulphides are characteristically replacements of carbonized plant remains.   (13) They were deposited only under special conditions of sedimentation (1933, pp. 482–483).

It is abundantly clear—and it must be significant—that while the ore minerals unquestionably occur in preferred sedimentary units and, within these, are localized in specific lenses, other sedimentary structures, and near-surface solution features, *they are not stratiform*.   Almost every feature of their occurrence indicates that *they have been deposited in their present positions distinctly later than sedimentation*. In their present form they are therefore unquestionably products of subsurface solution activity.   Upon this all are agreed.   What is *not* known, however, is the precise nature or source of these solutions.

The principal hypotheses are

1. The ore-forming solutions were ground water, and the host rocks were artesian and subartesian aquifers along which such ground water moved.   These near-surface solutions carried metal ions, together with SO_4^{--}, leached from associated strata, particularly tuffs.   The ore minerals were precipitated on encountering local reducing environments—and bacterial sulfate reduction —related to decaying vegetation in the sediments.
2. The ore-forming solutions were deep-seated and derived from igneous rocks that have not yet been bared by erosion.   The solutions moved up along fractures and were fed from these into the sedimentary hosts, where they were precipitated biologically or otherwise.
3. The ore-forming solutions were mixtures of near-surface ground waters and deep-seated hydrothermal solutions.

Before attempting to discriminate among these, it is appropriate to look at some of the geochemical characteristics of the three principal metals.

It is known that in igneous processes uranium tends to concentrate in the later, more siliceous fractions.   Its average concentration in granite is about 4 ppm, but considerably higher concentrations are known in individual intrusions. Concentrations in siliceous tuffs are, if anything, a little higher than in the corresponding intrusives.   Vanadium in magmatic rocks is highest in the iron-bearing oxide minerals but ranges up to about 100 to 150 ppm in siliceous igneous rocks. Copper has a distinct preference for mafic rocks, and granites, rhyolites, and related types have quite a low copper content—of the order of 20 ppm.

Uranium has six valence states, of which U^+, U^{++}, U^{3+}, and U^{5+} are unknown or unimportant in nature. U^{4+} (uranous) and U^{6+} (uranyl) are the abundant ions. $U^{4+}O_2$ (uraninite) is only very slightly soluble in water and is highly stable under reducing conditions. U^{4+} in aqueous solution is readily oxidized to U^{6+} or, rather, the uranyl ion $(U^{6+}O_2)^{++}$.   As pointed out by McKelvey et al. (1955) it can be said in a general way that U^{4+} is stable under the same conditions as H_2S, HS^-, and S^{--}, whereas $(U^{6+}O_2)^{++}$ coexists with SO_4^{--} and HSO_4^-. $(U^{6+}O_2)^{++}$ does not bond strongly with most anions and hence forms few insoluble compounds; those that are known as minerals result from combination with minor elements such as P, V, and As rather than with the more abundant "rock-forming" elements.   On the other hand, the ion readily forms the stable sulfate complex $(U^{6+}O_2)SO_4$ so that the solubility of uranium is high in sulfate solutions, almost independent of pH.   Vanadium forms V^{++}, V^{3+}, and V^{5+}, and the most important process in the weathering of vanadium compounds is the production of the vanadate ion $(VO_4)^{3-}$.   This combines with water to form vanadic acid, which may in turn combine with appropriate cations to form vanadates or be reduced to compounds such as montroseite.   Vanadium is thus precipitated by double decomposition as insoluble heavy metal vanadates or, by reduction, as lower oxides and hydroxides in the presence of organic matter, particularly plant remains. Copper is, of course, derived principally by the oxidation of sulfide to sulfate.

The transport of uranium and vanadium together in a reduced state appears to be difficult except at very low pH values. Apparently, however, there is some feature of natural solutions that permits movement together at a not too low pH. Whatever these solutions were in the case of the sandstone ores, they were undoubtedly high in sulfate. On encountering reducing environments created by decaying plant remains, uranium and vanadium were precipitated as lower oxides and copper (and associated silver) as sulfides as a result of associated bacterial sulfate reduction. Some of the uranium was also precipitated by absorption on the plant material itself.

With this brief review of the principal chemical considerations and the general incidence of the three metals in rock types—felsic tuffs and granite-derived arkoses —associated with the Colorado Plateau deposits, we may look briefly at the three hypotheses.

Hypothesis 1 The ore-forming solutions were derived by leaching of associated sediments, transportation in sulfate solutions, and deposition by reduction and sulfide formation. In addition to the large amounts of tuff in the Permian-Mesozoic sequence of the Colorado Plateau, Gruner (1956) has pointed out that there are enormous quantities of granite and related arkosic rocks in the underlying Precambrian basement. Of the uranium he notes:

> The quantity of rock either granitic or tuffaceous that was available for leaching is enormous. There is no way of computing the volume of granite eroded but a crude attempt can be made to find the area that was affected on the basis of available maps and the sediments deposited since Pennsylvanian time. If one includes only one-fourth of Arizona, one-fourth of New Mexico, one-half of Utah, one-third of Colorado, and three-fourths of Wyoming, one obtains an area of about 210,000 square miles. If one conservatively assumes that in this area only a thickness of 500 feet of Precambrian rocks had been eroded and incorporated into the sediments in existence at one time or another since Pennsylvanian time, a minimum of 21,000 cubic miles with over 800 million tons of uranium would have been available. If only one part in 1000 of this metal has been concentrated, it would have made 200 deposits of a million tons each with a tenor of 0.4 per cent uranium. Checking these figures for areal distribution one finds that statistically there could be one such deposit, including all the hidden ones, in 1050 square miles. This figure seems too large as it gives a 10,000-ton deposit for about every ten square miles, if a more common size of ore body is assumed. Of course, there has been a large loss of deposits by erosion in late Tertiary and Recent times. This uranium would be widely scattered now but could also have formed new deposits if climatic factors had been favorable. Just because the volcanic tuffs have not been included in the calculations does not mean that they are unimportant (1956, p. 513).

$CaSO_4$ is an abundant associate and there therefore seems no question that sulfate was present in the ore solutions. Movement of uranium as carbonate complexes in some cases, coupled with the insolubility of copper carbonate, might also account for the general separation of copper from uranium-vanadium.

The association of the metals with plant remains accords well with the suggestion that precipitation was brought about by reduction. This is supported by the fact that although many of the associated sediments are "red beds" (containing abundant Fe_2O_3), the *ore beds* are not red, as suggested by Garrels (see Jensen, 1958, p. 615). Organic reduction is clearly indicated by the lightness (negative $\delta^{34}S$) and extreme variability of $^{32}S/^{34}S$ ratios in the sulfide sulfur. According to Garrels and Jensen this noncoloration of the ore beds may be due to related reduction of preexisting red-bed material by both H_2S and the hydrogen produced by the reaction

$$4H_2S + Fe_2O_3 \rightarrow 2FeS_2 + H_2 + 3H_2O$$

Hypothesis 2 The ore-forming solutions were of deep-seated igneous-hydrothermal derivation and transport and were precipitated by "favorable bed" replacement and cavity filling. This is supported by (1) the broad association in some instances of ore distribution and major faults or volcanic centers and (2) the fact that the uranium-lead ages of the ores can be interpreted as approximating (50 to 90 $\times$ 10^6 years) to those of uraninites of hydrothermal veins of the Colorado Front Range (60 $\times$ 10^6 years).

The following points, emphasized particularly by Finch (1933) and Gruner (1956), seem to be important here:

1. Although veinlets are common within the deposits, larger "feeder" veins are extremely rare and apparently absent from the vast majority of occurrences.
2. In one place where a hydrothermal vein *is known* to have intersected a permeable pyroclastic unit (Marysvale, Utah; Gruner et al., 1951), the hydrothermal solutions have clearly been rapidly chilled, and have precipitated most of their load, at the intersection. Localization of this kind is not general in deposits of this class.
3. Even if the metals *had* remained in solution as they passed from conduit to aquifer, they:

   . . . would have had to move laterally great distances for it can hardly be expected that hydrothermal veins are situated under or in the neighborhood of each of the thousands of uranium occurrences of the western states. The originally *concentrated* uranium would have been decidedly *scattered* in this process only to be reconcentrated later in certain spots. This is possible but is it plausible? If we consider that the black ores are entirely independent of the present surface topography, we must concede that only a small proportion of the ore bodies can possibly have been discovered and that most of them never will be seen by man. Is it conceivable that uranium emanations were so prolific over 200,000 to 300,000 square miles that they contributed practically all their metal to the sediments? On the other hand, they left the intrusive rocks around which the deposits are supposed to be zoned, namely the La Sals, Abajos, Carrizos, and Ute Mountains, virtually barren of uranium, even of other veins except insignificant ones. If these U-bearing emanations joined the groundwater and run-off, how much *reconcentration* of the uranium could we expect? We are used to thinking in billions of tons of copper ore when we talk about the largest hydrothermal deposits of this metal. It would take a great many hydrothermal uranium deposits comparable in size to supply enough metal to the sedimentary rocks in order that ore bodies of the number and size known could be reconstituted (Gruner, 1956, p. 511).

4. Apparent regional associations of ore with major structures and/or igneous centers while known are certainly not general.

Hypothesis 3   The ore-forming solutions were derived from and transported by mixing of deep-seated with meteoric solutions, with deposition by local reduction, replacement, etc.   This is simply a blurred combination of hypotheses 1 and 2. Its principal virtues are that it combines the good points of the two hypotheses while avoiding their difficulties—particularly those of 2.   While the theory cannot easily be disproved, it is equally difficult to establish it.   Perhaps, in some cases, individual ore bodies do result from a combination of these two processes.

While recognizing that these problems are rarely clear cut and that various mechanisms may contribute, here and there, in endlessly varying proportions, the author considers that sequences of events such as those proposed in hypothesis 1 are probably the most important in the development of deposits of the sandstone–uranium-vanadium-copper association.   However, for the reader who wishes to obtain more information on the controversy and form his own opinion, the accounts of McKelvey et al. (1955), Gruner (1956), Jensen (1958), and Adler (1964) form an excellent starting point.

THE CONGLOMERATE–GOLD-URANIUM-PYRITE ASSOCIATION OF WITWATERSRAND TYPE

Deposits of this group have frequently been linked with those of the sandstone–uranium-vanadium-copper association, and indeed the two have a number of similarities.   There are some differences, however, and in view of these and the great importance of two of the Witwatersrand-type occurrences, these deposits are treated as a separate group.   At the same time the reader should be alive to the possibility that the two associations may have some fundamental connection.

DISTRIBUTION, FORM, AND SETTING

While there may well be a number of smaller analogs, the type is represented by three well-known occurrences—two of them of enormous commercial importance —occurring in three widely separated parts of the world.   These deposits are those of Witwatersrand in South Africa, Blind River in Canada, and Jacobina in Brazil. The first two are much the more important—the Rand as a producer of gold and uranium and Blind River as a producer of uranium.   Jacobina is essentially a producer of gold.

All three deposits are Precambrian.   Precise determination of ages by uranium-lead and lead-lead methods (using galena lead) is not possible because of almost certain differential movement of uranium and lead following original deposition and also because of the inevitable anomalous nature of the leads.   However, trace lead from zircon and monazite indicate that the age of the containing sedimentary rocks at Blind River is about 2500×10^6 years.   Lead-lead determinations on the Witwatersrand deposits indicate for the mineralization an age of not less than about

2200×10^6 years. The Jacobina sedimentary rocks are known only as being probably Proterozoic (Bateman, 1958, p. 417).

Deposits of this present type have often been referred to as *bankets*, or *conglomerate reefs*. They characteristically occur as disseminated bodies in beds and lenses of coarse conglomerate, which in turn occur as members of distinctly arenaceous successions. As with the "sandstone" group of ores just discussed, the host rock has often formed as a stream-channel accumulation so that the ore body it contains has one large dimension and two smaller ones. The form of the ore body may thus be simple, long, and sinuous. In other instances, however, the ore is apparently related to braided streams and to vertically bifurcating pebble beds, in which case the form of a deposit's outline may be quite complex.

The outstanding feature of the setting of these three deposits is their occurrence in strikingly similar quartz-pebble conglomerates. Not only are the ore-bearing beds or "reefs" similar within deposits, but they are remarkably similar from one deposit to another. The similarity of the Rand and Blind River conglomerates can only be described as uncanny and is now an established part of the folklore of economic geology. These conglomerates appear to be of fluviatile to shallow-water deltaic type.

At *Witwatersrand* the gold-uranium ores form part of four Precambrian units, known (in ascending order) as the Dominion Reef, Witwatersrand, Ventersdorp, and Transvaal Systems. Each is composed of a variety of sedimentary and volcanic rocks, the latter consisting of both lavas and tuffs and being in large part andesitic. In each of the systems the mineralization occurs in conglomerates, some of which are basal, others not. The Upper Witwatersrand ore-bearing beds are a good example of the "Rand type" occurrence. These beds have clearly been deposited in shallow water. Environments suggested are closed or continental basins in an arid to semiarid climate, marine beach, marine or lacustrine delta, and glacial or fluvioglacial. The outer periphery of the curved regional syncline in which the beds are now found appears to coincide quite closely with the old shoreline of the basin of deposition, and as one proceeds from the edge of this basin toward its center, the number and coarseness of conglomerate beds decrease. In sympathy with this, concentrations of gold and uranium dwindle and then disappear with increasing distance from the shoreline.

The rock types immediately associated with the conglomerates are those to be expected in the sedimentary environment concerned—they are gritty and sandy sediments (both quartzose and arkosic), with some shales. The conglomerates themselves are composed of glassy quartzose pebbles closely packed in a quartzitic to arkosic matrix containing some chlorite and sericite. The ore minerals occur almost entirely in the matrix. Scour-and-fill structures, current bedding, and braided channels are typical of the conglomerate beds. Ore-bearing conglomerates are related to several features of stratigraphy:

1. *Regional unconformities.* Some of the conglomerates—notably those of the Transvaal and Ventersdorp Systems—occur at regional unconformities.
2. *Intraformational stratigraphic breaks.* The conglomerates were deposited on

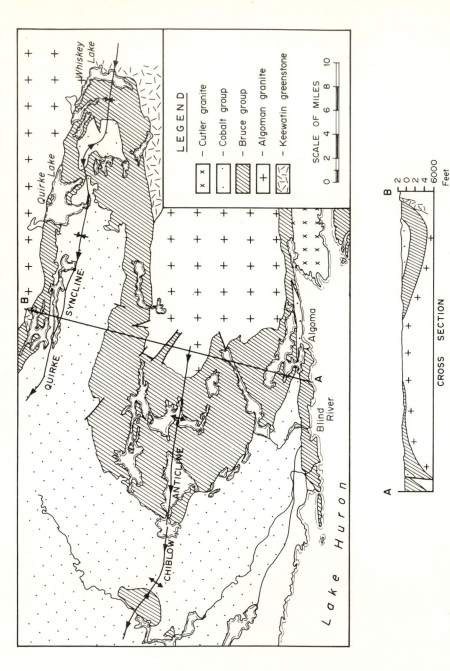

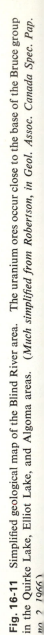

Fig. 16-11 Simplified geological map of the Blind River area. The uranium ores occur close to the base of the Bruce group in the Quirke Lake, Elliot Lake, and Algoma areas. (*Much simplified from Robertson, in Geol. Assoc. Canada Spec. Pap. no. 2, 1966.*)

essentially flat erosion surfaces constituting minor disconformities at different horizons in the Upper Witwatersrand succession and, to a lesser extent, in the Dominion Reef System.

3. *Stream-channel fillings.*   As already noted, conglomerates occur along sinuous or braided stream channels, in local erosional hollows, and in wide depressions created by large-scale scouring.

Within these environments the mineralization is sharply confined to the conglomerate layers, tailing out where the conglomerate grades into gritty layers. Mineralization does not pass out of conglomerate layers across the stratification except in isolated instances of small-scale, late-stage veining.

The *Blind River* deposits occur in a succession of sandstones, arkoses, grits, and conglomerates of Huronian age, apparently remarkably similar in both petrology and environment of accumulation to those of the Rand.   The principal units in the area are given in Fig. 16-11 (after Robertson, 1966, pp. 124–125).   The basement of sedimentation was composed of Keewatin volcanic and sedimentary rocks and granites intrusive into these.   The ore minerals are clearly associated with basal Bruce Group (Mississagi) conglomerates that in turn are related to basement topography and Mississagi stream channels and that have been later folded (Fig. 16-11).   It appears that the succession is of fluvial to deltaic type, modified by progressive northward submergence of the shoreline and overlain by marine or lacustrine siltstones and shales.   According to Derry (1960, p. 913) the principal deposits are localized in three channels that appear to be river beds in a delta.   These three channels have been interpreted as a single sinuous course by Bain (1960, p. 719), as shown in Fig. 16-12.   The three mineralized areas, suspected to be distinct courses by Derry, are shown by full outlines in Fig. 16-12.

Regarding the host conglomerates at Blind River, Derry notes:

There is a close similarity between all ore-bearing conglomerates from the Pronto mine in the south to the Algom-Quirke mine in the north.   In all cases the conglomerates consist largely of quartz pebbles with some pebbles of black chert, and rarely "foreign" pebbles, embedded in a darker matrix.   The matrix consists of fine-grained quartz, feldspar, and sericite and has a variable but considerable content of pyrite, generally between 15 and 25 per cent of the matrix in ore-grade conglomerate.   It carries such heavy minerals as zircon and monazite, and the uranium minerals are invariably restricted to this matrix.   The thickness of uraniferous conglomerate beds varies up to 12 feet or so but in certain cases—mainly by the merging of several conglomerate bands— a thickness of 20 to 30 feet may be reached.

The relationship of the ore-bearing conglomerate to the basement varies from the Pronto mine, where it is almost everywhere a basal conglomerate, to the situation on both sides of the syncline where the ore-bearing conglomerate may lie 100 feet or so above the basement but may approach or touch it locally where the original basement floor formed topographic "highs."

The number of conglomerate beds containing a sufficient proportion of uranium to be considered as millfeed varies from the single basal conglomerate at Pronto to the four or five distinct horizons on the north side of the syncline.   Many other conglomerate beds, generally narrower or less consistently developed, carry sub-commercial proportions of uranium (1960, p. 912).

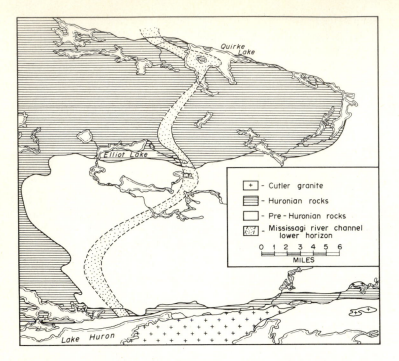

Fig. 16-12   Uraniferous river channel in the Lower Mississagi sediments, as interpreted by Bain.   (*Redrawn from Bain, Econ. Geol., 1960.*)

The *Jacobina* environment appears similar again.   The deposits here occur in the early Proterozoic Jacobina Series, which rests directly on Archean granite and granite gneiss.   The Jacobina Series has been divided into two units—a lower *Canus Group* of thin-bedded and cross-bedded quartzites containing a number of conglomerate horizons and an overlying *Serra Group* of massive quartzites.   The ore-bearing "reefs" are confined to the Canus Group and generally lie distinctly above the basal beds.   This sedimentary unit extends over a distance of at least 20 miles and contains many conglomerate horizons (Bateman, 1958).   [More recently Cox (1967) has divided the Jacobina Series into four units, of which the lowermost is the *Bananiera*; this is overlain by the ore-bearing conglomerate-quartzite unit, which is named the *Serra do Corrego Formation*.]   Within the ore-bearing unit there are two types of conglomerate reef—Chabu type (white quartz pebbles in an arenaceous matrix) that is essentially barren and Piritoso type (pyrite bearing) that contains ore.   Bateman notes:

> In the Piritoso-type reefs reddish-pink glassy quartz pebbles, most of which are in the range of $1-1\frac{1}{2}$ inches in diameter, account for 60 per cent of the reef.   The matrix consists largely of green chlorite and sericite liberally sprinkled with finely divided pyrite.   Some pyrite is observed filling fractures in the matrix.   Gold-bearing ore shoots and small, but significant, amounts of uranium occur in such reefs, there clearly being a

quantitative relationship between the gold, uranium, and pyrite. The quartzite enclosing the reef is a light greenish color and, in this respect, resembles the quartzite containing the uraniferous conglomerates at Blind River (1958, pp. 421–422).

The general nature of the Jacobina occurrences is illustrated in Fig. 16-13.

CONSTITUTION

The three principal areas of occurrence constitute a neat spectrum of gold-uranium compositions. Blind River is important virtually solely for its uranium and contains only sporadic small quantities of gold; the Jacobina occurrence is currently a source only of gold, and uranium, while always present, is very minor; and the great Witwatersrand deposits are important both for gold and uranium—the world's greatest single source of gold and among the first three as a source of uranium.

Witwatersrand. The principal "metallic" constituents of the Rand ores are pyrite, gold, uraninite, thucholite (a uranium oxide-hydrocarbon aggregate), osmiridium, and minor amounts of cobalt, nickel, copper, lead, zinc, and arsenides. Sericite, chlorite, and chloritoid are important in addition to quartz in the matrices. The pyrite is generally fine and constitutes about 2 to 10 per cent of the rock. Uraninite is also fine and gold is extremely fine grained (being rarely visible in hand specimens). Thucolite occurs as rounded grains, colloform sheets, and veinlets. The osmiridium is also very fine and at least partly intergrown with the gold. The gold characteristically contains silver that varies in amount from about 5 to about 16 per cent and averages about 10 per cent; i.e., the gold averages about 900 fine. Chromite and zircon are the two most important associated detrital heavy minerals.

The ore minerals occur mainly as pore-space fillings in the sandy matrix of the conglomerate. However, all of them are occasionally found as fine veinlets. Metal contents appear to be highest where the conglomerates are thickest and the pebbles coarse and well sorted. The highest gold and uranium values do not necessarily occur at any particular level, e.g., the base of a conglomerate reef. As emphasized by Bain (1960, pp. 715–716) the conglomerates generally underwent assortment until pebbles touched—usually at the bottom of the reef as it now exists. Where a few pebbles occur toward the top of individual beds, they are usually dispersed in quartz sand. Occasionally, however, well-sorted pebble layers appear locally at some distance above the basal pebble zone. The pattern of vertical distribution of gold and uranium—which show a strong tendency to occur together—shows clearly that they have concentrated in the lower part of the well-sorted pebble layers, whether these are at the base, middle, or near the top of any given reef.

Variation in the silver content of the gold is systematic to a degree and appears to show at least a semblance of a pattern:

1. Within a single reef, silver/gold decreases with depth in the reef. The silver content is lowest at the base of a reef and, since the highest gold values tend

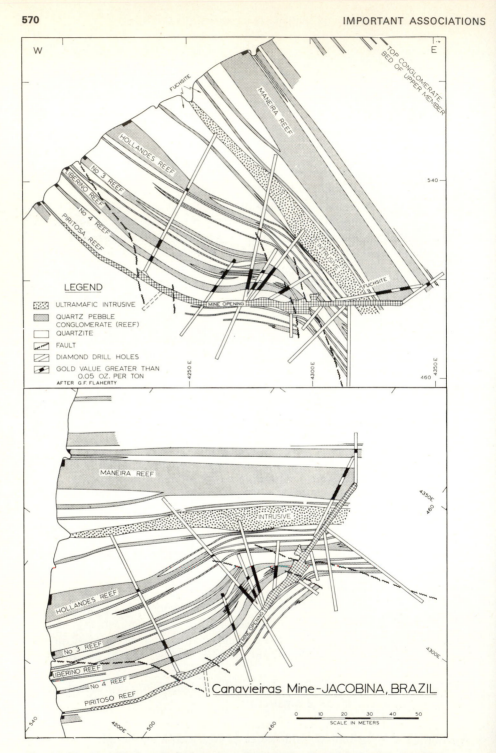

Fig. 16-13 Simplified geological sections through the Canavieiras mine, Jacobina, Brazil. (*From Gross, Econ. Geol., 1968.*)

to occur at the base, the *ratio* Ag/Au tends to decrease with increase in the *absolute abundance* of the gold.

2. Within reefs there is no systematic change in Ag/Au with direction of transport of the containing sediments. There is apparently some semblance of radial patterns of Ag/Au, but according to Hargraves (1963, p. 966), these do not appear to relate to patterns of sedimentation.

3. There do not appear to be clear variations in Ag/Au ratios between reefs.

Blind River. Uranium is the only metal of economic interest here; the gold, while present, is of no or negligible economic importance. The uranium content (as U_3O_8) of the ore zone is about 0.10 per cent over about a 10-foot thickness in the principal bed.

The radioactive minerals here are brannerite [probably $(U,Ca,Fe^{++},Y,Th)_3$-$(Ti,Si)_5O_{16}$; U = 27.9 to 43.6 per cent, Th = 0.26 to 4.4 per cent; most U present as U^{6+} but some as U^{4+}—Heinrich, 1958, p. 49], uraninite, uranothorite, thucholite, and an obscure type of radioactive mixture with monazite as a conspicuous component. Sulfides include pyrite, pyrrhotite, sphalerite, molybdenite, and galena. Other "heavy" minerals are ilmenite, magnetite, chromite, rutile, zircon, tourmaline, garnet, greenalite, grunerite, hematite, hornblende, chamosite, diopside, and orthopyroxene, together with epidote, fluorite, apatite, biotite, and carbonate (Robertson and Steenland, 1960, p. 679, referring to unpublished identifications by V. G. Milne, 1958). Gold is very low apart from sporadic "highs."

Apart from the quartz and feldspar of the matrix, pyrite is by far the most conspicuous mineral, occurring in amounts up to 15 per cent of the total rock and 15 to 25 per cent of the matrix in well-packed conglomerates. It appears as euhedral, subhedral, and rounded grains and is a constituent of both beds and cross-cutting veinlets. The stratified material occupies simple beds, cross-bedding, and scour channels and is often "sized"—i.e., appears to show graded bedding. Where it is found as veinlets it occurs as aggregates and films rather than as discrete grains, and the veinlets cut both matrix and quartz pebbles. The brannerite appears to be metamict and occurs as small rounded to subrounded grains. Uraninite occurs as angular to subrounded grains. Two analyses (Roscoe, 1959) of uraninite from Quirke Lake mine give 6.3 per cent ThO_2–64 per cent U_2O_3 and 5.9 per cent ThO_2–60 per cent U_2O_3, showing that the uraninite is quite notably thorium-rich. Open cracks in the ore zone often contain hydrocarbon mixtures. Some of these are of very low radioactivity, but others are highly radioactive and are true thucholite. This occurs in seams up to about $\frac{1}{4}$ inch thick, almost invariably accompanied by pyrrhotite.

Jacobina. The mineralogy here appears to be simple: pyrite, gold, and very small quantities of a uranium mineral thought to be uraninite. No hydrocarbons have been detected in the ore, and thorium is present in no more than trace amounts. Gold is present in small amounts in most of the conglomerates, and in the ores it rises to about 0.45 oz/ton. Uranium varies sympathetically with the gold; the general level in the conglomerates is of the order of 0.001 per cent expressed as U_3O_8, but in the ore this rises to about 0.015 per cent. Pyrite is fine and ranges up

Table 16-2   Principal features of ores from Witwatersrand, Blind River, and
Jacobina.   (From Bateman, 1958, p. 424)

	Witwatersrand	Blind River	Jacobina
Gold (oz/ton)	0.2–0.8	Trace to erratic highs	0.25–0.45
Uranium (% U_3O_8)	0.01–0.06	0.09–0.13	0.02–0.04
Thorium (% ThO_2)	Trace	0.05	Trace
Pyrite (%)	2–10 (fine grained)	5–12 (coarse grained)	2–5 (fine grained)
Pebbles	Glassy quartz	Glassy quartz	Glassy quartz (pink)
Pebble diameter (av. inch)	1	$1\frac{1}{4}$–$1\frac{3}{4}$	1–$1\frac{1}{2}$
Pebble-matrix ratio	70:30	65:35–40:60	60:40
Sericite and/or chlorite (%)	Present	15–25	15–25
Carbon	Significant	Negligible	Negligible
Accessory minerals	Marcasite	Marcasite	No mineralogical
	Pyrrhotite	Pyrrhotite	studies undertaken
	Ilmenite	Magnetite	
	Cobalt arsenide	Cobaltite	
	Galena	Galena	
	Sphalerite	Sphalerite	
	Monazite	Monazite	
	Zircon	Zircon	
	Osmiridium	Chalcopyrite	
	Diamond	Molybdenite	
	Chromite	Rutile	
	Tourmaline	Anatase	
	Rare earths	Rare earths	

to about 5 per cent.   Some, but certainly not all, of the gold is associated with this.

The economic concentrations of gold tend to occur in the foresets of current-bedded conglomerate-quartzite.   Most of the gold and uraninite are present as grains and pore-space fillings, though there are some veinlets.   In contrast to Blind River, though like some of the Witwatersrand ores, the gold at Jacobina tends to concentrate toward the top of conglomerate levels.   While the fineness of the gold at Jacobina ($\simeq 900$) is about the same as that of the Rand, no sign of systematic spatial variations of Ag/Au has been detected at Jacobina.

Comparatively little has been done on isotope abundances in these ores, a somewhat surprising state of affairs in view of the importance of the ore type and the intensity of the controversy surrounding its origin.

As would be expected the lead isotope abundances reflect a variably substantial radiogenic component; the leads are *anomalous* and in this case are of no great interest from either the geochemical or geochronological points of view. The sulfur, however, appears to show rather specific isotope abundance features that may well be important genetic indicators.

The major part of the the sulfur isotope data available so far is for the Witwatersrand, though extensive unpublished measurements by M. L. Jensen confirm that the principal features of the Rand sulfur also hold for Blind River and Jacobina. All of several hundred measurements on pyrite sulfide from the three occurrences show that $^{32}S/^{34}S$ ratios are extremely constant throughout and very close to, but usually just slightly lower than, the meteoritic value. By far the largest spread so far reported for Rand material is that of pyrite from the Vaal Reef of the Klerksdorp area, for which Jensen and Dechow (1964) found $\delta^{34}S$ to vary between $+0.2$ and $+6.6\%$. Measurements on 17 samples from the Elsburg and Basal Reefs of the Upper Witwatersrand System by Hoefs, Nielsen, and Schidlowski (1968) gave a range of $+1.0$ to $+4.0\%$, with a very small spread about the mean of approximately $+2.5\%$. These results also showed that there was negligible isotopic difference between the sulfide closely associated with carbonaceous matter (mean $+2.57\%$) and that not associated with such carbonaceous matter (mean $+2.51\%$). There was also very little difference between the sulfide of "original detrital" pyrite and that of "recrystallized" pyrite, though the latter *might* be just very slightly depleted in ^{34}S with respect to the "detrital" pyrite. Generally similar results have been obtained more recently by Chukhrov, Vinogradov, and Ermilova (1970). Although detailed information is not currently available on the isotopic constitution of the Blind River and Jacobina pyritic sulfur, Jensen's measurements have shown this to be very constant and close to $\delta^{34}S = 0$.

ORIGIN

A comparison of the principal features of the three ores is given in Table 16-2 (after Bateman, 1958, p. 424). Their settings are, as we have seen, very similar indeed.
There are three principal theories of "origin":

1. *Hydrothermal.* The gold and uranium, and some of the other constituents, have been introduced from some outside source—e.g., from a granitic intrusion or by lateral secretion from elsewhere—by warm aqueous solutions of some kind.
2. *Placer.* The principal components have been derived by erosion from adjacent Precambrian basements, carried by streams, and deposited, in association with conglomerates, as placers.
3. *Modified placer.* The principal components have been deposited first as placers and then redistributed on a small scale *within* the parent reef, with accompanying textural modification.

It is probably true to say that few if any geologists support a "pure" placer theory; virtually all "placerists" consider that there has been at least some post-depositional redistribution and textural modification. The rival hypotheses are thus the "hydrothermal" and the "modified placer" theories. The latter has many advocates—particularly among those geologists immediately concerned with the mining fields—but if the hydrothermal theory has fewer supporters it cannot be said that they are any less tenacious than the placerists. Both views are held strongly, and indeed the resulting debate has often been described as acrimonious.

Once again it has to be admitted that the problem is not yet solved and that there appear to be points of merit on both sides.

In support of their theory the "modified placerists" point to:

1. The remarkable association of the ore minerals with features of erosion and sedimentation
2. Rounding (i.e., detrital appearance) of some of the ore minerals
3. The involvement of some of the ore minerals in sedimentary structures, e.g., the development of what appears to be graded bedding in pyrite-rich bands
4. The very minor nature of vein structures and lack of obvious fault conduits for hydrothermal solutions

This evidence is regarded as doubtful or ambiguous by the hydrothermalists, who counter with:

1. The association of ore minerals with conglomerates and grits lying in erosion channels and basement depressions does not necessarily require placer deposition; it may simply be that these erosion-sedimentation structures and coarse rock types provided good conduits for the passage of—and deposition from—hydrothermal solutions.
2. The ore minerals are not invariably concentrated at the base of reefs as those of undoubted placer accumulations usually are.
3. The rounded shape of some of the uraninite may be the result of colloidal deposition rather than erosional attrition.
4. Uraninite and pyrite could not have survived oxidation during erosional transport unless the contemporary atmosphere had been quite unlike that of today.
5. If these minerals were alluvial, they would have been accompanied by large quantities of titaniferous iron ores; these, however, are lacking.
6. Modern gold placers contain little or no uraninite. However, they do contain substantial ThO_2, giving high Th/U ratios—of the order of 10:1. If such ratios are typical of sedimentary deposits, the very low Th/U of the Witwatersrand-Blind River-Jacobina ores indicate that these deposits are not sedimentary. [This, however, has been challenged by Friedman (1958, p. 889), who has shown that at Blind River, Th/U ratios range up to at least 13:1—well within the postulated sedimentary range.]
7. Ag/Au ratios are the opposite of what might have been expected from the usual ground-water leaching of silver in placers; this would have produced silver-poor gold in the *upper* parts of the reefs, whereas the upper parts of the Witwatersrand reef are silver-rich, the lower silver-poor.

Additional evidence now lies in the very recently acquired sulfur isotope data, and indeed this ore type may well be one for which isotope abundances provide critical geochemical information. The remarkable consistency of $\delta^{34}S$ and its close approach to the meteoritic value strongly suggest a fundamental magmatic source. However, the very constancy of values argues against extensive hydrothermal migration and long-drawn-out accretional deposition, both of which processes provide extensive opportunity for fractionation and hence ultimate variability of $\delta^{34}S$. The same constancy also indicates that there was no opportunity for fractionation by oxidation-reduction reactions such as those common in erosional and sedimentation processes occurring on the earth's surface at the present day. The sulfur isotope evidence seems to point quite unambiguously to a compound origin for the pyrite and hence, indirectly, for the gold, uranium, and associated minor components. The widespread constancy of $\delta^{34}S$ and its near-meteoritic value suggest that:

1. The pyrite sulfur was initially derived from a deep igneous source.
2. The pyrite itself formed and was first deposited in an igneous (either subsurface or volcanic) environment with negligible fractionation or contamination of the sulfur.
3. The pyrite was eroded, transported, and sedimented without going into solution; i.e., it was apparently not significantly exposed to oxygen so that sulfide sulfur was not converted to sulfate sulfur and then back again to sulfide with the oxidation-reduction fractionation that this would entail.
4. On sedimentation the pyrite was buried and protected from postsedimentational oxidation.
5. Postdepositional migration involved in the development of veinlets and outgrowths involved negligible fractionation and probably took place over short distances in entirely reducing environments.

The most likely explanation for the lack of weathering of the pyrite (and accompanying uraninite) seems to be a very rapid accumulation and burial of the whole clastic assemblage not far from its erosional source at a time when the oxygen content of the earth's atmosphere and the prevailing surface temperatures were low.

The problem is not yet resolved. The weight of opinion—and particularly the opinion of those who are directly concerned with the application of geological principles to the immediate problems of finding and following ore—rests with the modified placer hypothesis, though, of course, this does not necessarily mean that it is the correct one. The author leans toward the modified placer theory, largely because of the remarkable—and remarkably widespread—tie between the incidence of ore minerals, sedimentary structure, and sedimentary petrography and because of the very telling new evidence of sulfur isotopes. It has to be agreed, however, that there are some formidable geochemical problems yet to be solved.

CONCLUDING STATEMENT ON THE THREE STRATABOUND ORE ASSOCIATIONS

All three associations are of wide occurrence and each seems quite definitely to constitute a "type." For the most part the ores are not stratiform or "bedded" as are the ore types of Chap. 15. However, there is no doubt that within individual districts the ores have distinct stratigraphic affiliations, and from one province to another—all over the world—each is affiliated with its own specific suite of rock types, each in turn the product of a distinctive environment of sedimentation. That the ores of all three form veins, crusts, pore-space fillings, replacements, and other features indicative of *introduction* shows quite clearly that in their present form the ores are not sediments.   That they favor specific horizons in specific rock suites formed in specific sedimentary environments does suggest, however, that they may have originally been deposited in those sediments and later reorganized *within* them.

There seem to be two particularly remarkable things about the limestone–lead-zinc ores: the relative absence of copper and the small within-deposit spread of $^{32}S/^{34}S$ ratios.   The near absence of copper seems extraordinary since copper occurs in much the same quantity as lead and zinc in any potential source one cares to consider.   Strangely, the copper-rich ores of the *stratiform* lead-zinc association tend strongly to occur as veins in conspicuously carbonate-rich sedimentary rocks—as at Mount Isa.   Is there some connection?   This is an intriguing question and one that invites investigation.   The heaviness and the small within-deposit spread of $\delta^{34}S$ in the limestone–lead-zinc ores are highly reminiscent of the stratiform ores, but in the present case this seems to suggest Barton's methane reaction (as opposed to bacterial sulfate reduction) rather than an immediate volcanic source.

The sandstone–uranium-vanadium-copper and conglomerate–gold-uranium associations occur in rather similar settings and, of course, do not have the obvious marine affiliations shown by limestone–lead-zinc.   However, it appears that the "sandstone" ores are generally post-Precambrian, whereas the "conglomerate" ores are all Precambrian.   In addition the "conglomerate" ores occur in conglomerate almost exclusively—very little ore occurs in associated silt—whereas the "sandstone" ores, while favoring coarse grits and sands, occur in a range of clastic sediments from conglomerate to shale.   There are sharp differences in composition, too; in one case the uranium is associated with significant vanadium and copper and in the other case with gold.   Also the uranium is conspicuously higher—by an order of magnitude—in the younger sandstone deposits than in the older conglomerate ores.

The lightness and high within-deposit variability of $^{32}S/^{34}S$ ratios in the sandstone–uranium-vanadium-copper ores is in striking contrast to the near-meteoritic values, and remarkable between- and within-deposit uniformity, of $^{32}S/^{34}S$ in the conglomerate–gold-uranium-pyrite ores.   Whatever their derivation, the "sandstone" ores have probably had quite complex migrational histories in terms of oxidation-reduction reactions, and a major organic influence in their deposition seems almost certain.   The "conglomerate" ores, on the other hand, have apparently passed through only a brief and sheltered migrational phase—in particulate rather than ionic form and out of significant contact with oxygen—and their deposition and subsequent minor modification has been essentially abiological.   Although the "sandstone" and "conglomerate" associations have some features in common, one suspects that they are rather different entities in principle.

RECOMMENDED READING

A. Limestone–lead-zinc association

Brown, J. S. (ed.): Genesis of stratiform lead-zinc-barite-fluorite deposits (Mississippi Valley type deposits), *Econ. Geol. Monograph* no. 3, 1967.

Jackson, S. A., and F. W. Beales: An aspect of sedimentary basin evolution; the concentration of Mississippian Valley-type ores during late stages of diagenesis, *Bull. Can. Petrol. Geol.*, vol. 15, pp. 383–433, 1967.

B. Sandstone–uranium-vanadium-copper association

Adler, H. H.: The conceptual uranium ore role and its significance in uranium exploration, *Econ. Geol.*, vol. 59, pp. 46–53, 1964.

Finch, J. W.: Sedimentary copper deposits in the Western States, in "Ore Deposits of the Western States," Lindgren Volume, pp. 481–487, American Institute of Mining and Metallurgical Engineers, New York, 1933.

Fischer, R. P.: The uranium and vanadium deposits of the Colorado Plateau region, in "Ore Deposits of the United States 1933/67," J. D. Ridge (ed.), pp. 735–746, American Institute of Mining, Metallurgical and Petroleum Engineers, Inc., New York, 1968.

Gruner, J. W.: Concentration of uranium in sediments by multiple migration-accretion, *Econ. Geol.*, vol. 51, pp. 495–520, 1956.

Jensen, M. L.: Sulfur isotopes and mineral genesis, in H. L. Barnes (ed.), "Geochemistry of Hydrothermal Ore Deposits," pp. 143–165, Holt, Rinehart & Winston, Inc., New York, 1967.

McKelvey, V. E., D. L. Everhart, and R. M. Garrels: Origin of uranium deposits, *Econ. Geol. 50th Anniv. Vol.*, pp. 464–533, 1955.

C. Conglomerate–gold-uranium-pyrite association

Bain, G. W.: Patterns to ores in layered rocks, *Econ. Geol.*, vol. 55, pp. 695–731, 1960.

Bateman, J. D.: Uranium bearing auriferous reefs at Jacobina, Brazil, *Econ. Geol.*, vol. 53, pp. 417–425, 1958.

Gross, W. H.: Evidence for a modified placer origin for auriferous conglomerates, Canavieiras Mine, Jacobina, Brazil, *Econ. Geol.*, vol. 63, pp. 271–276, 1968.

Haughton, S. H. (ed.): "The Geology of Some Ore Deposits in Southern Africa," vol. I, The Geological Society of South Africa, Johannesburg, 1964.

Robertson, J. A.: The relationship of mineralization to stratigraphy in the Blind River area, Ontario, *Geol. Assoc. Can. Spec. Pap.* 3, 1966.

17
Ores of Vein Association

Ores occurring as veins have had an enormous influence on ore-genesis theory. Indeed it may be said that vein phenomena have had an all-pervading and dominating influence on attitudes to "ores" vis-à-vis "rocks" for almost 400 years—a dominance that possibly reached its greatest heights in the mid-twentieth century and that is diminished only a little even now. The nature of this preoccupation with veins, and its reasons, has already been discussed in Chap. 2. In its modern form it undoubtedly began with the researches of Agricola in the latter half of the sixteenth century and his famous statement that it was only the ignorant who could hold that veins were of the same age as the rocks that enclosed them.

While vein phenomena are unquestionably of great importance, there seems little doubt that such importance has been exaggerated and that this is being increasingly recognized as the second half of the twentieth century progresses. The most important factor in this change of attitude is, as we have already noted, changing ideas as to what constitutes ore. Until the nineteenth century, only rich material (ca. 3 per cent and higher copper, for example) could be satisfactorily worked so that, apart from the richer bedded deposits such as Rammelsberg and parts of the Kupferschiefer, it was only veins that *were* ores. However, with greatly improved mining, concentrating, and smelting methods, masses of rock containing 0.4 per cent copper and even less can now be economically worked so that almost every kind of rock occurrence provides its ores, and veins no longer hold their dominance.

However, the fact that they have lost their dominance does not mean that they are no longer important, and some vein occurrences and associations are very significant indeed. The great vein systems of Butte, Montana, still produce enormous quantities of copper. Much gold and silver still come from veins of the precious metal telluride type, much tin and tungsten come from veins of granitoid association—and there are many other examples. It is with these that the present chapter is concerned.

GENERAL CONSIDERATIONS

KINDS OF VEINS

As a class, vein ores are extremely heterogeneous, varying enormously in constitution, size, form, and geological environment—and, presumably, in origin. Some are clearly of pegmatitic type and represent a late stage of silicate-oxide-sulfide differentiation. Some are of sublimate type, related particularly to volcanic activity. Probably most veins result principally from aqueous solution activity, though the nature and environment of this covers a very wide range from deep-seated, high-temperature–high-pressure igneous hydrothermal processes to near-surface, low-temperature–low-pressure ground-water phenomena. In this general connection the considerations of Chaps. 5, 6, and 7 are of fundamental significance.

Constitutionally, the sulfides are of dominating importance overall, though here and there the native metals and oxides are the more abundant and significant constituents. Complexity of constitution and microstructure also varies greatly. Some veins consist of only one or two minerals; the ubiquitous, almost mono-mineralic, stibnite veins and many of the limestone-type galena-sphalerite veins are examples. On the other hand, some vein types display extensive assemblages, sometimes involving over 30 minerals. Occurrences of "Cobalt" type and some precious metal–base metal sulfide veins fall into this category. Similarly micro-structures range from coarse to fine, from simple random aggregates to the most complex precipitation, oriented growth, and transformation structures, and from simple undeformed encrustations and fillings to highly deformed and annealed aggregates with unresolvably complex geological histories. More than any of our other associations, vein deposits as a class demand consideration of all possible factors for the elucidation of their histories—solution derivation and migration phenomena, mineral stabilities, crystal growth processes, and the development of structures during deformation and annealing.

ZONING

Much emphasis has been placed on the zoning of veins. Zoning has been regarded by many investigators as a rather special characteristic of veins as opposed to other types of deposits, and it has been supposed that it might have very substantial relevance in the elucidation of origin and conditions of formation. The greatest modern exponent of this field of study is C. F. Park, who has published extensively on the subject (cf. particularly Park, 1955). That many veins display, or are involved in, zoning has been recognized for a long time and has been regarded as being of scientific relevance for at least most of the twentieth century. Such zoning may take the form of systematic constitutional changes within *individual* veins or the geographical or constitutional zoning of *groups* of veins about igneous foci or other prominent geological features.

These observations, however, must be seen in perspective and in the context of ore occurrence in general. While much has been made of zoning in veins, there are really very few ore deposits, or groups of them, *of any kind* that do not show zoning of one sort or another. Layered chromite and titaniferous oxide deposits show both within- and between-deposit zoning, as do some of the nickel sulfide occurrences (cf. the Lunnon Shoot at Kambalda, described in Chap. 11). Sedimentary iron and manganese deposits show well-developed zoning resulting from facies patterns. Stratiform sulfide ores almost invariably show systematic, gradational, changes in mineralogy and metal ratios, and we have already seen that the Roan copper deposits and the Sullivan lead-zinc ore body are very clear examples of this. Stratabound ores may behave likewise; limestone–lead–zinc ores may occur in zoned patterns relative to limestone banks and reefs, and certain stratabound uranium and gold ores may occur as groups "zonally" related to old shorelines and deltas. There is nothing extraordinary about the development of zonal patterns; indeed, if one considers the whole spectrum of ore occurrence, it may be seen that zoning of one kind or another is virtually ubiquitous, and a lack of zoning is probably a much more noteworthy feature than its presence!

Some of the best—and earliest known—examples of *within-deposit* zoning appear in some of the vein ores of Cornwall, phenomena first recognized by de la Bêche in 1839. Many Cornish veins show a general, progressive change in composition with depth, the general nature of which is indicated in Fig. 17-1 (after Dewey, 1925). Changes in opaque mineral assemblages are accompanied by somewhat less pronounced changes in nonopaque groupings. This zoning has been attributed to progressive fall in temperature with increasing distance from the presumed underlying granitic sources; tin and tungsten oxides were deposited first in the hottest, deepest region of each vein, followed by copper at a slightly higher level, and so on.

Another well-established example of vein zoning, in this case largely of *between-deposit* type, is that of Butte, Montana. This occurrence consists of a large number of veins that have formed in extremely complicated fashion within the mass of the *Butte Quartz Monzonite*.† Zoning, which involves the whole complex of veins, is expressed both horizontally and vertically. The horizontal aspect of the pattern is the more prominent and was first shown by Sales (1913) to have three principal components: (1) a *central zone* or core of altered quartz monzonite in which the copper ores are essentially free of sphalerite and manganese minerals,

† The interested reader is particularly directed to a very fine account of the Butte occurrence by C. Meyer et al. in J. D. Ridge (ed.), "Ore Deposits of the United States 1933/67," American Institute of Mining, Metallurgical and Petroleum Engineers, Inc., New York, 1968.

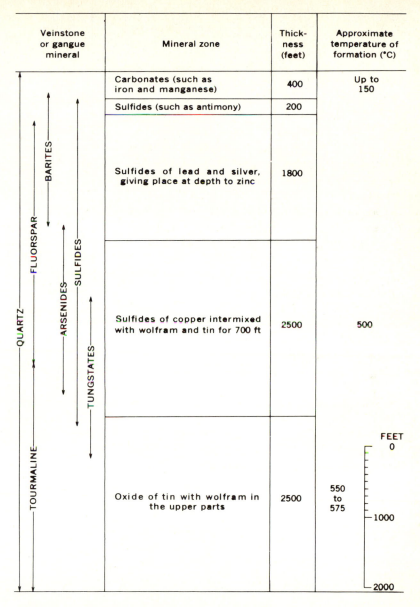

Veinstone or gangue mineral	Mineral zone	Thick-ness (feet)	Approximate temperature of formation (°C)
	Carbonates (such as iron and manganese)	400	Up to 150
	Sulfides (such as antimony)	200	
	Sulfides of lead and silver, giving place at depth to zinc	1800	
	Sulfides of copper intermixed with wolfram and tin for 700 ft	2500	500
	Oxide of tin with wolfram in the upper parts	2500	550 to 575

Fig. 17-1 Simplified, diagrammatic representation of zoning in the Cornish veins. (*From Sainsbury, U.S. Geol. Surv. Bull. 1301, after Dewey, Proc. Geologists' Assoc., London, 1925.*)

(2) an *intermediate zone* consisting of copper ore containing conspicuous sphalerite, and (3) a *peripheral zone* in which zinc and manganese are the characteristic and abundant economic metals. The general disposition of these zones, and the positions within them of some of the more important mining leases, are shown in

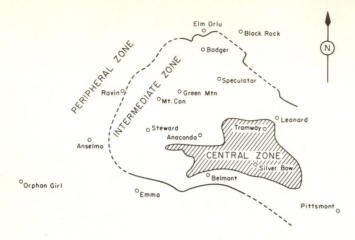

Fig. 17-2 Plan at altitude 4600 feet showing principal features of ore-mineral zoning at Butte. (*From Park, Econ. Geol. 50th Anniv. Vol., 1955.*)

Fig. 17-2 (from Park, 1955).   A more recent sectional view is shown in Fig. 17-3 (from Meyer et al., 1968).

A third well-documented example of zoning—like Butte, involving a large number of veins distributed through a small mineralized district—is that of the oxide-sulfide-carbonate mineralization of Zeehan in western Tasmania.   This was first documented in 1910 by Twelvetrees and Ward, to whom Fig. 17-4 is due, and has recently been substantiated by an exhaustive investigation by Both and Williams (1968), who on the basis of a very detailed chemical and mineralogical investigation, confirmed the existence of four zones that, progressing outward from the interior of the Heemskirk granite mass, are

1. *Cassiterite zone.* This includes the tin deposits within and adjacent to the granite.
2. *Pyritic zone.* Sulfides, including sphalerite, chalcopyrite, and tetrahedrite, appear here, with pyrite the dominant iron-bearing "gangue" mineral; sideritic carbonate appears only very occasionally.
3. *Intermediate (sideropyritic) zone.* The pyritic zone grades into one in which pyrite and sideritic carbonate are of comparable abundance; in addition to sphalerite, chalcopyrite, etc., this zone contains relatively large amounts of lead sulfosalts and tetrahedrite rich in silver.
4. *Sideritic zone.* The increase in the proportion of carbonate continues outward, and pyrite becomes relatively rare.

Both and Williams (1968, pp. 234–242) have also noted quite a number of subsidiary mineralogical changes accompanying this broad zonation: e.g., sphalerite tends to diminish outward, as does its iron content; arsenopyrite is not abundant

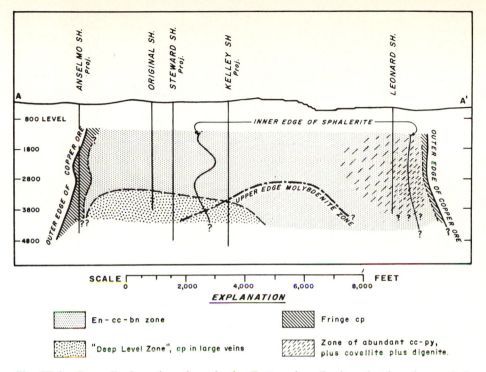

Fig. 17-3 Generalized section through the Butte mineralization showing the vertical disposition of metal sulfide zones. (*From Meyer et al., in "Ore Deposits of the United States," 1933/67.*)

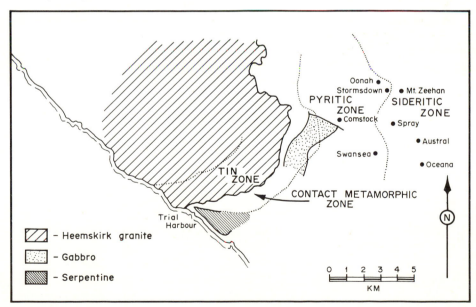

Fig. 17-4 Generalized plan view of the geology and related mineralogical zoning near Zeehan, Tasmania. (*Adapted from Twelvetrees and Ward, Tasmanian Geol. Surv. Bull., 1910.*)

in the inner zone, but increases going outward and then almost disappears in the sideritic zone; and coarse-grained pyrrhotite is confined to the pyritic zone.

These Cornish, Montanan, and Tasmanian occurrences are no more than good examples of a common and widespread phenomenon; as already noted, zoning in vein ores—as in most other associations—is very much the rule rather than the exception.

WALL-ROCK ALTERATION

Since vein ores represent the filling of openings—actual or incipient—they represent the addition of new material to the environment. Such addition naturally involves transport, which is able to take place as a result of high temperatures of the substances concerned—temperatures which endow the latter with high mobilities as gases, vapors, melts, or highly plastic solids—or as the result of the presence of a solvent. Whichever is the case—high temperatures, solutions, or both—migration of the new substances through, and their deposition in, rock openings usually leads to at least some modification—i.e., "alteration"—of the walls of the openings concerned. It is this phenomenon, fundamental and simple in principle, that is referred to as *wall-rock alteration*. It may involve addition to, subtraction from, or isochemical modification of the wall rock. Where addition occurs, there is usually some subtraction also, and this is generally accompanied by some reorganization of original components relative to each other. Alteration may be quite superficial or it may extend substantial distances into the body of the wall rocks.

Wall-rock alteration has already been touched on briefly in Chap. 12 in connection with porphyry copper deposits. Here it takes the form of contact metamorphism of the wall rocks and associated autometamorphism of the intrusions themselves (cf. Lowell and Guilbert, 1970). Wall-rock alteration is an important accompaniment of most *introduced* ore bodies, and although we are concerned with it here in relation to veins, it is not restricted to ores of this type, and contact metamorphic effects are *ipso facto* variants of it. Considerable attention has been paid to the phenomenon not only because of its scientific interest but particularly because diagnostic wall-rock alteration enlarges, effectively, the targets for which the mining geologist explores.

The principal types of alteration about the American porphyry coppers have already been outlined in Table 12-9 (after Creasey, 1966). The following more general treatment, which has substantial application to vein deposits, is largely according to the work of Meyer and Hemley (1967). The principal types of alteration recognized by these investigators are as follows.

Argillic alteration. As the name implies, this is characterized particularly by the development of clay minerals and affects principally intermediate and calcic plagioclases in intermediate igneous rocks, e.g., in andesites, diorites, etc. At lower levels of intensity the principal clay minerals are dickite, kaolinite, halloysite, and metahalloysite, which, with their common approximate chemical composition $Al_2Si_2O_5(OH)_4$, represent the subtraction of Ca and Na principally from the relevant feldspars. K feldspar and biotite may be minor associates, and there is usually small-scale development of sulfides. At more advanced stages of alteration, characteristic minerals are dickite, kaolinite, and pyrophyllite, usually with sericite and quartz and often with alunite, pyrite, tourmaline, and topaz. Alunitization, for example, may be brought about by the reaction of muscovite with H^+ and SO_4^{--} (Harvey and Vitaliano, 1964):

$$KAl_3Si_3O_{10}(OH)_2 + 4H^+ + 2SO_4^{--} \rightarrow KAl_3(SO_4)_2(OH)_6 + 3SiO_2 \quad (17\text{-}1)$$

    Muscovite   Alunite

Sericitic alteration. Sericite, quartz, and pyrite are the characteristic minerals of this type of alteration; that is, H^+ is added, breaking down minerals such as orthoclase and chlorite to sericite, and introduced sulfur combines with iron of iron-bearing silicates to form pyrite. According to Meyer and Hemley (1968, p. 172), sericitic alteration is very often closely associated with sulfide ore. Typical reactions (Meyer and Hemley, 1967) are

$$3KAlSi_3O_8 + 2H^+ \rightarrow KAl_2AlSi_3O_{10}(OH)_2 + 2K^+ + 6SiO_2 \quad (17\text{-}2)$$

    Orthoclase   Sericite

and, of a more complex kind,

$$2Al(Mg,Fe)_5AlSi_3O_{10}(OH)_8 + 5Al^{3+} + 3Si(OH)_4 + 3K^+ + 2H^+ \rightarrow$$

      Chlorite

$$3KAl_2AlSi_3O_{10}(OH)_2 + 10(Fe,Mg)^{++} + 12H_2O \quad (17\text{-}3)$$

            Sericite

The large amount of iron liberated in Eq. (17-3) would, of course, be available for combination with introduced sulfur for the formation of pyrite.

Propylitic alteration. This is a common and widespread type of alteration in andesites and related rocks, in which it was described by Becker as long ago as 1882. Assemblages here include variable but notable quantities of epidote (including zoisite and clinozoisite), albite, chlorite, and carbonate. Sericite and pyrite are also usually present; less common are other sulfides, zeolites, and montmorillonite. Chloritization of albite may take the following course:

$$2NaAlSi_3O_8 + 4(Mg,Fe)^{++} + 2(Fe,Al)^{++} + 10H_2O \rightarrow$$

    Albite

$$(Mg,Fe)_4^{++}(Fe,Al)_2^{3+}Si_2O_{10}(OH)_8 + 4SiO_2 + 2Na^+ + 12H^+ \quad (17\text{-}4)$$

            Chlorite

and sodium and silica so liberated may be used in reactions such as the conversion of andesine to clinozoisite:

$$2NaCaAl_3Si_5O_{16} + 2SiO_2 + Na^+ + H_2O \rightarrow$$

    Andesine

$$Ca_2Al_3Si_3O_{12}(OH) + 3NaAlSi_3O_8 + H^+ \quad (17\text{-}5)$$

            Clinozoisite

In some areas, or in some *zones* of certain areas, alteration may have given rise to particularly notable amounts of one or more minerals of the propylitic association. In such cases terminology may be more specific, and according to whichever mineral happens to be prominent, the alteration is referred to as *albitization*, *carbonatization*, or *chloritic, zeolitic,* or *pyritic* alteration.

Potassium silicate alteration. This is also referred to simply as *potassium alteration* and is common about hot springs, near-surface volcanic deposits, and porphyry coppers. Potassium feldspars and micas are the essential minerals here, and among the associated sulfides, pyrite, chalcopyrite, and molybdenite are most common. Anhydrite seems to be a prominent component of this type of alteration in some localities.

Silicification. This involves an increase in the proportion of silica (as quartz, chalcedony, or opaline silica) in the wall rock. The silica may be added from materials migrating through the adjacent fissures or may be produced *in situ* by chemical breakdown of one or more of the rock constituents. The latter type of process is illustrated by the reactions of Eqs. (17-1), (17-2), and others. Silicification is the type of alteration most commonly associated with sulfide ore deposition.

Overall, wall-rock alteration involves movement chiefly of Na, K, Ca, and Si into or out of the body of the host. Alteration, like the contents of the veins themselves, may be zonal, and different zones may overlap, or be telescoped. The nature of wall-rock alteration depends not only on the nature of the primary rocks themselves but, quite clearly, on the nature and state of the substances causing or catalyzing the alteration. Critical factors are oxygen, hydrogen, and CO_2 and sulfur partial pressures, total pressures, temperatures, and pH. The very process of alteration, with the exchange reactions implicit in it, must change the ore-carrying media, particularly where these are aqueous solutions. Such changes may substantially alter the capacity of fluids to transport ore.

Also—most importantly from the point of view of our understanding of ore-forming processes—wall-rock alteration may give important clues concerning the nature of the gases and liquids from which the different vein ores have been deposited.

STRUCTURAL AFFILIATIONS

Since veins represent the fillings of openings, they are *ipso facto* tied to geological structure in a most fundamental way. Usually associated structures are tectonic, though they are not invariably so. Important limestone–lead-zinc occurrences are well known to have been deposited in limestone solution cavities, and if "vein" and "structure" are seen in their widest sense, one must include such deposits as pore-space fillings in coarse sandstones, amygdales in vesicular lavas, and linings on minor dehydration and cooling cracks.

Most veins, however, occur in what are quite clearly tectonic breaks. These range from shallow, small-scale faults associated with individual volcanic centers to large deep-seated fractures associated with earth movement on a regional scale. Some are simple faults, others have a large shear component, and others may represent tension fracturing associated with folding. Veins may be confined to or notably enriched at the intersections of one set of faults with another or at the intersections of faults or joints with chemically unstable wall rocks such as limestones, basic lavas, or ultramafic intrusions. Such association between structure and ore is very pronounced and highly systematic in some vein provinces, in which case a thorough knowledge of local geological structure is of the greatest importance in mineral exploration.   Recognition of this has led to the development of classifications of vein ores based not on their compositions or general geological environments but on the nature of the structures in which they occur.   Undoubtedly the outstanding classification of this kind is that of Bateman (1942), which gives an extremely neat and comprehensive synthesis of the morphology and structural associations of "fissure fillings."†

GANGUE MINERALS

We have already considered the nature and stabilities of the *ore minerals* in Chaps. 4 and 5 and the chemistry of ore-forming media in general in Chap. 6.   No special reference has been made, however, to *gangue mineral*‡ formation, although in some of the early chapters it was emphasized that these are almost certainly as relevant from the point of view of understanding ore-forming processes as are the ore minerals themselves.   The gangue minerals are particularly prominent and spectacular in vein ores as a class, and as a result the more common ones have been the subject of substantial hydrothermal experimentation.   The principal minerals are quartz and calcite, though others such as fluorite, barite, gypsum, anhydrite, strontianite, celestite, and dolomite are often important.   Experimental investigations of gangue assemblages were begun by the French experimentalists in the middle of the nineteenth century and were brought to a high state of refinement in the mid-twentieth century by people such as G. C. Kennedy and G. W. Morey. Much of this work has recently been extended and skillfully correlated by H. D. Holland.   The following four examples of results from Kennedy and Holland are intended to provide no more than the briefest glimpse of this important and expanding field of investigation.

The solubility of quartz in water as a function of temperature and pressure is shown in Fig. 17-5 (from Kennedy, 1950, p. 639).   It must be kept in mind that this represents simply the system $SiO_2–H_2O$; the diagram does no more than show the *general nature* of relations, which would be considerably modified by the more complex mixtures of most natural environments.

It was noted in Chap. 6 that some modern ore solutions, and many of those older ones preserved as fluid inclusions, are highly saline.   Although a number of "salts" are usually present, NaCl is almost invariably the most abundant.   Figure

---

† The interested reader is particularly referred to chap. 5, pp. 107–137, in A. M. Bateman, "Economic Mineral Deposits," 2d ed., John Wiley & Sons, New York, 1950, for an outstandingly clear treatment of the structural associations of vein ores.

‡ The term *gangue mineral* is a very old and well-established one that is now becoming somewhat out of date and that might perhaps best fall into complete disuse—with all due respect to its hardy Cornish originators.   In the first place a number of minerals formerly regarded as useless, and hence "gangue," are becoming quite valuable as a result of changing values and new and improved technologies.   Secondly the use of the terms *ore mineral* and *gangue mineral* has tended to induce scientific as well as economic separateness, or distinctiveness—an unnatural separation which the author feels should be eliminated entirely from scientific considerations.

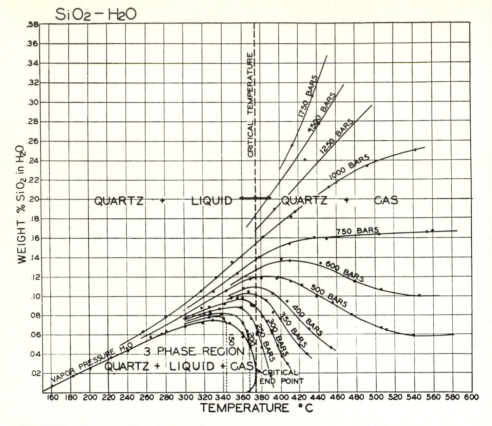

Fig. 17-5 The solubility of quartz in water at temperatures up to 560°C and pressures up to 1750 bars. (*From Kennedy, Econ. Geol., 1950.*)

17-6*a* and *b* (Holland, 1968) shows the effect of NaCl concentration (and temperature) on the solubility of the common gangue minerals fluorite and calcite. Figure 17-7 (also from Holland, 1968) shows very clearly the shape of the solubility surface of calcite in the system $CaCO_3$–$NaCl$–CO_2–H_2O between 50 and 300°C, between 0 and 70 atm pressure of CO_2, and at an NaCl concentration of approximately 1 *m*.

VEIN ASSOCIATIONS

Vein deposits as a group include some highly distinctive ore types, and a number of these occur in quite specific geological associations. Indeed, in spite of their "introduced" nature, a number of vein ore types appear to be just as closely tied to environments as do any of the groups of ores we have already considered.

Although a great deal of geological effort has been devoted to the study of veins—their distribution, affiliations, shapes, sizes, and mineralogical features—

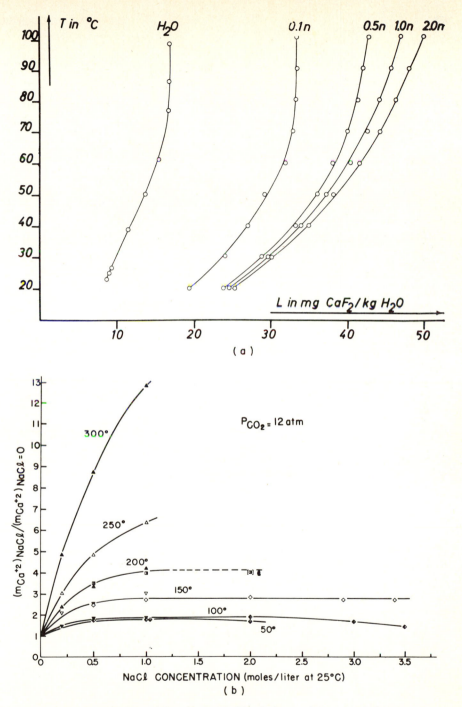

Fig. 17-6 (*a*) Solubility of fluorite in NaCl solutions. (*From Strubel, Neues Jahrb. Mineral., 1965.*) (*b*) Effect of NaCl on the solubility of calcite in water at a CO_2 pressure of 12 atm. [*From Holland, in H. L. Barnes (ed.), "Geochemistry of Hydrothermal Ore Deposits." Copyright © 1967 by Holt, Rinehart and Winston, Inc. Reprinted by permission of Holt, Rinehart and Winston, Inc., Publishers, New York.*]

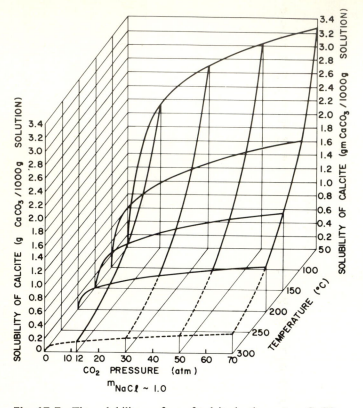

Fig. 17-7 The solubility surface of calcite in the system $CaCO_3$–$NaCl$–CO_2–H_2O between 50 and 300°C, between 0 and 70 atm pressure CO_2, and at an NaCl concentration of approximately 1 *m*. [*From Holland, in H. L. Barnes* (ed.), *"Geochemistry of Hydrothermal Ore Deposits." Copyright © 1967 by Holt, Rinehart and Winston, Inc. Reprinted by permission of Holt, Rinehart and Winston, Inc., Publishers, New York.*]

our knowledge of them is still not highly systematized. Only a few types of veins have been well documented as categories. Other groups are recognized, but our knowledge of them has not been brought together in a really systematic way. Other categories probably await recognition. The following three groups— *precious metal telluride ores of volcanic association, base metal sulfide veins of igneous association,* and *native silver: cobalt-nickel arsenide ores of "Cobalt type"*— are given as examples of the *kinds* of category into which at least many vein deposits probably fall. This is referred to again at the end of the chapter.

THE VOLCANIC PRECIOUS METAL TELLURIDE ASSOCIATION

One of the most conspicuous and distinctive vein associations is that of gold and silver telluride with fractures in volcanic rocks. This is not, of course, to say that all precious metal tellurides are of obviously volcanic association; they are not, and we have already noted in Chap. 11 the presence of hessite at Sudbury and tellurides in the southern African layered intrusions. Telluride mineralization is also known in sedimentary rocks, though these are usually closely associated with volcanic units, and near the contacts of subvolcanic intrusions. Precious metal telluride deposits of the present kind usually contain sulfides as well and may grade off into, or be related to, precious metal veins containing little or no telluride. In spite of these complexities and apparent small divergences, the volcanic precious metal telluride ores constitute the most important type of telluride occurrence and are one of the most clearly distinguished of the vein associations.†

DISTRIBUTION

The distribution of this type of ore occurrence is quite sharply defined, both in space and time. Two categories of occurrence are clear:

1. *Circum-Pacific belt.* The largest group of deposits occurs sporadically along the "Pacific girdle of fire," i.e., around the circum-Pacific volcanic belt. The more notable occurrences are in Fiji (the Tavua gold field), the Philippines, Japan, and California-Colorado-Nevada-Mexico. All are of Tertiary age.
2. *Australian-Canadian Precambrian shields.* These deposits, although widely separated geographically and possessing no direct tectonic connection, are very similar and may be grouped together. Both are closely tied to Precambrian volcanic "greenstone" belts, in each case of the order of 2500 million years old. This group includes the famous deposits of Kalgoorlie in Australia and Kirkland Lake and Porcupine in Canada.

Lesser occurrences are known in Europe, Russia, and elsewhere but are of much less importance than the above two groups.

FORM AND SETTING

In their *form* these deposits are usually rather irregular. Where they occur in fairly large, continuous breaks—as with some of the older deposits—veins are fairly continuous, uniform, and predictable. However, even in the case of the Australian and Canadian deposits, substantial quantities of the ore occur in small irregular fractures related to, or clearly separate from, the major veins of the localities concerned. The "main break" or Kirkland Lake fault is more than 10,000 feet long and 7000 feet deep. Associated with this, however, are a number

---

† The interested reader is particularly referred to an old but outstanding account of this and related gold-silver associations given under the heading "The young gold-silver lodes" in Beyschlag, Vogt, and Krusch, "Ore Deposits," vol. 2, pp. 515–600, Macmillan & Co., Ltd., London, 1916.

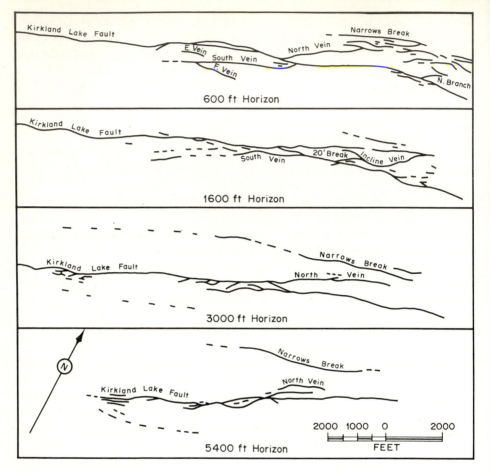

Fig. 17-8 The Kirkland Lake fault, or "main break," and some of its subsidiary fractures at Kirkland Lake, Ontario. (*From Ward et al. in "Structural Geology of Canadian Ore Deposits," a symposium arranged and published by the Canadian Institute of Mining and Metallurgy, 1948.*)

of lesser ore-bearing breaks, as indicated in Fig. 17-8, and in turn there are innumerable smaller fractures far too small to be shown in the diagram. Generally similar situations are found at Kalgoorlie and in some of the larger zones of precious metal telluride mineralization in the southwestern United States.

Many of the Tertiary deposits occur as fillings of small, irregular faults and tension fractures around volcanic centers and in breccia pipes. Shapes of individual veins and the configurations of ore bodies are often quite unsystematic and very difficult to predict.

In all cases the ores are *substantially* fissure fillings, though there is often some subsidiary replacement of vein walls. As fissure fillings they frequently show clear accretionary features such as crustification and the development of well-formed crystals. In some cases, however—particularly in the older deposits —post-ore movement has led to deformation. Where subsequent annealing has occurred, such deformed ores become granular and lose at least much of their original accretionary structure.

The *setting* of the ores is by definition volcanic, though as already pointed out there are some minor apparent deviations from this. The Tertiary circum-Pacific occurrences are naturally the ones of most obvious volcanic affinity, since many of them are demonstrably related not only to volcanic rocks but to present-day volcanic physiography and surface structure. For example, the Tavua mineralization of Fiji is clearly related to a volcanic center and caldera that still stands as a prominent feature of the local scenery. The incidence of ore within the caldera is shown in Figs. 17-9 and 17-10. Some of the Japanese deposits show comparably clear relations with Tertiary to Modern volcanic centers. The relation between the incidence of ore and volcanic structures is not, of course, so obvious in the Precambrian examples, though careful field work in the Canadian areas is beginning to show that this exists in some areas at least. Although, superficially, the Tertiary and Precambrian groups may appear to show some differences of setting, it seems very likely that these are not of fundamental significance and represent erosional, and later deformational, modification of the older environments rather than any real difference of kind.

While most of the deposits are found in faults and joints in lavas and tuffs and in agglomerate piles and breccia plugs, some, including representatives of the Tertiary group, occur close to or within "plutonic" intrusions. At Acupan in the Philippines the ore occurs principally in a quartz diorite and to only a minor extent in associated andesite and volcanic breccia (Callow and Worley, 1965). In Japan, the veins of the Oya mine, Miyagi Prefecture, occur partly in sedimentary rocks within the contact metamorphic aureole of a granodiorite, while ores of the Kinkei mine, Nagano Prefecture, occur in sericitic, graphitic, and chloritic schists not far from a granodioritic intrusion (Watanabe, 1952). However, the association of ore with such rocks is certainly not always as "anomalous" as it might seem; in some cases the "plutonic" rock may in fact be quite shallowly "subvolcanic" and in others the plutonic rock may be much older, having simply been partly or wholly transected by the (later) volcanic pipes with which the ores have their real association.

While we have up to this point emphasized the volcanic association generally —the occurrence of the ores in calderas, pyroclastic plugs and piles, and lavas— we have not paused to consider the compositional types of the volcanic rocks concerned. Constitutions appear to vary greatly and no definite tie between this type of mineralization and a particular lava composition is yet apparent. Basalts and andesites are more frequently associated than more felsic types, particularly among the Precambrian occurrences, but felsic rocks are quite common. In the western and southwestern Pacific, andesites (and their fragmental equivalents) are the overwhelmingly dominant associates, though basaltic rocks are also present in Fiji, and liparites are quite common about the Japanese occurrences. In some of the western United States deposits, on the other hand, rocks as felsic as quartz latites and rhyolites are associated, while at Cripple Creek, Colorado, the rocks are quite alkaline and include phonolites and alkaline basaltic rocks.   If there is some systematic relation between ore occurrence and some feature of volcanic rock composition, this has yet to be recognized.

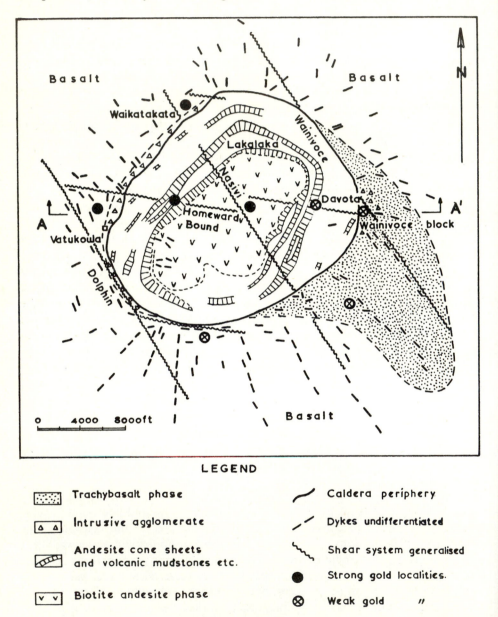

LEGEND

Trachybasalt phase		Caldera periphery	
Intrusive agglomerate		Dykes undifferentiated	
Andesite cone sheets and volcanic mudstones etc.		Shear system generalised	
Biotite andesite phase		Strong gold localities.	
		Weak gold ″	

Fig. 17-9 Geological sketch map of the Vatukoula caldera showing the spatial relation of ore occurrence to the principal geological features.   (*From Denholm, Aust. Inst. Mining Met. Proc., 1967.*)

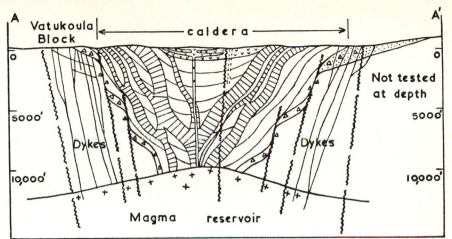

Fig. 17-10 Diagrammatic section through *AA'* of Fig. 17-9. (*From Denholm, Aust. Inst. Mining Met. Proc., 1967.*)

CONSTITUTION

In addition to gold and silver tellurides, ores of this class contain variable quantities of other tellurides, occasional selenides, elemental gold, silver, tellurium, and occasional minor selenium, and a variety of base metal sulfides and associated sulfosalts. Sulfides and sulfosalts are sometimes easily the most abundant constituents. Markham (1960, pp. 1469–1470) has suggested that in the Tertiary deposits native tellurium tends to be abundant and free gold rare, whereas in the Precambrian ores native tellurium is rare and free gold abundant.

The general nature of assemblages may be illustrated by reference to those of some of the major occurrences.

1. Tertiary deposits

   Tavua (*Vatukoula*), *Fiji* (Stillwell, 1949; Denholm, 1967, and others): krennerite, sylvanite, empressite, hessite, petzite, coloradoite, melonite, native tellurium, native gold, tetrahedrite, chalcopyrite, bornite, sphalerite, galena, stibnite, pyrrhotite, pyrite, quartz, calcite (often Mn bearing).

   Acupan, Philippines (Callow and Worley, 1965): petzite, hessite, altaite, sylvanite, calaverite, coloradoite, native gold and silver, electrum, tetrahedrite, chalcopyrite, bornite, famatinite, pyrargyrite, proustite, sphalerite, galena, stibnite, cinnabar, hematite, pyrite, marcasite, arsenopyrite, quartz, calcite (including Mn bearing), rhodochrosite, rhodonite, barite, anhydrite, gypsum.

   Téiné mine, Date, Hokkaido (Watanabe, 1952): sylvanite, krennerite (or calaverite), rickardite, native tellurium, native gold, tetrahedrite, luzonite, enargite, stibnite, bismuthinite, orpiment, chalcopyrite, sphalerite, galena, pyrite, marcasite, quartz, calcite, barite.

Table 17-1 Chemical compositions of some precious metal telluride ores from Japan.† (From Watanabe, Tohoku University Science Report, 1952)

	1	2	3	4	5	6	7	8
SiO_2	83.25	48.32	54.01	24.07	29.87	71.58	57.42	50.20
Al_2O_3	1.55	0.50	1.08	0.30	0.58	3.36	1.97	1.11
MgO	0.20	0.10	0.10	0.25	0.02	0.02	0.03	0.02
CaO	0.08	Trace	0.35	Trace	0.01	0.27	0.21	0.24
$BaSO_4$	0.32	Trace	14.57	60.13	4.18		0.06	0.11
Fe	4.27	1.22	1.83	1.84	1.32	0.65	7.53	4.15
Cu	1.79	20.66	10.74	0.09	27.54	Trace	0.06	0.16
Zn	0.78	1.64	0.80	0.05	1.69	Trace	0.31	0.36
Pb	Trace	Trace	Trace	Trace	Trace		Trace	
Sb	1.12	10.15	5.95	0.30	14.68			
As	0.05	2.71	1.17	Trace	3.20	Trace	Trace	Trace
Bi	Trace	0.35	0.04	0.01	0.37	3.03	0.42	1.64
S	4.43	12.44	7.10	0.71	16.83	0.06	4.13	3.01
Se	0.02	0.09	0.06	1.73	0.34	Trace	0.24	0.48
Te	0.08	0.33	0.83	10.06	0.10	18.44	17.70	29.44
Au	1.43×10^{-3}	0.35×10^{-3}	0.67×10^{-3}	4.63×10^{-3}	0.30×10^{-3}	0.15	6.15	4.61
Ag	0.07	0.07	0.14	0.01	0.18	0.33	0.06	3.27
Total	97.94	98.58	98.77	99.55	100.91	97.89	96.29	98.80

† Analyses 1–5, quartzose tetrahedrite-rich ores from the Téiné mine, Date, Hokkaido; analysis 6, Rendaiji mine; and analyses 7 and 8, Suzaki mine (both of these mines are on the Idzu Peninsula, near Tokyo).

Analyses of a range of ore from the Téiné mine (Watanabe, 1952) are given in Table 17-1.

Cripple Creek, Colorado (Loughlin and Koschmann, 1935; Koschmann and Bergendahl, 1968): calaverite, sylvanite, krennerite, petzite, hessite, silver-copper telluride, pyrite, sphalerite, chalcopyrite(?), galena, tetrahedrite, cinnabar, fluorite, quartz (and chalcedony), calcite, roscoelite, dolomite, ankerite, celestite.

2. Precambrian deposits

Kalgoorlie, Western Australia (Stillwell, 1931; Baker, 1958; Markham, 1960; Cabri, 1967): hessite, petzite, sylvanite, krennerite, calaverite, coloradoite, empressite, altaite, tetradymite, nagyagite, melonite, weissite, native tellurium, gold, arsenic, umangite, eucairite, berzelianite, clausthalite(?), tetrahedrite-tennantite, bournonite-seligmannite, chalcopyrite, bornite, covellite, sphalerite, galena, pyrite, pyrrhotite, galena, quartz, calcite and mixed carbonates, sericite, chlorite, tourmaline, ilmenite, magnetite, hematite, rutile.

Porcupine, Ontario (Keys, 1940; Thompson, 1949): coloradoite, petzite, hessite, tetradymite, altaite, calaverite, sylvanite, native gold, chalcopyrite,

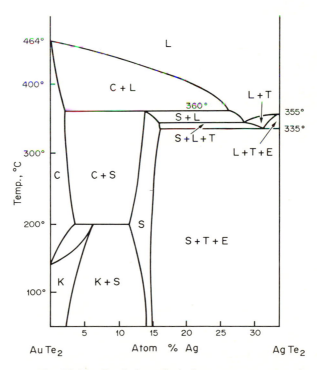

Fig. 17-11 Partly hypothetical temperature-composition relationships along the $AuTe_2$–$AgTe_2$ tie in the Au–Ag–Te system depicted in Fig. 5-28. (*Redrawn from Markham, Econ. Geol., 1960.*)

sphalerite, galena, pyrrhotite, pyrite, arsenopyrite, quartz, calcite, ankerite, tourmaline, scheelite, albite, sericite.

Kirkland Lake, Ontario (tellurides only, after Thompson, 1949; sulfides and others are also present): hessite, petzite, calaverite, altaite, coloradoite, melonite, tetradymite.

Noranda, Quebec (tellurides only, as for Kirkland Lake): hessite, petzite, calaverite, krennerite, altaite, tellurbismuth, rickardite(?), tetradymite.

ORIGIN

This seems quite unequivocal, at least in the case of the Tertiary to Modern deposits localized within and about well-preserved volcanic centers. The tellurides and associated minerals here have been deposited from solutions or vapors constituting part of the volcanic activity itself. Temperature probably covered a wide range from $100°C < T < 600°C$, and pressures would, for such near-surface activity, have been quite low. A partly hypothetical view of phase relations in the Au–Ag–Te system (after Markham, 1960, and Kullerud, 1964) was given in Chap. 5, and a likely set of relationships along the $AuTe_2$–$AgTe_2$ tie is shown in Fig. 17-11 (after Markham, 1960, p. 1164). The nature of the solutions and vapors doubtless varied fairly substantially, though the general presence of calcite indicates that the pH of the relevant hydrothermal solutions could not have been low.

As already suggested, it seems very likely that the Precambrian deposits are closely analogous to the Tertiary occurrences. If one allows for the effects of deformation and mineralogical modification accompanying metamorphism and then later effects of retrogressive metamorphism—including changes in equilibrium assemblages and the onset of recrystallization and annealing generally—it is not difficult to see the older deposits as fairly simple analogs of the younger ones.

BASE METAL SULFIDE VEINS OF IGNEOUS ASSOCIATION

Clearly this association is a very broad one, and there is virtually no doubt that it could be readily divided into quite a number of smaller groupings. To begin with, it includes veins of both plutonic and volcanic affiliation. These two have not been separated because, while some veins are clearly of volcanic connection and others are clearly of plutonic, there are many whose identity in this respect is by no means clear. In addition there are probably more than a few igneous masses currently thought of as "plutonic" that will in time come to be recognized as being quite shallowly "subvolcanic." Added to this broad spread of association there is an enormous heterogeneity of form and constitution—constitutional heterogeneity here is in marked contrast to the comparative (for vein deposits) homogeneity of the precious metal telluride deposits we have just considered. Clearly there is little question that the base metal sulfide veins of igneous association constitute, as a class, the most heterogeneous of all groups of ore deposits.

DISTRIBUTION

This is wide in both space and time. The ores occur abundantly on all continents —and on many islands, some of which are quite small—and in rocks of all geological ages.   By their very nature they are developed almost exclusively in orogenic belts, and if one excludes the veins associated with carbonatites, the "orogenic" association is virtually universal.   In addition to the fact that they *occur* in rocks of all ages, there does not seem to be any significant differences in the *frequency* of occurrence in rocks of different ages or any clear systematic change in composition with age.   Also, there does not seem to be any clear change in frequency of incidence or in composition with depth of erosion of the orogen concerned.   Young, barely eroded belts seem to have no fewer and no more than do more substantially, or very deeply, eroded provinces.   Similarly it seems quite impossible to sustain any suggestion of a broad tendency for change in composition with increase in depth within the fold belt.

Some of the great "vein provinces" of the world are the Precambrian and Paleozoic veins of Scandinavia, the Paleozoic Pennine and Cornwall provinces of England, the Paleozoic Harz province of southeastern Germany, the Precambrian of Canada, the Precambrian to Tertiary of western North and South America, and the Paleozoic Eastern Highlands Belt of Australia.   Many other provinces are, of course, known, and among these that of the Mesozoic to Tertiary belts of the western and southwestern Pacific is becoming rapidly more important as exploration there gathers pace.

FORM

Like other vein deposits, those of this group range from small irregular veinlets to large entities extending, or regularly repetitive, for thousands of feet.   The larger deposits represent single large breaks or intersections of them, and in addition associated smaller veins or veinlets—usually related to conjugate fractures of one kind or another—are fairly general.   Base metal sulfide veins as a class probably attain larger sizes than any other group of vein deposits.

In addition to those ore bodies localized by single fractures or small groups of related breaks, a significant number are in the form of stockworks or volcanic breccia plug fillings.   The stockworks—masses of small, irregular veins occupying cooling cracks and other small tension and shear fractures—are most common in the marginal portions of discordant intrusions, and in this respect the present class of deposits merges with the porphyry coppers.   In many cases—as with many porphyry coppers of the southwestern United States—the association is not obviously, if at all, volcanic.   However, in others, such as Rio Tinto in Spain, large numbers of veinlets of stockwork type are localized in prominent volcanic centers and are clearly of quite shallow subvolcanic formation.

Some very important vein deposits are in the form of fillings in volcanic breccia plugs, including explosion and collapse breccias, and coarse pyroclastic material blown out of, and fallen back into, calderas.   This "breccia pipe" type of deposit is important in Mexico and the Andean region of South America and is now beginning to be recognized in other parts of the world, particularly the Precambrian of Canada and Western Australia and the Southwest Pacific.   The great Cananea deposit in Mexico; Toquepala, Peru; Braden, Chile; Mons Cupri, Western Australia; and a number of deposits of the Noranda area in Canada are of this type.

SETTING

This has already been partly dealt with under the last two headings.   On a broad scale the setting is orogenic, extending from early eugeosynclinal rocks right through to those developed during folding, postorogenic uplift, and intrusion.   The deposits themselves may occur within volcanic centers, apparently as a contemporaneous part of the overall volcanic process.   They may occur as epigenetic features in lava flows, pyroclastic rocks, and other volcanic and associated sedimentary rocks.   They are very common around and within subvolcanic to plutonic intrusive masses and, very frequently, occur within orogenic belts many miles from volcanic centers or from outcropping plutonic rocks.

While spatial relations with their igneous associates are thus variable and sometimes not obvious, there does seem to be some tie between igneous rock composition and the occurrence of ores of this type.   Whether in plutonic or volcanic setting, the association is usually with rocks in the *compositional* range diorite-quartz diorite-quartz monzonite-granodiorite.   Thus in volcanic areas the ores are for the most part associated with rocks in the andesite-quartz latite range, with dacitic rocks perhaps the most frequently occurring of the rock types involved. While there is, of course, a substantial overlap, rock compositions are therefore rather more SiO_2-rich than with the precious metal telluride ores, which generally favor andesitic types.   Compositions, however, are very similar to those we have noted in connection with porphyry coppers.   Occasionally ores of the present type are associated with gabbroic compositions at one extreme or granitic at the other, but this tends to be the exception.   It is perhaps worth noting that ores in volcanic or shallowly subvolcanic terrains are more likely to be associated with compositional extremes than those in plutonic environments; in addition to occurring in the above "intermediate" types, base metal sulfide veins—particularly of cupriferous type—are quite common in basaltic rocks and not at all uncommon in quartz latites and rhyolites.

CONSTITUTION

This is highly variable.   Iron, copper, lead, and zinc are the principal sulfide components, though the proportions of these and the nature and incidence of their associates vary enormously.

Neglecting the iron sulfides, veins may be nearly monometallic in any one of copper, zinc, or lead.   Copper-rich (essentially chalcopyrite) veins occur in association with all rock compositions from mafic (Canadian "greenstones") to felsic (Rio Tinto rhyolites), though they are most frequent in dioritic-andesitic areas.   Copper-zinc veins have a generally similar association, but lead-zinc and highly lead-rich veins tend to have a more specifically felsic association.   As with the stratiform ores of Chap. 15, copper and lead tend to be antipathetic, though this feature is greatly subdued in the vein ores (Fig. 17-12).   Most of the copper, zinc, and lead is normally present in the common sulfides chalcopyrite, sphalerite, and galena, but some veins contain substantial quantities of sulfosalts.   In such cases the precious metals, particularly silver, may become quite abundant.   Indeed in some districts the base metal sulfides assume a subordinate role, and sulfosalts,

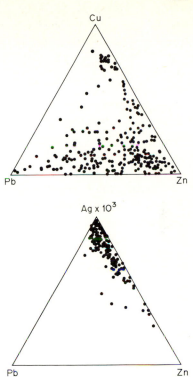

Fig. 17-12 Cu–Zn–Pb and Ag–Zn–Pb relations (atoms per cent) in the Conrad vein, Tingha, New South Wales. (*Redrawn from Richards, Aust. Inst. Mining Met. Proc., 1963.*)

other precious metal compounds, and the native elements (particularly gold and electrum) themselves are the dominant economic components. Zoning—by mineralogy, chemical constitution, and wall-rock alteration—is extremely common among members of this association.

The structures of the ores themselves are rather variable but often highly entertaining for the mineralogist. Most of them show the usual gross features of crustification and development of crystal faces, but in addition a wide variety of growth, precipitation, and deformation structures are common. The primary growth structures include the development of crystal faces and zones in single crystals and crustification and impingment boundaries where aggregates are involved. It is in deposits of this type that exsolution textures (the *precipitation structures* of Chap. 9) obtain their greatest development and in which the majority of the pairs listed in Table 9-4 were originally found. Deformation in ores of this type has received little attention until very recently but is probably much more widespread than has been generally recognized. Grain boundaries are often far from those of equilibrated random aggregates, and their lack of maturity probably reflects simple growth impingement in some cases and unannealed deformation in others. Many of the now undeformed ores of this type undoubtedly owe their microstructures to annealing rather than to primary growth. Although a great deal of microscopic work has been done on ores of this general association, much more remains and there appears to be considerable scope for the application of the principles discussed in Chaps. 9 and 10.

ORIGIN

The *general* nature of the origin of deposits of this group is beyond question: they are deposited from solutions, vapors, and gases that are dominantly of igneous derivation.   As we have noted earlier, the observation of ore-mineral deposition in volcanic fissures is commonplace and has been recognized as such for well over 100 years.   There is every reason to believe—as did de Beaumont and his immediate predecessors in the middle of last century—that analogous processes go on in subvolcanic and deeper environments.   Interaction of igneous solutions with ground water is doubtless nearly universal and must play a significant—and in some cases critical—part in the localization of many ores of this type, particularly those formed at intermediate and greater depths.   Some ores of this association may well be derived from later leaching of igneous materials—particularly pyroclastic beds—rather than from emanations derived directly from the igneous rocks during the actual intrusive or extrusive phase.   "Ore solutions" similar to those of the Salton Sea and Red Sea brines (cf. Chaps. 7 and 15) might well have obtained much of their heavy metal content as they percolated through buried volcanic formations.

NATIVE SILVER:COBALT-NICKEL ARSENIDE ORES OF "COBALT" TYPE

This is one of the most spectacular and clearly defined of all ore types.   It has been referred to, particularly in Chap. 4, as the principal habitat of many of the more exotic arsenide minerals and their assemblages.   It will have been noted, however, that the ores are not referred to now as part of an association.   The reason for this is simple: while the ore type displays remarkably constant characteristics *in itself*, it seems to show little consistency in its geological associations.

DISTRIBUTION

Cobalt-type ores are not of common occurrence, though one could certainly not go so far as to say they are rare.   They are sporadically distributed over a wide area in Europe; the deposits of the Erzgebirge, Germany, are by far the best known here, but there are also occurrences in the Taver copper mine in Northern Sweden, at Kongsberg in Norway, and in Cornwall, Portugal, France, and Sardinia.   The outstanding group of occurrences is, of course, that of the Cobalt district, northeast of Sudbury, Ontario.   Other deposits in Canada occur at Great Bear Lake, N.W.T.   A further group of deposits occurs in southwestern North America, in the Arizona-New Mexico-Mexico province.   Minor deposits are also known in Argentina and South Africa.   The age of the deposits ranges from Precambrian to Mesozoic and perhaps Tertiary, and in the present state of knowledge there is no indication that the ore type, or any recognizable variants of it, has any particular affiliation with any particular geological period.

FORM AND SETTING

The ore bodies are accretionary fillings of faults and joints. They therefore have two larger dimensions and a very much smaller one, thickness commonly being of the order of a few inches to 2 feet and usually not more than about 5 feet. Veins commonly pinch and swell and form "horsetails," where single veins fray out into several minor ones before terminating. Variable impregnation of the wall rocks by ore minerals and associated gangue is common.

The *setting* of the ores is highly variable. The great Canadian province is about 60 miles north-south by 70 miles east-west and contains large numbers of veins cutting Precambrian slates, quartzites, graywackes, and conglomerates. Veins almost invariably occur not too far from quartz-diabase sills—a fact that has been taken by numerous investigators to suggest a genetic connection. The famous veins of the Erzgebirge of Germany and Czechoslovakia occur in Precambrian schists where these have been intruded by Carboniferous granites and related porphyries. The Cornish deposits occur in Paleozoic sedimentary rocks in association with upper Paleozoic granitic stocks and their porphyritic phases. Some of the Hungarian deposits occupy fractures in a dioritic stock intrusive into Devonian schists, and veins in Sardinia occur in association with intrusions of granite, aplite, and porphyry into Silurian black schists. Veins in the Batopilas district of Mexico occur in faulted diorite and andesite that have been intruded by granite, and in the Sabinal district they occur in Lower Cretaceous sediments, including limestones, that have been intruded by a large mass of alaskite (quartz-sodium orthoclase rock). The ores of Wickenburg, Arizona, occur in faults within granite gneiss that has been intruded by dikes of granite pegmatite and fine diabase. In short, if there is some consistency in the setting of this ore type, such consistency is certainly not obvious.

CONSTITUTION

The two outstanding features of the ores themselves are their constitutional complexity and the prevalence in them of delicate zonal and dendritic structures.

The mineral assemblages in the more notable occurrences are given in Table 17-2 (after Bastin, 1939). Apart from the common sulfides such as pyrite, arsenopyrite, chalcopyrite, and galena, the principal opaque minerals are silver, bismuth, niccolite, smaltite, and chloanthite. Clearly the native metal, arsenide, sulfarsenide-antimonide, sulfosalt, and sulfide assemblages are also substantial when the ore type is viewed as a whole. In addition, a number of occurrences—notably the great Cobalt district itself—*individually* exhibit extensive opaque assemblages. The Cobalt area has yielded almost 40 different opaque species.

Apart from quartz, the carbonates are the most abundant gangue minerals, and among them calcite—often a pale-pink manganese-bearing variety—is often conspicuous.

The microstructures of these ores are most spectacular, and their delicacy and complexity are the ore mineralogist's delight. Simple accretionary structures are often beautifully developed, but the really outstanding features of the ores are their very fine dendritic growths. These principally involve native silver and the Ni–Co

Table 17-2   Mineralogical compositions of some native silver: cobalt-nickel arsenide ores.   (After Bastin, 1939)

	Joahanngeorgenstadt, Germany	Joachimsthal, Germany	Annaberg, Germany	Schneeberg, Germany	Freiberg, Germany	Dobschau, Hungary	Turtmannstal, Switzerland	Chalanches, France	Sarrabus, Sardinia	Cobalt, Ontario	So. Lorraine, Ontario	Gowganda, Ontario	Silver Islet, Ontario	Great Bear Lake, N.W.T., Canada	Bullard's Peak, New Mexico, U.S.A.	Wickenberg, Arizona, U.S.A.	Balmoral, So. Africa	Batopilas, Mexico
Elements, etc.																		
Alemontite								×										
Amalgam										×								
Antimony								×										
Arsenic		×		×	×													
Bismuth	×	×	×	×	×		×	×		×	×	×		×				
Gold								×										
Mercury								×										
Silver	×	×	×	×	×			×	×	×	×	×	×	×	×	×		×
Sulfides																		
Argentite	×	×	×	×	×			×	×	×	×	×	×	×		×		×
Berthierite							×											
Bismuthinite	×	×	×	×	×		×											
Bornite		×				×	×			×			×					
Boulangerite																		
Chalcocite		×	×	×					×	×			×					
Chalcopyrite	×	×		×		×	×	×	×	×	×	×	×	×		×	×	
Cinnabar								×										
Cosalite										×								
Covellite													×					
Galena	×	×	×	×	×			×	×	×	×	×	×	×				×
Linnaeite																		
Marcasite	×	×	×	×	×					×			×					
Millerite	×	×	×							×								
Molybdenite																	×	
Orpiment																		
Pentlandite																		
Pyrite	×	×	×	×		×	×	×	×	×	×	×	×	×		×		×
Pyrrhotite		×		×					×	×								
Realgar																		
Sphalerite	×	×	×	×				×	×	×	×			×	×			×
Sternbergite	×	×		×														
Stibnite				×				×										
Stromeyerite										×				×				
Ullmanite																		

Table 17-2 (continued)

	Joahanngeorgenstadt, Germany	Joachimsthal, Germany	Annaberg, Germany	Schneeberg, Germany	Freiberg, Germany	Dobschau, Hungary	Turtmannstal, Switzerland	Chalanches, France	Sarrabus, Sardinia	Cobalt, Ontario	So. Lorraine, Ontario	Gowganda, Ontario	Silver Islet, Ontario	Great Bear Lake, N.W.T., Canada	Bullard's Peak, New Mexico, U.S.A.	Wickenberg, Arizona, U.S.A.	Balmoral, So. Africa	Batopilas, Mexico	Sabinal, Mexico
Tellurides																			
Hessite														×					
Arsenides and sulfarsenides																			
Algodonite																			
Arsenopyrite	×	×		×		×	×		×	×	×			×				×	
Chloanthite	×	×	×	×	×	×	×	×		×			×	×		×			
Cobaltite				×		×	×		×	×	×		×						
Domeykite													×						
Gersdorffite										×									
Lollingite						×				×	×								
Maucherite										×									
Niccolite	×	×	×	×	×	×	×	×		×	×	×	×	×		×			
Rammelsbergite		×	×	×			×		×	×								×	×
Safflorite				×					×	×								×	×
Skutterudite										×			×		×				
Smaltite	×	×	×	×	×	×			×	×		×	×	×			×		
Antimonides and sulfantimonides																			
Breithauptite									×	×	×								
Dyscrasite								×		×									
Ullmannite									×	×									
Sulfosalts																			
Bournonite				×	×				×										
Emplectite										×									
Enargite																×			
Epigenite																			
Jamesonite					×														
Klaprothite																			
Matildite										×									
Polybasite		×		×	×				×	×								×	×
Proustite	×	×	×	×	×				×	×						×		×	
Pyrargyrite		×	×	×	×			×	×	×								×	×
Pyrostilpnite					×														
Rittingerite		×																	
Stephanite		×	×	×	×				×	×									
Tennantite																×			

Table 17-2 (continued)

	Joahanngeorgenstadt, Germany	Joachimsthal, Germany	Annaberg, Germany	Schneeberg, Germany	Freiberg, Germany	Dobschau, Hungary	Turtmannstal, Switzerland	Chalanches, France	Sarrabus, Sardinia	Cobalt, Ontario	So. Lorraine, Ontario	Gowganda, Ontario	Silver Islet, Ontario	Great Bear Lake, N.W.T., Canada	Bullard's Peak, New Mexico, U.S.A.	Wickenberg, Arizona, U.S.A.	Balmoral, So. Africa	Batopilas, Mexico
Sulfosalts (*continued*)																		
Tetrahedrite		×								×	×		×	×				
Wittichenite																		
Xanthoconite		×								×								
Haloids																		
Fluorite	×	×	×	×					×	×			×	×				
Oxides																		
Magnetite														×				
Opal																		
Quartz	×	×	×	×		×	×	×	×	×	×	×	×	×	×	×	×	×
Specularite	×	×											×	×				
Carbonates																		
Ankerite						×												
Aragonite										×								
"Brown spar"	×	×	×				×	×										
Calcite	×	×	×	×		×		×	×	×	×		×			×	×	×
Dolomite	×	×		×						×			×					
Rhodochrosite													×					
Siderite				×		×		×								×	×	
Sulfates																		
Anhydrite																		
Barite	×		×			×			×	×		×					×	×
Celestite																		
Gypsum		×						×										
Strontianite																		
Uranates																		
Pitchblende	×	×	×	×										×				
Others																		
Tourmaline						×												

arsenides, particularly the safflorites. The native silver, which sometimes contains intergrown native bismuth, develops cubic dendritic patterns that in turn yield cruciform shapes in cross sections. Frequently the crosses are composed of silver, or silver-plus-calcite, cores that are overgrown by arsenides. The whole dendritic mass may then be embedded in calcite. Quite often the cruciform shapes are rounded and are better referred to as rosettes. Not infrequently colloform structures are developed and involve silver, bismuth, native arsenic, and other components. In many cases the initial accretionary structures are slightly deformed and disrupted, permitting late-stage veins to develop in earlier ore and gangue minerals and to extend into the walls. Vein deposition at all stages may be followed by transitory periods of replacement, and so a whole gallery of accretionary, precipitation, deformation, and replacement structures may develop in a single vein or system of veins. This, with the very extensive assemblages that so often develop, gives great scope for the determination of long, complicated paragenetic sequences—that for the Silverfields area (about 2 miles south of the town of Cobalt, Ontario) shown in Fig. 17-13 (from Petruk, 1968) is typical.

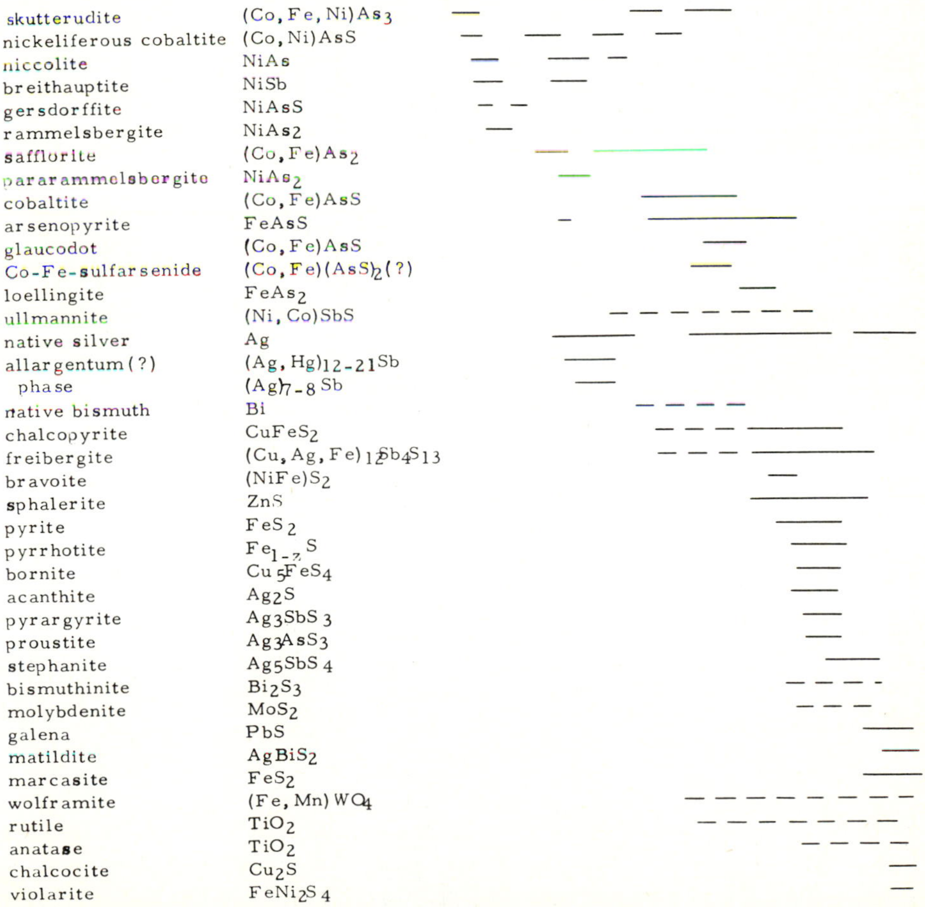

skutterudite	$(Co, Fe, Ni)As_3$
nickeliferous cobaltite	$(Co, Ni)AsS$
niccolite	$NiAs$
breithauptite	$NiSb$
gersdorffite	$NiAsS$
rammelsbergite	$NiAs_2$
safflorite	$(Co, Fe)As_2$
pararammelsbergite	$NiAs_2$
cobaltite	$(Co, Fe)AsS$
arsenopyrite	$FeAsS$
glaucodot	$(Co, Fe)AsS$
Co-Fe-sulfarsenide	$(Co, Fe)(AsS)_2(?)$
loellingite	$FeAs_2$
ullmannite	$(Ni, Co)SbS$
native silver	Ag
allargentum(?)	$(Ag, Hg)_{12-21}Sb$
phase	$(Ag_{7-8}Sb$
native bismuth	Bi
chalcopyrite	$CuFeS_2$
freibergite	$(Cu, Ag, Fe)_{12}Sb_4S_{13}$
bravoite	$(NiFe)S_2$
sphalerite	ZnS
pyrite	FeS_2
pyrrhotite	$Fe_{1-x}S$
bornite	Cu_5FeS_4
acanthite	Ag_2S
pyrargyrite	Ag_3SbS_3
proustite	Ag_3AsS_3
stephanite	Ag_5SbS_4
bismuthinite	Bi_2S_3
molybdenite	MoS_2
galena	PbS
matildite	$AgBiS_2$
marcasite	FeS_2
wolframite	$(Fe, Mn)WO_4$
rutile	TiO_2
anatase	TiO_2
chalcocite	Cu_2S
violarite	$FeNi_2S_4$

Fig. 17-13 Mineral assemblage and approximate sequence of mineral deposition in the Silverfields deposit, Cobalt, Ontario. (*From Petruk, Econ. Geol., 1968.*)

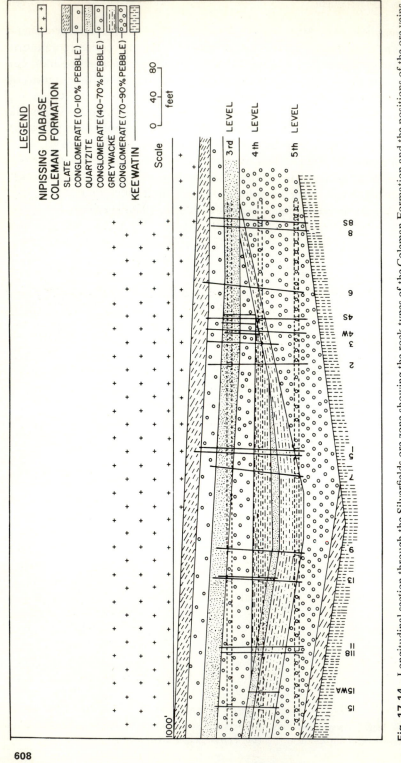

Fig. 17-14 Longitudinal section through the Silverfields ore zone showing the rock types of the Coleman Formation and the positions of the ore veins, which are represented by the near-vertical lines. *(From Petruk, Econ. Geol., 1968.)*

ORIGIN

Just as ideas on the origin of primary nickel ores were for many years dominated by theories developed around the ores of Sudbury (cf. Chap. 11), so ideas on the origin of ores of the present group have been heavily influenced by observations in the Cobalt area.

There appears to have been little doubt that the primary portions of these ores, wherever they occur, have been deposited from solutions derived from igneous intrusions. In the Cobalt province it happens that the only intrusions are composed of diabase, with small quantities of associated aplite. In the Cobalt area itself this diabase—a representative of the Nipissing diabase that is such a prominent feature of the eastern Canadian Precambrian—is in the form of a gently dipping sheet, about 1000 feet thick, that has been injected more or less parallel to, and just above, the uncomformable contact of the Cobalt series of sedimentary rocks and the underlying Keewatin basement (cf. Fig. 17-14). Most of the veins occur in the Cobalt rocks, between the Keewatin-Coleman Formation unconformity and the sill, with minor quantities in joints in the sill itself. While it may easily be said that the spatial association between veins and diabase here is a somewhat fortuitous, structural one, there is also clearly the possibility of a genetic connection—that the mineralizing solutions come from (1) the diabase itself at a late stage in the latter's cooling history and/or (2) basaltic magma chambers lying deep beneath the present zone of interest, which gave rise to the diabase magma of the sills and, essentially independently, to the ore solutions. Apart from minor aplites apparently closely related to the Nipissing diabase sills, no other intrusive igneous rocks of appropriate age have been found. The evidence as it appears in the Cobalt province—by far the most important province in the world for this ore type—has therefore been taken by many people to indicate the likelihood of a general genetic link between *basic* magmas and native silver:Co–Ni arsenide veins.

This has been questioned, however, by Bastin (1939), who has drawn attention to the igneous associations found elsewhere. Clearly for the world as a whole, intrusions of granitic affinity—granites, granitic aplites, granite porphyries, etc.—are almost ubiquitous associates. In some cases, notably Sweden, Cornwall, Germany, Switzerland—and even in parts of the Ontario province—the Ag–Co–Ni veins merge into or are intimately associated with base metal veins similar in all respects to those of normal granodioritic-granitic affinity. Also, as is emphasized by Bastin (1939, pp. 20–23), bismuth is a characteristic component of the Ag–Co–Ni veins, and uranium is sometimes present in notable amounts. Where these two metals occur independently of Ag–Co–Ni mineralization, their affinities are conspicuously with *felsic* igneous rocks. All this seems to suggest the real affinity of the Ag–Co–Ni ores may well be a felsic igneous one and that the Cobalt province, important though it is, is "anomalous." Bastin suggests, however, that even this anomaly may be no more than apparent. In the Bruce mines area, on the north shore of Lake Huron and hence southwest of Cobalt, there occurs Ag–Co–Ni mineralization in rocks similar to those about Cobalt and containing diabase sills. However, granite also occurs here. Apparently the ore veins in this locality developed after the intrusion of the granite, which in turn has occurred after the emplacement of the Nipissing diabase sills. The evidence is not un-equivocal, but Bastin (1939, p. 38) considers it more than possible that the connec-tion is with the granite and that by analogy the similar and not too distant Cobalt veins are derived from a related granite as yet unexposed.

Keeping in mind this doubt, evidence of worldwide association seems to indicate that the affiliation of this ore type is a felsic, granitic, one.

CONCLUDING STATEMENT

It has clearly not been possible in one short chapter to cover more than a very small part of the large and complex category of vein ores. To some readers this will doubtless appear as a fault of emphasis on the part of the author—that he, more concerned with other ore types, has neglected the important group of vein ores to which far more space should have been devoted. This may, of course, be correct, and in his defense the author can only plead that he had in mind the declin-ing relative importance of veins as sources of metals. Such a point of view may be regarded as unworthy by the purists, though to the author it is a matter of bowing to reality.

Having regard to features of form, setting, composition, and origin, it is probably true to say that vein ores as a category cover the whole spectrum from the simplest to the most complex of all mineral occurrences. There could be no sim-pler an entity than a monomineralic, diagenetic, pyrite or sphalerite vein and prob-ably no more complex an entity than a polymineralic, polygenetic vein of a severely deformed and intruded metasedimentary environment. It is thus hardly surprising that there are many vein associations, that many such associations are blurred and grade into each other, and that many vein occurrences cannot be readily fitted into any neat pigeonhole at all.

The three vein associations considered in this chapter are, of course, no more than a very small group of examples, and it must be kept in mind that there are in fact a wealth of associations of other kinds. As we have already noted, the great group of base metal sulfide veins of igneous affiliation—the second associa-tion of this chapter—undoubtedly includes several associations. Some of the more conspicuous among those not referred to earlier are the stibnite-quartz, stibnite-quartz-gold, and stibnite-quartz-scheelite-gold deposits that appear, with remark-ably consistent features, in intruded terrains here and there all over the world. Quartz-molybdenite, quartz-chalcopyrite-molybdenite, and quartz-molybdenite-bismuth-bismuthinite veins associated with felsic granitoids are another, if less frequent, category. Similarly there are a number of conspicuous oxide-bearing associations. Pegmatitic cassiterite and wolframite-bearing veins have been sub-stantial sources of tin and tungsten; quartz-cassiterite, quartz-wolframite, and a variety of more complex variants of these are important and well known. Among the native metal-bearing veins are those containing gold-quartz, gold-pyrite-quartz, and gold-calcite-pyrite-quartz assemblages, all of which are among the best known of all ore types and have yielded large quantities of gold in many parts of the world. Native antimony and native bismuth also occur in quite substantial quantities in "igneous" veins—sometimes as the sole constituents but more frequently as components of base metal sulfide veins.

Taking the category of vein deposits as a whole, it may be said that most native metal, and most oxide-bearing types, have quite clear igneous affiliations. The same is the case with the tellurides, and with the sulfides associated with them. However, while many other sulfide veins are probably also of direct igneous derivation, a great number are not. These range from the simplest sedimentary-diagenetic-burial metamorphic veins through to the most complexly polymetamorphic types occurring in highly deformed, high-grade metamorphic terrains. Such sulfide veins form under a range of conditions from the lowest temperature-pressure aqueous conditions of diagenetic environments right through to the highest temperatures and pressures of regional metamorphism. Involved in this is the whole gamut of hydrous and dry diffusional processes—the lateral secretion of Sterry Hunt and Sandberger—taking place during metamorphism. And this is far from all; the history of most veins has not ended with the final deposition of the ore minerals—*formation* has often been accompanied and followed by *deformation*. The shape, microstructure, and even the mineralogical constitution are often so heavily modified by such deformation that present features reflect barely at all the source and early history of the ores concerned. As a result it is far from easy to see any general relations between the complexities of ores and the complexities of their histories. Certainly many of those with a simple history are simple in themselves, but some with quite complex histories appear quite simple also. Similarly some ores which appear to have had quite simple histories are just as complex mineralogically as others known to have passed through a long history of events. Vein ores as a class are indeed an elusive and a difficult group, and they will doubtless provide problems for many generations of ore petrologists yet to come.

RECOMMENDED READING

Barnes, H. L. (ed.): "Geochemistry of Hydrothermal Ore Deposits," Holt, Rinehart and Winston, Inc., New York, 1967.

Bateman, A. M.: "Economic Mineral Deposits," 2d ed., John Wiley & Sons, Inc., New York, 1950 (see especially pp. 94–163).

Helgeson, H. C.: "Complexing and Hydrothermal Ore Deposition," Pergamon Press, New York, 1964.

Krauskopf, K. B.: "Introduction to Geochemistry," McGraw-Hill Book Company, New York, 1967.

Park, C. F.: The zonal theory of ore deposits, *Econ. Geol. 50th Anniv. Vol.*, pp. 226–248, 1955.

Skinner, B. J., D. E. White, and R. E. Mays: Sulfides associated with the Salton Sea geothermal brine, *Econ. Geol.*, vol. 62, pp. 316–330, 1967.

18
Ore Deposits of Metamorphic Affiliation

Many ore bodies occur in metamorphic terrains and among those that do are representatives of all the types and associations we have considered so far. Indeed if one includes all metamorphosed host rocks down to those of the albite-epidote-hornfels facies of contact metamorphism and the quartz-albite-muscovite-chlorite subfacies of regional metamorphism, it can be said that the majority of the world's ore occurrences are contained in metamorphic rocks. This is particularly true of base metal sulfide ores and to a lesser extent those of the precious metals.

Ores in metamorphic rocks may be related to metamorphism in two principal ways. They may be metamorph*ic* in the sense that the valuable constituents have been concentrated into their present positions by metamorphic processes. On the other hand they may be metamorph*osed* in the sense that metamorphism has simply changed the form and/or chemical arrangement of a preexisting deposit, having had no or negligible effect on concentration. Both categories of ore occurrence are of course metamorph*ic* in the sense in which that term is normally applied to common silicate and carbonate rocks.

Of the two main primary classes of rocks—igneous and sedimentary—it is the sedimentary rocks that are most frequently involved in, and most intensely modified by, metamorphic processes. As it happens it is also the ores in sedimentary rocks that seem to be most frequently—and most closely—related to metamorphism.   While representatives of any of the igneous associations we have considered—ultramafic-nickel, carbonatite-rare earth, porphyry-copper, and the others—may be found deformed or involved in metamorphism in one way or another, such cases are probably the exception rather than the rule.   On the other hand it may almost be said that it is rare to find ores in sedimentary rocks without clear manifestations of associated metamorphism.

TYPES AND FACIES OF METAMORPHISM

For present purposes we may follow Turner and Verhoogen's definition of metamorphism: it "is the mineralogical and structural adjustment of solid rocks to physical or chemical conditions which have been imposed at depths below the surface zones of weathering and cementation and which differ from the conditions under which the rocks in question originated" (1951, p. 450).   We may then subdivide metamorphism in general into four principal types:

1. *Contact metamorphism*.   That developed as an essentially local phenomenon in aureoles about intrusive bodies of igneous rock.
2. *Regional metamorphism*.   That developed over areas of the order of thousands of square miles—i.e., of "regional" extent—in association with burial and folding.
3. *Metasomatic metamorphism*.   That developed as chemical changes brought about by migrating liquids, vapors, or gases; a preexisting *host rock* is attacked by such fluids and is wholly or partly *replaced* by new material referred to as a *metasome*; metasomatic changes may develop as a part of contact or regional metamorphism or quite independently.
4. *Dislocation metamorphism*.   That developed in restricted zones of movement— such as faults and shears—and of intense disruption and deformation.

Since contact and regional metamorphism result from increase in temperature and pressure, it follows that progressive increase in either or both of these variables leads to the development of a series of *grades* of each of the two types of metamorphism.   For present purposes we may follow nomenclatures proposed by Turner and Verhoogen (1951).   Each grade, or *facies*, is defined by the presence of a particular index mineral or a particular equilibrium mineral assemblage.

CONTACT METAMORPHISM

This appears to develop over a wide range of temperatures but within a fairly narrow range of pressures.   Turner and Verhoogen (1951, p. 509) suggest temperatures of 300 to 800°C (very occasionally up to 1000°C) and load pressures from about 100 to 3000 atm.   Their principal *facies* of contact metamorphism in order of increasing temperature (at constant water pressure) or decreasing water pressure (at constant temperature) are

1. *Albite-epidote-hornfels*.   This is developed on the outer margins of contact-metamorphic aureoles or as the principal component of a weakly developed

aureole; in the former case it passes inward to the hornblende-hornfels facies.

2. *Hornblende-hornfels.*   This is commonly extensively developed; it passes inward to pyroxene-hornfels facies or may abut the intrusion in cases of weaker metamorphism.

3. *Pyroxene-hornfels.*   This forms the innermost zones of well-developed aureoles.

4. *Sanadinite.*   This is developed in xenoliths and in localized areas about shallow intrusions (low confining pressure allows the escape of H_2O and other volatiles, leading to the development of anhydrous, high-temperature phases).

REGIONAL METAMORPHISM

This also appears to develop over a wide range of both temperature and pressure, though temperatures in the lower half of the range, i.e., below about 350°C, are overwhelmingly the more common.   As already noted, sedimentary—including pyroclastic—and basic volcanic rocks are the materials principally affected by regional metamorphism.   With increase in load and temperature, diagenesis shades subtly into the low grades of regional metamorphism; mineral assemblages change and with them the physical structure of the rock, yielding progressively slates, schists, and then gneisses.   The recognized facies of regional metamorphism and their principal subfacies, approximately in order of increasing temperature, are given by Turner and Verhoogen (1951, p. 531) as:

1. Zeolite
2. Greenschist.   Three subfacies are
    a. Quartz-albite-muscovite-chlorite
    b. Quartz-albite-epidote-biotite
    c. Quartz-albite-epidote-almandine
3. Glaucophane-schist.   This facies probably corresponds to the same temperature range as that of the greenschist facies but higher pressures.
4. Almandine-amphibolite.   Four subfacies are
    a. Staurolite-almandine
    b. Kyanite-almandine-muscovite
    c. Sillimanite-almandine-muscovite
    d. Sillimanite-almandine-orthoclase
5. Granulite.   Two subfacies are
    a. Hornblende-granulite
    b. Pyroxene-granulite
6. Eclogite

The principal equilibrium mineral assemblages are indicated by the subfacies, and it is unnecessary to consider them in further detail here.   Metamorphism of *zeolitic* facies may be said to commence at temperatures between 100 and 200°C and pressures between 1000 and 2000 atm.   That of the *greenschist* facies takes place between the approximate limits 300 to 500°C and 3000 to 8000 atm pressure. Rocks of the *glaucophane-schist* facies seem to be developed by the metamorphism of geosynclinal suites of basalts, tuffs, graywackes, and cherts; their temperature range is probably essentially the same as that for the greenschist facies, though pressure is probably rather higher. The conditions of metamorphism in the *almandine-amphibolite* facies are in turn more intense, and Turner and Verhoogen (1951, p. 553) indicate temperatures in the general range 550 to 750°C, with pressures of 4000 to 8000 atm. Granulites (i.e., gneisses) are the distinctive product of the *granulite* facies of metamorphism, which appears to involve high temperatures (probably 700 to 800°C), high but rather variable total pressures, and low partial pressures of water. The *eclogite* facies constitutes the highest grade of regional metamorphism and involves high pressures, as indicated by the high densities of eclogites (3.4 to 3.6) as compared with the chemically analogous gabbros ($\simeq 3.0$). Rocks of this group are, however, comparative rarities, and while they are of great interest in silicate petrogenesis, they are of little or no importance as associates of ore.

METASOMATIC METAMORPHISM

As already noted, *metasomatism* or, synonymously, *replacement*, may proceed as a component process of contact and regional metamorphism. Indeed this is probably the case in almost all major contact and regional metamorphic events. For example, the development of calc-silicate rocks is usually partly a manifestation of metasomatism accompanying contact metamorphism, and regional scapolitization may result from widespread metasomatism accompanying regional metamorphism. On the other hand, it may proceed without the development of other metamorphic features: the conversion of peridotites to serpentine and talc, the replacement of fossils by silica or pyrite, and the replacement of rocks and fossils by oxide and sulfide ores usually proceeds without the development of other metamorphic modification of the materials concerned.

Concerning the intensity of metasomatism, one can do little more than give a qualitative estimate such as "slight," "moderate," "extensive," and so on. In other words the measure of replacement is its extent or intensity—not variation in type or "rank." It varies, from one example to another, from barely perceptible to 100 per cent replacement of the host by the metasome.

DISLOCATION METAMORPHISM

This involves intense structural modification of rocks and, like metasomatic metamorphism, is not readily divisible into grades; it can be described only by a qualitative estimate of its intensity and by reference to any specific structures developed.

ASSOCIATION OF ORES WITH TYPES OF METAMORPHISM

A wide variety of relations between ore and metamorphism has been recognized. Since it is in sedimentary rocks that the majority of large ore deposits occur and since regional metamorphism is the modifying process most frequently undergone by sediments, it follows that it is in regionally metamorphosed sedimentary terrains that most ores of metamorphic association occur. Ores associated with contact-metamorphosed rocks are much fewer in number, though they are by no means uncommon.   Many ores—particularly those in veins—are associated with dislocation metamorphism.   Metasomatic metamorphism has been regarded as an overwhelmingly important process in the formation of nonvein ores, though as we shall see, modern investigations suggest that it may be a good deal less important than previously supposed.   Apart from metasomatic change, the ore minerals of any given deposit may be related to the associated metamorphism in any one of three different ways:

1. They may have been *modified* by the metamorphic process, their original concentration representing a *prior* phenomenon.
2. They may have been *concentrated* by the metamorphic process, and hence constitute a *contemporary* phenomenon.
3. They may have been *superimposed* (by metasomatism, injection, etc.) on distinctly older, earlier metamorphosed, host rocks, their concentration constituting a *subsequent* phenomenon.

ORES OF CONTACT-METAMORPHIC ASSOCIATION

It seems likely that most deposits of contact zones represent either of the first two possibilities—they have been formed by contact metamorphism or modified by it. Which of these two relations holds in any particular case may not be easy to determine, and it would not be surprising if a number of deposits now diagnosed as having been *produced* by contact metamorphism are found in the future to have been only *modified* by it.

DEPOSITS PRODUCED BY CONTACT METAMORPHISM

There are four principal aspects of the production of contact-metamorphic ores.

1. Source rocks Contact deposits are virtually exclusively associated with large plutonic intrusions, i.e., batholiths, stocks, and bosses.   Minor concentrations may develop adjacent to dikes and sills and along the underside of large flows, but such deposits are usually of insignificant size and are comparative rarities.   The most common source types are the notably hydrous intermediate to felsic-intermediate rocks—quartz diorites, quartz monzonites, granodiorites, and their close relatives.   Comparatively dry basic and ultrabasic magmas—gabbros, norites, and peridotites—rarely yield contact deposits of note.   Granites as a class do not seem to constitute notable sources, though occasionally they have produced quite important deposits of this type.   The reason for the preferential association of contact deposits with the quartz diorite-granodiorite group rather than with gabbros and peridotites seems to be simply that the more siliceous types evolve much more volatile matter—the medium by which the ore materials are transferred from source to host.   Following this line of reasoning, however, it is not easy to see why granites should be such modest sources of contact ores.   This is clearly a matter for further study.

2. Conditions of transfer Initial temperatures are of necessity high and may reach the vicinity of 1000°C. As the magma cools and solidifies, temperatures fall in accordance with the thermal conductivity and thickness of the overlying rock. Ultimately the temperature equilibrates at that dictated by the local thermal gradient.

By analogy with the gases observed to evolve from volcanoes and from the evidence of the mineral assemblages of many contact aureoles, the principal compounds involved in the transfer process are water, halogens, metallic halides, sulfur and sulfur compounds, including the sulfides, CO_2, and perhaps nitrogen compounds. Phosphorus and boron are also conspicuous in many assemblages and undoubtedly arrive as compounds of igneous volatiles. During the period of gas transfer, materials are probably deposited as a result of heterogeneous reactions, for example,

$$Fe_3O_4 + 9Cl \rightleftharpoons 3FeCl_3 + 2O_2$$

and by reaction with the invaded rock, for example,

$$2FeCl_3 + 3CaCO_3 \rightarrow Fe_2O_3 + 3CaCl_2 + 3CO_2$$

As the intrusion and its contact cool, solution activity undoubtedly begins to take over from gas transfer and eventually becomes the principal mechanism of movement. Because contact-metamorphic generation of ores is in large part a high-temperature gas-liquid solution process, its products are often referred to as *pyrometasomatic*, or *pneumatolytic*, deposits.

3. Host rocks There is no question that certain rock types are far more frequently hosts to contact-metamorphic deposits than are others. As might have been expected it is the more highly reactive and relatively soluble rock types that constitute the most receptive hosts, and among these the carbonate rocks stand out. Most of these are impure, and the initial heating brought about by the intrusion tends to cause early development of calc-silicate assemblages appropriate to the facies of contact metamorphism being developed. Solution, especially of the carbonate fraction of the rock, makes way for the deposition of new compounds derived from the intrusion and compounds formed by reaction between igneous and host-rock materials. All carbonate types—limestone, dolomitic limestone, and dolomite—seem highly susceptible to contact-metamorphic addition, though in particular districts one type often seems more susceptible than others.

Noncarbonate sedimentary rocks, igneous rocks, and older metamorphic types are not receptive and rarely appear as hosts. Of these the more basic igneous rocks, with their high content of unstable mafic minerals, are the most likely to be affected. The Buena Vista iron ores of Churchill County, Nevada, appear to be examples of selective contact metasomatism of this type of material. Here, in a sequence containing shales, sandstones, cherts, limestones, dolomites, and a variety of intermediate to felsic lavas and tuffs, contact metasomatism has picked out layers of hornblende andesite for the deposition of large quantities of magnetite and the iron-vanadium spinel, coulsonite.

4. General statement The *distribution* of deposits of this kind is wide—they occur throughout the world in rocks of all ages. Since, however, they are generally developed about the roof areas of the relevant intrusions, they are more prominent in less deeply eroded terrains and hence are more conspicuous in Paleozoic, Mesozoic, and Tertiary provinces than in the Precambrian. By their very nature they are distributed here and there throughout belts of folding and plutonic intrusion and hence along the sites of old geosynclines. Within major loci of this type their distribution is largely controlled by the coincidence of quartz diorite-granodiorite plutons and carbonate-bearing sedimentary units. Some of the porphyry copper deposits provide good examples of contact mineralization. In Chap. 12 it was noted that while porphyry coppers *sensu stricto* are located in the parent stock, there are some cases where the mineralization extends into the immediately adjacent intruded rocks and even into rocks at some distance from the contact but still within the contact-metamorphic aureole. Such deposits seem to be genuinely derived from the central stock. The Bingham mineralization, with its central orthomagmatic copper and peripheral contact-metamorphic pyritic and sphalerite-galena deposits, provides a good example. It is noteworthy that in deposits of this kind the contact-metamorphic ores usually adhere to the general principle that ore favors carbonate, or carbonate-bearing, sedimentary rocks. Many contact ores are not, of course, related to porphyry copper occurrence and simply occur adjacent to, or as a separate part of the aureole of, an apparently "normal" plutonic mass. Their *form* is controlled largely by the configuration of, and distribution of inhomogeneities in, the host rock. One boundary of a deposit is normally defined by the intrusive contact, though the ore may extend a few feet back into the intrusive in the form of stockworks and disseminations. From the contact, the ore usually fingers out along, and within the confines of, the host rock. Where the host is, for example, an impure limestone unit, the extent of ore mineral deposition is often largely controlled by the composition of individual beds, leading to the development of a highly digitated outer margin (Fig. 18-1). A few contact ore bodies are large (major dimension of the order of thousands of feet) but most are small and of doubtful economic significance. Aspects of *setting* have been largely dealt with under *distribution* and *form*. We may summarize, however, by listing the main features of environment. In decreasing order of scale:

a. Here and there along the length of fold mountain belts
b. Against, or close to (always within the contact-metamorphic aureole) plutonic intrusions, usually of quartz diorite to granodiorite composition and most commonly of subsequent type
c. Within carbonate rocks of varying purity and, very occasionally, in other rock types such as igneous rocks rich in unstable mafic minerals

Composition varies from simple to highly complex and represents a combination of metamorphic minerals formed from the host matrix itself with those formed by reaction with, and simple deposition from, the igneous emanations. Of the introduced substances, silica is usually the most important. This often combines

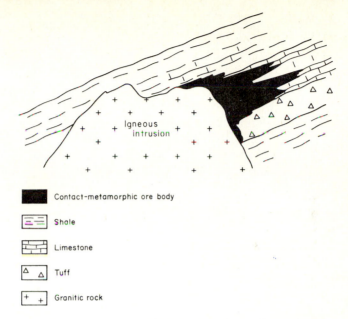

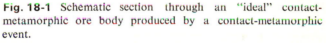

■ Contact-metamorphic ore body

Shale

Limestone

△ △ Tuff

+ + Granitic rock

Fig. 18-1 Schematic section through an "ideal" contact-metamorphic ore body produced by a contact-metamorphic event.

with the carbonate host rock to form calc-silicates and also forms quartz veins, which may be abundant. Addition of the alkali metals leads to reactions yielding feldspars and micas; addition of fluorine leads to the formation of minerals such as fluorite, topaz, and fluorapatite; addition of boron yields borosilicates such as tourmaline and axinite. Of the "ore elements," iron is the most frequently, and usually the most abundantly, introduced. It is often accommodated in silicates such as andradite garnet and hedenbergitic pyroxene, but its main mass is usually in the form of magnetite. Hematite, pyrite, arsenopyrite, and pyrrhotite are lesser, though sometimes important, iron minerals. Such assemblages of andradite, hedenbergite, and iron oxides and/or sulfides formed by contact-metamorphic addition of silica and iron to carbonate rocks are often referred to as *skarns*. Quite substantial deposits of iron ore may have formed in this way.

In some cases sulfides, arsenides, and related minerals occur abundantly in contact zones. Common minerals are arsenopyrite, pyrite, pyrrhotite, chalcopyrite, sphalerite, and galena, and one or more sulfosalts are often present. While some of these sulfide concentrations are of ore-deposit size and grade, they are not important as a class.

Contact-metamorphic ores occur as irregular masses, as isolated "splashes" in calc-silicates, and as veins and disseminations. They are often coarse, and particularly where brightly colored sulfides and calc-silicates are associated, they may be of quite spectacular appearance. For the most part their microstructures are not well known; in view of their high-temperature history it might be expected that they would be well annealed and that exsolution structures might often be preserved.   Any later movement, however, would modify grain-boundary configurations and might cause such complications as the addition of deformation twins to others earlier formed by annealing.

The *origin* of these deposits is not altogether clear; their components probably have two derivations, which in many cases may well have been compounded. The materials may be primary, in the sense that they have reached their final concentration entirely through igneous differentiation processes, or they may represent material engulfed by the pluton at an earlier, deeper stage of its intrusion—material which was melted, assimilated, and then regurgitated, so to speak, in the final stages of magma consolidation.   Many ores probably result from compounding of two such lineages.   At the present time there is no known way of discriminating between these various possibilities, though lead isotope studies may well yield a method.

DEPOSITS MODIFIED BY CONTACT METAMORPHISM

Occurrences of this type have not received wide recognition as a class, probably because almost any deposit at a major igneous contact has been assumed to have been *derived from* the adjacent igneous rock.   Again there is a problem of discrimination.   What could be criteria for distinguishing between products of *generation* on the one hand and those of *modification* on the other?

The problem is undoubtedly made easier where there are a number of deposits in the district concerned.   If the majority of these are independent of intrusive contacts, it is possible that the one or two occurring at contacts do so simply because they have been encroached upon by the relevant intrusion.   More specifically, if the deposits close to the contacts show the same stratigraphic and structural affiliations as do the deposits distant from the contacts, the former would probably represent preexisting deposits *modified* by the metamorphism.   But if their geological setting were different entirely, contact-metamorphic *generation* would be indicated.   In addition, appropriate evidence of contamination of the igneous rock in the vicinity of the ore body, or of substantial "back veining," should help to indicate the relation between intrusion and adjacent ore.

The best documented examples of contact-metamorphosed ores seem to be those involving the intrusion of large ore bodies by dikes and other minor igneous masses.   Probably the most famous instance of this is the intrusion of massive pyritic sphalerite-chalcopyrite ore by diabase dikes at the Horne mine, at Noranda in Quebec.   This has been investigated on several occasions during the past 40 years and until recently has been thought to represent the preservation of *pre-ore* dikes during the replacement of surrounding rhyolitic rocks by sulfides.   However, a very thorough investigation by Mookherjee and Suffel (1968) has shown conclusively that the dikes are younger than the ores.   Among a number of features it has been noted that:

1. On the sulfide side of the contact, pyrite is converted to magnetite, there is a marked increase in the proportion of pyrrhotite of monoclinic type and chal-

copyrite blebs in sphalerite, and there is the development of sulfide-sulfide and sulfide-silicate reaction zones.

2. The chilled margin of the diabase is *glassy* where in contact with sulfides but *microcrystalline* where in contact with silicate wall rock.

The modification of the sulfide mineralogy adjacent to the contact seems quite clearly to be a thermal effect associated with the dike intrusion, and the sharp association of glassy selvages where the dikes abut sulfides appears to be due to the more rapid chilling along these contacts—in turn due to the high heat conductivity of the sulfide "wall rock" as compared with the rhyolite.

A number of other examples of minor intrusion into sulfides are known. At the Eustis mine, in the Eastern Townships of Quebec (Stevenson, 1937, p. 335), a 40-foot camptonite dike has intruded a cupriferous (chalcopyrite) pyritic mass, producing a pyrrhotite-chalcopyrite-cubanite-anthophyllite assemblage in the contact zone and also developing pyrrhotite and pyrrhotite-chalcopyrite pseudomorphs after pyrite. Further examples of the intrusion of sulfide ore bodies by doleritic dikes are known at the Quemont, Normetal, and Mattagami Lake mines in Quebec and in the Wilroy and Geco mines at Manitouwadge, Ontario. There is also a notable occurrence at the Machia Magra Hill, Rajasthan, India, where a lead-zinc sulfide ore is cut by a 40-foot-thick doleritic intrusive that is partly a sill and partly a dike (see Mookherjee, 1970). Sulfur loss near intrusions is common, and where it is most pronounced the iron sulfide goes to magnetite. In some ores containing both pyrite and pyrrhotite, the iron sulfides convert to magnetite closest to the contact, the sulfur so released combining with pyrrhotite a little farther from the contact to form second-generation pyrite porphyroblasts. Contact metamorphism of sulfides by larger, granitoid, masses is less well documented, though examples have been described from the Canadian Shield, British Columbia (Muraro, 1966), Scandinavia (Antun, 1967), Japan (Watanabe, 1960), Britain (Neumann, 1950), and elsewhere.

In general, and in addition to thermally activated grain growth, contact metamorphism leads to sulfur loss from pyrite and its conversion to pyrrhotite or magnetite, conversion of pyrrhotite to its monoclinic modification or to magnetite, reaction of chalcopyrite with iron sulfides to form cubanite, reequilibration of sphalerite with iron minerals to form ZnS of higher iron content, and appropriate textural readjustment of galena. Contact metamorphism of oxide concentrations (iron, manganese) often leads to the reduction of some of the compounds concerned (e.g., hematite to magnetite) and usually leads to the development of extensive oxide-silicate-carbonate assemblages by the reaction of the oxides with the associated silicate and carbonate rocks—and new materials introduced from the intrusions—under contact-metamorphic conditions.

General statement The *distribution*, *form*, and *setting* of contact-*metamorphosed* deposits would clearly be expected to be of enormous variety; virtually any ore type and association might well—through coincidences of original position and the vagaries of plutonic intrusion—be encountered and modified by a mass of rising magma. As we have already noted, deposits of contact-metamorph*osed* type have not been widely recognized and worked upon, and so there is little documentation of them. It follows that the same is the case concerning mineralogical and chemical *constitution*, including microstructures; the potential variety is enormous, but the facts—apart from those noted above—have not yet been obtained. The microstructures of the ores may, of course, be expected to parallel those of associated silicate and carbonate rocks; that is, the distil parts of the deposit may be expected to show "original" features, which, however, progressively change as the contact is approached. Such changes include grain growth to yield notably coarser grain sizes, secondary grain growth (i.e., the development of "porphyroblasts"), grain-boundary equilibration, the formation of inversion and annealing twins, and so on. Galena, and to a lesser extent chalcopyrite, sphalerite, and pyrrhotite, often migrate on a small scale and become concentrated as minor fracture fillings within and adjacent to the parent sulfide mass. This is especially common in the case of galena. In the later stages of the metamorphic event the same sulfides—and again particularly galena—may "back-vein" the intrusive rock itself.

Chemical changes are likely to be more complicated and less easy to predict than the physical ones and to depend very much on whether the ores are of sulfide or oxide type and whether or not the locale has been a closed or an open system. When the system is open, i.e., when the vapor pressure of the ore exceeds the confining pressure, substances such as S, CO_2, the halogens, and water are likely to be lost. Sulfides decompose: high-sulfur sulfides go to ones of lower sulfur content, for example, $FeS_2 \rightarrow FeS + S \uparrow$, and/or the sulfides in general may lose all their sulfur, the metals involved going to the oxides or recombining to form silicates. Where the system is closed, sulfides and carbonates do not decompose but adjust their compositions to conform with sulfide-carbonate-silicate-oxide equilibrium requirements at the higher temperatures.

The *origin* of contact-metamorphic ores, like their form, geological setting, and constitution, involves almost limitless variety. Probably the most frequently affected primary ores are those of sedimentary origin, though doubtless older vein deposits, igneous ores, and preexisting regionally metamorphosed deposits are involved in contact metamorphism from place to place.

ORES OF REGIONAL-METAMORPHIC ASSOCIATION

Again we have the possibility of two genetic categories—ores that have been concentrated by regional-metamorphic processes and ores that have been deposited prior to metamorphism and later modified by it. There is, however, an additional problem with the ores of regional-metamorphic affiliation: at what stage of the geological history of a given area can regional metamorphism be said to have commenced? This is, of course, an old problem in silicate metamorphism, but it is just as relevant with the sulfides as with the silicates. Clearly we are concerned here with diagenesis and its role in concentrating and modifying sulfide and oxide concentrations. A further problem, already discussed in Chap. 15, is the possible importance of replacement. Some ores in regionally metamorphosed rocks may have been in no way involved in the regional metamorphism of the area; they may be much later superimpositions, deposited as pseudomorphs of part of the metamorphic sequence in which they now occur. This, however, constitutes *metasomatic metamorphism* and is therefore considered a little further on.

DEPOSITS PRODUCED BY REGIONAL METAMORPHISM

For convenience we shall regard metamorphism as commencing at the diagenetic stage. Although relatively little work has been done on the behavior of sulfides and oxides during diagenesis (cf., however, Amstutz et al., 1964), there seems no doubt that processes of this stage are capable of forming new ore minerals and of modifying, moving, and concentrating those already present. Likely results of diagenetic processes must therefore be taken well into account in considering the development of ores by regional metamorphism.

The diagenetic separation of iron and manganese oxides, and the concentration of the latter, has already been discussed in Chap. 14. According to Strakhov (1966) the higher solubility of manganese oxides at low pH and their lesser readiness to oxidize and reprecipitate allows manganese to migrate more readily than iron in many lake muds and silts, leading to diagenetic enrichment of manganese in particular areas of some lakes and confined seas. Concentration of iron oxides during diagenesis is probably slight at most, but there may be considerable fixation of iron as the sulfide. It is now well established (see, e.g., Kaplan, Emery, and Rittenberg, 1963) that the iron of sediment pore water is readily precipitated by bacterial reduction of sulfate—with consequent generation of S^{--}—below the mud-water interface in appropriate marine and lacustrine environments. In general this yields only low concentrations of iron—usually sparsely disseminated crystals and crystalline aggregates of pyrite or marcasite. Much richer concentrations of iron sulfide may possibly form by the pyritization of sedimentary iron oxide concentrations, though, of course, it must be kept in mind that there is no diagenetic concentration of iron—the iron has accumulated in its present position during *sedimentation* and has acted simply as a precipitant of sulfur from migrating H_2S that is presumably generated in and leaked from underlying organic sediments. Such iron sulfide-rich sediments would thus represent diagenetic concentration of sulfur. It seems likely that at least some of the sulfide in sulfide-rich sedimentary rocks (e.g., Wabana) has formed in this way, and Ramdohr (1958) considers that some of the pyrite of the Witwatersrand gold-uranium ores has developed by late-stage pyritization of magnetite and goethite.

It has also been suggested (e.g., Stanton, 1955) that compactional movement of pore waters may be important in the diagenetic development of some base metal sulfide ores of "manto" and stratiform type (see Chap. 15). In this case a situation as shown in Fig. 18-2 is postulated. A unit of volcanic material, largely pyroclastic, underlies a unit containing lenses of sedimentary iron sulfide. The pyroclastic unit is presumed, on the basis of observations about modern volcanoes, to contain substantial traces of heavy metal halides (chiefly chlorides) sublimed in vesicles and fractures in the tuffs, breccias, and lavas concerned. Most of the halide

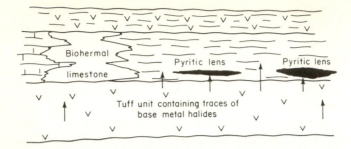

Fig. 18-2 Possible mechanism of formation of some pyritic base metal ore bodies during diagenesis and very early regional (burial) metamorphism; heavy metal halides in the pore waters of a pyroclastic unit may encounter pyritic lenses when forced upward by compaction.

material is dissolved in the pore waters and, with increasing burial and compaction, those solutions are expressed and moved upward.   As they rise, most of the metallic halide is dispersed and lost.   Some, however, inevitably encounters overlying iron sulfide lenses, and metals such as copper, lead, and zinc displace some of the iron from the sulfide, thus precipitating as sulfides themselves.   The result is a pyritic base metal deposit, the nonferrous metal fraction of which may be regarded as being the result of diagenetic–low-grade-metamorphic processes.

In addition to affecting concentration, diagenetic–low-grade-(burial) meta-morphic processes also yield new microstructures (Amstutz et al., 1964).   The most common is the development of pyrite porphyroblasts in shales and carbonate rocks.   In many cases the mechanism of formation of such pyrite cubes is not clear, though in others there is no doubt that the crystals form through some kind of "retexturing" of concretions or by the coalescence of closely packed groups of smaller crystals.   In some ores (e.g., New Brunswick; Stanton, 1959) pyrite con-cretions contain parallel zones of chalcopyrite, sphalerite, and galena.   With slight metamorphism the curved outlines of the zones, and of the concretions as a whole, begin to take up a suggestion of a cubic shape.   All stages can be traced from this right through to the point where the pyrite has been transformed into a single idiomorphic crystal containing concentric zones of the other sulfides.   The zoned pyrites at Mount Isa figured by Grondijs and Schouten (1937, e.g., figs. 69, 74) look to be good examples of an advanced stage in this process.

While there seems no doubt that some concentration and characteristic microstructural development take place in the interval between burial and the onset of low-grade metamorphism, the continuation of these processes at higher metamorphic grades is not established.   Indeed there does not seem to be good evidence of metamorphic *concentration* of ores once the rocks have ceased to contain free water.   There have been many suggestions that high-grade metamorphism might cause the expulsion of trace metals from mafic minerals and then somehow induce the metals to reprecipitate as sulfides in high local concentration.   Several investigators have, for example, postulated such an origin for the Broken Hill ore body.   However, while there is no doubt that many large volumes of rock—particularly those containing significant proportions of basic sills, lavas, and pyroclastic rock—possess sufficient metal to yield a large ore deposit, there is no indication of processes that might have permitted large-scale migration and concentration.

General statement   We may therefore conclude that sulfide and oxide ore bodies *produced* by regional metamorphism probably constitute a rather small and not very well-defined category.   It appears that it is the diagenetic and very early stages of metamorphism that are most important and that metamorphism above greenschist facies is not important in inducing concentration.   We would expect the *distribution* of ores of this type to be sparse, though probably wide in space and time.   The *forms* of the concentrations appear to be variable—sparse disseminations, lens-like concentrations in sediments, infillings along bedding planes, mud cracks and stylolites, and so on.   Their *setting* is principally sedimentary or volcanic-sedimentary and usually shows some relation to reduced zones of sedimentation.   Their *constitution* is usually fairly simple: manganese and iron oxides and hydroxides and the common sulfides pyrite, marcasite, pyrrhotite, chalcopyrite, sphalerite, and galena.   Arsenides and others may occur as traces.   In their *origin*, the valuable materials are probably almost always sedimentary and/or volcanic.

DEPOSITS MODIFIED BY REGIONAL METAMORPHISM

This group looks as if it may well be the most interesting and important category among the ores of metamorphic affiliation.   It includes many of the ores of metasedimentary rocks discussed in Chap. 15, many of the large deposits of the Canadian Precambrian Shield, and the Rammelsberg and other deposits of Europe and elsewhere.   It was, of course, ores of this type in the northern Appalachians that were investigated by Hitchcock and Emmons.   The fact that they were of metamorphic affiliation was largely ignored between 1920 and 1950, but this was reversed very sharply by the findings of Ramdohr (1950) and King and his colleagues (1953) on Broken Hill (cf. Chaps. 2 and 15).   The group includes ores of igneous, sedimentary, and vein associations, but those of sedimentary and volcanic-sedimentary association are by far the most important.

Before proceeding further it must be emphasized that these deposits—particularly those in metasedimentary rocks—continue to be the subject of considerable controversy, and their mode of emplacement must be regarded as still unresolved.

1. Diagenetic stage

Clearly it is not easy to separate deposits concentrated by diagenesis from those simply modified—in place—by diagenetic processes. The principal effects are coarsening on a microscale and the development of crystal faces when interfacial free-energy requirements dictate it.   Pyrite tends to form sharply developed cubes, other forms appearing very occasionally.   Some framboid development may continue into this stage.   Pyrrhotite in sedimentary rocks usually aggregates into small, flat, lensoidal bodies that appear as short spindles in section. The larger dimensions of these are almost always parallel to the bedding planes, and they are usually polycrystalline. Any accompanying chalcopyrite usually occurs in the body of a spindle and has grain size and shape similar to that of the pyrrhotite. Sphalerite-chalcopyrite aggregates often form extremely fine lattice intergrowths of chalcopyrite in sphalerite. The chalcopyrite follows the octahedral planes of the sphalerite, for the most part, and the structure as a whole is highly reminiscent of exsolution. However, in the present case the structure appears to be one of low-temperature segregation resulting from the slight coarsening of an extremely fine (microcrystalline) primary aggregate. Sphalerite-galena aggregates show a fine, wispy distribution of galena intergranular with respect to sphalerite. Where sphalerite is the minor phase it occurs as numerous tiny near-spherical bodies—a shape apparently resulting from the characteristic dihedral angle as discussed in Chap. 9—in a matrix of very fine galena.

Some of the silicate and oxide minerals of sedimentary iron and manganese deposits (cf. Chaps. 13 and 14) may be affected by diagenetic and very low-grade burial-metamorphic processes. For example, in iron formations, greenalite appears to be transferred into minnesotaite and perhaps thuringite by extremely low-grade metamorphic processes. According to Hewitt, there appears to be a distinct tendency for manganese oxides to "age":

> Recent work by the writer in a few districts indicates that the number of manganese oxides originally deposited as sediments is small and that following burial, especially under thick cover and with deformation, the number of oxide minerals steadily increases. The oxide minerals that are formed after burial depend largely upon the chemical elements deposited in the original oxides, unless the beds are involved in regional metamorphism.
>
> In modern bogs, it appears that the deposited material is largely uncrystallized hydrous oxides. In several deposits of early Pliocene age (Nevada; Baja California) most of the material still consists of uncrystallized oxides, the specific gravity of which ranges from 1.3 to 1.6. In each of these districts, however, there are local bodies of crystallized oxides, such as cryptomelane, pyrolusite, hollandite, and coronadite; these have probably been formed after burial, however.
>
> In the lower Oligocene deposits of the Chiatura district, south Russia, the most abundant mineral is pyrolusite, which seems to have formed before burial, but in some of the explored areas, an earthy hydrous oxide, manganite (brown belta), is still present. In the Eocene deposits of Cuba, the most widespread and abundant oxide is todorokite, followed by pyrolusite and manganite. In Morocco, in beds of Lias age and younger, the principal oxides are psilomelane, pyrolusite, and coronadite. In the beds of Permo-Trias age and older, however, braunite and hausmannite are common, but psilomelane is also present. In India, the most widespread oxide in the Precambrian stratified deposits is apparently braunite, but psilomelane and even pyrolusite are common. This brief review indicates significant evolution of the manganese oxide minerals with age, depth of burial, and deformation (1966, pp. 453 and 456).

2. Metamorphic stage   Clearly the diagenetic stage grades imperceptibly into that of metamorphism.

That ores have in some cases suffered regional metamorphism was first recognized through schistose and gneissic structures and particularly through the schistose intermingling of ore minerals and country rock along the margins of some deposits (Emmons, 1909, on the Maine deposits).   It has also been recognized for quite a long time (e.g., Carstens, 1931; Foslie, 1938; King, 1958) that the grain size of the ore minerals in massive base metal ore bodies generally increases with increase in the grade of metamorphism of the containing rocks.   For example, host rocks of the MacArthur River ores (see Chap. 15) have barely passed the diagenetic stage, and the ores are extremely fine.   Grain size is measured in microns; the galena, for example, is generally "fine grained, occurring as ragged, wispy grains and aggregates averaging less than five microns, usually one to three microns in diameter" (Cotton, 1965, p. 199).   Ores in rocks of greenschist grade, such as those of Captain's Flat, New South Wales, and Bathurst, New Brunswick, are appreciably coarser (grain size of the order of 0.01 to 1.0 mm).   Ores in higher grade terrains are coarser again; those of Broken Hill, in sillimanite-grade host rocks, have an average grain diameter of at least 3.0 mm, and here and there patches of ore range up to 8.0-cm grain size.   The many massive sulfide ore bodies in the Caledonian geosynclinal belt of Norway include virtually the whole range of this relationship between sulfide grain size and metamorphic grade of the enclosing rocks (Vokes, 1968).   Thus the grain size of the ore minerals runs closely parallel to the metamorphic grain size of the enclosing rocks, suggesting that the ore minerals have been involved with the silicates and others in the regional-metamorphic process.   (It might, of course, be argued that this correspondence in grain size is due to metasomatic pseudomorphism.   However, as noted in Chap. 15, the latter possibility is usually discounted by the fact that the ore minerals do not appear in the form of those country-rock minerals that are idiomorphic, e.g., micaceous minerals, garnet, sillimanite.   A further test in some cases would be to see whether host-mineral dihedral angles are inherited by the ore minerals.)

As in the case of contact metamorphism, regional metamorphism leads to modification of both the chemical and physical characteristics of the sulfide and oxide minerals concerned—modifications that parallel those in "ordinary" silicate and carbonate minerals.   As a generalization—and it must be emphasized that it is a generalization to which all sorts of interesting and significant exceptions will be found in the future:

1. Oxide concentrations are more notable for their chemical changes—since they react so readily with accompanying silicate and carbonate—and are less notable for their physical changes, usually persisting as essentially equigranular aggregates and doing little more than coarsening and developing crystal faces when appropriate.
2. Sulfide concentrations are less notable for their chemical changes—since they do not react with accompanying silicate and carbonate—and are more notable for their physical changes, which follow from their ready deformation and high capacity for annealing and grain growth.

The two most important groups of *oxide* concentration in the present context are, of course, the sedimentary iron and manganese deposits.   Stratiform zinc oxide and silicate concentrations, alpine-type chromite ores, and perhaps anorthosite-type iron-titanium concentrations are additional, though much lesser, categories.

The regional metamorphism of sedimentary *iron* has been studied fairly extensively and has involved a range from slightly ferriferous pelites (Chinner, 1960) through to high-grade concentrations of iron oxides. Iron deposits containing minor manganese and/or titanium have also been investigated.

Metamorphism of iron formation *per se*, as represented by a number of the Lake Superior "iron ranges," has been studied by James (1955). James' study is a particularly interesting one since it involves the interaction of facies of metamorphism with facies of iron sedimentation. James' location of the metamorphic isograds is shown in Fig. 18-3, and the mineralogical constitution of the various sedimentary facies within each metamorphic facies is given in Table 18-1. In each case the metamorphic grade is indicated by the mineralogies of the associated pelitic rocks. James considers that the major part of the magnetite here is not, as has commonly been supposed, due to metamorphic breakdown of sideritic carbonate but is probably essentially primary (i.e., an original component of magnetite-bearing sedimentary zones). The grunerite appears to result from carbonate-silicate reaction and hence is derived from the carbonate-silicate sedimentary facies.

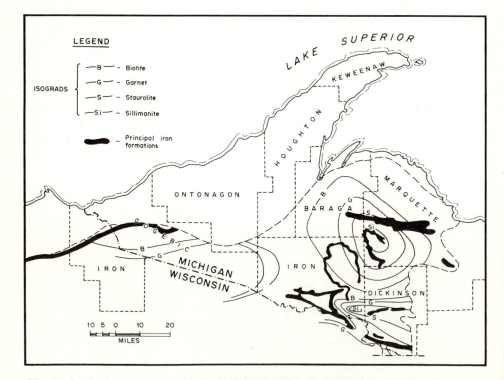

Fig. 18-3 Regional metamorphic isograds in part of the Lake Superior iron ore (banded iron formation) district. (*After James, Bull. Geol. Soc. Amer., 1955.*)

Table 18-1 Interplay of sedimentary and metamorphic facies in the Lake Superior iron ore province. (From James, 1955, p. 1475)

Major constituents are shown in uppercase letters

Facies		Composition†	Low grade‡	Intermediate¶	High grade§
			Sedimentary	*Metamorphic*	
Sulfide		Ferrous sulfide Organic carbon "Clay"	PYRITE CARBON (amorphous) QUARTZ Sericite	PYRITE GRAPHITE QUARTZ Micas Garnet (rare)	PYRITE GRAPHITE Pyrrhotite(?) Micas-garnet
Carbonate		Carbonate Chert Greenalite	CARBONATE QUARTZ Stilpnomelane Minnesotaite	GRUNERITE QUARTZ Magnetite Carbonate	GRUNERITE QUARTZ Magnetite Pyroxene
Silicate	Nonclastic	Greenalite Chert Carbonate Magnetite	MINNESOTAITE STILPNOMELANE QUARTZ CARBONATE Magnetite	GRUNERITE QUARTZ Magnetite	GRUNERITE QUARTZ Magnetite Pyroxene
	Partly clastic	Iron-rich clay Chert Carbonate Magnetite(?)	CHLORITE STILPNOMELANE QUARTZ CARBONATE MAGNETITE Biotite	GRUNERITE QUARTZ MAGNETITE EPIDOTE GARNET Carbonate Micas	GRUNERITE QUARTZ MAGNETITE Garnet Hornblende Pyroxene
Oxide	Magnetite-banded	Magnetite Carbonate Greenalite Chert	MAGNETITE STILPNOMELANE MINNESOTAITE CARBONATE QUARTZ	MAGNETITE GRUNERITE QUARTZ Garnet	MAGNETITE GRUNERITE QUARTZ Pyroxene Garnet
	Hematite-banded	Ferric oxide Chert Magnetite Calcite	HEMATITE QUARTZ Magnetite Calcite	SPECULAR HEMATITE QUARTZ Magnetite Calcite	SPECULAR HEMATITE QUARTZ Magnetite Calcite
			Less than 0.10 mm	0.1 to 0.20 mm	More than 0.20 mm
				Quartz grain size††	

† Inferred.
‡ Chlorite and biotite zones.
¶ Garnet and staurolite zones.
§ Sillimanite zone, essentially.   Inferred for some rocks.
†† Diameter of typical grains in relatively pure layers of chert.

From the microstructural point of view, James makes several interesting observations:

1. Grain size increases in a general way with increase in metamorphic grade. Using quartz of the more pure chert layers as an index, sizes are
   Chlorite zone: <0.05 mm, average 0.03 mm
   Biotite zone: 0.05 to 0.10 mm
   Garnet and staurolite zones: 0.10 to 0.20 mm
   Sillimanite zone: >0.20 mm
2. Oolitic structures in hematite-bearing material are preserved up to and including the sillimanite grade of metamorphism. In some oxide-rich sediments

Table 18-2 Stratigraphic sequence in the Wabush iron ore district. (From Klein, 1966, p. 250)

Pleistocene and Recent		Glacial, lake, and stream deposits
	Unconformity	
Proterozoic	Shabogamo Formation	Gabbro
	Intrusive contact	
	?	Amphibolites
	Upper	Quartz-specularite schist Quartz-magnetite rock
	Wabush Iron Formation	
	Lower	Quartz-carbonate-iron silicate schist Quartz-carbonate rock
	Carol Formation	Orthoquartzite and quartz-muscovite-kyanite-staurolite-garnet schist
	Duley Formation	Dolomite and calcite marble
	Katsao Formation	Garnet-kyanite-biotite-muscovite-quartz-feldspar schists and gneisses
	Unconformity	
Archean	Ashuanipi Complex	Granitic and granodioritic gneisses, schists, granite intrusives and possibly graphitic schists and gneisses, formerly classified as the "Nault Formation"

Table 18-3 Generalized stratigraphic sequence of members of the Wabush Iron Formation and the principal rock types within each member. (From Klein, 1966, p. 254)

The order of the rock types for each member has no stratigraphic significance.

Upper Wabush Formation	Quartz-specularite member	Banded specularite-magnesioriebeckite-Mn cummingtonite-rhodonite rock Banded specularite-rhodonite-rhodochrosite-Mn aegirine-calderite rock Banded quartz-specularite-rhodonite-kutnahorite-calderite rock Quartz-specularite-Mn cummingtonite schist Quartz-specularite-aegirine-augite-riebeckite-tremolite-calcite rock Quartz-specularite-anthophyllite schist Quartz-specularite schist
	Quartz-magnetite-specularite member	Banded quartz-magnetite-specularite rock
	Quartz-magnetite member	Banded quartz-magnetite-riebeckite-tremolite rock Banded quartz-magnetite-actinolite rock Banded to massive quartz-magnetite rock Banded quartz-magnetite-carbonate rock Banded quartz-magnetite-grunerite rock
Lower Wabush Formation	Quartz-silicate member	Quartz-grunerite-garnet-magnetite-ferrodolomite schist Grunerite-actinolite-diopside-magnetite rock Banded quartz-grunerite-ferrosalite rock Banded quartz-carbonate-grunerite-eulite rock Quartz-grunerite schist
	Quartz-carbonate-silicate member	Banded quartz-carbonate-grunerite rock
	Quartz-limonite-goethite member	Banded quartz-limonite and quartz-limonite-goethite rock
	Quartz-carbonate member	Banded quartz-carbonate-magnetite rock Banded quartz-carbonate rock

hematite forms platy porphyroblasts up to 1 inch in length in thick magnetite layers.

3. In some of the high-grade rocks hematite forms plates having a pronounced preferred orientation parallel to the axial planes of the folds.

Klein (1966) has studied the metamorphism of sedimentary *iron* and *iron-manganese* formations in the Wabush area of the Labrador Trough (see also Chap.

13).   As indicated by the associated pelitic schists, the prevailing grade of metamorphism is almandine-staurolite-kyanite.   The principal elements of the stratigraphic sequence as a whole are given in Table 18-2 and of the Wabush Iron Formation itself in Table 18-3.   Table 18-3 also includes the mineral assemblages corresponding to sedimentary layers of different compositions.   Postulated equi-

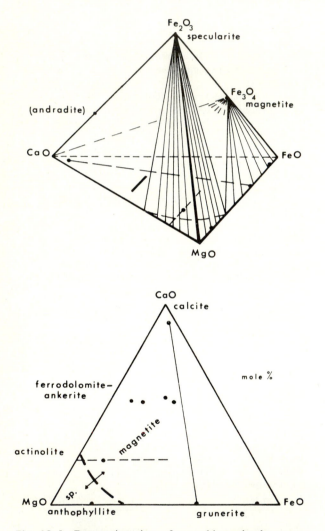

Fig. 18-4 Perspective view of assemblages in the system $CaO–FeO–Fe_2O_3–MgO–SiO_2–H_2O–CO_2$. Quartz is also present in all assemblages shown, and H_2O and CO_2 are regarded as perfectly mobile components.   On the base of the tetrahedron (shown as the lower of the two diagrams) are outlined the areas in which either magnetite or specularite is the iron oxide stable in the assemblage. (*From Klein, J. Petrol., 1966.*)

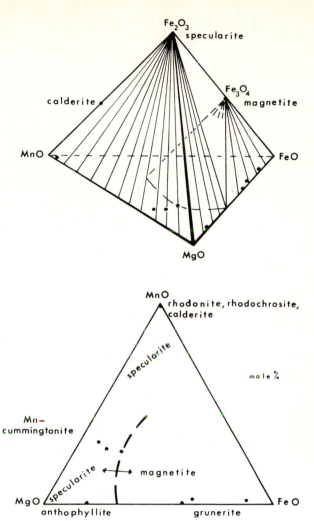

Fig. 18-5 Perspective view of assemblages in the system $FeO–Fe_2O_3–MgO–MnO–SiO_2–H_2O–CO_2$. Quartz is also present in all assemblages shown, and H_2O and CO_2 are regarded as perfectly mobile components. Regions of magnetite and specularite stability on the base of the tetrahedron are shown in the lower diagram. (*From Klein, J. Petrol., 1966.*)

librium diagrams for assemblages in two of the more important Wabush systems are given in Figs. 18-4 and 18-5. It should be noted here that all the assemblages shown contain quartz. Also CO_2 here is regarded by Klein as a perfectly mobile component, although he notes:

The relatively widespread occurrence, especially in the Carol Mine area, of quartz-ankerite and quartz-ferrodolomite assemblages that show no trace of reaction between the quartz and the carbonate to form ferromagnesian silicates, indicates that in large parts of the iron formation the CO_2 component was not free to diffuse out; in other words CO_2 seems to have behaved as an inert component (1966, p. 296).

However, the partial reaction of quartz and siderite in some areas to form grunerite and orthopyroxene suggests that in these cases the system was locally open to CO_2 to at least a minor extent. The frequent occurrence of the assemblages quartz-grunerite-calcite, actinolite-grunerite-calcite, and quartz-ferrosalite-grunerite-calcite, with calcite present only to the extent of 1 to 2 per cent, suggests that some CO_2 was able to *escape* from the system. In this general connection O_2 does not appear to have been particularly mobile either:

> The very frequent occurrence of quartz-magnetite-specularite rocks in which thin bands of magnetite alternate with thin bands of specularite indicates that the activity of the O_2 component was determined by the original assemblage, and not by its metamorphic history; as such O_2 cannot be considered as a perfectly mobile component (Klein, 1966, pp. 295–296).

The regional metamorphism of sediments *rich in manganese but containing little iron* received its first detailed attention in the work of Fermor (1909), who investigated a number of the metamorphosed sedimentary manganese deposits of India. Since that time similar occurrences, notably in the Minas Gerais area of Brazil, have also been investigated.

We have already noted that the manganese oxides appear to "age" through a succession of mineral species during the diagenetic stage. Increasing metamorphism leads to the development of Mn garnet (spessartite), rhodonite, rhodochrosite, and a wide variety of silicates and carbonates carrying manganese as a minor cation.

Fermor suggested that the Indian occurrences resulted from the regional metamorphism (ranging from low to high grade) of sediments containing unusually high proportions of manganese oxides and, probably, carbonates. Some of the more common assemblages described by Fermor are spessartite garnet, spessartite-quartz (to which Fermor gave the rock name *gondite*), spessartite-quartz-amphibole, spessartite-quartz-rhodonite, spessartite-rhodonite, and spessartite-rhodonite-magnetite.

The Brazilian occurrences show many similarities to those of India and have been described in great detail by Dorr, Soares, and Horen (1956). These investigators have divided the primary protores (i.e., the unweathered primary materials) of the Minas Gerais district (the most important of the Brazilian manganese provinces—see also Chap. 14) into three main types: manganese silicate-carbonate, marble-itabirite, and clastic, including tuffaceous rocks.

> The manganese-silicate-carbonate protore was formed by the regional dynamic metamorphism of manganiferous sediments believed to have originally consisted of syngenetic rhodochrosite, mudstone, and chert. These sediments were deposited under reducing conditions, possibly in a large silled basin or more probably in a series of smaller ones that may have been interconnected. There is no evidence pointing toward deposition

of syngenetic manganese oxide, and the ubiquitous presence of graphite and of alabandite in the protores proves that primary sedimentation could have occurred in this type of protore only under unusual environmental conditions.

The regional metamorphism caused the manganese carbonate to react with the chert and mudstone to form gondite and queluzite, eliminating the carbonate where it was not abundant and forming spessartite, rhodonite, and other manganiferous silicates and, where original chert was predominant, quartz. Where original manganese carbonate was in excess, rhodochrosite remains; in the Merid mine it forms as much as 45 per cent of the protore. Contact metamorphism caused by later intrusions raised the grade of metamorphism (normally in the epidote-amphibolite zone) and caused the formation of tephroite, rhodonite, bementite, neotocite, and the growth of other manganese silicate minerals to relatively large size. The net effect of the contact metamorphism was to retard ore-formation by creating minerals resistant to supergene enrichment. This type of metamorphism is local in distribution, whereas the dynamic metamorphism is regional.

The manganese silicate-carbonate protore is found only in a group of rocks of regional distribution which is thought to be of middle Precambrian age and which is older than the Minas series. The other two types of protores are found only in the Minas series or possibly in small part in the post-Minas Itacolumi series, both of presumed late Precambrian age.

The marble-itabirite protores are found in the middle group of the Minas series. These rocks, originally limestone, dolomite, and oxide-facies iron formation, form an intergrading series of chemical sediments deposited under oxidizing conditions in an oscillating epicontinental sea without sealed basins. There is no evidence of significant volcanic contributions to these rocks.

Manganese oxide was deposited synchronously with iron oxide in certain zones in the oxide-facies iron formation. Manganese also substitutes for Ca and Mg in the calcite and dolomite molecules up to about 4 per cent MnO in some marbles, forming manganoan calcite and manganoan dolomite.

Certain clastic rocks, including quartzites, meta-tuffs(?), and phyllites, contain small but locally high-grade manganese oxide deposits. Owing to depth of weathering, it is not possible to be sure of the source of manganese in these rocks, presumably it was derived from the meta-tuffs(?) or deposited as a primary chemical sediment with the clays from which the phyllites were formed by the low grade metamorphism to which the Minas series has been subjected (1956, pp. 280, 281).

By far the largest deposit in the Minas Gerais district is that of the Merid mine, where the ore has formed by the weathering enrichment of silicate-carbonate protore. The latter has developed from the original manganiferous sediments primarily as a result of regional metamorphism, though some contact-metamorphic effects are superimposed locally on the latter. The mineralogy is quite complex:

The most common minerals identified in the protore of the Merid mine were rhodochrosites of varying composition as well as manganoan calcite; spessartite garnet (zoned in some cases near intrusive contacts, with the percentage of Mg increasing toward the periphery); rhodonites of varying composition; asbestiform manganoan cummingtonite; graphite; quartz, tephroite; pyrophanite; thulite; neotocite; pyroxmangite; bementite; apatite and manganoan apatite; and manganiferous chert. The following sulfides and sulfarsenides were identified in the protore, disseminated in very fine grains (less than 0.005 mm.): pyrite; pyrrhotite; chalcopyrite; cobaltite; skutterudite; siegenite(?); bornite; covellite; and alabandite. With the exception of vein pyrite, which is coarser, none of the sulfide minerals are megascopically visible.

The sulfides and sulfarsenides are associated with the garnet, rhodonite, tephroite, or matrix rhodochrosite. The sulfides of the matrix are distributed in irregular bands or grouped aggregates in the matrix rock, which are interpreted as reflecting the original bedding. The pyrite in veins occurs with botryoidal or crystalline rhodochrosite and with asbestiform manganoan cummingtonite in addition to rhodochrosite. It is related in this type of occurrence to intrusive bodies in the protore (1956, pp. 290, 291).

Table 18-4 (after Dorr, Soares, and Horen, 1956, p. 328) gives the chemical composition of Merid protores, ranging from silicate-rich–carbonate-poor to silicate-poor–carbonate-rich.

Annerston (1968) was also concerned with the effects of oxygen activity on the composition of oxides and silicates during iron-formation metamorphism—in this case in the *titaniferous* Gällivare iron formation of northern Sweden. Table 18-5 gives the mineral assemblages and indicates that metamorphism has been of biotite-amphibolite grade. CO_2, O_2, and H_2O appear to have acted as perfectly mobile components. Ilmenite and sphene are the titanium-bearing minerals. Figure 18-6 shows hematite-magnetite-ilmenite-rutile relations as functions of T and P_{O_2}. This is, of course, modified in the Gällivare rocks by the presence of calcium, which gives $CaTiO_3$ (sphene). Annerston suggests that there was a close approach to chemical equilibrium in the metamorphism of this iron formation because element distribution between phases and Mg/Fe and Ti/(Mg + Fe + Ti) ratios of the ferromagnesian silicates in the presence of magnetite and hematite are all systematic.

Table 18-4   Chemical composition of a range of primary ore types of the Merid manganese ore body, Brazil. (From Dorr, Soares, and Horen, 1956, p. 329)

	1	2	3	4	5
SiO_2	13.17	18.01	21.77	22.85	35.89
Al_2O_3	3.52	4.47	5.32	4.77	10.66
TiO_2			0.44	0.26	0.38
Fe_2O_3	2.43	2.24	0.76	2.23	2.53
FeO	3.84	3.26	2.67	2.24	3.73
MnO	43.50	43.81	44.14	41.95	34.70
MnO_2					
CaO	2.47	2.64	5.57	2.32	3.80
MgO	3.75	3.91	4.67	2.79	2.53
K_2O					
Na_2O					
S				0.004	0.006
CO_2	19.39	15.53	13.15	16.02	4.66
C	3.19	1.09	0.07		
P_2O_5	0.68	0.80		0.17	0.14
H_2O^+	0.22	0.65	0.82	0.76	0.40
CoO					
NiO					
Total	96.16	96.31	99.38	96.36	99.48

Table 18-5 Mineral assemblages in the metamorphosed titaniferous Gällivare iron formation, northern Sweden. (From Annersten, 1968, p. 382)

Sample number	Hematite	Magnetite	Ilmenite	Biotite	Ca amphibole	Diopside	Quartz	Sphene	Calcite	Apatite	Scapolite	Microcline	Plagioclase
1		p†		p	p	p	p			p			An_8
2		p		p	p		p	p					An_{10}
3		p		p			p			p			An_{10}
4		p		p	p		p					p	An_6
5		p		p	p		p				p		An_{10}
6		p		p	p		p	p		p			An_5
7		p	p		p		p						An_{14}
8		p		p	p		p			p			An_{12}
9		p		p	p		p	p		p	p		$An_?$
10		p		p	p	p	p	p					An_6
11		p		p	p								An_8
12		p		p	p		p				p		An_8
13		p		p	p		p						$An_?$
14	p	p	p	p	p		p				p	p	An_{10}
15	p	p		p	p		p	p		p	p	p	An_2
16	p	p		p	p		p	p		p		p	An_5
17	p			p	p		p			p	p	p	$An_?$
18	p		p	p	p			p		p	p		An_4
19	p	p		p	p		p	p		p		p	An_2
20	p		p	p	p		p	p				p	An_6
21	p		p	p	p		p	p				p	An_4
22	p			p			p					p	An_{15}

† p = mineral present.

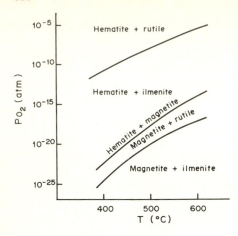

Fig. 18-6 Theoretical stability fields for the system Fe_3O_4–TiO_2 (rutile)–$FeTiO_3$–Fe_2O_3 as functions of oxygen pressure between 400 and 600°C. (*Redrawn from Annersten, Lithos, 1968.*)

The regional metamorphism of the *sulfides* is not nearly so well understood as that of the oxides, though it has recently begun to receive greatly increased attention.   Sulfide-sulfide (arsenide, etc.) equilibria at high pressures and tempera-tures have, of course, been well worked out, as was shown in Chap. 5.   Sulfide-oxide systems are less well known but have received some attention.   In contrast, sulfide-silicate systems, which are of vital concern in the chemical evolution of metamorphosed sulfide ores, are barely known.   What investigations have been carried out have been chiefly concerned with commercial smelting operations and with ores in igneous rocks.

Among the sulfides themselves, little can be learned of natural metamorphic assemblages because of lack of quenching.   As was noted in Chap. 5 some of the sulfide systems are quite unquenchable, and most reequilibrate fairly quickly with fall in temperature.   While, among the latter, some compositions may possibly be quenched in contact-metamorphic rocks (possible though unlikely), it seems quite impossible for quenching to have occurred in regionally metamorphosed materials.   We cannot therefore discuss metamorphic assemblages of sulfides in terms of sulfide compositions.   On the other hand, sulfide-oxide and sulfide-silicate systems may well retain their high-temperature features, just as do many silicate systems.   Studies of equilibria in these two general types of system look potentially highly interesting.   However, at the time of this writing there is very little information on the chemistry of regional metamorphism of sulfide-bearing rocks, apart from the suggestion that increasing metamorphism may lead to sulfur loss and perhaps to increasing complexity of sulfide-mineral assemblages.   Sulfur loss involves the conversion of pyrite to pyrrhotite and, in some cases, to magnetite and seems fairly well substantiated.   Greater mineralogical complexity has been observed in some of the more highly metamorphosed stratiform ores, though whether this really reflects genuine metamorphic modification or merely original constitutional differences between the deposits concerned is by no means clear. McDonald (1968) has suggested that the three large Australian stratiform ore bodies given below (see also Chap. 15) may illustrate the phenomenon of increase in mineralogical complexity with increase in grade of metamorphism. The Mac-Arthur River deposit is quite unmetamorphosed, Mount Isa has suffered only very slight metamorphism, and Broken Hill appears to have been metamorphosed to sillimanite grade.

MacArthur River opaque assemblage	Mount Isa opaque assemblage	Broken Hill opaque assemblage	
Pyrite	Pyrite	Pyrite	
Sphalerite	Sphalerite	Sphalerite	
Galena	Galena	Galena	
Chalcopyrite	Chalcopyrite	Chalcopyrite	
	Marcasite	Marcasite	
	Pyrrhotite	Pyrrhotite	
	Valleriite	Valleriite	
	Tetrahedrite	Tetrahedrite	
	Polybasite	Polybasite	
	Jamesonite	Jamesonite	
	Arsenopyrite	Arsenopyrite	
	Gold	Gold	
	Pyrargyrite	Pyrargyrite	
	Sternbergite	Sternbergite	
	Proustite ⎰ Not identified		
	Pentlandite ⎱ at Broken Hill	Wurtzite	Löllingite
		Cubanite	Willyamite
		Argentite	Dyscrasite
		Stromeyerite	Smaltite
		Molybdenite	Niccolite
		Linnaeite	Wolframite
		Breithauptite	Gudmundite
		Enargite	Bournonite
		Covellite	Miargyrite
		Boulangerite	Meneghinite
		Berthierite	Cobaltite
		Cassiterite	Stannite
		Magnetite	Ilmenite
		Ullmanite	Silver
		Antimony	

In addition, where highly metamorphosed deposits such as that of Broken Hill undergo later, low-temperature deformation, further mineralogical modification may take place as sulfides undergo *retrogressive metamorphism*. A late (ca. 500×10^6 years) retrograde episode associated with local small-scale deformation is now well recognized at Broken Hill, and silicates and sulfides have undergone retrogressive metamorphism together. Garnet, for example, has broken down to biotite and chlorite, and concurrently, pyrrhotite has been replaced by marcasite and tetrahedrite by chalcopyrite-gudmundite and related intergrowths (Richards, 1966).

In contrast to this dearth of chemical knowledge there is now a considerable body of information on the *physical* effects of regional metamorphism on sulfide ores.   Following on from diagenesis (where the sulfides occur in sediments), early changes are probably those of grain growth.   The first effect of this is the modification of depositional and diagenetic structures, usually of concretionary kind, composed of large numbers of fine grains.   The features developed following the onset of grain growth are essentially those described in Chap. 9.   Interaction of the crystal structures of adjacent grains leads to grain shapes and sizes yielding the minimum total interfacial free energy attainable under the conditions prevailing.   Among the allotriomorphic particles, grain shape follows from the size of the dihedral angles.   Other interfaces achieve energy minima by the development of crystal faces on either of the two substances in contact.   Because of this, some minerals have a greater propensity for the development of crystal faces than others, leading to the development of a sulfide crystalloblastic sequence analogous to that of silicates and carbonates.   The crystalloblastic sequence involving some of the common sulfides, silicates, and carbonates is garnet, magnetite-arsenopyrite, pyrite, dolomite, tremolite, muscovite, chlorite, pyrrhotite, sphalerite, chalcopyrite, galena.   To a first approximation this sequence probably holds for the other members of the sulfide-structure groups represented by the minerals given, e.g., others of the pyrite group, such as cobaltite, sperrylite, and gersdorffite, also have a strong tendency to idiomorphism and would probably assume a position close to that of pyrite where they occurred.

In this connection it must be kept in mind that shape results from the interaction of the growing particle with the medium in which growth proceeds.   For this reason a given substance will develop different shapes in different growth media. For example, the scattered pyrite frequently found in shales adjacent to ore bodies is usually sharply cubic, while that of the ore itself is often no more than subidiomorphic.   Similarly pyrite in sphalerite usually shows better crystal faces than pyrite in galena, and so on.

Onset of significant directed pressure, however, soon leads to deformation. At high confining pressures the deformation is plastic, leading to the flattening of grains and the development of much lattice bending and twinning.   As noted in Chap. 10, slow deformation in the form of creep may lead to the development of subgrains at this stage.   Hard minerals may be drawn out into eyes ("augen") in matrices of softer minerals, yielding sulfide schists and gneisses.   Frequently the augen are pyrite, this often in a pyrrhotite or sphalerite matrix.   Such pyrite augen often suffer rotation, the extent and sense of movement being indicated by spirally arranged inclusions of other sulfides within the pyrite grains.   In lead-zinc ores, sphalerite forms the augen, and galena forms the matrix in which such augen are set (Fig. 18-7).   The matrices are deformed intensely, particularly in the zones immediately surrounding augen.   Such deformation usually decreases with distance into the matrix from the phase boundary.   The augen themselves are, of course, usually deformed, though to a lesser extent than the soft matrix and usually by granulation rather than by plastic changes.   In many cases softer minerals, particularly galena, may be squeezed out of their matrices and into small cracks developing in adjacent wall rock, yielding veinlets that usually have a simpler composition and a coarser grain size than the parent sulfide mass.

The pyritic copper-lead-zinc deposits of Rammelsberg are probably the most famous of ores modified by regional metamorphism. That these ores had undergone metamorphism was clearly recognized by German geologists well back in the nineteenth century and was accepted unequivocally by Lindgren and Irving in 1911. Many other ore bodies—notably those of the Appalachians (Bathurst, New Brunswick; Ducktown, Tennessee, etc.), the Caledonides of Scandinavia, the Copperbelt of Rhodesia and Katanga, and the Precambrian and Paleozoic belts (Broken Hill, Mount Isa, Rosebery, etc.) of eastern Australia—are now recognized as possessing microstructures similar in principle to those of Rammelsberg and as having undergone regional metamorphism.

If metamorphic temperatures have been high, it is likely that metamorphic conditions would be essentially those of "hot-working" in metals and that the sulfides would anneal almost as they deformed. At lower temperatures any pause in the regional metamorphic event would, if those temperatures exceeded 250 to 300°C, permit at least some annealing. Single-phase triple-junction angles and dihedral angles, thrown out of equilibrium during deformation, would tend to readjust by grain-boundary movement. Some recrystallization may take place, as observed in the chalcopyrite of Buchans, Newfoundland, by Newhouse and Flaherty (1930). In some cases—particularly commonly with galena—strain may be eliminated by polygonization and the development of subgrains. Where single-phase aggregates have been severely flattened and disparate grain sizes developed, larger grains grow sideways at the expense of smaller grains leading to

1. The resumption of approximately equidimensional grain shapes with straight or gently curved boundaries and equilibrium triple-junction angles
2. Concomitant increase in overall grain size

Clearly a number of deformations, each separated by a pause allowing grain growth, may induce very substantial increases in grain size. This may well explain the coarseness of ores such as those of Broken Hill. The ore bodies in this case—whatever their mode of deposition—are now known to have been subjected to several periods of deformation. Successive strain-anneal-strain . . . events would inevitably lead to the grain sizes now observed.

In some cases the sulfide aggregate is not *quite* of a single phase; i.e., a second and perhaps other phases occur as grain-boundary impurities. Where the distribution of such impurity particles is not quite regular, and there are patches here and there where grain boundaries are devoid of them, annealing may lead to inhibition-dependent coarsening and the development of a porphyroblastic structure.

In many cases, as we have already noted in Chap. 10, regional metamorphism leads to the development of preferred orientations in the sulfides analogous to those of regionally metamorphosed silicate and carbonate rocks. This may be detected by optical anisotropism, as in the case of pyrrhotite, by twin traces, as in sphalerite, by reflectance of cleavage pits, as in galena, or by shape, as in pyrite. Preferred orientations of pyrrhotite are sometimes so good that they can readily be detected simply by scanning between crossed nicols. In some well-bedded, relatively slightly deformed ores—for example, those of Sullivan, British Columbia—the pyrrhotite shows good preferred orientation, with the c axis tending to be normal to the bedding planes. Kanehira (1959) has made a comparative study of preferred orientations in associated silicates, carbonates, sulfides, and oxides in the ore of the Chihara mine, Japan. The minerals studied were quartz, muscovite, chlorite, calcite, pyrite, sphalerite, chalcopyrite, and hematite. In the main body of schistose material (i.e., neglecting occurrences in veins, vugs, pressure shadows, etc.):

1. *Quartz* (c crystallographic axes) shows good preferred orientation, with the development of an ac girdle.
2. *Calcite* shows marked shape elongation parallel to the lineation of the rock, and the optic axes are distributed on an ac girdle, with a maximum near the c axis.
3. *Chlorite* and *muscovite* (poles to the basal planes) develop ac girdles, with a sharp concentration near the c axis.
4. *Sphalerite*, other than that of inclusions, exhibits twinning (the exact nature of this twinning is not clear from Kanehira's account), with twin planes showing a slight tendency to parallelism with the ab tectonic plane.
5. Much of the *pyrite* shows marked elongation parallel to the lineation, as in the case of calcite.
6. The basal planes of the *hematite* crystals have a clear tendency to lie parallel to the schistosity (ab) plane, their poles being concentrated on an ac girdle in a similar way to that observed in muscovite.

Kanehira sums up by noting:

. . . the metamorphic fabrics are clearly found in some ores as well as in schists, and not only the large-scale structures of the ore deposits are in harmony with the structures of the country rocks, but also the fine fabrics of the minerals forming the ore deposits are generally concordant with those of the minerals of the prevailing schists (1959, p. 334).

Preferred orientation of this kind may have important implications in the development of new microstructures during the annealing stage. Where an ore mineral undergoes full recrystallization during annealing, the resulting aggregate may be random or it may acquire a new preferred orientation, as pointed out in Chap. 10. Where recrystallization is followed by extensive grain growth, variation in the *degree* of such preferred orientation in single-phase aggregates may lead to *secondary grain growth*, or *coarsening*, also as described in Chap. 10. The result is the development of anhedral to subhedral porphyroblasts. These are widespread in much of the chalcopyrite of the Mount Isa copper ore, for example, and no doubt they will eventually be recognized as a common feature in many ores.

General statement Although there is a great deal of investigation yet to be done, it is tempting to postulate that ores *modified* by regional metamorphism constitute a common and widespread category. Perhaps, indeed, it is not unreasonable to suggest that the proportion of regionally metamorphosed ores among all ores is much the same as the proportion of regionally metamorphosed silicate and carbonate rocks among all "ordinary" rocks.

Recognizing that some speculation is involved, it may therefore be suggested that, in their *distribution*, ores of this category are widespread in space and time. By analogy with "ordinary" rocks it may be expected that the relative incidence of these ores increases with increase in the age of the terrain concerned; i.e., most Precambrian ores have probably suffered some regional metamorphism, whereas many of Tertiary age would not.

The *form* of the deposits must vary enormously. The primary factor influencing form must, of course, be the configuration in which the ore was originally deposited. This primary form is then modified by stress. Layers of orthomagmatic chromite and magnetite-ilmenite are often squeezed into lenses or "pods." Layered ores in sedimentary rocks are folded, puckered, and in some cases, apparently concentrated in fold crests. Clark (1967) has drawn attention to the likely large-scale effects of deformation on stratiform sulfides. He points out that where movement involves an extensive shear component, a massive sulfide concentration would yield as an incompetent unit relative to the harder enclosing rocks. Thus movement would be largely concentrated along the sulfide layer, smearing out the sulfide, inducing the development of mildly transgressive contacts, and causing thinning and elongation of the deposit as a whole. Clark suggests that in higher grade metamorphic terrains, originally "stumpy" stratiform lenses may suffer extreme elongation and attenuation parallel to the prevailing mineral lineation. This has clear consequences in mineral exploration. Like deposits in sediments, those in veins may be extensively folded and, in some cases, sheared. Repeated shearing along a mineralized fracture often leads to a "shearing out" of the ore body concerned, so that what was originally a continuous sheet is converted into a series of pods. This, too, has obvious implications in mineral exploration and exploitation.

Since ores of all types may be regionally metamorphosed and since any individual deposit may be metamorphosed to any one of a number of grades, variability of form is paralleled by variability of *setting*. The primary nature of the environment may thus be igneous or sedimentary. Superimposed on this are the elements of the metamorphism. Added to this there may have been more than one episode of regional metamorphism, or regional metamorphism may be the final episode in a series of events that includes an earlier metamorphism of different type (e.g., a contact metamorphism). The potential variation in setting *as now observed* is thus enormous. On the other hand, it may be recalled that some rocks—through their chemical and physical nature and/or their time of formation relative to the tectonic cycle—are more likely to suffer regional metamorphism than others. Those most frequently involved in regional metamorphism are geosynclinal sedimentary and pyroclastic rocks and associated mafic lavas. It is therefore the stratiform and related ores in these formations that are most frequently regionally metamorphosed. Chromite deposits and their associates in early alpine-type intrusions are also frequently affected.   Orthomagmatic deposits in later plutonic (granitic-dioritic) intrusions and those in and associated with later basic dikes and layered sills are, as one might expect, neither commonly nor severely affected.   Ores in sedimentary rocks laid down over stable basements in epicontinental seas are likewise not notably affected.

The *composition* of regionally metamorphosed deposits is as variable as their form and setting.   It is, of course, heavily governed by primary constitutions, though the final chemical and mineralogical makeup will also depend on whether the mass acted as a closed or as an open system during metamorphism.   The likelihood of water, CO_2, and sulfur loss must be kept in mind.   Pronounced changes in mineralogy are likely in both closed and open systems, particularly where the ore minerals are oxides.   The equilibria developed in sulfide-silicate and related systems may well be significant but await exhaustive investigation.

The *origin* of the deposits is compound and varies in a way analogous with that of regionally metamorphosed silicate and carbonate rocks.

ORES OF METASOMATIC ASSOCIATION

As with those of contact- and regional-metamorphic association, ores showing metasomatic affiliations may have been *produced* by metasomatism or simply *modified* by it.   Since metasomatism is a process that must usually involve considerable time and since it apparently often proceeds in stages, it is probably also true to say that most ores formed by metasomatism have also, in the protracted history of their development, been continuously modified by it.

As we noted in Chap. 2 and again in Chap. 15, many modern investigators have regarded metasomatic metamorphism as the most important of all ore-forming processes.   There is no question that metasomatism does indeed proceed in a wide variety of environments.   Fossil cephalopods now preserved in beautiful form as pyrite are clearly the result of metasomatic replacement.   Similarly silica pseudomorphs after corals in limestone, and after wood in shales or stream gravels, result from metasomatism.   In ores exposed to weathering, "secondary" sulfides may be observed—at this very day—replacing older "primary" sulfides, and the replacement of magnetite by hematite is a common phenomenon.   Alteration of the wall rocks of veins—usually by the introduction of one or more of H_2O, K, SiO_2, and S— is also a well-established result of metasomatic metamorphism.

However, in spite of this there is increasing doubt of the overall importance of metasomatism in ore formation.   Sulfides (and oxides) often occur in and take the form of sedimentary beds, suggesting the possibility of gross pseudomorphism of a number of adjacent strata (e.g., Broken Hill, Mount Isa, Sullivan).   However, when one examines the ore microscopically, evidence for pseudomorphism on a fine scale is often very hard to find.   At Broken Hill, for example, the host rocks contain abundant garnet and sillimanite—two minerals that characteristically occur as idiomorphic to subidiomorphic crystals.   Well-defined crystal faces are common here, and both minerals have clearly recognizable habits.   However, in spite of intensive microscopical work during the last 50 years or so, no pseudomorphs of sulfide after garnet or sillimanite have been recorded. This seems to indicate quite clearly that however the sulfides were precipitated in the first place, their *present form and position are not due to metasomatic metamorphism.* Similarly neither Mount Isa nor Sullivan—both of which are beautifully bedded and which are hence possibly grossly pseudomorphous after bedding of carbonate or silicate —show any evidence of replacement of specific nonsulfide minerals on a microscopic scale.

Thus while metasomatism is undoubtedly a geological process and while it is almost certainly the process by which *some* ores have formed, evidence for it in most of the world's great ore bodies is either unclear or highly equivocal. It seems likely that its operation has been invoked too loosely and indeed quite uncritically in some cases. Metasomatism implies pseudomorphism, and the obtaining of evidence of the latter must be an important objective wherever the possibility of metasomatic metamorphism is being investigated.

General statement Since ideas on the importance of metasomatism in ore formation are currently very much in a state of flux, it is not easy to make a reliable general statement concerning ore that results from it. Deposits that in the 1940s were confidently regarded as examples *par excellence* of metasomatic concentration (e.g., the Roan, Mount Isa, Broken Hill, Sullivan) are now, as we saw in Chap. 15, strongly suspected of being formed by quite different processes. Indeed, doubt has been cast on the importance of metasomatism almost wherever it has been applied to important sulfide and oxide ores, and until the problem is resolved only the most tentative and generalized statements about their features can be made.

Keeping in mind that any subsurface liquids or gases—not only those of immediate igneous derivation—can lead to replacement, the *distribution* of this group of ores is probably widespread in terms of space, time, and geological environment. Those resulting from diagenetic change and the activity of pore-space fluids in compacting sediments might be expected to appear wherever appropriate sedimentary rocks occurred. Those derived from igneous hydrothermal fluids would be expected to lie within belts of plutonic intrusion and metamorphism. The *form* of the deposits would be expected to be highly variable. If they replaced the host rock *in toto*, their outlines would be those of the host in the locality concerned. Clearly the potential for variation in form here is very great indeed. Since they may not replace the entire host in many cases, the potential variety of shape is even greater. If, for example, the extent of replacement of a sedimentary unit varies significantly from one small bed to another (due to small stratigraphical changes in the physical and/or chemical nature of the host sediment), the extremities of a deposit may have a digitated form, as is the case in many contact-metamorphic (i.e., contact-metasomatic or *pyrometasomatic*) deposits.

The geological *setting* of deposits of this kind is also variable, though they most frequently occur in sedimentary rocks, and within these, in carbonate rocks, shales, carbonate-bearing shales, siltstones, and in pyroclastic horizons. These may or may not be folded. Metasomatic deposits derived from granites or related "plutonic" intrusions would be expected to occur in orogenic belts, not too far from intrusive bodies, and in calcareous sediments or pyroclastic rocks adjacent to faults which acted as conduits for the flow of solutions from source to the site of metasomatism.

The *constitution* of the metasomatic group of ores appears to vary between wide limits. Sulfides, arsenides, etc., oxides, and silicates may all develop by replacement, and a wide variety of cations appear to occur in a wide variety of proportions. Certainly there appear to be no characteristic compositions (as in many of our other associations) or ranges of composition.

As regards *origin* it may be said that the metasomatic group of ores has a wide derivation—granitic intrusions, older sedimentary and pyroclastic rocks of all kinds—and that they may be transported in a wide variety of aqueous solutions and gases. As noted in Chaps. 7 and 17 those sources of origin of which we have greatest knowledge are the gases and hydrothermal solutions associated with volcanism and the hot concentrated brines associated with permeable sediments.

ORES ASSOCIATED WITH DISLOCATION METAMORPHISM

As we have already noted, this type of metamorphism involves intense structural modification of rocks, usually on a very local scale. The most common loci of dislocation metamorphism are large faults, shear zones, and layers of sedimentary (or igneous) rocks whose composition and attitude are particularly conducive to movement along them.

All types of ore occurrences may be affected by dislocation metamorphism, and the whole or only part of a particular deposit may be involved. This type of metamorphism is, of course, essentially a *modifying* process and can have little or no effect on the *concentration* of ores. Some large ore bodies, such as Broken Hill, may be intersected by faults having a large shear component, leading to the development of localized dislocation metamorphism (Fig. 18-7) and, in some cases, accompanying retrograde metamorphism. In such cases the proportion of the total ore body involved in dislocation is usually quite small. Vein ore bodies are also frequently affected, particularly those that are structurally controlled. In this case some kind of opening is first developed, most commonly a fault or fault intersection. While deposition of any sulfides necessarily follows the initial creation of such an opening, there is no reason why movement along the faults concerned should cease with ore deposition. Indeed it very frequently does not, and movement accompanying or following deposition may severely modify the ore concerned. Frequently in such cases the *whole* ore body is deformed.

Dislocation metamorphism takes the form of brittle fracture of the harder minerals and plastic flow of the softer ones. Where both are present, the fragments of the harder minerals often become strung out in the flowing matrices provided by the softer minerals. Vein deposits of the Coeur d'Alene district of Idaho provide outstanding examples of the dislocation metamorphism of almost monomineralic galena. The latter shows pronounced flattening of grains and schistosity in hand specimen, and under the microscope it displays abundant bending of cleavage and subgrain formation (cf. Chap. 10). Similar deformed

Fig. 18-7 Galena-sphalerite-garnet gneiss from Broken Hill. Sphalerite (medium gray) and garnet (dark gray, best developed in the lower half of the photograph) form the augen. (*From Stanton and Willey, Econ. Geol., 1970.*)

galena veins are well known about Mount Farrell, Tasmania, and there are numerous other examples. Stibnite veins, in which the stibnite is usually contained in a matrix of quartz and/or carbonate, is very frequently—indeed almost universally—affected by dislocation metamorphism. In hand specimen, individual crystals, in the elongated blade form characteristic of this mineral, can readily be seen to be bent, kinked, and granulated. Under the microscope, the deformation features of such stibnite are by far the most spectacular seen in common sulfides. Kink bands abound, and there are usually numerous deformation twins and translation lamellae. In many cases the stibnite is partly recrystallized (in the strict sense, as defined in Chap. 10), in which case it might be said—from a purely physical point of view—that sulfide retrogressive metamorphism has occurred.

Many polymetallic veins are deformed by late movement along the structures that localized them. Indeed it is very difficult to find any fully filled vein that does not, with careful scrutiny, exhibit abundant evidence of deformation. Broken Hill provides excellent examples of the effects of localized shearing in a "nonvein" ore body. Indeed, as the reader will now have realized, this deposit is a compound case—a spectacular example of a regionally metamorphosed ore body that has later suffered local dislocation, and associated retrograde, metamorphism. The latter, first described in detail by Richards (1966), has led to the formation of schists and breccias. In both classes of material the galena, which provides the plastic matrix in most cases, shows abundant slip, kinking, twinning, cleavage distortion, and subgrain formation. Sphalerite is twinned, kinked, bent, and drawn out and in some cases has shattered severely. Chalcopyrite—a very minor constituent of the Broken Hill sulfide schists and breccias—has deformed by the development of spindle-shaped deformation twins, kinks, and microfaults. The remaining principal sulfide, pyrrhotite, has deformed by bending, twinning, and kinking and also shows much very conspicuous subgrain development. An interesting feature of these dislocated ores is that the disposition of small bodies of minor phases is often markedly influenced by deformation structures in the major (host) phases. For example Richards (1966) notes that small bodies of pyrrhotite and chalcopyrite in sphalerite are often localized along sphalerite deformation twins, at intersections of deformation and annealing twins, and at intersections of deformation twins with grain boundaries. As Bateman (1925) had noted earlier in connection with the Coeur d'Alene ores, it is clear that the present textures of the Broken Hill ores give little or no indication of the original process of deposition. Indeed there seems no doubt that original textures of the ores have been completely obliterated.

General statement As with the other types of ore deposit of metamorphic affiliation, those related to dislocation metamorphism exhibit considerable variety in most of their features. Since they are modifications of all other types of ore occurrences, including many associated with earlier metamorphisms, their *distribution* is necessarily as broad—and as limited—as that of all other ore occurrences together. Since they result from faulting and shearing of earlier formed ore, their *form* is often more or less planar. Where they have developed at the intersections of faults, or of faults with folds, they may take the form of pipes or of two-dimensional saddles or basins. Clearly there is some potential for variation here. Their geological *setting* covers the full span of all the associations we have examined so far and of a variety of minor ones we have not been able to consider. The only environmental characteristic common to all ores associated with dislocation metamorphism is that they occur in a region of faulting and that some of these faults intersect earlier formed deposits of ore. Their *constitution* is limited only by that of ores in general. Their *origin*, like that of all ores of metamorphic affiliation, is compound: their initial deposition probably encompasses every primary ore-forming process, they may then be affected by any or all of the three foregoing types of metamorphism, and they are then involved in dislocation. Clearly even this is not necessarily the end of their geological history.

RECOMMENDED READING

Kalliokoski, J.: Metamorphic features in North American massive sulfide deposits, *Econ. Geol.*, vol. 60, pp. 485–505, 1965.

Stanton, R. L., and H. Gorman: A phenomenological study of grain boundary migration in some common sulfides, *Econ. Geol.*, vol. 63, pp. 907–923, 1968.

Stanton, R. L., and H. G. Willey: Natural work-hardening in galena, and its experimental reduction, *Econ. Geol.*, vol. 65, pp. 182–194, 1970.

Vokes, F. M.: A review of the metamorphism of sulphide deposits, *Earth Sci. Rev.*, vol. 5, pp. 99–143, 1968.

19
Ore Type and the
Tectonic Cycle

or
Relations between the Evolution
of Ore Type and the Evolution
of Geological Environments
and the Continental Crust

The basic philosophy of this book has been that ores are rocks and, therefore, that in most cases they are likely to be intrinsic parts of the environments in which they occur. While it has not been possible to examine *all* the different kinds of ore bodies that are known, we have considered quite a wide variety of them and in doing so have been led to consider a similarly wide variety of geological environments. Examination of the evidence generally confirms our original suspicion— most deposits do indeed appear to be fundamentally related to their surroundings. But not only this; different kinds of ore have different kinds of environment. It appears therefore that we may be more specific: ores are not merely *part* of their environment—*each is a characteristic part of a particular kind of environment.*

This leads to a second simple but vital principle. The geological environment existing in any segment of the earth's crust or surface does not remain static —it is constantly changing. In most cases the changes concerned are progressive and systematic; one kind of environment gradually gives way to another, the succession of environmental metamorphoses following an evolutionary pattern. The broad process and its result are, of course, what is commonly referred to as *crustal evolution,* and in its mature stages any segment of continental crust bears evidence of a long succession of transitory environments. Most of these environments are those that we have found to contain ore deposits, in many cases of their own particular, characteristic kind. It follows from this that *any given pattern of evolution of the crust is accompanied by a parallel pattern of evolution of ore type.*

A GENERALIZED PATTERN OF CRUSTAL EVOLUTION

It is not possible nor desirable here to consider crustal evolution in detail nor to dwell upon controversies surrounding it. Problems such as those of continental drift and the origin and evolution of magmas are clearly highly relevant to ore genesis but involve far too much evidence and argument for adequate—or appropriate—treatment here.

The author has always been attracted to the idea that the continents have developed largely by accretional growth: that each has developed from an early volcanic nucleus or nuclei and that these have been joined and added to by the development of peripheral volcanic arcs, the accumulation of volcanic matter, and the products of developing erosion. This theme is doubtless complicated by the development of later fractures and relative movement—perhaps drift—of the "continental crust" so formed, but the basic pattern remains there all the same. A process of outward growth, somewhat discontinuous but with each period of growth conforming to an evolutionary pattern, seems to form a plausible framework for the succession of geological environments we have just referred to.

In highly simplified form we may visualize the nucleus beginning as a broad swell on the ocean floor. Swells of this kind, clearly early stages in the development of volcanic islands, have been detected by seismic measurements along the modern arc of the Solomon Islands in the Southwest Pacific, and they are probably developing in similar situations elsewhere at the present day. The material involved is the oceanic basaltic crust. As the swell, or groups of swells, develops it begins to undergo block faulting, and by this time there has been at least some extrusion of basic lava. The blocks of the axial portions of the faulted swells eventually protrude above the sea, forming islands composed of basic pillow lavas, principally basalts poor in olivine, together with spilites and keratophyres. Pyroclastic rocks are virtually absent at this stage, but the first erosional products are formed.

Volcanism continues, but it changes to basaltic andesite, then gradually and progressively to andesites, dacites, and rhyolites. With the onset of the andesitic stage, pyroclastic activity becomes conspicuous, and this increases progressively with increase in the felsic nature of the volcanism. By this time the island is quite large—of the order of 100 miles in length—and new swells are developing and evolving, either in linear or arcuate arrangement with respect to the earlier swells. The older islands are now undergoing extensive erosion and much detrital and chemical sediment is accumulating about them. At some stage this sediment begins to move under the influence of gravity—it may gradually slough off the sides of volcanic piles and down the sides of sea-floor basins, or it may sag, or drape, over blocks of basaltic basement as these move in accordance with continued block faulting. The sedimentary rocks may become heavily faulted and quite intensely folded as a result of the simple combination of near-vertical block faulting and slow gravity collapse.

At about this time the earliest "plutonic" rocks—rocks of granitoid texture —appear.   These are often in pipe or stock form, in which case they are almost certainly of fairly shallow subvolcanic nature, in spite of their coarse granitoid texture.   Other masses are larger and less regular in shape, though they are usually somewhat elongated, with their long dimensions essentially parallel to the axis of the island or the island chain.   Whether these granitoids are products of magmatic differentiation or some kind of transformation of some of the now deeply buried pyroclastic rocks or both processes is by no means clear.   Their composition falls generally in the range diorite-granodiorite.

By this time the oldest of our original swells have become quite large islands —of the order of the size of Java and Sumatra at the present day—and they and smaller intervening swells begin to coalesce.   Where events have taken place in relatively isolated positions, as for example, the Solomon Islands at present or the "protocontinents" (Goodwin, 1966) of the Canadian Shield some 2500 million years ago, coalescence leads to the development of a new youthful continental mass.   Where the island chain extends outward from a continent (e.g., the Aleutians), coalescence leads to the development of a new protuberance such as that of the Malayan Pennisula and the Indonesian islands.   Where the island chain has developed parallel and close to a continental margin, infilling leads to the coalescence of the islands among themselves and the chain as a whole with the continent itself.   This appears to be illustrated today by Japan and its associated island chains developing as increments to the Asian continental mass and in the past by the development of volcanic belts such as those of the Appalachians in North America and the Eastern Highlands Belt of Australia.   Using current tectonic parlance, it would be said that the first two situations involved *eugeosynclinal* accumulation, whereas the third involved *miogeosynclinal* accumulation between the continental shore and the offshore arc and *eugeosynclinal* accumulation in the trough on the outer side of the arc (see Jacobs, Russell, and Wilson, chap. 14, for further details of these aspects).

Following their consolidation as a new "continental" mass, or as an addition to a preexisting continent, the segment of rocks continues to evolve.   Further plutonic intrusion of granodioritic to granitic kind—i.e., of rather more siliceous nature—occurs.   The terrain suffers compressional folding and faulting, varying degrees of metamorphism, and eventually becomes part of a "continental shield."   Extensive erosion of mountain ranges leads to peneplanation, the broad flat areas so developed becoming susceptible to inundation whenever sea level rises.   In this way broad epicontinental seas may develop over large areas of stable continent, giving rise to the development of extensive deposits of detrital sediments, numerous large carbonate reef complexes, and the deposition of evaporites and other products of chemical sedimentation.   Where the platform is relatively smooth and only gently sloping, very minor changes in sea level may result in quite extensive lateral movements of the shorelines.   This in turn may lead to widespread merging and interdigitation of shallow-water marine and fluviatile sediments—particularly sediments of the lower reaches of large braided streams, deltas, and outwash fans.

THE PATTERN OF CRUSTAL EVOLUTION
AND ITS RELATION TO CHANGING ORE TYPE

The sequence of crustal evolution just outlined has admittedly been painted with rather broad strokes.   However, although crude, it does seem to accord quite well with observational evidence, and it provides a reasonable approximation to the "normal" sequence of events without being dependent on any tectonic and petro- logical theories that happen to be favored at any particular time.   Keeping in mind the many simplifications involved, we shall take it as a framework and ex- amine the ways in which the various types of ore occurrences appear to fit the evo- lutionary pattern.

EARLY VOLCANIC STAGE

This is the stage of early, basic volcanism, and as we have already seen, it involves the period of development of sea-floor swelling, early faulting, and troughs between swells and fault blocks.

In the basic lavas themselves, native copper and associated sulfides may be precipitated as orthomagmatic disseminations and vesicular fillings in the basalts. The native copper and chalcopyrite occurring as disseminated specks in very mafic lavas of some of the Solomon Islands are examples.   Also in the basic lavas (and associated volcanic sills) there may develop notable concentrations of nickel and associated sulfides.   Possible examples of this are found in the deposits of Kam- balda and elsewhere in Western Australia, and perhaps at Thompson in Manitoba.

Precious metal concentrations, particularly telluride-bearing types, may form at this stage, though the exact timing here is hard to determine in the case of the Precambrian deposits.   Certainly the Fiji occurrences suggest that formation may occur quite early in the tectonic cycle, though the timing in old deposits such as those of the Kalgoorlie and Kirkland Lake areas is not clear.   Deformation has imposed an effective camouflage over most of the Precambrian deposits.

Some ore-forming chemical sedimentation is also characteristic of this stage, though these processes come to their peak a little later.   Manganese (of the "green- stone" association of Shatskiy) and iron compounds are precipitated as chemical sediments, and there may be substantial amounts of jasper formed with them or on its own.   Manganese seems to be rather more important at this stage, though some of the Lake Superior-Ontario-Quebec iron formations appear to be associated with the early basaltic stages of "protocontinent" formation.

EARLY ALPINE-TYPE ULTRAMAFIC STAGE

This appears to follow very closely the early basic lava stage and to involve both intrusions and minor, rather less ultramafic, lavas and pyroclastic rocks.   As already noted the earliest ultramafic intrusion appears to occur in major faults in the lava pile, and while such ultramafic rocks have been widely suggested to be primary mantle material, there is also some evidence to suggest that they may merely be subvolcanic "residues" related to some of the basaltic rocks and derived andesitic types.

The principal ores here are the alpine-type chromite deposits, which occur in quantity only in the intrusions. Some of the nickel sulfide concentrations already referred to as part of the earlier stage *may* in fact have accumulated in some of the lavas of the present period. Nickel also occurs in nickel-rich olivine, which may be later converted to ore by weathering. Chromite deposits of alpine type occur in Paleozoic to Tertiary "serpentine belts" throughout the world, as noted in Chap. 11.

DEVELOPING EUGEOSYNCLINAL STAGE

By this time volcanic islands are well established and enlarging by bodily uplift, volcanic accretion, and the accumulation of sediment. The volcanic products are becoming more felsic and the pyroclastic component, as compared with the lavas, is beoming more prominent.

Although, as we have seen all too clearly in Chap. 13, it is impossible at present to generalize about the origin of banded iron formations, it appears that at least very significant development of this ore type occurs at this stage of continental development. The Precambrian banded iron formations of Ontario and western Quebec have been shown very clearly by Goodwin (cf. Chap. 13) to have developed as volcanic chemical sediments—as a result of sea-floor exhalative activity—in the increasingly felsic stages of the Precambrian volcanic arcs of that region. Presumably deposition took place in troughs (i.e., eugeosynclines and their smaller, satellitic, basins) in interisland regions of the arcs concerned. The precise provenance and timing of development of some of the other great banded iron formation provinces is still not clear, though the observations by La Berge of tuff bands within several of them tempt one to speculate that the Ontario situation may be fairly general. Considerable jasper may be deposited at this stage, and also some manganese, as at Cuyuna.

Modern analogs of the Precambrian processes appear to be operating in the Kuriles (Ebeko), the Solomons (Simbo), and elsewhere.

This stage of continental development also appears to be the most important for the formation of stratiform sulfide deposits of marine and marine-volcanic association. While a few of these ores (e.g., Cyprus and some of the Japanese deposits) are associated with basaltic and more mafic lavas, the vast majority are associated with andesitic and dacitic rocks—particularly the pyroclastic equivalents of these compositions. Small banded iron formations, and often minor concentrations of manganese (and barite), are frequent associates. Most of the deposits are found not far from the margins of the eugeosynclines, and biological activity in the environment—even as far back as 2500 million years or more—is almost always indicated. Many of the Ontario base metal ores, Mount Isa and MacArthur River in the Precambrian, Bathurst, New Brunswick, and eastern Australia in the Paleozoic, and many Japanese deposits of Tertiary age are examples.

These stratiform ores probably include chemical sediments, sulfide pyroclastic concentrations, and perhaps, in some cases, lavas. All are, however, essentially sea-floor volcanic accumulations. Related to them, and still well preserved in Mesozoic to Tertiary terrains that have suffered only moderate erosion, are sulfide-bearing subvolcanic intrusions, breccia plugs, and related shallow igneous bodies of andesitic to rhyolitic composition. These are the "porphyry coppers," which are probably often related to the development of stratiform concentrations at this stage of tectonic-sedimentary evolution.

A word should perhaps be said at this point concerning the "Steinmann Trinity"—the long recognized association of serpentine belts, mafic lavas, and radiolarian cherts and jaspers. In fact the association is much more than a trinity. Graywackes, particularly of andesitic derivation, are also almost always present and represent redistributed andesitic pyroclastic material. In addition, there are a number of associated *ores*. Volcanic manganese concentrations are almost ubiquitous, and the aforementioned jaspers may grade into iron concentrations. Stratiform pyritic lenses are very common and usually contain one or more base metal sulfides. Of these, copper is probably the most frequently occurring nonferrous metal, and cupriferous pyritic lenses are a widespread feature of these environments. Indeed it seems better now to refer to the *Steinmann Association* rather than the Steinmann Trinity—*an association in which certain ore types form an integral and sometimes very important part*.

ADVANCED EUGEOSYNCLINAL AND MIOGEOSYNCLINAL STAGE

Although the volcanic festoons that are isolated from continents or extend outward from the latter are characterized by igneous sedimentation (with, in appropriate climates, reef sedimentation) throughout their earlier history, we have already noted that those peripheral to the continents see a mixing of their own volcanic materials with the products of continental erosion. Thus a hybrid sedimentation develops on the continental side of the arc—the miogeosyncline—while volcanic sedimentation continues on the outer flank and the seaward trough—the eugeosyncline. During this stage the increasingly felsic volcanism gives rise to continuing deposition of marine-volcanic sulfide ores (stratiform and otherwise) and banded iron formations in the eugeosyncline. For some reason as yet unknown, *major* iron formations are not associated with the base metal sulfide deposits. Some iron formations have—as we have seen in Chap. 13—substantial iron sulfide facies, and many large base metal sulfide masses have minor bands of iron formation associated with them, but major deposits of the two kinds do not seem to occur together. This is undoubtedly significant, though the reasons for it remain unknown.

On both sides of the arc, but perhaps particularly on the miogeosynclinal side, volcanic base metals are contributed to reef and off-reef environments, eventually to be concentrated by sedimentary, diagenetic, and later processes to form an early category of limestone–lead-zinc deposits. It appears that the miogeosynclinal and Usinsk (eugeosynclinal) types of manganese deposits also develop at this stage and in these environments.

EARLY CONTINENTAL AND INTRUSIVE STAGE

By this stage the volcanic islands are welded to each other, and to the continental margin in the case of arcs developed in proximity to a continent. Further intense fault movement, compressive folding, the formation of more felsic plutonic intrusions, the waning of volcanism, and the continuation of rapid erosion are characteristic. The pattern of "plutonic" igneous activity here is not easy to unravel, but the earliest part of it may be the formation of highly siliceous granitic rocks and related pegmatites. To these are related disseminated granitic, contact-metamorphic, and pegmatitic cassiterite deposits; quartz-cassiterite veins; quartz-wolframite, quartz-scheelite, and quartz-gold veins; stibnite and stibnite-scheelite-gold veins; and base metal sulfide veins, often accompanied by arsenopyrite and characterized by increasing proportions of lead and zinc as compared with copper. Some of the base metal sulfide veins here may result from the destruction of earlier formed eugeosynclinal stratiform ores; i.e., some of the sulfide ores here may represent *dissipation* rather than the generally accepted *concentration*, as suggested in Chap. 2.

Several other types of intrusion and ore formation may occur at about this stage. The plutonic type of anorthositic iron-titanium oxide association may develop at about this time in the deep roots of the new crustal segment. As we noted in Chap. 12, the type of metamorphism developed here suggests substantial depth of formation and considerable late-stage movement and dislocation. Perhaps a little later in the sequence of events the large layered oxide (minor sulfide)-bearing intrusives of gabbroic composition develop. These are emplaced largely into volcanic segments of the crust, but since faulting—and particularly folding—is relatively minor, intrusion would appear to have taken place after substantial tectonic stability had developed. In the more mafic parts of these intrusions, e.g., in the Bushveld complex, the principal oxide is chromite, whereas in the harzburgitic to anorthositic layers, e.g., Bushveld, Stillwater, Duluth, the oxides are ilmenite, magnetite, and other (minor) Fe–Ti–O minerals. Whether some major nickel-copper sulfide deposits such as those of Sudbury are also to be identified with this tectonic stage is not known, though they may be.

A mineralization that *continues* its emplacement into this stage is that of the alpine-type ultramafic association. The ultramafic rocks, by this time well serpentized and still within or close to the original large fault systems in which they had earlier formed, continue—with their associated chromite—to be squeezed and moved about in their greasy, plastic state, taking up new positions as each succeeding earth movement dictates. The shape and lithological and structural association of the chromite ore bodies become more and more complicated as this history of movement progresses.

SHELF AND SHIELD STAGE

This is an important period from the point of view of ore formation and involves a variety of both igneous and sedimentary ores. Those deposits formed during the earlier stages of the evolutionary cycle may have already undergone substantial metamorphism by this time, and in many cases they continue to do so as the "maturing" of the new continental segment proceeds.

The outpouring of flood basalts, probably a fairly early event after the establishment of a new portion of a continent, does not seem to have had great importance in the formation of mineral deposits. The copper-bearing lavas of the Precambrian basaltic sequence of the Keweenaw Peninsula, Lake Superior, do appear to be of "flood" type, however, and probably constitute an important exception.

Perhaps the most important deposits are those developed under shallow-water marine and near-shore fluviatile conditions along shallow continental shelves and around the margins of shallow epicontinental seas. This group of deposits includes:

1. Later-stage limestone–lead-zinc deposits, often associated with oil-bearing strata and evaporite beds as at Pine Point in Canada and elsewhere.
2. Nonvolcanic sedimentary manganese deposits (orthoquartzite-glauconite-clay association) such as those of the Tertiary estuarine beds of Nikopol' and Chiatura in Russia; also, in arid areas, manganese deposits of Moroccan type.
3. Ironstones of the Clinton, Lorraine, and English Mesozoic type, again associated with near-shore, often estuarine or lagoonal, sedimentation.
4. "Sandstone-type" copper-uranium-vanadium ores such as those of the Colorado Plateau, formed in rather coarse sediments of outwash fans, near-shore braided stream systems, deltas, and shoreline sediments of epicontinental seas. As pointed out in Chap. 16, there has undoubtedly been substantial solution activity and redistribution of the ore minerals here, but the primary association among tectonics, lithology, and ore occurrence seems quite clear.
5. Gold-uranium deposits of Witwatersrand-Blind River-Jacobina type, localized in coarse conglomerates and grits of braided stream channels, in turn probably representing the lower reaches of large stream systems.
6. Base metal sulfide deposits of apparently nonvolcanic provinces. Like the limestone–lead-zinc deposits, these may be quite closely associated in the stratigraphic column with evaporites. Two sulfide provinces that, though not altogether alike, appear to be of this general type are the Kupferschiefer-Marl Slate of Europe and England and the Copperbelt of Zambia.
7. A very minor category—the bog or marsh iron ores such as those of the present Northern Hemisphere.

The igneous ores of this final stage of continental evolution are much less important than the sedimentary groups and may be somewhat later—i.e., they seem to have a closer affinity with the older, more distinctly shield-type portions of the continents. The principal ore types are those of the *carbonatite association*, which appear to be associated with very late, tensional features in the old shields.

ORE TYPES IN SPACE AND TIME

The very fact that ore type evolves in parallel with the evolution of continental masses carries with it the implication that particular ores—and, as a corollary, particular metals—should have been conspicuously concentrated in certain places at certain times. That some areas of the continents are indeed "mineralized" to a notable degree and that certain periods of geological time have been more prolific in this respect than others have long been recognized and expressed in the concepts of *metallogenetic epochs* and *metallogenetic regions* or *provinces*. We may define a metallogenetic epoch as *a geological period during which notable ore formation took place*. A metallogenetic region† is less simply defined. To some geologists it is *a region in which notable ore formation took place during a particular geological period, i.e., metallogenetic epoch*, while to others it is *a region in which notable ore formation has taken place during one or more metallogenetic epochs*. This distinction appeared to have quite definite significance in the period when it was thought that most ore bodies were of subsurface emplacement and immediate plutonic derivation. However, since it is now being recognized that ores have much wider origins and follow a finite and often protracted evolutionary pattern just as all other continental rocks do, the distinction is losing its relevance. A given metallogenetic region may contain ores of many epochs, though the latter are not necessarily distinct and indeed usually grade into each other simply as a succession of periods of continental evolution.

Although great emphasis has been placed here on the parallelism between the evolution of the "ordinary rocks" and the "ores" of a continental segment, it must be made clear that not all members or features of these two categories are necessarily developed—or developed to a comparable degree—in the appropriate parts of all continents, i.e., during *all* evolutionary sequences. In some cases certain ore types, as with certain rock types, seem to have been missed altogether, and in other cases particular types seem to have developed to an extraordinary degree. This has led to the apparent development of specific metal enrichments—so that various areas have become known as "copper provinces," "iron provinces," and so on. We shall return to this shortly.

Since they appear to be among the earliest ores of the "ideal" evolutionary cycle, it is perhaps appropriate to look first at the distribution of banded iron formations in space and time. We may then consider some of the more important ore types that succeed them in the sequence, noting particularly those regions and epochs in which unusual development of specific ore types occurred.

1. BANDED IRON FORMATIONS

These are well developed on all the world's present shield areas. Their development has therefore been wide in space, and the presence of an "iron province" is far from being an exclusive feature of any particular continent.

They do, however, seem to show quite a remarkable restriction in time; in spite of possible minor exceptions, they are restricted to the Precambrian. Within

† The terms *metallogenetic region* and *metallogenetic province* are used synonymously here.

this, there is a remarkable clustering of deposits about an age close to 2200 million years—a very distinct "iron epoch."

2. SULFIDE NICKEL ORES OF MAFIC-ULTRAMAFIC ASSOCIATION

So far these appear to be confined almost entirely to three distinct provinces—the Superior-Ungava area of Canada, the Western Australian shield, and the Kola Peninsula-Siberia region of Scandinavia-Russia (Figs. 11-7, 11-12, 11-8, respectively). Most nickel ores of this kind are Precambrian, the oldest known being about 2500 million years in age. A few are Paleozoic to Mesozoic. There thus appear to be a very restricted number of nickeliferous regions but no distinct "nickeliferous epochs."

3. STRATIFORM SULFIDE DEPOSITS OF VOLCANIC AFFINITY

These are widespread and are again well represented in all continents—around old volcanic nuclei and along volcanic arcs and eugeosynclines (Fig. 15-1). In contrast to the banded iron formations they are well spread through geological time, from about 2700 million years to Modern. There do, however, appear to be two apparent "bursts"—at about 1500 to 1700 $\times$ 10^6 years (e.g., Sullivan, Broken Hill, Mount Isa) and at about 300 to 500 $\times$ 10^6 years (e.g., northern Appalachians, Caledonides, eastern Australia). These two bursts also appear to be notably lead-rich. Both of these features *may*, of course, be more apparent than real and require further investigation.

4. PRECIOUS METAL TELLURIDE DEPOSITS OF VOLCANIC ASSOCIATION

Although isolated minor deposits occur elsewhere (e.g., eastern Europe, Armenia), these ores occur in three outstanding provinces and apparently fall into two epochs. Two provinces—those of Ontario-Quebec and Western Australia—are about 2500 $\times$ 10^6 years old and form part of the old volcanic nuclei. The other—that of the circum-Pacific belt—is substantially Tertiary and related to clearly recognizable volcanic arcs and extensions of these. Thus deposits of this kind, while constituting a fairly general ore type, show particularly notable development in three quite clearly defined metallogenetic regions and in two similarly defined metallogenetic epochs.

5. IRON-TITANIUM OXIDE ORES OF ANORTHOSITIC ASSOCIATION

These occur in relatively minor amount in Africa (the Bushveld complex), Australia, Scandinavia, and elsewhere, but the really major development occurs along the length of southern Quebec and into southern Ontario and the Adirondack area of New York—i.e., in the Grenville province of the North American Precambrian. The ore type also appears—in layered form—about the western end of Lake Superior (e.g., Duluth) and in Montana (Stillwater complex). There is thus a clear titaniferous province through central North America (Fig. 19-1), a province constituting part of a metallogenetic epoch about 1000 to 1400 million years old.

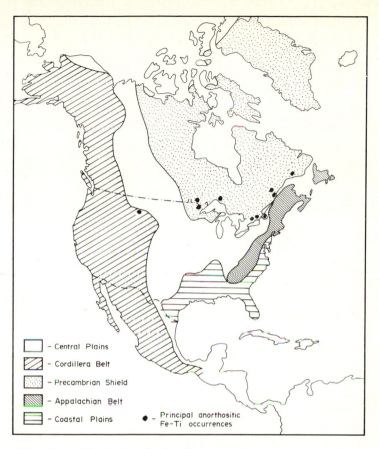

Fig. 19-1 The magmatic titaniferous province (anorthositic iron-titanium oxide deposits) of middle North America.

6. THE "PORPHYRY COPPERS"

These are moderately widespread in the circum-Pacific region, as we have seen. The great province of this ore type is, however, that of the western Americas—a somewhat discontinuous belt extending from Chile through Central America, Mexico, Arizona, and Nevada to northern British Columbia (Fig. 12-11). The metallogenetic epoch concerned is essentially a Mesozoic one. Another belt—the Southwest Pacific province—including the Philippines, New Guinea, and the Solomon Islands is now also becoming apparent, and this appears to be essentially Tertiary in age.

7. "PLUTONIC" VEIN DEPOSITS

This is a highly complex and widespread group and is developed in fold belts of all ages. Usually any given metallogenetic province of this type contains a multiplicity of metals and is not readily identified by a particular metal or metals.

Exceptions are such metals as tin, tungsten, and cobalt, which often do occur in such a way as to give rise to quite distinct provinces.   The tin provinces of eastern Australia, Southeast Asia (Figs. 19-2 and 19-3), and Ireland-England-France-Italy (Skinner, 1969, p. 52) and the tungsten province of western North America and the tin-tungsten province of Kwangsi-Kwangtung-Hunan in China (Lamey,

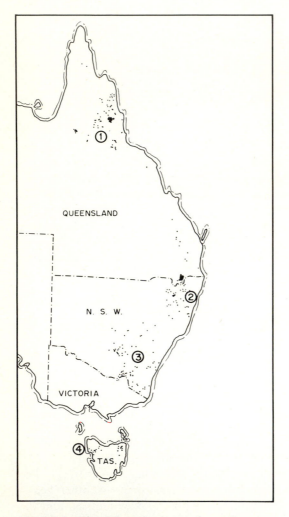

Fig. 19-2   The Paleozoic tin province of eastern Australia.   (*After Hills, Fifth Empire Mining Met. Congr. Trans., Melbourne, 1953.*)

1. Herberton-Cooktown province
2. New England province
3. Southeastern New South Wales-northern Victorian province
4. Northern Tasmanian province

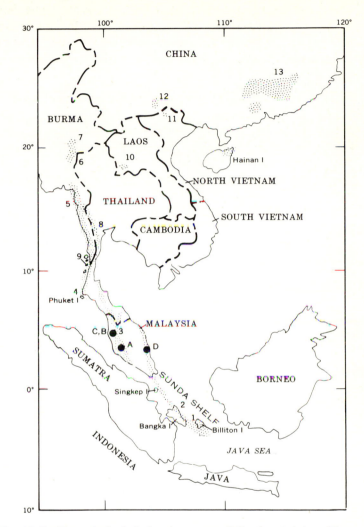

Fig. 19-3 The principal tin-bearing areas of Southeast Asia. (*From Sainsbury, U.S. Geol. Surv. Bull. no. 1301, 1969.*)

1. Billiton (Belitung) Island, Indonesia
2. Bangka Island, Indonesia
3. Malaysia
   A. Gunong Bakau mines, Selangor
   B. Lahat pipe, Kinta Valley, Perak
   C. Beatrice mine, Selabin
   D. Pahang mines, Pahang
4. Phuket Island, Thailand
5. Maulmein area, Burma
6. Mawchi area, Burma
7. Byingyi district, Burma
8. Tavoy district, Burma
9. Mergui Islands, Burma
10. Nam Pha Tene mine, Laos
11. Pia Oak Mountains, Tonkin, Vietnam
12. Kochiu area, China
13. Kwangtung-Kwangsi-Hunan provinces, China

1966, p. 165) are good examples.   The Ontario (Cobalt) native silver:cobalt-nickel arsenide province is another beautifully defined and distinctive example of a plutonic(?) vein province.

8. LIMESTONE–LEAD-ZINC ORES

Although these are products of both geosynclinal and shelf-shield sedimentation, the latter seem to be by far the more important *in toto* and provide a most beautiful example of a metallogenetic region—a very extensive one but well defined all the same.   This is the North American lead-zinc belt, closely associated with Paleozoic limestone reefs and banks and extending discontinuously from New York to Texas and to the Canadian North West Territories (Fig. 19-4).   Another very conspicuous, if not quite so spectacular, province is that of Europe-North Africa, in this case of Upper Paleozoic and Mesozoic age.

It is interesting to note that this ore type is virtually entirely post-Precambrian

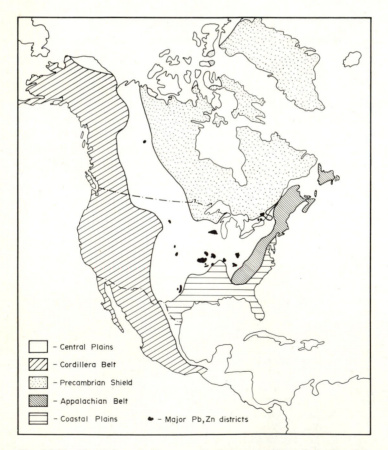

Legend:
- □ – Central Plains
- ▨ – Cordillera Belt
- ⣿ – Precambrian Shield
- ▤ – Appalachian Belt
- ▤ – Coastal Plains
- ● – Major Pb, Zn districts

Fig. 19-4   The Paleozoic limestone lead-zinc province of the Central Plains region of the United States and Canada.

—it is essentially a Paleozoic-Mesozoic phenomenon—and is almost completely confined to the Northern Hemisphere.

9. MANGANESE DEPOSITS OF SEDIMENTARY AFFILIATION

The two most important genetic groups of pre-Modern terrains are the volcanic, and some of the shelf, associations.

The volcanic deposits are well represented on all continents and many modern volcanic islands and in all geological periods. As with the banded iron formations they are most conspicuous in the Precambrian, but they are also quite common in Paleozoic rocks (where they often show up very clearly as members of our Steinmann Association) and in Mesozoic to Modern volcanic accumulations.

In contrast, the shelf association (i.e., the orthoquartzite-glauconite-clay association of Nikopol'-Chiatura type) seems to be remarkably restricted in both space and time. This might not be notable were it not for the fact that these ores are quantitatively so important. As noted in Chap. 14 virtually all these deposits —including some very large ones—are of Lower and Middle Oligocene age and occur along the northern hinterland of the Black Sea. This probably constitutes the most restricted and clearly defined of all the important metallogenetic epochs and provinces so far recognized.

10. THE CARBONATITE ASSOCIATION

We have seen that the development of carbonatites has been widespread in space and time and that the current very active search for them progressively emphasizes this ubiquity. Throughout, their preferred habitat appears to be that of tensional features in shield areas. However, as in the case of so many of the more important types, carbonatites do seem to have had their time and place of *particular* development. For the world as a whole, the great *epoch* of carbonatite formation has been that of the Mesozoic-Tertiary, and the outstanding *province* has been that of East Africa.

SPECULATIONS ON THE FACTORS THAT MAY INFLUENCE THE DISPOSITION OF METALLOGENETIC EPOCHS AND REGIONS

While it is clear that each of the main ore types fits into its "normal" position or positions in the general pattern of continental evolution, it is equally clear—as we have already noted—that particular ore types, like other rock types, are not equally developed in appropriate stages of the development of *all* continents. Superimposed on the general pattern of metallogenetic epochs and regions there has been—from time to time and from place to place—quite extraordinary and conspicuous development of a particular member of our series of ore types. It is such "aberrations" that have produced the world's great "iron," "tin," "lead-zinc," "copper," and other "provinces."

Why should such aberrations develop? Their incidence has had enormous economic consequences and is a scientific problem of obvious importance. It has been widely suggested that they may simply reflect regional variations in the metal content of the crust and subcrust, but this is questionable and difficult to sustain. The present situation is that the answer is just not known, and the problem continues to be one of the most intriguing avenues of research in the geology of ores.

Enigmatical though these aberrations are, there is, perhaps, enough evidence to justify at least some preliminary speculation on the matter. We have seen that the formation of ores results from processes going on both within and at the surface of the earth. Notable variation in the intensity of ore formation thus probably results from fluctuations in the intensities of these processes and from fluctuations in the nature and intensity of the *interaction* of such processes. Such a suggestion is usually sufficient to bring horror to those of deep uniformitarian convictions, but it is well to remember that the principle of uniformitarianism holds that *the present is the key to the past*—not that *the past was identical to the present*. From our knowledge of evolutionary patterns and the incidence of catastrophic events among celestial bodies, it would be surprising indeed if no progressive evolutionary changes, and occasional "sudden events," did not take place on earth. If this view is correct—and there is very good reason to think that it is—the development of metallogenetic provinces by the interaction of various trains of evolutionary change, and of these with particular bursts of geological activity, takes on quite a normal and expectable aspect.

There are at least three ore associations that appear to illustrate this point of view very well. These are the banded iron formations, the limestone–lead-zinc deposits, and the manganese deposits of the orthoquartzite-glauconite-clay association so abundantly developed along the Black Sea hinterland. The space-time relationships of these ore types are quite distinctive and illustrate the kinds of approach that might well be used to gain a better understanding of metallogenetic epochs and provinces in general.

As the earth has aged it has continually lost its contained gas, the continents have built up as volcanic scar tissue and its erosional derivatives, the hydrosphere and atmosphere have accreted and matured, and the biosphere has nucleated and evolved. It may be that, in the first place, the rate of evolution of gas, and its gross composition, have changed with time. It seems likely that rate of degassing would have tended to decrease, and this immediately brings us to a first example of the possible effect of changes in intensities of processes on the development of "irregularities" in the incidence of ores in space and time. It is known (Martin and Piwinskii, 1969) that iron in rocks moves quickly to a hydrous phase if this is present. Perhaps early, rapid degassing led not only to the supply of most of the water of the oceans but also to the extensive stripping of iron from the subcrust. In this way the evolution and accumulation of large quantities of iron would have been a natural accompaniment of the evolution and accumulation of large quantities of water—a type of event that continued as the earth evolved but could never be repeated on the scale of the initial loss. Perhaps this is the beginning of the explanation of the incidence in space and time of the larger *banded iron formations*— their remarkably widespread distribution in space as opposed to their rather restricted distribution in time.

It is strange that the Paleozoic should see such a tremendous development of *limestone–lead-zinc* ores. It might be said that their sudden appearance is apparent only here and that their absence from Precambrian rocks is an erosional feature. A little reflection shows this to be impossible to sustain. Lead appears to have escaped from the crust from early Precambrian to Recent times, as is evidenced by the incidence of the stratiform ores, so the problem does not seem to be *basically* one of supply. Since the deposits are so closely tied to reef structures and to environments of such conspicuously biological aspect, the possibility arises that primary deposition may have been heavily influenced by some new phase of biological evolution that manifested itself in the warm waters of broad epicontinental seas. The remarkable paucity of copper points to a highly efficient process of metal segregation, which may also suggest a biological factor. Perhaps these limestone–lead-zinc ores result from an interaction of both subterranean and surface factors: a subterranean one of *somewhat* unusually abundant supply (also manifested in the particularly conspicuous development of stratiform lead-zinc-bearing deposits of the eugeosynclines of the same era) and a surface one of climatic and oceanographic environment conducive to the onset of a new biological evolutionary burst.

While acknowledging the distinctiveness and importance of the Mesozoic ironstones, porphyry coppers, and carbonatites, perhaps one of the most extraordinary metallogenetic events of post-Precambrian times has been the development

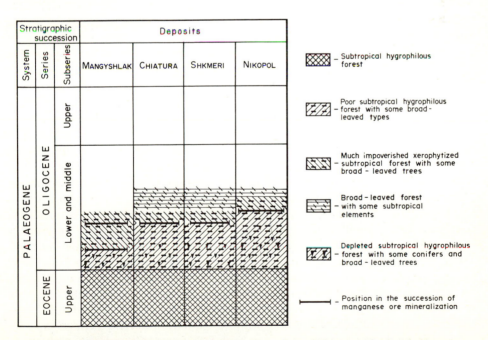

Fig. 19-5 Diagram illustrating sharp restriction of the Chiatura and related manganese deposits of Russia to climatic changes during the Lower to Middle Oligocene period.

of *manganese ores* around the northern hinterland of the Black Sea. Although there *may* have been subterranean hot springs involved, the Russian work suggests that the history of these ores is essentially a surface one: weathering, stream transport, and estuarine deposition. A most interesting observation made by the Russians (see Shterenberg et al., 1965) is that principal deposition of the manganese always seems to have coincided with a sharp change in climate and vegetation. This is illustrated in Fig. 19-5. From their pollen and spore studies, Shterenberg and his colleagues deduced that the manganese ores formed as the climate changed from humid subtropical to one of cooler, drier, temperate kind. This may have affected the derivational weathering processes, or it may have wrought some critical change in the depositional environment. Judging from the remarkable degree of segregation of manganese from iron that has occurred, it seems most likely that it was the depositional process that was the most critical one. Perhaps in this case the common processes of normal minor manganese deposition in shallow-water sediments have been facilitated to an extraordinary degree by the coincidence of a suitable manganese-bearing source terrain, the evolutionary appearance of organisms capable of extremely efficient and selective fixation of manganese, and the sudden onset of climatic conditions particularly conducive to the explosive development of the organisms concerned. Such an explanation involves no departure from "normal" processes and does not require the particular area to have been notably rich in total manganese. All it requires is the coincidence of auspicious climatic and physiographic conditions with a particular phase of biological evolution. The situation seems no different in principle from the present coincidence of the modern climate, the proliferation of mankind, and the explosive pollution of the atmosphere during the past 200 years.

CONCLUDING STATEMENT

The main purpose of this final chapter has been to "draw the threads together." In the preceding pages we have looked at the principles governing the materials of ores—the nature and stabilities of ore minerals and the mechanisms by which such minerals grow, deform, and anneal—and then we have examined and considered the settings, or environments, in which the different kinds of ores occur. There seems no question that most ore bodies are intrinsic parts of their environments and that particular kinds of environments are characterized by particular kinds of ore. Certain vein ore types, being migrant and hence superimposed entities, may not have quite such close ties with their geological surroundings, though there are many others—such as the volcanic precious metal telluride veins—that most certainly do. Because the genetic connection between ore and environment is so general, it naturally follows—as the author has tried to demonstrate in this chapter —that the ore types of any given area change with time in parallel with changes in the assemblages of other rock types and hence that such changes in ore type follow an evolutionary pattern conforming with the pattern of development of the continental segment concerned.

This brings us back to our original hypothesis: that ores are rocks and should be regarded, investigated, and sought by the methods of petrology—*ore petrology*. Ores, like any other rocks, are part of the stuff of which continents are made; tiny, it is true, but continental building blocks nonetheless. It follows from this that, contrary to much twentieth-century opinion, the understanding of most ores is to be found *in their surroundings as much as in themselves*. Looking back on the history of the study of ores, one is tempted to suggest that, for a while at least, we might well study the ores themselves a little less and their surroundings—particularly their *regional* settings—a little more. Clearly such environmental considerations must be basically lithological and must range from the grandest to the smallest scale. At one extreme there lies the relation of regional patterns of ore occurrence to lithology and of lithology to the tectonic evolution of large segments of the continental crust. At the other extreme there is the detail of the primary structures of the rocks themselves and of their modification by local deformation and metamorphism. A whole spectrum of environmental features lies between. Such an approach demands great geological perspective and balance, but it seems likely to yield many prizes both in scientific understanding and mineral discovery. For rising generations of ore geologists, this huge field of environmental studies presents great opportunity and an exciting challenge. As Thomas Crook so clearly appreciated in the early part of this century, each of the great family of ore types has its special place in earth history, and there is perhaps a no more important or intriguing part of petrology—using that term in its best and widest sense—than ore petrology.

RECOMMENDED READING

Bilibin, Yu. A.: "Metallogenic Provinces and Metallogenic Epochs," Queens College Press, Flushing, New York, 1968.

Goodwin, A. M.: Archaean protocontinental growth and mineralization, *Can. Mining J.*, vol. 87, pp. 57–60, 1966.

Krauskopf, K. B.: Source rocks for metal-bearing fluids, in H. L. Barnes (ed.), "Geochemistry of Hydrothermal Ore Deposits," pp. 1–33, Holt, Rinehart and Winston, Inc., New York, 1967.

Pereira, J., and C. J. Dixon: Evolutionary trends in ore deposition, *Trans. Inst. Mining Met.*, vol. 74, pp. 505–527, 1965.

Turneaure, F. S.: Metallogenetic provinces and epochs, *Econ. Geol. 50th Anniv. Vol.*, pp. 38–98, 1955.

Bibliography†

Adams, F. D.: An experimental investigation into the action of differential pressure on certain minerals and rocks, employing the process suggested by Professor Kick, *J. Geol.*, vol. 18, pp. 489–525, 1910.

——— and J. T. Nicholson: An experimental investigation into the flow of marble, *Phil. Trans. Roy. Soc. London, Ser. A*, vol. 195, pp. 363–401, 1901.

Adler, H. H.: The conceptual uranium ore roll and its significance in uranium exploration, *Econ. Geol.*, vol. 59, pp. 46–53, 1964.

Ahlfeld, F.: Ueber Zinnkies (Beiträge zur Geologie und Mineralogie Boliviens. Nr. 5), *Neues Jahrb. Mineral.*, sec. A, vol. 68, pp. 268–287, 1933.

Allen, E. T.: Chemical aspects of volcanism with a collection of the analyses of volcanic gases, *J. Franklin Inst.*, vol. 193, pp. 29–80, 1922.

——— and E. G. Zies: A chemical study of fumaroles of the Katmai region, *Geophys. Lab. Carnegie Inst. Wash. Pap.* no. 485, pp. 75–155, 1919.

——— and ———: A chemical study of the fumaroles of the Katmai region, *Tech. Pap. Natl. Geogr. Soc.*, vol. 1, no. 2, pp. 75–155, 1923.

Amstutz, G. C. (ed.): "Sedimentology and Ore Genesis," Elsevier Publishing Company, Amsterdam, 1964.

Anderson, A. L.: Some pseudo-eutectic ore textures, *Econ. Geol.*, vol. 29, pp. 577–589, 1934.

Anderson, C. A.: Massive sulfides and volcanism, *Econ. Geol.*, vol. 64, pp. 129–146, 1969.

Anger, G., H. Nielsen, H. Puchelt, and W. Ricke: Sulfur isotopes in the Rammelsberg ore deposit (Germany), *Econ. Geol.*, vol. 61, pp. 511–536, 1966.

Annersten, H.: A mineral chemical study of a metamorphosed iron formation in Northern Sweden, *Lithos*, vol. 1, pp. 374–397, 1968.

Antun, P.: Sedimentary pyrite and its metamorphism in the Oslo region, *Norsk Geol. Tidsskr.*, vol. 47, pp. 211–236, 1967.

Arnold, R. G.: Equilibrium relations between pyrrhotite and pyrite from 325°C to 743°C, *Econ. Geol.*, vol. 57, pp. 72–90, 1962.

Aston, R. L.: The tensile deformation of large aluminium crystals at crystal boundaries, *Proc. Cambridge Phil. Soc.*, vol. 23, pp. 549–560, 1927.

Ault, W. U.: Isotopic fractionation of sulfur in geochemical processes, in P. H. Abelson (ed.), "Researches in Geochemistry," pp. 241–260, John Wiley & Sons, Inc., New York, 1959.

——— and J. L. Kulp: Isotopic geochemistry of sulphur, *Geochim. Cosmochim. Acta*, vol. 16, pp. 201–235, 1959.

——— and ———: Sulfur isotopes and ore deposits, *Econ. Geol.*, vol. 55, pp. 73–100, 1960.

Aust, K. T., and B. Chalmers: The specific energy of crystal boundaries in tin, *Proc. Roy. Soc. London, Ser. A*, vol. 201, pp. 210–215, 1950.

Baas Becking, L. G. M., I. R. Kaplan, and D. Moore: Limits of the natural environment in terms of pH and oxidation-reduction potentials, *J. Geol.*, vol. 68, pp. 243–284, 1960.

Bailey, D. K.: Carbonatite volcanoes and shallow intrusions in Zambia, in O. F. Tuttle and J. Gittins (eds.), "Carbonatites," pp. 127–154, Interscience Publishers, Inc., New York, 1966.

Bain, G. W.: Patterns to ores in layered rocks, *Econ. Geol.*, vol. 55, pp. 695–731, 1960.

Bainbridge, K. T., and A. O. Nier: Relative isotopic abundances of the elements, *Natl. Res. Council (U.S.), Nuclear Science Ser., Prelim. Rep.* 9, 1950.

Baker, G.: Tellurides and selenides in the Phantom Lode, Great Boulder Mine, Kalgoorlie, in "F. L. Stillwell Anniversary Volume," pp. 15–40, The Australasian Institute of Mining and Metallurgy, Melbourne, 1958.

Barnes, H. L.: Ore solutions, *Carnegie Inst. Wash., Ann. Rept. Dir. Geophys. Lab. Year Book* 59, pp. 137–141, 1960.

† This bibliography is intended to give a comprehensive but not exhaustive cover of relevant literature. Details of less important and more obscure works cited are all obtainable from the appropriate listed references.

————: In E. Roedder, Report on S.E.G. symposium on the chemistry of ore-forming fluids, August–September, 1964, *Econ. Geol.*, vol. 60, pp. 1380–1403, 1965.

———— and C. K. Czamanske: Solubilities and transport of ore minerals, in H. L. Barnes (ed.), "Geochemistry of Hydrothermal Ore Deposits," pp. 334–381, Holt, Rinehart and Winston, Inc., New York, 1967.

Barrett, C. S.: "Structure of Metals: Crystallographic Methods, Principles and Data," 2d ed., McGraw-Hill Book Company, New York, 1952.

———— and T. B. Massalski: "Structure of Metals: Crystallographic Methods, Principles and Data," 3d ed., McGraw-Hill Book Company, New York, 1966.

Barton, P. B.: Some limitations on the possible composition of the ore-forming fluid, *Econ. Geol.*, vol. 52, pp. 333–353, 1957.

————: The chemical environment of ore deposition and the problem of low-temperature ore transport, in P. H. Abelson (ed.), "Researches in Geochemistry," John Wiley & Sons, Inc., New York, 1959.

————: Possible role of organic matter in the precipitation of the Mississippi Valley ores, in J. S. Brown (ed.), "Genesis of Stratiform Lead-Zinc-Barite-Fluorite Deposits," *Econ. Geol. Monograph* no. 3, pp. 371–378, 1967.

———— and B. J. Skinner: Sulfide mineral stabilities, in H. L. Barnes (ed.), "Geochemistry of Hydrothermal Ore Deposits," pp. 236–333, Holt, Rinehart and Winston, Inc., New York, 1967.

———— and P. Toulmin: Sphalerite phase equilibria in the system Fe–Zn–S between 580°C and 850°C, *Econ. Geol.*, vol. 58, pp. 1191–1192, 1963.

———— and ————: The electrum-tarnish method for the determination of the fugacity of sulphur in laboratory sulfide systems, *Geochim. Cosmochim. Acta*, vol. 28, pp. 619–640, 1964.

———— and ————: Phase relations involving sphalerite in the Fe–Zn–S system, *Econ. Geol.*, vol. 61, pp. 815–849, 1966.

Bastin, E. S.: The nickel-cobalt-native silver ore type, *Econ. Geol.*, vol. 34, pp. 1–40, 1939.

Bateman, A. M.: Silver-lead deposits of Slocan District, British Columbia, *Econ. Geol.*, vol. 20, pp. 554–572, 1925.

————: "Economic Mineral Deposits," 2d ed., John Wiley & Sons, Inc., New York, 1950.

————: The formation of late magmatic oxide ores, *Econ. Geol.*, vol. 46, pp. 404–426, 1951.

———— and S. G. Lasky: Covellite-chalcocite solid solution and ex-solution, *Econ. Geol.*, vol. 27, pp. 52–86, 1932.

Bateman, J. D.: Uranium bearing auriferous reefs at Jacobina, Brazil, *Econ. Geol.*, vol. 53, pp. 417–425, 1958.

Bauer, H. L., R. A. Breitrich, J. J. Cooper, and J. A. Anderson: Porphyry copper deposits in the Robinson Mining District, Nevada, in S. R. Titley and C. L. Hicks (eds.), "The Geology of the Porphyry Copper Deposits: Southwestern North America," pp. 233–244, University of Arizona Press, Tucson, 1966.

Bear, L. M.: The Mineral Resources and Mining Industry of Cyprus Bull. no. 1, 1963.

Beck, P. A.: Annealing of cold-worked metals, *Advan. Phys.*, vol. 3, pp. 245–324, 1954.

Bennett, E. M.: Lead-zinc-silver and copper ore deposits of Mount Isa, in J. McAndrew (ed.), "Geology of Australian Ore Deposits," pp. 233–246, Eighth Commonwealth Mining and Metallurgical Congress, Australia and New Zealand, 1965.

Bergenfelt, S.: Om förekomsten av selan: Skelleftefältets sulfidmalmer, *Geol. Fören. Stockholm Förh.*, vol. 75, pp. 327–359, 1953.

Berry, L. G.: Studies of mineral sulpho-salts: IV–Galeno-bismutite and "lillianite," *Amer. Mineralogist*, vol. 25, pp. 726–734, 1940.

———— and B. Mason: "Mineralogy: Concepts, Descriptions and Determinations," W. H. Freeman and Company, San Francisco, 1959.

Beyschlag, F., J. H. L. Vogt, and P. Krusch (trans. S. J. Truscott), "The Deposits of the Useful Minerals and Rocks," vol. 2, pp. 515–1249, Macmillan & Co., Ltd., London, 1916.

Bilibin, Yu. A.: "Metallogenic Provinces and Metallogenic Epochs," pp. 1–35, Queens College Press, Flushing, New York, 1968.

Bischoff, J. L.: Red Sea geothermal brine deposits: their mineralogy, chemistry and genesis, in E. T. Degens and D. A. Ross (eds.), "Hot Brines and Recent Heavy Metal Deposits in the Red Sea," pp. 368–401, Springer-Verlag New York Inc., New York, 1969.

Boldt, J. R., and P. Queneau: "The Winning of Nickel: Its Geology, Mining and Extractive Metallurgy," The International Nickel Company of Canada, Limited and Methuen & Co., Ltd., London, 1967.

Bonatti, E., and O. Joensuu: Deep-sea iron deposit from the South Pacific, *Science*, vol. 154, pp. 643–645, 1966.

Borchert, H.: Ueber Entmischungen im System Cu–Fe–S und ihre Bedeutung als "geologische Thermometer," *Chemie Erde*, vol. 9, pp. 145–172, 1934.

———: Genesis of marine sedimentary iron ores, *Trans. Inst. Mining Met.*, vol. 69, pp. 261–279, 1960.

Bose, P. N.: Geology of the Lower Narbada Valley between Nimáwar and Káwant, *Mem. Geol. Surv. India*, Mem. 21, pp. 1–72, 1884.

Bostrom, K., and H. T. Peterson: Precipitates from hydrothermal exhalations on the East Pacific Rise, *Econ. Geol.*, vol. 61, pp. 1258–1265, 1966.

Both, R. A., and K. L. Williams: Mineralogical zoning in the lead-zinc ores of the Zeehan Field, Tasmania. Part II: Paragenetic and zonal relationships, *J. Geol. Soc. Australia*, vol. 15, pp. 217–243, 1968.

Boue, A.: Mémoire géologique sur l'Allemagne, *J. de Physique*, 1822.

Bowen, N. L.: The broader story of magmatic differentiation, briefly told, in "Ore Deposits of the Western States," Lindgren Volume, pp. 106–128, American Institute of Mining and Metallurgical Engineers, New York, 1933.

Brandes, H.: Zur Theorie des Kristallwachstums, *Z. phys. Chem.*, vol. 126, pp. 196–210, 1927.

Brathwaite, R. L.: The geology of the Rosebery ore deposits, Ph.D. thesis, University of Tasmania, 1969.

Bravais, M. A.: Etudes crystallographiques, *J. de l'Ecole Polytechnique*, vol. 167, 1851.

Bray, J. M.: Ilmenite-hematite-magnetite relations in some emery ores, *Amer. Mineralogist*, vol. 24, pp. 162–170, 1939.

Breislak, S.: "Introduzione alla Geologia," 1811.

Brögger, W. C.: Die Eruptivgesteine des Kristianiagebiefes, IV. Das Fengebeit in Telemark, *Norsk Vidensk. Selsk. Skriftu. I, Math. Nat. Kl.* (1920) no. 9, pp. 1–408, 1921.

Brown, A. S.: Mineralization in British Columbia and the copper and molybdenum deposits, *Trans. Can. Inst. Mining Met.*, vol. 72, pp. 1–15, 1969.

Brown, G. M.: The layered ultrabasic rocks of Rhum, Inner Hebrides, *Phil. Trans. Roy. Soc. London, Ser. B*, vol. 240, pp. 1–53, 1956.

Brown, J. S.: "Ore Genesis: A Metallurgical Interpretation, an Alternative to the Hydrothermal Theory," Thomas Murby & Co., London, 1950.

———: Oceanic lead isotopes and ore genesis, *Econ. Geol.*, vol. 60, pp. 47–68, 1965.

———: Isotopic zoning of lead and sulphur in Southeast Missouri, in J. S. Brown (ed.), "Genesis of Stratiform Lead-Zinc-Barite-Fluorite Deposits," *Econ. Geol. Monograph* no. 3, pp. 410–426, 1967.

Buckley, H. E.: "Crystal Growth," John Wiley & Sons, Inc., New York, 1951.

Buddington, A. F.: Correlation of kinds of igneous rocks with kinds of mineralization, in "Ore Deposits of the Western States," Lindgren Volume, pp. 350–385, American Institute of Mining and Metallurgical Engineers, New York, 1933.

——— and D. H. Lindsley: Iron-titanium oxide minerals and synthetic equivalents, *J. Petrol.*, vol. 5, pp. 310–357, 1964.

Buerger, M. J.: The plastic deformation of ore minerals, *Amer. Mineralogist*, vol. 13, pp. 1–17 and pp. 35–51, 1928.

———: Translation-gliding in crystals, *Amer. Mineralogist*, vol. 15, pp. 45–64, 1930.

———: Translation-gliding in crystals of the NaCl structural type, *Amer. Mineralogist*, vol. 15, pp. 174–187, 1930.

———: The significance of "block structure" in crystals, *Amer. Mineralogist*, vol. 17, pp. 177–191, 1932.

————: A common orientation and a classification of crystals based upon a marcasite-like packing, *Amer. Mineralogist*, vol. 22, pp. 48–56, 1937.

————: The relative importance of the several faces of a crystal, *Amer. Mineralogist*, vol. 32, pp. 593–606, 1947.

Buerger, N. W.: The unmixing of chalcopyrite from sphalerite, *Amer. Mineralogist*, vol. 19, pp. 525–530, 1934.

Bunn, C. W.: Adsorption, oriented overgrowth and mixed crystal formation, *Proc. Roy. Soc. London, Ser. A*, vol. 141, pp. 567–593, 1933.

Burst, J. F.: Mineral heterogeneity in "glauconite" pellets, *Amer. Mineralogist*, vol. 43, pp. 481–497, 1958.

Bushendorf, F., H. Nielsen, H. Puchelt, and W. Ricke: Schwefel-Isotopen-Untersuchungen am Pyrit-Sphalerit-Baryt-Lager Meggen/Lenne (Deutschland) und an verschiedenen Devon-Evaporiten, *Geochim. Cosmochim. Acta*, vol. 27, pp. 501–523, 1963.

Cabrera, N., and R. V. Coleman: Theory of crystal growth from the vapor, in J. J. Gilman (ed.), "The Art and Science of Growing Crystals," pp. 3–28, John Wiley & Sons, Inc., New York, 1963.

Cabri, L. J.: Note on the occurrence of calaverite and petzite in the Phantom lode, Great Boulder Mine, Kalgoorlie, *Proc. Australasian Inst. Mining Met.*, no. 222, p. 95, 1967.

Callahan, W. H.: Some spatial and temporal aspects of the localization of Mississippi Valley-Appalachian type ore deposits, in J. S. Brown (ed.), "Genesis of Stratiform Lead-Zinc-Barite-Fluorite Deposits," *Econ. Geol. Monograph* no. 3, pp. 14–19, 1967.

Callow, K. J., and B. W. Worley: The occurrence of telluride minerals at the Acupan Gold Mine, Mountain Province, Philippines, *Econ. Geol.*, vol. 60, pp. 251–268, 1965.

Cameron, E. N.: "Ore Microscopy," John Wiley & Sons, Inc., New York, 1961.

————: Structure and rock sequences of the critical zone of the eastern Bushveld Complex, *Mineralog. Soc. Amer., Spec. Pap. 1*, pp. 93–107, 1963.

————: Chromite deposits of the Eastern part of the Bushveld Complex, in S. H. Haughton (ed.), "The Geology of Some Ore Deposits in Southern Africa," vol. II, pp. 131–168, The Geological Society of South Africa, Johannesburg, 1964.

Campbell, N.: Tectonics, reefs and stratiform lead-zinc deposits of the Pine Point area, Canada, in J. S. Brown (ed.), "Genesis of Stratiform Lead-Zinc-Barite-Fluorite Deposits," *Econ. Geol. Monograph* no. 3, pp. 59–70, 1967.

Carpenter, H. C. H., and M. S. Fisher: A metallographic investigation of native silver, *Trans. Inst. Mining Met.*, vol. 41, pp. 382–403, 1932.

Carroll, D.: The role of clay minerals in the transportation of iron, *Geochim. Cosmochim. Acta*, vol. 14, pp. 1–28, 1958.

Carruthers, D. S.: An environmental view of Broken Hill ore occurrence, in J. McAndrew (ed.), "Geology of Australian Ore Deposits," pp. 339–351, Eighth Commonwealth Mining and Metallurgical Congress, Australia and New Zealand, 1965.

———— and R. D. Pratten, The stratigraphic succession and structure in the Zinc Corporation Ltd. and New Broken Hill Consolidated Ltd., Broken Hill, N.S.W., *Econ. Geol.*, vol. 56, pp. 1088–1102, 1961.

Carstens, C. W.: Oversigt over Trondhjemsfeltets bergbygning, *Norske Vidensk. Selsk. Förh.*, 1919.

————: Die Kiesvorkommens im Porsangergebiet, *Norsk Geol. Tidsskr.*, vol. 12, pp. 171–177, 1931.

————: Zur Geochemie einiger norwegischen Kiesvorkommen, *Norske Vidensk. Selsk. Förh.*, vol. 14, no. 10, p. 36, 1941.

Castano, J. R., and R. M. Garrels: Experiments on the deposition of iron with special reference to the Clinton iron ore deposits, *Econ. Geol.*, vol. 45, pp. 755–770, 1950.

Chadwick, C. A.: Eutectic alloy solidification, *Progr. Mater. Sci.*, vol. 12, pp. 97–182, 1963.

Chalmers, B.: The influence of the difference of orientation of two crystals on the mechanical effect of their boundary, *Proc. Roy. Soc. London, Ser. A*, vol. 162, pp. 120–127, 1937.

Chapman, E. P., Jr.: Geology of the Brenda molybdenum copper deposit, *Bull. Can. Inst. Mining Met.*, 70th Annual General Meeting, Vancouver, 1968.

Chinner, G. A.: Pelitic gneisses with varying ferrous/ferric ratios from Glen Clova, Angus, Scotland, *J. Petrol.*, vol. 1, pp. 178–217, 1960.

Chukhrov, F. V., V. I. Vinogradov, and L. P. Ermilova: On the isotopic sulfur composition of some Precambrian strata, *Miner. Deposita*, vol. 5, pp. 209–222, 1970.

Clark, L. A.: Variations in stratiform sulfide ores as functions of differing conditions of metamorphism, Unpublished paper, Department of Geological Sciences, McGill University, 1967.

——— and G. Kullerud: The sulfur-rich portion of the Fe–Ni–S system, *Econ. Geol.*, vol. 58, pp. 853–885, 1963.

Clarke, F. W.: The data of geochemistry, 5th ed., *U.S. Geol. Surv. Bull.* 770, 1924.

Cloud, P. E., Jr.: Significance of the Gunflint (Precambrian) flora, *Science*, vol. 148, pp. 27–35, 1965.

Coche, L., D. Dastillon, M. Deudon, and P. Emery: Complements à l'étude du bassin ferrifère de Lorraine, Le Bassin de Landres-Amermont, Centre documentution sidérurgique, Paris, 1954.

Cochrane, C. W., and A. B. Edwards: The Roper River oolitic ironstone formations, *Commonwealth Scient. Ind. Res. Org.*, *Aust.*, *Mineragraphic Investigations, Tech. Pap.* 1, 1960.

Coleman, R. C., and M. Delevaux: Occurrence of selenium in sulfides from some sedimentary rocks of the Western United States, *Econ. Geol.*, vol. 52, pp. 499–527, 1957.

Collins, W. H., T. T. Quirke, and E. Thomson: Michipicoten Iron Ranges, *Mem. Geol. Surv. Can.*, Mem. 147, 1926.

Colony, R. J.: Ore mineral sequence, in E. E. Fairbanks (ed.), "The Laboratory Examination of Ore Minerals," New York, 1928.

Cooper, J. A., and J. R. Richards: Lead Isotope measurements on sediments from Atlantis II and Discovery Deep Areas, in E. T. Degens and D. A. Ross (eds.), "Hot Brines and Recent Heavy Metal Deposits in the Red Sea," pp. 499–511, Springer-Verlag New York Inc., New York, 1969.

Cotton, R. E.: H.Y.C. lead-zinc-silver ore deposit, MacArthur River, in J. McAndrew (ed.), "Geology of Australian Ore Deposits," pp. 197–200, Eighth Commonwealth Mining and Metallurgical Congress, Australia and New Zealand, 1965.

Cousins, C. A.: The platinum deposits of the Merensky Reef, in S. H. Haughton (ed.), "The Geology of Some Ore Deposits in Southern Africa," vol. II, pp. 225–238, The Geological Society of South Africa, Johannesburg, 1964.

Cox, D. P.: Regional environment of the Jacobina auriferous conglomerate, *Econ. Geol.*, vol. 62, pp. 773–780, 1967.

Creasey, S. C.: Hydrothermal alteration, in S. R. Titley and C. L. Hicks (eds.), "The Geology of the Porphyry Copper Deposits, Southwestern North America," pp. 51–74, University of Arizona Press, Tucson, 1966.

Crook, T.: The genetic classification of ore deposits, *Mineral. Mag.*, vol. 17, pp. 55–85, 1914.

———: "History of the Theory of Ore Deposits," Thomas Murby & Co., London, 1933.

Croxford, N. J. W.: "The Mineralogy of the No. 7 Lead-Zinc Ore Body, Mt. Isa, and Its Interpretation," Ph.D. thesis, University of New England, New South Wales, 1962.

———: A mineralogical examination of the MacArthur lead-zinc-silver deposit, *Proc. Australasian Inst. Mining Met.*, no. 226, pp. 97–108, 1968.

Cullen, D. J.: Tectonic implications of banded ironstone formations, *J. Sed. Petrol.*, vol. 33, pp. 387–392, 1963.

Czamanske, C. K.: In E. Roedder, Report on S.E.G. symposium on the chemistry of the ore-forming fluids, August–September, 1964, *Econ. Geol.*, vol. 60, pp. 1380–1403, 1965.

———, E. Roedder, and F. C. Burns: Neutron activation analysis of fluid inclusions for copper, manganese and zinc, *Science*, vol. 140, pp. 401–403, 1963.

Daly, R. A.: "Igneous Rocks and the Depths of the Earth," McGraw-Hill Book Company, New York, 1933.

Dana, J. W.: "The System of Mineralogy of James Dwight Dana and Edward Salisbury Dana, Yale University, 1837–1892," 7th ed., entirely rewritten and greatly enlarged by C. Palache, H. Berman, and C. Frondel, vol. I, John Wiley & Sons, Inc., New York, 1944.

Darken, L. S., and R. W. Gurry: "Physical Chemistry of Metals," McGraw-Hill Book Company, New York, 1953.

Davidson, C. F.: Oolitic ironstones of fresh-water origin, *Mineral. Mag.*, vol. 104, pp. 158–159, 1961. (Review of A. L. Yanitsky, Oligocene oolitic iron ores of northern Turgai and their genesis, *Tr. Inst. Geol. Rudn. Mestorozhd., Petrofr., Mineralog. i Geokhim.*, Trudy, vol. 37, 1960.)

Davies, R.: Experimental investigation of chalcocite: annealing and plastic deformation at elevated temperatures, *Can. J. Earth Sci.*, vol. 2, pp. 98–117, 1965.

Davis, G. R.: The origin of the Roan Antelope Copper Deposit of Northern Rhodesia, *Econ. Geol.*, vol. 49, pp. 575–615, 1954.

Dawson, J. B.: Oldoinyo Lengai—an active volcano with sodium carbonatite lava flows, in O. F. Tuttle and J. Gittins (eds.), "Carbonatites," pp. 155–168, Interscience Publishers, Inc., New York, 1966.

Deans, T.: Economic mineralogy of African carbonatites, in O. F. Tuttle and J. Gittins (eds.), "Carbonatites," pp. 385–413, Interscience Publishers, Inc., New York, 1966.

De Beaumont, E.: Note sur les emanations volcaniques et metallifères, *Bull. Soc. Geol. France*, vol. 4, p. 1249 (Séance du 5 Juillet), 1847.

Deer, W. A., R. A. Howie, and J. Zussman: "Rock-Forming Minerals," vol. 3, "Sheet Silicates," Longmans, Green & Co., Ltd., London, 1962.

Degens, E. T., and D. A. Ross (eds.): "Hot Brines and Recent Heavy Metal Deposits in the Red Sea," Springer-Verlag New York Inc., New York, 1969.

Delahaye, P., M. Pourbaix, and P. Van-Rysselberche: Diagrammes d'équilibre potential—pH de quelques eléments, *Compt. Rend. Troiszieme Réunion Comité thermodynam. Cinét. Electrochim. Berne 1951*, pp. 15–19, 1952.

Denholm, L. S.: Lode structures and ore shoots at Vatukoula, Fiji, *Proc. Australasian Inst. Mining Met.*, no. 222, pp. 73–84, 1967.

Derry, D. R.: Evidence of the origin of the Blind River uranium deposits, *Econ. Geol.*, vol. 55, pp. 906–927, 1960.

Desborough, G. A.: The significance of accessory magmatic sphalerite in basic rocks to the origin of nickeliferous pyrrhotite ores, *Econ. Geol.*, vol. 61, pp. 370–375, 1966.

Dewey, H.: The mineral zones of Cornwall, *Proc. Geologists' Assoc. (Engl.)*, vol. 36, pp. 107–134, 1925.

Dietz, R. S.: Sudbury structure as an astrobleme, *J. Geol.*, vol. 72, pp. 412–434, 1964.

Djurle, S.: An x-ray study on the system Cu–S, *Acta Chem. Scand.*, vol. 12, pp. 1415–1426, 1958.

Donnay, J. D. H., and D. Harker: A new law of crystal morphology extending the law of Bravais, *Amer. Mineralogist*, vol. 22, pp. 446–467, 1937.

Dorr, J. V., I. Soares, and A. Horen: The manganese deposits of Minas Gerais, Brazil, *Int. Geol. Congr., 20th, Mexico, Manganese Symp.*, vol. III, pp. 279–346, 1956.

Dunham, K. C.: Neptunist concepts in ore genesis, *Econ. Geol.*, vol. 59, pp. 1–21, 1964.

Edwards, A. B.: Some ilmenite micro-structures and their interpretation, *Proc. Australasian Inst. Mining Met.*, no. 110, pp. 39–58, 1938.

———: A note on some tantalum-niobium minerals from Western Australia, *Proc. Australasian Inst. Mining Met.*, no. 120, pp. 731–744, 1940.

———: The copper deposits of Australia, *Proc. Australasian Inst. Mining Met.*, no. 130, pp. 105–171, 1943.

———: Solid solution of tetrahedrite in chalcopyrite and bornite, *Proc. Australasian Inst. Mining Met.*, no. 143, pp. 141–155, 1946.

———: Mineralogy of the Middleback iron ores, in A. B. Edwards (ed.), "Geology of Australian Ore Deposits," pp. 464–472, Fifth Empire Mining and Metallurgical Congress, Australia and New Zealand, 1953.

————: "Textures of the Ore Minerals and Their Significance," Australasian Institute of Mining and Metallurgy, Melbourne, 1947; revised ed. 1954.

————: The composition of the Peko Copper Orebody, Tennant Creek, *Proc. Australasian Inst. Mining Met.*, no. 175, pp. 55–82, 1955.

———— and G. C. Carlos: The selenium content of some Australian sulphide deposits, *Proc. Australasian Inst. Mining Met.*, no. 172, pp. 32–63, 1954.

Ehrenberg, C. G.: Vorläufige Mittheilungen ueber das wirklige Vorkommen fossiler Infusorien und ihre grosse Verbreitung, *Poggendorff's Annalen*, band 38, pp. 213–227, 1836.

Ehrenberg, H., A. Pilger, and F. Schröder: Das Schwefelkies-Zinkblende-Schwerspatlager von Meggen (Westfalen); (Hannover: Niedersächsisches Landesamt für Bodenforschung), *Monographien der Deutschen Blei-Zink-Erzlagerstätten*, no. 7, 1954.

Ellison, S. P.: Economic applications of paleoecology, *Econ. Geol. 50th Anniv. Vol.*, pp. 867–884, 1955.

Emmons, W. H.: Some regionally metamorphosed ore deposits and the so-called segregated veins, *Econ. Geol.*, vol. 4, pp. 755–781, 1909.

————: On the mechanism of the deposition of certain metalliferous lode systems associated with granitic batholiths, in "Ore Deposits of the Western States," Lindgren Volume, pp. 327–349, American Institute of Mining and Metallurgical Engineers, New York, 1933.

Erd, R. C., H. T. Evans, Jr., and D. H. Richter: Smythite, a new iron sulfide, and associated pyrrhotite from Indiana, *Amer. Mineralogist*, vol. 42, pp. 309–333, 1957.

Evans, R. C.: "An Introduction to Crystal Chemistry," 2d ed., University Press, Cambridge, 1964.

Fenner, C. N.: Pneumatolytic processes in the formation of minerals and ores, in "Ore Deposits of the Western States," Lindgren Volume, pp. 58–105, American Institute of Mining and Metallurgical Engineers, New York, 1933.

Fermor, L. L.: The manganese ore deposits of India, *Mem. Geol. Surv. India, Mem.* 37, 1909.

Finch, J. W.: Sedimentary copper deposits of the Western States, in "Ore Deposits of the Western States," Lindgren Volume, pp. 481–487, American Institute of Mining and Metallurgical Engineers, New York, 1933.

Findlay, A.: "The Phase Rule and Its Applications," 9th ed., revised and enlarged by A. N. Campbell and N. O. Smith, Dover Publications, Inc., New York, 1951.

Fleischer, M. F., and W. E. Richmond: The manganese oxide minerals: a preliminary report, *Econ. Geol.*, vol. 38, pp. 269–286, 1943.

Fletcher, N. H.: Heterogeneous nucleation of ice crystals, *J. Aust. Inst. Metals*, vol. 10, pp. 101–105, 1965.

Folinsbee, R. E., R. Krouse, and A. Sasaki: Sulfur isotopes and the Pine Point lead-zinc deposits, *Econ. Geol.*, vol. 61, p. 1307, 1966.

Foslie, S.: Discussion on lecture by Th. Vogt, *Norsk Geol. Tidsskr.*, vol. 17, pp. 214–216, 1938.

Frank, F. C., and H. H. Wills: Crystal growth and dislocations, *Advan. Phys.*, vol. 1, pp. 91–109, 1952.

Freeze, A. C.: On the origin of the Sullivan orebody, Kimberley, B.C., *Can. Inst. Mining Spec. Vol.* 8, pp. 263–294, 1966.

French, B. M.: Sudbury structure, Ontario: Some petrographic evidence for an origin by meteorite impact, *Meteoritic Soc.*, 1966.

Friedman, G. M.: On the uranium-thorium ratio in the Blind River, Ontario, uranium-bearing conglomerate, *Econ. Geol.*, vol. 53, pp. 889–890, 1958.

Frondel, C.: Exsolution growths of zincite in manganosite and of manganosite in periclase, *Amer. Mineralogist*, vol. 25, pp. 534–538, 1940.

Fyfe, W. S.: Isomorphism and bond type, *Amer. Mineralogist*, vol. 36, pp. 538–542, 1951.

————: "Geochemistry of Solids: An Introduction," McGraw-Hill Book Company, New York, 1964.

Garlick, W. G.: The Syngenetic theory, in F. Mendelsohn (ed.), "Geology of the Northern Rhodesian Copperbelt," pp. 146–165, Roan Antelope Copper Mines Limited and MacDonald & Co. (Publishers) Ltd., London, 1961.

————: Association of mineralization and algal reef structures on Northern Rhodesian Copperbelt, Katanga, and Australia, *Econ. Geol.*, vol. 59, pp. 416–427, 1964.

Garrels, R. M.: "Mineral Equilibria at Low Temperature and Pressure," Harper & Brothers, New York, 1960.

Gill, J. E.: Experimental deformation and annealing of sulfides and interpretation of ore textures, *Econ. Geol.*, vol. 64, pp. 500–508, 1969.

Girifalco, L. A.: "Atomic Migration in Crystals," Blaisdell Publishing Company, New York, 1964.

Gjelsvik, T.: Geochemical and mineralogical investigations of titaniferous iron ores, West Coast of Norway, *Econ. Geol.*, vol. 52, pp. 482–498, 1957.

Gold, D. P.: Average chemical composition of carbonatites, *Econ. Geol.*, vol. 58, p. 988, 1963.

Goldschmidt, V. M.: In A. Muir (ed.), "Geochemistry," Oxford University Press, London, 1954.

———— and L. Strock: Zur Geochemie der Selen. II, *Nachr. Ges. Wiss. Gottingen, math-nat. Kl.*, vol. 1, pp. 123–142, 1935.

Goodspeed, C. E.: Mineralization related to granitization, *Econ. Geol.*, vol. 47, pp. 146–168, 1952.

Goodwin, A. M.: Facies relations in the Gunflint iron formation, *Econ. Geol.*, vol. 51, pp. 565–595, 1956.

————: Some aspects of Archean structure and mineralization, *Econ. Geol.*, vol. 56, pp. 897–915, 1961.

————: Structure, stratigraphy and origin of iron formations, Michipicoten area, Algoma District, Ontario, Canada, *Geol. Soc. Amer. Bull.*, vol. 73, pp. 561–586, 1962.

————: Geochemical studies at the Helen Iron Range, *Econ. Geol.*, vol. 59, pp. 684–718, 1964.

————: Mineralized volcanic complexes in the Porcupine–Kirkland Lake–Noranda region, Canada, *Econ. Geol.*, vol. 60, pp. 955–971, 1965.

————: Archean protocontinental growth and mineralization, *Can. Mining J.*, vol. 87, pp. 57–60, 1966.

Govett, G. J. S.: Origin of banded iron formations, *Bull. Geol. Soc. Amer.*, vol. 77, pp. 1191–1212, 1966.

Gray, A.: The correlation of the ore-bearing sediments of the Katanga and Rhodesian copper belt, *Econ. Geol.*, vol. 25, pp. 783–804, 1930.

Green, J., and A. Poldervaart: Some basaltic provinces, *Geochim. Cosmochim. Acta*, vol. 7, pp. 177–188, 1955.

Greig, J. W., E. Posnjak, H. E. Merwin, and R. B. Sosman: Equilibrium relationships of Fe_3O_4, Fe_2O_3 and oxygen, *Amer. J. Sci.*, vol. 230, pp. 239–316, 1935.

Grigor'yev, D. P.: Three types of plastic deformation in galena, *Mineralog. Sb., L'vovsk. Geol. obshchest.*, vol. 12, pp. 129–143, 1958. (Russian text; English translation by Royer & Royer Inc.)

Grogan, R. M., and J. C. Bradbury: Origin of the stratiform fluorite deposits of Southern Illinois, in J. S. Brown (ed.), "Genesis of Stratiform Lead-Zinc-Barite-Fluorite Deposits," *Econ. Geol. Monograph* no. 3, pp. 40–51, 1967.

Grondijs, H. F., and C. Schouten: A study of the Mount Isa ores, *Econ. Geol.*, vol. 32, pp. 407–450, 1937.

Gross, G. A.: Geology of iron deposits in Canada, vols. I, II, and III, *Economic Geology Report* no. 22, Geological Survey of Canada, 1967.

Gross, K. A.: X-ray line broadening and stored energy in deformed and annealed calcite, *Phil. Mag.*, vol. 12, pp. 801–813, 1965.

Gross, W. H.: Evidence for a modified placer origin for auriferous conglomerates, Canavieiras Mine, Jacobina, Brazil, *Econ. Geol.*, vol. 63, pp. 271–276, 1968.

Gruner, J. W.: Organic matter and the origin of the Biwabik Iron-bearing Formation of the Mesabi Range, *Econ. Geol.*, vol. 17, pp. 407–460, 1922.

————: Crystal structure types, *Amer. Mineralogist*, vol. 14, pp. 173–187, 1929.

————: Structures of sulphides and sulphosalts, *Amer. Mineralogist*, vol. 14, pp. 470–481, 1929.

————: The structure and chemical composition of greenalite, *Amer. Mineralogist*, vol. 21, pp. 405–425, 1936.

————: The composition and structure of minnesotaite, a common iron silicate in iron formations, *Amer. Mineralogist*, vol. 29, pp. 363–372, 1944.

————: "Mineralogy and Geology of the Mesabi Range," St. Paul, Minnesota, Iron Range Resources and Rehabilitation Comm., 1946.

————: Concentration of uranium in sediments by multiple migration-accretion, *Econ. Geol.*, vol. 51, pp. 495–520, 1956.

————, W. G. Fetzer, and I. Rapaport: The uranium deposits near Marysvale, Piute County, Utah, *Econ. Geol.*, vol. 46, pp. 243–251, 1951.

Guild, F. N.: A microscopic study of the silver ores and their associated minerals, *Econ. Geol.*, vol. 12, pp. 297–353, 1917.

Guinier, A.: Substructures in crystals, in W. Shockley (ed.), "Imperfections in Nearly Perfect Crystals," pp. 402–440, John Wiley & Sons, Inc., New York, 1952.

Hall, A. L.: The Bushveld igneous complex of the Central Transvaal, *Mem. Geol. Surv. Div. Un. S. Afr.*, Mem. 28, 1932.

Hallimond, A. F.: Iron ores–Bedded ores of England and Wales—Petrography and Chemistry, *Mem. Geol. Surv. U.K. Spec. Rep. Mineral Resources*, vol. 29, 1925.

Hammond, P.: Allard Lake ilmenite deposits, *Econ. Geol.*, vol. 47, pp. 634–649, 1952.

Hanson, G.: Manganese in Canada, *Int. Geol. Congr., 20th, Mexico, Manganese Symp.*, vol. III, pp. 9–14, 1956.

Harder, E. C.: Iron-depositing bacteria and their geologic relations, *U.S. Geol. Surv. Prof. Pap.* 113, 1919.

Harder, H.: Beitrag zur Petrographie und Genese der Hämatiterze des Lahn-Dill Gebietes, *Heidelberger Beitr. Mineral. Petrog.*, vol. 4, nos. 1–2, pp. 54–66, 1954.

Hargraves, R. B.: Silver-gold ratios in some Witwatersrand conglomerates, *Econ. Geol.*, vol. 58, pp. 952–970, 1963.

Harker, A.: "Metamorphism," Methuen & Co., Ltd., London, 2d ed., 1939.

Harris, J. F.: Summary of the geology of Tanganyika. pt. IV: Economic Geology, *Mem. Geol. Surv. Tanganyika*, Mem. 1, 1961.

Harvey, R. D., and C. J. Vitaliano: Wall rock alteration in the Goldfield district, Nevada, *J. Geol.*, vol. 72, pp. 564–579, 1964.

Haughton, S. H. (ed.): "The Geology of Some Ore Deposits in Southern Africa," vol. II, The Geological Society of South Africa, Johannesburg, 1964.

Hawley, J. E.: The Sudbury ores; Their mineralogy and origin, *Can. Mineralogist*, vol. 7, pt. 1, 1962.

———— and D. F. Hewitt: Pseudo-eutectic and pseudo-exsolution intergrowths of nickel arsenides due to heat effects, *Econ. Geol.*, vol. 43, pp. 273–279, 1948.

———— and I. Nicol: Selenium in some Canadian sulphides, *Econ. Geol.*, vol. 54, pp. 608–628, 1959.

———— and R. L. Stanton: The facts: the ores, their minerals, metals and distribution, in The Sudbury ores: their mineralogy and origin, by J. E. Hawley, *Can. Mineralogist*, vol. 7, pt. 1, pp. 30–128, 1962.

————, R. L. Stanton, and A. Y. Smith: Pseudo-eutectic intergrowths in arsenical ores from Sudbury, *Can. Mineralogist*, vol. 6, pt. 5, pp. 555–575, 1961.

Hayes, A. O.: The Wabana ores of Newfoundland, *Mem. Can. Geol. Surv.*, Mem. 66, 1915.

Hegeman, F.: Über Sedimentäre Lagerstätten mit Submarines Vulkanischen Stoffzufuhr, *Fortschr. Mineral.*, vol. 27, pp. 54–55, 1950.

Heinrich, E. W.: "Mineralogy and Geology of Radioactive Raw Materials," McGraw-Hill Book Company, New York, 1958.

————: "The Geology of Carbonatites," Rand McNally & Company, Chicago, 1966.

Helgeson, H. C.: "Complexing and Hydrothermal Ore Deposition," Pergamon Press, New York, 1964.

Hess, F. L.: Uranium, vanadium, radium, gold, silver and molybdenum sedimentary deposits, in "Ore Deposits of the Western States," Lindgren Volume, pp. 450–480, American Institute of Mining and Metallurgical Engineers, New York, 1933.

Hewitt, D. F.: Stratified deposits of the oxides and carbonates of manganese, *Econ. Geol.*, vol. 61, pp. 431–461, 1966.

———— and M. Fleischer: Deposits of the manganese oxides, *Econ. Geol.*, vol. 55, pp. 1–55, 1960.

Hewitt, R. L., and G. M. Schwartz: Experiments bearing on the relation of pyrrhotite to other sulphides, *Econ. Geol.*, vol. 32, p. 1070, 1937.

Hey, M. H.: A new review of the chlorites, *Mineral. Mag.*, vol. 30, pp. 277–292, 1954.

Heyl, A. V.: Minor epigenetic, diagenetic and syngenetic sulfide, fluorite, and barite occurrences in the Central United States, *Econ. Geol.*, vol. 63, pp. 585–594, 1968.

Hitchcock, E.: "Geology of Massachusetts," 1833 (from W. H. Winchell and H. V. Winchell: Iron Ores of Minnesota, *Bull. Geol. Nat. Hist. Surv. Minn.*, vol. 6, 1891).

Hobbs, B. E.: Recrystallization of single crystals of quartz, *Tectonophysics*, vol. 6, pp. 353–401, 1968.

————, D. M. Ransome, R. H. Vernon, and P. F. Williams: The Broken Hill Orebody, Australia. A review of recent work, *Miner. Deposita*, vol. 3, pp. 293–316, 1968.

Hoefs, J., H. Nielsen, and M. Schidlowski: Sulfur isotope abundances in pyrite from Witwatersrand conglomerates, *Econ. Geol.*, vol. 63, pp. 975–977, 1968.

Högbom, A. C.: Uber das nephelinsyenit auf der Insel Alnö, *Geol. Fören. Stockholm Förh.*, vol. 17, pp. 100–160 and 214–256, 1895.

Holland, H. D.: Gangue minerals in hydrothermal deposits, in H. L. Barnes (ed.), "Geochemistry of Hydrothermal Ore Deposits," pp. 382–436, Holt, Rinehart and Winston, Inc., New York, 1967.

Hollingworth, S. E., and J. H. Taylor: The Northampton sand ironstone—stratigraphy, structure and reserves, *Mem. Geol. Surv. U.K.*, 1951.

Horikoshi, E.: Volcanic activity related to the formation of the Kuroko-type deposits in the Kosaka district, Japan, *Miner. Deposita*, vol. 4, pp. 321–345, 1969.

Hough, J. L.: Fresh-water environment of deposition of Precambrian banded iron-formations, *J. Sed. Petrol.*, vol. 28, pp. 414–430, 1958.

Hu, H., and C. S. Smith: The formation of low-energy interfaces during grain growth in alpha and alpha-beta brasses, *Acta Met.*, vol. 4, pp. 638–646, 1956.

Huber, N. K.: Some aspects of the origin of the Ironwood Iron-Formation of Michigan and Wisconsin, *Econ. Geol.*, vol. 54, pp. 82–118, 1959.

Hunt, T. S.: The geognostical history of the metals, *Trans. Amer. Inst. Mining Engrs.*, vol. 1, pp. 331–342, 1873.

Hutchinson, R. W.: Genesis of Canadian massive sulphides reconsidered by comparison to Cyprus deposits, *Can. Mining Met. Bull.*, vol. 68, pp. 972–986, 1965.

Hutton, J.: Theory of the Earth, *Trans. Roy. Soc. Edinburgh*, 1788.

Jackson, E. D.: Primary textures and mineral associations in the ultramafic zone of the Stillwater Complex, Montana, *U.S. Geol. Surv. Prof. Pap. 358*, 1961.

————: Stratigraphic and lateral variation of chromite composition in the Stillwater Complex, *Mineral. Soc. Amer. Spec. Pap. 1*, pp. 46–54, 1963.

Jackson, S. A., and F. W. Beales: An aspect of sedimentary basin evolution: the concentration of Mississippi Valley-type ores during late stages of diagenesis, *Bull. Can. Petrol. Geol.*, vol. 15, pp. 383–433, 1967.

Jacobs, J. A., R. D. Russell, and J. T. Wilson: "Physics and Geology," McGraw-Hill Book Company, New York, 1959.

James, H. L.: Iron formation and associated rocks in the Iron River district, Michigan, *Bull. Geol. Soc. Amer.*, vol. 62, pp. 251–266, 1951.

————: Sedimentary facies of iron formation, *Econ. Geol.*, vol. 49, pp. 235–293, 1954.

————: Zones of regional metamorphism in the Precambrian of Northern Michigan, *Bull. Geol. Soc. Amer.*, vol. 66, pp. 1455–1488, 1955.

————: Chemistry of the iron-rich sedimentary rocks, *U.S. Geol. Surv. Prof. Pap.* 440, chap. W, 1966.

Jeffries, Z., and R. S. Archer: "Science of Metals," McGraw-Hill Book Company, New York, 1924.

Jensen, M. L.: Sulfur isotopes and the origin of sandstone type uranium deposits, *Econ. Geol.*, vol. 53, pp. 598–616, 1958.

————: Sulfur isotopes and hydrothermal mineral deposits, *Econ. Geol.*, vol. 54, pp. 374–394, 1959.

————: Sulfur isotopes and mineral genesis, in H. L. Barnes (ed.), "Geochemistry of Hydrothermal Ore Deposits," pp. 143–165, Holt, Rinehart and Winston, Inc., New York, 1967.

Kanasewich, E. R., and R. M. Farquhar: Lead isotope ratios from the Cobalt-Noranda area, Canada, *Can. J. Earth Sci.*, vol. 2, pp. 361–384, 1965.

Kanehira, K.: Geology and ore deposits of the Chihara Mine–Ehime Prefecture, Japan, *J. Fac. Sci. Univ. Tokyo Sect. II*, vol. 11, pp. 309–338, 1959.

Kaplan, I. R., K. O. Emery, and S. C. Rittenberg: The distribution and isotopic abundance of sulfur in recent marine sediments off Southern California, *Geochim. Cosmochim. Acta*, vol. 27, pp. 297–331, 1963.

———— and S. C. Rittenberg: Microbiological fractionation of sulphur isotopes, *J. Gen. Microbiol.*, vol. 34, pp. 195–212, 1964.

————, R. E. Sweeney, and A. Nissenbaum: Sulfur isotope studies on Red Sea geothermal brines and sediments, in E. T. Degens and D. A. Ross (eds.), "Hot Brines and Recent Heavy Metal Deposits in the Red Sea," pp. 474–498, Springer-Verlag New York Inc., New York, 1969.

Kemp, J. F.: The role of igneous rocks in the formation of veins, *Trans. Amer. Inst. Mining Engrs.*, vol. 31, pp. 169–197, 1901.

————: Ore deposits at the contacts of intrusive rocks and limestones; and their significance as regards the general formation of veins, *Econ. Geol.*, vol. 2, pp. 1–13, 1907.

Kennedy, G. C.: A portion of the system silica-water, *Econ. Geol.*, vol. 45, pp. 629–653, 1950.

Keys, M. R.: Paragenesis in Hollinger veins, *Econ. Geol.*, vol. 35, pp. 611–628, 1940.

King, H. F.: Notes on ore occurrences in highly metamorphosed Precambrian rocks, in "F. L. Stillwell Anniversary Volume," pp. 143–168, The Australasian Institute of Mining and Metallurgy, Melbourne, 1958.

———— and E. S. O'Driscoll: The Broken Hill Lode, in A. B. Edwards (ed.), "Geology of Australian Ore Deposits," pp. 578–600, Fifth Empire Mining and Metallurgical Congress, Australia and New Zealand, 1953.

———— and B. P. Thomson: Geology of the Broken Hill District, in A. B. Edwards (ed.), "Geology of Australian Ore Deposits," pp. 533–577, Fifth Empire Mining and Metallurgical Congress, Australia and New Zealand, 1953.

Kingery, W. D.: "Introduction to Ceramics," John Wiley & Sons, Inc., New York, 1960.

Kinkel, A. R., Jr.: The Ore Knob Massive Sulphide Deposit, North Carolina: An example of recrystallized ore, *Econ. Geol.*, vol. 57, pp. 1116–1122, 1962.

————: Massive pyritic deposits related to volcanism, and possible methods of emplacement, *Econ. Geol.*, vol. 61, pp. 673–694, 1966.

Klein, C.: Mineralogy and petrology of the metamorphosed Wabush Iron Formation, Southwestern Labrador, *J. Petrol.*, vol. 7, pp. 246–305, 1966.

Knight, C. L.: Ore genesis—the source bed concept, *Econ. Geol.*, vol. 52, pp. 808–817, 1957.

Knopf, A.: Petrology, *Geol. Soc. Amer. 50th Anniv. Vol.*, pp. 333–363, 1941.

Koschmann, A. H., and M. H. Bergendahl: Principal gold-producing districts of the United States, *U.S. Geol. Surv. Prof. Pap.* 610, pp. 1–283, 1968.

Kraume, E., F. Dahlgrun, P. Ramdohr, and A. Wilke: Die Erzlager des Rammelsberges bei Goslar, *Monographein der Deutschen Blei-Zink-Erzlagerstätten*, no. 8, 1955.

Krauskopf, K. B.: Dissolution and precipitation of silica at low temperatures, *Geochim. Cosmochim. Acta*, vol. 10, pp. 1–26, 1956.

———: The heavy metal content of magmatic vapor at 600°C, *Econ. Geol.*, vol. 52, pp. 786–807, 1957.

———: Separation of manganese from iron in sedimentary processes, *Geochim. Cosmochim. Acta*, vol. 12, pp. 61–84, 1957.

———: "Introduction to Geochemistry," McGraw-Hill Book Company, New York, 1967.

Krieger, P.: Bornite-klaprotholite relations at Conception del Oro, Mexico, *Econ. Geol.*, vol. 35, pp. 687–697, 1940.

Krumbien, W. C., and R. M. Garrels: Origin and classification of chemical sediments in terms of pH and oxidation-reduction potentials, *J. Geol.*, vol. 60, pp. 1–33, 1952.

Kullerud, G.: The FeS–ZnS system: a geological thermometer, *Norsk Geol. Tidsskr.*, vol. 32, pp. 61–147, 1953.

———: The upper stability curve of covellite, *Carnegie Inst. Wash., Ann. Rept. Dir. Geophys. Lab. Year Book* 56, pp. 195–197, 1957.

———: Phase relations in the Fe–S–O system, *Carnegie Inst. Wash., Ann. Rept. Dir. Geophys. Lab. Year Book* 56, pp. 198–200, 1957.

———: The Fe–Ni–S system, *Carnegie Inst. Wash., Ann. Rept. Dir. Geophys. Lab. Year Book* 62, pp. 175–189, 1963.

———: Review and evaluation of recent research on geologically significant sulphide-type systems, *Fortschr. Mineral.*, vol. 41, pp. 221–270, 1964.

——— and H. S. Yoder, Jr.: Pyrite stability relations in the Fe–S system, *Econ. Geol.*, vol. 54, pp. 533–572, 1959.

——— and ———: Sulfide-silicate relations, *Carnegie Inst. Wash., Ann. Rept. Dir. Geophys. Lab. Year Book* 62, pp. 215–218, 1963.

——— and ———: Sulfide-silicate relations, *Carnegie Inst. Wash., Ann. Rept. Dir. Geophys. Lab. Year Book* 63, pp. 218–222, 1964.

——— and R. A. Yund: Polydymite stability relations, *Carnegie Inst. Wash., Ann. Rept. Dir. Geophys. Lab. Year Book* 60, pp. 176–178, 1961.

——— and ———: The Ni–S system and related minerals, *J. Petrol.*, vol. 3, pp. 126–175, 1962.

La Bêrge, C. L.: Altered pyroclastic rocks in South African iron-formations, *Econ. Geol.*, vol. 61, pp. 572–581, 1966.

Lamey, C. A.: "Metallic and Industrial Mineral Deposits," McGraw-Hill Book Company, New York, 1966.

Lamplugh, G. W., C. B. Wedd, and J. Pringle: Iron ores–Bedded ores of the Lias, Oolites and later formations in England, *Mem. Geol. Surv. U.K., Spec. Rep. Mineral Resources*, vol. 24, 1920.

Lausen, C.: Graphic intergrowth of niccolite and chalcopyrite, Worthington Mine, Sudbury, *Econ. Geol.*, vol. 25, pp. 356–364, 1930.

Le Conte, J.: Discussion on Pošepný's paper, *Trans. Amer. Inst. Mining Engrs.*, vol. 24, pp. 996–1006, 1895.

Lepp, H., and S. S. Goldich: Origin of Precambrian iron formations, *Econ. Geol.*, vol. 59, pp. 1025–1060, 1964.

Lindgren, W.: "Ore Deposits," McGraw-Hill Book Company, New York, 1911.

———: Pseudo-eutectic textures, *Econ. Geol.*, vol. 25, pp. 1–13, 1930.

———: Differentiation and ore deposition, Cordilleran Region of the United States, in "Ore Deposits of the Western States," Lindgren Volume, pp. 152–180, American Institute of Mining and Metallurgical Engineers, 1933.

——— and J. D. Irving: Origin of the Rammelsberg ore deposits, *Econ. Geol.*, vol. 6, pp. 303–313, 1911.

Lister, G. F.: The composition and origin of selected iron-titanium deposits, *Econ. Geol.*, vol. 61, pp. 275–310, 1966.

Ljunggren, P.: Differential thermal analysis and x-ray examination of iron and manganese bog ores, *Geol. Fören. Stockholm Förh.*, vol. 77, pp. 135–147, 1955.

Loughlin, G. F., and A. H. Koschmann: Geology and ore deposits of the Cripple Creek district, Colorado, *Colorado Sci. Soc. Proc.*, vol. 13, pp. 217–435, 1935.

Lowell, J. D., and J. M. Guilbert: Lateral and vertical alteration-mineralization zoning in porphyry ore deposits, *Econ. Geol.*, vol. 65, pp. 373–408, 1970.

Lusk, J.: Base metal zoning in the Heath Steele B-1 orebody, New Brunswick, Canada, *Econ. Geol.*, vol. 64, pp. 509–518, 1969.

——— and J. H. Crockett: Sulfur isotope fractionation in coexisting sulfides from the Heath Steele B-1 orebody, New Brunswick, Canada, *Econ. Geol.*, vol. 64, pp. 147–155, 1969.

Lyall, K. D.: The origin of mechanical twinning in galena, *Amer. Mineralogist*, vol. 51, pp. 243–247, 1966.

——— and M. S. Paterson: Plastic deformation of galena (lead sulphide), *Acta Met.*, vol. 14, pp. 371–383, 1966.

McCarthy, K. A., and B. Chalmers: Energies of grain boundaries in silver chloride crystals, *Can. J. Phys.*, vol. 36, pp. 1645–1651, 1958.

McDonald, J. A.: Metamorphism and its effects on sulphide assemblages, *Miner. Deposita*, vol. 2, pp. 200–220, 1967.

McKelvey, V. E., D. L. Everhart, and R. M. Garrels: Origin of uranium deposits, *Econ. Geol. 50th Anniv. Vol.*, pp. 464–533, 1955.

McKie, D.: Goyazite and florencite from two African carbonatites, *Mineral. Mag.*, vol. 33, pp. 281–297, 1962.

McLean, D.: "Grain Boundaries in Metals," Oxford University Press, London, 1957.

MacLeod, W. N.: Banded iron formations of Western Australia, in J. McAndrew (ed.), "Geology of Australian Ore Deposits," pp. 113–117, Eighth Commonwealth Mining and Metallurgical Congress, Australia and New Zealand, 1965.

Macnamara, J., W. Fleming, A. Szabo, and H. C. Thode: The isotopic constitution of igneous sulphur and the primordial abundance of the terrestrial sulphur isotopes, *Can. J. Chem.*, vol. 30, pp. 73–76, 1952.

Macnamara, P. M.: Rock types and mineralization at Panguna porphyry copper prospect, Upper Kaverong Valley, Bougainville Island, *Proc. Australasian Inst. Mining Met.*, no. 228, pp. 71–80, 1968.

Maksimov, A. A.: Types of manganese and iron-manganese deposits in Central Kazakhstan, *Int. Geol. Rev.*, vol. 2, pp. 508–521, 1960.

Malan, S. P.: Stromatolites and other algal structures at Mufulira, Northern Rhodesia, *Econ. Geol.*, vol. 59, pp. 397–415, 1964.

Margolin, H. (ed.): Recrystallization, grain growth and textures, Amer. Soc. Metals Seminar, 1966.

Markham, N. L.: Synthetic and natural phases in the system Au–Ag–Te, pt. I, *Econ. Geol.*, vol. 55, pp. 1148–1178, 1960.

Martin, R. F., and A. J. Piwinskii: Experimental data bearing on the movement of iron in an aqueous vapour, *Econ. Geol.*, vol. 64, pp. 798–803, 1969.

Mason, B.: System Fe_2O_3–Mn_2O_3; comments on the names bixbyite, sitaparite and partridgeite, *Amer. Mineralogist*, vol. 29, pp. 66–69, 1944.

Maucher, A., and H. J. Schneider: The Alpine lead-zinc ores, in J. S. Brown (ed.), "Genesis of Stratiform Lead-Zinc-Barite-Fluorite Deposits," *Econ. Geol. Monograph* no. 3, pp. 71–89, 1967.

Mendelsohn, F. (ed.): "The Geology of the Northern Rhodesian Copper Belt," Roan Antelope Copper Mines Limited and MacDonald & Co. (Publishers) Ltd., London, 1961.

Mero, J. L.: Ocean-floor manganese nodules, *Econ. Geol.*, vol. 57, pp. 747–767, 1962.

———: "The Mineral Resources of the Sea," Elsevier Publishing Company, Amsterdam, 1965.

Merwin, H. E., and R. H. Lombard: The system Cu–Fe–S, *Econ. Geol.*, vol. 32, pp. 203–284, 1937.

Meyer, C., and J. J. Hemley: Wall rock alteration, in H. L. Barnes (ed.), "Geochemistry of Hydrothermal Ore Deposits," pp. 166–235, Holt, Rinehart and Winston, Inc., New York, 1967.

————, E. P. Shea, C. C. Goddard, L. G. Seihen, J. M. Guilbert, R. N. Miller, J. F. McAleer, C. B. Brox, R. C. Ingersoll, Jr., G. J. Burns, and T. Wigal: Ore deposits at Butte, Montana, in J. D. Ridge (ed.), "Ore Deposits in the United States 1933/67," Graton-Sales Volume, pp. 1373–1416, The American Institute of Mining, Metallurgical, and Petroleum Engineers, Inc., New York, 1968.

Michener, C. E.: "Minerals Associated with Larger Sulphide Bodies of the Sudbury Type," Unpublished Ph.D. thesis, University of Toronto, 1940.

Miller, A. R., C. D. Densmore, E. T. Degens, J. C. Hathaway, F. T. Manheim, P. F. McFarlin, R. Pocklington, and A. Jokela: Hot brines and recent iron deposits in deeps of the Red Sea, *Geochim. Cosmochim. Acta*, vol. 30, pp. 341–350, 1966.

Miller, D. S., and J. L. Kulp: Isotopic study of some Colorado Plateau ores, *Econ. Geol.*, vol. 53, pp. 937–948, 1958.

Mitcham, T. W., and C. G. Evensen: Uranium ore guides, Monument Valley district, Arizona, *Econ. Geol.*, vol. 50, pp. 170–176, 1955.

Mookherjee, A., and G. Suffel: Massive sulphide–late diabase relationships, Horne Mine, Quebec: genetic and chronological implications, *Can. J. Earth Sci.*, vol. 5, pp. 421–432, 1968.

Moore, E. S.: Observations on the Broken Hill lode, N.S.W., *Econ. Geol.*, vol. 11, pp. 327–348, 1916.

Morey, C. W.: The solubility of solids in gases, *Econ. Geol.*, vol. 52, pp. 225–251, 1957.

———— and J. M. Hesselgeser: The solubility of some minerals in superheated steam at high pressures, *Econ. Geol.*, vol. 46, pp. 821–835, 1951.

Mott, N. F.: Slip at grain boundaries and grain growth in metals, *Proc. Phys. Soc. London*, vol. 60, pp. 391–394, 1948.

Muraro, T. W.: Metamorphism of zinc-lead deposits in southeastern British Columbia, *Can. Inst. Mining Met. Spec. Vol.* 8, pp. 239–241, 1966.

Naldrett, A. J., and E. L. Gasparrini: Archean nickel sulfide deposits in Canada: their classification, geological setting and genesis with some suggestions as to exploration, *Geol. Soc. Aust., Spec. Pub.* no. 3, 1971.

———— and G. Kullerud: Investigations of the nickel-copper ores and adjacent rocks of the Sudbury District, Ontario, *Carnegie Inst. Wash., Ann. Rept. Dir. Geophys. Lab. Year Book* 64, pp. 177–188, 1965.

Necker, A. L.: An attempt to bring under general geological laws the relative position of metalliferous deposits, with regard to the rock formations of which the crust of the earth is formed, *Proc. Geol. Soc. London*, vol. 1, p. 392, 1832.

Neumann, H.: Notes on the mineralogy and geochemistry of zinc, *Mineral. Mag.*, vol. 28, pp. 575–581, 1949.

————: Pseudomorphs of FeS after pyrite in the Balachulish slates, *Mineral. Mag.*, vol. 29, pp. 234–238, 1950.

Newhouse, W. H.: The equilibrium diagram of pyrrhotite and pentlandite and their relations in natural occurrences, *Econ. Geol.*, vol. 22, pp. 284–299, 1927.

————: The composition of vein solutions as shown by liquid inclusions in minerals, *Econ. Geol.*, vol. 27, pp. 419–436, 1932.

————: A pyrrhotite-cubanite-chalcopyrite intergrowth from the Frood Mine, Sudbury, Ontario, *Amer. Mineralogist*, vol. 16, pp. 334–337, 1931.

————: Opaque oxides and sulphides in common igneous rocks, *Bull. Geol. Soc. Amer.*, vol. 47, pp. 1–51, 1936.

———— and C. F. Flaherty: The texture and origin of some banded or schistose sulphide ores, *Econ. Geol.*, vol. 25, pp. 600–620, 1930.

Nishimura, M.: Statistical study of the contents of iron and manganese in the mineral springs in Japan, *J. Chem. Soc. Japan*, sec. 73, pp. 749–753, 1952.

Nissen, A. E., and S. L. Hoyt: On the occurrence of silver in argentiferous galena ores, *Econ. Geol.*, vol. 10, pp. 172–179, 1915.

Nockolds, S. R.: Average chemical compositions of some igneous rocks, *Bull. Geol. Soc. Amer.*, vol. 65, pp. 1007–1032, 1954.

Northcott, L.: Veining and sub-boundary structures in metals, *J. Inst. Metals*, vol. 59, pp. 225–253, 1936.

Oftedahl, C.: A theory of exhalative-sedimentary ores, *Geol. Fören. Stockholm Förh.*, vol. 80, pp. 1–19, 1958.

Ohashi, R.: On the origin of Kuroko of the Kosaka mine, *J. Geol. Soc. Tokyo*, vol. 26, pp. 107–132, 1919.

————: On the stratified Kuroko, *J. Geol. Soc. Tokyo*, vol. 26, pp. 341–346, 1919.

————: On the origin of the Kuroko of the Kosaka copper mine, Northern Japan, *J. Akita Min. College*, vol. 2, pp. 11–18, 1920.

Olsen, J. C., D. R. Shawe, L. C. Pray, W. N. Sharp, and D. F. Hewitt: Rare earth mineral deposits of the Mountain Pass District, San Bernardino County, California, *U.S. Geol. Surv. Prof. Pap. 261*, 1954.

Orcel, J., and S. T. Pavlovitch: Les caractères microscopiques des oxydes de manganèse et des manganites naturels, *Bull. Soc. Franc. Mineral.*, vol. 54, pp. 108–179, 1931.

Paltridge, I. M.: An algal biostrome fringe and associated mineralization at Mufulira, Zambia, *Econ. Geol.*, vol. 63, pp. 207–216, 1968.

Park, C. F.: The zonal theory of ore deposits, *Econ. Geol. 50th Anniv. Vol.*, pp. 226–248, 1955.

Paterson, M. S.: The melting of calcite in the presence of water and carbon dioxide, *Amer. Mineralogist*, vol. 43, pp. 603–606, 1958.

Pecora, W. T.: Carbonatites: A review, *Bull. Geol. Soc. Amer.*, vol. 67, pp. 1537–1556, 1956.

Pereira, J., and C. J. Dixon: Evolutionary trends in ore deposition, *Trans. Inst. Mining Met.*, vol. 74, pp. 505–527, 1965.

Peters, W. C., A. H. James, and C. W. Field: Geology of the Bingham Canyon porphyry copper deposit, in S. R. Titley and C. L. Hicks (eds.), "The Geology of the Porphyry Copper Deposits: Southwestern North America," pp. 165–176, University of Arizona Press, Tucson, 1966.

Petruk, W.: Mineralogy and origin of the Silverfields silver deposit in the Cobalt area, Ontario, *Econ. Geol.*, vol. 63, pp. 512–531, 1968.

Pettijohn, F. J.: "Sedimentary Rocks," Harper & Brothers, New York, 1949.

Philpotts, A. R.: Origin of the anorthosite-mangerite rocks in Southern Quebec, *J. Petrol.*, vol. 7, pp. 1–64, 1966.

————: Origin of certain iron-titanium oxide and apatite rocks, *Econ. Geol.*, vol. 62, pp. 303–330, 1967.

Poldervaart, A.: Crust of the earth (A Symposium), *Geol. Soc. Amer. Spec. Pap. 62*, 1955.

Porrenga, D. H.: Glauconite and chamosite as depth indicators in the marine environment, *Marine Geol.*, vol. 5, pp. 495–501, 1967.

Pošepný, F.: The genesis of ore-deposits, *Trans. Amer. Inst. Mining Engrs.*, vol. 23, pp. 197–369, 1894.

Powell, J. L., P. M. Hurley, and H. W. Fairbairn: The strontium isotopic composition and origin of carbonatites, in O. F. Tuttle and J. Gittins (eds.), "Carbonatites," pp. 365–378, Interscience Publishers, Inc., New York, 1966.

Quirke, T. T., Jr.: Geology of the Temiscamie iron-formation, Lake Albanel iron range, Mistassini territory, Quebec, Canada, *Econ. Geol.*, vol. 56, pp. 299–320, 1961.

Ramberg, H.: Titanic iron ore formed by dissociation of silicates in granulite facies, *Econ. Geol.*, vol. 43, pp. 553–570, 1948.

Ramdohr, P.: Beobachtungen an Magnetit, Ilmenit, Eisenglanz und Ueberlegungen ueber das System $FeO-Fe_2O_3-TiO_2$, *Neues Jahrb. Mineral. Geol. Paläontol.*, vol. 54, sec. A, pp. 320–379, 1926.

————: Ueber Schapbachit, Matildit und den Silber–und Wismutgehalt mancher Betiglanze, *Sb. Preuss. Akad. Wiss., Phys-Math.* Kl. VI, pp. 71–91, 1938.

————: Die Erzmineralien in gewöhnlichen magmalischen Gesteinen, *Abhandl. Preuss. Akad. Wiss., Jahrgang*, pp. 1–43, 1940.

————: The ore deposit of Broken Hill in New South Wales in the light of new geological knowledge and ore microscopic investigations: Translation (by Consolidated Zinc Pty. Ltd.) of the German text of the paper in *Heidelberger Beitr. Mineral. Petrog.*, vol. 2, p. 25, 1950.

————: Die Lagerstätte von Broken Hill in New South Wales, im Lichte der neuen geologischen Erkenntnisse und erzmikrosckpischer Untersuchungen, *Heidelberger Beitr. Mineral. Petrog.*, vol. 2, pp. 291–333, 1950.

Ramsdell, L. S.: Studies on silicon carbide, *Amer. Mineralogist*, vol. 32, pp. 64–82, 1947.

Read, W. T., Jr.: "Dislocations in Crystals," McGraw-Hill Book Company, New York, 1953.

Revelle, R. R.: Marine bottom samples collected on the Pacific Ocean by the *Carnegie* on its seventh cruise, *Carnegie Inst. Wash. Pub.* 556, pt. 1, 1944.

Richards, J. R.: Lead isotopes at Dugald River and Mount Isa, Australia, *Geochim. Cosmochim. Acta*, vol. 31, pp. 51–62, 1967.

Richards, S. M.: The abundance of copper, zinc, lead and silver in a discordant hydrothermal ore body of the Conrad Mine, N.S.W., *Proc. Australasian Inst. Mining Met.*, no. 208, pp. 43–53, 1963.

————: Mineragraphy of fault-zone sulphides, Broken Hill, N.S.W., *Commonwealth Scient. Ind. Res. Org. Aust., Mineragraphic Investigations Tech. Pap.* 5, 1966.

Ridge, J. D. (ed.): "Ore deposits of the United States 1933/67," vols. I and II, The American Institute of Mining, Metallurgical and Petroleum Engineers, Inc., New York, 1968.

Robertson, J. A.: The relationship of mineralization to stratigraphy in the Blind River area, Ontario, *Geol. Assoc. Can., Spec. Pap.* 3, 1966.

Robertson, D. S., and N. C. Steenland: On the Blind River uranium ores and their origin, *Econ. Geol.*, vol. 55, pp. 659–694, 1960.

Roedder, E.: Environment of deposition of stratiform (Mississippi Valley Type) ore deposits, from studies of fluid inclusions, in J. S. Brown (ed.), "Genesis of Stratiform Lead-Zinc-Barite-Fluorite Deposits," pp. 371–378, *Econ. Geol. Monograph* no. 3, 1967.

————: Fluid inclusions as samples of ore fluids, in H. L. Barnes (ed.), "Geochemistry of Hydrothermal Ore Deposits," pp. 515–574, Holt, Rinehart and Winston, Inc., New York, 1967.

Romberger, S. B.: See H. L. Barnes (ed.), "Geochemistry of Hydrothermal Ore Deposits," p. 380, Holt, Rinehart and Winston, Inc., New York, 1967.

Roscoe, S. M.: On thorium-uranium ratios in conglomerate and associated rocks near Blind River, Ontario, *Econ. Geol.*, vol. 54, pp. 511–512, 1959.

Rose, A. W., and W. W. Baltosser: The porphyry copper deposit at Santa Rita, New Mexico, in S. R. Titley and C. L. Hicks (eds.), "The Geology of the Porphyry Copper Deposits: Southwestern North America," pp. 205–220, University of Arizona Press, Tucson, 1966.

Roseboom, E. H.: The $CoAs_2$–$NiAs_2$–$FeAs_2$–As system, *Carnegie Inst. Wash., Ann. Rept. Dir. Geophys. Lab. Year Book* 56, pp. 201–204, 1957.

————: Skutterudites (Ca,Ni,Fe)As_{3-x}: composition and cell dimensions, *Amer. Mineralogist*, vol. 47, pp. 310–327, 1962.

————: Co–Ni–Fe diarsenides: compositions and cell dimensions, *Amer. Mineralogist*, vol. 48, pp. 271–299, 1963.

————: An investigation of the system Cu–S and some natural copper sulphides between 25° and 700°C, *Econ. Geol.*, vol. 61, pp. 641–672, 1966.

———— and G. Kullerud: The solidus in the system Cu–Fe–S between 400° and 800°C, *Carnegie Inst. Wash., Ann. Rept. Dir. Geophys. Lab. Year Book* 57, pp. 222–227, 1958.

Rosenfeld, I., and O. A. Beath: "Selenium: Geobotany, Biochemistry, Toxicity and Nutrition," Academic Press, Inc., New York, 1964.

Rosenhain, W., and J. C. W. Humfrey: The tenacity, deformation, and fracture of soft steel at high temperature, *J. Iron Steel Inst. (London)*, vol. 87, pp. 219–271, 1913.

Ross, V.: Geochemistry, crystal structure and mineralogy of the sulfides, *Econ. Geol.*, vol. 52, pp. 755–774, 1957.

Royer, L.: Recherches expérimentales sur l'épitaxie ou orientation mutuelle des cristaux d'espèces différentes, *Bull. Soc. Fran. Min.*, vol. 51, pp. 7–159, 1928.

Rubey, W. W.: The geologic history of sea water, *Bull. Geol. Soc. Amer.*, vol. 62, pp. 1111–1147, 1951.

Russell, H. D., S. A. Hiemstra, and D. Groeneveld: The mineralogy and petrology of the carbonatite at Loolekop, eastern Transvaal, *Trans. Geol. Soc. S. Afr.*, vol. 57, pp. 197–208, 1954.

Russell, R. D., and R. M. Farquhar: "Lead Isotopes in Geology," Interscience Publishers, Inc., New York, 1960.

Sainsbury, C. L.: Tin resources of the world, *U.S. Geol. Surv. Bull. no.* 1301, pp. 1–55, 1969.

Sakai, H.: Isotopic properties of sulfur compounds in hydrothermal processes, *Geochem. J. (Nagoya)*, vol. 2, pp. 29–49, 1968.

——— and H. Nagasawa: Fractionation of sulfur isotopes in volcanic gases, *Geochim. Cosmochim. Acta*, vol. 15, pp. 32–39, 1958.

Sakamoto, T.: The origin of the Pre-Cambrian banded iron ores, *Amer. J. Sci.*, vol. 248, pp. 449–474, 1950.

Sales, R. H.: Ore deposits at Butte, Mont., *Trans. Amer. Inst. Mining Engrs.*, vol. 46, pp. 3–109, 1913.

Sampson, E.: Chromite deposits, in W. H. Newhouse (ed.), "Ore Deposits as Related to Structural Features," pp. 110–125, Princeton University Press, Princeton, N.J., 1942.

Sandberger, F.: Untersuchungen über Erzgänge, vol. 1, 1882.

Sasaki, A., and H. R. Krouse: Sulfur isotopes and the Pine Point lead-zinc mineralization, *Econ. Geol.*, vol. 64, pp. 718–730, 1969.

Sawkins, F. J., A. C. Dunham, and D. M. Hirst: Iron-deficient low-temperature pyrrhotite, *Nature*, vol. 204, pp. 175-176, 1964.

Schachner-Korn, D.: Ein metamorphes Erzgefüge, *Heidelberger Beitr. Mineral. Petrog.*, vol. 1, pp. 407–426, 1948.

Schmitt, H. A.: The porphyry copper deposits in their regional setting, in S. R. Titley and C. L. Hicks (eds.), "The Geology of the Porphyry Copper Deposits: Southwestern North America," pp. 17–34, University of Arizona Press, Tucson, 1966.

Schneiderhöhn, H.: Die genetische Einteilung der Gesteine und Minerallagerstätten, *Zeitschr. prakt. Geol.*, vol. xl, pp. 168–172, 1932.

———: The formation of ore deposits and geotectonics, Translation from German of the paper in *Erzbergb. Metallhisttwes.*, 1953.

——— and P. Ramdohr: "Lehrbuch der Erzmikroskopie," vol. II, Berlin, 1931.

Scholtz, D. L.: The magmatic nickeliferous ore deposits of East Griqualand and Pondoland, *Trans. Proc. Geol. Soc. S. Afr.*, vol. 39, pp. 81–210, 1936.

Schwartz, G. M.: Intergrowths of chalcopyrite and cubanite: experimental proof of the origin of intergrowths and their bearing on the geologic thermometer, *Econ. Geol.*, vol. 22, pp. 44–61, 1927.

———: Experiments bearing on bornite-chalcocite intergrowths, *Econ. Geol.*, vol. 23, pp. 381–397, 1928.

———: Relations of chalcocite-stromyerite-argentite, *Econ. Geol.*, vol. 30, pp. 128–146, 1935.

——— and C. F. Park, Jr.: Pseudo-eutectic textures, *Econ. Geol.*, vol. 25, pp. 658–663, 1930.

Schwellnus, C. M., and J. Willemse: Titanium and vanadium in the magnetic iron ores of the Bushveld Complex, *Trans. Proc. Geol. Soc. S. Afr.*, vol. 46, pp. 23–38, 1943.

Sederholm, J. J.: The average composition of the earth's crust in Finland, *Bull. Comm. Geol. Finl.*, vol. 12, no. 70, 1925.

Shatskiy, N. S.: On manganiferous formations and the metallogeny of manganese, Paper I. Volcanogenic-sedimentary manganiferous formations, *Int. Geol. Rev.*, vol. 6, pp. 1030–1056, 1964.

Shcherbina, V. V., and R. N. Zer'yan: Paragenesis of silver and gold tellurides as solid phases in the system Ag–Au–Te, *Geochem. Int.*, no. 4, pp. 653–657, 1964.

Shockley, W., and W. T. Read: Quantitative predictions from dislocation models of crystal grain boundaries, *Phys. Rev.*, vol. 75, p. 692, 1949.

Shterenberg, L. E., L. A. Kozyar, V. G. Morozova, G. P. Gapochka, and G. Yu. Butuzova: The age of Chiaturi manganese deposit and its place among other deposits in the South of the European part of the Soviet Union, *Econ. Geol. USSR*, vol. 1, nos. 5–6, pp. 9–24, 1965.

Siemes, H.: Zum Rekristallisationverhalten von natürlich verformten Bleiglanzen, *Neues Jahrbuch Miner. Abh.*, vol. 102, pp. 1–30, 1964.

Sindeeva, N. D.: "Mineralogy and Types of Deposits of Selenium and Tellurium," Interscience Publishers, Inc., New York, 1964.

Skinner, B. J.: "Earth Resources," Prentice-Hall, Inc., Englewood Cliffs, N.J., 1969.

——— and D. L. Peck: An immiscible sulfide melt from Hawaii, *Econ. Geol. Monograph* no. 4, pp. 310–322, 1969.

———, D. E. White, H. J. Rose, and R. E. Mays: Sulfides associated with the Salton Sea geothermal brines, *Econ. Geol.*, vol. 62, pp. 316–330, 1967.

Slawson, W. F., and R. D. Russell: Common lead abundances, in H. L. Barnes (ed.), "Geochemistry of Hydrothermal Ore Deposits," pp. 77–108, Holt, Rinehart and Winston, New York, 1967.

Smith, C. S.: Grains, phases and interfaces: an interpretation of microstructure, *Trans. Amer. Inst. Mining Met. Engrs.*, vol. 175, pp. 15–51, 1948.

———: Interphase interfaces, in W. Shockley (ed.), "Imperfections in Nearly Perfect Crystals," pp. 377–401, John Wiley & Sons, Inc., New York, 1952.

———: "Metal Interfaces," American Society for Metals, Cleveland, Ohio, 1952.

———: Some elementary principles of polycrystalline microstructure, *Met. Rev.*, vol. 9, no. 33, pp. 1–48, 1964.

Smith, F. C.: Structure of zinc sulphide minerals, *Amer. Mineralogist*, vol. 40, pp. 658–675, 1955.

Smitheringale, W. G., and M. L. Jensen: Sulfur isotopic composition of the Triassic igneous rocks of eastern United States, *Geochim. Cosmochim. Acta*, vol. 27, pp. 1183–1208, 1963.

Smyth, F. H., and L. H. Adams: The system calcium oxide-carbon dioxide, *Amer. Chem. Soc. J.*, vol. 45, pp. 1169–1184, 1923.

Sorby, H. C.: On the microscopic structures of crystals, indicating the origin of minerals and rocks, *Quart. J. Geol. Soc. London*, vol. 14, pp. 453–500, 1858.

Souch, B. E., T. Podolsky, et al.: The sulphide ores of Sudbury: Their particular relationship to a distinctive inclusion-bearing facies of the nickel irruptive, *Econ. Geol. Monograph* no. 4, pp. 252–261, 1969.

Spurr, J. E.: "The Ore Magmas," McGraw-Hill Book Company, New York, 1923.

Stanton, R. L.: "Lower Palaeozoic Mineralization and Features of Its Environment Near Bathurst, Central Western New South Wales," Ph.D. thesis, University of Sydney, 1954.

———: The genetic relation between limestone, volcanic rocks and certain ore deposits, *Aust. J. Sci.*, vol. 17, no. 5, pp. 173–175, 1955.

———: Lower Palaeozoic mineralisation near Bathurst, N.S.W., *Econ. Geol.*, vol. 50, pp. 681–714, 1955.

———: Studies of polished surfaces of pyrite, and some implications, *Can. Mineralogist*, vol. 6, pp. 87–118, 1957.

———: Mineralogical features and possible mode of emplacement of the Brunswick Mining and Smelting Orebodies, Gloucester County, N.B., *Bull. Can. Inst. Mining Met.*, vol. 52, no. 570, pp. 631–643, 1959.

————: Geological theory and the search for ore, *Mining & Chem. Eng. Rev.*, vol. 53, no. 7, pp. 48–55, 1961.

————: Elemental constitution of the Black Star Orebodies, Mount Isa, and its interpretation, *Trans. Inst. Mining Met.*, vol. 72, pp. 69–124, 1962.

————: Mineral interfaces in stratiform ores, *Trans. Inst. Mining Met.*, vol. 74, pp. 45–79, 1964.

————: Compositions of stratiform ores as evidence of depositional processes, *Trans. Inst. Mining Met., Sec. B*, vol. 75, pp. B75–B84, 1966.

———— and H. Gorman: A phenomenological study of grain boundary migration in some common sulfides, *Econ. Geol.*, vol. 63, pp. 907–923, 1968.

———— and H. G. Willey: Natural work-hardening in galena, and its experimental reduction, *Econ. Geol.*, vol. 65, pp. 182–194, 1970.

———— and T. A. Rafter: The isotopic constitution of sulphur in some stratiform lead-zinc sulphide ores, *Miner. Deposita*, vol. 1, pp. 16–29, 1966.

———— and ————: Sulfur isotope ratios in co-existing galena and sphalerite from Broken Hill, New South Wales, *Econ. Geol.*, vol. 62, pp. 1088–1091, 1967.

———— and R. D. Russell: Anomalous leads and the emplacement of lead sulfide ores, *Econ. Geol.*, vol. 54, pp. 588–607, 1959.

Steiner, A., and T. A. Rafter: Sulfur isotopes in pyrite, pyrrhotite, alunite and anhydrite from steam wells in the Taupo Volcanic Zone, New Zealand, *Econ. Geol.*, vol. 61, pp. 1115–1129, 1966.

Stevens, R. E.: Composition of some chromites of the Western Hemisphere, *Amer. Mineralogist*, vol. 29, pp. 1–34, 1944.

Stevenson, J. S.: Mineralization and metamorphism at the Eustis mine, Quebec, *Econ. Geol.*, vol. 32, pp. 335–363, 1937.

Stillwell, F. L.: Observations on the mineral constitution of the Broken Hill lode, *Proc. Australasian Inst. Mining Met.*, no. 64, pp. 1–76 and 97–172, 1926.

————: The occurrence of telluride minerals at Kalgoorlie, *Proc. Australasian Inst. Mining Met.*, no. 84, pp. 116–190, 1931.

————: Occurrence of tellurides at Vatukoula, Fiji, *Proc. Australasian Inst. Mining Met.*, no. 154–155, pp. 3–28, 1949.

Stout, W.: The iron-bearing formations of Ohio, *Bull. Geol. Surv. Ohio*, Ser. 4, vol. 45, 1944.

Strahan, A., W. Gibson, J. C. Cantrill, R. L. Sherlock, and H. Dewey: Iron ores–Pre-Carboniferous bedded ores of England and Wales, *Mem. Geol. Surv. U.K., Spec. Rep. Mineral Resources*, vol. 13, 1920.

Strakhov, N. M.: Types of manganese accumulation in present-day basins: their significance in understanding of manganese mineralization, *Int. Geol. Rev*, vol. 8, pp. 1172–1196, 1966.

Stranski, I. N.: Zur Theorie des Kristallwachstums, *Z. phys. Chem.*, vol. 136, pp. 259–278, 1928.

Stringham, B.: Igneous rock types and host rocks associated with porphyry copper deposits, in S. R. Titley and C. L. Hicks (eds.), "The Geology of the Porphyry Copper Deposits: Southwestern North America," pp. 35–40, University of Arizona Press, Tucson, 1966.

Sullivan, C. J.: Ore and granitization, *Econ. Geol.*, vol. 43, pp. 471–498, 1948.

————: Metallic melting point and ore deposition, *Econ. Geol.*, vol. 49, pp. 555–574, 1954.

Suzuki, J., and H. Ohmachi: Manganiferous iron ore deposits in the Tokoro district of northeastern Hokkaido, Japan, *Int. Geol. Congr., 20th, Mexico, Manganese Symp.*, vol. 4, pp. 199–204, 1956.

Tatsumi, T.: Sulfur isotopic fractionation between co-existing sulfide minerals from some Japanese ore deposits, *Econ. Geol.*, vol. 60, pp. 1645–1659, 1965.

Tatsumi, T. (ed.): "Volcanism and Ore Genesis," University of Tokyo Press, Tokyo, 1970.

Taylor, J.: Report on the state of knowledge respecting mineral veins, *Rep. Br. Assoc.*, 3d Mtg., 1834.

Taylor, J. H.: Petrology of the Northampton Sand Ironstone Formations, *Mem. Geol. Surv. U.K.*, 1949.

Temple, K. C., and N. W. Le Roux: Syngenesis of sulfide ores: sulfate-reducing bacteria and copper toxicity, *Econ. Geol.*, vol. 59, pp. 271–278, 1964.

Thayer, T. P.: Preliminary chemical correlation of chromite with the containing rocks, *Econ. Geol.*, vol. 41, pp. 202–217, 1946.

———: Some critical differences between alpine-type and stratiform peridotite-gabbro complexes, *Int. Geol. Congr., 21st, Copenhagen, 1960, Rept. Session, Norden*, pt. 13, pp. 247–259, 1960.

———: Principal features and origin of podiform chromite deposits, and some observations on the Guleman-Soridag District, Turkey, *Econ. Geol.*, vol. 59, pp. 1497–1524, 1964.

Thode, H. G.: Variations in abundances of isotopes in nature, *Research (London)*, vol. 2, pp. 154–161, 1949.

———, H. Kleerekoper, and D. McElcheran: Isotope fractionation in the bacterial reduction of sulphate, *Research (London)*, vol. 4, pp. 581–582, 1951.

———, J. Monster, and H. B. Dunford: Sulphur isotope geochemistry, *Geochim. Cosmochim. Acta*, vol. 25, pp. 159–174, 1961.

Thomas, W. N.: "Broken Hill Ore Occurrence Reinterpreted," Consolidated Zinc Pty. Ltd. Report (unpublished), 1960.

Thompson, R. M.: The telluride minerals and their occurrence in Canada, *Amer. Mineralogist*, vol. 34, pp. 342–382, 1949.

Titley, S. R., and C. L. Hicks (eds.): "Geology of the Porphyry Copper Deposits: Southwestern North America," University of Arizona Press, Tucson, 1966.

Trendall, A. F.: Three great basins of Precambrian banded iron formation deposition: a systematic comparison, *Geol. Soc. Amer. Bull.*, vol. 79, pp. 1527–1544, 1968.

Tudge, A. P., and H. G. Thode: Thermodynamic properties of isotopic compounds of sulfur, *Can. J. Res. Sec. B*, vol. 28, pp. 567–578, 1950.

Turek, A.: "Rubidium-Strontium Isotopic Studies in the Kalgoorlie-Norseman Area," Ph.D. thesis, Australian National University, 1966.

Turner, F. J., and J. Verhoogen: "Igneous and Metamorphic Petrology," McGraw-Hill Book Company, New York, 1951; 2d ed., 1960.

Tuttle, O. F., and J. Gittins (eds.): "Carbonatites," Interscience Publishers, Inc., New York, 1966.

Twelvetrees, W. H., and L. K. Ward: The orebodies of the Zeehan Field, *Geol. Surv. Bull. Tasmania*, No. 8, 1910.

Uglow, W. L.: Gneissic galena ore from the Slocan district, British Columbia, *Econ. Geol.*, vol. 20, pp. 573–586, 1917.

Ulmer, G. C.: Experimental investigations of chromite spinels, *Econ. Geol. Monograph* no. 4, pp. 114–130, 1969.

——— and W. B. White: The existence of chromous ion in the spinel solid solution series $FeCr_2O_4$–$MgCr_2O_4$, *J. Amer. Ceram. Soc.*, vol. 49, pp. 50–51, 1966.

Vaasjoki, O.: On the natural occurrence of manganese in Finland, *Int. Geol. Congr., 20th, Mexico, Manganese Symp.*, 1956.

Van der Veen, R. W.: "Mineragraphy and ore deposition," The Hague, 1925.

Van Hise, C. R.: Some principles controlling the deposition of ores, *Trans. Amer. Inst. Mining Engrs.*, vol. 30, pp. 27–177, 1900.

———: Treatise on metamorphism, *U.S. Geol. Surv. Monograph* 47, 1904.

——— and W. S. Bayley: The Marquette Iron-bearing district of Michigan, *U.S. Geol. Surv. Monograph* 28, 1897.

——— and C. K. Leith: The geology of the Lake Superior region, *U.S. Geol. Surv. Monograph* 52, 1911.

Varentsov, I. M.: "Sedimentary Manganese Ores," Elsevier Publishing Company, Amsterdam, 1964.

———: Contribution to the establishment of the conditions in which the Nikopol' and other deposits of the Southern Ukrainian manganese ore field formed, *Econ. Geol. USSR*, vol. 1, nos. 5–6, pp. 25–39, 1965.

Verma, A. R.: "Crystal Growth and Dislocations," Butterworth Scientific Publications, London, 1953.

Verwoerd, W. J.: South African carbonatites and their probable mode of origin, *Ann. Univ. Stellenbosch*, vol. 41, ser. A, no. 2, 1966.

Vinogradov, A. R.: S^{32}/S^{34} isotopic composition of meteorites and of the earth, U.N.E.S.C.O./N.S./R.I.C./62 Int. Conf. on Radioisotopes in Scientific Research, 1957.

Vokes, F. M.: A review of the metamorphism of sulphide deposits, *Earth-Sci. Rev.*, vol. 5, pp. 99–143, 1968.

Volmer, M.: Zum Problem des Kristallwachstums, *Z. phys. Chem.*, vol. 102, pp. 267–275, 1922.

—— and W. Schultze: Kondensation an Kristallen, *Z. phys. Chem.*, Ser. A, vol. 156, pp. 1–22, 1931.

Von Eckermann, H.: Alnökalkens genesis i belysning av diamantborrningar och ny detalj-kartering, *Geol. Fören. Stockholm. Förn.*, vol. 62, pp. 102–105, 1940.

——: The alkaline district of Alnö Island, *Sveriges Geol. Undersokn, Arsbok*, Ser. Ca, no. 36, 1948.

——: Progress of research on the Alnö carbonatite, in O. F. Tuttle and J. Gittins (eds.), "Carbonatites," pp. 30–32, Interscience Publishers, Inc., New York, 1966.

——: The petrogenesis of the Alnö alkaline rocks, *Bull. Geol., Inst. Univ. Uppsala*, vol. 40, pp. 25–36, 1961.

Wager, L. R., E. A. Vincent, and A. A. Smales: Sulphides in the Skaergaard Intrusion, East Greenland, *Econ. Geol.*, vol. 52, pp. 855–903, 1957.

Wagner, P. A.: The iron deposits of the Union of South Africa, *Mem. Geol. Surv. Div. Un. S. Afr.*, Mem. 26, 1928.

Waldschmidt, W. A.: Deformation in ores, Coeur d'Alene District, Idaho, *Econ. Geol.*, vol. 20, pp. 573–586, 1925.

Walker, F.: Differentiation of the Palisade diabase, *Bull. Geol. Soc. Amer.*, vol. 51, pp. 1059–1106, 1940.

Ward, W., O. S. Perry, K. Griffin, G. H. Charlewood, H. Hopkins, G. MacIntosh, and S. P. Ogryzlo: The gold mines of Kirkland Lake, in "Structural Geology of Canadian Ore Deposits," pp. 644–653, Canadian Institute of Mining and Metallurgy, Inc., Montreal, 1948.

Watanabe, M.: Modes of occurrences of tellurium bearing minerals in Japan, *Sci. Rep. Tohoku Univ. Ser. III*, vol. 4, pp. 45–80, 1952.

Watanabe, T.: Characteristic features of ore deposits found in contact-metamorphic aureoles in Japan, *Int. Geol. Rev.*, vol. 2, pp. 946–966, 1960.

Werner, A. G.: "Neue Theorie von der Entestehung de Gänge," 1791.

White, D. E.: Environments of generation of some base-metal ore deposits, *Econ. Geol.*, vol. 63, pp. 301–335, 1968.

Whitney, J. D.: "Metallic Wealth of the United States," 1854.

Willemse, J.: A brief outline of the geology of the Bushveld Igneous Complex, in S. H. Haughton (ed.), "The Geology of Some Ore Deposits in Southern Africa," vol. II, pp. 91–130, The Geological Society of South Africa, Johannesburg, 1964.

——: The geology of the Bushveld Igneous Complex, the largest repository of magmatic ore deposits in the world, *Econ. Geol. Monograph* no. 4, 1969.

Williams, D.: Genesis of sulphide ores, *Proc. Geol. Assoc.*, vol. 71, pp. 245–311, 1960.

——: Further reflections on the origin of the porphyries and ores of Rio Tinto, Spain, *Trans. Inst. Mining Met.*, vol. 71, pp. 265–266, 1962.

Wilson, H. D. B. (ed.): Magmatic ore deposits: a symposium, *Econ. Geol. Monograph* no. 4, 1969.

Winchell, N. H., and H. V. Winchell: The iron ores of Minnesota, Bulletin no. 6, Geological and Natural History Survey of Minnesota; Minneapolis: Harrison and Smith, State Printers, 1891.

Woodall, R., and G. A. Travis: The Kambalda Nickel Deposits, Western Australia, Ninth Commonwealth Mining and Metallurgical Congress, London, 1969.

Wyllie, P. J.: Experimental studies of carbonatite problems: The origin and differentiation of carbonatite magmas, in O. F. Tuttle and J. Gittins (eds.), "Carbonatites," pp. 311–352, Interscience Publishers, Inc., New York, 1966.

—— and O. F. Tuttle: Experimental verification for the magmatic origin of carbonatites, *Int. Geol. Congr., 21st, Copenhagen, 1960, Rept. Session, Norden*, pt. 13, pp. 310–318, 1960.

—— and ——: The system $CaO-CO_2-H_2O$ and the origin of carbonatites, *J. Petrol.*, vol. 3, pp. 238–243, 1960.

Zapffe, C.: Deposition of manganese, *Econ. Geol.*, vol. 26, pp. 799–832, 1931.

Zelenov, K. K.: On the discharge of iron in solution into the Okhotsk Sea by thermal springs of the Ebeko volcano (Paramuchir Island), *Dokl. Akad. Nauk. SSSR*, vol. 120, pp. 1089–1092, 1958.   (In Russian; English translation published by Consultants Bureau, Inc., New York, pp. 497–500, 1959.)

——: Transportation and accumulation of iron and aluminium in volcanic provinces of the Pacific, *Izv. Akad. Nauk. SSSR, Ser. Geol.*, pp. 47–59, 1959.

Zies, E. G.: The Valley of Ten Thousand Smokes–I.   The fumarolic incrustations and their bearing on ore deposition, *Tech. Pap. Natl. Geogr. Soc.*, vol. 1, no. 4, pp. 1–61, 1929.

——: The Valley of Ten Thousand Smokes–II.   The acid gases contributed to the sea during volcanic activity, *Tech. Pap. Natl. Geogr. Soc.*, vol. 1, no. 4, pp. 62–79, 1929.

Zurbrigg, H. F.: Thompson Mine Geology, *Bull. Can. Mining Met.*, vol. 66, pp. 227–236, 1963.

Zwicky, F.: On the imperfections of crystals, *Proc. Natl. Acad. Sci. U.S.*, vol. 15, pp. 253–259, 1929.

Name Index

Subject Index